Interface Science and Composites

INTERFACE SCIENCE AND TECHNOLOGY
Series Editor: Arthur Hubbard

In this Series:

Vol. 1: Clay Surfaces: Fundamentals and Applications
Edited by F. Wypych and K.G. Satyanarayana

Vol. 2: Electrokinetics in Microfluids
By Dongqing Li

Vol. 3: Radiotracer Studies of Interfaces
Edited by G. Horányi

Vol. 4: Emulsions: Structure Stability and Interactions
Edited by D.N. Petsev

Vol. 5: Inhaled Particles
By Chiu-sen Wang

Vol. 6: Heavy Metals in the Environment: Origin, Interaction and Remediation
Edited by H.B. Bradl

Vol. 7: Activated Carbon Surfaces in Environmental Remediation
Edited by Teresa J. Bandosz

Vol. 8: Tissue Engineering: Fundamentals and Applications
By Yoshito Ikada

Vol. 9: Particles at Interfaces: Interactions, Deposition, Structure
By Zbigniew Adamczyk

Vol. 10: Interface Science in Drinking Water Treatment: Theory and Applications
Edited by G. Newcombe and D. Dixon

Vol. 11: Surface Complexation Modelling
Edited by Johannes Lützenkirchen

Vol. 12: Theory of Colloid and Interfacial Electric Phenomena
By Hiroyuki Ohshima

Vol. 13: Sorbent Deformation
By A.V. Tvardovskiy

Vol. 14: Advanced Chemistry of Monolayers at Interfaces: Trends in Methodology and Technology
Edited by Toyoko Imae

Vol. 15: Macromolecules in Solution and Brownian Relativity
By Stefano A. Mezzasalma

Vol. 16: The Properties of Water and their Role in Colloidal and Biological Systems
By Carel Jan van Oss

Vol. 17: Chemistry on Modified Oxide and Phosphate Surfaces: Fundamentals and Applications
By Robson Fernandes de Farias

Vol. 18: Interface Science and Composites
By Soo-Jin Park and Min-Kang Seo

INTERFACE SCIENCE AND TECHNOLOGY – VOLUME 18

Interface Science and Composites

Soo-Jin Park and Min-Kang Seo

Department of Chemistry,
Inha University, South Korea

ELSEVIER

Amsterdam • Boston • Heidelberg • London • New York • Oxford
Paris • San Diego • San Francisco • Singapore • Sydney • Tokyo
Academic Press is an imprint of Elsevier

Academic Press is an imprint of Elsevier
Radarweg 29, PO Box 211, 1000 AE Amsterdam, The Netherlands
Linacre House, Jordan Hill, Oxford OX2 8DP, UK
32, Jamestown Road, London, NWI 7BY, UK
225 Wyman Street, Waltham, MA 02451, USA
525 B Street, Suite 1900, San Diego, CA 92101-4495, USA

First edition 2011

ISBN: 978-0-08-097605-1
ISSN: 1573-4285

Contents

Preface

The goal of 'interface science and composites' is to facilitate the manufacture of technological materials with optimized properties on the basis of a comprehensive understanding of the molecular structure of interfaces and their resulting influence on composite materials processes. Of course, establishing such a complete understanding would be a monumental undertaking. Consider the parameters which govern interface structure and behavior: besides the crystallographic and compositional variables which define a polycrystalline system, the external conditions, temperature, stress, electric/magnetic fields, radiation and chemical environment, prevailing historically and currently, need to be taken into account.

From the early development of composites of various natures, the optimization of the interface has been of major importance. While there are many reference books available on composites, few of them deal specifically with the science and mechanics of the interface of materials and composites. Further, many recent advances devoted solely to research in composite interfaces have been scattered in a variety of published literature and have yet to be assembled in a readily accessible form. To this end this book is an attempt to bring together recent developments in the field, both from the materials science and mechanics perspective, in a single convenient volume. The central theme of the book is tailoring the interface science of composites to optimize the basic physical principles than just on blind use of materials as well as mechanical performance and structural integrity of composites with enhanced strength/stiffness and fracture toughness (or specific fracture resistance). It also deals mainly with interfaces in advanced composites made from high-performance fibers, such as glass, carbon, aramid and some inorganic fibers, and matrix materials encompassing polymers, carbon, metals/alloys and ceramics.

The broad spectrum of interface science and technology includes several possible combinations of solid, liquid and gas phases. Of these, four are of practical importance: gas on solid, liquid on solid and solid on solid. The understanding of the interface in composites is the central point in the high-performance advanced composites. To cover all these combinations in a medium-sized book to a reasonable degree of detail is almost impossible. Knowledge of the basic principles and theories of this branch of interface science and composites has certainly become important, if not indispensable, for researchers and students who are working in the areas of both science and engineering. The book contains ten chapters of discussions on interface science, composite interfaces and microstructure-property relationships with underlying fundamental physical and mechanical principles. Chapter 1 reviews the physical or intermolecular interaction as well as the concepts of acid-base interaction. Chapter 2 explains the long-range forces to surface dynamics of solid-gas interfaces. Chapter 3 introduces the intermolecular forces, describing their role in determining the solid surface properties. Adhesion at solid-solid interfaces involving interfacial phenomena is presented in Chapter 4, along with the correlations among adhesion, London dispersive and/or acid-base interactions. Interface applications in micro- and nanosystems affecting modern societal needs are presented in Chapter 5. Chapter 6 is devoted to reviewing two essential parameters and processing techniques for manufacturing the composites. The properties and applicability of composites associated with adopting matrix types are interpreted and compared in Chapter 7. The evaluation techniques that have been developed to assess the reinforcement fibers, matrix resins, reinforcement-matrix interface and final composites are presented in Chapter 8, along with the techniques for non-destructive evaluation. Chapter 9 describes the experimental fiber-matrix interfacial properties on a microscopic scale. The roles of the interface and its effects on the mechanical performance of the fiber-reinforced composites are addressed from several viewpoints. Recent research efforts to augment the applications of composites by means of nanoscaled materials are presented in Chapter 10. Especially, the development of a nanolevel dispersion of graphene particles in a polymer matrix has opened a new and interesting area in materials science in recent years.

The authors are grateful to everyone who has contributed to this book.

Soo-Jin Park
Min-Kang Seo

Chapter 1

Intermolecular Force

1.1. INTRODUCTION

The dawn of the 19th century brought new ways of observing phenomena at the molecular level. The work of crystallographers and chemists elucidated the arrangement of atoms within matter and laid the foundation of our present knowledge. This story, familiar to every physicist, was written by Young [1], van der Waals [2], Keesom [3], Debye [4, 5] and, finally, London [6, 7], who developed a rigorous quantum mechanical description of intermolecular forces.

These forces result from the electromagnetic interactions between the electrons and nuclei forming molecules, and thus their calculation requires solving the Schrödinger equation for a system of interacting particles. This can be done approximately, on the basis that electrons move much more rapidly than nuclei, so electronic and nuclear motions can be separated. The energy associated with electronic motion is the potential energy for the motion of the nuclei, and can be regarded as the intermolecular interaction potential.

At the most fundamental level, all atomistic interactions are electromagnetic. In spite of this unifying and underlying fundamental principle, various types of atomic and molecular interactions show sufficient specificity either in the underlying theories or in their relative strength within different regimes of interatomic or intermolecular separations.

Our objectives in this book are to look into the interface science, and establish relations for scale from the molecular-level forces to forces between microscopic substances, including the surface dynamics, the surface free energy at a given temperature, and the surface energy. It should be noted that interface science has been extremely broadened in various applications, providing essentially limitless areas for investigation.

1.2. LONG-RANGE FORCE

The fundamental importance of bonding energies between bodies is traditionally divided into two broad classes: chemical bonds or short-range forces, and physical forces (or intermolecular bonds or long-range forces).

Interface Science and Composites, Volume 18
ISSN: 1573-4285, doi: 10.1016/B978-0-12-375049-5.00001-3

These features point to the key role of the two-phase and three-phase surface phenomena in various technological processes. For most materials under laboratory conditions, the properties are replaced by a composite system comprising of the solid materials and liquid interlayer or capillary bridges. The properties of the system as a whole are determined by the properties of not only the solid phase, but also the liquid interlayer, the character of the interaction between the liquid and the solid particles and finally, the character of the interaction between the particles separated by the liquid interlayer. The presence of liquid interlayer can play a decisive role in, e.g., the self-organization of particles, their flow ability their dispensability, and so on, creating a comprehensive framework and the language of these forces in science as well as identifying the strategies to exploit them for the design of new materials and devices. The most topical study areas at different interfaces are listed in Table 1-1 [8].

The energies are largely dependent on the distance at which one body feels the presence of the other. Usually, the long-range force is called a 'Lennard-Jones potential [9], and has a minimum value at a certain distance.

Table 1-1 Topical Study Areas at Different Interfaces

Interfaces	Designation	Topical areas
Solid-Gas	Adsorption	Solid aerosol, adsorption, catalysis, corrosion, diffusion, surface energy, thin films, permeation, osmosis, filtration, oxidation, charge transfer, condensation, and nucleation
Solid-Liquid	Wettability	Sol, gel, colloidal suspension, solid emulsion, wetting, spreading, surface tension, friction, lubrication, diffusion, pervaporation, capillarity, electrochemistry, galvanic effects, corrosion, cleaning, filtration, ion electro migration, optical properties, charge transfer, nucleation, and growth
Solid-Solid	Adhesion	Solid suspension, adhesion, cohesion, corrosion, passivation, epitaxial growth, wear, friction, diffusion, thin films, delamination, creep, mechanical stability, durability, solid state devices, blend and alloy, charge transfer, nucleation, and growth abrasion

For the long-range force, it is generally accepted that the distance between two bodies is nearly always greater than about 0.3 nm, at which the resulting configuration is taken to be an equilibrium one [10–12], as shown in Fig. 1-1.

The same can be done in the graphite lattice, as shown in Fig. 1-2. The bonding force acting between two neighboring atoms can be directly demonstrated as a function of inter atomic separation, resulting in anisotropic properties. The bond energy in the *c* direction is commonly called van der Waals bond

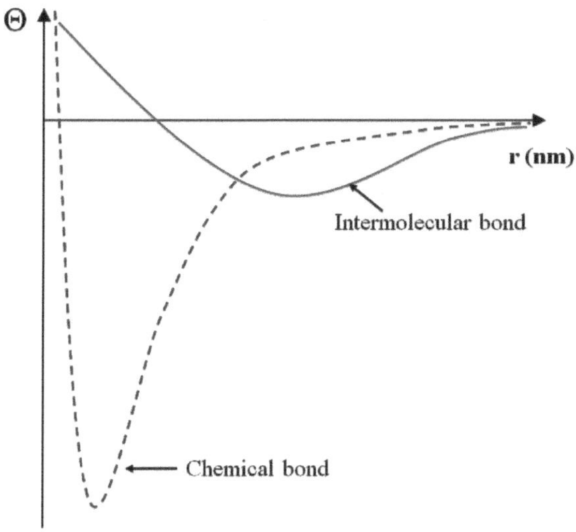

Fig. 1-1 Potential energy *vs.* distance curve.

or π electron interaction and is estimated to be 17 ~ 33 kJ/mol between the planes, as compared to about the 100 ~ 750 kJ/mol of the chemical covalent nature or σ-bond within the planes [9, 13].

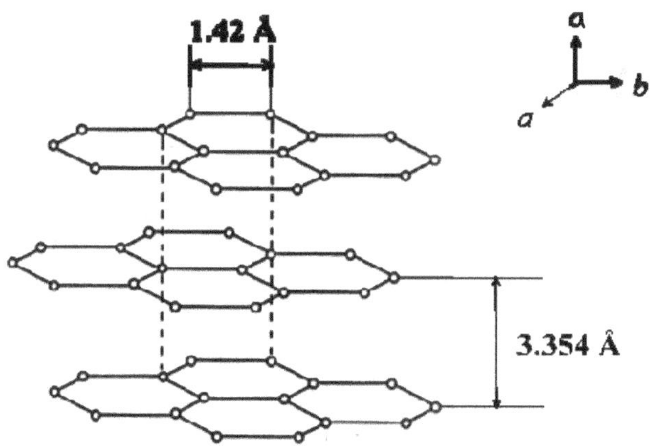

Fig. 1-2 Anisotropy of the physical and chemical bonds in the graphite lattice.

We begin with the short-range force between two point charges, q_1 and q_2, separated by a distance, x, in a vacuum, which is, from Coulomb's law:

$$f = \frac{q_1 q_2}{x^2} \tag{1-1}$$

The potential energy of interaction $U = -\int f dx$ is then:

$$U = \frac{q_1 q_2}{x} \tag{1-2}$$

where U is in ergs if q is in electrostatic units and x in centimeters. The electric field a distance x from a charge is:

$$E = \frac{q}{x^2} \tag{1-3}$$

such that the force is given by the product qE. The sign of E follows that of q; the interaction energy is negative if attractive and positive if repulsive.

We next consider a molecule having a dipole moment $\mu = qd$, that is, one in which charges q^+ and q^- are separated by a distance d. A dipole aligned with a field experiences a potential energy, $U = \mu E$, where again U is in ergs if μ is in esu/cm. The conventional unit of a dipole moment is the *Debye*, $1\text{D} = 1 \times 10^{-18}$ esu/cm or 3.336×10^{-30} C/m, corresponding to unit electronic charges 0.21 Å apart.

At distances far from the dipole, the length d becomes unimportant and the dipole appears as a 'point dipole'. The potential energy for a point dipole in the field produced by the charge in Eq. (1-3) is:

$$U = \frac{\mu q}{x^2} \tag{1-4}$$

The field produced far from a dipole ($x >> d$) is:

$$E = \frac{\mu}{x^3}(3\cos^2\theta + 1)^{1/2} \tag{1-5}$$

where θ is the angle between the position vector x and the dipole direction. Along the dipole direction ($\theta = 0$) this field becomes simply, $E = 2\mu/x^3$. A dipole interacts with the field of a second dipole to give an interaction potential energy:

$$U = \frac{\mu_1 \mu_2}{x^3}[2\cos\theta_1\cos\theta_2 - \sin\theta_1\sin\theta_2\cos\phi] \tag{1-6}$$

where θ_i is the angle between the center-to-center line x and the dipoles and ϕ is the azimuthal angle as illustrated in Fig. 1-3. The maximum attraction occurs with the aligned dipoles $\theta_1 = \theta_2 = 0$; thus for identical dipoles:

$$U_{\max} = -\frac{2\mu^2}{x^3} \tag{1-7}$$

whereas the maximum repulsion will be of the same magnitude when the dipoles are aligned in the opposite direction ($\theta_1 = \theta_2 = 180$). In a gas or a liquid, thermal agitation tends to rotate the dipoles into random orientations while the interaction potential energy favors alignment. The resulting net interaction potential energy (determined by Keesom in 1912) is:

$$U_{av} = -\frac{2\mu^4}{3kTx^6} \tag{1-8}$$

This orientation interaction thus varies inversely with the sixth power of the distance between dipoles. Remember, however, that the derivation has assumed separations largely compared with *d*.

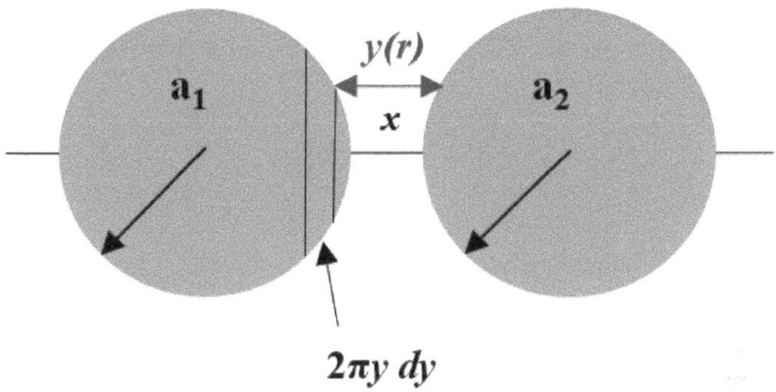

Fig. 1-3 Geometry for the interaction between two dipoles.

Another interaction involving dipoles is that between a dipole and a polarizable molecule. A field induces a dipole moment in a polarizable molecule or atom:

$$\mu_{ind} = \alpha_0 E \tag{1-9}$$

where α is the polarizability and has units of volume in the cgs system. It follows from $U = \mu$E that:

$$U(\alpha_0 E) = \mu_{ind} E = -\frac{\alpha E^2}{2} \tag{1-10}$$

where the negative sign implies attraction and the factor of 0.5 arises because we integrate from zero field or infinite separation to the field or position of interest $\int_0^E \mu_{\text{ind}} dE$. The induced dipole is instantaneous on the time scale of molecular motions and the potential energy is independent of temperature and is averaged over all orientations to give:

$$U(\alpha_0 \mu) = -\frac{\alpha \mu^2}{x^6} \tag{1-11}$$

This is the result worked out by Debye in 1920 and referred to as the Debye or induction interaction.

As an exercise, it is not difficult to show that the interaction of a polarizable molecule with a charge *q* is:

$$U(\alpha_0 q) = -\frac{\alpha_0 q^2}{2r^4} \tag{1-12}$$

We have two interaction potential energies between uncharged molecules that vary with distance to the minus sixth power as found in the Lennard-Jones potential. Thus far, none of these interactions accounts for the general attraction between atoms and molecules that are neither charged nor possess a dipole moment. After all, CO and N are similarly sized, and have roughly comparable heats of vaporization and hence the molecular attraction, although only the former has a dipole moment.

In general, the long-range forces, $\Theta(r)$, may be represented by the sum of the two separate potentials:

$$\Theta(r) = \frac{\Theta_R}{r^n} - \frac{\Theta_A}{r^6} \tag{1-13}$$

where Θ_R, Θ_A, n are the positive constants and n is a number usually taken to be between 8 and 16, with the subscripts R and A indicating repulsive and attractive, respectively. This equation was first proposed by Mie [14], and was extensively investigated by Lennard-Jones [9], when $n = 12$.

The Lennard-Jones potential (the so-called 6-12 equation) commonly holds for nonpolar molecules having no permanent dipole moment such as helium, argon, and methane [7, 15, 16]. Nevertheless, this potential can be expected to give an accurate description of the long-range forces only for sufficiently long distances between the two bodies [10, 17].

There are many different types of van der Waals attractive forces; these forces involve the inverse sixth-power, and are always negative in Eq. (1-13). These forces may be classified as follows, depending on the type of interaction.

1. London dispersive force: induced dipole-induced dipole interaction or deformation polarizability-deformation polarizability interactions [7, 15].
2. Debye inductive force: induced dipole-permanent dipole interaction [18, 19].
3. Keesom orientational force: permanent dipole-permanent dipole interaction [3].

In addition to dipole moments in London force such as hydrogen, ethylene, and carbon dioxide [15], it is possible for molecules to have quadrupole or higher multipole moments; these multiple moments are due to the concentration of electric charge at four (or higher multipole) separate points in the molecules, giving rise to similar interactions proportional to r^{-8} in dipole-quadrupole and to r^{-10} in quadrupole-quadrupole interactions. For long-range distances the r^{-8} in the interaction term is in any case smaller than the r^{-6} in the Lennard-Jones potential, and the effect of the higher multipole moments seems always to be negligible [15]. Avgul and Kiselev [20] produced a study on the adsorption energy of a variety of gases on the basal graphite plane in which dipole-quadrupole and quadrupole-quadrupole were found to contribute to the r^{-6} term in orders of 10 percent and less than $1 \sim 2$ percent respectively. However, the effects of the quadrupole or higher multipoles, such as octapoles and hexadecapoles, cannot be underestimated in short-range forces between two bodies [15] or in the repulsive contribution to the potential investigated by Israelachvili *et al.* [21–23].

Table 1-2 shows the approximate values for the Keesom (μ–μ), Debye or induction (μ–α_0), and London or dispersion (α_0–α_0) interactions for several molecules. Even for highly polar molecules, the last is very important. The first two interactions are difficult to handle in the condensed systems since they are sensitive to the microscopic structure through the molecular orientation. It will be seen that all these interactions give rise to an attraction varying with the inverse sixth power of the intermolecular distance van der Waals interactions. This is the dependence indicated by the a/V^2 terms in the van der Waals equation of state for a nonideal gas [7]:

$$\left(P + \frac{a}{V^2}\right)(V - b) = RT \qquad (1\text{-}14)$$

where V is the volume per mole, and a and b are constants, the former giving a measure of the attractive potential and the latter the actual volume of a mole of molecules.

Table 1-2 Contributions to van der Waals' Interaction Between Neutral Molecules

Molecule	$10^{24}\alpha_0^*$ (cm^3)	$h\nu^*$ (eV)	$10^{18}\mu^*$ (esu/cm)
He	0.2	24.7	0
Ar	1.6	15.8	0
CO	1.99	14.3	0.12
HCl	2.63	13.7	1.03
NH_3	2.21	16	1.5
H_2O	1.48	18	1.84

$^*\alpha_0$ is the deformation polarizability h the Plank's constant, v the quantized harmonic oscillator of frequency or electronic vibrational frequency in the ground state, and μ dipole moment.

For the first order, the dispersion (α_0–α_0) interaction is independent of the structure in a condensed medium and should be approximately pairwise additive. Qualitatively, this is because the dispersion interaction results from a small perturbation of electronic motions so that many such perturbations can add up without serious mutual interaction. Because of this simplification and its ubiquity in colloid and surface science, dispersion forces have received the most significant attention in the past half-century. Tables 1-3 and 1-4 list the key equations in cgs/esu units and SI units for long-range interactions.

Table 1-3 Conversions Between cgs/esu and SI Constants and their Units

Function	cgs/esu	SI
Potential	$Volt_{esu}$	300 volts [V]
Ionization energy, $h\nu_0$	$eV = 1.6 \times 10^{-12}$ [erg]	$eV = 1.6 \times 10^{-19}$ joule [J]
Charge	q [esu]	$\frac{q}{\sqrt{4\pi\varepsilon_0}}$ [coulomb (C)]
Polarizability	α [cm^3]	$\frac{\alpha}{4\pi\varepsilon_0}$ [m^3]
Dipole moment	μ, $D = 10^{-18}$ [esu/cm]	$\frac{\mu}{\sqrt{4\pi\varepsilon_0}}$, $D = 3.336 \times 10^{-30}$ [C/m]
Electronic charge	$e = 4.803 \times 10^{-10}$ [esu]	$e = 1.602 \times 10^{-19}$ [C]
Permittivity	1	$\varepsilon_0 = 8.854 \times 10^{-12}$ [C^{2}J^{-1}m^{-1}]
Boltzmann constant	$k = 1.38 \times 10^{-16}$ [erg/K]	$k = 1.38 \times 10^{-23}$ [J/K]

Table 1-4 Interaction Potential Energies in cgs/esu and SI

Function	cgs/esu	SI
Coulomb's law	$U = \frac{q_1 q_2}{x}$	$U = \frac{q_1 q_2}{4\pi\varepsilon_0 x}$
Keesom interaction	$U = -\frac{2\mu^4}{3kTx^6}$	$U = -\frac{\mu^4}{3kT(4\pi\varepsilon_0)x^6}$
Debye, induction interaction	$U = -\frac{\mu^2\alpha}{x^6}$	$U = -\frac{\mu^2\alpha}{(4\pi\varepsilon_0)^2 x^6}$
London, dispersion interaction	$U = -\frac{3h\nu_0\alpha_1\alpha_2}{4x^6}$	$U = -\frac{3h\nu_0\alpha_1\alpha_2}{4(4\pi\varepsilon_0)^2 x^6}$

The importance of long range interactions in the synthesis, design, and manipulation of materials at the nanometer scale was thus recognized from the very beginning of nanoscience. However, it is only recently that the intricacies of not only van der Waals forces, referred to by Feynman, but also all long range interactions have emerged in unexpectedly many research areas. These areas include the quantum field theory, the quantum and classical density functional theories, various mean-field and strong-coupling statistical mechanical formulations, liquid state integral equations, and computer simulations. These theoretical repercussions have led to novel experimental designs and methods with concomitant novelty and prospects in technology.

The role of long range interactions in self-assembling active devices constructed of heterogeneous components is fundamental. These interactions govern the stability of component clusters which are essential for the design of nanodevices and nanoactuators. The new technological paradigms that might be developed as a consequence of these fundamental studies promise new ways of thinking that bring old problems close to solution.

1.2.1. van der Waals Interaction

1.2.1.1. Concept of van der Waals Interaction

The net energy of the intermolecular interaction or internal physical energy, Θ, between two separated bodies, is the result of both attractive and repulsive, effects. The repulsive interaction is created between two neighboring molecules so that they avoid occupying the same space. Thus, this interaction rises very steeply to high positive values when the intermolecular separation falls below a certain distance. It otherwise has little effect on the internal energy.

Table 1-5 Various Binding Energies and Equilibrium Distances [8]

Bond type	Binding energy (kJ/mol)	Equilibrium distance (Å)
Primary or Chemical Bond		
Ionic	550 ~ 1100	1 ~ 2
Covalent (including coordinate)	60 ~ 750	1 ~ 2
Metallic	100 ~ 400	1 ~ 2
Secondary or Intermolecular (Physical) Bond		
London	≤ 45	3 ~ 5
Debye	≤ 3	3 ~ 5
Keesom	≤ 25	3 ~ 5
Hydrogen	≤ 55	2.4 ~ 3.1

As can be seen in Table 1-5 there are several possible attractive interactions, collectively called van der Waals forces (London, Debye, and Keesom effects) characterized by the same inverse sixth power dependence on equilibrium distance and hydrogen bonding. So, the attraction of the long-range force (subscript A), Θ_A, becomes:

$$\Theta_A = \Theta_A^L + \Theta_A^D + \Theta_A^K + \Theta_A^H \qquad (1\text{-}15)$$

where the superscripts L, D, K, and H refer to London, Debye, Keesom and hydrogen bond, respectively.

In the van der Waals attraction, the most important thing is the dielectric constant, ε, dependent on the frequency at which the alternating electric field varies. This is the name given to the factor by which the capacitance of a parallel plate condenser is increased upon the insertion of an insulating material because the net charges appear on the surface of the dielectric between the plates [24]. Under the electric field, dielectric molecules are polarized, so that an electric dipole moment can be induced. These polarized charges are referred to as the (total) polarizability, α. Under these conditions the polarizability of the medium may be classified by three contributions [24–26].

1. The deformation of temporary or induced dipoles of the molecules due to the displacement of the average positions of the electrons relative to the nuclei of the molecules (electronic polarizability or mean molecular electron polarizability, α_e, or so-called deformation polarizability, α_0).
2. The deformation of the nuclear skeleton of the molecules (atomic polarizability, α_a, or mean molecular vibrational polarizability, α_v). This polarizability is independent of temperature.
3. The partial (because of the disorientating effect of thermal motion) alignment of permanent dipoles (orientation polarizability, α_μ). This contribution accounts for the temperature-dependence of the dielectric constant, since increased thermal agitation tends to scramble the permanent dipoles. Their orientation polarizability, α_μ, is defined according to the strength of the (permanent) dipole moment, μ, of the molecule

$$\alpha_\mu = \frac{\mu^2}{3k_B T} \tag{1-16}$$

where k_B and T are the Bolzmann constant and Kelvin temperature, respectively.

In this classification, it is essential to note that the deformation polarizability α_0, of the molecule, comprises the electronic polarizability and atomic polarizability, where:

$$\alpha_0 = \alpha_e + \alpha_a \tag{1-17}$$

In Eq. (1-17), the atomic polarizability is generally so small compared to the electronic polarizability that it can be neglected [27]; that is, $\alpha_0 \approx \alpha_e$. Therefore, the total polarizability, α, of the molecule may be indicated by the sum of two contributions of polarizability (the so-called Debye–Langevin equation), such as:

$$\alpha = \alpha_0 + \frac{\mu^2}{3k_B T} \tag{1-18}$$

Thus, it can be summarized that the main contributors to the van der Waals attraction are the deformation polarizability independent of the presence of the permanent dipole and the orientation polarizability that is the average effect of the rotation of the molecule, depending on the temperature in the electric field.

1.2.1.2. London Dispersive Force

Dispersion or London force related to the nonpolar properties of the van der Waals attraction exists between all the adjacent pairs of atoms or molecules. As mentioned above, the origin of this force is the instantaneous charge redistribution in one molecule, which polarizes the electron clouds in adjacent molecules, resulting in nonpolar (or nonspecific) intermolecular attraction.

In the 1930s, London [7,15] showed the dispersive force as depending on the deformation polarizability, α_0. The first ionization potential, *I*, of molecules, on the basis of quantum mechanics, gives out energy, *E*, for the system:

$$E = \frac{1}{2}hv \tag{1-19}$$

where *h* is the Plank's constant and *v* the quantized harmonic oscillator of frequency or electronic vibrational frequency in the ground state.

Using Eq. (1-19), the magnitude of the London dispersive force in 1 mole between two identical molecules may be expressed by the following Eq. (1-20) in SI units, in which $4\pi\varepsilon_0 = 1.11265 \times 10^{-10}$ C/m/V is used as a conversion factor:

$$\Theta_A^L = -\frac{3}{4}N_A \cdot hv\left(\frac{\alpha_0}{4\pi\varepsilon_0}\right)^2\left(\frac{1}{r}\right)^6 \tag{1-20}$$

where N_A is the Avogadro's number, and ε_0 the permittivity of vacuum.

Meanwhile, in the London dispersive force (Eq. (1-20)), the characteristic electronic vibrational frequency, *v*, is directly related to the deformation polarizability, α_0, of the molecule, as shown by [26, 28, 29]:

$$v = \frac{1}{2\pi}\sqrt{\frac{e^2}{\alpha_0 m_e}} \tag{1-21}$$

where *e* and m_e are elementary charge (1.602×10^{-19} Coulomb) and mass of electron (9.019×10^{-31} kg), respectively.

For the nonidentical molecules, the individual frequencies and deformation polarizabilities are taken into account in Eq. (1-20):

$$\Theta_A^L = -\frac{3}{2}N_A \cdot h\left(\frac{v_1 v_2}{v_1 + v_2}\right)\frac{\alpha_{0,1}\alpha_{0,2}}{(4\pi\varepsilon_0)^2}\left(\frac{1}{r_{1-2}}\right)^6 \tag{1-22}$$

For the calculation of the magnitude of the London dispersive force in 1 mole, the quantity *hv* in Eq. (1-22) may be regarded as being energy-equivalent and is sometimes approximated by the first ionization potential, *I*:

$$\Theta_A^L \approx -\frac{3}{2}N_A \cdot \left(\frac{I_1 I_2}{I_1 + I_2}\right)\frac{\alpha_{0,1}\alpha_{0,2}}{(4\pi\varepsilon_0)^2}\left(\frac{1}{r_{1-2}}\right)^6 \tag{1-23}$$

Furthermore, to obtain a simple equation for two interacting molecules of gas (or liquid, subscript *L*) and solid (*S*) having the respective characteristic electronic vibrational frequency (or quantized harmonic oscillator of frequency) ν_Lnd ν_S, in Eq. (1-22), we can substitute the geometric mean for the harmonic mean, as below [26, 30]:

$$\frac{\nu_L \nu_S}{\nu_L + \nu_S} = \frac{\sqrt{\nu_L \nu_S}}{2} \tag{1-24}$$

This approximation has accuracy with an error rate of less than 4% [26]. It is then retained in (Eqs. (1-25) and (1-26)) to describe the London dispersive component of the Gibbs potential free energy of interaction, $-\Delta G_{\mathrm{A}}^{\mathrm{L}}$, between two nonidentical molecules based on the Lennard-Jones potential:

$$\Theta_{\mathrm{A}}^{\mathrm{L}} = -\Delta G_{\mathrm{A}}^{\mathrm{L}} \tag{1-25}$$

$$\Theta_{\mathrm{A}}^{\mathrm{L}} = -\frac{3}{4}\frac{\mathrm{N_A}}{(4\pi\varepsilon_0)^2}\left(\frac{1}{r_{S-L}}\right)^6 \{(h\nu_S)^{1/2}\cdot\alpha_{0,S}\}\cdot\{(h\nu_L)^{1/2}\cdot\alpha_{0,L}\} \tag{1-26}$$

If the equilibrium distance of adsorbent-adsorbate interaction, r_{S-L}, can be assumed to be a constant of 0.3 nm for all the probes studied, when the heat of vaporization, which may be shown to equal the long-range force, has an order of magnitude of about 27 to 45 kJ mol^{-1}[31–33], it can be possible to rewrite Eq. (1-26) as:

$$\Theta_{\mathrm{A}}^{\mathrm{L}} = \mathrm{K}\cdot\{(h\nu_S)^{1/2}\cdot\alpha_{0,S}\}\cdot\{(h\nu_L)^{1/2}\cdot\alpha_{0,L}\} \tag{1-27}$$

where

$$K = -\frac{3}{4}\frac{N_A}{(4\pi\varepsilon_0)^2}\left(\frac{1}{r_{S-L}}\right)^6 \tag{1-28}$$

The relation shown above reveals that the London dispersive component of the Gibbs potential free energy of interaction, $-\Delta G_{\mathrm{A}}^{\mathrm{L}}$, of a solid is a function of the characteristics of a liquid, $\{(h\nu_L)^{1/2}(\alpha_{0,L})\}$, [or a function of $\alpha_{0,L}^{3/4}$, since $h\nu$ is also a function of α_0 in Eq. (1-21)]. Therefore, Eq. (1-27) allows us to calculate the London dispersive component of the adsorbate-adsorbent interaction for a given liquid when the quantity $\{(h\nu_L)^{1/2}(\alpha_{0,L})\}$ is defined as a characteristic of the probe considered from the basis of the polarizability of molecules, as listed in Table 1-6.

Van der Waals forces document the important new directions of development in the field, achieved both in experiment and in theory, and should give an idea of future potential. Why are van der Waals interactions so important? It is well known that these weak interactions play an important role in chemistry, physics, and, in particular, all the biodisciplines. Since all life on earth may be viewed as a matter of supramolecular chemistry, with van der Waals forces playing a central role, the understanding of these interactions is important for any progress in the targeted synthesis of new drugs. Even in this age of combinatorial chemistry it remains true that the progress in designing highly specific drugs is strongly accelerated by intimate knowledge of the intermolecular forces that control the specificity of interaction and the binding constants of complexes such as those among drugs and the DNA, RNA, or proteins.

Table 1-6 Characteristics and Percentages of the London, Debye, and Keesom Contributions to the van der Waals Attraction Between Various Molecules [8]

Molecules (abbrev.)	$\alpha_0\ 10^{40a}$ ($C\ m^2\ V^{-1}$)	μ^b (Debye)	$h\nu^c$ (eV)	London[d] (%)	Debye[d] (%)	Keesom (%)
n-Pentane (C5)	11.15	0.05	3.30	100	0	0
n-Hexane (C6)	13.19	0.00	3.03	100	0	0
n-Heptane (C7)	15.24	0.085	2.83	100	0	0
n-Octane (C8)	17.29	0.0.	2.63	100	0	0
n-Nonane (C9)	19.34	-	2.49	100	0	0
Carbon tetrachloride (CCl_4)	11.66	0	3.23	100	0	0
Benzene (Bz)	11.58	0	3.20	100	0	0
Toluene (To)	13.70	0.36	2.99	99.4	0.6	0
Xylene (Xy)	15.88	0	2.79	100	0	0
Chloroform ($CHCl_3$)	9.47	1.013	3.40	89.5	5.3	5.2
Diethylether (Et_2O)	9.92	1.15	3.54	86.5	6.1	7.4
Ethyl acetate (EtOAc)	9.82	1.78	3.36	58.6	11.0	30.4
Tetrahydrofuran (THF)	8.77	1.63	3.73	63.1	9.8	27.1
Pyridine (Py)	10.62	2.19	3.40	44.2	10.9	44.9
Acetone (AC)	7.14	2.88	4.14	14.2	7.4	78.4
Nitromethane (NM)	5.51	3.46	4.703	4.4	4.6	91.0
Formamide (Fa)	4.68	3.73	5.39	3.2	3.5	93.3
Acetonitrile (An)	4.88	3.92	4.94	2.8	3.3	93.9
Water (H_2O)	1.45	1.85	8.68	9.7	4.4	85.9

[a] *Computed from refractive index, n, (at 20 °C) in Ref. 5 according to Lorentz-Lorenz-Debye equation [Ref. 26, 30], which is*

$$\left(\frac{n^2-1}{n^2+2}\right)\frac{M}{\rho}=\frac{N_A\alpha_0}{3\varepsilon_0} \tag{1-29}$$

where M, ρ, and ε_0 are relative molecular mass, molecular density, and permittivity of vacuum, respectively.

[b] *From Ref. 7, where 1 Debye = 3.33564×10^{-30} C m.*

[c] *Computed from Eq. (1-21) and from the values of α_0, computed from the first column in this table, where 1 eV = 1.6022×10^{-19} J.*

[d] *Computed from the first three columns of this Table.*

1.2.1.3. Specific Force

Induction (or Debye) and orientation (or Keesom) forces Θ_A^{D+K}, which are the specific (or polar) properties of the van der Waals attraction, exist in the presence of the dipole moment and (total) polarizability, resulting in specific (or polar) intermolecular attraction.

Debye [5, 19] showed that an electrical field induces a dipole in a nearby dipolar molecule and the magnitude of (Permanent) dipole moment, μ, is proportional to the electrical field, E_F:

$$\mu = \alpha E_F \tag{1-30}$$

where α is the total polarizability of the dipolar molecule.

As shown by Debye, the potential energy, or intermolecular force (in this work, Θ_A^{D+K}) in one mole may be expressed by the following Eq. (1-31) in SI units:

$$\Theta_{\mathrm{A}}^{\mathrm{D+K}} = -2\mathrm{N_A}\alpha\mu^2 \frac{1}{(4\pi\varepsilon_0)^2}\left(\frac{1}{r}\right)^6 \tag{1-31}$$

Combining Eqs. (1-28) and (1-31) gives the general result:

$$\Theta_{\mathrm{A}}^{\mathrm{D+K}} = -2\mathrm{N_A}\mu^2 \frac{1}{(4\pi\varepsilon_0)^2}\left(\frac{1}{r}\right)^6\left(\alpha_0 + \frac{\mu^2}{3k_{\mathrm{B}}\mathrm{T}}\right) \tag{1-32}$$

For a pair of identical molecules, it should be noted that in Eq. (1-32), the first term determined with regard to the deformation polarizability is the so-called 'Debye inductive force', and the second term is generally called a 'Keesom orientational force' between the molecules when the dipole moment is considered in the intermolecular attractive system.

As has been already noted for nonidentical molecules, Eq. (1-32) is divided by two terms, for the Debye inductive force, $\Theta_{\mathrm{A}}^{\mathrm{D}}$:

$$\Theta_{\mathrm{A}}^{\mathrm{D}} = -\frac{\mathrm{N_A}}{(4\pi\varepsilon_0)^2}\left(\alpha_{0,1}\mu_2^2 + \alpha_{0,2}\mu_1^2\right) \times \left(\frac{1}{r_{1-2}}\right)^6 \tag{1-33}$$

For the Keesom orientational force, $\Theta_{\mathrm{A}}^{\mathrm{K}}$:

$$\Theta_{\mathrm{A}}^{\mathrm{K}} = -\frac{2}{3}\frac{\mathrm{N_A}}{(4\pi\varepsilon_0)^2 k_{\mathrm{B}}\mathrm{T}}\mu_1^2\mu_2^2\left(\frac{1}{r_{1-2}}\right)^6 \tag{1-34}$$

Eqs. (1-20) and (1-31) can be combined to give the van der Waals attraction (superscript LDK) for a pair of identical molecules in 1 mole, $\Theta_{\mathrm{A}}^{\mathrm{L+D+K}}$:

$$\Theta_{\mathrm{A}}^{\mathrm{L+D+K}} = -\frac{\mathrm{N_A}}{(4\pi\varepsilon_0)^2}\left(\frac{1}{r}\right)^6\left(\frac{3}{4}h\nu\alpha_0^2 + 2\alpha_0\mu^2 + \frac{2}{3}\frac{\mu^4}{k_{\mathrm{B}}\mathrm{T}}\right) \tag{1-35}$$

The fractional contributions of the van der Waals attraction (namely, London, Debye and Keesom) can be determined in the intermolecular attractive system in Eq. (1-35), since all the three contributions show the same functional dependence on the equilibrium distance of molecules, to such an extent that they can be conveniently considered together.

Table 1-6 shows these fraction contributions calculated for a variety of useful molecules for surface dynamics. As can be seen in Table 1-6, the London dispersive component acts as a major contribution to the van der Waals attraction in certain cases, except those cases of high polar molecules, which strictly depend on the high dipole moment. In the case of water, hydrogen bonding is strong enough to contribute additionally to the interaction.

Thus, the real London dispersive contribution is even less than the value shown in Table 1-6.

1.2.1.4. Hydration Force

In this section, we introduce the hydration force in recent advances as the secondary long-range force, a weakly intramolecular force.

When two surfaces are brought into contact, repulsive forces at about the 1 nm range have been measured in aqueous electrolytes between a variety of surfaces: clays, mica, silica, alumina, lipids, DNA, and surfactants. Because of the correlation with the low (or negative) energy of wetting of these solids with water, the repulsive force has been attributed to the energy required to remove the water of hydration from the surface, or the surface adsorbed species, presumably because of the strong charge-dipole, dipole–dipole, or H-bonding interactions. These forces have been termed as hydration forces [22, 34–36].

Even now, the origin of hydration forces is not clear and several effects are being discussed. Certainly the fact that one layer of water molecules is bound to solid surfaces is important. The hydration force, however, extends over more than only two water layers. Israelachvili and Wennerstöm point out that the effect of the first water layer should not even be called a hydration force because it is caused by the interaction between the water molecules and the solid surface and not by water–water interactions (See Fig. 1-4) [35].

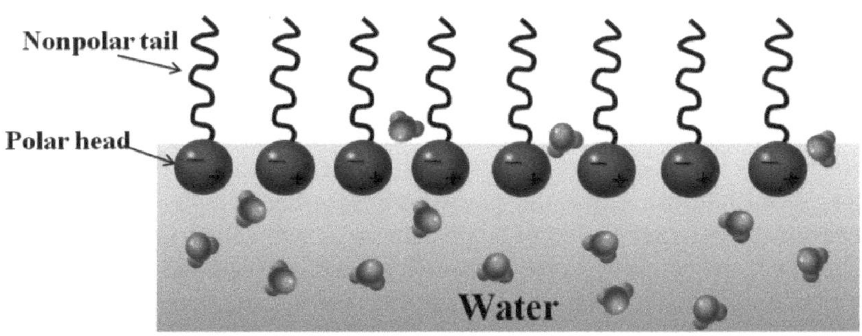

Fig. 1-4 Surfactant molecules arranged on an air-water interface by driving of hydration forces.

We focus here on the aspects of phase stability of the surfactant solutions or colloidal microcrystals when the stability, coexistence, or swelling is due to a hydration force, and is not of immediate electrostatic origin. In such situations, the absence of an identified 'electrostatic effect' such as a link between the Debye lengths and phase limits is due either to the absence of charge or to an effect independent of the presence of added salt [37]. It may seem paradoxical to attribute a long range to the hydration force, which can persist only for a length in which the drive for structural alignment of the solvent around the solute can overcome the effects of Brownian motion [38]. This force is only 'long' when compared to the hydrogen bonding, complexation, and the other nearest neighbor interactions considered in the chemistry of colloids. For good model systems, in the absence of salt, the hydration force can be detected by applied osmotic pressure as low as a few hundred Pa, with typical distances between surfactant aggregates of up to $\sim$ 3 nm [39]. At $\sim$ 1 nm, the hydration pressure can grow to hundreds of atmospheres between the planar surfaces.

The distance dependence characterizing the exclusion of small solutes from the macromolecular surfaces follows the same exponential behavior as that of the hydration force between the macromolecules at close spacings. Similar repulsive forces are seen for the exclusion of nonpolar alcohols from the highly charged DNA and of salts and small polar solutes from the hydrophobically modified cellulose [40]. The exclusion magnitudes for different salts follow the Hofmeister series, which has long been thought to be connected with water structuring [41].

One feature is the intriguing connection with the distribution of salts in the thin liquid films on ice. The connection between hydration effects in water and the Bjerrum defect distribution in ice has been noted before [42] and is due to the structuring of water molecules close to the macroscopic surfaces. In ice this is described by a redistribution of orientational Bjerrum defects, whereas in water it is usually discussed within water solvation or hydration models. In both cases, however, ion redistribution couples with hydration patterns.

Solvation of the interacting macromolecular surfaces, and modulation of this solvation by cosolutes such as salts exquisitely regulates the equilibria of specific association in chemistry and biology. Depending on whether the cosolute is preferentially excluded from, or attracted to, the surfaces of the macromolecules, a cosolute can either increase or decrease the complex stability [43]. However, the dynamic action of a cosolute on complexation is not yet understood, and there is no way to predict which kinetic constant, the 'on rate' or the 'off rate' has greater impact.

Between hydrophobic surfaces a completely different interaction is observed. Hydrophobic surfaces attract each other (See Fig. 1-5) [44]. This attraction is called London dispersive interaction. The interaction between the solid hydrophobic surfaces of about 45 kJ/mol in van der Waals attraction was determined by Park and Israelachvili [8, 45, 46]. With the surface force apparatus they observed an exponentially decaying attractive force between the two mica surfaces with an adsorbed monolayer of the cationic surfactant cetyltrimethylammoniumbromide (CTAB). Since then the hydrophobic force has been investigated by different groups and its existence is now generally reported [44]. The origin of the hydrophobic force is discussed in the 1.2.1.2.

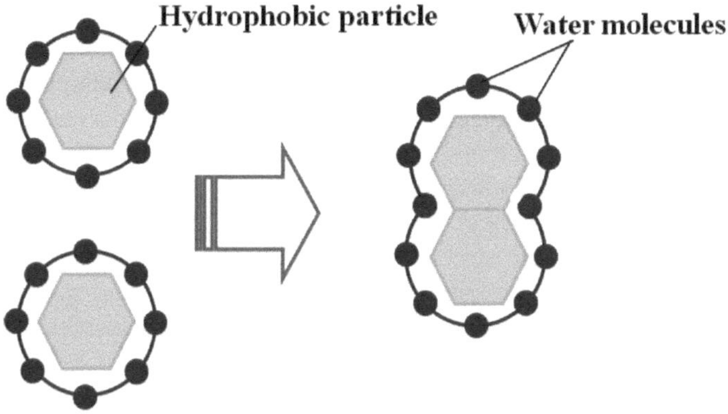

Fig. 1-5 Hydrophobic interactions.

Usually, two components of the attraction are observed [47]. One is the long-range and decays roughly exponentially with a decay length of typically 1 ~ 2 nm. This can be attributed to a change in the water structure when the two surfaces approach each other. The second component is more surprising: it is very long-ranged and extends out to 100 nm in some cases. Its origin is not understood. One hypothesis is that this attraction is due to the gas bubbles that form spontaneously [48]. This is called cavitation. Estimations of the rate of cavitation, however, result in values that are much too low. Another hypothesis is that there are always some gas bubbles residing on the hydrophobic surfaces. Once these gas bubbles come into contact they fuse and cause a strong attraction due to the meniscus force. An open question remains: how these bubbles can be stable, since the reduced vapor pressure inside a bubble and the surface tension should lead to immediate collapse. Non-DLVO forces also occur when the aqueous medium contains surfactants, which form micelles, or poly electrolytes. For the detailed discussion of this complex interaction, Claesson *et al.* [49] have reported on it.

Research in superhydrophobicity recently accelerated with a letter that reported man-made superhydrophobic samples produced by allowing the alkylketene dimer (AKD) to solidify into a nanostructured fractal surface [50]. Many papers have since presented fabrication methods for producing the superhydrophobic surfaces, including particle deposition, vapor deposition [51], sol-gel techniques [52], plasma treatments [53], and casting techniques [54]. Current opportunities for research lie mainly in the fundamental research and practical manufacturing [55].

Debates have recently emerged concerning the applicability of the Wenzel and Cassie-Baxter models. It has become clear that both the static and dynamical properties can be controlled *via* surface patterning. Superhydrophobicity is perhaps the prime example: by making the surface rough, the contact angle of a hydrophobic surface can be increased to close to 180° [56, 57]. The two possible states, i.e., Wenzel (collapsed) [58] and Cassie–Baxter (suspended) [59], exhibit clear differences in drop mobility.

In an experiment designed to challenge the surface energy perspective of the Wenzel and Cassie–Baxter model and promote a contact line perspective, water drops were placed on a smooth hydrophobic spot in a rough hydrophobic field, a rough hydrophobic spot in a smooth hydrophobic field, and a hydrophilic spot in a hydrophobic field [60]. These tests showed that the surface chemistry and geometry at the contact line affected the contact angle and contact angle hysteresis, but that the surface area inside the contact line had no effect. An argument that increased jaggedness in the contact line and enhances the droplet mobility has also been proposed [61].

The coverage of solid or liquid surfaces with atoms and, more recently, with organic or inorganic molecules is an area of broad scope that has received the attention of the scientific community during the recent years. This interest is based on the fact that the presence of the molecules usually modifies the surface properties resulting in new materials with enhanced properties suitable for the preparation of devices in molecular electronics or for the study of emerging science and technology. In this regard, the design and development of the coated surfaces showing unprecedented optoelectronic properties require a detailed understanding of the phenomena occurring at the atomistic scale at the interface.

Thus, the 2D arrangement is a result of a combination of weak noncovalent intermolecular forces (such as van der Waals or dispersive forces) with molecule-substrate interactions, in which the crystalline symmetry of the surface plays a leading role. Therefore, the interface is an important tool for the development and understanding of the emergent from macroscience to nanoscience at the surfaces.

1.2.2. Hydrogen Bonding

The most common physicochemical effect encountered in the strong long-range nature of surface dynamics [62] is the result of hydrogen bonding, in which the hydrogen atoms serve as bridges linking together two atoms of high electronegativity, such as FH–F, NH–F, NH–N, NH–O,CH–N, OH–N, CH–O, and OH–O. As to the general shape of the Lennard-Jones potential, it is generally in the range 0.24 to 0.31 nm, which values are substantially smaller than the van der Waals radii [63]. That is, the strength of the bond becomes favorable to a maximum of about 55 kJ/mol [64–66], much bigger than when only van der Waals forces are involved.

The structure of the hydrogen atom with acceptable electrons as the acceptor can interact strongly with the nearby electronegative atoms. Thus, hydrogen bonding is expected to play a role in the interaction between two bodies bearing the functional groups, such as hydroxyl, carbonyl, carboxyl, amino, and similar groups. In the 1970s, Kamlet and Taft [66, 67] introduced the ideas about the role of acid-base interactions of a solvent in hydrogen bonding in the sense of the

Lewis acid–base theory [68]. In this chapter, we consider the role of hydrogen bonding of solid surfaces for the specific force of the long-range contribution to the surface dynamics.

Table 1-5 (see 1.2.1.1) gives the accumulated data from the literature [69–72] on the binding energies and the equilibrium distances of the primary (or chemical) and secondary (or intermolecular) forces.

In particular, long-range forces are shown to be the sum of the van der Waals force and hydrogen bonding. Van der Waals forces are also expressed as the total sum of the London dispersive forces (apolar–apolar), Debye induction forces (apolar–polar), Keesom orientational forces (polar–polar), and repulsive forces.

As a rule, it is then possible to note that the intermolecular interaction or the Gibbs free energy, $-\Delta G_A$, of a solid surface can be considered as the sum of two components: a dispersive (or nonpolar, superscript L) component, i.e., attributable to the London force, and a specific (or polar, SP) component owing to all other types of interactions (Debye, Keesom, hydrogen bonding (H), and other weakly polar effects):

$$-\Delta G_A = \Theta_\mathrm{A} \tag{1-36}$$

$$-\Delta G_\mathrm{A} = \Theta_\mathrm{A}^{van\ der\ Waals} + \Theta_\mathrm{A}^\mathrm{H} \tag{1-37}$$

$$-\Delta G_\mathrm{A} = \Theta_\mathrm{A}^\mathrm{L} + \Theta_\mathrm{A}^\mathrm{SP} \tag{1-38}$$

where

$$\Theta_\mathrm{A}^\mathrm{SP} = \Theta_\mathrm{A}^\mathrm{D} + \Theta_\mathrm{A}^\mathrm{K} + \Theta_\mathrm{A}^\mathrm{H} + \cdots \tag{1-39}$$

1.3. ACID-BASE INTERACTIONS

1.3.1. Concept of Acid-Base Interactions

The idea of acids and bases has been a concept of great importance in chemistry since the earliest times, in some cases helping to correlate large amounts of data and in others leading to new predictive ideas. Jensen [73] describes a useful approach in the preface to his book on the Lewis acid–base concept:

Acid-base concepts occupy a somewhat nebulous position in the logical structure of chemistry. They are, strictly speaking, neither facts nor theories and are, therefore, never really 'right' or 'wrong'. Rather they are classificatory definitions or organizational analogies. They are useful or not useful. ... The study of their historical evolution...clearly shows that the acid-base definitions are always a reflection of the facts and theories current in chemistry at the time of their formulation and that they must, necessarily, evolve and change as the facts and theories themselves evolve and change.... the older definitions...generally represent the most powerful organizational analogy consistent with the facts and theories extant at the time.

Practical acid-base chemistry known in ancient times, developed gradually during the time of the alchemists, and was first satisfactorily explained in molecular terms after Ostwald and Arrhenius established the existence of ions in aqueous solution in 1880–1890. During the early development of the acid–base theory, the experimental observations included the sour taste of acids and the bitter taste of bases, indicator color changes caused by the acids and bases, and the reactions of acids with bases to form salts. Partial explanations included the idea that all acids contained oxygen (oxides of nitrogen, phosphorus, sulfur, and the halogens, all form acids in water), but by the early nineteenth century many acids that did not contain oxygen were known. By 1838, Liebig [74] defined acids as 'compounds containing hydrogen, in which the hydrogen can be replaced by a metal', a definition that still works well in many instances.

1.3.2. Arrhenius Concept

The Arrhenius definition [75] of acid-base reactions is a development of the hydrogen theory of acids, devised by Svante Arrhenius, whose work was used to provide a modern definition of acids and bases that followed from his work in 1884 with Friedrich Wilhelm Ostwald in establishing the presence of ions in aqueous solution, and led to Arrhenius receiving the Nobel Prize in Chemistry in 1903, for the 'recognition of the extraordinary services, ... rendered to the advancement of chemistry by his electrolytic theory of dissociation'.

In this concept, the 'Arrhenius acids form hydrogen ions (or hydronium ions, H_3O^+) in aqueous solution, Arrhenius bases form hydroxide ions in solution', and the reaction of hydrogen ions and hydroxide ions to form water is the universal aqueous acid-base reaction. The ions accompanying the hydrogen and hydroxide ions form a salt, so the overall Arrhenius acid-base reaction can be written:

$$\text{acid} + \text{base} \rightarrow \text{salt} + \text{water} \tag{1-40}$$

For example,

$$\text{hydrochloric acid} + \text{sodium hydroxide} \rightarrow \text{sodium chloride} + \text{water} \tag{1-41}$$

$$H^+ + Cl^- + Na^+ + OH^- \rightarrow Na^+ + Cl^- + H_2O \tag{1-42}$$

This explanation works well in the aqueous solutions, but is inadequate for nonaqueous solutions and for gas and solid phase reactions in which H^+ and OH^- may not exist, and for which later definitions by Brønsted-Lowry and Lewis are more appropriate for general use.

1.3.3. Brønsted-Lowry Concept

Brønsted [76] defined an acid as a species with a tendency to lose a proton and a base as a species with a tendency to add a proton. These definitions expanded the Arrhenius list of acids and bases to include the gases HCl and NH_3, along with many others. This definition also introduced the concept of conjugate acids and bases differing only in the presence or absence of a proton, and described all reactions as occurring between a stronger acid and base to form a weaker acid and base:

$$\underset{\text{acid 1}}{H_3O^+} + \underset{\text{base 2}}{NO_2^-} \rightarrow \underset{\text{base 1}}{H_2O} + \underset{\text{acid 2}}{HNO_2}$$

Conjugate acid–base pairs :

Acid	Base
H_3O^+	H_2O
HNO_2	NO_2^-

(1-43)

In water, HCl and NaOH react as the acid H_3O^+ and the base OH^- to form water, which is the conjugate base of $H_3O,^+$ and the conjugate acid of OH^-. Reactions in nonaqueous solvents having ionizable protons parallel those in water. An example of such a solvent is liquid ammonia, in which NH_4Cl and $NaNH_2$ react as the acid NH_4^+ and the base NH_2^-, to form NH_3, which is both a conjugate base and a conjugate acid:

$$NH_4^+ + Cl^- + Na^+ + NH_2^- \rightarrow Na^+ + Cl^- + 2NH_3 \quad (1\text{-}44)$$

with the net reaction:

$$\underset{\text{acid}}{NH_4^+} + \underset{\text{base}}{NH_2^-} \longrightarrow \underset{\text{conjugated base and conjugated acid}}{2NH_3} \quad (1\text{-}45)$$

1.3.4. Solvent System Concept

Aprotic nonaqueous solutions require a similar approach, but with a different definition of acid and base. The solvent system definition [77] applies to any solvent that can dissociate into a cation and an anion (auto dissociation), where the cation resulting from auto dissociation of the solvent is the acid and the anion is the base. The Arrhenius reaction:

$$\text{acid} + \text{base} \rightarrow \text{salt} + \text{water} \quad (1\text{-}40)$$

and the Brønsted acid-base reaction:

$$\text{acid 1} + \text{base 2} \rightarrow \text{base 1} + \text{acid 2} \quad (1\text{-}46)$$

can then become:

$$\text{acid} + \text{base} \rightarrow \text{solvent (both acid and base)} \quad (1\text{-}47)$$

In the solvent BrF_3, for example, the dissociation takes the form:

$$2BrF_3 \leftrightarrow BrF_2^+ + BrF_4^- \quad (1\text{-}48)$$

and the acid + base reaction is the reverse:

$$2BrF_3 \leftrightarrow BrF_2^+ + BrF_4^- \quad (1\text{-}48)$$

with BrF_2^+ the acid and BrF_4^- the base. Solutes, that increase the concentration of the acid BrF_2^+ are classified as acids, and those that increase the concentration of BrF_4^- are classified as bases. For example, SbF_5 is an acid in BrF_3:

$$SbF_5 + BrF_3 \rightarrow BrF_2^+ + SbF_6^- \quad (1\text{-}49)$$

Table 1-7 Properties of Solvents [78]

Solvent	Acid cation	Base anion	pK ion (25 °C)	Boiling point (°C)
Protic Solvents				
Ammonia, NH_3	NH_4^+	NH_2^-	27	−33.38
Sulfuric acid, H_2SO_4	$H_3SO_4^+$	HSO_4^-	3.4 (10 °C)	330
Acetic acid, CH_3COOH	$CH_3COOH_2^+$	CH_3COO^-	14.45	118.2
Hydrogen fluoride, HF	H_2F^+	HF_2^-	~12 (0 °C)	19.51
Methanol, CH_3OH	$CH_3OH_2^+$	CH_3O^-	18.9	64.7
Water, H_2O	H_3O^+	OH^-	14	100
Aprotic Solvent				
Solvent	Boiling point (°C)			
Dinitrogen tetroxide, N_2O_4	21.15			
Sulfur dioxide, SO_2	−10.2			
Pyridine, C_5H_5N	115.5			
Acetonitrile, CH_3CN	81.6			
Diglyme, $CH_3(OCH_2CH_2)_2OCH_3$	162.0			
Bromine trifluoride, BrF_3	127.6			

Ionic fluorides such as KF are bases in BrF_3:

$$F^- + BrF_3 \longrightarrow BrF_4^- \tag{1-50}$$

Of course, autoionizing protonic solvents such as H_2O and NH_3, also satisfy the solvent system definition: the solutes that increase the concentration of the cation (H_3O^+, NH_4^+) of the solvent are considered acids, and solutes that increase the concentration of the anion (OH^-, NH_2^-) are considered bases.

Table 1-7 gives some of the properties of common solvents.

Caution is needed in interpreting acid-base reactions and indeed, any reaction. For example, $SOCl_2$ and SO_3^{2-} react as acid and base, in SO_2 solvent:

$$SOCl_2 + SO_3^{2-} \leftrightarrow 2SO_2 + 2Cl^- \tag{1-51}$$

It was at first believed that $SOCl_2$ dissociated and the resulting SO^{2+} reacted with SO_3^{2-}:

$$SOCl_2 \leftrightarrow SO^{2+} + 2Cl^- \tag{1-52}$$

$$SO^{2+} + SO_3^{2-} \leftrightarrow 2SO_2 \tag{1-53}$$

However, the reverse reactions should lead to oxygen exchange between SO_2 and $SOCl_2$, but none is observed [78, 79]. The details of the $SOCl_2 + SO_3^{2-}$ reaction are still uncertain.

1.3.5. Lewis Concept

Lewis [80] defined a base as an electron-pair donor and an acid as an electron-pair acceptor. This definition further expands the list to include the metal ions

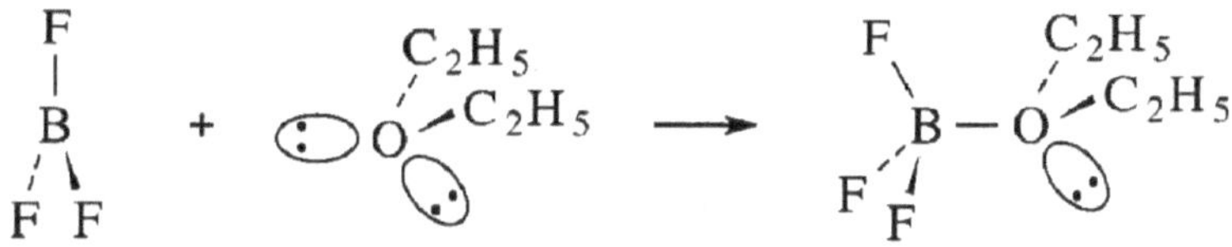

Fig. 1-6 Boron trifluoride ether adduct.

and other electron-pair acceptors as acids and provides a handy framework for the nonaqueous reactions. Most of the acid-base descriptions in this book will use the Lewis definition, which encompasses the Brønsted and solvent system definitions. In addition to all the reactions above, the Lewis definition includes reactions such as:

$$Ag^{+} + 2 : NH_3 \longrightarrow H_3N : Ag : NH_3^{+} \tag{1-54}$$

with silver ion (or other cation) as an acid and ammonia (or other electron-pair donor) as a base. In reactions such as this, the product is often called an adduct, a product of the reaction of a Lewis acid and base to form a new combination. Another example of a Lewis acid-base adduct is a common reagent in synthesis, the boron trifluoride-diethyl ether adduct, $BF_3O(C_2H_5)_2$. Since fluorine is the most electronegative element, the boron atom in BF_3 is quite positive. Lone pairs on the oxygen of the diethyl ether are attracted to boron; the result is that one of the lone pairs bonds to boron, changing the geometry around B from planar to nearly tetrahedral, as shown in Fig. 1-6. As a result, BF_3, with a boiling point of −99.9 °C, and diethyl ether, with a boiling point of 34.5 °C, form an adduct with a boiling point of 125 °C to 126 °C (at which temperature it decomposes into its two components).

Table 1-8 Chemical Phenomena Subsumed by the Category of Lewis Acid-base (Acceptor-Donor) Reactions

(A)	Systems covered by the Arrhenius, solvent system, Lux-Flood, and proton acid-base definitions
(B)	Traditional coordination chemistry, and "nonclassical" complexes
(C)	Solvation, solvolysis, and ionic dissociation phenomena, in both aqueous and nonaqueous solutions
(D)	Electrophilic and nucleophilic reactions, in organic and organometallic chemistry
(E)	Charge-transfer complexes, so-called molecular addition compounds, weak intermolecular forces, H-bonding, etc.
(F)	Molten salt phenomena
(G)	Various miscellaneous areas such as chemiadsorption of closed-shell species, intercalation reactions in solids, so-called ionic metathesis reactions

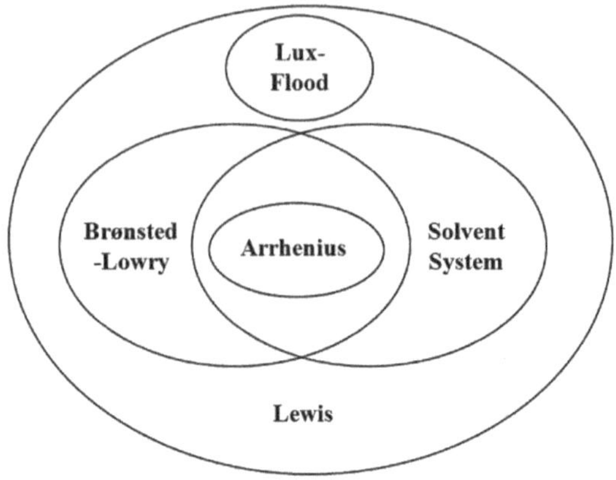

Fig. 1-7 Venn diagram showing the relationship between the various chemical systems classified as acid-base by the five major acid-base definitions [88].

Lewis acid-base adducts involving metal ions are coordination compounds. The rest of this chapter will develop the Lewis concept, in which adduct formation is common.

Other acid-base definitions have been proposed. While they are useful in particular types of reactions, none has been widely adopted for general use. The Lux-Flood definition [81–83] is based on the oxide ion, O^{2-}, as the unit transferred between the acids (oxide ion acceptors) and bases (oxide ion donors). The Usanovich [84] definition proposes that any reaction leading to a salt (including oxidation-reduction reactions) should be considered an acid-base reaction. This definition could include nearly all the reactions and has been criticized for this all-inclusive approach. The Usanovich definition is rarely used today. The electrophile-nucleophile approach of Ingold [85] and Robinson [86], widely used inorganic chemistry, is essentially the Lewis theory with terminology related to the reactivity (electrophilic reagents are acids, nucleophilic reagents are bases).

Table 1-8 lists the major classes of the chemical phenomena that are subsumed under the general category of Lewis acid-base reactions. The relevance of the Lewis concepts to each of these areas is for the most part self-evident so that it is only necessary to comment briefly on each and to indicate where the reader can find a more detailed treatment. Discussions of the relationship between the Lewis definitions and the more restricted Arrhenius, Lux-Flood, solvent-system, and proton definitions have been given by several authors, the most thorough being that of Day and Selbin [87]. Fig. 1-7 summarizes these relationships by means of a Venn diagram.

Finally, Table 1-9 summarizes these acid-base definitions.

Table 1-9 Comparison of Acid-base Definitions

	Definitions		Examples	
	Acid	Base	Acid	Base
Lavoisier	Oxide of N, P, S	Reacts with acid	SO_3	NaOH
Liebig	Replaccable H	Reacts with acid	HNO_3	NaOH
Arrhenius	Hydronium ion	Hydroxide ion	H^+	OH^-
Br½nsted	Proton donor	Proton acceptor	H_3O^+	H_2O
			H_2O	OH^-
			NH_4^+	NH_3
Solvent system	Solvent cation	Solvent anion	BrF_2^+	BrF_4^-
Lewis	Electron-pair acceptor	Electron-pair donor	Ag^+	NH_3
Usanovich	Electron acceptor	Electron donor	Cl_2	Na

1.3.6. Pearson's Hard and Soft Acids and Bases

The hard and soft acids and bases (HSAB) concept was developed by Ralph Pearson [89, 90] as an explanation of the data concerning the reactions of metal ions and anions; the concept has since been expanded to include many other reactions and has recently been placed on a more mathematical foundation [90–92].

For many years, chemists tried to explain experimental observations such as the insolubility of the silver halides and other salts that can be used to separate metal ions into groups for identification in the qualitative analysis schemes. Fajans [93] proposed that the insolubility of a salt in water was a consequence of the degree of covalent bonding in these compounds. Fajans proposed the following correlations:

1. Covalent character increases with increase in the size of the anion and decrease in size of the cation.
2. Covalent character increases with increasing charge on either ion.
3. Covalent character is greater for cations with non-noble gas electronic configurations.

For example, $Fe(OH)_3$ is much less soluble than $Fe(OH)_2$ (rule 2), AgS is much less soluble than AgO (rule 1), FeS is much less soluble than $Fe(OH)_2$ (rules 1, and 2), Ag_2S is much less soluble than AgCl (rule 2), and salts of the transition metals in general are less soluble than those of the alkali and alkaline earth metals (rule 3). These rules are helpful in predicting the behavior of the specific cation-anion combinations in relation to the others, although they are not sufficient to explain all such reactions. The HSAB concept provides a more general approach that covers some of the exceptions.

Ahrland, Chatt, and Davies [94] classified some of the same phenomena (as well as others) by dividing the metal ions into class (a) ions, including most metals, and class (b) ions, a smaller group including Cu^{+}, Pd^{2+}, Ag^{+}, Pt^{2+}, Au^{+}, Hg_2^{2+}, Tl^{+}, Tl^{3+}, Pb^{2+}, and heavier transition metal ions. The members of class (b) are located in a small region in the periodic table at the lower right side of the transition metals. The class (b) ions form halides whose solubility is, in order, F > Cl > Br > I the reverse of the solubility order of class (a) halides. The class (b) metal ions also have a larger enthalpy of reaction with phosphorous donors than with nitrogen donors, again the reverse of the class (a) metal ion reactions. In the periodic Table shown in Fig. 1-7, the elements that are always in class (b) and those that are commonly in class (b) when they have low or zero oxidation states are identified. In addition, the transition metals have class (b) character in compounds in which their oxidation state is zero.

Ahrland, Chatt, and Davies [94] explained the class (b) metals as having *d* electrons available for π bonding. Therefore, the high oxidation states of elements to the right of the transition metals have more class (b) in their reactions with halides, but Tl (III) shows stronger class (b) character because Tl (I) has two 6*s* electrons that screen the 5*d* electrons and keep them from being fully available for π bonding. Elements farther left in the table have more class (b) character in the low or zero oxidation states when more *d* electrons are present (See Fig. 1-8).

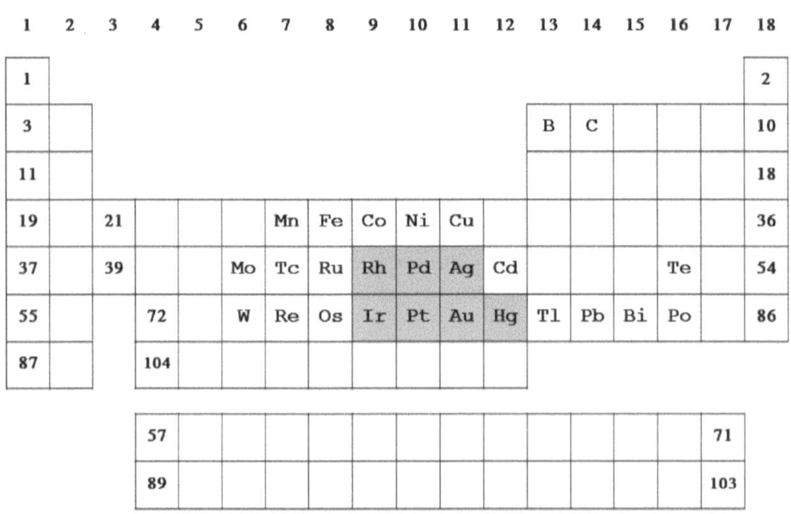

Fig. 1-8 Location of class (b) metals in the periodic table. Those in the outlined region are class (b) acceptors. Others indicated by their symbols are borderline elements, whose behavior depends on their oxidation state, and the donor. The remainder (blank) are class (a) acceptors [94].

Donor molecules or ions that have the most favorable enthalpies of reaction with class (b) metals are those that are more readily polarizable and have vacant *d* or π^* orbitals available for π bonding.

Pearson has designated the class (a) ions as hard acids and class (b) ions as soft acids. Bases are also classified as hard or soft. For example, the halide ions range from F^-, a very hard base, through less hard Cl^- and Br^- to I^-, a soft base. Reactions are more favorable for the hard-hard and the soft-soft interactions than for a mix of hard and soft in the reactants. Much of the hard-soft distinction depends on polarizability, the degree to which the molecules form slightly polar species that can then combine with the other molecules. Hard acids and bases are relatively small, compact, and nonpolarizable, while soft acids and bases are larger and more polarizable (therefore softer). The hard acids are therefore any cations with large positive charge (3+ or larger) or those whose *d* electrons are relatively unavailable for π bonding. Soft acids are those whose *d* electrons or orbitals are readily available for π bonding. In addition, the more massive the atom, the softer it is likely to be, because the large number of inner electrons shield the outer ones and make the atom more polarizable. This description fits the class (b) ions well: they are primarily 1+ and 2+ ions with filled or nearly filled *d* orbitals, and most are in the second and third rows of the transition elements, with 45 or more electrons. Table 1-10 lists bases and acids in terms of their hardness or softness.

The trends in the bases are even easier to see, with $F^- > Cl^- > Br^- > I^-$ the hardness order of the halides. Again, more electrons and larger sizes lead to softer behavior. In another example, S^{2-} is softer than O^{2-} because it has more electrons spread over a slightly larger volume, making S^{2-} more polarizable. Within a group, such comparisons are easy; as the electronic structure and size changes, and comparisons become more difficult but are still possible.

More detailed comparisons are possible, but another factor, called the inherent acid-base strength, must also be kept in mind in these comparisons.

An acid or a base may be either hard or soft and at the same time either strong or weak. The strength of the acid or base may be more important than the hard-soft characteristics; both must be considered at the same time. For example, if two soft bases are in competition for the same acid, the one with more inherent base strength may be favored unless there is a considerable difference in the softness. Such comparisons require care; seldom is one factor totally responsible for the reaction, and the reaction is nearly always a competition between acid-base pairs. As an example, consider the following reaction. Two hard-soft combinations react to give a hard-hard and a soft-soft combination, although ZnO is composed of the strongest acid (Zn^{2+}) and the strongest base (O^{2-}).

$$\underset{\text{soft-hard}}{ZnO} + \underset{\text{hard-soft}}{2LiC_4H_9} \longrightarrow \underset{\text{soft-soft}}{Zn(C_4H_9)_2} + \underset{\text{hard-hard}}{Li_2O} \qquad (1\text{-}55)$$

In 1963, Pearson unified the conclusions from his earlier study of the Edwards equation with those deduced from the study of the aqueous stability

Table 1-10 Hard and Soft Acids and Bases [95]

Hard	Border line	Soft
Acid		
H^+, Li^+, Na^+, K^+ (Rb^+, Cs^+)	Fe^{2+}, Co^{2+}, Ni^{2+}, Cu^{2+}, Zn^{2+},	$Co(CN)_5^{3-}$, Pd^{2+}, Pt^{2+}, Pt^{4+}
Be^{2+}, $Be(CH_3)_2$, Mg^{2+}, Ca^{2+}, Sr^{2+} (Ba^{2+}),	$B(CH_3)_3$, GaH_3,	BH_3, $Ga(CH_3)_3$, $GaCl_3$,$GaBr_3$, GaI_3, Tl^+, $Tl(CH_3)_3$
Sc^{3+}, Ga^{3+}, Gd^{3+}, Lu^{3+}, Th^{4+}, U^{4+}, UO_2^{2+}, Pu^{4+}, Ti^{4+}, Zr^{4+}, Hf^{4+}, VO^{2+}, Cr^{3+}, Cr^{6+}, MoO^{3+}, WO^{4+}, In^{3+}, La^{3+}, Mn^{2+}, Mn^{7+}, Fe^{3+}, Co^{3+}, Ce^{4+},	Rh^{3+}, Ir^{3+}, Ru^{3+}, Os^{2+},	Cu^+, Ag^+, Au^+, Cd^{2+}, Hg^+, Hg^{2+}, CH_3Hg^+
BF_3, BCl_3, $B(OR)_3$, Al^{3+}, $Al(CH_3)_3$, $AlCl_3$, AlH_3	R_3C^+, $C_6H_5^+$, Sn^{2+}, Pb^{2+}	CH_2, carbenes
CO_2, RCO^+, CH_3Sn^{3+}, $(CH_3)_2Sn^{2+}$,Si^{4+}, Sn^{4+},	NO^+, Sb^{3+}, Bi^{3+}, SO_2	Br_2, Br+, I_2, I+, ICN, etc.
N^{3+}, RPO_2^+, $ROPO_2^+$, As^{3+}		HO^+, RO^+, RS^+, RSe^+, Te^{4+}, RTe^+,
SO_3, RSO_2^+, $ROSO_2^+$		O, Cl, $RO\cdot$, $RO_2\cdot$
Ions with oxidation states of 4 or higher		Metals with zero oxidation oxidation state
HX (hydrogen-bonding molecules)		π acceptors: trinitrobenzene, choroanil, quinines, tetracyanoethylene, etc.
Base		
F^-, (Cl^-),	Br^-,	H^-,
H_2O, OH^-, O^{2-}, ROH, RO^-, R_2O,	NO_2^-, SO_3^{2-},	I^-,
CH_3COO^-, NO^{3-}, ClO^{4-}	N_3^-, N2, $C_6H_5NH_2$, C_5H_5N	R^-, C_2H_4, C_6H_6, CN^-, RNC, CO,
CO_3^{2-}, SO_4^{2-}, PO_4^{3-}		RSH, RS^-, R_2S
NH_3, RNH_2, N_2H_4		SCN^-, R_3P, $(RO)_3P$, R_3As, $S_2O_3^{2-}$, R_2S, RS^-

Table 1-11 Correlations Subsumed by the HSAB Principle

Substrates, correlating with pK_a of base (high β or b); class A or (a) acceptor, H^+-like ions or lithophiles	→	Hard acids
Substrates, correlating with E_n or P_n of base (high α or a); class B or (b) acceptor, Hg^{2+}-like ions or chalcophiles	→	Soft acids
Bases, with large pK_a values; donors high on the class (a) affinity series	→	Hard bases
Bases, with large E_n of P_n values; donors high on the class (b) affinity series	→	Soft bases

constants and made the identifications listed in Table 1-11. He also proposed the following rules to summarize the experimental data [96–100]:

Rule 1. Equilibrium: Hard acids prefer to associate with the hard bases and soft acids with soft bases.
Rule 2. Kinetics: Hard acids react readily with the hard bases and soft acids with soft bases.

The idea, that the kinetics and thermodynamics of a series of reactions follow the same correlation is valid only to the extent that the reactions obey the noncrossing rule. A selection of typical hard and soft acids and bases is given in Table 1-10.

As described in these general rules, the hard-hard combinations are more favorable energetically than the soft-soft combinations. When in doubt, this explanation may be helpful in deciding the determining factor in a reaction. Also, either the hard-hard or the soft-soft combination can lead to insoluble salts, but such cases show that the rules have limitations. Some cations considered hard will precipitate under the same conditions as others that are clearly soft. For this reason, any predictions based on the HSAB must be considered tentative, and solvent and other interactions must be considered carefully.

1.3.7. Drago's *E, C* Equation

A quantitative system of acid-base parameters proposed by Drago and Wayland [101–104] uses the equation:

$$-\Delta H = E_A E_B + C_A C_B \tag{1-56}$$

where ΔG is the enthalpy of the reaction $A + B \rightarrow AB$ in the gas phase or in an inert solvent, and E and C are parameters calculated from experimental data. Drago has separated the enthalpy into two components, where E is a measure of the capacity for electrostatic interactions and C a measure of the tendency to form covalent bonds. The subscripts refer to the values assigned to the acid and base, with I_2 chosen as the reference acid and *N,N*-dimethylacetamide and diethyl sulfide as reference bases. The defined values (in units of kcal/mol) are:

	E_A	C_A	E_B	C_B
I_2	1.00	1.00		
N,N-dimethylacetamide			1.32	
Diethyl sulfide				7.40

Values of E_A and C_A for the selected acids and E_B and C_B for selected bases are given in Table 1-12. Combining the values of these parameters for acid-base pairs gives the enthalpy of reactions in kcal/mol; multiplying by 4.184 J/cal converts to joules (although we use joules in this book, these numbers were originally derived for calories and we have chosen to leave them unchanged).

Examination of the table shows that most acids have lower C_A values and higher E_A values than I_2. Since I_2 has no permanent dipole, it has little electrostatic attraction for bases and, therefore, has a low E_A. On the other hand, it has a strong tendency to bond with some other bases, accounted for by a relatively large C_A. Because 1.00 was chosen as the reference value for both parameters for I_2, C_A values are mostly below 1 and E_A values are mostly above 1. For C_B and E_B, this relationship is reversed.

The example of iodine and benzene shows how these tables can be used.

$$\underset{\text{acid}}{I_2} + \underset{\text{base}}{C_6H_6} \longrightarrow I_2 \cdot C_6H_6 \tag{1-57}$$

$$-\Delta H = E_A E_B + C_A C_B \tag{1-58}$$

or

$$\Delta H = -(E_A E_B + C_A C_B) \tag{1-59}$$

$$\begin{aligned} \Delta H &= -(1.00 \times 0.681 + 1.00 \times 0.525) \\ &= -1.206\ \text{kcal/mol or} - 5.046\ \text{kJ/mol} \end{aligned} \tag{1-60}$$

The experimental value of ΔH is − 1.3 kcal/mol, or − 5.5 kJ/mol, 10 percent larger [106]. This is a weak adduct (other bases combining with I_2 have enthalpies as exothermic, as −12 kcal/mol, or −50 kJ/mol), and the calculation does not agree with experiment as well as many. Because there can be only one set of numbers for each compound, Drago has developed statistical methods for averaging

Table 1-12 E_A, C_A, E_B, and C_B Values [105]

Acid	E_A	C_A	Base	E_B	C_B
I_2	0.50	2.00	NH_3	2.31	2.04
H_2O	1.54	0.13	CH_3NH_2	2.16	3.12
SO_2	0.56	1.52	$(CH_3)_2NH$	1.80	4.21
HF	2.03	0.30	$(CH_3)_3N$	1.21	5.61
HCN	1.77	0.50	$C_2H_5NH_2$	2.35	3.30
CH_3OH	1.25	0.75	$(C_2H_5)_3N$	1.32	5.73
H_2S	0.77	1.46	$HC(C_2H_4)_3N$	0.80	6.72
HCl	3.69	0.74	C_5H_5N	1.78	3.54
C_6H_5OH	2.27	1.07	$4\text{-}CH_3C_5H_4N$	1.74	3.93
$(CH_3)_3COH$	1.36	0.51	$3\text{-}CH_3C_5H_4N$	1.76	3.72
$HCCl_3$	1.49	0.46	$3\text{-}ClC_5H_4N$	1.78	2.81
CH_3COOH	1.72	0.86	CH_3CN	1.64	0.71
CF_3CH_2OH	2.07	1.06	$CH_3C(O)CH_3$	1.74	1.26
C_2H_5OH	1.34	0.69	$CH_3C(O)OCH_3$	1.63	0.95
$1\text{-}C_3H_7OH$	1.14	0.90	$CH_3C(O)OC_2H_5$	1.62	0.98
PF_3	0.61	0.36	$HC(O)N(CH_3)_2$	2.19	1.31
$B(OCH_3)_3$	0.54	1.22	$(C_2H_5)_2O$	1.81	1.63
AsF_3	1.48	1.14	$O(CH_2CH_2)_2O$	1.86	1.29
$Fe(CO)_5$	0.10	0.27	$(CH_2)_4O$	1.64	2.18
CHF_3	1.32	0.91	$(CH_2)_5O$	1.70	2.02
$B(C_2H_5)_3$	1.70	2.71	$(C_2H_5)_2S$	0.24	3.92
H^+	45.00	13.03	$(CH_3)_2SO$	2.40	1.47
CH_3^+	19.70	12.61	C_5H_5NO	2.29	2.33
Li^+	11.72	1.45	$(CH_3)_3P$	1.46	3.44
K^+	3.78	0.10	$(CH_3)_2O$	1.68	1.50
NO^+	0.10	6.86	$(CH_3)_2S$	0.25	3.75
NH_4^+	4.31	4.31	CH_3OH	1.80	0.65
$(CH_3)_2NH_2^+$	3.21	0.70	C_2H_5OH	1.85	1.09
$(CH_3)_4N^+$	1.96	2.36	C_6H_6	0.70	0.45
$C_5H_5NH^+$	1.81	1.33	H_2S	0.04	1.56
$(C_2H_5)_3NH^+$	2.43	2.05	HCN	1.19	0.10
$(CH_3)_3NH^+$	2.60	1.33	H_2CO	1.56	0.10
H_3O^+	13.27	7.89	CH_3Cl	2.54	0.10
$(H_2O)_2H^+$	11.39	6.03	CH_3CHO	1.76	0.81
$(H_2O)_3H^+$	11.21	4.66	H_2O	2.28	0.10
$(H_2O)_4H^+$	10.68	4.11	$(CH_3)_3COH$	1.92	1.22
$(CH_3)_3Sn^+$	7.05	3.15	C_6H_5CN	1.75	0.62
$(C_5H_5)Ni^+$	11.88	3.49	F^-	9.73	4.28
$(CH_3)NH_3^+$	2.18	2.38	Cl^-	7.50	3.76
			Br^-	6.74	3.21
			I^-	5.48	2.97
			CN^-	7.23	6.52
			OH^-	10.43	4.60
			CH_3O^-	10.03	4.42

Table 1-13 Acids and Bases with Parallel Changes in *E* and *C*

Acids	C_A	E_A
m-$CF_3C_6H_4OH$	0.53	4.48
$B(CH_3)_3$	1.70	6.14
Bases	C_B	E_B
C_6H_6	0.681	0.525
CH_3CH	1.34	0.886
$(CH_3)_2CO$	2.33	0.987
$(CH_3)_2SO$	2.85	1.34
CH_3	3.46	1.36

experimental data from many different combinations. In many cases, the agreement between calculated and experimental enthalpies is within 5 percent.

One phenomenon not well accounted for by other approaches is seen in Table 1-13 [107]. It shows a series of four acids and five bases in which both *E* and *C* increase. In most descriptions of bonding as the electrostatic (ionic) bonding increases, covalent bonding decreases, but these data show both increasing at the same time. Drago argues that this means that the *E* and *C* approach explains acid-base adduct formation better than one alternative, the HSAB theory described in the next section.

Drago's system emphasizes the two factors involved in acid-base strength (electrostatic and covalent) in the two terms of his equation for the enthalpy of reaction, while Pearson's puts more obvious emphasis on the 'covalent' factor. Pearson has proposed the equation $\log K = S_A S_B + \sigma_A \sigma_B$, with the inherent strength S modified by a softness factor σ. Larger values of strength and softness then lead to larger equilibrium constants or rate constants [108, 109].

Although Pearson attached no numbers to this equation, it does show the need to consider more than just hardness or softness in working with acid-base reactions. Both systems (Pearson's HSAB, Drago's *E* and *C* parameters) are useful, but neither covers every case, and it is usually necessary to make judgments about the reactions for which information is incomplete. With *E* and *C* numbers available, quantitative comparisons can be made. When those numbers are not available, the HSAB approach can provide a rough guide for predicting reactions. Examination of the tables also shows little overlap of the examples chosen. Neither approach is completely satisfactory; both can be of considerable help in classifying reactions and predicting which reactions will proceed and which will not.

1.3.8. Gutmann's Donor and Acceptor Numbers (*DN-AN*)

Gutmann introduced the Donor Number (*DN*) [110] and Acceptor Number (*AN*) [111] to describe the AB properties of the Lewis definitions. *DN* were developed in order to correlate the behavior of a solute (such as its solubility, redox potential, or degree of ionization) in a variety of donor solvents with a given solvent's coordinating ability, that is, with its basicity or donicity. *DN* is defined in terms of the molar exothermic heat of mixing of the candidate solvent with a reference acid, antimony pentachloride ($SbCl_5$), in a dilute (10^{-3} M) solution in dichloroethane (*D*) (A relative measure of the basicity of a solvent *D* is given by the Gibbs free energy of its reaction with an arbitrarily chosen reference acid).

$$DN = -\Delta G \cdot (\mathrm{SbCl_5} : \text{ Base solvent}, D) \quad (\mathrm{kJ/mol}) \tag{1-61}$$

It is assumed (and graphically tested by plotting log *K* of the *D*-$SbCl_5$ reaction *vs.* *DN*) that entropy effects are constant and that one-to-one adducts are formed so that the *DN* is a reflection of the inherent *D*-$SbCl_5$ bond strength.

The most important assumption of the *DN* approach, however, is that the order of the base strengths established by the $SbCl_5$ scale remains constant for all the other acids (solutes), the value of the ΔG formation of a given adduct being linearly related to the *DN* of the base (solvent) *via* the equation:

$$-\Delta G_{\mathrm{D \cdot A}} = \mathrm{a} \cdot \mathrm{DN}_{\mathrm{D \cdot SbCl_5}} + b \tag{1-62}$$

where a and b are constants characteristic of the acid.

Graphically this means that a plot of the *DN* for a series of donor solvents *vs.* the $-\Delta G$ formation of their adducts with a given acid will give a straight line. Example plots are shown in Fig. 1-9. By experimentally measuring the ΔG formation of only two adducts for a given acid, one can predict, *via* the resulting characteristic line of the acid, the ΔG formation of its adducts with any other donor solvent for which the *DN* is known.

Since in this case the van der Waals interactions of the tested molecules and those of the reference acid molecules are confined almost entirely to the solvent, the measured heats of mixing have no contribution, owing to changes in the van der Waals interactions with the neighbor molecules [112]. *DN* is therefore solely determined by the AB interaction between the tested solvent and $SbCl_5$. $SbCl_5$ is chosen as the reference probe because it forms 1:1 adducts with all the donor molecules such that their AB properties can be evaluated on a per molecule or a per mole basis.

Furthermore, Gutmann introduced an analogous *AN* scale for the correlation of the solute-solvent interactions in acidic solvents. The adduct formation leads to a change in its structure from a bipyramidal to a distorted octahedral configuration, thus involving similar hybridization energies. The higher the *DN*, the stronger is the basicity. *AN* was originally defined as the relative ^{31}P-NMR downfield shift ($\Delta\delta^{AB}$) induced in $(C_2H_5)_3PO$, when dissolved in a pure candidate acidic solvent. *AN* values are scaled to an arbitrarily chosen value of 100 for the shift produced by the 1:1 $(C_2H_5)_3PO$-$SbCl_5$ adduct in dichloroethane. *AN* values have been measured for 34 solvents. Again, it is assumed that the relative solvent acceptor order established by the $(C_2H_5)_3PO$ scale remains

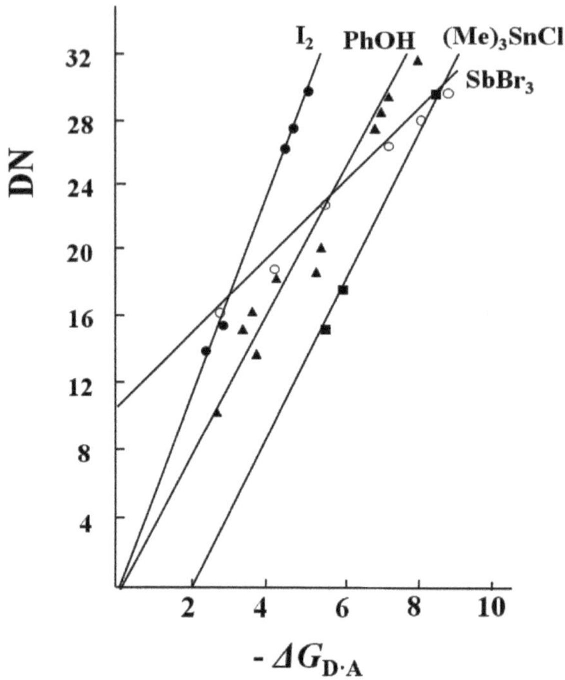

Fig. 1-9 Typical donor number plots showing that the heats of formation of the adducts between a given acid, and a series of bases are linearly proportional to the donor numbers of the bases involved. To simplify the graph, data points have not been labeled with the corresponding bases.

constant for all the other basic solutes. A selection of typical *AN* values is shown in Table 1-14 [113]. For strongly amphoteric solvents one must, of course, consider the relative importance of both the donor number and acceptor number simultaneously.

The Gutmann *DN* and *AN*, are measures of the strength of solvents, as Lewis acids or bases. The *DN* is based on the ^{31}P-NMR chemical shift of triethylphophine oxide ($(C_2H_5)_3PO$) in the solvent. The *AN* is based on the heat of the reaction between the solvent and $SbCl_5$ in dichloroethane.

Gutmann suggested that the enthalpy of AB adduct formation be written as:

$$-\Delta G^{AB} = \frac{AN_A DN_B}{100}, \tag{1-63}$$

which, unfortunately, had only limited success in predicting enthalpies of untried acid-base pairs. It was later found that the ^{31}P NMR spectrum was appreciably shifted downward not only by acid-base interactions, but also by van der Waals interactions [114]. An improved Acceptor Number, AN*, was

Table 1-14 Gutmann's *DN* and *AN* Values [119]

Solvent	*DN*	*AN*
1,2-dichloroethane	0	16.7
2-methyl-2-propanol	21.9	27.1
2-propanol	21.1	33.8
2-propanone	17	12.5
4-methyl-2-oxo-1,3-dioxolane	15.1	18.3
acetonitrile	14.1	18.9
benzonitrile	11.9	15.5
butanol	19.5	36.8
dichloromethane	1	20.4
dimethyl sulfoxide	29.8	19.3
ethanol	19.2	37.9
formamide	24	39.8
hexamethylphosphoramide	38.8	10.6
hexane	0	0
methanol	19	41.5
N,N-dimethylacetamide	27.8	13.6
N,N-dimethylformamide	26.6	16
N-methylformamide	27	32.1
N-methylpyrrolidon	27.3	13.3
nitrobenzene	4.4	14.8
nitromethane	2.7	20.5
propanol	19.8	37.3
sulfolane	14.8	19.2
tetrachloromethane	0	8.6
trichloromethane	4	23.1
water	18	54.8

then defined as the enthalpy of AB adduct formation of the probe of interest with reference to the base Et_3PO:

$$AN^* = -\Delta G(Et_3PO : \text{Acid solvent}) \quad (\text{kJ/mol}), \tag{1-64}$$

The value of AN^* is thus solely determined by the AB interaction between the solvent of interest and Et_3PO. AN^* is found to be proportional to $AN\text{-}AN^d$:

$$\text{AN}^* = 0.288\left(\text{AN} - \text{AN}^{\,\text{d}}\right) \tag{1-65}$$

where AN^d is the van der Waals contribution to AN and can be calculated from surface tension measurements. The higher the AN^*, the stronger the acidity.

Gutmann's approach recognizes the bifunctionality of most materials. However, it ignores the fact that the covalent (soft) and electrostatic (hard) interactions have independent contributions to the acidity and basicity. The most important assumption of Gutmann's approach is that the relative order of basicity and acidity established by the $SbCl_5/Et_3PO$ scale remains unchanged for all the other acids or bases, which may not always be the case. This is illustrated by

Pearson's HSAB principle which states that 'Hard acids prefer to associate with hard bases, and soft acids prefer to associate with soft bases' [115].

When the surface free energy is forced on the interface of two adherents, the contribution of the two components of surface free energy of each adherent is realized, as discussed in this chapter, as the hydrophilic and hydrophobic moieties. Then, it is easily possible to say that the hydrophilic or specific component, including hydrogen bonding, is transformed by the Gutmann's electron acceptor-donor system at a given temperature. According to this definition, the ability of the acid-base interaction may be characterized by means of the total hydrophilic component of a solid surface, whereas Pearson's and Drago's approaches include the hydrophobic or London dispersive component of a surface free of a solid surface, as the basis for the enthalpy of formation. Moreover, it greatly helps to determine the characteristic acidity (K_A) and basicity (K_D) of the solid surface on the basis of the specific component of surface free energy of the two adherents, at the interfaces.

Acid-base forces are known to contribute significantly to interactions in the macromolecular systems. The evolution of the acid-base concepts is traced from their early stages, as represented by the work of Arrhenius, Brønsted-Lowry, Lewis, Pearson, Drago, and Gutmann, to their current, complex state, with particular reference to their application, to the interface science in various applications.

1.4. DLVO (Derjaguin, Landau, Verwey, and Overbeek) Theory

1.4.1. Concept of DLVO (Derjaguin, Landau, Verwey, and Overbeek) Theory

In 1923, the first successful theory for ionic solution was developed by Debye and Hückel [116]. The framework of linearized Debye-Hückel theory was applied to describe the colloidal dispersions. After that, Levine and Dube [117, 118] found that between colloidal particles there were both a medium-range strong repulsion and a long-range strong attraction, but Levine and Dube could not describe the stability and instability of the colloidal dispersion. In 1941, Derjaguin and Landau provided an initial theory of combination of the attraction and repulsion forces [119]. 7 years later, Verwey and Overbeek got the same answer [120]. Both groups obtained their results independently. In their work, they corrected the defect of the Levine-Dube theory for colloidal systems and formulated the classical standard theory of colloidal dispersions, which successfully described the irreversible process of coagulation of the colloidal particles [121, 122].

In the case of ionic fluids and particles having charged surfaces, the stability of a liquid film is to a great extent governed by the electrostatic forces due to the overlap of the electric double layers in the liquid interlayer. The main difficulty in calculating these forces consists of the need to determine the spatial arrangement and concentration of ions in the interlayer with allowance for the deformation of the ionic atmospheres as the surfaces approach each other.

The DLVO theory (named after Derjaguin, Landau, Verwey, and Overbeek) in its simplest form looked at the two main forces acting on the charged colloidal particles in a solution. The two forces are: (1) electrostatic repulsion and (2) van der Waals attractive force.

The electrostatic stabilization of particles in a suspension is successfully described by the DLVO theory. The interaction between two particles in a suspension is considered to be the result of a combination of the van der Waals attraction potential and the electric repulsion potential. There are some important assumptions in the DLVO theory:

1. Infinite flat solid surface,
2. Uniform surface charge density,
3. No re-distribution of the surface charge, i.e., the surface electric potential remains constant,
4. No change of concentration profiles of both the counter-ions and the surface charge determining ions, i.e., the electric potential remains unchanged, and
5. Solvent exerts influences via the dielectric constant only, i.e., no chemical reactions between the particles and solvent.

The DLVO theory provides a good explanation of the interaction between the two approaching particles. The theory states that colloidal stability is determined by the potential energy of the particles (Φ_T) summarizing two parts: potential energy of the attractive interaction due to van der Waals force (Φ_A), and potential energy of the repulsive electrostatic interaction (Φ_R):

$$\Phi_T = \Phi_A + \Phi_R \tag{1-66}$$

Van der Waals interaction between the two nanoparticles is the sum of the molecular interaction for all pairs of molecules composed of one molecule in each particle, as well as for all pairs of molecules with one molecule in a particle and one in the surrounding medium, such as the solvent. Integration of all the van der Waals interactions between two molecules over the two spherical particles of radius, r, separated by a distance, S, as illustrated in Fig. 1-10, gives the total interaction energy or attraction potential [123]:

$$\Phi_A = -\frac{A}{6}\left\{\frac{2r^2}{(S^2+4rS)} + \frac{2r^2}{(S^2+4rS+4r^2)} + \ln\left[\frac{(S^2+4rS)}{(S^2+4rS+4r^2)}\right]\right\} \tag{1-67}$$

where the negative sign represents the attraction nature of the interaction between two particles, and A is a positive constant termed the Hamaker constant, which has a magnitude on the order of $10^{-19} \sim 10^{-20}$ J, and depends on the polarization properties of the molecules in the two particles and in the medium that separates them.

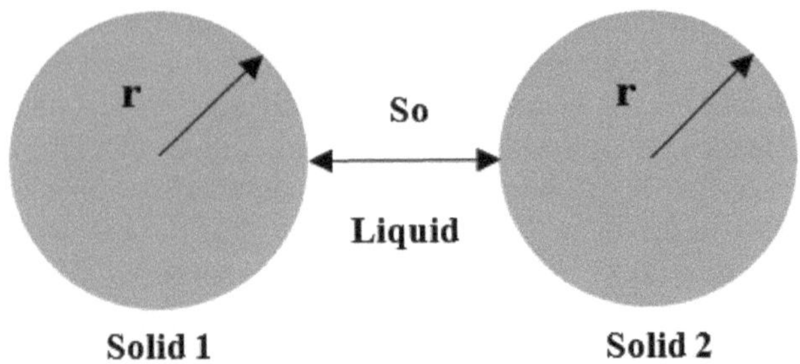

Fig. 1-10 The total interaction energy or attraction potential between two molecules.

It should be noted that in solving the problem of electrostatic interaction of the charged surfaces separated by an ionic interlayer the boundary conditions are first of all determined by the charging mechanism. In particular, if surface charging occurs by the adsorption of ions from the solution and the entropy contribution to the free energy of the system is independent of the amount of the adsorbed ions, the charged surfaces will approach each other under the constant-potential boundary condition. If the charging occurs by complete dissociation of the surface groups the constant-charge boundary conditions are applied. In the

Table 1-15 Hamaker Constants for some Common Materials

Materials	A_i (10^{-20} J)
Metal	16.2 ~ 45.5
Gold	45.3
Oxides	10.5 ~ 15.5
Al_2O_3	15.4
MgO	10.5
SiO_2 (fused)	6.5
SiO_2 (quartz)	8.8
Ionic crystals	6.3 ~ 15.3
CaF_2	7.2
Calcite	10.1
Polymers	6.15 ~ 6.6
Polyvinyl chloride	10.8
Polyethylene oxide	7.5
Water	4.35
Acetone	4.20
Carbon tetrachloride	4.78
Chlorobenzene	5.89
Ethyl acetate	4.17
Hexane	4.32
Toluene	5.40

case of incomplete dissociation of the ionized groups or if the entropy depends on the concentration of the adsorbed ions, the charge-regulation boundary condition is used [124, 125]. Yet another type of boundary conditions correspond to the interaction of the 'soft' ion-permeable particles or polyelectrolyte particles [126]. If the particle size greatly exceeds the Debye wavelength, the potential in the bulk of the particle is always equal to the Donnan potential.

Table 1-15 lists some Hamaker constants for a few common materials [127]. Eq. (1-67) can be simplified under various boundary conditions. For example, when the separation distance between two equal sized spherical particles is significantly smaller than the particle radius, i.e., S/r << 1, the simplest expression of the van der Waals attraction can be obtained:

$$\Phi_A = -\frac{Ar}{12S} \tag{1-68}$$

Table 1-16 Simple Formulas for the van der Waals Attraction Between Two Particles

Particles	Φ_A
Two spheres of equal radius, r*	$-\frac{Ar}{12S}$
Two spheres of unequal radii, r_1 and r_2*	$-\frac{Ar_1r_2}{6S(r_1+r_2)}$
Two parallel plates, with thickness of δ, interaction per unit area	$-\frac{A}{12\pi\left(S^{-2}+(2\delta+S)^{-2}+(\delta+S)^{-2}\right)}$
Two blocks, interaction per unit area	$-\frac{A}{12\pi S^2}$

*r, r_1 and r_2 >> S.

Table 1-16 lists the other simplified expressions of the van der Waals attraction potential. From this table, it should be noticed that the van der Waals attraction potential between the two particles is different from that between two flat surfaces. Furthermore, it should be noted that the interaction between the two molecules is significantly different from that between the two particles. Van der Waals interaction energy between the two molecules can be simply represented by:

$$\Phi_A \propto -S^{-6} \tag{1-69}$$

As shown in Fig. 1-11, although the nature of the attraction energy between two particles is the same as that between two molecules, integration of all the interaction between molecules from two particles and from the medium results in a totally different dependence of force on the distance. The attraction force between two particles decays much more slowly and extends over distances of nanometers. As a result, a barrier potential must be developed to prevent agglomeration. Two methods are widely applied to prevent the agglomeration of particles: electrostatic repulsion and steric exclusion.

Also, electric repulsive potential energy is presented as:

$$V_R = 2\pi\varepsilon\varepsilon_0 r\zeta^2 e^{-kx} \tag{1-70}$$

where, ε is the dielectric constant of the solvent, ε_0 vacuum permittivity, ζ zeta potential, and k a function of the ionic concentration (k^{-1} is the characteristic length of the Electric Double Layer).

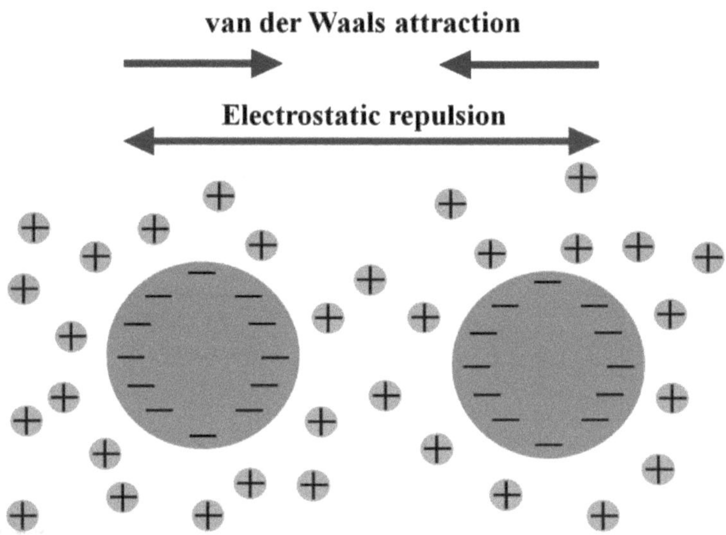

Fig. 1-11 The potential energy of the interaction between two particles.

The minimum of the potential energy determines the distance between two particles corresponding to their stable equilibrium. The two particles form a loose aggregate, which can be easily re-dispersed. A strong aggregate may be formed at a shorter distance corresponding to the primary minimum of the potential energy (not shown in the picture). In order to approach the distance of the primary minimum the particle should overcome the potential barrier.

Fig. 1-12 shows the van der Waals attraction potential electric repulsion potential and the combination of the two opposite potentials as a function of distance from the surface of a spherical particle [128]. At a distance far from the solid surface, both the van der Waals attraction potential and electrostatic repulsion potential reduce to zero. Near the surface there is a deep minimum in the potential energy produced by the van der Waals attraction. A maximum is located a little farther away from the surface, as the electric repulsion potential

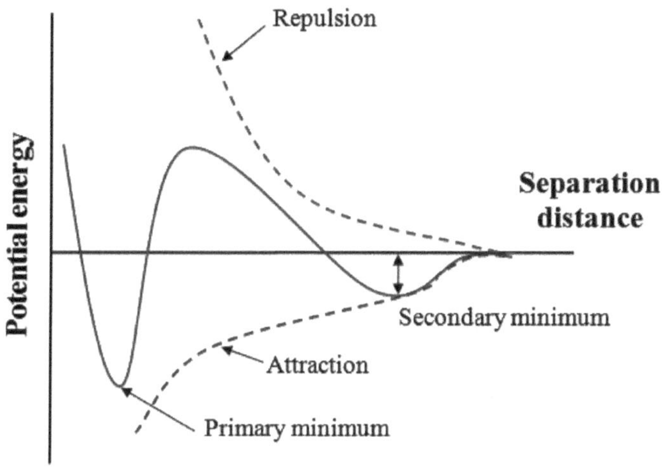

Fig. 1-12 Interaction potential energy as a function of the intermolecular distance.

dominates the van der Waals attraction potential. The maximum is also known as the repulsive barrier. If the barrier is greater than $\sim$ 10 kT, where k is Boltzmann constant, the collisions of two particles produced by Brownian motion will not overcome the barrier and agglomeration will not occur. Since the electric potential is dependent on the concentration and valence state of the counter-ions as given in Eqs. (1-71) and (1-72), and the van der Waals attraction potential is almost independent of the concentration and valence state of the counter-ions, the overall potential is strongly influenced by the concentration and valence state of the counter-ions.

The electric potential drops approximately according to the following equation:

$$E \propto e^{-\kappa(h-H)} \tag{1-71}$$

where $h \geq H$, which is the thickness of the Stern layer, $1/\kappa$ is known as the Debye-Hückel screening strength and is also used to describe the thickness of the double layer, and κ is given by:

$$\kappa = \left\{ \frac{(F^2 \Sigma_i C_i Z_i^2)}{\varepsilon_r \varepsilon_0 R_g T} \right\}^{1/2} \tag{1-72}$$

where F is Faraday's constant, ε_0 the permittivity of vacuum, ε_r the dielectric constant of the solvent, and C_i and Z_i are the concentration and valence of the counter-ions, of type i. This equation clearly indicates that the electric potential in the vicinity of the solid surface decreases with increased concentration and valence state of counter-ions, and increases exponentially with an increased dielectric constant of the solvent.

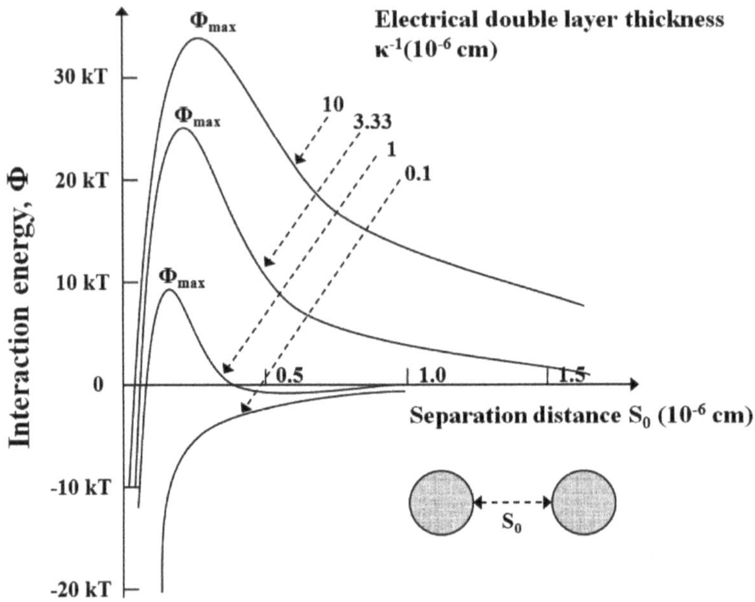

Fig. 1-13 An increase in concentration and valence state of counter-ions results in a faster decay of the electric potential.

The DLVO potential (primary minimum in Fig. 1-12) is obtained by adding hard-core, screened coulombic and van der Waals potentials. The height of the repulsive barrier indicates how stable the system is. The values of minima and maxima as well as their position depend on the solution characteristics [129].

An increase in the concentration and valence state of the counter-ions results in a faster decay of the electric potential as schematically illustrated in Fig. 1-13 [130]. As a result, the repulsive barrier is reduced and its position is pushed toward the particle surface. The secondary minimum in Fig. 1-13 does not necessarily exist in all the situations, and is present only when the concentration of counter-ions is high enough. If a secondary minimum is established, particles likely associate with each other, which is known as flocculation.

The mechanisms of the van der Waals forces and the electrostatic interactions are considered the theory of stability of lyophobic colloids. In this theory of lyophobic surfaces and colloids, the particles are treated as interacting weakly with the dispersion medium, which usually permits a correct description of the stability of the interlayers (Fig. 1-14), the kinetics of coagulation and the destabilizing effect of the electrolyte additives using the two approaches to calculations of surface forces.

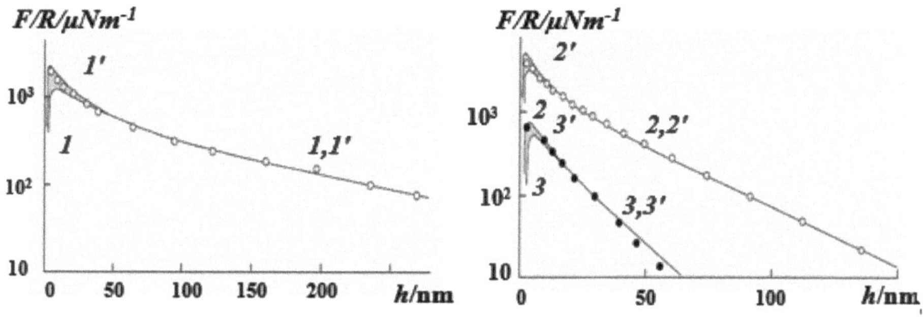

Fig. 1-14 Reduced forces of interaction (*F/R*) between two crossed cylindrical mica surfaces.

Additionally, there is an important occurrence in the flocculation of aqueous colloids. A suspension of charged particles experiences both double-layer repulsion and dispersion attraction, and the balance between these determines the ease and hence the rate at which the particles aggregate. Verwey and Overbeek [131, 132] considered the case of two colloidal spheres and calculated the net potential energy versus distance curves of the type illustrated in Fig. 1-15 for the case of $\psi_0 = 25.6$ mV (i.e., $\psi_0 = kT/e$ at 25 °C). At low ionic strength, as measured by κ, the double-layer repulsion is overwhelming except at very small separations, but as κ increases, a net attraction at all distances is finally attained. There is a critical region of κ such that a small potential minimum of about $kT/2$ occurs at a distance of separation *s* about equal to a particle diameter. This minimum is known as the secondary minimum and can lead to weak, reversible aggregation under certain conditions of the particle size, surface potential, and Hamaker constant [133].

We believe that the last stage of the evolution of the DLVO theory in its classical form should be dated to the early 1960s; at that time, the macroscopic theory of the van der Waals forces was already elaborated and various theories for the calculation of electrostatic interactions between same-type and different-type particles separated by uniform liquid or solvent interlayers were developed. However, the development of the DLVO theory was accompanied by an accumulation of the experimental data that not only showed poor quantitative agreement, but in some cases qualitatively contradicted the predictions of the classical DLVO theory. Examples are provided by recent studies on the stability of foam films [134, 135], wetting films and interlayers between the solid surfaces [134–140].

The presence of a large repulsive potential barrier between the secondary minimum and contact prevents flocculation. One can thus see why increasing ionic strength of a solution promotes flocculation. The net potential per unit area between the two planar surfaces is given by Ref. [141],

$$U(x)_{net} = \frac{64 n_0 kT}{\kappa}\left(\frac{e^{y_0/2}-1}{e^{y_0/2}+1}\right)^2 e^{-2\kappa x} - \frac{A}{12\pi x^2} \tag{1-73}$$

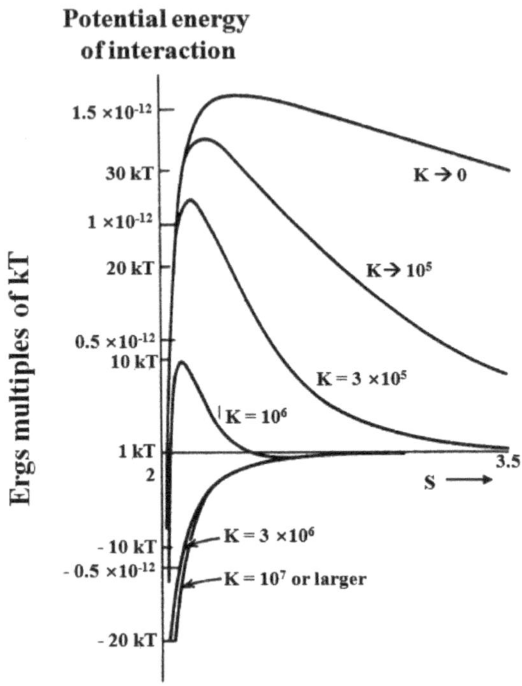

Fig. 1-15 The effect of electrolyte concentration on the interaction potential energy between two spheres where K is κ in cm^{-1}.

where $y_0 = ze\psi_0/kT$. If we assume that flocculation will occur when no barrier exists, we require that $U(x) = 0$ and $dU(x)/dx = 0$ at some value of x. This defines a critical electrolyte concentration,

$$n_0 = \frac{1152}{exp(4)} \frac{\varepsilon^3 (kT)^5}{e^6 A^2 z^6} \left(\frac{e^{y_0/2} - 1}{e^{y_0/2} + 1} \right)^4 \tag{1-74}$$

which varies as $1/z^6$, as shown in the Schulze-Hardy rule. Thus, for a z-z electrolyte, equivalent flocculation concentrations would scale as $1 : \left(\frac{1}{2}\right)^6 : \left(\frac{1}{3}\right)^6$ or 100: 1.6: 0.13, for a 1-1, 2-2, and 3-3 electrolyte, respectively. Actually, the higher-valence ions have an increased tendency for specific adsorption, so the flocculation effectiveness becomes a matter of reduction of ψ_0, as well as a matter of reduction of the double layer thickness [142].

Quantitative measurements of flocculation rates have provided estimates of Hamaker constants in qualitative agreement with theory. One assumes diffusion-limited flocculation when the probability to aggregate decreases with the exponential of the potential energy barrier height, as illustrated in Fig. 1-15. The barrier height is estimated from the measured flocculation rate; other measurements [143] give the surface (or zeta) potential leaving the Hamaker constant to be determined from Eq. (1-73) [144–146]. Complications arise from the assumption of constant surface potential during aggregation, from double-layer relaxation during aggregation [145–149], and from nonuniform charge distribution on the particles [150–152]. In studies of the stability of ZnS sols in NaCl and $CaCl_2$, Durán and co-workers [153] found they had to add the Lewis acid-base interactions, developed by van Oss [154] to the DLVO potential to model their measurements. Alternatively, the initial flocculation rate may be measured at an ionic strength such that no barrier exists. By this means PWP was found to be about 0.7×10^{-13} erg for the aqueous suspensions of polystyrene latex [155]. The hydrodynamic resistance between the particles in a viscous fluid must generally be recognized to obtain the correct flocculation rates [133].

Direct force measurements on the SFA by Israelachvili and co-workers and others also confirm the DLVO theory for many cases [156–159]. An example of a force measurement is shown in Fig. 1-16 as a plot of force over radius, *F/R*, *vs.* surface separation for the lipid bilayer-coated surfaces in two salt solutions at two ionic strengths [160]. Generally the DLVO potential works well until one gets to separations on the order of the Stem layer [161], at which the hydrated ions can eliminate the primary minimum [162].

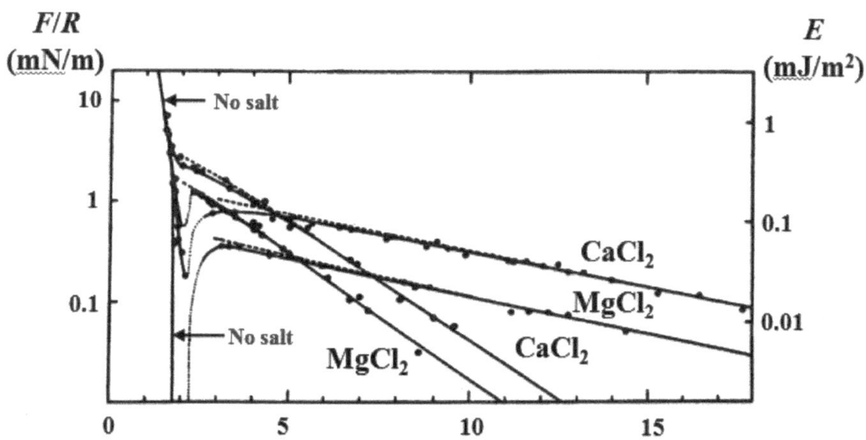

Fig. 1-16 The force between two crossed cylinders coated with mica and carrying adsorbed bilayers of phosphatidylcholine lipids at 22 °C. The solid symbols are for 1.2 mM salt, while the open circles are for 10.9 mM salt. The solid curves are the DLVO theoretical calculations.

1.4.2. Electric Double Layer

An electric double layer is a phenomenon that plays a fundamental role in the mechanism of the electrostatic stabilization of colloids. Colloidal particles gain negative electric charge when negatively charged ions of the dispersion medium are adsorbed on the particle surface. A negatively charged particle attracts the positive counterions surrounding the particle. As shown in Fig. 1-17, an electric double layer is the layer surrounding a particle of dispersed phase, including the ions adsorbed on the particle surface and a film of the countercharged dispersion medium. The electric double layer is electrically neutral.

An electric double layer consists of three parts:

1. Surface charge: charged ions (commonly negative) adsorbed on the particle surface.
2. Stern layer: counterions (charged opposite to the surface charge), attracted to the particle surface and closely attached to it by the electrostatic force.
3. Diffuse layer: a film of the dispersion medium (solvent) adjacent to the particle. Diffuse layer contains free ions with a higher concentration of the counterions. The ions of the diffuse layer are affected by the electrostatic force of the charged particle.

The electrical potential within the electric double layer has a maximum value on the particle surface (Stern layer). The potential drops with the increase of distance from the surface and reaches 0 at the boundary of the electric double layer.

When a colloidal particle moves in the dispersion medium, a layer of the surrounding liquid remains attached to the particle. The boundary of this layer is called the slipping plane (shear plane).The value of the electric potential at the slipping plane is called the zeta potential, which is a very important parameter in the theory of interaction of colloidal particles.

Although many important assumptions of the DLVO theory were not satisfied in real colloidal systems, in which small particles dispersed in a diffusive medium, the DLVO theory was still found to be valid and was widely applied in practice, as long as the following conditions are met:

1. Dispersion is very dilute, so that the charge density and distribution on each particle surface and the electric potential in the proximity next to each particle surface are not interfered with by other particles.
2. No other force is present besides the van der Waals force and the electrostatic potential; i.e., gravity is negligible or the particle is significantly small, and there exist no other forces, such as magnetic fields.
3. Geometry of particles is relatively simple, so that the surface properties are the same over the entire particle surface, and, thus surface charge density and distribution as well as the electric potential in the surrounding medium are the same.
4. The double layer is purely diffusive, so that the distributions of counter-ions and charge determining ions are determined by all the three forces: electrostatic force, entropic dispersion, and Brownian motion.

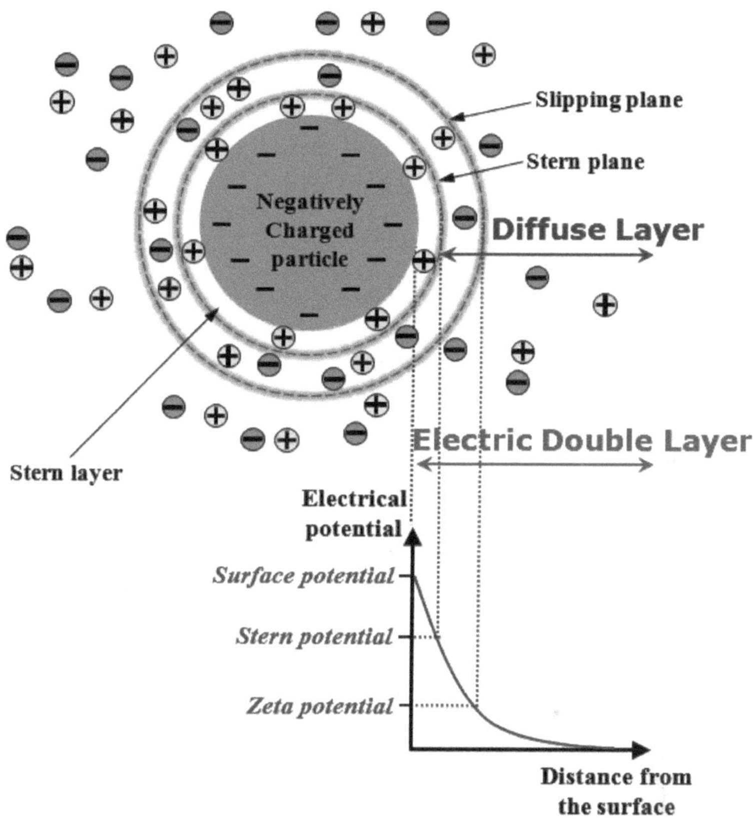

Fig. 1-17 Diagram of electric double layer.

However, it should be noted that electrostatic stabilization is limited by the following facts:

1. Electrostatic stabilization is a kinetic stabilization method.
2. It is only applicable to dilute systems.
3. It is not applicable to electrolyte sensitive systems.
4. It is almost impossible to re-disperse the agglomerated particles.
5. It is difficult to apply to multiple phase systems, since, in a given condition, different solids develop different surface charges and electric potentials.

It is very clear that some of the assumptions are far from the real picture of two particles dispersed in a suspension. For example, the surface of particles is not infinitely flat, and the surface charge density is likely to change when two charged particles get very close to each other. However, in spite of the assumptions, the DLVO theory works very well in explaining the interactions between two approaching particles that are electrically charged, and thus the theory is widely accepted in the research community of colloidal science.

Furthermore, it is found that the quantitative analysis of the surface forces is based on the Gouy-Chapman [163, 164] theory of diffuse ionic atmospheres and on London's theory of molecular forces. These two approaches underlie the DLVO theory of the stability of lyophobic colloids. In the framework of the DLVO theory, the total interaction energy is determined by the dispersion (van der Waals) and electrostatic mechanisms.

1.4.3. Zeta Potential

The stability of many colloidal solutions depends critically on the magnitude of the electrostatic potential (ψ_0) at the surface of the colloidal particles. One of the most important tasks in colloid science is therefore to obtain an estimate of ψ_0 under a wide range of electrolyte conditions. In practice, one of the most convenient methods for obtaining ψ_0 uses the fact that a charged particle will move at some constant, limiting velocity under the influence of an applied electric field. Even quite small particles (i.e., $< 1\ \mu$m) can be observed using a dark-field microscope, and in this way their velocity can be directly measured. This technique is called micro-electrophoresis, and what is measured is the electromobility (μ) of a colloid, which is its speed (u) divided by the applied electric field (E) [165].

So, from now on, we will determine if an estimate of ψ_0 can be obtained from the measured electro mobility of a colloidal particle. It turns out that it can be obtained simply, through analytic equations only for the cases of very large and very small particles. Thus, if a is the radius of an assumed spherical colloidal particle, we can obtain direct relationships between the electro mobility and the surface potential if either $\kappa a > 100$ or $\kappa a < 0.1$, where κ^{-1} is the Debye length of the electrolyte solution. Let us first look at the case of small spheres (where $\kappa a < 0.1$), which leads to the Hückel equation.

1.4.4. Hückel Equation (κa < 0.1)

The spherically symmetric potential around a charged sphere is described by the Poisson-Boltzmann equation:

$$\frac{1}{r^2}\frac{\mathrm{d}}{\mathrm{d}r}\left[r^2\frac{\mathrm{d}\Psi}{\mathrm{d}r}\right] = -\frac{\rho(r)}{\varepsilon_0 \mathrm{D}} \tag{1-75}$$

where $\rho(\mathrm{r})$ is the charge density and Ψ the potential at a distance r away from a central charge. This equation can be simplified using the Debye-Hückel or linear approximation valid for low potentials [116]:

$$\frac{1}{r^2}\frac{\mathrm{d}}{\mathrm{d}r}\left[r^2\frac{\mathrm{d}\Psi}{\mathrm{d}r}\right] = \kappa^2\Psi \tag{1-76}$$

which has the simple, general solution:

$$\Psi = \frac{A\exp(\kappa r)}{r} + \frac{B\exp(-\kappa r)}{r} \tag{1-77}$$

The constant A must equal zero for the potential Ψ to fall to zero at a large distance away from the charge; the constant B can be obtained using the second boundary condition, in which $\psi = \psi_0$ at $r = a$, where a is the radius of the charged particle and ψ_0 the electrostatic potential on the particle surface. Thus, the following result is obtained:

$$\Psi_0 = \frac{B\,\exp(-\kappa r)}{r} \tag{1-78}$$

and, therefore,

$$\Psi_0 = \frac{\Psi_0 a\,\exp[-\kappa(r-a)]}{r} \tag{1-79}$$

The relationship between the total charge q on the particle and the surface potential is obtained using the fact that the total charge in the electrical double-layer around the particle must be equal to and of opposite sign to the particle charge, that is:

$$q = -\int_a^\infty 4\pi r^2 \rho(r) dr \tag{1-80}$$

where $\rho(\mathrm{r})$ is the charge density at a distance r from the center of the charged particle. The value of $\rho(\mathrm{r})$ can be obtained from a combination of Eqs. (1-75) and (1-76), assuming the linear approximation is valid, and, hence:

$$q = 4\pi\varepsilon_0 \mathrm{D}\kappa^2 \int_a^\infty r^2 \Psi \mathrm{d}r \tag{1-81}$$

Now, using Eq. (1-79) for Ψ,

$$q = 4\pi\varepsilon_0 \mathrm{D}\kappa^2 a\Psi_0 \int_a^\infty r\,\exp[-\kappa(r-a)]\mathrm{d}r \tag{1-82}$$

Integration using Leibnitz's theorem gives:

$$q = 4\pi\varepsilon_0 \mathrm{D} a\Psi_0(1+\kappa a) \tag{1-83}$$

Rearranging this equation leads to a useful physical picture of the potential around a sphere, thus:

$$\Psi_0 = \frac{q}{4\pi \mathrm{D}\varepsilon_0 a} - \frac{q}{4\pi \mathrm{D}\varepsilon_0 (a+\kappa^{-1})} \tag{1-84}$$

This result corresponds to a model of a charged particle with a diffuse layer charge (of opposite sign) at a separation of $1/\kappa$, as illustrated in Fig. 1-18.

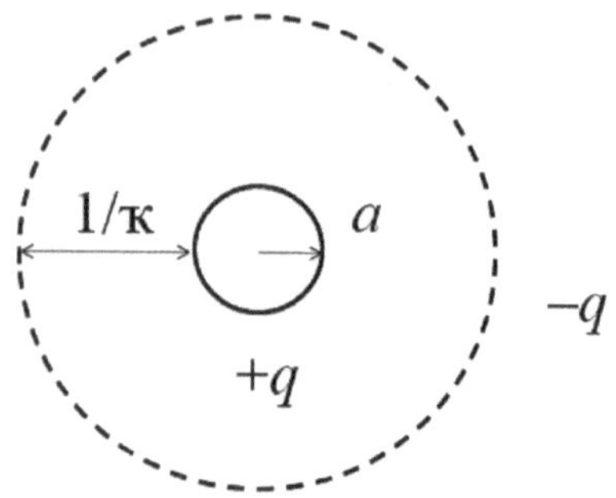

Fig. 1-18 Diagram of the diffuse electrical double-layer around a small, charged colloid.

Since we now have Eq. (1-83), which relates the charge on the particle to the surface potential, we can combine this with the forces acting on a moving particle in an applied electric field. Thus, when the particle is moving at a constant velocity (u), the electrostatic force on the particle (qE) must equal the drag force, which may be assumed (for laminar, steady fluid flow) to be that given by Stoke's Law (i.e., $F_{drag} = 6\pi a u \eta$). Using Eq. (1-83) and the fact that we have defined the electro mobility (μ) of a particle as u/E, we obtain the result that:

$$\Psi_0 = \frac{3\mu\eta}{2\varepsilon_0 D(1 + \kappa a)} \qquad (1\text{-}85)$$

which for $\kappa a \ll 1$ becomes:

$$\Psi_0 = \frac{3\mu\eta}{2\varepsilon_0 D} = \zeta \qquad (1\text{-}86)$$

In this result, the condition of the small particles means that the actual size of the particles (which is often difficult to obtain) is not required. For reasons to be discussed later, we will call the potential obtained by this method the zeta potential (ζ) rather than the surface potential. In the following section, we will consider the alternative case of large colloidal particles, which leads to the Smoluchowski equation.

1.4.5. Smoluchowski Equation ($\kappa a > 100$)

From now on, the case of large colloidal particles is considered an alternative derivation for, situations in which the particle radius is much larger than the Debye length (i.e., $\kappa a > 100$). This situation is best described by the schematic diagram given in Fig. 1-19, in which the surface of the large particle is assumed to be effectively flat relative to the double-layer thickness. It is also assumed, in this approach, that the fluid flows past the surface of the particle in parallel layers of increasing velocity according to the distance from the surface.

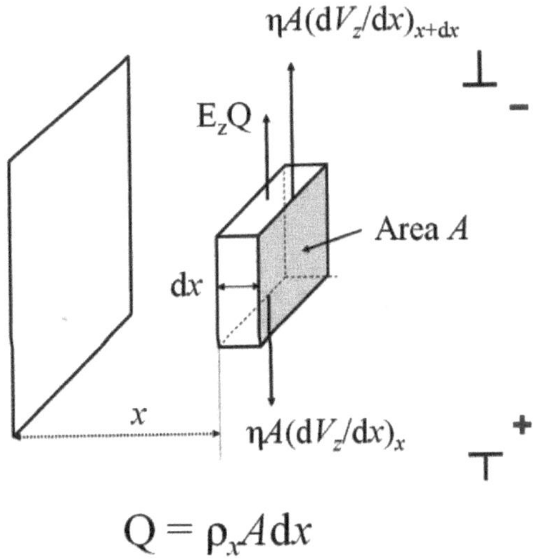

Fig. 1-19 Schematic diagram of the balance in forces acting on a fluid element close to the surface of a large colloidal particle.

At the surface the fluid has zero velocity (relative to the particle), and at a large distance away, the fluid moves with the same velocity as the particle, but in the opposite direction. It is also assumed that the flow of the fluid does not alter the ion distribution in the diffuse double-layer (i.e., in the x direction). Under these conditions the mechanical equilibrium can be considered in a fluid element, between x and $x + \mathrm{d}x$, when the viscous forces acting in the z direction on the fluid element due to the velocity gradient in the x direction, are precisely balanced by the electrostatic body force acting on the fluid due to the charge contained in it. Thus, we obtained the mechanical equilibrium condition that:

$$E_z\rho_x \mathrm{A}\mathrm{d}x = \eta \mathrm{A}\left(\frac{\mathrm{d}V_z}{\mathrm{d}x}\right)_x - \eta \mathrm{A}\left(\frac{\mathrm{d}V_z}{\mathrm{d}x}\right)_{x+\mathrm{d}z} \tag{1-87}$$

or

$$E_z\rho_x \mathrm{A}\mathrm{d}x = -\eta \mathrm{A}\left(\frac{\mathrm{d}^2V_z}{\mathrm{d}x^2}\right)\mathrm{d}x \tag{1-88}$$

We can then relate the charge density, ρ_x, to the electrostatic potential using the one-dimensional Poisson-Boltzmann equation,

$$\frac{\mathrm{d}^2\Psi}{\mathrm{d}x^2} = -\frac{\rho_x}{\varepsilon_0 \mathrm{D}} \tag{1-89}$$

Thus, in Eq. (1-88)

$$E_z\varepsilon_0 \mathrm{D}\frac{\mathrm{d}^2\Psi}{\mathrm{d}x^2}\mathrm{d}x = \eta\left(\frac{\mathrm{d}^2V_z}{\mathrm{d}x^2}\right)\mathrm{d}x \tag{1-90}$$

which on integration gives

$$E_z \varepsilon_0 \mathrm{D} \frac{\mathrm{d}\Psi}{\mathrm{d}x} = \eta \left(\frac{\mathrm{d}V_z}{\mathrm{d}x} \right) + c_1 \tag{1-91}$$

Since $d\,\Psi/dx = 0$ when, $d\mathrm{V}_z/dx = 0$, the integration constant, c_1, must be equal to zero and a second integration,

$$\int_{\Psi=0}^{\Psi=\zeta} E_z \varepsilon_0 \mathrm{D} \frac{\mathrm{d}\Psi}{\mathrm{d}x} \mathrm{d}x = \int_{V_z}^{0} \eta \left(\frac{\mathrm{d}V_z}{\mathrm{d}x} \right) \mathrm{d}x \tag{1-92}$$

produces the result that:

$$E_z \varepsilon_0 \mathrm{D} \zeta = -\eta V_z \tag{1-93}$$

if it is assumed that $\mathrm{D} \neq f(x)$ and $\eta \neq f(x)$ (i.e., that the fluid is Newtonian). Since $-\mathrm{V}_z$ refers to the fluid velocity, this term can be easily converted to particle velocity (i.e., $\mathrm{V}_p = -\mathrm{V}_z$) and, from our definition of electro mobility, (μ), it follows that:

$$\zeta = \frac{\mu \eta}{\varepsilon_0 \mathrm{D}} \tag{1-94}$$

This important result is called the 'Smoluchowski equation' and, as before, the zeta potential is directly related to the mobility and does not depend on either the size of the particle or on the electrolyte concentration [166].

In summary, for the two extreme cases:

$$\zeta = \frac{3\mu\eta}{2\varepsilon_0 D}, \quad \text{for } \kappa a \ll 1(<0.1) \tag{1-95}$$

$$\zeta = \frac{\mu\eta}{\varepsilon_0 D}, \quad \text{for } \kappa a \gg 1(>100) \tag{1-96}$$

1.5. Summary

The main purpose of this chapter is primary to review the physical or intermolecular interaction, including the van der Waals interaction and hydrogen bonding. Moreover, it is transformed in the London force as a hydrophobic term, and the Debye, Keesom, and hydrogen bonding as a hydrophilic term. The importance of intermolecular interaction in the synthesis, design, and manipulation of materials from macroscale to nanoscale gives unexpectedly many research areas.

It is also reviews the concepts of the acid-base interaction about Arrhenius, Brønsted-Lowry, Lewis, Pearson, and Drago's studies in terms of interface sciences, and especially about Gutmann's approach in the hydrophilic element of the excess surface free energy. Furthermore, the DLVO (Derjaguin, Landau,

Verwey, and Overbeek) theory is also proposed in relation to the interface science.

The author hopes that this book of modern theoretical approaches to determine the surface forces of different nature acting in interface science technology, will help the researchers to reasonably choose procedures and treatments of systems, containing particles, composites, and so on, in order to attain the desired results.

REFERENCES

[1] T. Young, Philosophical Transactions of the Royal Society 95 (1867) 65.
[2] J.D. Van der Waals, In over de Continuiteit van den Gas- en. Vloeistoftoestand, Thesis. Leiden Univ, Leiden (1873).
[3] W.H. Keesom, Phys. Z 22 (1921) 129.
[4] P. Debye, Phys. Z 13 (1912) 295.
[5] P. Debye, Phys. Z 21 (1920) 178.
[6] F. London, Phys. Z 63 (1930) 245.
[7] F. London, Z. Phys. Chem. Abt. B 11 (1930) 222.
[8] S.J. Park, Van der Waals Interaction at Surfaces. In Encyclopedia of Surface and Colloid Science, pp 5570. Taylor & Francis, New York (2006).
[9] J.E. Lennard-Jones, B.M. Dent, Transactions of the Faraday Society 24 (1928) 92.
[10] P.C. Hiemenz, Principles of Colloid and Surface Chemistry, 2nd edit, pp 611. Marcel Dekker, New York (1986).
[11] J.B. Donnet, S.J. Park, Carbon 29 (1991) 995.
[12] R.C. Reid, J.M. Prausnitz, T.K. Sherwood, The Properties of Gases and Liquids, 3rd edit, pp 181. McGraw-Hill, New York (1977).
[13] E. Fitzer, W. Hüttner, Journal of Physics D: Applied Physics 14 (1981) 347.
[14] G. Mie, Ann. Phys. (Leipzig) 11 (1903) 657.
[15] F. London, Transactions of the Faraday Society 33 (1937) 8.
[16] J.M. Prausnitz, R.N. Lichtenthaler, E.G. de Azevedo, Molecular Thermodynamics of Fluid-Phase Equilibria, 2nd edit, pp 48. Prentice-Hall, Englewood Cliffs (1986).
[17] M.J. Kim, J.C. Holste, K.R. Hall, J.C. Slattery, Journal of Colloid and Interface Science 158 (1993) 488.
[18] A.W. Adamson, Physical Chemistry of Surfaces, 5th edit. John Wiley & Sons, New York (1990).
[19] P. Debye, Phys. Z 22 (1921) 302.
[20] N.N. Avgul, A.V. Kiselev, Chemistry and Physics of Carbon, pp 1. Marcel Dekker, New York (1970).
[21] J.N. Israelachvili, R.M. Rashley, Nature 306 (1983) 249.
[22] J.N. Israelachvili, Accounts of Chemical Research 20 (1987) 415.
[23] J.N. Israelachvili, S.J. Kott, Journal of Chemical Physics 88 (1988) 7162.
[24] R.A. Alberty, F. Daniels, Physical Chemistry, 5th edit, pp 310. John Wiley & Sons, New York (1979).
[25] R. Schmid, V.N. Sapunov, Non-Formal Kinetics, pp 123. Verlag-Chemie, Weinheim (1982).
[26] J.B. Donnet, S.J. Park, H. Balard, Chromatographia 31 (1991) 434.
[27] J.N. Isrealachvili, Intermolecular and Surface Forces, 2nd edit. Academic Press, London (1991).

[28] J.C. Slater, J.G. Kirkwood, Physics Review 37 (1931) 682.
[29] S.J. Park, E. Papirer, J.B. Donnet, Journal of Chemical Physics 91 (1994) 203.
[30] J.B. Donnet, S.J. Park, Carbon 29 (1991) 955.
[31] J.B. Donnet, S.J. Park, M. Brendlé, Carbon 30 (1992) 263.
[32] R.C. Reid, J.M. Prausnitz, T.K. Sherwood, The Properties of Gases and Liquids, 3rd edit. McGraw-Hill, New York (1977).
[33] D.R. Lide, Handbook of Chemistry and Physics, 71st edit. CRC Press, Boca Raton (1990).
[34] J.A. Riddick, W.B. Bunger, Organic Solvents. Wiley-Interscience, New York (1970).
[35] J. Israelachvili, H. Wennerström, Nature 379 (1996) 219.
[36] S. Leikin, V.A. Parsegian, D.C. Rau, R.P. Rand, Annual Review of Physical Chemical 44 (1993) 369.
[37] J.C. Hao, T. Zemb, Current Opinion in Colloid & Interface Science 12 (2007) 129.
[38] B. Todd, C. Stanley, N.Y. Sidorova, D.C. Rau, Hydration Forces: Water and Biomolecules. Wiley, New York (2007).
[39] D. Carrièrea, M. Page, M. Dubois, T. Zemb, H. Cölfen, A. Meister, L. Belloni, M. Schönhoff, H. Möhwald, Colloids and Surfaces A 303 (2007) 137.
[40] C. Bonnet-Gonnet, S. Leikin, S. Chi, D.C. Rau, V.A. Parsegian, Journal of Physics Chemistry B 105 (2001) 1877.
[41] B. Todd, V.A. Parsegian, A. Shirahata, T. Thomas, D.C. Rau, Biophysical Journal 94 (2008) 4775.
[42] D. Gruen, S. Marcelja, Journal of the Chemical Society Faraday Transactions 2 79 (1983) 211.
[43] D. Harries, J. Rosgen, Methods in Cell Biology 84 (2008) 679.
[44] H.K. Christenson, P.M. Claesson, Advances in Colloid Interface Science 91 (2001) 391.
[45] J. Israelachvili, R.M. Pashley, Nature 300 (1982) 341.
[46] J. Israelachvili, R.M. Pashley, Journal of Colloid and Interface Science 98 (1984) 500.
[47] P.M. Claesson, C.E. Blom, P.C. Herder, B.W. Ninham, Journal of Colloid and Interface Science 96 (2002) 1.
[48] P. Attard, Advances in Colloid Interface Science 104 (2003) 75.
[49] V. Bergeron, P.M. Claesson, Advances in Colloid Interface Science 96 (2002) 1.
[50] T. Onda, S. Shibuichi, N. Satoh, K. Tsujii, Langmuir 12 (1996) 2125.
[51] M. Miwa, A. Nakajima, A. Fujishima, K. Hashimoto, T. Watanabe, Langmuir 16 (2000) 5754.
[52] A.N.J. Shirtcliffe, G. McHale, M.I. Newton, C.C. Perry, Langmuir 19 (2003) 5623.
[53] D.O.H. Teare, C.G. Spanos, P. Ridley, E.J. Kinmond, V. Roucoules, J.P.S. Badyal, S.A. Brewer, S. Coulson, C. Willis, Chemistry of Material 14 (2002) 4566.
[54] J. Bico, C. Marzolin, D. Quéré, Europhysics Letters 47 (1999) 743.
[55] C. Extrand. In Self-Cleaning Surfaces: An Industrial Perspective, MRS Bulletin (2008).
[56] D. Quere, Physica A 313 (2002) 32.
[57] D. Quere, Reports on Progress Physics 68 (2005) 2495.
[58] R.N. Wenzel, Industrial & Engineering Chemistry 28 (1936) 988.
[59] A.B.D. Cassie, S. Baxter, Transactions of the Faraday Society 40 (1944) 546.
[60] L.C. Gao, T.J. McCarthy, Langmuir 23 (2007) 3762.
[61] W. Chen, A.Y. Fadeev, M.C. Hsieh, D. Öner, J. Youngblood, T.J. McCarthy, Langmuir 15 (1999) 3395.
[62] P.A. Small, Journal of Applied Chemistry 3 (1953) 71.
[63] A. Bondi, Journal of Physics Chemistry 68 (1964) 441.
[64] J.M. Prausnitz, R.N. Lichtenthaler, E.G. de Azevedo, Molecular Thermo Dynamics of Fluid-phase Equilibria, 2nd edit. Prentice-Hall, Englewood Cliffs (1986).

[65] O. Olabisi, L.M. Robeson, M.T. Shaw, Polymer-Polymer Miscibility. Academic Press, New York (1979).
[66] M.J. Kamlet, R.W. Taft, Journal of the American Chemical Society 98 (1976) 377.
[67] R.W. Taft, M.J. Kamlet, Journal of the American Chemical Society 98 (1976) 2886.
[68] G.N. Lewis, Valence and the Structure of Atoms and Molecules. Chemical Catalog Co, New York (1923).
[69] S.J. Park, Long-range Force Contributions to Surface Dynamics. In Interfacial Forces and Fields: Theory and Applications. Marcel Dekker, New York (1990).
[70] C. Gatti, P.J. MacDougall, R.F.W. Bader, Journal of Chemical Physics 88 (1988) 3792.
[71] D. R. Lide. In Handbook of Chemistry and Physics, 71st edit, CRC Press: Boca Raton, 9.1. (1990).
[72] G. Vidali, G. Ihm, H.Y. Kim, M.K. Cole, Surface Science Reports 12 (1991) 133.
[73] W.B. Jensen, The Lewis Acid-Base Concepts, pp 511. Wiley, New York (1980).
[74] J. Liebig, Annales Pharmaceutiques Francaise 26 (1838) 113.
[75] S. Arrhenius, Philosophical Magazine 41 (1896) 237.
[76] J.N. Brønsted, Rec. Trav. Chim. 42 (1923) 718.
[77] A.F.O. Germann, Journal of the American Chemical Society 47 (1925) 2461.
[78] W.L. Jolly, The Synthesis and Characterization of Inorganic Compounds. Prentice Hall, Englewood Cliffs (1970).
[79] R.E. Johnson, T.H. Norris, J.L. Huston, Journal of the American Chemical Society 73 (1951) 3052.
[80] G.N. Lewis, Valence and the Structure of Atoms and Molecules. Chemical Catalog Co, New York (1923).
[81] H. Lux, Electrochem, Z 45 (1939) 303.
[82] H. Flood, T. Förland, Acta Chemica Scandinavica 1 (1947) 592.
[83] W.B. Jensen, The Lewis Acid-Base Concepts. Wiley, New York (1980).
[84] M. Usanovich, Zh. Obshch. Khim 9 (1939) 182.
[85] C.K. Ingold, Chemical Reviews 15 (1934) 225.
[86] R. Robinson, Outline of an Electrochemical (Electronic) Theory of the Course of Organic Reactions. Institute of Chemistry, London (1932).
[87] C.M. Day, J. Selbin, Theoretical Inorganic Chemistry, 2nd edit. Reinhold, New York (1969).
[88] W.B. Jensen, Chemical Reviews 78 (1978) 1.
[89] R.G. Parr, R.G. Pearson, Journal of the American Chemical Society 105 (1983) 7512.
[90] R.G. Pearson, Journal of the American Chemical Society 107 (1985) 6801.
[91] R.G. Pearson, Journal of the American Chemical Society 108 (1986) 6109.
[92] R.G. Pearson, Inorganic Chemistry 27 (1988) 734.
[93] K. Fajans, Naturwissenschaften 11 (1923) 165.
[94] S. Ahrland, J. Chatt, N.R. Davies, Quarterly Reviews of the Chemical Society 12 (1958) 265.
[95] R.G. Pearson, Journal of Chemical Education 45 (1968) 581.
[96] R.G. Pearson, J. Songstad, Journal of the American Chemical Society 89 (1967) 1827.
[97] R.G. Pearson, Advances in Linear Free Energy Relationships (Chapman, N. B. & Shorter, J., eds.). Plenum Press, New York (1972).
[98] R.G. Pearson, Science 151 (1966) 172.
[99] R.G. Pearson, Chemistry in Britain 3 (1967) 103.
[100] R.G. Pearson, Surv. Prog. Chem. 5 (1969) 1.
[101] R.S. Drago, B.B. Wayland, Journal of the American Chemical Society 87 (1965) 3571.

[102] R.S. Drago, G.C. Vogel, T.E. Needham, Journal of the American Chemical Society 93 (1971) 6014.
[103] R.S. Drago, Structure Bonding 15 (1973) 73.
[104] R.S. Drago, L.B. Parr, C.S. Chamberlain, Journal of the American Chemical Society 99 (1977) 3202.
[105] R.S. Drago, Journal of the American Chemical Society 112 (1990) 8953.
[106] R.M. Keefer, L.J. Andrews, Journal of the American Chemical Society 77 (1955) 2164.
[107] R.S. Drago, Journal of Chemical Education 51 (1974) 300.
[108] R.G. Pearson, Journal of the American Chemical Society 85 (1963) 3533.
[109] R.G. Pearson, Hard and Soft Acids and Bases. Dowden, Stroudsburg, Pennsylvania (1973).
[110] V. Gutmann, A. Steininger, E. Wychera, Monatsh Chemical 97 (1966) 460.
[111] U. Mayer, V. Gutmann, W. Genger, Monatsh Chemical 106 (1975) 1235.
[112] F.M. Fowkes, J. Adhes, Sci. Tech 4 (1990) 669.
[113] V. Gutmann, Coordination Chemistry Reviews 18 (1976) 225.
[114] F.L. Riddl, F.M. Fowkes, Journal of the American Chemical Society 112 (1990) 3259.
[115] R.G. Pearson, Hard and Soft Acids and Bases. Dowden, Stroudsburg, Pennsylvania (1973).
[116] P. Debye, E. Hückel, Physalische Zeitschrift 24 (1923) 185.
[117] S. Levine, Proceedings of the Royal Society London A 170 (1939) 145.
[118] S. Levine, G.P. Dube, Transactions of the Faraday Society 35 (1940) 1125.
[119] B. Derjaguin, L. Landau, Acta Physicochimica. URSS 14 (1941) 633.
[120] E.J.W. Verwey, J.T.G. Overbeek, Theory of the Stability of Lyophobic Colloids. Elsevier, Amsterdam (1948).
[121] N. Ise, I.S. Sogami, Structure Formation in Solution: Ionic Polymers and Colloidal Particles. Springer, New York (2005).
[122] S. Bhattacharjee, M. Elimelech, M. Borkovec, Croatica Chemica Acta 71 (1998) 883.
[123] P.C. Hiemenz, Principles of Colloid and Surface Chemistry. Marcel Dekker, New York (1977).
[124] J. Lyklema, J.F.L. Duval, Advances in Colloid Interface Science 114 (2005) 27.
[125] H. Ohshima, Y. Inoko, T. Mitsui, Journal of Colloid and Interface Science 86 (1982) 57.
[126] H. Oshima, Colloid Stability-The Role of Surface Forces I, Tadros. Wiley-VCH, Weinheim (2007).
[127] A.C. Pierre, Introduction to Sol-Gel Processing. Kluwer, Norwell, Miami (1998).
[128] G. D. Parfitt. In Dispersion of Powders in Liquids with Special Reference to Pigments, Appl. Sci., London, 1. (1981).
[129] G. Malescio, Nature Materials 2 (2003) 501.
[130] J.T.G. Overbeek, Journal of Colloid and Interface Science 58 (1977) 408.
[131] E.J.W. Verwey, J.T.G. Overbeek, Theory of the Stability of Lyophobic Colloids. Elsevier, Amsterdam (1948).
[132] E.J.W. Verwey, J.T.G. Overbeek, Journal of the Chemical Society Faraday Transactions 1 42B (1946) 117.
[133] W.B. Russel, D.A. Saville, W.R. Schowalter, Colloid Dispersions. Cambridge University Press (1989).
[134] R. Cohen, G. Ozdemir, D. Exerowa, Colloids and Surfaces B 29 (2003) 197.
[135] C. Stubenrauch, R. Strey, Journal of Physics Chemistry B 109 (2005) 19798.
[136] B. Diakova, C. Filiatre, D. Platikanov, A. Foissy, M. Kaisheva, Advances in Colloid Interface Science 96 (2002) 193.
[137] M. Giesbers, J.M. Kleijn, M.A. Cohen Stuart, Journal of Colloid and Interface Science 252 (2002) 138.

[138] M.M. Kohonen, H.K. Christenson, The European Physical Journal E 6 (2001) 315.
[139] A. Mukhopadhyay, B.M. Law, Physical Review Letters 83 (1999) 772.
[140] P.M. Claesson, M. kjellin, K.J. Rojas, C. Stubenrauch, Physical Chemistry Chemical Physics 8 (2006) 5501.
[141] A.W. Adamson, A.P. Gast, Physical Chemistry of Surfaces, 6th edit, pp 232. John-Wiley & Sons, New York (1997).
[142] J.T.G. Overbeek, Pure and Applied Chemistry 52 (1980) 1151.
[143] A.W. Adamson, A.P. Gast, Physical Chemistry of Surfaces, 6th edit, pp 182. John-Wiley & Sons, New York (1997).
[144] R.H. Ottewill, A. Watanabe, Kolloid-Z 173 (1960) 7.
[145] G.D. Parfitt, N.H. Picton, Journal of the Chemical Society Faraday Transactions 1 64 (1968) 1955.
[146] C.G. Force, E. Matijević, Kolloid-Z 224 (1968) 51.
[147] S.S. Dukhin, J. Lyklema, Faraday Discussions of the Chemical Society 90 (1990) 261.
[148] J. Kijlstra, H. van Leeuwen, Journal of Colloid and Interface Science 160 (1993) 424.
[149] S.Y. Shulepov, S.S. Dukhin, J. Lyklema, Journal of Colloid and Interface Science 171 (1995) 340.
[150] H. Kihira, N. Ryde, E. Matijević, Journal of the Chemical Society Faraday Transactions 1 88 (1992) 2379.
[151] H. Kihira, E. Matijević, Langmuir 8 (1992) 2855.
[152] J.A. Maroto, F.J. de las Nieves, Progress in Colloid and Polymer Science 98 (1995) 89.
[153] J.D.G. Durán, M.C. Guinde, A.V. Delgado, F. González-Caballero, Langmuir 11 (1995) 3648.
[154] C.J. van Oss, Colloids Surfaces A 78 (1993) 1.
[155] J.W.T. Lichtenbelt, C. Pathmamanoharan, P.H. Wiersma, Journal of Colloid and Interface Science 49 (1974) 281.
[156] J.N. Israelachvili, B.M. Ninham, Journal of Colloid and Interface Science 58 (1977) 14.
[157] J.N. Israelachvili, G.E. Adams, Nature 262 (1976) 772.
[158] R.M. Pashley, Journal of Colloid and Interface Science 83 (1981) 531.
[159] J. Marra, Journal of Physics Chemistry 90 (1986) 2145.
[160] J. Marra, J.N. Israelacchvili, Biochemistry 24 (1985) 4608.
[161] O. Stern, Z. Elektrochem 30 (1924) 508.
[162] J.N. Israelachvili, Intermolecular and Surface Forces. Pp 79. Academic, San Diego (1992).
[163] G. Gouy, Journal of Physics 9 (1910) 457.
[164] G. Gouy, Annals of Physics 7 (1917) 129.
[165] R.M. Pashley, M.E. Karaman, Applied Colloid and Surface Chemistry, pp 102. John-Wiley & Sons (2004).
[166] M. Smoluchowski, Annals of Physics 326 (1906) 756.

Chapter 2

Solid-Gas Interaction

2.1. INTRODUCTION

Solid interface science is the study of physical and chemical phenomena that occur at the interface of two phases, including the solid-gas interfaces, solid-liquid interfaces (Chapter 3), and solid-solid interfaces (Chapter 4). This chapter deals with various aspects of a very important type of situation, in which some technique or theory is distributed between a solid phase and a gaseous one. The importance of solid-gas interface in science has promoted great interest in areas such as adsorbents, catalysis, magnetism, microelectronics, optoelectronics, pharmacy, and polymer composite materials. In particular, the final mechanical characteristics and the stability of composite materials depend largely on the degree of interaction, achieved at interfaces between a matrix and the reinforcing materials in the practical application of composites. The interaction of reinforcing materials and the matrix can be determined, by the adsorption of the gas used on the solid surface under study.

Generally, gases below their critical temperature, tend to adsorb with a solid surface as a result of general Vander Waals interactions. The molecules that disappear from the gas either enter the inside of the solid, or remain on the outside attached to the surface. The former phenomenon is termed absorption (or dissolution) and the latter adsorption. The solid that takes up gas, is called the adsorbent, and the gas or vapor taken up on the surface, is called the adsorbate. It is not always easy to tell whether the gas is inside the solid, or merely at the surface because most of the practical absorbents are very porous bodies, with large internal surfaces. In this case of physical adsorption, as it is called interest centers on the size and nature of the adsorbent-adsorbate interactions and on those between adsorbate molecules. The attractive force is due to correlated charge fluctuations, in the two bonding partners, i.e., between mutually induced dipole moments. In molecular physics, where these dipoles can be considered as 'point' dipoles, the attractive potential is that between the attracting dipoles.

The adsorption on the solid-gas interfaces has been studied theoretically for most of this century and the simplest of the resulting theories provide the insight

Interface Science and Composites, Volume 18
ISSN: 1573-4285, doi: 10.1016/B978-0-12-375049-5.00002-5

needed for most of the applications. The last few decades have seen a very rapid extension of our ability to explore the composition and structure of solid surfaces in the presence of gases, and the interaction of solid-gas has been long treated by many scientists (Langmuir, Freundlich, Fowler, Brunauar, etc.). [1–8]. They have reported most of the adsorption equations, such as the Langmuir [1, 2], the Van Bemmelen-Freundlich [3, 4], the Fowler-Guggenheim [5], and the Brunauer–Emmett–Teller (BET) [6] models, while others are derived from a solid-gas adsorption hypothesis. These adsorption equations are widely used for determining the surface area and pore size distribution, of a variety of different solid materials by gas adsorption measurements. This section will investigate two adsorption equations, one attributed to the Langmuir's equation, and one to BET's equation, and applies their equations to experimental data. The measurement of adsorption at the solid-gas interfaces also forms an essential part of many fundamental and applied investigations on the nature and behavior of solid surfaces [9, 10]. In addition, this section focuses on the introduction of the reversible molecular force of a solid surface measured by the solid-gas chromatographic technique, which is one of the adsorption analyses based on the deformation polarizability, and the dipole moment of molecules. Two methods, which were discussed in the chromatographic study in [10], the inverse gas chromatography at infinite dilution and at finite concentration, are largely considered in this chapter.

2.2. STRUCTURE AND CHEMICAL COMPOSITION OF SOLID SURFACES

The structure and surface chemistry of the solid as a whole is crucially dependent upon the nature and type of surfaces exposed on the crystallites. We have described briefly the importance of a solid adsorbent, its specific surface area, porosity, and pore size distribution in the 2.1. In principle, therefore, we can understand the surface properties of any solid if we know the amount of each type of surface exposed and have detailed knowledge of the properties of each and every type of surface plane.

A chemically pure, molecularly smooth, uniformally arranged planar solid surface is seldom, if ever, found in nature and is difficult to realize under laboratory conditions except over tiny areas. The first problem is how to describe the real structure, and the second, even more important is how to determine it. Fig. 2-1 shows some of the principal types of departure from crystallographic ideality that affects the properties of the surface. Most of the surface atoms do lie on smooth terraces, but these are not of a large extent and are separated from each other by steps, generally one atom high but not necessarily so limited. Isolated atoms (adatoms) may sit atop terraces or against steps. Vacancies (missing atoms) may occur in comparable locations. The riser of a step will have kinks, slightly different from vacancies. The influence of these imperfections stems from the fact, that the atoms in various locations will have

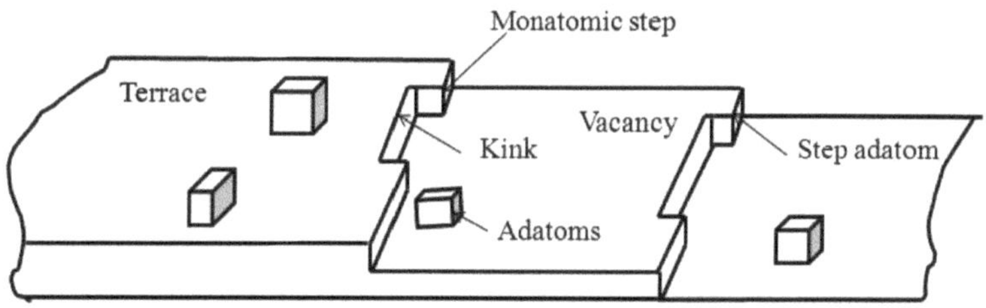

Fig. 2-1 Schematic representation of the types of heterogeneities found on an atomic scale.

different numbers of the nearest regularity, overlying crystals with assorted defects (vacancies, interstitial atoms, dislocations, etc.), and will be expected to have different surface energies [11].

In addition, it is important to know as much as possible about the microscopic structure of the surface and surface spectroscopic and diffraction techniques. The most commonly employed technique is to prepare macroscopic (i.e., size ~ cm) single crystals of metals, and then to deliberately cut them in a way that exposes a large area of the specific surface of interest. Scanning electron microscope (SEM), transmission electron microscopy (TEM), scanning tunneling microscopy (STM), FT-Raman spectra, and atomic force microscopy (AFM), are now widely used to obtain the structure of surfaces. The composition of the surface atoms or molecules is very important, especially in the field of adsorbents and catalysis, where mixed surface compositions are common. The determination of the valence state of surface atoms is important, such as by X-ray photoelectron spectroscopy (XPS), energy-dispersive X-ray spectroscopy (EDS or EDX), X-ray diffraction (XRD), and X-ray absorption fine structure (EXAFS), for chemical analysis.

Most solids have foreign atoms or molecular groupings on their surfaces, that are so tightly held that they do not really enter into adsorption-desorption equilibrium and so can be regarded as part of the surface structure. Generally speaking, the adsorption capacity of adsorbents depends on the pore structure and nature of the solid. Various physico-chemical modifications of solid have been developed and successfully used for the adsorption of gases. For example, the surface modification of the carbon materials has been mentioned as having an important effect on their adsorptive behaviors; depending on conditions, the modified surface may be acidic or basic, and the structure of carbon materials may be affected.

As one example, electrochemical treatment is a well known method of introducing surface functional groups containing oxygen onto carbon surfaces. The effects of different surface functional groups on the surface structure of modified porous carbons are investigated in acidic solution under different current densities [12]. As shown in Fig. 2-2, in the XPS results, the oxygen content of the carbons increased with an increasing current density. It is seen that

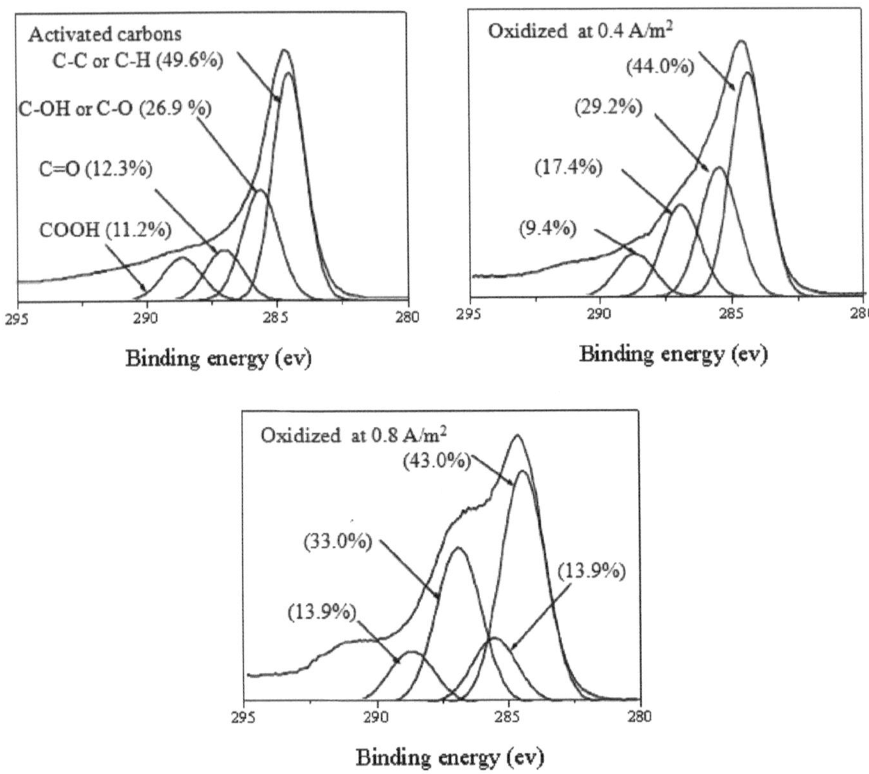

Fig. 2-2 High-resolution C1s peaks of electrochemically oxidized activated carbons as a function of current density.

the carbon surfaces were composed of various surface functional groups, including phenolic hydroxyl groups (C–OH or –C–O–), carbonyl groups (C=O), or carboxylic acid groups (–COOH). Usually, C–OH or –C–O– and –COOH groups, have an especially strong acidity, whereas the C=O group has a weak acidity (or is neutral), in the aspect of its acid–base characteristics. From this point of view, the growth of the C–OH or –C–O– groups and the enhancement of the C=O group growth at 0.8 A/m^2 of current density probably results in the high adsorption characteristics of the carbon surfaces in this electrochemical oxidation system.

Like other oxidation treatments, the electrochemical treatment can endow carbons with a number of polar functional groups but at the same time cause the loss of some of their advantages, such as the specific surface area or total pore volume. Table 2-1 shows that the specific surface area, total pore volume, and micro pore volume of oxidized activated carbons, decreases with an increasing current density [12].

Table 2-1 Textural Properties of the Activated Carbons as a Function of Current Density

Samples	S_{BET}[a]	V_T[b]	V_M[c]	F_M[d]
Activated carbons	1195	0.51	0.49	96.1
Oxidized at 0.4 A/m^2	955	0.41	0.37	90.2
Oxidized at 0.8 A/m^2	925	0.39	0.35	89.7

[a] *Specific surface area (m^2/g).*
[b] *Total pore volume (cm^3/g).*
[c] *Micropore volume (cm^3/g).*
[d] *Fraction of micropore (%).*

To study the influence of surface treatment on the polar gas adsorption of the activated carbons, an ammonia gas removal test was previously reported [13]. Activated carbons were modified by ozone treatment, to enhance the efficiency of the removal of ammonia gas, over the activated carbons. The variation of ammonia adsorption versus the reaction time of each specimen is shown in Fig. 2-3 [13]. It shows that the ozone-treated activated carbons reveal much higher removal efficiency than the pristine activated carbons. This is due to the acid-base interaction of each functional group containing oxygen on the surfaces of activated carbons, which plays the key role in the determination of the gas adsorption capacity.

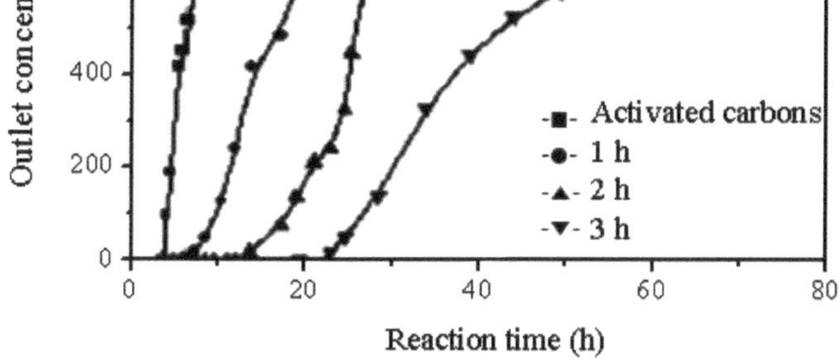

Fig. 2-3 Ammonia removal efficiency of activated carbons as a function of ozone treatment time.

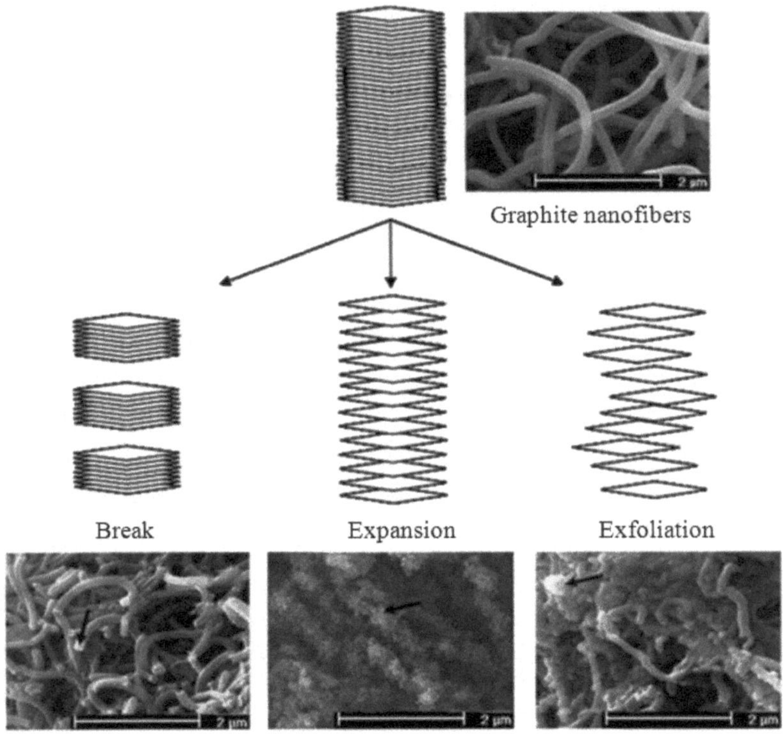

Fig. 2-4 Schematic diagrams of mechanisms of pore development of porous graphite nanofibers.

Another experimental study prepared porous graphite nanofibers *via* KOH activation at different temperatures [14]. The information on the mechanisms of pore development of graphite nanofibers is shown in Fig. 2-4. According to the types of pore development, there can be three mechanisms: breaking, expansion, and exfoliation. Breaking means just breaking the fibers and forming shorter fibers. Expansion indicates the opening of the pores so this mechanism does not cause the demolition of fibrous shapes of graphite nanofibers. Exfoliation means the co-reaction of both the separation of layers and the breaking of the fibers, resulting in the collapse of the fiber shapes.

The textural properties of the porous carbons were analyzed, using N_2 adsorption isotherms at 77 K, and shown in Table 2-2 [14]. As can be seen, the textural properties, including the specific surface area and the pore volumes of graphite nanofibers are proportionally enhanced with an increasing activation temperature. The pore structure of porous graphite nanofibers is enhanced by a chemical activation process that involved increasing the activation temperature at a fixed KOH ratio and the pore development behaviors being determined by the expansion of graphite layers up to 1000 °C.

Table 2-2 Textural Properties of Porous Graphite Nanofibers with Various Activation Temperatures

Samples	S_{BET}[a]	V_T[b]	V_M[c]	F_M[d]
Graphite nanofibers	320	0.44	0.31	28.5
Activated at 900 °C	500	0.46	0.28	39.0
Activated at 1000 °C	580	0.52	0.31	40.5
Activated at 1100 °C	790	0.83	0.54	34.6

[a] *Specific surface area (m^2/g).*
[b] *Total pore volume (cm^3/g).*
[c] *Micropore volume (cm^3/g).*
[d] *Fraction of micropore (%).*

Carbon nanotubes, as a gas adsorbent have become one of the most interesting materials due to their novel structural characteristics, such as their nanometer hollow tubes and high surface area with a narrow pore distribution size. The surface structure and chemical nature of carbon nanotubes has been a subject of much discussion [15].

As another example, the surface structure of the activated multi-walled carbon nanotubes, with the well-developed pore structures and a highly specific surface area after CO_2 activation at different temperatures is discussed [15]. Fig. 2-5 shows the XRD patterns of the multi-walled carbon nanotubes at an activation temperature in the range of 500-1100 °C. The decrease in the (002) Bragg peak intensity of XRD results is due to an increase in the irregularity of the layer structures, while a widening of the peak range results in the collapse of the layer structures [15].

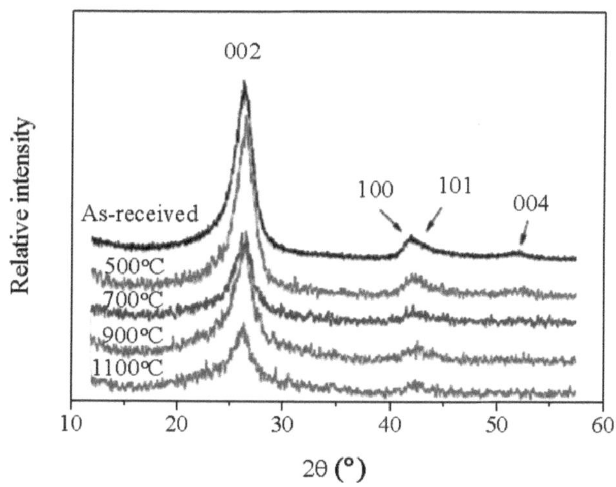

Fig. 2-5 XRD patterns of the activated carbon nanotubes at different activation temperatures.

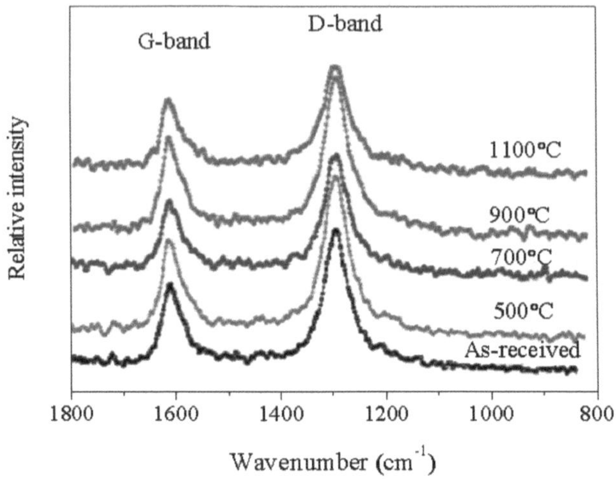

Fig. 2-6 FT-Raman spectra of the activated multi-walled carbon nanotubes at different activation temperatures.

The FT-Raman spectrum is as a very valuable tool in characterizing carbon-based nanostructures. Each activated carbon nanotube consists of two characteristic bands, namely the D-band at ~ 1280 cm^{-1} and the G-band at ~ 1600 cm^{-1}, as shown in Fig. 2-6 [15]. Note the frequencies at around 1600 cm^{-1}, which corresponded to the tangential C–C stretching vibration known as a graphite mode G-band. At around 1280 cm^{-1}, due to the local defects that originated from structural imperfections in the multi-walled carbon nanotubes it is named the defect mode D-band.

Measuring the ratios of the intensities of the D- and G-bands can reveal an idea of the disorder of the carbon nanotubes. As shown in Fig. 2-7, the ratio of I_D/I_G in 1100 °C sharply decreases which might be attributed to the fact that the disordered structure of the external surface of carbon nanotubes decomposes at a higher temperature [15]. The figure also shows the introduction of oxygen complexes on the carbon nanotubes surfaces during the CO_2 activation process. Fig. 2-7 also shows the surface pH of the activated multi-walled carbon nanotubes. The pH of the pristine activated multi-walled carbon nanotubes is around 7.5, which decreases with an increasing activation temperature. This is due to the introduction of oxygen complexes on the multi-walled carbon nanotubes surfaces during the CO_2 activation process [15].

Recently, one-atom-thick, two-dimensional materials, graphene nanosheets, have shown a high specific surface area. Yang *et al.* [16] reported that a preheated high-temperature environment is believed to be critical, for a chemical-exfoliation based production of graphene starting from graphite oxide, and the result is seen in Fig. 2-8.

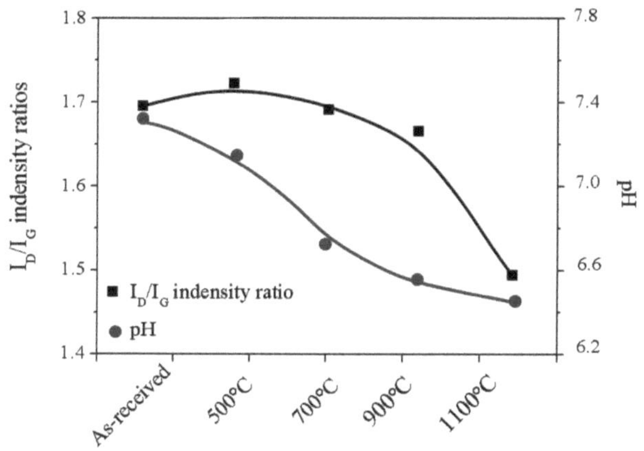

Fig. 2-7 I_D/I_G intensity ratios and pH of the activated carbon nanotubes at different activation temperatures.

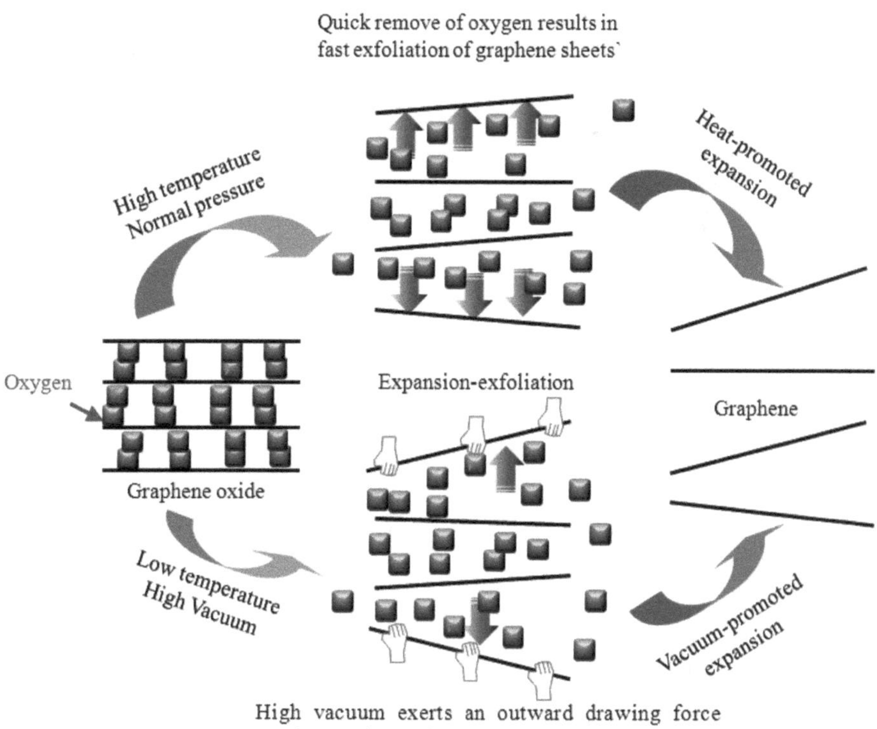

Fig. 2-8 Schematic representation of chemical exfoliation of graphenes. (top) High-temperature (above 1000 °C) exfoliation, under an atmospheric pressure; (bottom) low-temperature (as low at 200°C) exfoliation under high vacuum, where high vacuum introduces a negative pressure surrounding the graphene layers.

The high vacuum, accompanying the flowing out of oxygen at a low temperature, exerts an outward drawing force on the expanding graphene layers, which helps accelerate the expansion of graphene layers and results in an effective exfoliation of graphene layers (as schematically represented in Fig. 2-8) [16]. A SEM image of the low-temperature exfoliated graphene nanosheets (Fig. 2-9(a)), shows, that these layers are exfoliated to a very large extent. The specific surface area of the exfoliated graphene nanosheets is 382 m^2/g (Fig. 2-9(b)) [16].

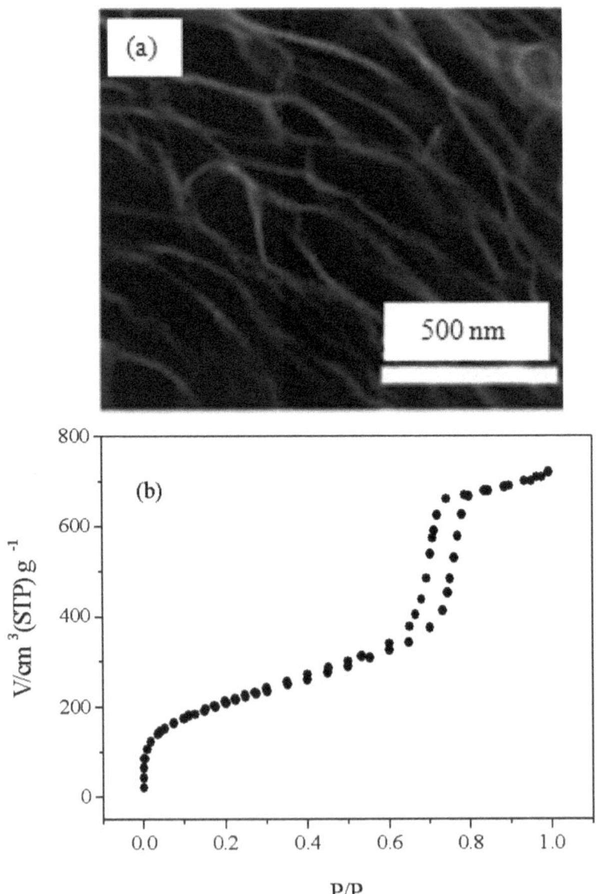

Fig. 2-9 SEM image (a) N_2/77 K adsorption isotherms and (b) of the prepared graphene nanosheets.

Zeolites, a class of porous crystalline aluminosilicates built of a periodic array of TO_4 tetrahedral (T = Si or Al), have been widely used in separation applications mainly because of their unique ability of molecular sieving [17–19]. The presence of aluminum atoms in these silicate-based molecular sieve materials, introduces negative framework charges that are compensated with exchangeable cations in the pore space (often alkali cations), and these structural characteristics of zeolites enable them to adsorb a wide variety of gas molecules, including acidic gas molecules, such as CO_2, SO_2, etc. For example, the CO_2 adsorption properties of zeolites are also influenced by the porous characteristics of the framework. Aguilar-Armenta *et al.* reported adsorption isotherms of three natural zeolites including erionite, mordenite, and clinoptilolite measured at 290 K [18]. The isotherms show variations of adsorption capacities according to the zeolite type, ranging from *ca.* 1.6 to 2.7 mmol/g at a pressure of 0.1 bar, as shown in Fig. 2-10 [18]. In addition, Tezel *et al.* carried out, an experimental screening study of various synthetic zeolite adsorbents, including 13X (SiO_2/Al_2O_3 ratio: 2.6–3.0), NaY (a commercial NaY zeolite with a SiO_2/Al_2O_3 ratio: > 20), HY-5 (SiO_2/Al_2O_3 ratio: 4.8 or 5.6), ZSM-5 (a commercial zeolite, based on the ZSM-5 structure with a large SiO_2/Al_2O_3 ratio: > 1000) [19]. The pure component CO_2 adsorption isotherms of these zeolites displayed a wide range of adsorption capacities from 1.2 mmol/g (HY-5) to 4.5 mmol/g (13X) at 295 K and 1 bar [19]. This is attributed to the different Si/Al ratios of the zeolites as well as the amount and type of charge-balancing cations.

As above experimental results, most techniques have been used for surface structural analysis as evidenced by several convincing studies of solid surfaces

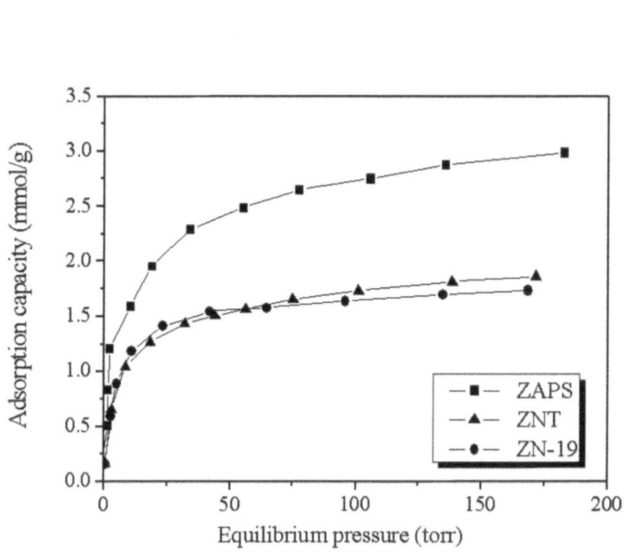

Fig. 2-10 Adsorption isotherms of CO_2 and CH_4 at 290 K on natural zeolites, including erionite (ZAPS), mordenite (ZNT), and clinoptilolite (ZN-19).

and chemical nature. And, the surface treatments influence the solid-gas interaction.

2.3. ADSORPTION ISOTHERMS

When a gas comes into contact with a solid surface, molecules of the gas will adsorb to the surface in quantities that are a function of their partial pressure in the solid. The measurement of the amount of gas adsorbed over a range of partial pressures at a single temperature, results in a graph known as an adsorption isotherm.

The amount of gas adsorbed per unit weight of adsorbent is a function of both pressure and temperature, besides, of course, the nature of the gas. The data may be represented as isotherms, V (P), at constant T, isobars, V (T) at constant P, or isosteres, P (T) at constant V. Isotherms are the closest to direct experiment. Isosterms are largely employed in calculating heats of adsorption from Eq. (2-1) [20]:

$$\ln\frac{P_2}{P_1} = \frac{\Delta H_{vap}(T_2 - T_1)}{RT_1T_2} \quad (2\text{-}1)$$

where ΔH_{vap} is the enthalpy of vaporization (or heat of vaporization).

It is often measured at the normal boiling point of a substance; although tabulated values are usually corrected to 298 K, the correction is often smaller than the uncertainty in the measured value. The heat of vaporization is temperature-dependent, though a constant heat of vaporization can be assumed for small temperature ranges and below the critical temperature ($T_r < 1.0$). The heat of vaporization diminishes with increasing temperature and it vanishes completely at $T_r < 1.0$, because above the critical temperature the liquid and vapor phases no longer co-exist [20].

2.3.1. IUPAC Classification of Adsorption Isotherms

Many different types of isotherms have been observed in the literature [6, 21–26]. These isotherms can have very different shapes depending on the type of adsorbent, the type of adsorbate, and intermolecular interactions between the gas and the surface of solid. The first systematic attempt to interpret adsorption isotherms for solid-gas equilibrium was introduced by Brunauer, Deming, Deming, and Teller (BDDT) in 1940 [21]. These authors classified isotherms into five types.

The BDDT classification has become the core of the modern IUPAC classification of adsorption isotherms; these BDDT isotherms and an additional one introduced much later by Sing [23], which completes the IUPAC classification [25, 26], are illustrated in Fig. 2-11.

In physical adsorption, adsorption isotherms can be classified as one of six types, as shown in Table 2-3. Adsorption occurs because it results naturally

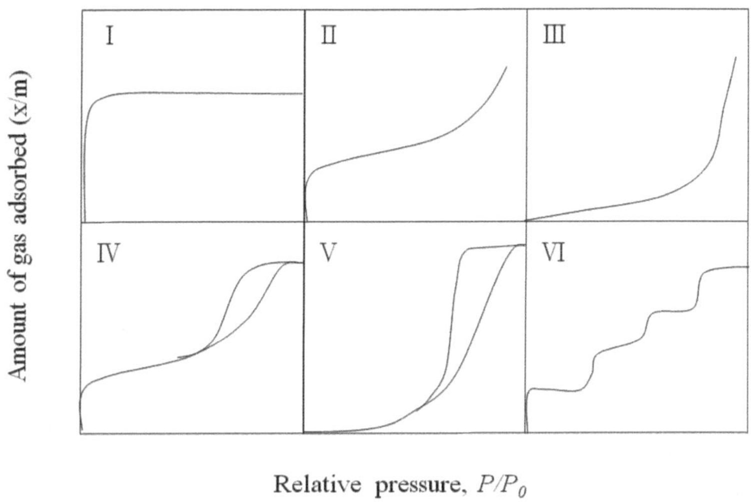

Fig. 2-11 IUPAC classification of adsorption isotherms.

Table 2-3 Features of Adsorption Isotherms

Type	Features: Interaction between sample surface and adsorbate	Features: Porosity	Sample-adsorptive example
I	Relatively strong	Micropores (<2 nm)	Activated carbon-N_2
II	Relatively strong	Nonporous	Oxide-N_2
III	Weak	Nonporous	Carbon-water vapor
IV	Relatively strong	Mesopore (2~50 nm)	Silica-N_2
V	Weak	Mesopore Micropore	Activated carbon-water vapor
VI	Relatively strong sample surface has an even distribution of energy	Nonporous	Graphite-Krypton

in a lower free energy for the system, Type I isotherms characterize microporous adsorbents. Types II and III describe adsorption on macroporous adsorbents, with strong and weak adsorbate-adsorbent interactions, respectively. Types IV and V represent adsorption isotherms with hysteresis. Finally, Type VI has steps. The BDDT and IUPAC classifications have two deficiencies: they are incomplete and they give the incorrect impression that adsorption isotherms are always monotonic functions of pressure. This is because the IUPAC classification only takes into account, adsorption at subcritical temperatures.

Until now, we have mainly treated adsorption onto non-porous surfaces. In reality, most of the industrial, and many natural materials, are porous: textiles, paper, bricks, sand, porous rocks, food products, zeolites, etc. We start our discussion with a classification of pores, according to their size, which is recommended by IUPAC:

1. Macropores have a diameter, larger than 50 nm. Macropores are so wide, that gases adsorb virtually to flat surfaces.
2. Mesopores are in the range of 2 $\sim$ 50 nm. Capillary condensation often dominates the filling of mesopores. Below the critical temperature, multi-layers arise. Pores, on one hand, limit the number of layers, but on the other hand, capillary condensation can occur.
3. Micropores are smaller than 2 nm. In micropores the structure of the adsorbed fluid is significantly different from its macroscopic bulk structure. Confined liquids are a highly active area of research because of their unique properties. An important example of microporous materials are zeolites, which are used for catalysis.

Type I isotherms are obtained for the micro-porous adsorbents such as active carbons or zeolites, for examples, the N_2 isotherms, for the untreated and the plasma-treated activated carbon fibers [27]. To study the changes of surface morphologies of the untreated and the plasma-treated activated carbon fibers, SEM images are used, at a resolution of 50,000, as shown in Fig. 2-12(a and b). It is observed, that many macropores appeared after the oxygen plasma by the destruction of the carbon surfaces, and this can lead to a decrease of the specific surface area and total pore volume. As shown in Fig. 2-12(c), the pore structure is approximately Type I, according to the IUPAC classification. It is evident that most of the pore volume of the samples is filled below a relative pressure of about 0.1, indicating that these samples are highly microporous. After a sharp increase up to relative pressure 0.1, the isotherms show very small increases in further adsorption. It also shows the specific surface area and total pore volume decrease somewhat with plasma treatment.

Examples of Type II adsorption isotherms are of adsorption of N_2, on multi-walled carbon nanotubes and seen in Fig. 2-13 [28]. Fig. 2-13(a) shows TEM images of Pt-loaded activated multi-walled carbon nanotubes after the CO_2 activation for 20 min. It shows the average diameter of multi-walled carbon nanotubes being around 25 nm. Moreover, the Pt particles are uniformity dispersed on the multi-walled carbon nanotubes. Fig. 2-13(b) shows the N_2 isotherm of the Pt-loaded activated multi-walled carbon nanotubes as a function of the CO_2 flow time. It shows the evolution of all the samples on adsorption isotherms, P/P_0 increases can be described as a Type II isotherm,

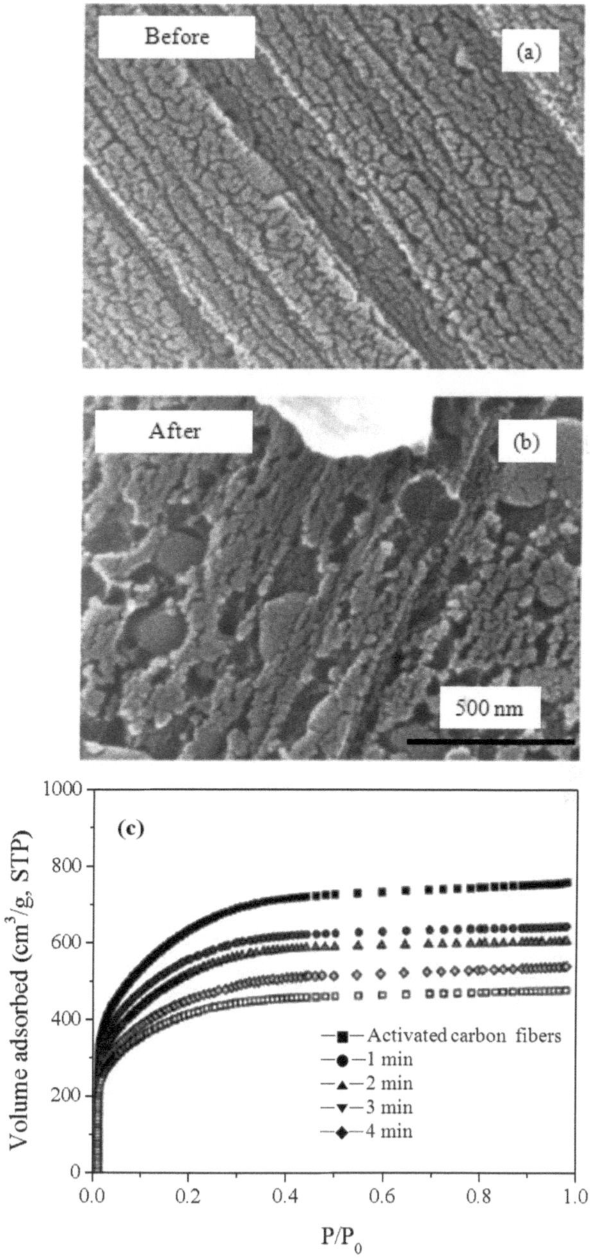

Fig. 2-12 SEM photographs (a and b), and N_2/77 K adsorption isotherms (c), of the activated carbon fibers, before and after oxygen plasma treatments.

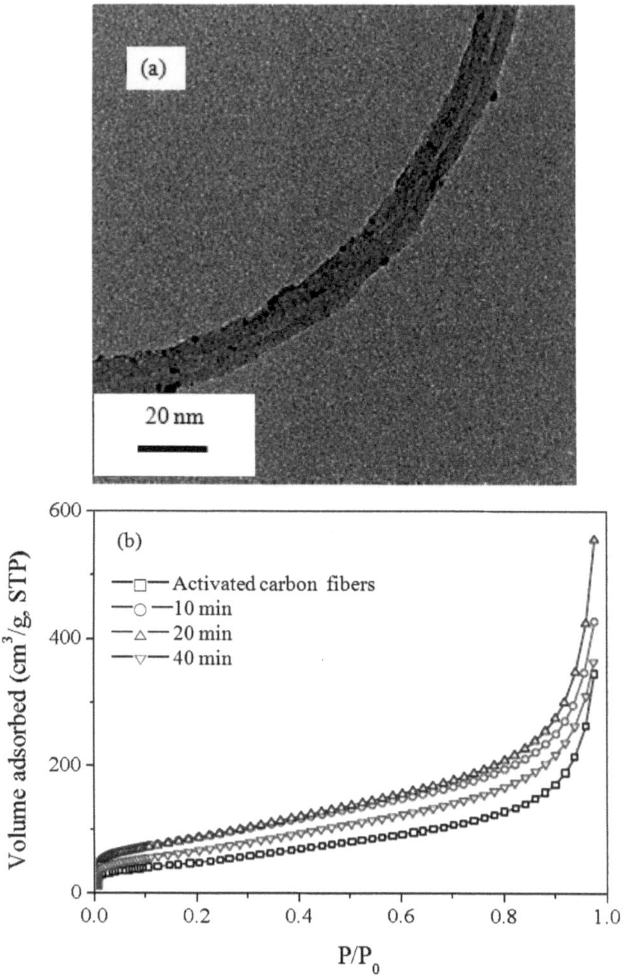

Fig. 2-13 TEM image (a) and N_2/77 K adsorption isotherms (b), of the Pt-loaded activated carbon nanotubes, as a function of CO_2 low time.

which is often associated with micropores (diameter: < 2 nm), and mesopores (diameter: 2–100 nm), based on the classification recommended by IUPAC [25, 26]. A higher activation temperature often leads to a larger specific surface area, and thus, contributes to a larger nitrogen adsorption capacity.

Examples of Types III and V adsorption isotherms are of adsorption of N_2 at 77 K on a nanoporous silicon carbide/nano-sized fiber complex [29]. The nanoporous silicon carbide/nano-sized fiber complex is made from precursors of various ratios, with SiO_2 as silicon source and polyacrylonitrile as the carbon source, by a simple electrospinning method, and economical carbothermal reduction. The SEM image of the stabilized electrospun polyacrylonitrile-SiO_2 complex (C/Si = 5/1 and 7/1 mol ratio) is shown representatively in Fig. 2-14 (a and b). The diameters of the stabilized electrospun complex fibers decreased with increasing C/Si mol ratios of the precursor solution in a degree. In particular, C/Si = 7/1 has a porous appearance not observed in the stabilized electrospun polyacrylonitrile-SiO_2 complex. It is believed that it is generated by vaporization of carbonaceous gas made from carbon and oxygen. The adsorption–desorption isotherms show as Type III adsorption isotherms when the mol ratio of carbon source/silicon source is 3/1, 4/1, and 5/1 (Fig. 2-14(a)). However, adsorption–desorption isotherms show Type V adsorption isotherms is 6/1 and 7/1 when the mol ratio of carbon source/silicon source is 6/1 and 7/1 (Fig. 2-14(b)). Generally, Type III adsorption isotherms are characteristic of weak gas-solid (adsorbate-adsorbent) interactions by non-porous or macroporous solids, and Type V adsorption isotherms are characteristic of weak gas-solid (adsorbate-adsorbent) interactions by mesoporous or microporous solids (Fig. 2-14 (c)).

Examples of Type IV adsorption isotherms are of adsorption of $N_2$77 K on mesoporous silica [30]. Among the best known mesoporous solids, MCM-41 (Mobile Crystalline Materials) is widely known as a hexagonal mesoporous material that has various applications in the fields of catalysis chemistry, adsorption, materials sciences, and the petrochemical industry, due to the large specific surface area and narrow and controllable pore size distributions. The TEM image shows a well-ordered hexagonal pore structure, as shown in Fig. 2-15(a) [30]. N_2 adsorption-desorption isotherms of Ni doped MCM-41 are typical Type IV isotherms with hysteresis loops caused by capillary condensation in the mesopores, which means that all samples have mesoporous structures, as shown in Fig. 2-15(b).

As one example, the adsorption isotherm for N_2/46.2 K to the basal plane of graphite is shown in Fig. 2-16 [31]. The isotherms show a sequence of several monolayer steps and approach a vertical line asymptotically at a pressure interpreted as P_0. It is a typical Type VI isotherm based on the classification recommended by IUPAC [25, 26].

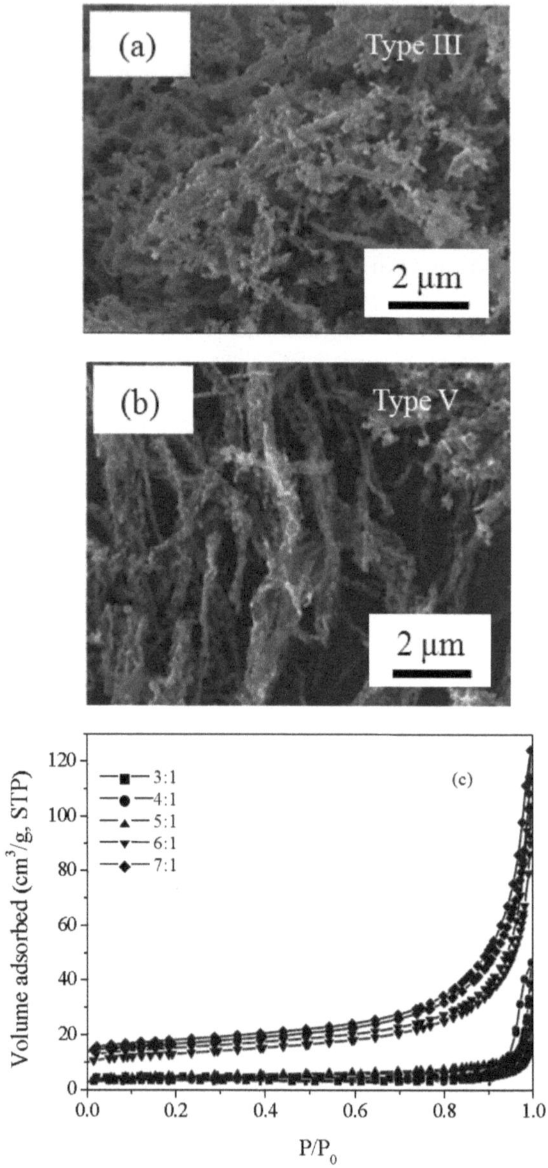

Fig. 2-14 SEM image ((a) = 5:1 and (b) = 7:1) and N_2/77 K adsorption isotherms (c), of the nanoporous silicon carbide nano-sized fiber complexes, as a function of polyacrylonitrile/SiO_2 mol ratio.

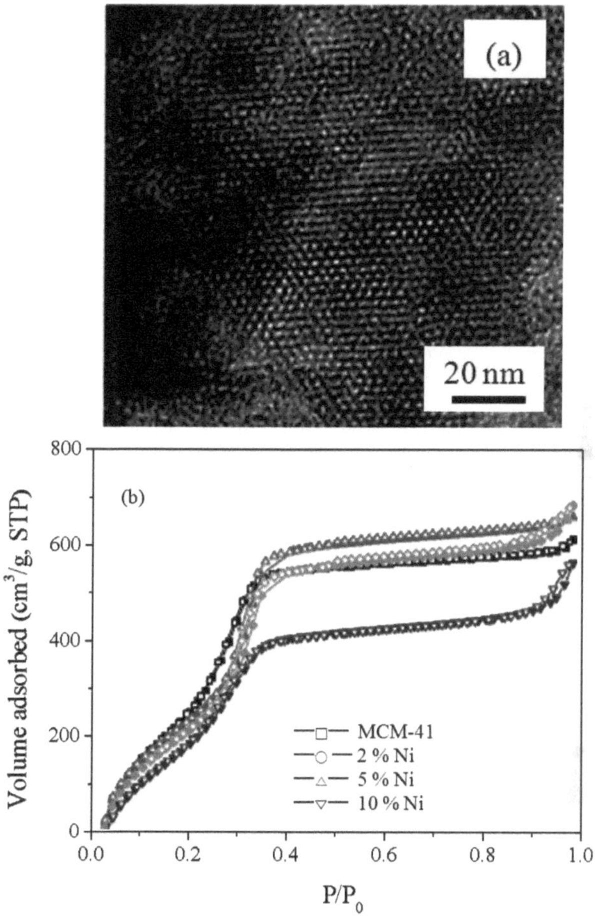

Fig. 2-15 TEM image (a) and N_2/77 K adsorption isotherms (b) of Ni doped MCM-41 as a function of Ni contents.

2.3.2. Langmuir Isotherm

In 1916, Langmuir proposed the monolayer adsorption theory [1]. The equation of the adsorption isotherm of the following assumption is noted.

1. Adsorption sites on adsorbent surface have the same adsorption energy.
2. One adsorption site is occupied by one adsorbed molecule.
3. There is no lateral interaction between adsorbed molecules.
4. Adsorption is complete when a mono-molecular layer or monolayer is formed.

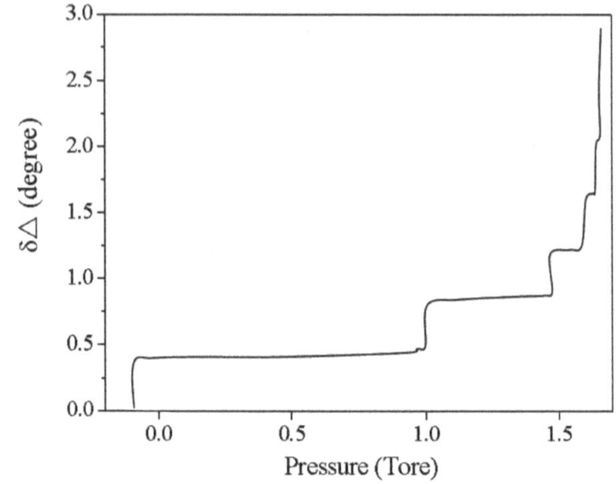

Fig. 2-16 Adsorption of N_2 to a single basal plane of graphite at a temperature of 46.2 K as determined by ellipsometry. Plotted is the change in the ellipsometric angle Δ *vs.* pressure. The subsequent adsorption of at least four layers, at defined pressures, can clearly be distinguished.

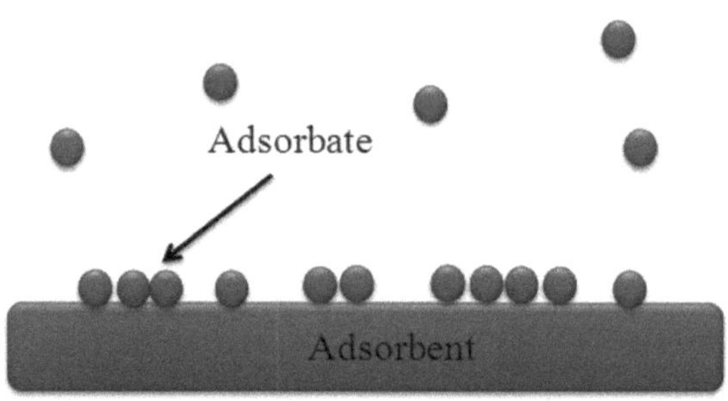

Fig. 2-17 Schematic drawing of the Langmuir adsorption model.

The typical Langmuir adsorption model assumed that on the surface there are a certain number of binding sites per unit area *S* (Fig. 2-17).

An isotherm for Type I data was obtained by Langmuir, in 1918 [2]; it is usually written to

$$x/m = a\mathrm{P}/(1 + b\mathrm{P}) \qquad (2\text{-}2)$$

where *x*/*m* is the quantity of material adsorbed per unit amount of adsorbent (often gas volume at STP per gram); *k*, *a*, *b*, and *n* are 'constants,' whose interpretation can be related to the derivations of the isotherms; and P is the gas pressure.

Two types of derivation yield the Langmuir isotherm – the kinetic and the thermodynamic. Kinetically it can be visualized that as the equality of the rates at which gas molecules join and leave the adsorbing surface. If the surface has S sites per gram, S_0 of which are occupied, the condensation rate is written $k_1(S - S_0)$ where k_1 can be derived from the kinetic theory as the number of impacts per unit time for the gas at the given pressure on an area equal to that of one site:

$$k_1 = \mathrm{A_0P}/(2\pi mk\mathrm{T})^{1/2} \tag{2-3}$$

The evaporation rate is written as k_2 where k_2 will depend on the tenacity with which the adsorbed molecule is held; i.e.,

$$k_2 = k' exp - q/k\mathrm{T} \tag{2-4}$$

Equating the two rates then leads to:

$$\frac{S_0}{S} = \theta \tag{2-5}$$

$$= \frac{\dfrac{\mathrm{A_0P}}{k'(2\pi\, mk\mathrm{T})^{1/2} exp - q/k\mathrm{T}}}{1 + \dfrac{\mathrm{A_0P}}{k'(2\pi\, mk\mathrm{T})^{1/2} exp - q/k\mathrm{T}}} \tag{2-6}$$

$$= \frac{K\mathrm{P}}{1 + K\mathrm{P}} \tag{2-7}$$

It is not difficult to show the equivalence of Eqs. (2-2) and (2-5), (2-6), and (2-7): θ is the fraction of the surface occupied by molecules of area A_0 on the surface mass is m, and residence time is $1/k'exp - q/kT$, all at gas pressure P.

Superficially, Eqs. (2-5), (2-6), and (2-7), look like a one-parameter equation, but a second unknown hides in the relation of x/m to θ.

The second procedure for deriving an adsorption isotherm makes use of the thermodynamic principle that the chemical potentials of the adsorbate as a gas and as an adsorbed layer must be the same. The chemical potentials are obtained by differentiating the Helmholtz free energies, which, in turn, are obtained from the appropriate partition functions [32].

Take the result for the gas directly from physical chemistry:

$$\mu_g = -k\mathrm{T}\, ln\, \mathrm{Z}^{(g)} \tag{2-8}$$

$$= -k\mathrm{T}\, ln\, \mathrm{Z}^{(g)}_{trans}\, \mathrm{Z}^{(g)}_{int} \tag{2-9}$$

For the adsorbed gas the situation is slightly different because it must take into account the number of ways our S_0 adsorbed molecules can be distributed among the S sites. Also, the three degrees of translational freedom for the gas molecules go over into two in the adsorbed state plus one vibration of the bond between the adsorbate and adsorbent. The partition function is:

$$\mathrm{Z}^{(s)} = \frac{\mathrm{S!}}{(\mathrm{S} - \mathrm{S_0})!\mathrm{S_0}!}\left(\mathrm{Z}^{(S)}_{trans}\, \mathrm{Z}^{(S)}_{vib}\, \mathrm{Z}^{(S)}_{int}\right) exp \frac{\mathrm{q}}{kT} \tag{2-10}$$

and then use Stirling's theory on the factorials in the form:

$$x! = (x/e)^x \tag{2-11}$$

and differentiating with respect to S_0, obtain for the chemical potential of the adsorbate.

$$\mu_S = kT\,ln\,\frac{S_0}{S - S_0} - kT\,ln\,Z^{(S)}_{trans}\,Z^{(S)}_{vib}\,Z^{(S)}_{int} \tag{2-12}$$

Before taking the step of equating μ_g and μ_s we examine the components of the partition function in more detail. For translation in one dimension the energy levels are

$$\varepsilon_n = n^2h^2/8\pi a^2 m \tag{2-13}$$

where a is the length of the 'box' in which the molecule moves.

The partition function then is obtained as

$$Z_{trans} = \sum_n exp - \frac{\varepsilon_n}{kT} \tag{2-14}$$

$$= \int_0^\infty exp - \frac{\varepsilon_n}{kT}dn \tag{2-15}$$

$$= \left(\frac{2\pi mkT}{h^2}\right)^{1/2} a \tag{2-16}$$

For the gas (three dimensions) a^3 is interpreted as volume, and replaced by kT/P so that:

$$Z^{(g)}_{trans} = (2\pi mkT/h^2)^{3/2}(kT/P) \tag{2-17}$$

For the adsorbed state we get:

$$Z^{(S)}_{trans} = (2\pi mkT/h^2)A_0 \tag{2-18}$$

where the area a^2 is written A_0.

The new vibration of the adsorbate against the surface has (in the harmonic oscillator approximation) energy levels $(n + 1/2)\,hv$ leading to

$$Z^{(S)}_{vib} = \frac{exp\,hv/kT}{exp(hv/kT) - 1} \sim \frac{kT}{hv} \tag{2-19}$$

where the expansion is justified, since hv/kT is likely much less than unity.

Now we insert these expressions into Eqs. (2-8) and (2-9), and equate chemical potentials to obtain:

$$\frac{\theta}{1-\theta} = \frac{A_0 P}{(2\pi\,mkT)^{1/2}}\cdot\frac{1}{v}exp\frac{q}{kT}\frac{Z^{(s)}_{int}}{Z^{(g)}_{int}} \tag{2-20}$$

which is easily recognized as Eqs. (2-5), (2-6), and (2-7) with the meaning for *K* of:

$$K = \frac{A_0 exp(q/kT) Z_{int}^{(s)}}{(2\pi\ mkT)^{1/2} Z_{int}^{(g)}} \quad (2\text{-}21)$$

Change in the internal component of *Z* due to adsorption has been extensively studied spectroscopically. A widespread use of the Langmuir isotherm is the estimation of the surface area of adsorbents from the limiting adsorption at high pressures (θ=1) and an acceptable A_0. The standard procedure is, using Eq. (2-2), to plot the reciprocal of the adsorption, *m/x*, against the reciprocal pressure. The intercept, *b/a*, is the reciprocal of the limiting adsorption. The principal lack in the preceding derivations is recognition that at fair θ lateral interaction may affect the adsorption [on either basis, through k_2 of Eq. (2-10), or $Z_{int}^{(s)}$ of Eq. (2-3).

The Langmuir equation is derived from the adsorption model on micro or nanoporous solid surfaces such as active carbons or zeolites, and gives a Type I isotherm (see Fig. 2-11). A Type I isotherm is illustrated in the study of activated carbon, as seen in Fig. 2-18 [33]. It is now widely accepted that the initial part of the Type I isotherm represents micropore filling and that the slope of the plateau a high relative pressure is due to multilayer adsorption on the nonmicroporous structure, i.e., in mesopores, in macropores, and on the external surfaces.

In such microporous adsorbents, the Type I isotherm can be explained as filling of the adsorbate molecules into micropores; that is, the initial sharp rise of the isotherm corresponds to the filling process of adsorbate into micropores, and

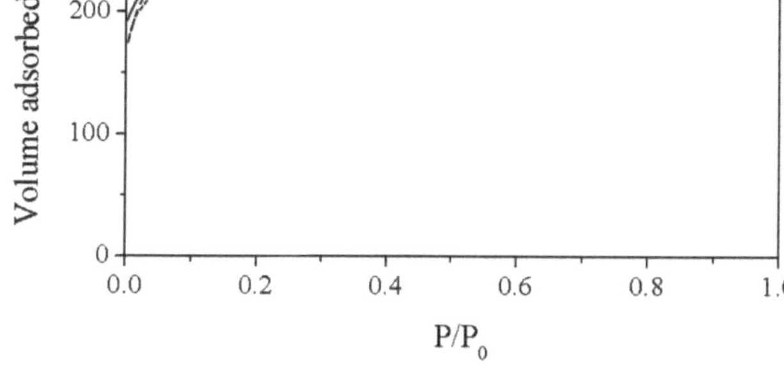

Fig. 2-18 Volume of N_2/77 K adsorbed, on the pristine activated carbons (ACs), and treated activated carbons *vs.* relative pressure.

Table 2-4 Textural Properties of Pt/Ru-Graphite Nanofibers (Pt/Ru-GNFs) Catalysts as a Function of the Activation Temperature

Samples	Langmuir Surface area (m^2/g)	BET Surface area (m^2/g)
Graphite nanofibers	392	259
Pt/Ru-GNF-700	472	267
Pt/Ru-GNF-800	541	307
Pt/Ru-GNF-900	517	305
Pt/Ru-GNF-1000	445	302

the flat region of the isotherm appears when the micropores are completely filled by adsorbate. It is reasonable to find that the monolayer volume of microporous adsorbent obtained from the Langmuir plot corresponds to the micro pore volume. The specific surface area of microporous active carbons, if it is calculated from the monolayer volume of the Langmuir plot reaches 3000 m^2/g. This value is larger than the maximum specific surface area 2630 m^2/g, the value calculated by BET's method in which both sides of the graphite layer in the active carbon are covered by a monolayer of adsorbate.

For example, the graphite nanofibers treated at various temperatures are used as carbon supports to improve the efficiency of PtRu catalysts [34]. In Table 2-4, the specific surface area of graphite nanofibers is calculated from the monolayer volume of the Langmuir plot and reaches 392 $\sim$ 541 m^2/g. Evidently, these values are larger, than those obtained from BET's technique, as seen in Table 2-4.

2.3.3. Brunauer-Emmett-Teller (BET) Isotherm

In 1938, Brunauer, Emmett, and Teller reported the multilayer adsorption theory (the BET theory) [6]. They extended Langmuir's monolayer theory to multilayer adsorption. The derived equation is named the BET equation after the first letter of each of their names. The BET's equation has been utilized in determining the monolayer volume of adsorbed gas and the specific surface area of the adsorbents.

The BET's equation is derived, on the following hypotheses:

1. Surface is energetically homogeneous; i.e., all adsorption sites on the bare surface have the same adsorption energy (E_1).
2. There is no lateral interaction, between adsorbed molecules.
3. The adsorption energies, in the second and all higher layers (E_2, E_3. . .), are equal to the condensation energy of the adsorptive; i.e., $E_2 = E_3 = \ldots = E_i = E_L$.

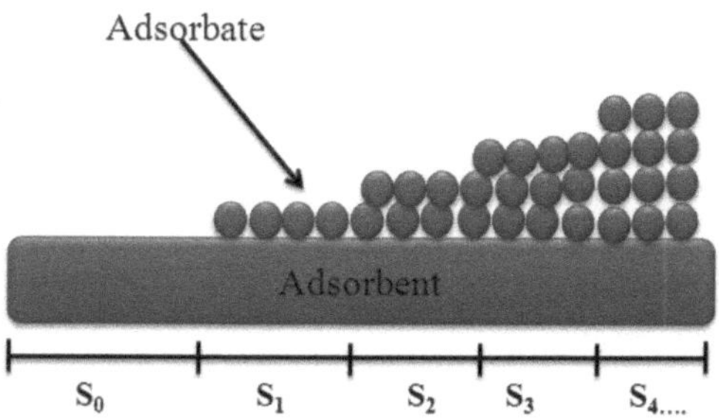

Fig. 2-19 BET model of adsorption.

It was recognized long ago, that adsorption limited to a unimolecular layer could not account for isotherms of Types II-V, but there is no reason to assume, that adsorption is so limited. The famous BET theory [6] uses the same approach as does the Langmuir idea but for a series of discrete layers (Fig. 2-19). That is, 'vertical' interaction occurs between molecules, in layer one and the bare surface, layer two and layer one, etc., but lateral interaction is ignored. In the upper layers the situation approaches that of the relation between a gas and its liquid condensate. For them, at least, ignoring lateral interaction while considering vertical interaction of the same molecules, seems very artificial. Nevertheless, BET's equation has continued to be successfully used. Its derivation, like that of the Langmuir isotherm, can be either kinetic or thermodynamic [6, 35], or give a kinetic one, while the thermodynamic approach may be found [36, 37], and gives some delightful reminiscences of its development.

The model is of S adsorption sites altogether, with S_1 having one adsorbed molecule, S_2 two molecules, etc., so that the total adsorption becomes the sum of S_i. The complete partition function, analogous to Eq. (2-10) is

$$Z^{(s)} = \frac{S!}{(S-S_1)!S_1!} \cdot \frac{S_1!}{(S_1-S_2)!S_2!} \cdot \frac{S_2!}{(S_2-S_3)!S_3!} \cdots (Z_1^{(s)})^{S_1} (Z_2^{(s)})^{S_2} (Z_3^{(s)})^{S_3} \cdots \tag{2-22}$$

Equilibrium, between successive layers requires that $d\ln Z^{(s)}/dS_i = 0$, from which it is not difficult to show that:

$$\frac{S_i - S_{i+1}}{S_{i+1} - S_{i+2}} = \frac{S_{i+1} - S_{i+2}}{S_{i+2} - S_{i+3}} \cdot \frac{Z_{i+2}^{(s)}}{Z_{i+1}^{(s)}} \tag{2-23}$$

where

$$Z_i^{(s)} = Z_{i\,trans}^{(s)} Z_{i\,int}^{(s)} exp\, q_i/kT \tag{2-24}$$

In the same way it can be shown (from equilibrium between the surface, and first and second layers) that,

$$\frac{S - S_1}{S_1 - S_2} = \frac{S_1 - S_2}{S_2 - S_3} \cdot \frac{Z_2^{(s)}}{Z_1^{(s)}} \tag{2-25}$$

Now, let us suppose that the top layer of adsorbed molecules is the *p*th. For this layer Eq. (2-25) becomes

$$\frac{S_{p-2} - S_{p-1}}{S_{p-1} - S_p} = \frac{S_{p-1} - S_p}{S_p} \cdot \frac{Z_p^{(s)}}{Z_{p-1}^{(s)}} \tag{2-26}$$

since $S_{p+1} = 0$.

This layer must have the same chemical potential as the gas so we can write (differentiating (Eq. (2-26)) with respect to S_p after putting in the Stirling approximations)

$$kT\, ln \frac{S_p}{S_{p-1} - S_p} - kT ln Z_p^{(s)} = \mu^0 + kT ln P \tag{2-27}$$

$Z_p^{(s)}$ is taken as similar to the partition function for the liquid; i.e.,

$$kT\, ln\, Z_L = \mu^0 + kT\, ln\, P \tag{2-28}$$

from which substitution gives:

$$\frac{S_p}{S_{p-1} - S_p} = \frac{P}{P_0} \frac{Z_p^{(s)}}{Z_L} \tag{2-29}$$

Progressive replacement of $(S_1 - S_2)/(S_2 - S_3)$ in Eq. (2-25), by its value from Eq. (2-23), leads to the set of equations:

$$(S_1 - S_2) = (S - S_1)xc_1 \tag{2-30}$$

$$(S_2 - S_3) = (S - S_1)x^2 c_1 c_2 \tag{2-31}$$

$$S_p = (S - S_1)x^P c_1 c_2 \ldots c_p \tag{2-32}$$

where $c_i = Z_i^{(s)}/Z_1$ and $x = P/P_0$.

The results of these lengthy manipulation are that we now can write a relation for our total adsorption, S_0, by summing the values of S_i obtainable from (Eqs. (2-30), (2-31), and (2-32)). This is done conveniently, using the identities

$$S_0 = (S_1 - S_2) - 2(S_2 - S_3) + 3(S_3 - S_4) + \cdots \tag{2-33}$$

$$S_0 = (S - S_1) + (S_1 - S_2) + (S_2 - S_3) + \cdots \quad (2\text{-}34)$$

Thus, we obtain as the immediate precursor to BET's isotherm

$$\theta = \frac{S_0}{S} \quad (2\text{-}35)$$

$$= \frac{xc_1(1 + 2xc_2 + 3x^2c_2c_3 + \cdots \mathrm{p}x^{\mathrm{p}-1}c_2 \cdots c_\mathrm{p})}{1 + xc_1 + x^2c_1c_2 + \cdots x^\mathrm{p}c_1c_2 \cdots c_\mathrm{p}} \quad (2\text{-}36)$$

The final assumption of the theory is that all c_i except c_1 are unity. Thus, Eqs. (2-35) and (2-36) reduce to:

$$\theta = \frac{xc_1(1 + 2x + 3x^2 + \ldots \mathrm{p}x^{\mathrm{p}-1})}{1 + x + x^2 + \ldots x^\mathrm{p}} \quad (2\text{-}37)$$

For both Eqs. (2-35) and (2-37), the numerator is seen to be the derivative of the denominator, but for Eq. (2-37), the latter is a geometric progression, so that where $c_1 = C$

$$\theta = \frac{xC[1 - (\mathrm{p}+1)x^\mathrm{p} + \mathrm{p}x^{\mathrm{p}+1}]}{(1 - x)(1 - x + xC - x^{\mathrm{P}+1}C)} \quad (2\text{-}38)$$

Now since $x < 1$ and (excepting for adsorption limited by pores) it is reasonable to let $p \to \infty$, we obtain BET's isotherm:

$$\theta = \frac{Cx}{(1 - x)(1 - x + Cx)} \quad (2\text{-}39)$$

Although we have retained the symbol θ, it can, of course, now far exceed unity and is more often written v/v_m, where v is the volume of gas adsorbed (reduced to STP), and v_m is the volume required for a monolayer. From the monolayer volume to the corresponding number of molecules to the area they cover is a direct route. The customary procedure is to cast Eq. (2-39) into the form;

$$\frac{x}{v(1 - x)} = \frac{1}{Cv_\mathrm{m}} + \frac{(C - 1)x}{Cv_\mathrm{m}} \quad (2\text{-}40)$$

here plot $x/v(1 - x)$ against x, and obtain C and v_m from the slope and intercept of the resulting straight line. For surface area determination the most commonly used gas is N_2 at low temperatures (77 K), where, except in special cases, its A_0 is 0.162 nm^2. For some high-energy solids, the lattice spacing of the solid, affects A_0, and it can be up to 0.2 nm^2 for N_2.

The curves in the above figure show BET's adsorption isotherm at the different C constants (see Fig. 2-20). When C is large (i.e., large adsorption heat), the isotherm is Type II, whereas when C is small (i.e., small adsorption heat), the

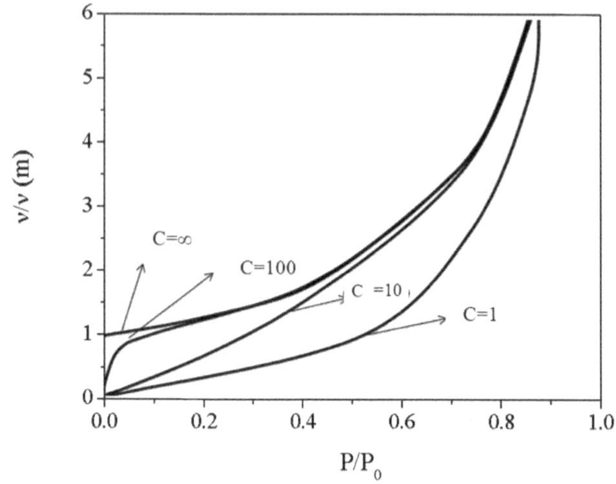

Fig. 2-20 BET isotherms various values of *C*.

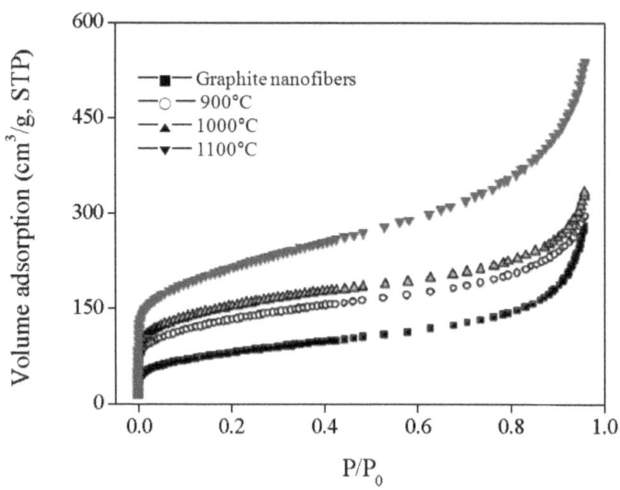

Fig. 2-21 N_2/77 K adsorption isotherms in porous graphite nanofibers, as a function of activation temperature.

adsorption isotherm is Type III (as shown in Fig. 2-11). The salient feature of BET's isotherm is that, despite the shortcomings of its theoretical base, it is capable of reproducing each of the first three types of isotherm illustrated in Fig. 2-11. For large C, it becomes the Langmuir isotherm, except as $x \rightarrow 1$. For intermediate C (from just over 1 to *ca.* 200), Type II isotherms with the point of inflection at reasonable values of the P/P_0 result. Values of C below 1 give

Table 2-5 Properties of Gases Using for Calculating Specific Surface Area

Gas	T[a] (K)	P_0[b] (Pa)	A[c] (nm^2/molecule)
N_2	77	1.013×10^5 at 77.4 K (liquid)	0.13 ~ 0.20 (0.162)
Ar	77	2.78×10^4 at 77.4 K (solid)	0.13 ~ 0.17 (0.15)
	87	1.013×10^5 at 87.3 K (liquid)	
Krypton	77	293 at 77.4 K (solid)	0.17 ~ 0.22 (0.20)
	77	338 at 77.4 K (liquid)	0.17 ~ 0.22 (0.20)
	90	2.74×10^3 at 90.2 K (solid)	0.17 ~ 0.22 (0.20)
Xenon	90	8.25 at 90.2 K (solid)	0.18 ~ 0.27 (0.23)
Methane	90	1.08×10^4 at 90.2 K (solid)	0.15 ~ 0.17 (0.16)
n-Butane	273	1.013×10^6 at 272.7 K (liquid)	0.32 ~ 0.57 (0.44)
CO_2	195	1.013×10^6 at 194.7 K (solid)	0.14 ~ 0.20 (0.20)

[a] *Typical adsorption temperature.*
[b] *Saturation vapor pressure at adsorption temperature.*
[c] *Area per adsorbed molecules in monolayer.*

Type III isotherms. In Table 2-5, the properties and ranges of useful gases are collected in Table 2-5, for calculating specific surface area of adsorbents [38].

As shown in Fig. 2-21, the porous graphite nanofibers by a KOH activation method were reported for manufacturing porous carbon nanofibers [39]. BET's equation is used to observe the specific surface areas (N_2/77 K). The curves of all of the samples show a middle form between Types I (the region of low relative pressure) and II (the region of middle and high relative pressure).

As seen in Table 2-6, BET's constant *C* increases with an increasing activation temperature up to 1000 °C, and then shows a diminution. The same trends can be found in the net heat of adsorption [39]. BET's constant can be strongly enhanced by increasing the micropore fraction. The increase of the *C* value, means the development of micropores in graphite nanofibers.

Table 2-6 Textural P roperties of Porous Graphite Nanofibers as a Function of Activation Temperature

Samples	C_{BET}[a]	E_n[b]	S_{BET}[c]	V_T[d]
Graphite nanofibers	161	3.264	320	0.439
Activated at 900 °C	182	3.343	500	0.464
Activated at 1000 °C	242	3.526	580	0.521
Activated at 1100 °C	237	3.513	790	0.830

[a] *BET's constant* $C = (a_1 v_2)/(a_2 v_1) e^{(q_1 - q_L)RT}$.
[b] *Net heat of adsorption (kJ/mol):* $\Delta E = q_1 - q_L = RT ln(C)(a_1 \cdot v_2 = a_2 \cdot v_1)$.
[c] *Specific surface area* (m^2/g).
[d] *Total pore volume* (cm^3/g).

2.4. MEASUREMENT OF ADSORPTION ISOTHERMS

Several experimental methods are applied to measure adsorption isotherms. The main problem is to determine the amount adsorbed. Because the sorbent needs to be regenerated for most commercial applications, adsorption processes are necessarily cyclic. A number of cyclic adsorption processes are available, depending on the way the sorbent is regenerated. These processes have been discussed extensively elsewhere. For the measurement of gas adsorption isotherms three different methods known as the volumetric gravimetric method and inverse gas chromatography, can be used.

2.4.1. Gravimetric Measurement

In the gravimetric method the amount of adsorbed gas at the equilibrium pressure is weighted with the aid of a microbalance. The adsorbent, usually in the form of a powder, is placed into a bulb and kept at the desired temperature. The bulb is mounted on a sensitive balance. Before the experiment, the bulb is evacuated. Then, the gas of interest is admitted into the bulb at a certain pressure. The increase in weight is measured. Dividing by the total surface area of the adsorbent, it will obtain the amount of adsorbed gas, on the surfaces of solid. The pressure is increased and the weight measurement is repeated. In this way a whole isotherm is recorded. Here it is not necessary to know the equation of state or the volume of the system. In addition, it is possible to continuously register the amount adsorbed.

Cooling or heating of the sample may be difficult, since direct contact with the sample is impossible [40]. The adsorption balance (see Fig. 2-22) consists essentially of a hollow beam (8) suspended between two strips of a flat spring (not shown in the figure) and counterweighted by an adjustable screw (6). The beam is a stainless steel tube, with an outer diameter of 6 mm and a wall thickness of 1 mm. A copper container (11) is connected to one side of the beam by means of a similar tube and a piece of silicon rubber (9). The gas to be adsorbed is admitted to the hollow beam by means of mercury filled seal (7), and placed in such a position that the axis of rotation of the beam coincides with the mercury surface. To prevent evaporation this surface is covered with a thin layer of oil. The balance is mounted in a vacuum-tight housing (5), which have an open connexion to the glass surrounding tube (10). During measurements at adsorption pressures above 0.2 mm Hg, the pressures in the beam and in the housing were kept at the same value by connecting tubes (1) and (2) with each other. The gas in tube (10), thus provides a thermal link, between the solid powder inside the container, and the cooling bath. At pressures below 0.2 mmHg, the tube (2) was closed, to prevent too low a pressure occurring in tube (10); the mercury seal can withstand this pressure difference. This displacement of the beam is transmitted to a strain gauge bridge (3) by means of a brass wire (4).

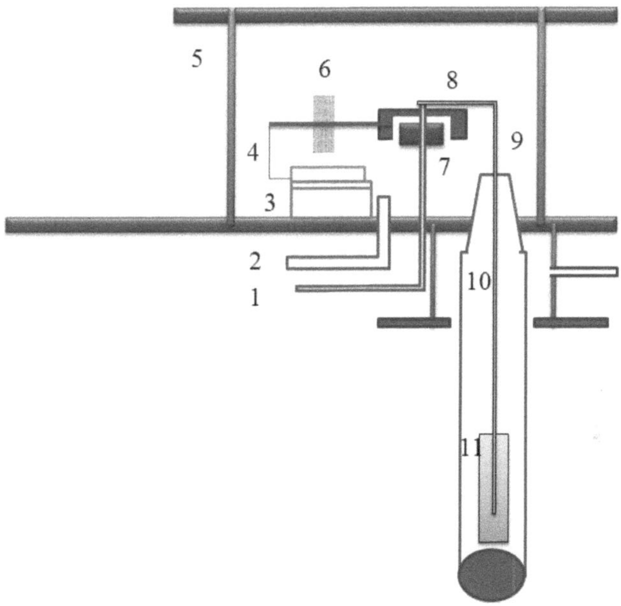

Fig. 2-22 Schematic drawing of the adsorption balance.

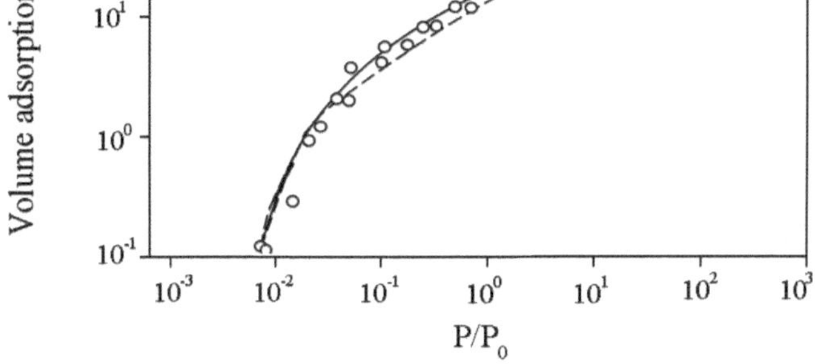

Fig. 2-23 Adsorption of H_2 at 77 K on the zeolite 5A.

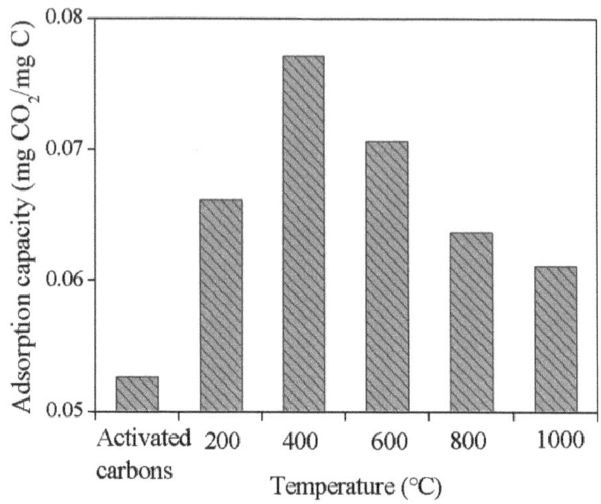

Fig. 2-24 CO_2 adsorption by pristine and ammonia-treated activated carbons as a function of the modification temperature after 300 min; adsorption temperature = 309.15 K, pressure = 1.025 bar, CO_2 flow = 25 ml/min.

As shown in Fig. 2-23, the H_2 adsorption isotherms at 20 K on zeolites 5A appear to be negligible as well. When the measurements are at pressures lower than 10^{-4} mmHg, the adsorption isotherms are in agreement with those obtained from the other volumetric measurements [40].

Jones *et al.* [41] and Przepiórski and co-workers [42] used thermogravimetric analysis (TGA) to compare the CO_2 adsorption of pristine activated carbons with those of carbons, prepared by ammonia treatments at different temperatures. They found, that there is an optimum temperature, for the ammonia treatment with the sample treated at 673 K exhibiting the highest CO_2 adsorption capacity of 1.73 mmol/g at 309 K and 1 bar, as shown in Fig. 2-24 [42]. Treatment at higher temperatures does not result in additional enhancement in adsorption capacity. A gradual decrease in the CO_2 adsorption capacities by heat treatments above 673 K is attributed to the blockage of micropores.

The gravimetric method is also used as the determination of the reversible absorption of CO_2 for practical applications. Barron *et al.* reported possible the CO_2 absorption applications of the polyethyleneimine modified single-wall carbon nanotubes by thermogravimetric analysis [43]. As can be seen from Fig. 2-25, once CO_2 is introduced the mass increases significantly until a stable maximum is reached [43]. Once the carrier gas is switched back to argon, a mass loss occurs back to the original mass for the polyethyleneimine (M_w: 25000) modified single-wall carbon nanotubes. Desorption of CO_2 is accomplished, by heating under argon at 75 °C. Under the conditions investigated the maximum absorption of 9.2% w/w was observed for polyethyleneimine modified single-wall carbon nanotubes at 27 °C. The ability of the polyethyleneimine modified single-wall

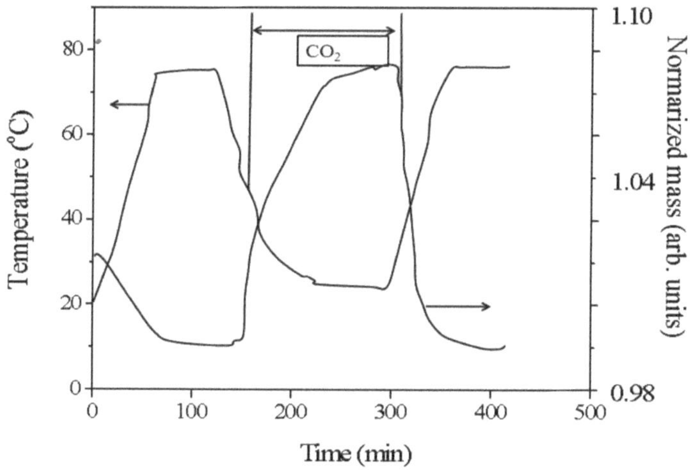

Fig. 2-25 Thermogravimetric analysis of polyethyleneimine (Mw: 25000) modified single-wall carbon nanotubes, showing the mass change, associated with the adsorption and desorption of CO_2. The mass has been normalized to the lowest mass recorded, which is equivalent to polyethyleneimine modified, single-wall carbon nanotubes.

carbon nanotubes to absorb CO_2, should be related to the number and the availability of NH_2 groups for reaction. This is because polyethyleneimine is known to react with CO_2 by means of the amines to form a carbamate [43].

Zelenak *et al.* reported the pure-component equilibrium isotherms for CO_2 using thermogravimetric analysis (TGA) on a mesoporous silica material [44]. The influence of the basicity of amine ligands on sorption of CO_2 over amine-modified mesoporous silica is investigated. The CO_2 sorption-desorption isotherms are performed at 25 °C, as shown in Fig. 2-26 [44]. Since the adsorption-desorption studies are performed using 10% CO_2/N_2 gas mixture, it is of interest to study the extent of coadsorption of CO_2 and N_2 on the materials. Therefore adsorption-desorption runs are repeated with pure CO_2 (99.999%) used during the sorption step and pure nitrogen used as an inert gas during the regeneration (desorption) step. Fig. 2-26 shows the adsorption-desorption cycles of carbon dioxide sorption, over SBA-12 (Santa Barbara Amorphous type materials)/amine sample. From Fig. 2-26, it is obvious that with pure CO_2 a similar uptake is observed during the experiment as in the study performed with 10% CO_2/N_2 gas mixture in adsorption step and are used in the regeneration step. This indicates that no significant co-adsorption occurred, which is in agreement with the results of other authors showing that N_2 sorption on functionalized silica is negligible and amine modified silica displays high CO_2/N_2 selectivity by acid-base interactions, among molecules of CO_2 as acidic gas, and the amines as basic centers.

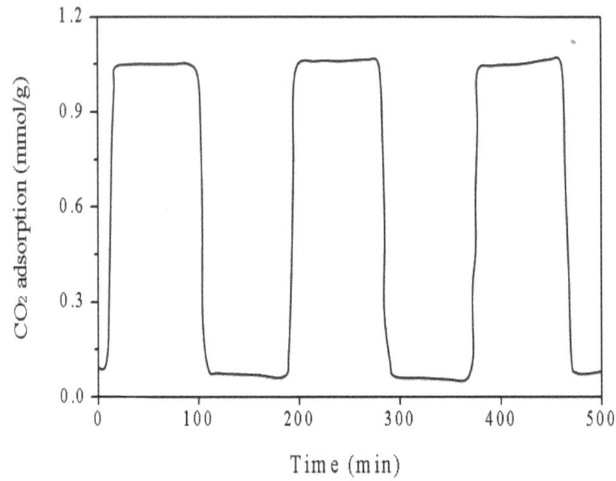

Fig. 2-26 Cycles of carbon dioxide sorption/desorption studied by thermogravimetric analysis at 25°C over aminopropyl modified SBA-12. The pure CO_2 (sorption) and N_2 (initial activation/desorption) were used during the experiment.

2.4.2. Volumetric Measurement

The volume of an adsorbed gas, at constant pressure and temperature is determined. First, we first determine the 'dead space' or volume of the bulb by admitting some nonadsorbing (or weakly adsorbing) gas such as Helium. Then, after evacuating the bulb, the gas of interest is admitted into the bulb. This is done at a constant pressure and temperature. The volume admitted into the bulb minus the dead space is the amount adsorbed. The advantages of this method are a relatively simple construction and a good heat contact between the adsorbent and the cooling (or heating) device.

2.4.2.1. Pressure Swing Adsorption (PSA)

For bulk separation, pressure swing adsorption (PSA) is more suitable. However, PSA is also used for drying of air and industrial gases. More applications continue to be discovered for PSA, including air prepurification (i.e., purification of air prior to cryogenic distillation, by removal of carbon dioxide, water, and hydrocarbons). Adsorption modelling, characterization of porous materials and industrial process design requires basic reliable adsorption data both for pure gases and mixtures. During the last decade, interest in high-pressure adsorption data has increased and it is now common to find such data in the literature. For example, considering the particular application of natural gas storage by adsorption, one must identify the best adsorbent and the most appropriate filling pressure that will lead to values of storage capacities that can be compared to the classical compressed natural gas storage. The experimental study should include the determination of adsorption isotherms at pressures up to 10^6 kPa [45].

Fig. 2-27 shows the general principle of the volumetric method and schematic diagram of the volumetric apparatus, respectively.

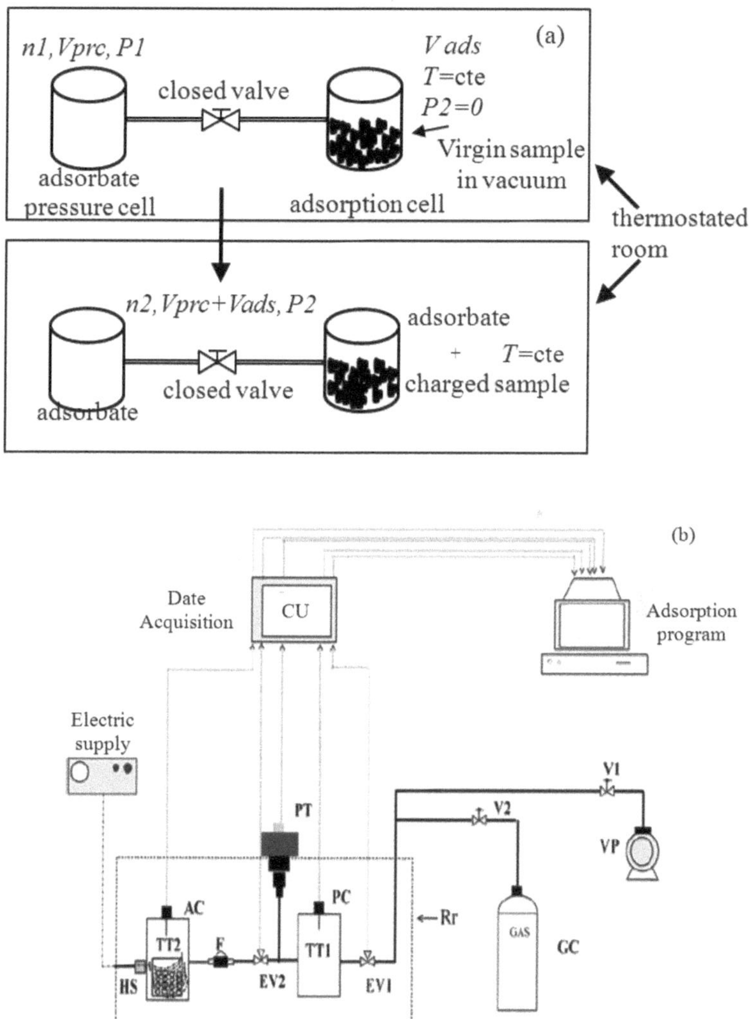

Fig. 2-27 General principle (a) and schematic diagram (b) of the volumetric apparatus.

Recently, hydrogen storage has attracted much attention as a promising energy resource and is expected to be applied to polymer electrolyte membrane fuel cell (PEMFC). Usually, the hydrogen storage capacities of the adsorbents

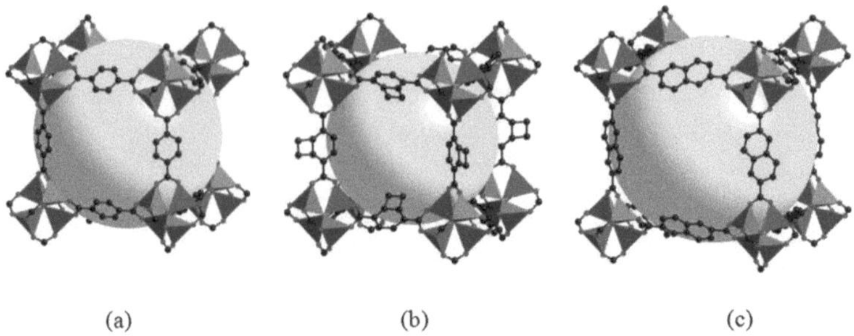

Fig. 2-28 Single-crystal X-ray structures of metal-organic frameworks-5 (MOF-5) (a), isoreticular metal-organic framework-6 (IRMOF-6) (b), and isoreticular metal-organic framework-6 (IRMOF-8) (c) illustrated for a single cube fragment, of their respective cubic three-dimensional extended structure. On each of the corners is a cluster $[OZn_4(CO_2)_6]$ of an oxygen-centered Zn_4 tetrahedron, that is bridged by six carboxylates of an organic linker (Zn, blue polyhedron; O, red spheres; C, black spheres). The large yellow spheres represent the largest sphere, that would fit in the cavities, without touching the van der Waals atoms of the frameworks. Hydrogen atoms have been omitted.

are evaluated using a high-pressure volumetric apparatus at a certain temperature. Yaghi *et al.* reported the very favorable hydrogen sorption properties obtained at 78 K, or an ambient temperature under safe pressures (up to 20 bar) with crystalline metal-organic frameworks (MOFs) having cubic cavities of uniform size and internal structure [46]. Fig. 2-28 shows inorganic $[OZn_4]^{6+}$ groups are joined to an octahedral array of $[O_2C\text{-}C_6H_4\text{-}CO_2]^{2-}$ (1, 4-benzenedicarboxylate, BDC) groups to form a robust and highly porous cubic framework. The metal-organic frameworks-5 structure motif and related compounds provide ideal platforms on which to adsorb gases, because the linkers are isolated from each other and accessible from all sides to sorbate molecules. The scaffolding-like nature, of metal-organic frameworks-5 and its derivatives lead to extraordinarily high apparent specific surface areas ($2500 \sim 3000\ m^2/g$) for these structures.

The hydrogen gas uptake is measured by metal-organic frameworks-5 at 78 K by introducing small amounts of H_2 into a chamber containing the guest free form of the material and monitoring the weight change as a function of increasing doses of H_2. The measured sorption isotherm shows type I behavior, in which the saturation is reached at low pressures, followed by a pseudo plateau, at higher pressure of H_2 with a maximum uptake of 45 mg of H_2 per gram of metal-organic frameworks-5 (4.5 wt.% = 17.2 H_2 per $Zn_4O\ (BDC)_3$ formula unit) and shown in Fig. 2-29(a) [46]. The observed sharp uptake of H_2 at lower pressure indicates favorable sorption interactions between the metal-organic frameworks-5 framework and H_2 molecules. In addition, similar to the reversible sorption of gases and organic vapors (N_2, Ar, CO_2, $CHCl_3$, CCl_4, C_6H_6, and C_6H_{12}) in metal-organic frameworks-5 [47, 48], adsorbed H_2 molecules can also be reversibly desorbed from the pores by reducing pressure.

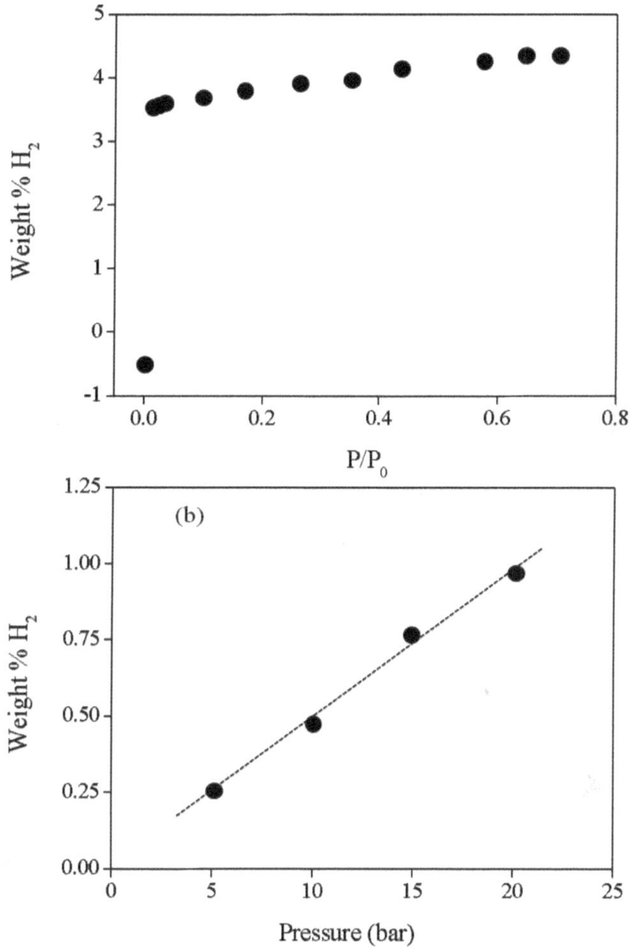

Fig. 2-29 Hydrogen gas, adsorption isotherm for metal-organic frameworks-5 at 78 K (a) and 298 K (b).

A different sorption apparatus is constructed, in which a 1.0 g sample of guest-free metal-organic frameworks-5 was charged with H_2 up to 20 bar and the weight change monitored as a function of H_2 uptake and release. Metal-organic frameworks-5 shows a substantial H_2 uptake that increased linearly with pressure, giving 1.0 wt.% at 20 bar, and shown Fig. 2-29(b). These findings demonstrate the potential of metal-organic frameworks for H_2 storage applications. The isotherm at an ambient temperature is expected to be approximately linear as observed because the material is noticeably under saturated with H_2

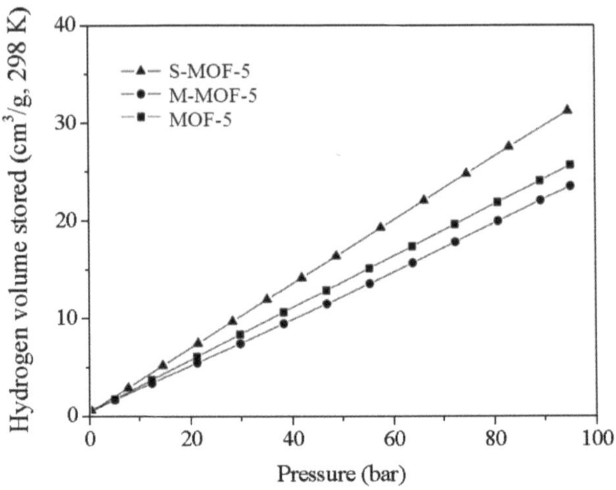

Fig. 2-30 Hydrogen storage capacity of metalorganic frameworks-5 (MOF-5), multi-walled carbon nanotubes/metal-organic frameworks-5 (M-MOF-5), and single-walled carbon nanotubes/metal-organic frameworks-5 (S-MOF-5).

in the pressure range explored and, in principle, at higher pressures can take up more H_2, up to at least the amount observed at 78 K (compare Fig. 2-29 (a and b)) [46].

Previously, the hydrogen storage behaviors of carbon nanotubes/metal-organic frameworks-5 hybrid composites (CNTs/MOF-5) were reported [49]. The hydrogen storage capacity, of single-carbon nanotubes/metal-organic frameworks-5 shows a higher value at 298 K and 100 bar than the other samples, and can be seen in Fig. 2-30. The carbon nanotubes provide abundant additional interface sites, with significant catalytic activity, with high-aspect-ratio characteristics, resulting from the bonds in association with carbon of graphitic sheets. In addition, the presence of carbon nanotubes on metal-organic frameworks-5 influences the enhancement of hydrogen storage, resulting from the modified physicochemical properties of carbon nanotubes/metal-organic frameworks-5 hybrid composites.

Another hydrogen storage behavior of porous graphite nanofibers decorated by Pt nanoparticles was investigated [50]. As shown in Fig. 2-31, the Pt particles were successfully loaded onto the carbon surfaces by using the TEM method. The size of the metal nanoparticles, varied from 1 to 20 nm. Pt nanoclusters of 3–6 nm, are observed in Fig. 2-31 (b), growing as large as $5 \sim 20$ nm, in the Fig. 2-31 (c) sample. Additionally, the Pt nanoclusters hyperactively grew on Pt-10/PGNF, resulting in a decrease of the total dispersion rate and increase of the average particle size.

The hydrogen storage behaviors of the Pt/porous graphite nanofibers were studied at 298 K and 100 bar [50], and are shown in Fig. 2-32. All

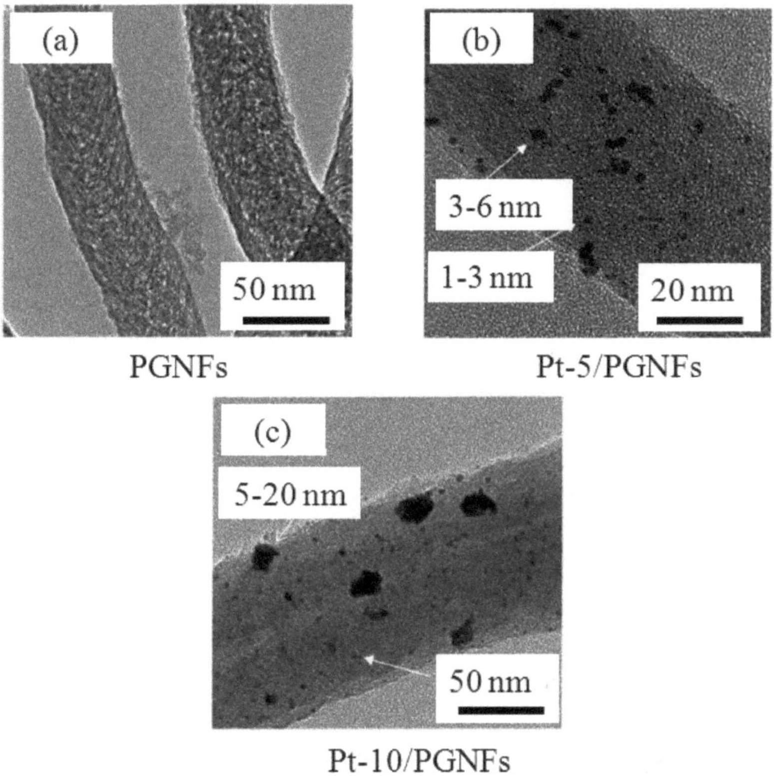

Fig. 2-31 TEM images of porous graphite nanofibers, and Pt supported porous graphite nanofibers, as a function of Pt content: (a) porous graphite nanofibers (PGNFs), (b) Pt-5/porous graphite nanofibers (Pt-5/PGNFs), and (c) Pt-10/porous graphite nanofibers (Pt-10/PGNFs).

of the Pt/porous graphite nanofibers, show a higher hydrogen uptake, than do the porous graphite nanofibers sample (0.25 mass %). This result clearly indicates, that the presence of Pt nanoparticles, enhances hydrogen uptake of porous graphite nanofibers, meaning that Pt plays the role of a catalyst in reducing the adsorption energy of hydrogen storage. Pt-5/porous graphite nanofibers show a higher hydrogen uptake than other samples. This result can be interpreted to mean, that some difference of specific surface area can be overcome by a higher amount of platinum, but the formation of large, poorly dispersed clusters and a severe decrease of specific surface area reduce the hydrogen storage capacity of carbon materials in this Pt/porous graphite nanofiber system.

The high pressure hydrogen storage behaviors of the Cu plated-activated carbon fibers, by a pressure swing adsorption (PSA) apparatus at 100 bar and

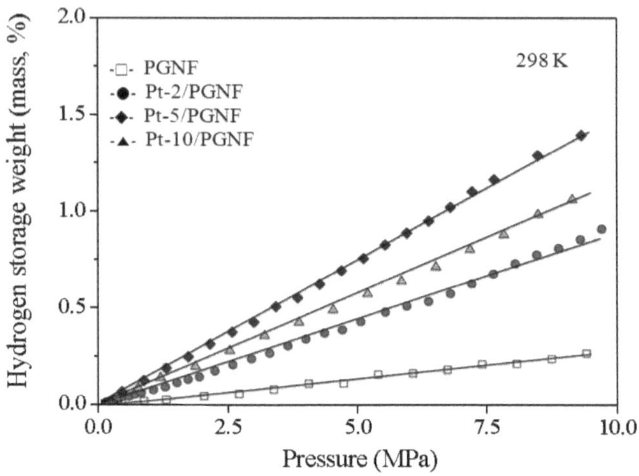

Fig. 2-32 Hydrogen storage behaviors of Pt decorated porous graphite nanofibers, as a function of metal content.

298 K were studied [51]. Park *et al.* proposed a dipole-induced model (or modified spill-over effects) in which the hydrogen molecules stored on the carbon surfaces (or into the carbon structures) by the dipole-induced effects at 298 K due to the presence of metal nanoparticles on carbon surfaces, and is shown in Fig. 2-33 [51], shows a schematic diagram of the modified spill-over effects on the interface of metal and H_2 atom. As illustrated in Fig. 2-33(a and b), the first hydrogen molecules adsorbed can be dissociated on the metal surfaces, and the surfaces of the metals are fully covered by these hydrogen atoms (as with Yang's mechanism [52]); after the formation of the hydrogen layer on the metal surfaces, the second hydrogen molecules can be physically adsorbed at high pressure, and can be changed partially to polar molecules due to the disequilibrium of electron clouds (or negatively charged) resulting from the strong electron acceptor characteristics of metals.

Meanwhile, the carbons near metal catalysts, have lost some of their electrons with rapidly and positively charged (or changed to electron acceptors) due to the strong electron-acceptor characteristics of near metallic catalysts. This can be exhibited in the formation of new hydrogen-friendly sites on carbon surfaces due to the electron-acceptor–donor interaction behaviors. The average binding energy of the physical bonds is 10 $\sim$ 50 kJ/mol, indicating that they can easily be broken. Thus, the third hydrogen molecules (see Fig. 2-33(d)) can be adsorbed at higher pressure, than indicated in Fig. 2-33(c). By adding the third hydrogen molecules, the hydrogen molecules physisorbed on the metal surfaces can spill over and finally be stored on the carbon surfaces that have electron-acceptor characteristics.

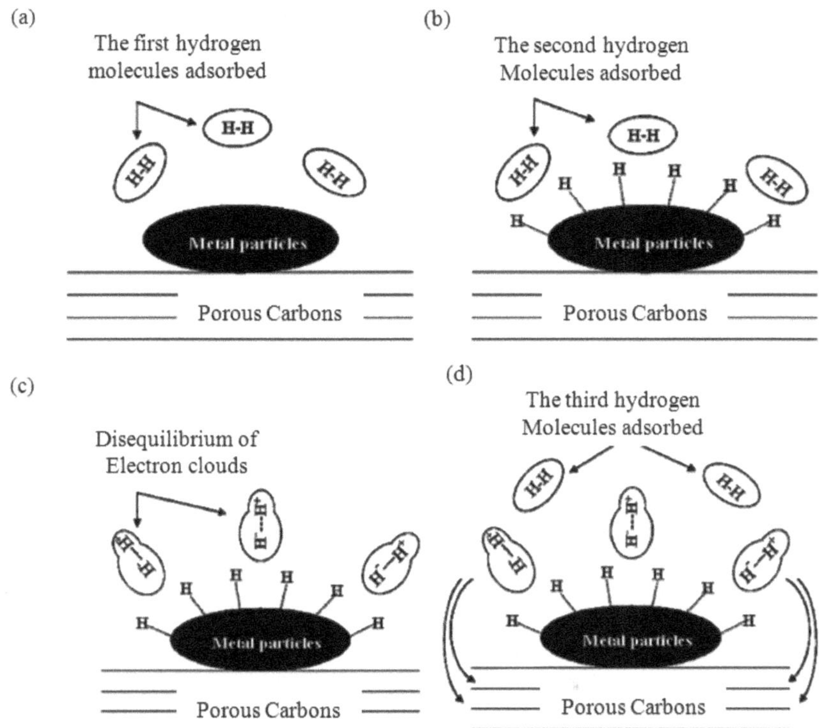

Fig. 2-33 A schematic diagram of the modified spill-over effects, on the interface of H_2 atoms and metal.

As shown in Fig. 2-34, the hydrogen-storage capacity was enhanced in the presence of Cu particles with a shorter plating time, due to the modified spill-over effects, and then began to decrease as a result of excessively plated Cu particles (at 60 min and 120 min) covering the activated carbon fiber surfaces [51].

The CO_2 storage capacity of carbon nanofibers at 298 K and 1 bar is investigated [53]. The carbon nanofibers are successfully expanded intercalating KOH followed by heat treatment in the temperature range of 700 ~ 1000 °C. As shown in Table 2-7, after heat treatment, the specific surface area, and total pore volume of carbon nanofibers increases up to 900 °C but then decrease at 1000 °C. It is interesting to note, that the micropore fractions of the treated carbon nanofibers also increased at 900 °C, whereas the mesopore fractions, decreased with increasing heat temperatures. The average pore diameter of treated carbon nanofibers was less than the pristine carbon nanofibers, which decreased with increasing temperature. This is due to the KOH activation being partially blocked by the porous structure of the carbon nanofibers at temperatures up to 1000 °C.

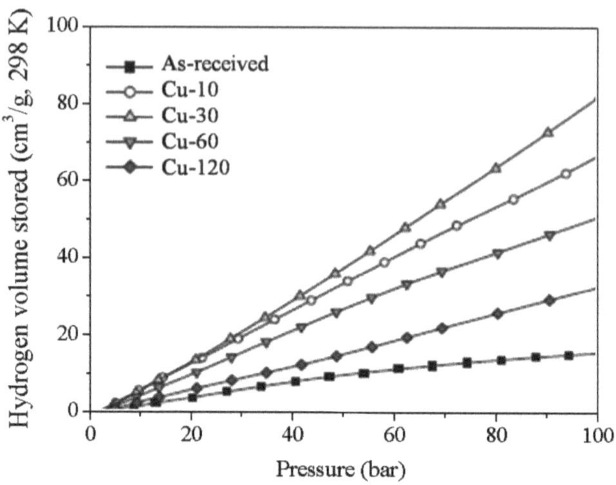

Fig. 2-34 Cu plated activated carbon fibers hydrogen-storage behaviors at 298 K as a function of plating time (10 ~ 120 min).

Table 2-7 Pore Structure Parameters for the Chemical Treated Carbon Nanofibers as a Function of Heat Temperature from N_2 and CO_2 Adsorption Isotherms

	N_2 (77 K)					CO_2 (298 K)			
Samples	S_{BET}[a]	V_T[b]	V_{Mi}[c]	V_{Me}[d]	D[e]	S_{BET}[a]	V[f]	D[g]	D[e]
Carbon nanofibers	178	0.250	0.060	0.190	57.0	28.1	0.017	6.2	28.9
Activated at 700 °C	249	0.458	0.012	0.446	73.5	39.3	0.026	5.7	23.5
Activated at 800 °C	256	0.468	0.064	0.404	73.1	40.1	0.030	5.2	29.3
Activated at 900 °C	567	0.708	0.274	0.434	49.9	69.5	0.053	5.7	30.1
Activated at 1000 °C	228	0.288	0.075	0.288	50.3	35.4	0.029	6.7	33.0

[a] *Specific surface area (m^2/g): BET equation ($P/P_0 = 0.05 \sim 0.1$).*
[b] *Total pore volume (cm^3/g): V_{ads} ($P/P_0 = 0.995$) 0.001547.*
[c] *Micropore volume (cm^3/g): Dubinin–Radushkevich equation.*
[d] *Mesopore volume (cm^3/g): BJH equations.*
[e] *Average pore diameter (Å): $2 \times S_{BET}/V_{ads}$.*
[f] *Micropore volume (cm^3/g): Dubinin–Astakhov equation.*
[g] *Average micropore diameter (Å): Horvath-Kawazoe equation.*

The aim was to improve the CO_2 adsorption capacity of the carbon nanofibers, by increasing the porosity of carbon nanofibers [53]. Fig. 2-35 shows that the activation temperature had a major influence on the CO_2 adsorption capacity and textural properties of the carbon nanofibers. The specific surface area, total pore volume, and mesopore volume of the carbon nanofibers increased after heat

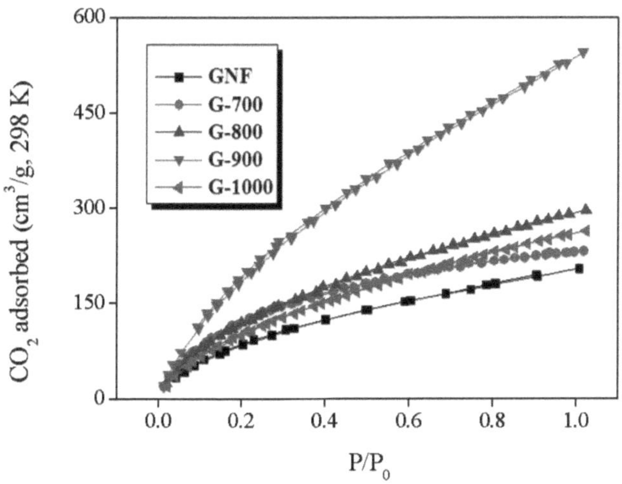

Fig. 2-35 CO_2 storage capacity of the chemical treated carbon nanofibers, as a function of heat temperature.

treatment. The CO_2 adsorption isotherms, showed the best CO_2 adsorption capacity with 59.2 mg/g (activated at 900 °C).

In addition, many adsorbents used the pressure swing adsorption (PSA) method to measure the capture properties of aim gases. For example, the adsorptive separation of gases (CO_2 or CH_4) by the PSA techniques has been most extensively studied [54–59].

2.4.2.2. Temperature Swing Adsorption (TSA)

For purification, temperature swing adsorption (TSA) is generally the process of choice. Regeneration of adsorbent in a temperature swing adsorption process is achieved by an increase in temperature. Fig. 2-36 schematically shows the effect of temperature on the adsorption equilibrium (Type I isotherm) of a single adsorbate [60].

For any given partial pressure of the adsorbate in the gas phase (or concentration in the liquid phase), an increase in temperature leads to a decrease in the quantity adsorbed. If the partial pressure remains constant at P_1, increasing the temperature from T_1 to T_2 will decrease the equilibrium loading from Q_1 to Q_2.

A relatively modest increase in temperature can affect a relatively large decrease in loading. It is therefore generally possible to desorb any components, provided that the temperature is high enough. However, it is important to ensure that the regeneration temperature does not cause degradation of the adsorbents.

A change in temperature alone is not used in commercial processes because there is no mechanism for removing the adsorbates from the adsorption unit once desorption from the adsorbents has occurred. A hot purge gas or steam

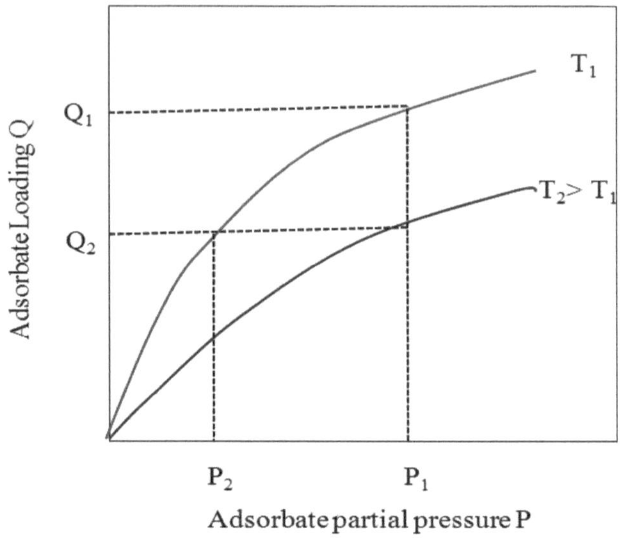

Fig. 2-36 Effect of temperature on the adsorption equilibrium (Type I adsorption isotherm).

passage through the bed to sweep out the desorbed components is almost always used in conjunction with the increase in temperature. A very important characteristic of temperature swing adsorption processes is that they are used virtually exclusively for treating feeds with low concentrations of adsorbates [60].

The extent of equilibrium adsorption at a given pressure diminishes with a rising temperature, so also the shape of the isotherm also changes, as illustrated in Fig. 2-37 with data obtained in 1910. As the figure shows, the amount of adsorption, at constant temperature increases while the curvature of the isotherms decreases [60].

As expected, the adsorption capacities of activated carbons, like those of other physisorbents such as zeolites, decrease rapidly as temperature increases [61, 62]. Do *et al.* measured single component CO_2 adsorption isotherms on activated carbon at pressures up to 0.2 bar and three different temperatures showing that the adsorption capacities dropped from about 0.75 to 0.1 mmol/g as the temperature increased from 298 to 373 K [61]. Na *et al.* measured the CO_2 adsorption isotherm on a commercial activated carbon at temperatures between 288 and 328 K using a volumetric apparatus, as shown in Fig. 2-38 [62]. The experimental data shows good agreement with predictions from the Langmuir-Freundlich model and illustrate that CO_2 uptake on this activated carbon decreases from about 3.2 to 1.5 mmol/g after a small temperature increase, from 288 to 328 K.

More recently, Bhatia *et al.* described the transport processes in bidisperse carbon particles by modelling the simultaneous mass transfer in both macropore

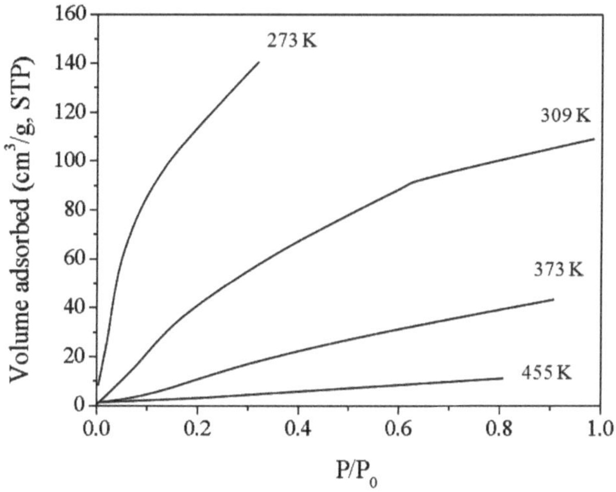

Fig. 2-37 Variation of adsorption isotherms on charcoal with rising temperature.

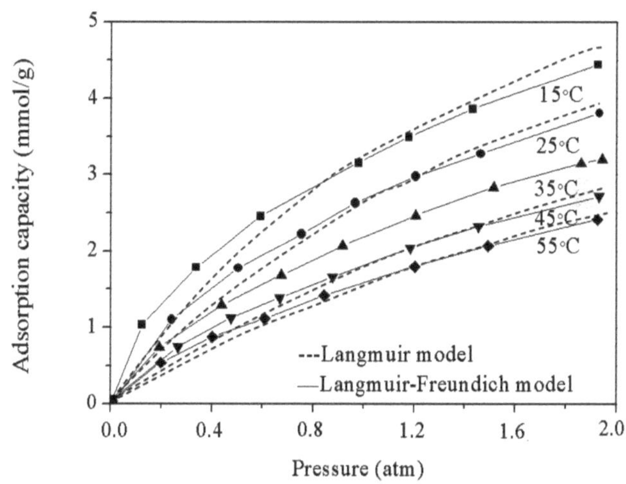

Fig. 2-38 CO_2 adsorption isotherms on a commercial activated carbon.

and micropore phases where the mass balance for the micropores is coupled with that of the macropores by a term that corresponded to the CO_2 flux at the intersection of both pores [63]. As an approximation for the activated carbon pores with a size distribution, the authors also proposed simplifying the system to have a uniform pore size that corresponded to the number-averaged micropore width, allowing identical CO_2 conductance through this pore. Predictions from this model fit the experimental data for the CO_2 adsorption-desorption kinetics measured on 6 mm slab particles of commercial Ajax carbon at three

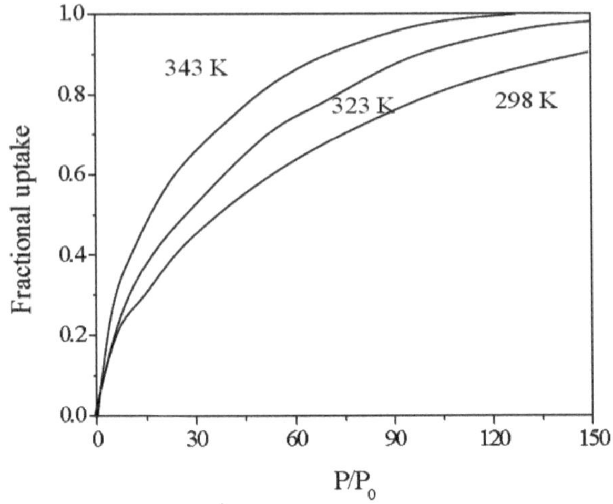

Fig. 2-39 Model fitting and experimental kinetics of CO_2 adsorption on a 6 mm slab particles of carbon at 20% mole fraction and different temperatures.

different temperatures, as shown in Fig. 2-39 [63]. The results show that despite a slight deviation in the fit of desorption kinetics, the model could successfully correlate the experimental data for adsorption kinetics. Overall, the adsorption kinetics of CO_2 on activated carbons are comparable to those on zeolites, which can approach their equilibrium capacity on time scale of minutes under many conditions. Pore structures of the activated carbons are typically represented by bidisperse pore models including micropores and macropores. In this sense, the CO_2 adsorption kinetics of activated carbons has been explained in terms of the diffusional mass transport through these pores between which the adsorbate molecules can be exchanged.

2.4.3. Gas Chromatography Mesurement

A simple method of obtaining gas-solid adsorption isotherms was introduced by James and Phillips [64] in 1954. In this method, pure carrier gas is initially passed through the column. At some time, the carrier gas is replaced by a continuous stream of a mixture of carrier gas and solute vapor at a constant concentration. This leads to a step change in the concentration-time profile of the solute, called the 'front boundary'. The boundary is carried by the mobile phase through the column. Due to the chromatographic principle, the solute is held back in the column. The time delay until its emergence at the column outlet, compared to the time pure carrier gas stays in the column, depends on the distribution coefficient of the solute, between the mobile phase (carrier gas) and the stationary phase (solid adsorbent).

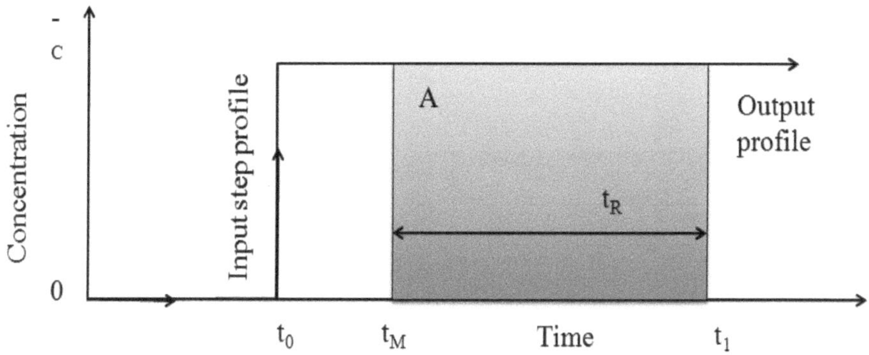

Fig. 2-40 Front boundary.

At a time t_1 the boundary reaches the end of the column. The term 'breakthrough', describes this phenomenon. After the breakthrough, the solute outlet concentration equals the solute inlet concentration. The time delay of the solute breakthrough, also t_1 called retention time, t_R, is obtained, by subtracting the travel time of pure carrier gas, t_M from the time the solute stays in the column. Fig. 2-40 shows the front boundary for an idealized case where the output is also a step.

The reduction of nitric oxide over Cu-plated activated carbon fibers is demonstrated in the copper electroplating [65]. The SEM image of the activated carbon fibers is covered with grain like nanoparticles and their amount increases with the plating time, as shown in Fig. 2-41.

Fig. 2-42, depicts that the NO conversion of the activated carbon fibers is studied. As a result, reveals the as-received sample has low activity, whereas the Cu-plated activated carbon fibers samples have high reduction activity. NO reduction efficiency curves maintain almost a fixed concentration, after 170 min of a reaction time, and increase with the copper content [65]. The Cu metals on the activated carbon fibers surfaces scavenged oxygen by oxidizing themselves into Cu_2O and finally CuO as a reductant. The Cu metals on the Cu-plated activated carbon fibers play a major role in the NO removal in this system [65].

As another example, the activated carbon fibers are produced by an oxyfluorination treatment to enhance the capacity of ammonia gas removal [66]. The specific surface area and total pore volume of activated carbon fibers decrease after oxyfluorination treatment, as shown in Table 2-8. The newly formed C-F bond and carbon–oxygen complexes have some influence on the blocking or filling of micropores on the activated carbon fibers surfaces. Meanwhile, the fraction of micropore volume slightly increases with the treatment. Thus some mesopores (near to micropore) are probably changed to

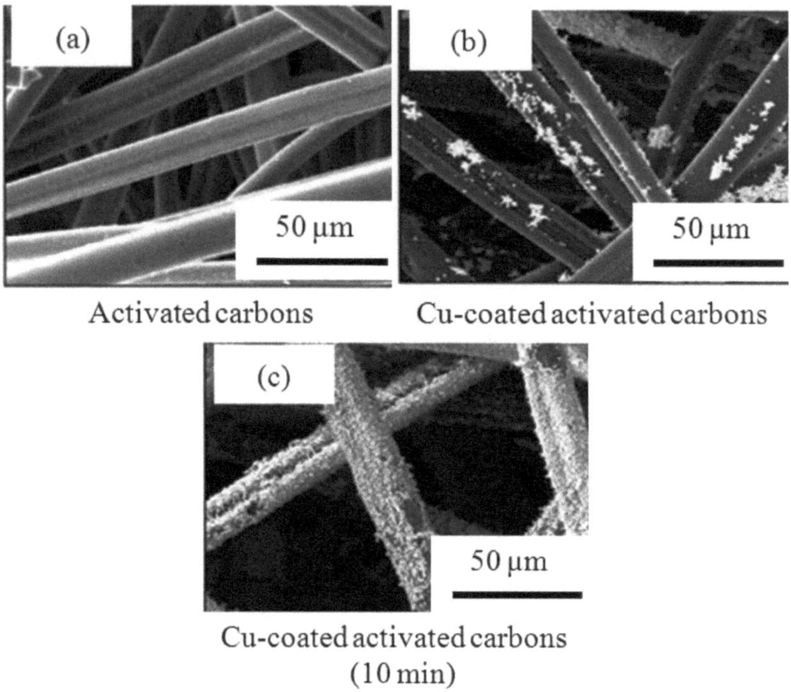

Fig. 2-41 SEM images of the electrolytically Cu-plated activated carbon fibers, as a function of the plating time.

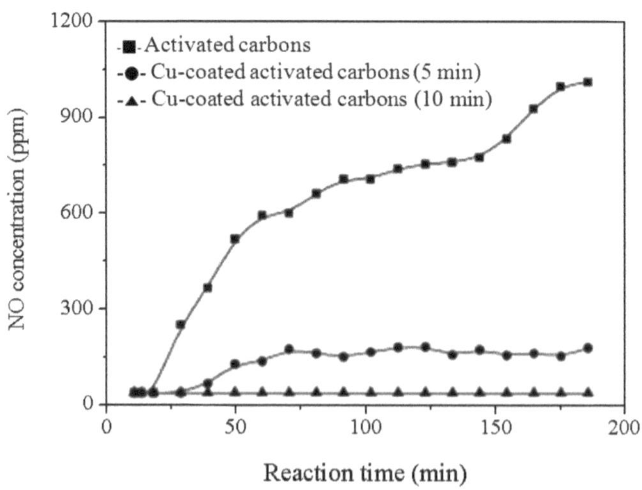

Fig. 2-42 NO reduction efficiency of electrolytically Cu-plated activated carbon fibers, as a function of the plating time.

Table 2-8 Textural Properties of the Oxyfluorinated Activated Carbon Fibers as a Function of the Treatment Time

	S_{BET}[a]	V_T[b]	V_M[c]	F_M[d]
Activated carbons	2120	1.22	1.15	94
2 min	1880	1.08	1.01	94
5 min	1730	0.95	0.90	95
10 min	1360	0.78	0.75	96

[a] *Specific surface area (m^2/g).*
[b] *Total pore volume (cm^3/g).*
[c] *Micropore volume (cm^3/g).*
[d] *Fraction of micropore (%).*

micropores by sticking some functional groups, onto the entrance of the pores, resulting in narrowing of the pore size.

The ammonia removal capacity of the activated carbon fibers before and after oxyfluorination is shown in Fig. 2-43 [66]. The capacities of all the oxyfluorinated activated carbon fibers are at a higher level than that of the as-received sample and the level increases with the treatment time. It can be seen that the ammonia-adsorption ability of the as-received sample, sharply decreases within a few hours, and reaches about 950 ppm after 10 h. These results prove that newly introduced surface functional groups containing carbon-oxygen or -fluorine complexes on activated carbon fiber surfaces play a major role in improving ammonia removal [66].

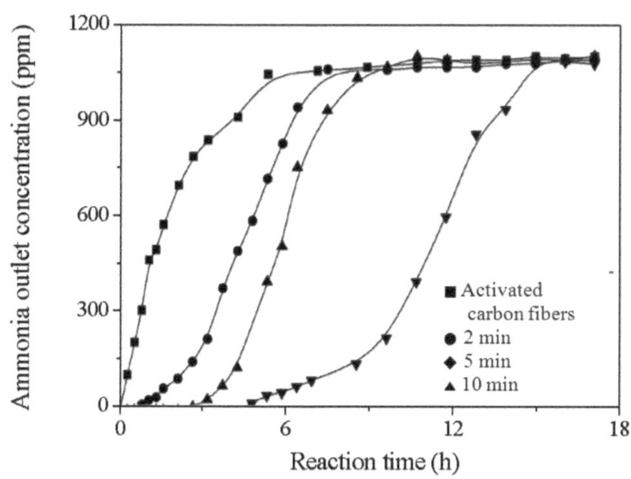

Fig. 2-43 Ammonia gas removal of the oxyfluorinated activated carbon fibers, as a function of treatment time.

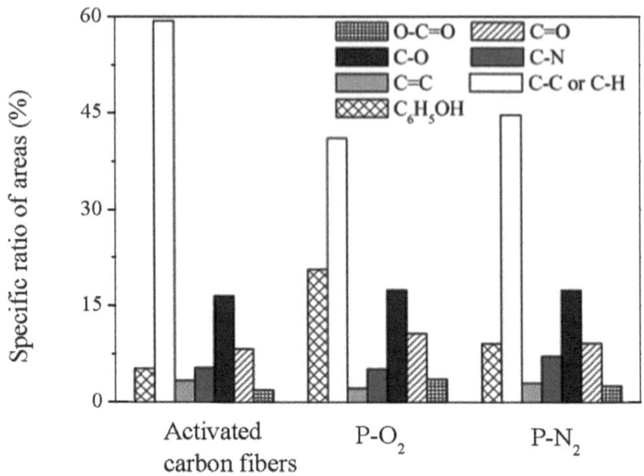

Fig. 2-44 Specific ratios of the areas of activated carbon fibers, before and after plasma treatments, in C1s XPS spectra (plasma treatment at O_2: P-O_2, plasma treatment at N_2: P-N_2).

The atmospheric pressure plasma treatments (Ar/O_2 and Ar/N_2) of activated carbon fibers are carried out to introduce hydrophilic functional groups on carbon surfaces in order to enhance the hydrogen chloride gas (HCl) adsorption [67]. As shown in Fig. 2-44, the oxygen and nitrogen contents on the activated carbon fibers are increased in O_2 and N_2, respectively [67].

The atmospheric pressure plasma with O_2 and N_2 gases increases the hydrophilicity of activated carbon fibers studied, resulting in an increase of hydrophilic HCl gas removal, as shown in Fig. 2-45 [67]. It is reasonable that the surfaces of P-N_2 have a basic nature due to the Ar/N_2 plasma treatment resulting in large nitrogen-containing functional groups, such as NH_4, –NH_2, and =NH. This basic nature of atmospheric pressure plasma with O_2 can lead to an increase in the removal efficiency of acidic gas, such as HCl. It is concluded, that the plasma treatment with active gases like O_2 and N_2 leads to the increase of HCl [67].

A nickel metal electroplating on the activated carbon fiber surfaces was carried out to remove the toxic hydrogen chloride gas [68]. As shown in Fig. 2-46 (a and b), the surface pore structures on activated carbon fibers undergo pore-blocking phenomena owing to deposition of nickel metal on the activated carbon fibers. As the electroplating proceeds, nickel metal nuclei migrate and aggregate, resulting in the formation of nickel particles and then the particles grow to a greater size. As shown in Fig. 2-46 (c), the sharp peaks around $2\theta = 44°$ and $51°$ correspond to the (111) and (200) planes of nickel on the surface of the Ni/activated carbon fibers. The intensity of the carbon peak in the Ni-plated activated carbon fibers decreased, whereas the nickel peaks and crystallinity in the Ni-plated activated carbon fibers are strengthened with an extended plating time. As the electroplating proceeds, nickel metal nuclei

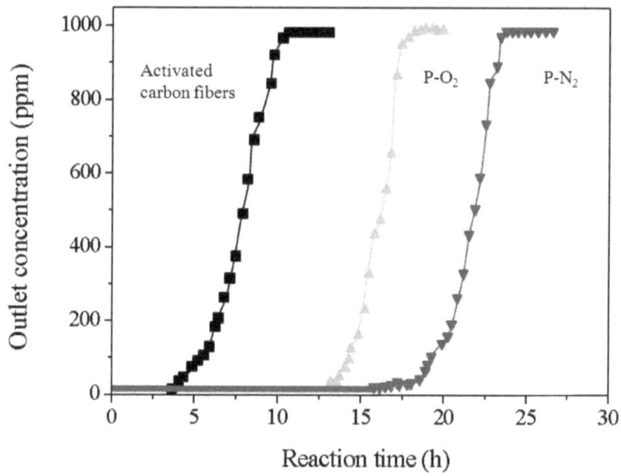

Fig. 2-45 Breakthrough curves of the activated carbon fibers before and after (plasma treatment at O_2:P-O_2 plasma treatment at N_2:P-N_2) plasma treated activated carbon fibers for HCl removal (Input HCl = 1013 ppm, feeding rate = 10 ml/min, and temperature = 25 °C).

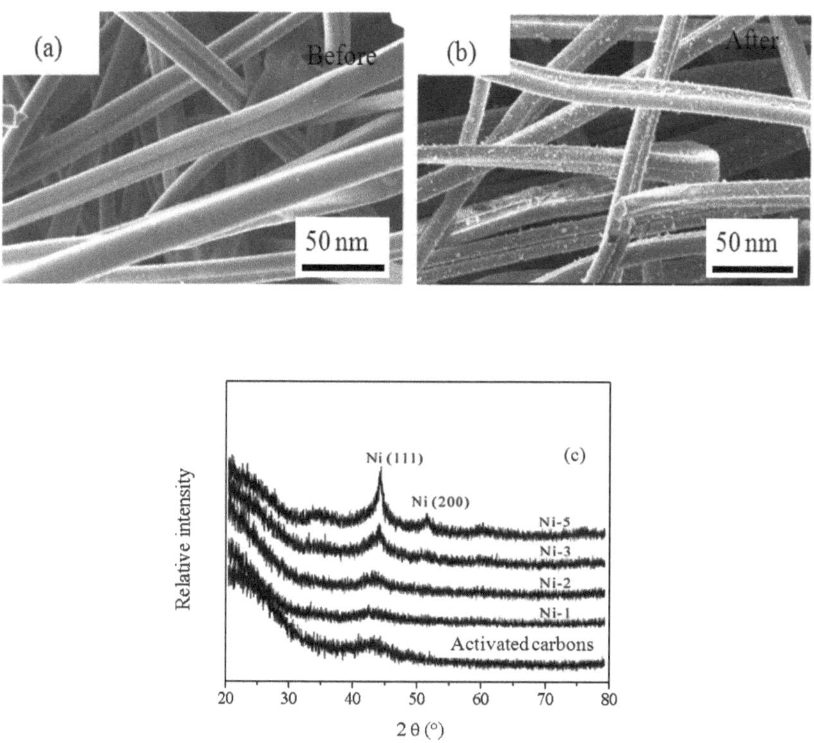

Fig. 2-46 SEM image (a, b) and XRD (c) patterns of the activated carbon fibers (a) and after electrolytic Ni-plated activated carbon fibers (b) as a function of Ni content.

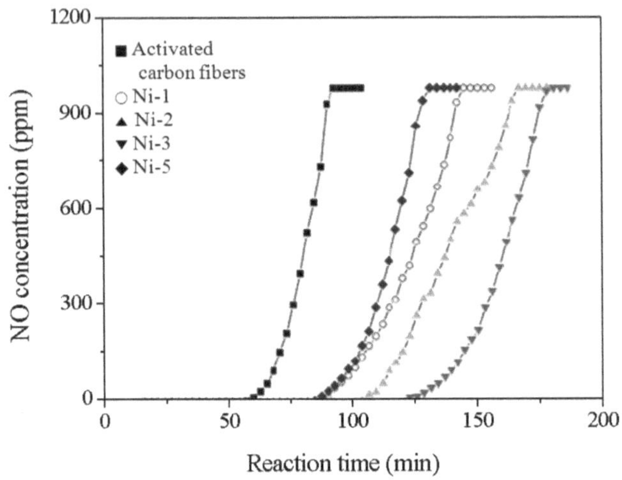

Fig. 2-47 HCl removal efficiency of the electrolytic Ni-plated activated carbon fibers as a function of Ni content.

migrate and aggregate resulting in the formation of nickel particles and then the particles grow to a greater size.

The electrolytically Ni-plated activated carbon fibers show much higher removal efficiency than that of untreated activated carbon fibers, and the 3.0 wt.% Ni-plated/activated carbon fibers sample shows the highest removal efficiency, as shown in Fig. 2-47 [68]. This result means that the removal efficiency depends on induced metal, but the specific surface area also affects it, as observed in the 5.0 wt.% Ni-plated activated carbon fibers sample. The catalytic reaction occurs by the strong interaction between Ni and Cl and formed ionic compound, such as $NiCl_2$, during the HCl gas flow through the packed sample. The nickel, as an active site, takes part in HCl removal, and enhances the removal efficiency [68].

In addition, Guliants *et al.* [69] determined the CO_2 capacity for polyethylenimine-substituted, chloropropyl silane-functionalized, and MCM-48, under dry conditions with gas chromatography. The amine efficiency for this material is 0.08 and very little pore volume existed after functionalization (0.02 cm^3/g), suggesting that many of the amines in the polyethylenimine are inaccessible due to nearly complete pore filling.

2.5. INFINITE AND FINITE CONCENTRATION

2.5.1. Solid-gas Interaction at Infinite Dilution

The technique of inverse gas chromatography at infinite dilution has been used for solid surface characterization with great success, especially for estimation of the dispersive component of surface energy by adsorption of n-alkanes [70–76].

It is generally accepted nowadays that the final mechanical characteristics and durability of a composite material depends largely on the degree of adhesion of the polymer matrix to the reinforcing filler that is achieved, which is the quality of the interface that is formed between the two solids [76].

An understanding of gas-solid interaction as a physical origin begins with an appreciation of the (total) polarizability of molecules, which is described as being related to both the deformation polarizability and dipole moment (or orientational polarizability) properties of two different substances at an interface. Park [10, 70, 76], examined the reversible molecular force of a solid surface measured by the gas-solid chromatographic technique at infinite dilution or at finite concentration, and described the energetic distribution of a solid surface with an increasing amount of probe molecule. This section describes the degree of gas-solid interaction for ignoring the interactions between the probe (or solute) molecules themselves.

2.5.1.1. Adsorption Gibbs free energy

In the inverse gas chromatography at infinite dilution or near zero surface coverage, the surface characterization of the solid is realized by injecting very small quantities of probes of known properties into a column packed with the solid. The retention time is the fundamental parameter characterizing the equilibrium distribution of the probe concentration, between the probes where all probe molecules behave independently. Therefore, the retention time is a function only of the adsorbate-adsorbent interaction [77].

Fig. 2-48 shows a chromatogram for linear elution chromatography. The amount of probe injected is set to the amount that corresponds to the limit of

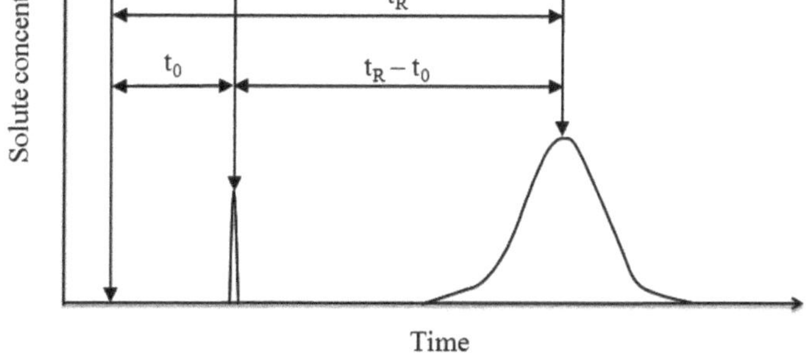

Fig. 2-48 Chromatogram for elution chromatography.

sensitivity of the detector. The retention time, t_R, is the time that the average molecule of a probe takes to travel the length of the column. The time is measured to be the midpoint of the symmetrical curve for non-polar probes and to be the position at which the total area of the asymmetrical curve for polar probes, is divided into equal areas. The time, t_0, is the zero retention time reference measured with a non adsorbing probe, such as methane, simply to pass through the column space from the inlet to the outlet.

Therefore, the net retention volume, V_N, may be calculated from

$$V_N = J_3^2 \cdot D \cdot (t_R - t_0) \tag{2-41}$$

where J_3^2 is the pressure-gradient correction factor taking into account the compressibility of the gas according to James and Martin [78], and D the corrected flow rate and J_3^2 and D are respectively defined by [77, 78]

$$J_3^2 = \frac{3}{2}\left[\frac{\left(\frac{\Delta P}{P_A}+1\right)^2 - 1}{\left(\frac{\Delta P}{P_A}+1\right)^3 - 1}\right] \tag{2-42}$$

and

$$D = D_S \frac{T_E}{T_C} \frac{P_A - P_{H_2O}}{P_A} \tag{2-43}$$

where P_A is atmospheric pressure (mmHg), ΔP the pressure deviation in the column (mmHg), D_S the flow rate with ΔP (cm^3/min), T_C and T_R, respectively (K), column and room temperatures, and P_{H_2O} the partial pressure of water vapor at room temperature (mmHg) [79, 80].

The equilibrium thermodynamic functions describing the retention process are essentially related to the net retention volume. For example, the intermolecular adsorption of Gibbs free energy, $-\Delta G_A$, for one mole of solute vapor from a reference gaseous state with the partial pressure, P_0, to a reference adsorbed state with the equilibrium spread pressure (or two-dimensional pressure), π_0a, is given as [81]

$$-\Delta G_A = RT\, ln\left(\frac{V_N \cdot P_0}{S_P \cdot W_g \cdot \pi_0}\right) \tag{2-44}$$

where S_P and W_g are respectively the specific surface area (m^2/g) and the weight (g) of the solid in the column.

In order to determine the intermolecular adsorption of Gibbs free energy, $-\Delta G_A$, in Eq. (2-44), two reference states and one intermolecular function state based on the polarizability of molecules are generally considered:

1. The reference state of Kembell and Rideal [82] where $P_0 = 1.013 \times 10^5$ Pa and $\pi_0 = 6.078 \times 10^{-5}$/Nm.

2. The reference state of De Boer and Kruyer [83] where $P_0 = 1.013 \times 10^5$ Pa and $\pi_0 = 3.38 \times 10^{-5}$/Nm.
3. The intermolecular function state, based on the polarizability of molecules at a given temperature in the column, as it was made before by Donnet and Park [84, 85].

Since the intermolecular adsorption of Gibbs free energy, $-\Delta G_A$, in Eq. (2-44), is independent of the specific surface area and weight of the solid in the column, except for the net retention volume, V_N, in the inverse gas chromatography at infinite dilution. Therefore $-\Delta G_A$ can be rewritten

$$-\Delta G_A = \mathrm{RT}\ ln V_N + C \tag{2-45}$$

where C is a constant dependent on the chosen reference states.

As mentioned before, it is generally accepted that the intermolecular adsorption of Gibbs free energy of an adsorbate-adsorbent system have been generally described by the sum of two terms [86, 87]; the London dispersive component, $-\Delta G_A^L$, and the specific component, $-\Delta G_A^{SP}$, as

$$-\Delta G_A = [-\Delta G_A^L] + [-\Delta G_A^{SP}] \tag{2-46}$$

with the superscript L and SP referring the London dispersive and specific components of Gibbs free energy, respectively.

Park and co-workers [84, 85] obtained a simple equation for two interacting molecules of gas (or liquid, subscript L) and solid (S) having the respective characteristic electronic vibrational frequency (or quantized harmonic oscillator of frequency) ν_L and ν_S in Eq. (1–22) [84, 85]. (See 1.2.1.1)

For the London dispersive component of the adsorbate-adsorbent interaction for a given liquid, the basis of the polarizability of molecules (in Table 1-6) are listed in Table 2-9.

$$-\Delta G_A^L = K \cdot \{(h\nu_S)^{1/2} \cdot \alpha_{0,S}\} \cdot \{(h\nu_L)^{1/2} \cdot \alpha_{0,L}\} \tag{2-47}$$

For polar probes, the additional or specific component of the Gibbs frees energy, $-\Delta G_A^{SP}$, resulting from polar interactions is then determined by the distance between the experimental points A and B of the same abscissa on the n-alkane lines as illustrated in Fig. 2-49.

This is Donnet-Park's approach [84, 85] based on the intermolecular function state that is derived from the polarizability of molecules in the adsorbate-adsorbent system for the experimental results of inverse gas chromatography at infinite dilution.

As an example, Table 2-9 shows the results of the adsorption of Gibbs free energy determined from the three present methods (two reference states and an intermolecular function state). In the comparison of these methods, the values obtained from the intermolecular function state are in good agreement with those from the heat of the vaporization of the liquid, which is assumed to be in the order of magnitude of long-range forces between two-identical molecules.

Table 2-9 Gibbs Free Energies (kJ/mol) for the Three Present Methods Between Probes and Carbon Fibers[a]

Probes	KR[b]	DB[c]	$-\Delta G_A$[d]	ΔH_V[e]
n-Hexane	22.57	18.19	34.74	31.56
n-Heptane	26.72	22.35	38.90	36.57
n-octane	30.88	26.50	43.05	41.49
Diethylether	20.37	15.60	32.55	27.10
Methyl acetate	21.93	17.56	35.10	32.29
Ethyl acetate	24.72	20.35	36.89	35.60
THF	23.66	19.29	35.84	31.99
Acetone	22.27	17.90	34.45	30.99
Nitromethane	25.22	20.85	37.39	38.27
Acetonitrile	25.29	20.92	37.46	32.94

Source: [10, 88].
[a] *High strength carbon fiber (T-300) electro-oxidized by the current density measured at 33.5 °C.*
[b] *Reference state of Kemball and Rideal [82].*
[c] *Reference state of de Boer and Kruyer [83].*
[d] *Intermolecular function state based on the polarizability of molecules [84, 85].*
[e] *Heat of vaporization measured at 25 °C [80].*

It is also generally accepted that the interaction energy of adsorbate-adsorbent may be greater than that of adsorbate-adsorbate, because their polarizabilities significantly increase with the molecular mass (or molecular volume) according to the Lorentz–Lorenz–Debye equation [10, 89, 90].

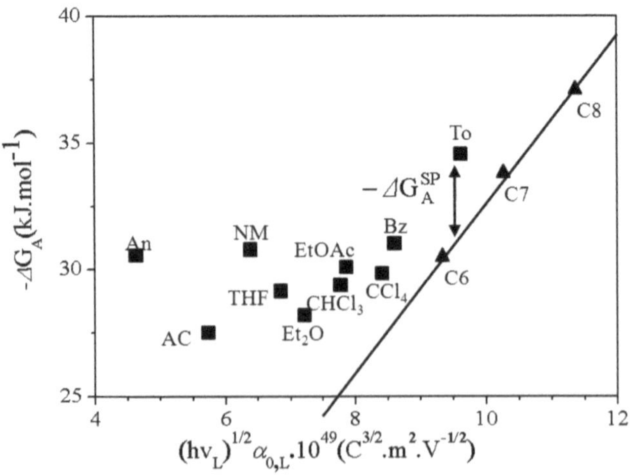

Fig. 2-49 Principle of determining the London dispersive and specific components of adsorption Gibbs free energy between untreated carbon fiber (supplied by Soficar, T-300), and all probes, measured at 54.5 °C.

Considering the interaction energy at a given experimental temperature, may lead to a very useful absolute value in the adsorbate-adsorbent system for determining the adsorption value of enthalpy in the infinite dilution region [91]. It also leads to an estimate of the maximum adsorption value on the highest active site of a solid surface when evaluating the distribution function with the amount adsorbed in a monolayer [70].

In the next section, the values obtained from the intermolecular function state are carefully discussed in terms of the specific component of surface free energy or acid-base interaction between adsorbate-adsorbent.

2.5.1.2. London Dispersive Component

The London dispersive component of the surface free energy, γ_S^L, of a solid is a predominant property for the prediction of behavior of nonpolar adsorbents such as polyolefins or of practically nonpolar adsorbents like some carbon materials (natural graphites and carbon fibers).

A simple approach for the determination of the γ_S^L using nonpolar probes such as n-alkanes in inverse gas chromatography at infinite dilution is demonstrated [91]. They also discussed the evaluation of the London dispersive component of the surface energy (or enthalpy, h_S^L), starting from the variation of the adsorption characteristics, of a series of long-chain n-alkanes molecules, with temperature.

The interaction of nonpolar probes, such as n-alkanes, with a nonpolar support may be studied by the work of adhesion, W_A^L, using by geometic mean.

$$W_{\mathrm{A}}^{\mathrm{L}} = 2\sqrt{\gamma_{\mathrm{S}}^{\mathrm{L}} \cdot \gamma_{\mathrm{L}}^{\mathrm{L}}} \tag{2-48}$$

where the subscripts S and L represent the solid and liquid (or probe), respectively, and $\gamma_{\mathrm{L}}^{\mathrm{L}}$ the London dispersive component, of the surface free energy of the probes.

Eq. (2-48) may be rewritten, in terms of the London dispersive component of adsorption Gibbs free energy, $-\Delta G_{\mathrm{A}}^{\mathrm{L}}$

$$W_{\mathrm{A}}^{\mathrm{L}} = \frac{-\Delta G_{\mathrm{A}}^{\mathrm{L}}}{\mathrm{N_A} \cdot a_{\mathrm{L}}} \tag{2-49}$$

and

$$-\Delta G_{\mathrm{A}}^{\mathrm{L}} = 2\mathrm{N_A} \cdot a_{\mathrm{L}} \cdot \sqrt{\gamma_{\mathrm{S}}^{\mathrm{L}} \cdot \gamma_{\mathrm{L}}^{\mathrm{L}}} \tag{2-50}$$

where $\mathrm{N_A}$ is Avogadro's number and a_{L} the sectional area of the adsorbed alkanes.

In the chromatographic process, three methods have been established for determining the London dispersive component of the surface free energy of a solid.

In the early 1980s, Gray *et al.* [71, 72, 92] developed a method for evaluating γ_S^L. They used a series of n-alkanes and determined the incremental adsorption free energy of a methylene group, i.e., $-\Delta G_A(CH_2)$. If a_{CH_2} is the cross-sectional area of an adsorbed CH_2 group, assumed by Gray *et al.* as 0.06 nm,^{2} and $\gamma_{CH_2}^L$ (or, simply γ_{CH_2} in the case of n-alkanes) is the surface free energy of a CH_2 group i.e., of a surface constituted of close packed CH_2 groups analogous to polyethylene, one obtains

$$-\Delta G_A(CH_2) = 2N_A \cdot a_{CH2} \cdot \sqrt{\gamma_S^L \cdot \gamma_{CH_2}} \tag{2-51}$$

where $-\Delta G_A(CH_2)$ is equal to

$$-\Delta G_A(CH_2) = RT \ln \frac{V_{n+1}(C_{n+1}H_{2n+4})}{V_n(C_nH_{2n+2})} \tag{2-52}$$

and γ_{CH_2} becomes [93, 94]

$$\gamma_{CH_2} = 35.6 - 0.058(t - 20) \text{ (in mJ/m}^2) \tag{2-53}$$

with *t* being the temperature in Celsius.

In Eq. (2-51), all terms are either known (N_A, a_{CH_2}, γ_{CH_2}) or measurable ($-\Delta G_A(CH_2)$), except the quantity of interest: γ_S^L. This procedure has been used for characterizing various solid surfaces by many authors [74, 95–98].

In a similar way, Schultz and co-workers [74, 99] proposed a relation for the calculation of γ_S^L, but based on a different variable, that is $[a_L \cdot (\gamma_L^L)^{1/2}]$, a_L the section of the adsorbed alkane, and γ_S^L the surface free energy of the injected n-alkanes (usually, C_5H_{12}-$C_{12}H_{26}$), obtained at room temperature [100, 101]. $[-\Delta G_A]$, may be calculated using the following equations

$$-\Delta G_A = RT \ln V_N + C \tag{2-45}$$

$$= 2N_A \cdot a_L \cdot \sqrt{\gamma_S^L \cdot \gamma_L^L} + C \tag{2-54}$$

where *C* is a constant dependent on the total surface area of the solid in the column and on the choice of a reference state for the adsorbed solute.

The term $[a_L \cdot (\gamma_L^L)^{1/2}]$ may be computed for a series of n-alkanes according to [91]

$$a_{n,L} \cdot (\gamma_{n,L}^L)^{1/2} = 1.57 + 36.25 \times C_n (\text{in } \mathring{A}^2 (\text{mJ/m}^2)^{1/2}) \\ n = 12, \quad R = 0.976, \quad SD = 0.47 \tag{2-55}$$

Here the subscript *n* is the number of carbon atoms of n-alkanes from C_5 to C_{12}. n is the number of n-alkanes probes included in regressions, *R* the correlation coefficient, and *SD* the standard deviation of the fit.

Eq. (2-54) can be transformed in Eq. (2-56), since the variation of $[-\Delta G_A]$ with $[a_L \cdot (\gamma_L^L)^{1/2}]$ is linear [91].

$$-\Delta G_A(CH_2) = N_A \cdot 2(\gamma_S^L)^{1/2} \times (36.25) \tag{2-56}$$

When measuring the quantity $[-\Delta G_A(CH_2)]$ according to Eq. (2-56) it is possible to determine the quantity of interest, γ_S^L. The values of γ_S^L obtained from the above method are similar to those obtained using either the method developed by Gray *et al.* [71, 72] or the wetting method [74, 102]. This method has also been used for characterizing various solid surfaces [74, 99, 103].

The analysis [91] is based on the deformation polarizability, $\{(h\nu L)^{1/2} \cdot \alpha 0, L\}$, of the liquid in Eq. (2-47) with the arithmetic mean instead of Fowkes' geometric mean in Eqs. (2-48) and (2-51).

In third method, the estimation of γ_{CH_2}, is estimated at 1.32 mJ/m^2 taking the difference between the surface free energies of n-alkanes having respectively n and $n+1$ of carbon atoms (e.g., C_5H_{12}-$C_{12}H_{26}$), instead of Eq. (2-53).

$$\gamma_{n,L} = 9.97 + 1.32 \times C_n \tag{2-57}$$

$n = 12$, R (correlation coefficient) = 0.976, SD (standard deviation) = 0.47

Owing to the very small value of γ_{CH_2}, the following first-order equation can be taken for the expression of the surface free energy, using an arithmetic mean, into as Zettlemoyer did [104]. It is not meaningful to apply the geometric or harmonic means given the important differences in the two considered parameters γ_S^L and γ_{CH_2} which is

$$W_A^L(\text{solid} - CH_2) = \gamma_S^L + \gamma_{CH_2} \tag{2-58}$$

One is more favorable for the determination of the two parameters, rather than. In the case of the adsorption system of a methylene group with a support, Eq. (2-51) can be transformed

$$-\Delta G_A(CH_2) = N_A \cdot a_{CH_2} \cdot \gamma_S^L + N_A \cdot a_{CH_2} \cdot \gamma_{CH_2} \tag{2-59}$$

As seen, this relation is satisfactorily verified, and Eq. (2-61) can be simplified, since the contribution of the quantity, $[N_A \cdot a_{CH_2} \cdot \gamma_{CH_2}]$ is usually less than 2% that of the solid [91].

$$-\Delta G_A(CH_2) = N_A \cdot a_{CH_2} \cdot \gamma_S^L \tag{2-60}$$

According to Gray *et al.* [71, 72, 93], the surface area of a methylene group, a_{CH_2} is equal to 0.06 nm^2, although different quantities have been estimated; e.g., 0.031 nm^2 was calculated from the spherical model in the most dense face-centered packing in the liquid state [in Eq. (2-61)] [105], 0.0524 nm^2 from Groszek's model being equal to the area of the hexagon in the graphite basal plane [106], 0.0611 nm^2 from Clint's crystalline model [107], and values up to

0.077 nm^2 (0.154 × 0.45 ± 0.05) recently from scanning tunneling microscopy (STM) on $C_{32}H_{66}$ [108].

$$a_L = \frac{6}{\sqrt{3}}\left(\frac{M}{4\sqrt{2}\cdot N_A\cdot \rho}\right)^{2/3} \tag{2-61}$$

where M is the molecular weight of the liquid, N_A Avogadro's number and ρ the density of the solidified liquid.

However, it should be noted that the determination by Eq. (2-60) is much less dependent on the variation of the surface area of a methylene group than those by Eqs. (2-51) and (2-54). In Eq. (2-60), all terms are either known or measurable, except the quantity of interest: γ_S^L Moreover, Eq. (2-60) is very interesting since it takes into account a small adsorbate, i.e., $-CH_2-$ group, whose surface area and surface free energy are slightly affected by temperature. This means that the variations in area and surface entropy of an adsorbed $-CH_2-$ group are negligible with temperature. Therefore, it is possible to determine the surface enthalpy derived from the Gibbs–Helmholtz equation

$$h_S^L = \gamma_S^L - T\left(\frac{\gamma_S^L}{dT}\right) \tag{2-62}$$

where h_S^L is the London dispersive component, of the surface enthalpy of a solid (mJ/m^2).

The plot of the natural logarithm of the net retention volume (or the adsorption free energy of the probes used [85]) against the number of carbon atoms of linear n-alkanes gives a slope from which, incremental adsorption free energy of a methylene group, $-\Delta G_A(CH_2)$, is determined, as illustrated in Fig. 2-50.

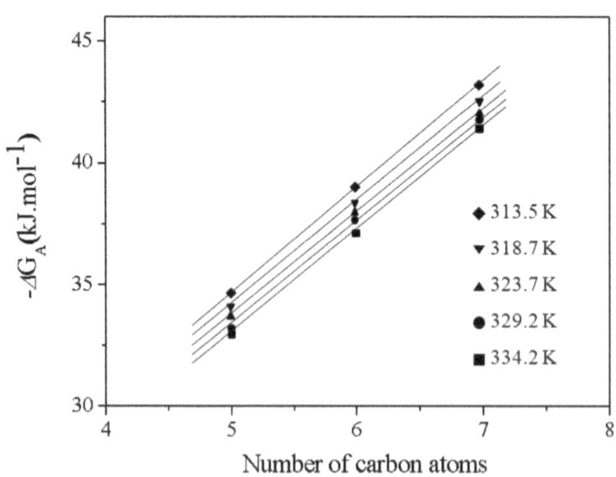

Fig. 2-50 Linear variation of the adsorption free energy, with the number of carbon atoms of n-alkanes at various chromatographic temperatures.

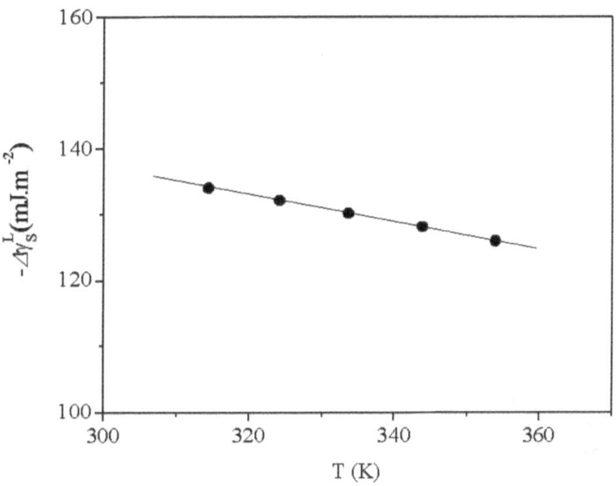

Fig. 2-51 Linear variation of γ^L_S, with temperature (K).

From the values thus obtained, at different temperatures, one finally obtains the London dispersive component of the surface free energy, γ^L_S, of the solid which can be extrapolated at room temperature, as shown in Fig. 2-51. The experimental values of γ^L_S evaluated from Gray, Schultz and our analytical methods are given in Table 2-10, with a 95% confidence level.

The method using Eq. (2-60) can also be used to evaluate surface enthalpy, h^L_S. The values for the studied carbonaceous materials are also listed in Table 2-10, with a 95% confidence level [91].

The BET surfaces, measured with N_2 at 77 K [6], indicate that these materials have very low specific surface areas having of 1.0 m^2/g, except sample V (500 m^2/g) [91]. In Table 2-10, the γ^L_S values determined from Eq. (2-60) may be observed:

The γ^L_S values obtained by Park's methods for the very low specific surface area (samples I ~ IV) are different from that measured by the BET technique, approximately 1.0 m^2/g. In addition, sample *V* has a moderately high γ^L_S value; i.e., only several times higher than with those measured for samples I-IV.

According to this observation, it is difficult to appreciate the maximum value of the surface free energy value of the surface free energy and surface enthalpy of a solid, especially in the case of microporous materials, which have widely efficient adsorption properties of the surface (Sample V). Therefore, for this material, more studies may be needed on the adsorption isotherm, spreading pressure, isosteric heat of adsorption, and even heterogeneities of solid surfaces. Park [91] was concerned with the finite concentration technique with an increasing amount adsorbed, which will be dealt to some extent in the next section. As shown in Table 2-10, the validity of this work is verified by the fact that the correlation coefficients (R) are largely constant and also the

Table 2-10 γ_S^L **Extrapolated at 25 °C Determined by Three Methods,** h_S^L **and** γ_S^L **Temperature Dependence of the Solid [**γ_S^L**: mJ/m**^{2}**,** h_S^L**: mJ/m**^{2}**,** $-\gamma_S^L/dT^f$**: mJ/m**^{2}**/K]**

Sample	I[a]	II[b]	III[c]	IV[d]	V[e]
Park's analysis	138 ± 1	151 ± 4	135 ± 1	99 ± 2	281 ± 14
	R[g] = −0.999	R = −0.992	R = −0.998	R = −0.996	R = −0.998
	SD[h] = 0.22	SD = 0.84	SD = 0.18	SD = 0.64	SD = 0.46
h_S^L	214 ± 8	239 ± 19	220 ± 12	256 ± 19	400 ± 39
$-\gamma_S^L/dT^f$	0.275	0.296	0.286	0.525	0.397

[a] *Natural graphite from Madagascar.*
[b] *Natural graphite from Madagascar heated to 800 °C and Ar plasma treated.*
[c] *Commercial synthetic graphites supplied by Lonza Co.*
[d] *Commercial SiC supplied by Aldrich Chemical Co.*
[e] *Pitch-based isotropic carbon fibers activated by steam at 900 °C.*
[f] *Values obtained from column 3 in this Table.*
[g] *Correlation coefficient.*
[h] *Standard deviation of the fit.*

standard deviations (SD) of the fit are quite low. The infinite dilution technique, which is useful for determining the incremental adsorption free energy of a methylene group, $-\Delta G_A(CH_2)$, permits us to evaluate the γ_S^L, and, moreover, provides a simple way to obtain another physical property of the surface, h_S^L.

2.5.1.3. Acid–Base Component

In order to determine the specific component, $-\Delta G_A^{SP}$ or acid–base component, $-\Delta G_A^{A-B}$ in Eq. (2-46), which causes the acid–base character of a solid surface, the method of intermolecular function state based on the polarizability of molecules was reported early [84, 85, 109].

As mentioned above, the specific component of Gibbs free energy, $-\Delta G_A^{SP}$, is also evaluated from the plot of $[-\Delta G_A]$ or $[\mathrm{RT}\ln V_N + C]$, in Eq. (2-46), *vs.* $\{(h\nu_S)^{1/2}\cdot\alpha_{0,L}\}$ obtained from the following Eq. (2-63).

$$-\Delta G_A^{SP} = \mathrm{RT}\,ln\frac{V_N(\text{polar} - \text{probes})}{V_N(\text{n} - \text{alkanes})} \tag{2-63}$$

From a physic-chemistry point of view, the specific component of Gibbs free energy, $-\Delta G_A^{SP}$, may be at least equal to zero for extremely nonpolar probes, and largely increases with their degree of dipole moments for polar probes according to Eq. (2-36).

Therefore, the experimental corresponding points to the polar probes always lie above the reference line of n-alkanes, and practically as a function of the degree of their polarity, δ, following the Stockmayer's equation [110, 111].

$$\delta = 1.94 \times \frac{\mu^2}{V_b T_b} \tag{2-64}$$

Table 2-11 Characteristics of the Probes

Probes	$(h\nu)^{1/2}\alpha_0 \cdot 10^{49}$ [a] $(C^{3/2}/m^2/V^{1/2})$	T_b [b] (K)	V_c [c] (cm^3/g/mol)	V_b [d] (cm^3/g/mol)	δ	Character
n-Pentane	8.1	309	304	114	0	Nonpolar
n-Hexane	9.2	342	370	140	0	
n-Heptane	10.3	372	432	165	0	
n-Octane	11.4	399	492	189	0	
n-Nonane	12.5	424	548	211	0	
CCl_4	6.2	350	276	103	0	
Benzene	8.6	353	259	96	0	
Toluene	9.5	384	316	119	0.01	Weakly Polar
Xylene	10.5	412	376	142	0.00	
Chloroform	7.8	335	239	89	0.07	
Diethylether	7.3	308	280	105	0.08	
Methyl acetate	6.2	330	228	84	0.21	
Ethyl acetate	7.9	350	286	107	0.16	
THF	6.8	340	224	83	0.18	Polar
Acetone	5.8	329	209	77	0.64	
Nitromethane	6.4	374	173	63	0.98	
Acetonitrile	4.7	355	173	63	1.33	

[a] *Computed from column in Table 1-6.*
[b] *From Ref. [10, 112].*
[c] *From Ref. [10, 111].*
[d] *Computed from the critical volume, V_c, according to Tyn and Calus [10, 113].*

where μ is the dipole moment (Debye), V_b the liquid molar volume at the normal boiling temperature (cm^3/g/mol), and T_b the boiling temperature in Kelvin (K).

Table 2-11 shows a few calculated values of the degree of polarity, for nonpolar and polar probes determined from Eq. (2-64) with the potential parameters, $\{(h\nu_S)^{1/2} \cdot \alpha_{0,L}\}$. Polar probes are normally defined as those which have $\delta \geq 0.07$ [110].

As an example, Park *et al.* measured the values of $[-\Delta G_A^L]$, $[-\Delta G_A^{SP}]$, and total Gibbs free energy, $[-\Delta G_A]$, of the studied polar probes for the untreated high-strength carbon fiber (namely T-300) at 54.5 °C, are listed in the Table 2-12 obtained from the results of the Fig. 2-49 [110].

Table 2-12 shows a few calculated values of the degree of polarity, for nonpolar and polar probes determined from Eq. (2-64) with the potential parameters, $\{(h\nu_S)^{1/2} \cdot \alpha_{0,L}\}$. Polar probes are normally defined as those which have $\delta \geq 0.07$ [110].

Based on the method derived from the intermolecular function state, the plotting of $[-\Delta G_A^{SP}]$ against the degree of polarity δ, of probes, indicates that all the experimental points corresponding to the weakly polar and polar probes are aligned, as seen in the Fig. 2-52 derived from the results of Fig. 2-49 [110].

Table 2-12 London Dispersive $[-\Delta G_A^L]$, Specific $[-\Delta G_A^{SP}]$, and Total Adsorption Gibbs Free Energy $[-\Delta G_A]$, in the Carbon Fibers Measured at 54.5 °C (in kJ/mol) According to Eqs. (2-46) and (2-47)

Probes	$-\Delta G_A^L$	$-\Delta G_A^{SP}$	$-\Delta G_A$
n-Hexane	30.7	0	30.7
n-Heptane	34.4	0	34.4
n-Octane	38.1	0	38.1
CCl_4	28.1	2.1	30.2
Benzene	28.7	2.4	31.1
Toluene	31.7	3.5	35.2
Chloroform	26.1	3.9	30.0
Diethylether	24.4	4.4	28.8
Ethyl acetate	26.4	4.1	30.5
Tetrahydrofuran	22.7	7.2	29.9
Acetone	19.4	9.1	28.5
Nitromethane	21.4	9.7	31.1
Acetonitrile	15.7	15.5	31.2

A study of the interaction of Lewis acids and bases (or electron acceptors and donors) in surface dynamics has led to new insight into interactions with various solid surfaces [109, 110, 114], as well as interactions at interfaces between two different substances. The acid-base interactions of Lewis, including the orientational properties of the charge transfer forces of Mulliken [115], occur

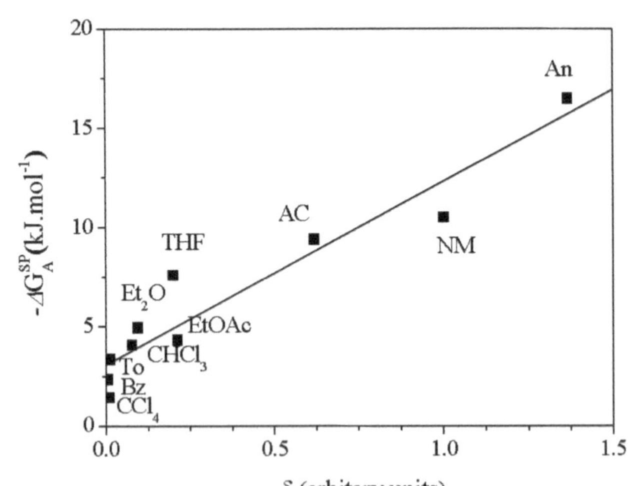

Fig. 2-52 Correlation between $[-\Delta G_A^{SP}]$ obtained from the results of Fig. 2-44 and degree of polarity **δ**, of the liquids used.

between specific (or polar) groups in substances. These interactions are quite dependent on the Stockmayer degree of polarity, δ, [116] as measured by dipole moment in Eq. (2-64). Furthermore, the concept that 'acids attract bases' may be substituted for the old rule that 'like attracts like' in the interaction of acids and bases. This is obvious in solubility and swelling studies with solid substances, where acidic surfaces interact strongly with basic solvents but not with acid solvents [117, 118]. Therefore, the Lewis concept of acids as electron acceptor (electrophilic, anion-solvating) and bases, as electron donors (nucleophilic, cation-solvating) can be taken into account in terms of specific interactions to include all molecules. The major characteristics of substances are essentially amphoteric and today, lots of efforts are still being made to exploit it. On the other hand, the acidity or basicity of organic molecules may be correlated by three empirical solvent parameter approaches.

Drago *et al.* [119] proposed a four-parameter equation, for predicting the calorimetric enthalpies of acid-base interactions, $-\Delta H^{ab}$, in neutral solvents, with spectroscopically determined shifts of the OH stretching frequency of phenol when interacting with a variety of bases of dilute solutions in carbon tetrachloride and tetrachloroethylene,

$$-\Delta H^{ab} = 0.0103\Delta\nu_{OH}(\mathrm{cm}^{-1}) + 3.08\cdot(\mathrm{kcal/mol}) \tag{2-65}$$

which further correlates the $\left[-\Delta H^{\mathrm{ab}}\right]$ values for a wide variety of acids and bases with the 'E' and 'C' equation

$$-\Delta H^{\mathrm{ab}} = C_{\mathrm{A}}C_{\mathrm{B}} + E_{\mathrm{A}}E_{\mathrm{B}} \tag{2-66}$$

where C and E stand for 'covalent contribution' and 'electrostatic contribution', respectively and subscripts A and B for acid and base, respectively.

According to Pearson's concept [120], acids and bases are euphemistically described by the terms 'soft' and 'hard', respectively.

In addition, Fowkes [114, 121] actually developed a technique for the determination of the work of adhesion, W_{A}, by applying a four-parameter equation for acid–base interaction in the system of solid-liquid using Eq. (2-48)

$$W_{\mathrm{A}} = 2\sqrt{\gamma_{\mathrm{S}}^{\mathrm{L}}\cdot\gamma_{\mathrm{L}}^{\mathrm{L}}} + f(C_{\mathrm{A}}C_{\mathrm{B}} + E_{\mathrm{A}}E_{\mathrm{B}}) \tag{2-67}$$

Linear salvation energy relationships (LSERs) using the solvatochromic parameters π^* (Debye inductive and Keesom orientational forces of the liquid), α (hydrogen-bond acidity) and β (hydrogen-bond basicity) was proposed by Kamlet-Taft and coworkers [122–127], in considering the hydrogen-bond acceptor and hydrogen-bond donor

$$E_{\mathrm{A}} = E_{\mathrm{A},0} + s\pi^* + a\alpha + b\beta \tag{2-68}$$

where E_{A} is the adsorption energy, s the Debye inductive and Keesom orientational forces, a the hydrogen-bond basicity, and b the hydrogen-bond acidity of the unlike molecules with respect to solvatochromic parameters π^*, α, and

Table 2-13 Characteristics of the Acido-Basicity of Probes

Probes	AN^a	DN^b	π^*	α	β
n-Pentane	0	0	−0.08	0	0
n-Hexane	0	0	−0.04	0	0
n-Heptane	0	0	−0.02	0	0
n-Octane	0	0	0.01	0	0
n-Nonane	0	0	–	–	–
CCl_4	8.6	0	0.28	0	0
Benzene	8.2	0.1	0.59	0	0.1
Toluene	–	–	0.55	0	0.11
Xylene	–	–	0.51	0	0.12
Chloroform	23.1	0	0.58	0.44	0
Diethylether	3.9	19.2	0.27	0	0.47
Methyl acetate	10.7	16.5	0.60	0	0.42
Ethyl acetate	9.3	17.1	0.55	0	0.45
THF	8.0	20.0	0.58	0	0.55
Acetone	12.5	17.0	0.72	0.08	0.48
Nitromethane	20.5	2.7	0.85	0.22	–
Acetonitrile	18.9	14.1	0.75	0.19	0.31

[a] *Acceptor numbers.*
[b] *Donor numbers.*

β, and the constant $E_{A,0}$ the values of E_A when all three terms in Eq. (2-68) are zero.

A few useful values of the solvatochromic parameters of the organic, in the chromatographic process, are listed in Table 2-13.

Doan and Drago [128, 129] reported in an extended work that Eq. (2-66) includes the spectral shifts obtained from the linear free energy relationships approach. They showed a four-parameter equation for a general set of electron donors (acceptors), in which the specific interaction or spectral shift with an electron acceptor (donors) can be fixed by linear free energy relationships.

Linear free energy relationships (LSERs) use the acceptor numbers (*AN*) and donor numbers (*DN*) of Gutmann's school [130–134]. According to Gutmann, donor numbers is defined as the negative of the enthalpy of adduct formation for the reaction of the electron donor (*D*) with $SbCl_5$ as a reference acceptor in a highly dilute (10^{-3} M) 1,2-dichloroethane solution.

$$DN = -\Delta H_{D-SbCl_5} \tag{2-69}$$

The electron acceptor numbers (*AN*) can be used as a measure of a solvent's ability, to take a share in the electron pairs from suitable donors. They are defined by means of a different experimental procedure on the basis of the NMR chemical shifts of phosphorus which are produced on transfer of Et_3PO through solvents. The electron acceptor numbers (*AN*) values are scaled from an arbitrarily chosen value of zero for the shift produced by hexane (or

1,2-dichloroethane) to one hundred for the shift produced by the 1: 1 $Et_3PO\text{-}SbCl_5$ adduct in 1,2-dichloroethane. An estimation of the coordinate bond energy is possible by using Gutmann's formula [131].

$$-\Delta H = \frac{DN \cdot AN}{100} \tag{2-70}$$

The necessity of having two independent terms, to describe the specific (or polar) interactions is relevant to the fact that most organic probes are 'amphoteric' and may as well act as electron donors or electron acceptors, as seen in Table 2-13. In Table 2-13, it should be noted that linear free energy relationships for a general set of electron acceptors or donors which is the specific component of adsorption enthalpy may be closely related to linear free energy relationships of Kamlet-Taft, resulting in the correlation coefficients being largely constant and the standard deviations being low, as below [88]

$$\begin{aligned} AN &= 0.73 + 14.76 \cdot \pi^* + 32.0 \cdot \alpha \\ \mathrm{n} &= 13, \mathrm{R} = 0.967, \mathrm{SD} = 1.53 \end{aligned} \tag{2-71}$$

$$\begin{aligned} DN &= 0.01 - 1.49 \cdot \pi^* + 39.97 \cdot \beta \\ \mathrm{n} &= 12, \mathrm{R} = 0.977, \mathrm{SD} = 1.55 \end{aligned} \tag{2-72}$$

where n is the number of probes included in regressions, R the correlation coefficient, and SD the standard deviation of the fit.

Therefore, the values estimated for the Lewis acids and bases of the probes derived from three different experimental methods (originally proposed by Drago, Kamlet-Taft, and Gutmann) are found to be in good agreement with each other. Among these methods, Park *et al.* particularly discussed Gutmann's method for the determination of the acido-basicity of the solid surface, because it leads to the coherent values of a solid for the specific component of the adsorption enthalpy, in chromatographic processes.

Thermodynamically, the specific components of adsorption enthalpy $\left[-\Delta H_{\mathrm{A}}^{\mathrm{SP}}\right]$ and adsorption enthalpy $\left[-\Delta S_{\mathrm{A}}^{\mathrm{SP}}\right]$ can be derived from the specific component of Gibbs free energy $\left[-\Delta G_{\mathrm{A}}^{\mathrm{SP}}\right]$ expressed as a function of the experimental temperatures in chromatographic processes, such as

$$-\Delta H_{\mathrm{A}}^{\mathrm{SP}} = \frac{\mathrm{d}(-\Delta G_{\mathrm{A}}^{\mathrm{SP}}/\mathrm{T})}{\mathrm{d}(1/\mathrm{T})} \tag{2-73}$$

$$-\Delta S_{\mathrm{A}}^{\mathrm{SP}} = \frac{\mathrm{d}(-\Delta G_{\mathrm{A}}^{\mathrm{SP}})}{d\mathrm{T}} \tag{2-74}$$

with T being the absolute temperature in Kelvin (K).

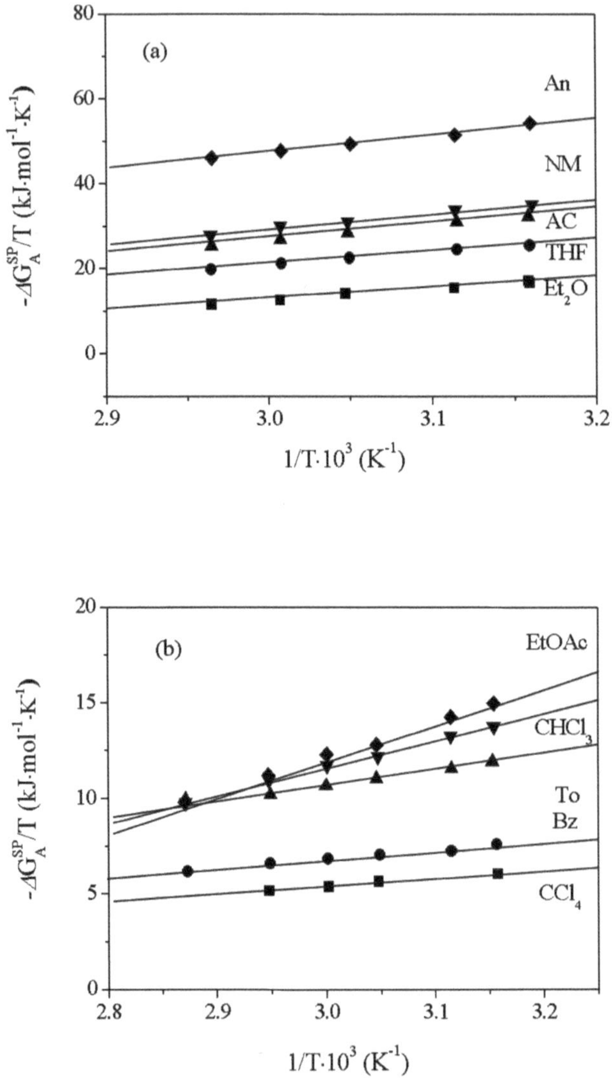

Fig. 2-53 Linear variation of $\left[-\Delta G_{\mathrm{A}}^{\mathrm{SP}}/T\right]$ against $\left[T^{-1}\right]$ between untreated carbon fiber (supplied by Soficar, T-300) and all probes, measured for the determination of $\left[-\Delta H_{\mathrm{A}}^{\mathrm{SP}}\right]$ and $\left[-\Delta S_{\mathrm{A}}^{\mathrm{SP}}\right]$.

For untreated carbon fibers, plotting of $\left[-\Delta G_{\mathrm{A}}^{\mathrm{SP}}/T\right]$ against $\left[T^{-1}\right]$, gives a straight line with slope $\left[-\Delta H_{\mathrm{A}}^{\mathrm{SP}}\right]$ and intercept $\left[-\Delta S_{\mathrm{A}}^{\mathrm{SP}}\right]$, as illustrated in Fig. 2-53, in the experimental temperature range of 35 ~ 75 °C. As an example, Table 2-14 shows $\left[-\Delta H_{\mathrm{A}}^{\mathrm{SP}}\right]$ and $\left[-\Delta S_{\mathrm{A}}^{\mathrm{SP}}\right]$ determined, by plotting

Table 2-14 The Specific Components of the Adsorption Enthalpy $[-\Delta H_A^{SP}]$ (in kJ/mol) and Entropy $[-\Delta S_A^{SP}]$ (J/mol/K) of the Carbon Fibers Used

	Fibers as-received[a]		Fibers modified[b]	
Probes	$[-\Delta H_A^{SP}]$	$[-\Delta S_A^{SP}]$	$[-\Delta H_A^{SP}]$	$[-\Delta S_A^{SP}]$
Diethylether	15.7	34	26.7	68
Ethyl acetate	17.2	40	30.1	71
Tetrahydrofuran	20.9	40	36.5	82
Acetone	21.5	38	36.8	78
Nitromethane	23.3	41	32.1	65
Acetonitrile	28.3	39	40.9	75

[a] *Untreated and unsized carbon fibers (T-300) supplied by Soficar.*
[b] *Electro-oxidized by the current density (1.4 A/m^2).*

the highly polar probes (in Table 2-11) and carbon fibers, untreated, and treated (by current density 1.4 A/m^2) [10].

According to Gutmann's definition, the ability of acid-base interaction may be characterized by the specific component of adsorption enthalpy, $\left[-\Delta H_A^{SP}\right]$ which can be determined from the following equations, as proposed by Papier *et al.* [135]

$$-\Delta H_A^{SP} = K_A \cdot DN + K_D \cdot AN \tag{2-75}$$

or

$$\frac{-\Delta H_A^{SP}}{AN} = K_A \cdot \frac{DN}{AN} + K_D \tag{2-76}$$

where K_A and K_D are respectively the characteristic acidity, and basicity of the solid surfaces.

In order to fit the chromatographic technique for the study of adsorption phenomena, it is not too much to emphasize that the choice of the probes should be judicious to obtain to data on its intermolecular capacity over the solid surface. For this work, some chromatographic probes displaying weakly polar properties (according to Stockmayer's polarity, δ), such as, CCl_4, benzene, toluene, xylene, and even chloroform, are not suitable for determining the acid-base character of a solid surface, since Gutmann's electron acceptor and donor numbers of the probes are too high as compared with their dipole moment, as seen in Table 1-16 [109]. This fact also confirms the validity of this analysis, since weakly polar probes are close to the reference line of n-alkanes, as early shown in Fig. 2-52.

The variation of $\left[-\Delta H_A^{SP}/AN\right]$ *vs.* $[DN/AN]$ of the probes shown in Table 10 are plotted in Fig. 2-54, which can give a straight line with slope K_A and K_D (K_A and K_D are respectively the characteristic acidity and basicity of the solid surfaces) are increased after electrochemical oxidation on carbon fibers. As a

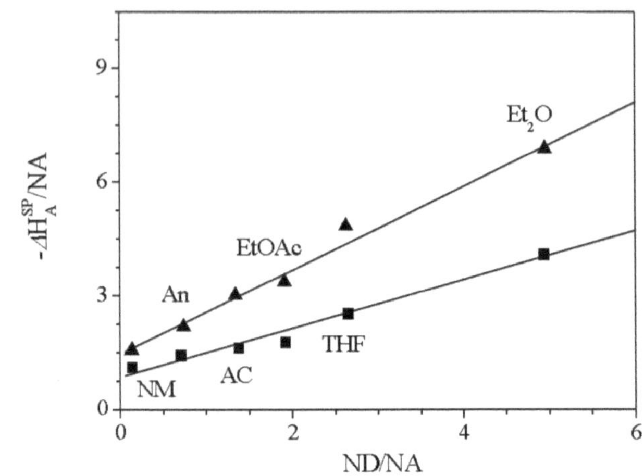

Fig. 2-54 Linear variation of $[-\Delta H_A^{SP}/AN]$ against [*DN*/*AN*] of the untreated (■) and electrochemically treated (▲) carbon fibers (supplied by Soficar, T-300), for the determination of the characteristic acidity and basicity of the fiber surfaces.

result, adhesion between the fibers and matrix in composite materials is more likely to be enhanced [110].

Another acid-base study on carbon fiber (supplied by Courtaulds Co.) surfaces produced the mechanical properties of composites as shown in Table 2-15, with a 90% confidence level in the arbitrary units [76].

In comparing the acido-basicity of the untreated and electrochemically or anodically treated fibers in Table 2-15, it is observed that the amine-treatments increase significantly the basicity of the fibers (Samples II and III), while commercial oxidation increases the acidy (IV), as expected. This behavior is probably explained in terms of the presence of any physisorbed species (such as, amine and oxygen functional groups for the fiber surfaces amined and oxidized, respectively). This can improve the degree of adhesion at interfaces of the composites for fiber surfaces incorporated with the epoxy matrix and play an important role in forming the fiber-matrix bond

2.5.2. Solid-Gas Interaction at a Finite Concentration

2.5.2.1. Equilibrium Spreading Pressure and Surface Free Energy

Surface free energy is a physical phenomenon caused by intermolecular interactions at an interface, i.e., London dispersive force, Debye inductive force, Keesom orientational forces, hydrogen bonding, Lewis acid–base interactions, and energetically homogeneous and heterogeneous interactions [10, 138–141]. In the early 1960s, the adsorption studies reported by Fowkes [142] and Graham [143] defined the equilibrium spreading pressure (π_e) and surface free energy (γ_s) of solids, using simple adsorbates such as nitrogen, argon, and hydrocarbon.

Table 2-15 The Acid-Base Character of Carbon Fiber Surfaces and the Degree of Adhesion Measured by Interfacial Shear Strength (IFSS) System on Monofilaments and Interlaminar Shear Strength (ILSS) on Composites

	Electro-oxidized liquids	K_A[a]	K_D[a]	$K_A \pm K_D$[a]	R[b]	SD[c]	n[d]	IFSS[e]	ILSS[f]
I	None	0.9 ± 0.1	1.4 ± 0.2	2.3 ± 0.3	0.996	0.02	7	88.5 ± 5	69.5 ± 5
II	Ethylene-diamine	0.9 ± 0.2	2.2 ± 0.3	3.1 ± 0.5	0.983	0.02	6	117 ± 5	87 ± 9
III	Triethylene-tetramine	1.0 ± 0.3	2.7 ± 0.4	3.7 ± 0.7	0.982	0.04	5	107 ± 3.5	103 ± 9
IV	Commercially standard	1.8 ± 0.2	1.6 ± 0.4	3.4 ± 0.6	0.995	0.10	7	≒ 125	115 ± 10

[a] *Values with 90% confidence level.*
[b] *Correlation coefficient.*
[c] *Standard deviation of the fit.*
[d] *Number or experimental polar probes included in regressions.*
[e] *Values of the interfacial shear strength (IFSS) measured by pull-out system (matrix: epoxy resin NARMCO 5208) reported by Desarmot et al. [137].*
[f] *Values of the interlaminar shear strength (ILSS) of the composites (matrix: epoxy resin NARMCO 5208, volume fraction of the fibers, $V_f = 70\%$, s/d = 3.5, c.h.s = 2 mm/min) reported by Desarmot et al. [136, 137].*

Park and Brendlé [91] showed, that an increase of the London dispersive component (γ_s^L) of surface free energy is one of the main factors in the adsorption capacity, along with the specific surface area of the solid.

As an example, the effects of polar acidic, polar basic, and nonpolar organic chemical surface oxidations of carbons on the surface and adsorption properties, microstructures, and surface energetics, are investigated by BET's isotherms. An interpretation based on the nitrogen amount adsorbed for filling a monolayer (a_0) was proposed for the determination of the Gibbs free energy of nitrogen adsorption, allowing evaluation of the equilibrium spreading pressure or London dispersive component of the surface free energy of the carbons studied. When the nitrogen adsorbate is considered in the adsorption studies, the equilibrium spreading pressure (π_e) can be determined by using the integrated form of the Gibbs equation [144]:

$$-\Delta G_A(N_2) = RT \int_0^1 \frac{a}{a_0} d(\ln P/P_0) \quad (2\text{-}77)$$

$$= N_A a_{N_2} \pi_e \quad (2\text{-}78)$$

where R is the ideal gas constant, T the absolute temperature, a the amount adsorbed at a relative pressure P/P_0, a_0 the amount adsorbed for filling a monolayer, N_A Avogadro's number, and a_{N_2} the cross-sectional area, of the adsorbed N_2.

Therefore, π_e can be calculated according to Eqs. (2-77) and (2-78):

$$\pi_e = \frac{RT}{N_A a_{N_2} a_0} \int_0^1 a d(\ln P/P_0) \quad (2\text{-}79)$$

Using the BET approach, it is then possible to determine the amount adsorbed, for filling a monolayer, a_0

$$\frac{P/P_0}{a(1 - P/P_0)} = \frac{1}{C_{BET} a_0} + \frac{(C_{BET} - 1)P/P_0}{C_{BET} a_0} \quad (2\text{-}80)$$

where C is the BET constant characteristic of the adsorbate-adsorbent.

Fig. 2-55 shows the adsorption isotherms of carbon studied before and after chemical treated with H_3PO_4, KOH, and C_6H_6. All of the adsorption isotherms are of Type III, according to the IUPAC classification.

This result shows that the adsorption behavior of the carbons treated with basic or non-polar chemical solution is similar to that of untreated carbon, for which, there is no significant change in the adsorption behavior as a function of relative pressure, P/P_0. From the plot of $(P/P_0)/a\ (1\text{-}P/P_0)$ versus $[P/P_0]$ ranging between 0.05 and 0.25, excellent straight lines are obtained (the correlation coefficient being greater than 0.999), which allows the determination of C_{BET}

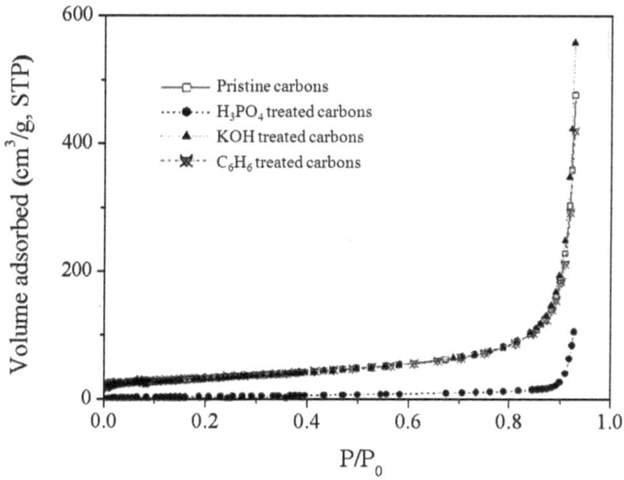

Fig. 2-55 N_2/77 K adsorption isotherms on the relative pressure of the carbons, before and after surface treatments.

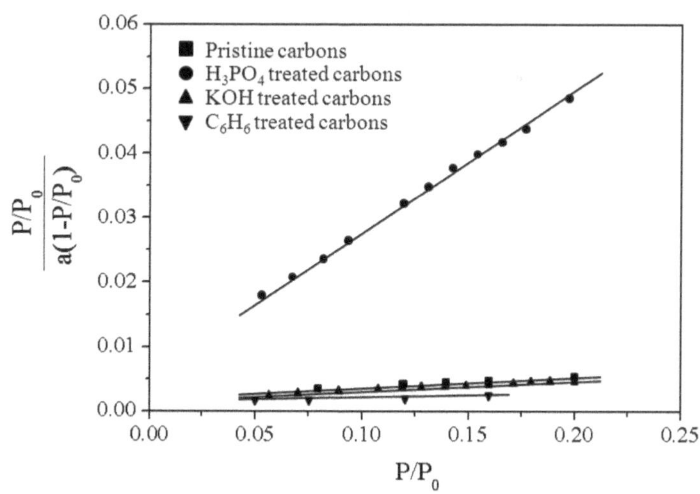

Fig. 2-56 BET distribution transform of the carbons, before and after surface treatments.

and a_0, as seen in Fig. 2-56. S_{BET}, C_{BET}, and a_0 can be calculated as usual, for BET measurements by Eq. (2-80); the results are listed in Table 2-16 [144].

It should be noted that π_e, based on a_0 and a_{N_2}, is considered from the adsorbate point of view, even if it can also be determined from the specific surface area (S_{BET}) of a solid surface [144], rather than from $[N_A a_{N_2} a_0]$ in Eq. (2-79).

Table 2-16 Surface and Adsorption Values of the Carbons Before and After Surface Treatments

Samples	S_{BET}[a]	C_{BET}[b]	a_0[c]
Un-treated carbon	112	112	1.17
Acidic-treated carbon	18	50	0.19
Basic-treated carbon	118	162	1.21
Nonpolar -treated carbon	114	134	1.15

[a]*Specific surface area (m^2/g).*
[b]*BET s constant.*
[c]*Amount adsorbed for filling a monolayer (mmol/g).*

From the values thus obtained from the nitrogen adsorbate, it is of interest to determine, the London dispersive component, γ_s^L, of surface free energy of the carbon investigated in the adsorption studies at 77 K. The relationship between γ_s^L and π_e for a solid surface is given by the following equation [145, 146]:

$$\gamma_s^L = \frac{(\pi_e + 2\gamma_{N_2})^2}{4\gamma_{N_2}^L} \tag{2-81}$$

where subscript s and superscript L represent the solid state and London dispersive component of surface free energy and $\gamma_{N_2}^L = \gamma_{N_2} = 10.5\ \mathrm{mJ/m^2}$ [142].

Fig. 2-57 shows the evolutions of the equilibrium spreading pressure and London dispersive component of surface free energy of the carbon studied. As

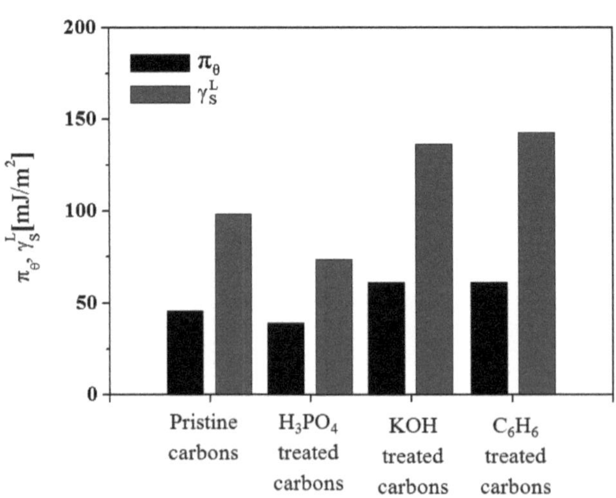

Fig. 2-57 Results of the equilibrium spreading pressure (π_e) and London dispersive component of surface free energy (γ_s^L) of the carbons before and after surface treatments.

an experimental result, the acidic treatment leads to decreases in both π_e and γ_s^L. However, π_e and γ_s^L increase in the case of basic-treated or non-polar-treated carbons. Acidic treatment led to significant decrease in adsorption amount, S_{BET}, and surface free energy of the carbons, due to aggregation of the microstructures and increasing weight of the swollen specimen in an equilibrium state. Polar basic and non-polar chemical treatments resulted in an increase of the equilibrium spreading pressure or London dispersive component of surface free energy of the carbon blacks, without significantly changing the surface and adsorption properties and microstructures [144].

For study, the surface free energy of carbon on the mechanical tearing results obtained from the rubber compound composites in Fig. 2-58 [144].

According to Kraus [147], the degree of adhesion between the filler surface and the rubber can be accessed from the swelling behavior of a sample in a solvent. The importance of tearing energy, G_{IIIC}, as a criterion of the interfacial adhesion relationship can be considered for study of the surface free energy of carbon on the mechanical tearing results obtained from the rubber compound composites.

The results of tearing energies measured in split cantilever beam tests are shown in Fig. 2-58. It indicates that the tearing energies of the rubber compounds filled with basic-treated or non-polar-treated carbon are higher than those of rubber compounds filled with acidic-treated carbon, or the pristine carbon. In the case of basic-treated or non-polar-treated carbon specimens, significant advantages accrued from the surface treatment, resulting in improved tearing energy due to an increase of the equilibrium spreading pressure of carbon. It is noted that modification of carbon surfaces appears to

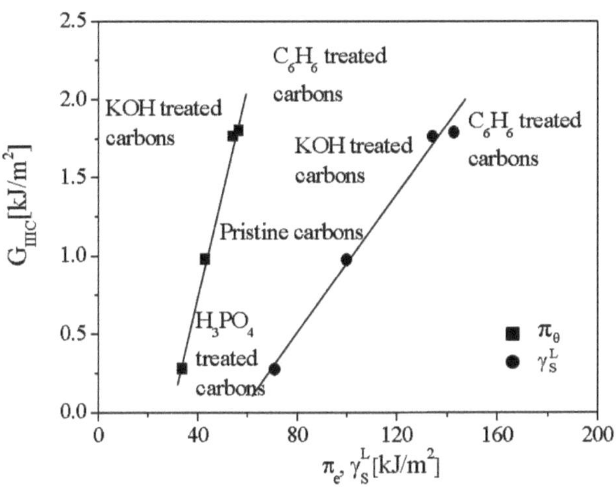

Fig. 2-58 Dependences of the tearing energy (G_{IIIc}) of the compoundings, on the equilibrium spreading pressure (π_e) and London dispersive component of surface free energy (γ_s^L) of the carbons before and after surface treatments.

increase the London dispersive component of surface free energy and the parameter of polymer–filler interaction describing the mechanical interfacial behavior of a nonpolar organic matrix, in a composite system.

2.5.2.2. *Inverse Gas Chromatography at a Finite Concentration*

The dependence of the inverse gas chromatography retention volume (or time) on the adsorbate concentration in the gas phase proved to be a very useful and fast way to determine the adsorption isotherms. Information on surface area, porosity, adsorbate-adsorbent interactions, can be obtained [148, 149].

For the access of the adsorption isotherm, the volumes injected are in very small quantities of less than 0.1 μL of gaseous probes for infinite dilution up to the saturation of the adsorbate, or up to the increase of net retention time when the quantity adsorbed increased, in the chromatography, on account of the sensitivity of the detector. In the chromatographic approach, the peak maxima method [81] is generally used to determine the net retention volumes, which are corrected by the compressibility of temperature as well as flow rate, as shown in Fig. 2-59.

The partial equilibrium pressure, $P(\mathrm{i})$, on the adsorbent can be determined from Eq. (2-82) [150].

$$P(\mathrm{i}) = \frac{\mathrm{RT}}{DV_0}\frac{V(\mathrm{i})}{\mathrm{A(i)}}H(\mathrm{i}) \quad (2\text{-}82)$$

where R is the gas constant (J/mole/K), T the temperature in Kelvin (K), D the flow rate taking into account gas compressibility (cm^3/min), V_0 the molecular mass of the adsorbate (g/mol), $V(\mathrm{i})$ the quantity injected for the injection

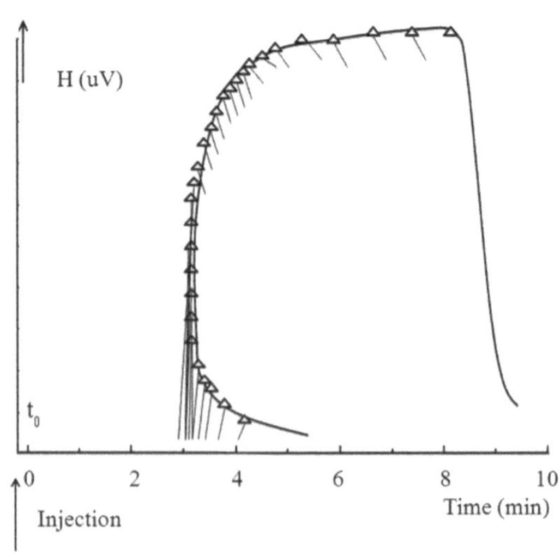

Fig. 2-59 Schematic diagram of the chromatographic peaks of a solute with the injection volume.

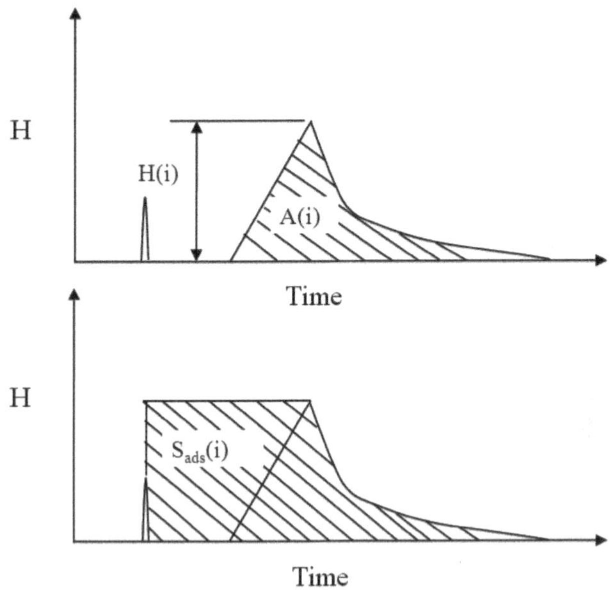

Fig. 2-60 Determinations of H (i), A (i), and S_{ads}(i) in the chromatographic peak.

volume, i the adsorbate (μV/s), H (i) the height of the chromatographic peak for the injection volume, and i the adsorbate (μV), as illustrated in Fig. 2-60.

If the surface of an adsorbent is energetically homogeneous, three well-known statistics are fundamentally considered for evaluating the distribution functions of the adsorption site energy of probability of a state of energy, $f(\varepsilon)$, in the adsorbate-adsorbent system [70].

1. Bose-Einstein distribution (B.E).

$$f(\varepsilon) = \frac{1}{exp\left(\frac{\varepsilon - \varepsilon_{f(\varepsilon)=1/2}}{RT}\right) - 1} \tag{2-83}$$

2. Maxwell-Boltzmann distribution (M.B).

$$f(\varepsilon) = \frac{1}{exp\left(\frac{\varepsilon - \varepsilon_{f(\varepsilon)=1/2}}{RT}\right)} \tag{2-84}$$

3. Fermi–Dirac distribution (F.D)

$$f(\varepsilon) = \frac{1}{exp\left(\frac{\varepsilon - \varepsilon_{f(\varepsilon)=1/2}}{RT}\right) + 1} \tag{2-85}$$

with $\varepsilon_{f(\varepsilon))=1/2}$ being the adsorption energy level measured at the half of adsorption isotherm (kJ/mol).

For the chromatographic process, we discuss Fermi–Dirac's distribution function for characterizing the energetics of a solid surface, because it is known, that this distribution is very powerful, especially, for conditions such as weakly van der Waals attraction and relatively low temperature systems, as compared with B.E (Bose–Einstein) and/or M.B (Maxwell–Boltzmann) distributions [151].

For the study of adsorption site energy, [$\varepsilon_{f(\varepsilon)=1/2}$] statistics, indicates the potential of the molecules in Eq. (2-85), such as Fermi's level, determining their tendency to move at an interface. In particular, as the temperature decreases, the equation shown in [70] is:

$$f(\varepsilon) = 0 \quad \text{if}\, \varepsilon > \varepsilon_{f(\varepsilon)=1/2} \tag{2-86}$$

$$f(\varepsilon) = 1 \quad \text{if}\, \varepsilon < \varepsilon_{f(\varepsilon)=1/2} \tag{2-87}$$

While evaluating the distribution function, the adsorption site energies, ε, may be determined by the experimental adsorption isotherm, according to Zeldowitch's original expression [150].

$$\varepsilon = \mathrm{RT} \cdot \ln\left(\frac{K}{\mathrm{P}}\right) \tag{2-88}$$

where K is the pre-exponential factor of Henry's constant and P the partial equilibrium pressure (Pa).

Therefore, the determination of the pre-exponential factor of Henry's constant is needed while using this definition, and the well-known methods proposed by Adamson [151] or by Hobson [152], or by Park and Donnet [70] could be considered for this purpose:

1. According to Adamson and Ling,

$$K = \mathrm{P}_0 \cdot exp\left(\frac{E_V}{\mathrm{RT}}\right) \quad \text{(in Pa)} \tag{2-89}$$

where P_0 is the saturation vapor pressure (Pa) and E_v the energy of vaporization of adsorbate (kJ/mol).

2. According to Hobson

$$K = 1.76 \times 10^4 (M\mathrm{T})^{1/2} \quad \text{(in torr)} \tag{2-90}$$

where M is the molecular weight of adsorbate (g/mol), T the temperature in Kelvin (K).

3. According to Park and Donnet [70]

$$K = exp\left(\frac{\varepsilon_1}{\mathrm{RT}}\right) \quad (\text{in Pa when P} = 1{\cdot}\text{Pa}) \tag{2-91}$$

where ε_1 is the adsorption of Gibbs free energy, $-\Delta G_{\mathrm{A}}$, determined from the infinite dilution which is related to the first value or the most active site value of the adsorbate-adsorbent system [70].

Hobson [152] originally proposed the pre-exponential factor of Henry's constant, K, as in Eq. (2-90) for the case of a simple gas (such helium, argon, and nitrogen) at a very low experimental temperature ($0 \sim 100$ K). This has been successfully studied in the heterogeneity distribution of adsorption site energies on solid surfaces [153–155]. The methods of Adamson-Ling and Hobson produce quite high values (50–80 kJ/mol) that are not consistent with the physical adsorption energy being equal to the heat of vaporization at ambient temperature. In addition, the starting adsorption free energy, ε_1, at a near zero coverage state obtained from Eqs. (2-92) and (2-93), is increased with temperature, which is not in agreement with the behavior of adsorption [70].

An analytical method, that can determine the adsorption free energy, ε_1 (or $-\Delta G_A$), can be proposed based on the deformation polarizability, α_0, and the dipole moment, μ, of a molecule of solute, during the adsorption study on the solid surfaces as measured from the inverse gas chromatography at infinite dilution [70]. When the first value, ε_1, of the adsorption free energy is equal to the energy measured at infinite dilution [i.e., the equilibrium concentration being extremely small, $P = 1$ Pa in Eq. (2-91)], the pre-exponential factor of Henry's constant, K, can be obtained, depending on the experimental temperatures.

$$-\Delta G_{\mathrm{A}} = [-\Delta G_{\mathrm{A}}^{\mathrm{L}}] + [-\Delta G_{\mathrm{A}}^{\mathrm{SP}}] \tag{2-46}$$

$$= -\frac{3}{4}\frac{\mathrm{N_A}}{(4\pi\varepsilon_0)^2}\left(\frac{1}{r_{S-L}}\right)^6\left\{(h\nu_{\mathrm{S}})^{1/2}\cdot\alpha_{0,\mathrm{S}}\right\}\cdot\left\{(h\nu_{\mathrm{L}})^{1/2}\cdot\alpha_{0,\mathrm{L}}\right\} + [-\Delta G_{\mathrm{A}}^{\mathrm{SP}}] \tag{2-92}$$

$$= \varepsilon_1 \tag{2-93}$$

$$= RT \cdot \ln K \text{ (in case of } P = 1 \text{ Pa)} \tag{2-94}$$

This approximation has an accuracy within an error of 5%, and then Park *et al.* utilize used Eq. (2-92) to describe the adsorption free energy, ε(i), as a function of the amount adsorbed calculated from the increase of equilibrium pressure [70].

Fig. 2-61 shows that the adsorption free energy, $-\Delta G_A$, of n-hexane (as non-polar probe) and ethyl acetate (polar probe) on the surface of carbon fibers (PAN-based high strength carbon fibers supplied by Soficar, T-300), were investigated by inverse gas chromatography at infinite dilution with changing experimental temperatures. The first adsorption value, ε_1, of the adsorption free energy at a given temperature for the adsorption isotherms at finite dilution (in this work, 30°C) can be determined by extrapolating. It enables the determination of the pre-exponential factor of Henry's constant, K, by Eq. (2-91).

For the study of the adsorption energy level, the molecular quantity adsorbed, a(i), on the adsorbent, is derived from Eq. (2-95).

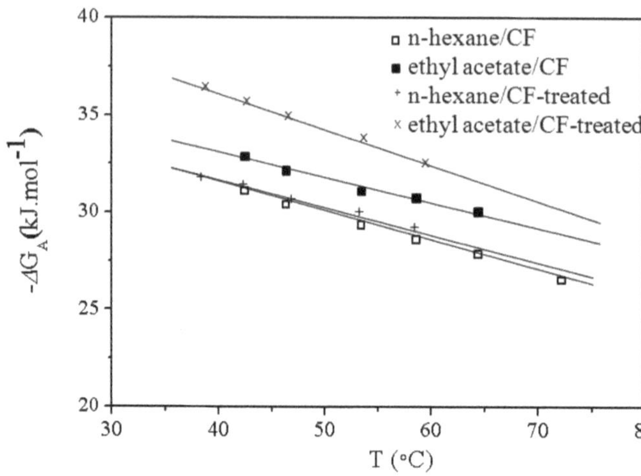

Fig. 2-61 Linear variation of adsorption free energy, $-\Delta G_A$, with experimental temperature (°C), measured at infinite dilution.

The pressure dependence on the quantity of injected adsorbate, is seen in Fig. 2-62 [148].

$$a(i) = \frac{1}{mV_0}\frac{V(i)}{A(i)}S_{\mathrm{ads}}(i) \tag{2-95}$$

where m is the weight of adsorbent (g), S_{ads} the adsorption area for the injection volume, i, the adsorbate (μV/s).

By the BET approach [6], it is possible to determine the amount adsorbed for filling a monolayer, a_0 , and to obtain

$$\frac{\mathrm{P/P_0}}{a(1-\mathrm{P/P_0})} = \frac{1}{C_{\mathrm{BET}}a_0} + \frac{(C_{\mathrm{BET}}-1)\mathrm{P/P_0}}{C_{\mathrm{BET}}a_{\mathrm{m}}} \tag{2-80}$$

where P_0 is the saturation pressure of the adsorbed, and C_{BET} a BET's constant characteristic of the adsorbate-adsorbent.

From the plot of [$P/P_0/a(1 - P/P_0)$] *vs.* [P/P_0] ranging between 0.05 and 0.25, excellent straight lines are obtained (the correlation coefficient being more than 0.99), which allows the determination of the C_{BET} and a_0 as seen in Fig. 2-62, and the adsorption isotherm, a/a_0, can be calculated in the usual BET way of Eqs. (2-80), and (2-95) as listed in Table 2-17 [70].

It is then possible to make the distribution curve of adsorption site energies, $f(\varepsilon)$, in Eq. (2-85), after determining the adsorption energy level, $\varepsilon_{f(\varepsilon)} = 1/2$, at the point of [$a/a_0 = 1/2$], as seen in Fig. 2-63. It obviously shows that the distribution curve of adsorption site energies, $f(\varepsilon)$, is very similar to the BET distribution transform, at low coverage ($\varepsilon > \varepsilon_{f(\varepsilon)=1/2}$), and behaves similarly to Langmuir's adsorption [1] towards the surface monolayer at high coverage.

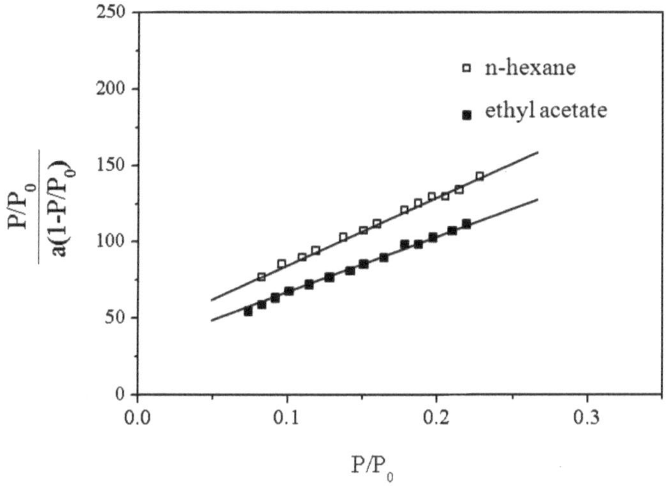

Fig. 2-62 BET distribution transform on the untreated carbon fibers (supplied by Soficar, T-300) measured at 30 °C.

The adsorption energy level or potential of the molecules, $\varepsilon_{f(\varepsilon)=1/2}$, for all the carbon fibers studied in Park's work, is presented in Table 2-17, which gives information on the eventual mobility of molecules at an interface [70].

The experimental temperature is taken as 30°C, due to the chromatographic limitation. It can however be noted that the result obtained from finite dilution is less sensitive to the temperature than those of infinite dilution, because of decreasing entropy of molecules with an increasing adsorbed amount.

Table 2-17 Results Obtained from Gas Chromatographic Method Measured at 30 °C

Samples	Untreated Carbon fibers		Treated carbon Fibers	
Probes	n-Hexane	Ethyl acetate	n-Hexane	Ethyl acetate
ε_1 (kJ/mol)	33.16	34.71	33.10	37.31
lnK (K in Pa)	13.16	13.77	13.13	14.80
a_o (μmol/g)	2.56	2.17	3.08	2.58
C_{BET}	10.76	11.25	16.23	14.52
$P/P_0{}^a$	0.24	0.39	0.21	0.55
a/a_m^a	1.05	1.45	1.10	1.97
$\varepsilon_{f(\varepsilon)=1/2}$ (kJ/mol)	14.7	17.0	15.6	20.3

[a] *Values measured up to the saturation of adsorption in the detector of chroatograph.*

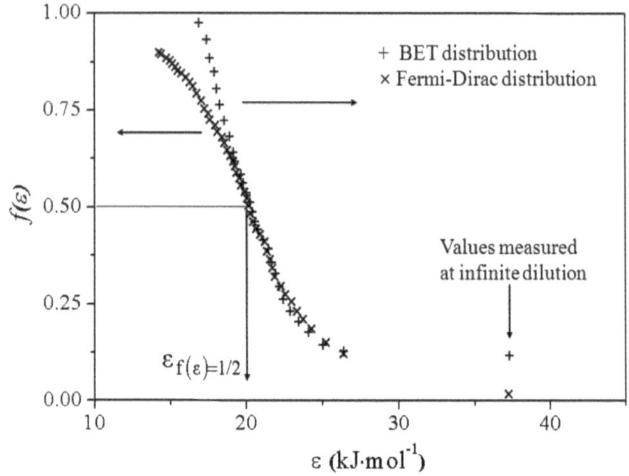

Fig. 2-63 Distribution functions of adsorption site energies based on the electrochemically treated carbon fibers (supplied by Soficar T-300) measured at 30 °C.

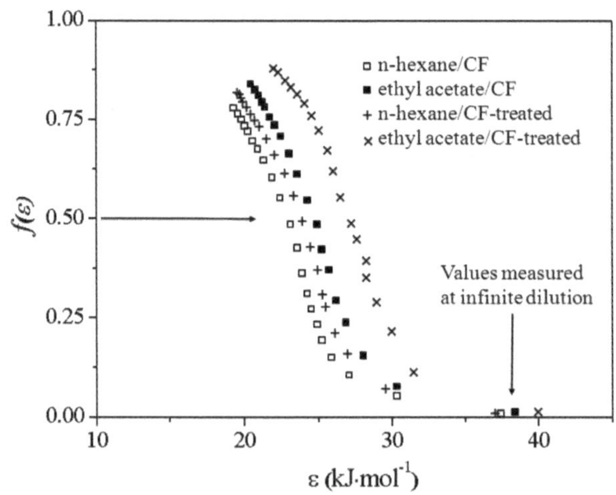

Fig. 2-64 Distribution functions of adsorption site energies based on Fermi-Dirac's law measured at 30 °C.

Fig. 2-63 shows all the experimental distribution functions of the studied adsorption site energies. In Fig. 2-64, the [$\varepsilon_{f(\varepsilon)=1/2}$] of the n-hexane and ethyl acetate adsorption are increased by electrochemical or anodic oxidation of the carbon fibers. In particular, the results of the ethyl acetate probe which is able to display not only the London dispersive but also the specific (or polar) component of the adsorption free energy on the modified fibers shows an increase of

$[\varepsilon_{f(\varepsilon)=1/2}]$, as well as distribution functions of adsorption site energies. This behavior is probably due to the effect of surface functional groups on adsorption energies [70].

A new interpretation based on the correlation between infinite and finite dilution of the solute is made for the physical adsorption of gases on the solid surfaces, determining the pre-exponential of Henry's constant in a given temperature. Fermi–Dirac's law is used for the evaluation of the distribution function of adsorption site energies. This adsorption behavior is similar to transform at high coverage in a monolayer. In addition, this adsorption permits us to determine the adsorption energy level, which may evaluate the tendency to move at an interface [70].

2.6. SUMMARY

The main purpose of this chapter was to review the long-range forces to surface dynamics of solid-gas interfaces. Against this background, theoretical and applied procedures are discussed to characterize the surfaces of solid. In principle, two kinds of intermolecular origin forces based on the deformation polarizability, α_0, and dipole moment, μ, of a molecule offer themselves: The London dispersive component and the specific (or acid-base) component of the adsorption Gibbs free energy. This was a new way to find surface free energy, surface energy (or surface enthalpy), acid-base character, degree of adhesion between solid-solid interfaces, adsorption isotherm, and energetic distribution in a monolayer. The study of distribution function characterization of a solid surface is still in its infancy for lack of heterogeneous isotherms, but, in principle, theories and techniques are available to foster the development of this important bridge between infinite states in solid surface chemistry.

REFERENCES

[1] I. Langmuir, Journal of the American Chemical Society 38 (1916) 2221.
[2] I. Langmuir, Journal of the American Chemical Society 40 (1918) 1368.
[3] J.M.V. Bemmelen, Landwirt. Versuch 35 (1888) 69.
[4] H.M.F. Freundlich, Zeitschrift für Physikalische Chemie 57 (1906) 385.
[5] R.H. Fowler, E.A. Guggenhcim, Statistical Thermodynamics. Cambridge University Press, London (1939).
[6] S. Brunauer, P.H. Emmett, E. Teller, Journal of the American Chemical Society 60 (1938) 309.
[7] O. Redlich, D.L. Peterson, Journal of Physical and Chemistry 63 (1959) 1024.
[8] P. Kisliuk, Journal of Physical and Chemistry Solids 3 (1957) 95.
[9] A.P. Henderson, L.N. Seetohul, A.K. Dean, P. Russell, S. Pruneanu, Z. Ali, Langmuir 25 (2009) 931.
[10] S.J. Park, Long-range Force Contributions to Surface Dynamics. Interfacial Forces and Fields: Theory and Applications. Marcel Dekker, New York (1990).
[11] A.W. Adamson, A.P. Gast, Physical Chemistry of Surfaces, 6th edit. John-Wiley & Sons, New York (1997).

[12] B.J. Kim, S.J. Park, Journal of Colloid and Interface Science 311 (2007) 311.
[13] S.J. Park, S.Y. Jin, Journal of Colloid and Interface Science 286 (2005) 417.
[14] B.J. Kim, Y.S. Lee, S.J. Park, Journal of Colloid and Interface Science 306 (2007) 454.
[15] S.Y. Lee, S.J. Park, International Journal of Hydrogen Energy 35 (2010) 6757.
[16] W. Lv, D.M. Tang, Y.B. He, C.H. You, Z.Q. Shi, X.C. Chen, C.M. Chen, P.X. Hou, C. Liu, Q.H. Yang, ACS Nano 3 (2009) 3730.
[17] K.S. Walton, M.B. Abney, M.D. LeVan, Microporous and Mesoporous Materials 91 (2006) 78.
[18] R. Hernandez-Huesca, L. Diaz, G. Aguilar-Armenta, Separation and Purification Reviews 15 (1999) 163.
[19] P. Harlick, F.H. Tezel, Microporous and Mesoporous Materials 76 (2004) 71.
[20] R.A. Alberty, F. Daniels, Physical Chemistry, 5th edit. Wiley, New York (1979).
[21] S. Brunauer, L. Deming, W. Deming, E. Teller, Journal of the American Chemical Society 62 (1940) 1723.
[22] J.T. Nolan, T.W. Mckeehan, R.P. Danner, Journal of Chemical Engineering Data 26 (1981) 112.
[23] S.J. Gregg, S.W. Sing, Adsorption Surface Area and Porosity, 2nd edit. Academic Press, London (1982).
[24] G.H. Findenegg, Fundamentals of Adsorption. Engineering Foundation, New York (1984).
[25] IUPAC Recommendations Pure Applied Chemistry 57 (1985) 603.
[26] IUPAC Recommendations Pure Applied Chemistry 66 (1994) 603.
[27] S.J. Park, B.J. Kim, Journal of Colloid and Interface Science 275 (2004) 590.
[28] S.Y. Lee, S.J. Park, Journal Solid State Chemistry 183 (2010) 2951.
[29] S.H. Lee, S.M. Yun, S.J. Kim, S.J. Park, and Y.S. Lee, Resonance Chemistry Intermediate in press (2010).
[30] S.Y. Lee, S.J. Park, Journal of Colloid and Interface Science 346 (2010) 194.
[31] J.W.O. Faul, U.G. Volkmann, Surface Science 227 (1990) 390.
[32] P.H. Hermans, Colloid Science. Elsevier, New York (1949).
[33] S.J. Park, K.D. Kim, Journal Colloid Interface Science 212 (2004) 186.
[34] S.J. Park, J.M. Park, M.K. Seo, Journal Colloid Interface Science 337 (2009) 300.
[35] S. Brunauer, The Adsorption of Gases and Vapors. Princeton University Press, Princeton (1943).
[36] C.T.L. Hill, An Introduction to Statistical Thermodynamics. Addison-Wesley, London (1960).
[37] S. Brunauer, Twenty Years of Colloid Surface Chemistry The Kendall Award Addresses. American Chemistry Society, Washington, D.C. (1973).
[38] S. J. Park, Unpublished results.
[39] B.J. Kim, S.J. Park, Nanotechnology 17 (2006) 4395.
[40] G.L.B.F. Flapper, H.M. Gijsman, Cryogenics 14 (1974) 150.
[41] S. Choi, J.H. Drese, C.W. Jones, ChemSusChem 2 (2009) 796.
[42] J. Przepiórski, M. Skrodzewicz, A.W. Morawski, Applied Surface Science 225 (2004) 235.
[43] E.P. Dillon, C.A. Crouse, A.R. Barron, ACS Nano 2 (2008) 156.
[44] V. Zelenak, D. Halamova, L. Gaberova, E. Bloch, P. Llewellyn, Microporous and Mesoporous Materials 116 (2008) 358.
[45] Y. Belmabkhout, M. Frère, G. De Weireld, Measurement. Science. Technology 15 (2004) 848.
[46] N.L. Rosi, J. Eckert, M. Eddaoudi, D.T. Vodak, J. Kim, M. O'keeffe, O.M. Yaghi, Science 300 (2003) 1127.
[47] H. Li, M. Eddaoudi, M. O'Keeffe, O.M. Yaghi, Nature 402 (1999) 276.
[48] M. Eddaoudi, H. Li, O.M. Yaghi, Journal American Chemical Society 122 (2000) 1391.

[49] S.J. Park, S.Y. Lee, Carbon Letters 10 (2009) 19.
[50] B.J. Kim, Y.S. Lee, S.J. Park, Journal Colloid Interface Science 318 (2008) 530.
[51] S.J. Park, B.J. Kim, Y.S. Lee, M.J. Cho, International Journal of Hydrogen Energy 33 (2008) 1706.
[52] A.J. Lachawiec, J.G. Qi, R.T. Yang, Langmuir 21 (2005) 11418.
[53] L.Y. Meng, S.J. Park, Journal Colloid Interface Science 352 (2010) 498.
[54] R.V. Siriwardane, M.S. Shen, E.P. Fisher, J.A. Poston, Energy Fuels 15 (2001) 279.
[55] R. Chatti, A.K. Bansiwal, J.A. Thote, V. Kumar, P. Jadhav, S.K. Lokhande, R.B. Biniwale, N. K. Labhsetwar, S.S. Rayalu, Microporous and Mesoporous Materials 121 (2009) 84.
[56] H. Yang, Z. Xu, M. Fan, R. Gupta, R.B. Slimane, A.E. Bland, I.J. Wright, Environmental Science Technology 20 (2008) 14.
[57] M.G. Plaza, C. Pevida, B. Arias, J. Fermoso, A. Arenillas, F. Rubiera, J.J. Pis, Journal of Thermal Analysis and Calorimetry 92 (2008) 602.
[58] C. Liu, L. Zhang, J. Deng, Q. Mu, H. Dai, H. He, Journal Physical Chemistry C 112 (2008) 19248.
[59] C.M. Tenney, C.M. Lastosskie, Environmental Progress and Sustainable Energy 25 (2006) 4343.
[60] F. Homfray, Chemistry 74 (1910) 129.
[61] D.D. Do, K. Wang, Carbon 36 (1998) 1539.
[62] B.K. Na, K.K. Koo, H.M. Eum, H. Lee, H.K. Song, Korean Journal Chemical Engineering 18 (2001) 220.
[63] L.P. Ding, S. Bhatia, K.F. Liu, Chemical Engineering Science 57 (2002) 3909.
[64] D.H. James, C.S.G. Phillips, Journal of Chemical Society., 1066.
[65] B.J. Kim, S.J. Park, Journal of Colloid and Interface Science 292 (2005) 493.
[66] S.J. Park, S.Y. Jin, Journal of Colloid and Interface Science 286 (2005) 417.
[67] S.J. Park, B.J. Kim, S.K. Ryu, Carbon Letter 5 (2004) 103.
[68] S.J. Park, S.Y. Jin, S.K. Ryu, Carbon Letter 5 (2004) 186.
[69] S. Kim, J. Ida, V.V. Guliants, J.Y.S. Lin, Journal of Physical Chemistry B 109 (2005) 6287.
[70] S.J. Park, J.B. Donnet, Journal of Colloid and Interface Science 200 (1998) 46.
[71] G.M. Dorris, D.G. Gray, Journal of Colloid and Interface Science 77 (1980) 353.
[72] S. Kalz, D.G. Gray, Journal of Colloid and Interface Science 82 (1981) 318.
[73] C. St-Flour, E. Papier, Industrial and Engineering Chemistry 21 (1982) 666.
[74] J. Schultz, L. Lavielle, C. Martin, Journal of Adhesion 23 (1987) 45.
[75] S. Dong, M. Brendlé, J.B. Donnet, Chromatographia 28 (1989) 10.
[76] S.J. Park, J.B. Donnet, Journal of Colloid and Interface Science 206 (1998) 29.
[77] J.R. Conder, C.L. Young, Physicochemical Measurement by Gas Chromatography. John Wiley & Sons, Chichester 25 (1979).
[78] A.T. James, A.J.P. Martin, Biochemistry Journal 50 (1952) 679.
[79] D.H. Desty, W.T. Swanton, Journal of Physical Chemistry 65 (1961) 766.
[80] D.R. Lide, In Handbook of Chemistry and Physics, 71st edit. CRC Press, Boca Raton, 6, 86 (1990).
[81] J.R. Conder, C.L. Young, Physicochemical Measurement by Gas Chromatography, pp. 430. John Wiley & Sons, Chichester (1979).
[82] C. Kemball, E.K. Rideal, Proceedings of the Royal Society of London A 187 (1946) 53.
[83] J.H. De Boer, S. Kruyer, Proceedings of the Koninklijke Nederlandse Akademie van Wetenschappen B 55 (1952) 451.
[84] J.B. Donnet, S.J. Park, Carbon 29 (1995) 955.

[85] J.B. Donnet, S.J. Park, H. Balard, Chromatographia 31 (1991) 434.
[86] F.M. Fowkes, Journal of Physical Chemistry 66 (1962) 382.
[87] F.M. Fowkes, Journal of Physical Chemistry 67 (1963) 2538.
[88] S.J. Park, Ph. D. dissertation, CNRS, France (1992).
[89] M.M. Selim, T.A. El-Nabarawy, Carbon 18 (1980) 287.
[90] F. Saura Calixto, A. Garcia Raso, Chromatographia 15 (1982) 521.
[91] S.J. Park, M. Brendlé, Journal of Colloid and Interface Science 188 (1997) 336.
[92] J. Anhang, D.G. Gray, Journal of Applied Polymer Science 27 (1982) 71.
[93] D.G. Legrand, G.L. Gaines Jr, Journal of Colloid and Interface Science 31 (1969) 162.
[94] G.L. Gaines Jr, Polymer Engineering Science 12 (1972) 1.
[95] E. Papirer, H. Balard, A. Vidal, European Polymer Journal 24 (1988) 783.
[96] H.P. Schreiber, F. St. Germain, Journal of Adhesion Science Technology 4 (1990) 319.
[97] M.J. Wang, S. Wolff, J.B. Donnet, Rubber Chemical Technology 64 (1991) 559.
[98] M.M. Chehimi, E. Pigois-Landureau, M. Delamar, Journal de Chimie Physique 89 (1992) 1173.
[99] J. Schultz, L. Lavielle, C. Martin, Journal de Chimie Physique 84 (1987) 231.
[100] J.A. Riddick, W.B. Bunger, Organic Solvents. Wiley-Inter science, New York (1970).
[101] J.J. Jasper, Journal of Physical Chemistry 59 (1955) 1019.
[102] J. Schultz, L. Lavielle, Inverse Gas Chromatography Characterization of Polymers and Other Materials. American Chemical Society, Washington, D.C. (1989).
[103] C. Bonnerup, P. Gatenholm, Journal of Adhesion Science and Technology 7 (1993) 247.
[104] A.C. Zettlemoyer, Hydrophobic Surfaces, pp. 1. Academic Press, New York (1969).
[105] P.H. Emmett, S. Brunauer, Journal of American Chemical Society 59 (1937) 1553.
[106] A.J. Groszek, Proceedings of Royal Society of London A 314 (1970) 473.
[107] J.H. Clint, Transactions of the Faraday Society 68 (1972) 2239.
[108] G.C. McGonigal, R.H. Bernhardt, D.J. Thomson, Applied Physics Letter 57 (1990) 28.
[109] J.B. Donnet, S.J. Park, M. Brendlé, Carbon 30 (1992) 263.
[110] S.J. Park, E. Papirer, J.B. Donnet, Journal de Chimie Physique 91 (1994) 203.
[111] R.C. Reid, J.M. Prausnitz, T.K. Sherwood, The Properties of Gases and Liquids, 3rd edit. McGraw-Hill, New York (1977).
[112] D.R. Lide, In Handbook of Chemistry and Physics, 71st edit. CRC Press, Boca Raton, 7. 2 (1990).
[113] M.T. Tyn, W.F. Calus, Processing 21 (1975) 16.
[114] F.M. Fowkes, Rubber Chemistry and Technology 57 (1983) 328.
[115] R.S. Mulliken, Journal of American Chemical Society 74 (1952) 821.
[116] R.C. Reid, J.M. Prausnitz, T.K. Sherwood, The Properties of Gases and Liquids, 3rd edit. McGraw-Hill, New York (1977).
[117] F.M. Fowkes, Surface and Interfacial Aspects of Biomedical Polymers. Plenum Press, New York (1985).
[118] O. Olabisi, L.M. Robeson, M.T. Shaw, Polymer-Polymer Miscibility. Academic Press, New York (1979).
[119] R.S. Drago, G.C. Vogel, T.E. Needham, Journal of American Chemical Society. 93 (1971) 6014.
[120] R.G. Pearson, Journal of American Chemical Society 108 (1986) 6109.
[121] F.M. Fowkes, D.W. Dwight, D.A. Cole, T.C. Huang, Journal of Non-Crystalline Solids 120 (1990) 47.
[122] M.J. Kamlet, R.W. Taft, Journal of American Chemical Society 98 (1976) 377.

[123] R.W. Taft, M.J. Kamlet, Journal of American Chemical Society 98 (1976) 2886.
[124] S. Ehrenson, Journal of Organic Chemistry 44 (1979) 1793.
[125] M.J. Kamlet, J.L.M. Abboud, M.H. Abraham, R.W. Taft, Journal of Organic Chemistry 48 (1983) 2877.
[126] M.J. Kamlet, R.M. Doherty, M.H. Abraham, P.W. Carr, R.E. Doherty, R.W. Taft, Journal of Physical Chemistry 91 (1987) 1996.
[127] M.J. Kamlet, R.M. Doherty, M.H. Abraham, Y. Marcus, R.W. Taft, Journal of Physical Chemistry 92 (1988) 5244.
[128] P.E. Doan, R.S. Drago, Journal of American Chemical Society 104 (1982) 4524.
[129] P.E. Doan, R.S. Drago, Journal of American Chemical Society 106 (1984) 2772.
[130] R. Schmid, V.N. Sapunov, Non-Formal Kinetics. Verlag-Chmie, Weinheim (1982).
[131] V. Gutmann, Electrochimica Acta 21 (1976) 661.
[132] A.J. Parker, U. Mayer, R. Schmid, V. Gutmann, Journal of Organic Chemistry 43 (1978) 1843.
[133] U. Mayer, Pure Applied Chemistry 51 (1979) 1697.
[134] V. Gutmann, Pure Applied Chemistry 51 (1979) 2197.
[135] C.S. Four, E. Papirer, Industrial& Engineering Chemistry Product Research and Development 21 (1982) 666.
[136] B. Barbier, M. Vilatte, G. Desarmot, In Proceedings of Third European Conference on Composite Materials, Bordeaux (1989).
[137] B. Barbier, J. Pinson, G. Desarmot, M. Sanchez, Journal of the Electrochemical Society 137 (1990) 1757.
[138] R. Tsunoda, Journal of Colloid and Interface Science 188 (1997) 224.
[139] E. Ruckenstein, Journal of Colloid and Interface Science 196 (1997) 313.
[140] M. Greiveldinger, M.E.R. Shanahan, Journal of Colloid and Interface Science 215 (1999) 170.
[141] S.B. Rho, M.A. Lim, Polymer Korea 23 (1999) 662.
[142] F.M. Fowkes, Industrial Engineering Chemistry 56 (1964) 40.
[143] D. Graham, Journal of Physical Chemistry 68 (1964) 2788.
[144] S.J. Park, J.S. Kim, Journal of Colloid and Interface Science 232 (2000) 311.
[145] P.C. Hiemenz, R. Rajagopalan, Principles of Colloid and Surface Chemistry, 3rd edit. Marcel Dekker, New York (1997).
[146] A.V. Tvardovski, A.A. Fomkin, Journal of Colloid and Interface Science 198 (1998) 296.
[147] G. Kraus, Journal of Applied Polymer Science 7 (1963) 861.
[148] A.V. Kiselev, Y.I. Yashin, In La Chromatographie Gaz-Solide Masson et Cie Paris (1969).
[149] J.D. Fast, 'Entropi', Bibliothèque Technique Philips, Paris (1961).
[150] J. Zeldowitsch, Acta Physicochimica URSS 1 (1935) 961.
[151] A.W. Adamson, I. Ling, Advances in Chemistry Series 33 (1961) 51.
[152] J.P. Hobson, Canadian Journal of Physics 43 (1965) 1941.
[153] D. Morel, H.F. Stoeckli, W. Rudzinski, Surface Science 114 (1982) 85.
[154] W. Rudzinski, L. Lajtar, A. Patrykiejew, Surface Science 67 (1977) 195.
[155] J. Cortes, P. Araya, Anal tica Quimica 85 (1989) 33.

Chapter 3

Solid-Liquid Interface

3.1. INTRODUCTION

The interaction of a liquid with a solid is characterized by the word 'Wetting'. It can be the spreading of a liquid over a solid surface, the penetration of a liquid into porous materials, or the displacement of one liquid by another. This phenomenon can help to characterize a surface, and to determine the interaction, between a solid and a liquid.

The solid/liquid interface plays a fundamental role in diverse fields and helps with an understanding of the physical phenomena and structural knowledge of the interface, at the atomic scale, for example in catalysis, crystal growth, lubrication, electrochemistry, colloidal system, and in many biological reactions. Unraveling the atomic structure at the solid/liquid interface is, therefore, one of the major challenges facing the surface science today to understand the physical processes in model systems.

One way to quantify a liquid's surface wetting characteristics is to measure the contact angle of a drop of liquid placed on the surface of an object. The contact angle is the angle formed by the solid/liquid interface and the liquid/vapor interface measured from the side of the liquid. Liquids wet surfaces when the contact angle is less than 90°. For a penetrant material to be effective, the contact angle should be as small as possible. In fact, the contact angle for most liquid penetrants is very close to 0°.

The wetting ability of a liquid is a function of the surface energy of the solid-gas interface, the liquid-gas interface, and the solid-liquid interface. The surface energy across an interface or the surface tension at the interface is a measure of the energy required to form the unit area of a new surface at the interface. The intermolecular bonds or cohesive forces between the molecules of a liquid cause surface tension. When the liquid encounters another substance, there is usually an attraction between the two materials. The adhesive forces between the liquid and the second substance will compete against the cohesive forces of the liquid. Liquids with weak cohesive bonds and a strong attraction to another material (or the desire to create adhesive bonds) will tend to spread over the material. Liquids

Interface Science and Composites, Volume 18
ISSN: 1573-4285, doi: 10.1016/B978-0-12-375049-5.00003-7

with strong cohesive bonds and weaker adhesive forces will tend to bead-up or form a droplet when in contact with another material.

Depending on the thermodynamic state or the hydrodynamic status of the liquid drop in which the contact angle is measured, two types of contact angles can be defined. If the contact angle is measured when either the liquid drop continues to spread or when its thermodynamic state conditions continue to change, the measured contact angle is termed the dynamic contact angle. However, if the contact angle is measured under conditions in which the liquid drop is stationary and the surrounding conditions are in the steady state, the measured contact angle is known as the static/equilibrium contact angle. Theoretically, the dynamic contact angle is related more to the hydrodynamic conditions, whereas the equilibrium contact angle depends on the surface properties of the solid-liquid-vapor system under the given thermodynamic conditions.

This chapter is focused on the surface/liquid interaction. We concentrate on the role of the intermolecular interactions at interfaces based on the dynamics, the surface free energy at a given temperature and the surface energy of a solid. The contact angle technique is chosen for studies of the wettability phenomena owing to its simplicity. The electrostatic force and self-assembly at the interfaces between a solid and a liquid are also discussed from the viewpoint of intermolecular interactions.

3.2. SURFACE ENERGETICS

Thermodynamically, the physical origin of the surface free energy is the excess Gibbs free energy of the matter at the interface. Atoms or molecules exposed at an interface are surrounded by fewer neighbors, such as solid, liquid, and gas phases, resulting in an anisotropic distribution of these neighbors, which is the characteristic of a surface. They must share some of the interaction energy with the neighboring molecules. Hence, the surface free energy, γ, represents the rate of change of the Gibbs free energy of the system with respect to the area, A, at a constant pressure and temperature [1].

$$\gamma = \left(\frac{\partial G}{\partial A}\right)_{T,P} \tag{3-1}$$

In principle, Eq. (3-1) can be used to calculate the surface tension of a condensed phase held together by the long-range forces.

Another notation that emphasizes the fact that the surface free energy at a given temperature is identical to the excess Gibbs free energy, G^E, can be expressed by a thermodynamic function:

$$G^E = \gamma = H^E - TS^E \tag{3-2}$$

here, the excess entropy, S^E, is

$$S^E = \left(\frac{\partial \gamma}{\partial T}\right) \tag{3-3}$$

Eq. (3-2) and Eq. (3-3) are combined to give the excess enthalpy, H^E:

$$H^E = \gamma - T\left(\frac{d\gamma}{dT}\right) \tag{3-4}$$

In Eq. (3-4), the excess surface enthalpy [2] is identical to the surface energy, h. Thus, Eq. (3-4) can be rewritten.

$$h = \gamma - T\left(\frac{d\gamma}{dT}\right) \tag{3-5}$$

The surface free energy at a given temperature and the surface energy (or enthalpy) for a solid system are shown in Eq. (3-5) [3].

3.3. CONTACT ANGLE AND SURFACE TENSION

3.3.1. Sessile Drop as a Force Balance

Contact angle (θ) measurements, as described by T. Young in 1805, remain at present the most accurate method for determining the interaction energy between a liquid (L) and solid (S) in a condensed state at the minimum equilibrium distance (ℓ_o) of S and L. This is defined geometrically as the angle formed by a liquid at the three-phase boundary, where a gas, liquid, and solid intersect. It is a measure of the result of the energy of cohesion of the liquid, in the guise of γ_L, and the energy of adhesion between a liquid and a solid.

Young [4] described the equilibrium contact angle at the three-phase boundary in terms of the vectorial sum, as shown in Fig. 3-1, resulting in the following equation of equilibrium force balance:

$$\gamma_{LV}\cos\theta = \gamma_{SV} - \gamma_{SL} \tag{3-6}$$

here, γ_i and γ_j are the interfacial tensions between phases i and j, with subscripts V, L, and S being the vapor, liquid and solid, respectively. In addition, θ is the equilibrium contact angle.

3.3.2. Spreading Pressure

In Young's expression, the solid-vapor interfacial tension, γ_{SV}, is the surface tension of the solid in equilibrium with the vapor of the wetting liquid. If γ_S is the surface tension of the solid against its own vapor or in a vacuum, then the following equation holds [5],

$$\gamma_S - \gamma_{SV} = \pi_\varepsilon \tag{3-7}$$

where π_e, is the equilibrium spreading pressure (or two-dimensional pressure) of the adsorbed film; this reflects the reduction of the surface tension of the solid due to vapor adsorption.

In general, π_e is small for moderately large values of θ, though it can become significant as θ drops below 10° [6, 7]. Fig. 3-1 shows, that low values of θ

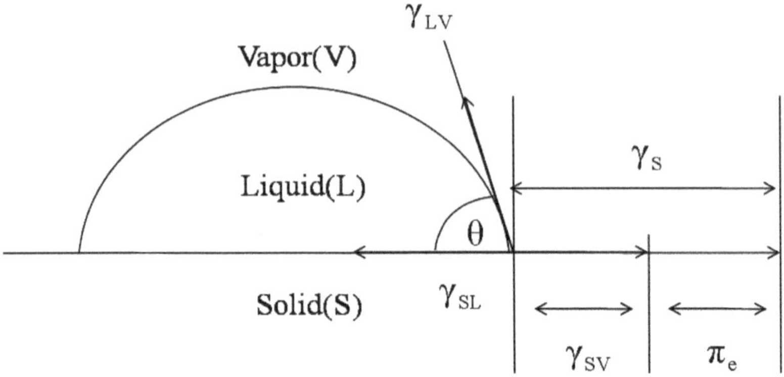

Fig. 3-1 Components of interfacial tension, used in deriving Young's equation.

indicate that the liquid spreads, or wets well, while high values indicate poor wetting. If the angle θ is less than 90°, the liquid is said to wet the solid. If it is greater than 90°, it is said to be non-wetting. A zero contact angle represents complete wetting.

The theoretical and applied aspects of the surface tension and the capillarity of a solid surface are widely elucidated in the literature [8, 9–15] and in other writings [16–18]. There are many methods available to measure the surface free energy (or surface tension) of liquid or solid surfaces, including the capillary rise, maximum bubble pressure, pendant drop, du Noüy ring, Wilhelmy slide, and other methods [8, 9–17]. Fowkes *et al.* [18] studied the possibility of spreading pressures arising with high-energy liquids deposited on the low-energy solids and found that this did not occur. On the other hand, when the vapor of a low-energy liquid interacts with a somewhat higher-energy solid surface, the effect of the resulting positive spreading pressure causes an increase in the contact angle of the water on that solid surface, which allows the determination of π_ε. Busscher *et al.* [19], on the other hand, contend that even when $\gamma_L > \gamma_S$, spreading pressures can have a considerable influence on the contact angle.

Only the liquid tensiometric method [20–23] will be discussed here, as it is a new and highly reproducible method that is capable of providing coherent values of a solid for the London dispersive and the specific components of the long-range force at the interfaces. In general, this method is effective when used with cylindrical solids with small diameters, such as carbon fibers (7 $\sim$ 8 μm), which cannot be properly assessed with the classical contact angle method [20].

3.3.3. Hysteresis of Contact Angle Measurement

The Young equation is held to be valid for contact angles by measured angle [24, 25]. Advancing contact angles (θ_a) are considered to be a measure of the apolar aspects of the surfaces while the retreating contact angles (θ_r) are for the polar aspects and the difference between θ_a and θ_r can be known by the contact angle hysteresis. Another important factor of contact angle hysteresis is the surface roughness. When the roughness radii of the surfaces are larger than 1 μm, correction factors are applied in this method to measure accurate contact angles. However, using smooth surfaces, which have small roughness radii (>1 μm) are by far more desirable. Other possible factors of contact angle hysteresis were discussed by Adamson [2].

Generally, the contact angle hysteresis is positive, i.e., $\theta_a \rangle \theta_r$, indicating complete or partial residual wetting of a solid surface, by the liquid. However, negative hysteresis ($\theta_a \rangle \theta_r$) can rarely take place when a component of the liquid is affect to a change in the properties of the solid surface. For example, the plurivalent counterions tend to neutralize the negatively charged solid surfaces and counteract the hydrophilicity of the surface by depressing the surface's electron donicity parameter (γ^-) [26–29].

The common positive contact angle hysteresis with polar liquids on the polar surfaces represents the insignificance of the equilibrium spreading pressure. Thus, in certain cases, a finite contact angle can be obtained in the condition of $\gamma_L \rangle \gamma_S$. Whereas, in all cases of $\theta_a \rangle \theta_r$, at least, the partially wetting of the liquid on the solid surface is possible. However, when the value of θ_a is much higher than that of θ_r, negligible wetting occurs. Thus, no quantitatively significant amount of wetting is caused by the evaporation of the liquid from the drop after the drop of the liquid on the dry surface is compared to the effect of a retreating angle.

The best means of measuring the contact angles most likely is by a direct observation of the drop, and examining the contact angles on either side of the drop. As mentioned above, the base of the drop is circular. The observation is done by viewing the drop with a telescope. A number of drops of obtained liquid should be measured and then the average value is taken. In addition, when the reflection of the liquid drop on the solid surface is visible, it becomes easier to examine the triple points, i.e., gas-liquid-solid, resulting in enhanced accuracy of the contact angle measurement [30]. Several other approaches to measurement of the contact angle, such as the Wilhelmy plate method, the tilting plate method, and the spherical body methods were described by Neumann and Good [31].

3.3.4. Surface Energy Measurements

The surface energy is generally defined as the work required to build a unit of area of a given surface [3], using the sessile drop technique, however, the surface energy is not quite as well defined. The surface energy obtained by the sessile drop technique depends not only on the solid samples, but the properties of used probe liquid, as well as the particular theory including various parameters.

There are numerous such theories developed by various researchers. These methods differ in several ways, in terms of derivation and convention for example. Most importantly, they differ in the number of components or parameters. Simpler methods containing fewer components simplify the system by the surface energy values into one number, whereas strict methods using more components are derived to distinguish various components of the surface energy. In addition, the total surface energy of solids and liquids depends on the different types of molecular interactions, such as the London dispersive and the polar or acid/base interactions, and is considered to be the sum of these independent components. Some theories account for more of these phenomena than do other theories. These distinctions are to be considered when deciding which method is appropriate for the experiment at hand. The following are a few commonly used contact angle methods to determine the most proper theory.

3.3.4.1. One-liquid Tensiometric Method

There are many methods available to measure the surface tension of a solid surface. The Zisman theory is the simplest and the most commonly used method to determine the surface energy, as it is a one-component theory and tends to be more useful in practice for surfaces with the lower energy levels. A technique developed by Zisman and Fox uses a series of homologous tests of liquids of a known surface tension [32]. This theory simply defines the surface energy, as being equal to the surface energy of the highest surface energy liquid that wets the solid completely, and the contact angles are thus recorded. It means that the droplet will disperse and then completely get wetted on the solid surface, for this liquid and for any other liquids with lower surface energies, but not for liquids with higher surface energies. In this method, this liquid can hypothetically be any liquid. The best way to determine the surface energy by the Zisman method is to acquire the data points of the contact angles for several liquids on the solid surface and then plot the cosine of that angle against the known surface energy of the liquids. From the Zisman plot, one can extrapolate the highest liquid surface energy, real or hypothetical, resulting in complete wetting of the sample with a contact angle of 0°.

3.3.4.2. Two-liquid Tensiometric Method

In the early 1960s, Fowkes [33, 34] introduced the concept of the surface free energy of a solid. The surface free energy is expressed using the sum of the two components: a dispersive component, γ_S^L, attributable to London attraction, and a specific component, γ_S^{SP}, owing to all other types of interactions, such as Debye, Keesom, hydrogen bonding, and other polar effects.

$$\gamma_S = \gamma_S^L + \gamma_S^{SP} \tag{3-8}$$

Similarly, for a liquid,

$$\gamma_L = \gamma_L^L + \gamma_L^{SP} \tag{3-9}$$

where the superscripts L and SP refer to the London dispersive and specific components, respectively, and subscripts S and L represent the solid and liquid phases, respectively.

Fowkes [35] and Good [36] also developed a novel technique for the determination of the work of adhesion, W_A, by employing the surface free energy obtained from the contact angle value in the system of a solid-liquid using the geometric mean [34]

$$W_A = W_A^L + W_A^{SP} \tag{3-10}$$

$$= \gamma_S + \gamma_L - \gamma_{SL} \tag{3-11}$$

$$= \gamma_L(1 + \cos\theta) \tag{3-12}$$

where

$$W_A^L = 2\sqrt{\gamma_S^L \cdot \gamma_L^L} \tag{3-13}$$

and, similarly

$$W_A^{SP} = 2\sqrt{\gamma_S^{SP} \cdot \gamma_L^{SP}} \tag{3-14}$$

combining Eqs. (3-10) to (3-14) gives

$$\gamma_{SL} = \gamma_S + \gamma_L - 2\sqrt{\gamma_S^L \cdot \gamma_L^L} - 2\sqrt{\gamma_S^{SP} \cdot \gamma_L^{SP}} \tag{3-15}$$

Assuming Young's expression, in Eq. (3-6), holds for a two-liquid system, that is, liquid (L_1)-liquid (L_2), Schultz *et al.* [37] proposed the equation

$$\gamma_{SL} = \gamma_{L_1L_2}\cos\theta_{SL_1L_2} + \gamma_{SL_1} \tag{3-16}$$

According to Eq. (3-16), it is then possible to describe γ_{SL_1} and γ_{SL_2} as

$$\gamma_{SL_1} = \gamma_S + \gamma_{L_1} - 2\sqrt{\gamma_S^L \cdot \gamma_{L_1}^L} - 2\sqrt{\gamma_S^{SP} \cdot \gamma_{L_1}^{SP}} \tag{3-17}$$

$$\gamma_{SL_2} = \gamma_S + \gamma_{L_2} - 2\sqrt{\gamma_S^L \cdot \gamma_{L_2}^L} - 2\sqrt{\gamma_S^{SP} \cdot \gamma_{S_2}^{SP}} \tag{3-18}$$

substituting Eq. (3-17) and Eq. (3-18) into Eq. (3-16) yields

$$\gamma_{L_1} - \gamma_{L_2} + \gamma_{L_1L_2}\cos\theta_{SL_1L_2} = \tag{3-19}$$

$$2\sqrt{\gamma_S^L}\left(\sqrt{\gamma_{L_1}^L} - \sqrt{\gamma_{L_2}^L} + \left(2\sqrt{\gamma_S^{SP} \cdot \gamma_{L_1}^{SP}} - 2\sqrt{\gamma_S^{SP} \cdot \gamma_{L_2}^{SP}}\right)\right)$$

When L_1 is formamide (subscript F) and liquid L_2 is n-alkanes (H), the term $\gamma_{L_2}^{SP}$ can be neglected because the surface free energy of n-alkanes consists of only the London dispersive force. Therefore, we can rewrite Eq. (3-19) as

$$\gamma_F - \gamma_H + \gamma_{FH}\cos\theta_{SFH} = 2\sqrt{\gamma_S^L}\left(\sqrt{\gamma_F^L} - \sqrt{\gamma_H} + \left(2\sqrt{\gamma_S^{SP}\cdot\gamma_F^{SP}}\right)\right) \quad (3\text{-}20)$$

where γ_F, γ_F^L, and γ_F^{SP} are the surface free energy of formamide and its London dispersive and specific components, respectively; γ_H is the surface free energy of the hydrocarbon; and γ_{FH} is the interfacial energy of the formamide/hydrocarbon.

If the contact angles for various n-alkanes are measured (typically C_6, C_8, C_{10}, C_{12}, and C_{16}), the plot of $(\gamma_F - \gamma_H + \gamma_{FH}\cos\theta_{SFH})$ against $\left(\sqrt{\gamma_F^L} - \sqrt{\gamma_H}\right)$ may result in a straight line with a slope of $2\sqrt{\gamma_S^L}$ and an intercept of $\left(2\sqrt{\gamma_S^{SP}\cdot\gamma_F^{SP}}\right)$.

Table 3-1 gives the values of parameters [38–45] in Eq. (3-20). The various experimental values and species of the polar liquids are listed in Table 3-1. It is

Table 3-1 **Surface Free Energy Characteristics of the Liquids, Measured at 20 °C [46]**

Liquid	γ_L^L (mJ/m²)	γ_L^{SP} (mJ/m²)	γ_L (mJ/m²)
n-Pentane	16.05	0	16.05
n-Hexane	18.40	0	18.40
n-Heptane	20.14	0	20.14
n-Octane	21.62	0	21.62
n-Nonane	22.85	0	22.85
n-Decane	23.83	0	23.83
n-Undecane	24.66	0	24.66
n-Dodecane	25.35	0	25.35
n-Tridecane	25.99	0	25.99
n-Tetradecane	26.56	0	26.56
n-Pentadecane	27.07	0	27.07
n-Hexadecane	27.47	0	27.47
Formamide	39.5	18.7	58.2
	18.1	40.24	58.34
	31.8	25.7	57.5
Diiodomethane	50.42	0.38	50.8
Water	21.8	51.0	72.8
	10.8	61.2	72.0
Ethylene glycol	18.91	29.29	48.2
	31.0	16.7	47.7
Glycerol	20.22	43.08	63.3
	33.9	29.8	63.7

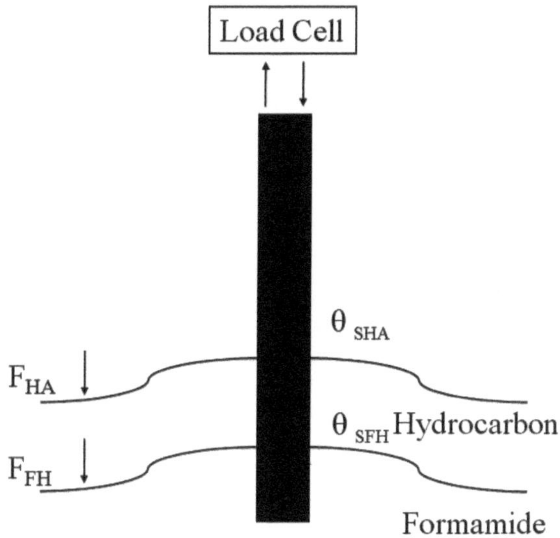

Fig. 3-2 Schematic diagram of the air-hydrocarbon-fiber and hydrocarbon-formamide-fiber interfaces.

also possible to plot the results using Eq. (3-20) to determine the London dispersive component, γ_S^L, and the specific component, γ_S^{SP}, of the solid surface.

Fig. 3-2 presents the basic principles of the two-liquid tensiometric method. The method involves the wetting of the solid by a solvent, generally formamide or water in the presence of hydrocarbons, as it was formulated before by Donnet and Schultz [22, 23]. In this method, a single fiber about 2 to 3 cm in length is attached from the arm of a micro-balance. During the immersion and emersion cycles, the change in force, F, due to the menisci of the liquid raised at the two interfaces is recorded.

At equilibrium, when the single fiber is immersed in the two liquids, the net increase in the force is contributed by three different forces: the apparent weight of the liquid raised at the air-hydrocarbon-fiber interface F_{HA}, the apparent weight of the liquid raised at the hydrocarbon-formamide-fiber interface F_{FH}, and the buoyancy acting on the length of the fiber immersed F_P, as shown in Fig. 3-2. When the diameter of the fiber is very small, the buoyancy contribution will be too small, to be neglected.

When the fiber is immersed in the hydrocarbon phase, the contribution F_{HA} can be given by

$$F_{HA} = C \cdot \gamma_H \cdot \cos\theta_{SHA} \tag{3-21}$$

where C is the circumstance of the fiber and θ_{SHA} is the contact angle of the hydrocarbon on the fiber in air or in hydrocarbon vapor.

When the contact angle θ_{SHA} is almost zero, the circumstance C is given by

$$C = \frac{F_{HA}}{\gamma_H} \tag{3-22}$$

and can be determined by measuring the force due to immersion in only the hydrocarbon phase.

Similarly, the contribution F_{FH} is given by an expression analogous to Eq. (3-21)

$$F_{FH} = C \cdot \gamma_{FH} \cdot \cos\theta_{SFH} \tag{3-23}$$

combining Eq. (3-22) and Eq. (3-23) gives

$$F_{FH} = \frac{F_{HA}}{\gamma_H} \cdot \gamma_{FH} \cdot \cos\theta_{SFH} \tag{3-24}$$

$$= \frac{F_{HA}}{\gamma_H} \cdot \gamma_{FH} \cdot (\sec\theta)^{-1}_{SFH} \tag{3-25}$$

where θ_{SFH} is the contact angle of formamide on the fiber in the presence of the hydrocarbon.

Considering Eq. (3-20), it is apparent that the major variable on the y-axis is $\gamma_{FH}\cos\theta_{SFH}$, which becomes identical to $\gamma_H F_{FH}/F_{HA}$ via Eq. (3-25). Connecting Eq. (3-20) with Eq. (3-25), a direct calculation of the contact angle, θ_{SFH}, is not required in this tensiometric method, as shown in Fig. 3-3.

When F_{HA} and F_{FH} are measured with a number of n-alkanes, it is possible to investigate the London dispersive γ_S^L and specific γ_S^{SP} components of the solid

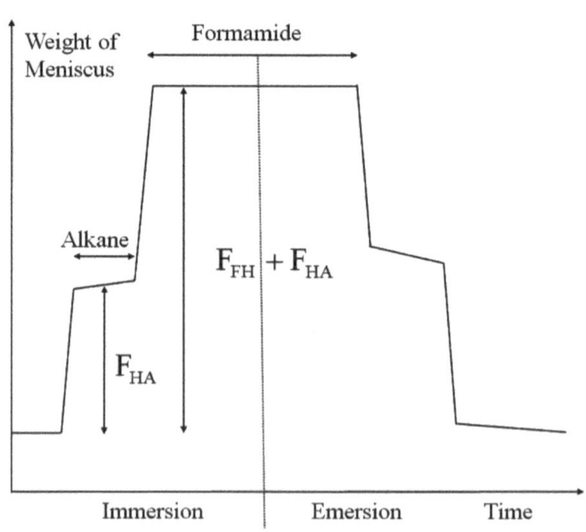

Fig. 3-3 Run showing readings in immersion-emersion on the recorder.

Table 3-2 Surface Free Energy Components of High Strength Carbon Fibers Studied by Two-liquid Tensiometric Method (in mJ/m^2)

Samples	γ_S^L	γ_S^{SP}	γ_S
Untreated[a]	26 ± 8	12 ± 3	38 ± 11
Plasma treated[b]	35 ± 2	25 ± 8	60 ± 3
Oxidized[c]	33 ± 6	18 ± 8	51 ± 7
Oxidized[d]	27 ± 6	28 ± 8	55 ± 7

[a] *Untreated and unsized carbon fiber (T-300) supplied by Soficar.*
[b] *By microwave generator (2.4 GHz) for 10 min.*
[c] *By commercially just anodic oxidation.*
[d] *By commercially fully anodic oxidation.*

surface using the following equation obtained after combining Eq. (3-20) and Eq. (3-25).

$$\gamma_F - \gamma_H\left(1 - \frac{F_{FH}}{F_{HA}}\right) = 2\sqrt{\gamma_S^L}\left(\sqrt{\gamma_F^L} - \sqrt{\gamma_H}\right) + \left(2\sqrt{\gamma_S^{SP} \cdot \gamma_F^{SP}}\right) \tag{3-26}$$

With this method, there are several comprehensive studies of the determination of the surface free energy of the high-strength and high-modulus carbon fibers before and after several surface treatments, including electrochemical [22, 41] and plasma [47–49] treatments. The surface free energy as well as its London dispersive γ_S^L and the specific γ_S^{SP} components are presented in Table 3-1 [20].

In Table 3-2, the surface free energy of the carbon fibers untreated and unsized is approximately 38 mJ/m.^{2} However, the surface free energy increases when the fibers are treated by either electrochemical or plasma methods. Particularly, the specific component of the surface free energy is enhanced by about 100 percent. It is interesting to note that the increase of the specific component of the surface free energy may be attributed to the introduction of a free radical or a functional group on the carbon surface. Thereby, the carbon surface becomes more active, finally leading to an increase in the degree of adhesion, for example by the fiber-matrix interaction, which is one of the determining parameters [50–54].

3.3.4.3. *Three-liquid Tensiometric Method*

In the Good and van Oss model, the surface energy is defined as [55, 56]:

$$\gamma = \gamma_S^L + 2\sqrt{\gamma_S^+\gamma_S^-} \tag{3-27}$$

where γ_S^L is dispersive component or Lifshitz-van der Waals interactions, and γ_S^+ and γ_S^- the polar components or the Lewis acid–base of a solid surface.

The van Oss theory separates the surface energy of solids and liquids into three components. It contains the dispersive surface energy and subdivides the polar component as being the sum of two specific components: the surface energy by acidic interactions (γ^+) and basic interactions (γ^-). The acid component theoretically describes the tendency of a surface to have polar interactions with a second surface that has the ability to act basic by donating electrons. Conversely, the base component represents the tendency of a surface to have polar interactions with another surface that acts acidic by accepting electrons. The relationship, between the surface energy components, liquid components, and the contact angle is defined as:

$$\gamma_L(1 + \cos\theta) = 2\left(\sqrt{\gamma_S^L \gamma_L^L} + \sqrt{\gamma_S^+ \gamma_S^-} + \sqrt{\gamma_L^+ \gamma_S^-}\right) \qquad (3\text{-}28)$$

Again, the best method for the determination of this theory is to use at least three different liquids depositing with known dispersive and polar components. One of the drops should have a dispersive component to its surface energy ($\gamma_L = \gamma_L^L$), one should have a dispersive and an acidic or basic component ($\gamma_L = \gamma_L^L + \gamma_L$), and the last liquid should have either a liquid with a dispersive and acidic or basic component, or a liquid with all three components ($\gamma_L = \gamma_L^L + \gamma_L^+ + \gamma_L^-$).

Being a three-component theory, it is naturally more robust than other theories and there is a great imbalance between the acid and base components of the polar surface energy. The van Oss theory is most suitable for testing the surface energies of the inorganics, organometallics, polymers, and surface-containing ions.

3.3.5. Contact Angle Measurements

3.3.5.1. Tilting Plate Method

The tilting plate method is a classical method to measure contact angles [2]. In this method, a wide solid plate is dipped into a liquid and the position of the solid plate is altered using an adjustable mount until the angle is formed, so that the liquid surface appears to remain perfectly flat right up to the surface of the solid. Nevertheless, there are at least three difficulties with this method; 1) It is difficult to adjust the solid plate at an angle such that the liquid surface is perfectly flat right up to the solid surface; 2) It is generally inconvenient to find the large sample of solid and the large volume of liquid required by the tilting plate method; 3) It is not always easy to prevent the surface and liquid contamination during the contact angle measurement.

The Wilhelmy plate/slide technique [57] is another classical method. The contact angle is examined by directly measuring the height of the capillary rise at a vertical flat plate. Obviously, an advantage of this technique is that the measurement of a contact angle becomes the measurement of a much more accurate height. Recently, a new method of measuring the capillary rise was

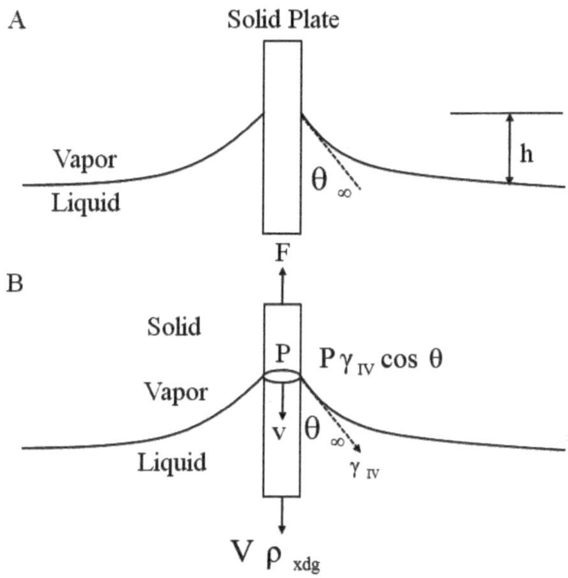

Fig. 3-4 (A) Schematic diagram of the Wilhelmy plate technique for contact angle measurement, (B) Schematic diagram of the Wilhelmy gravitational method for contact angle measurement.

developed by Budziak and Neumann [58] using the digital image analysis in conjunction with high-precision translation stages. As shown in Fig. 3-4, for a vertical flat plate brought into contact with a pool of liquid and then can rise on the plate above the undisturbed liquid surface. For such a capillary rise profile around a large vertical flat plate, the relationship between the contact angle and the height of the capillary rise can be obtained from a straightforward integration of the Laplace equation of capillarity:

$$\sin\theta = 1 - \frac{\Delta\rho_{lV} g h^2}{2\gamma_{lV}} \tag{3-29}$$

where $\Delta\rho_{lV} = \rho_l - \rho_V$ is the density difference between the liquid phase and the vapor phase, g the acceleration due to gravity, h the capillary rise height, and θ_∞ the contact angles of the liquid on the vertical plate.

Usually, the termination of the capillary rise profile is quite sharp under proper illumination and h can be accurately measured by a traveling microscope mounted on a high-precision translation stage. Another advantage of the Wilhelmy plate method is that it avoids the line tension effect on the measured contact angles, as the three-phase contact line in this case is a straight line. In addition, this method is particularly suited to obtain the temperature-dependence of the contact angle. However, it is generally inconvenient to find the large sample of the solid and the large volume of liquid required for the Wilhelmy slide technique.

Generally, both the tilting plate method and the Wilhelmy slide technique require a large flat solid surface and hence cannot be used to measure the contact angle of a curved surface such as a cylindrical fiber surface. The Wilhelmy gravitational method can be the only existing technique suitable for this type of contact angle measurement. In this method, shown in Fig. 3-4B, a solid hanged on an electrobalance is partially immersed in a test liquid, and then the contact angle is determined by measuring the force, F, required to balance the solid:

$$F = P_W \gamma_{lv} \cos\theta + V\Delta\rho 1_{Sl} g \tag{3-30}$$

where P_W is the wetted perimeter of the solid, V the liquid volume displaced by the solid or the volume of the solid immersed in the liquid, and $\Delta\rho_{Sl} = \rho_S - \rho_l$, the density difference between the solid phase and the liquid phase.

In the literature, the Wilhelmy gravitational method was also used to measure the temperature-dependence of the surface tension of bitumen. The contact angle between the platinum and the bitumen is zero, indicating complete wetting [59, 60]. However, the contact angle data are generally difficult to reproduce using the Wilhelmy gravitational method due to the non-constant diameters of the fibers and the penetration of the liquid into the fiber, leading to a change in the fiber's shape and density.

3.3.5.2. Wicking Method

Hard particles with a diameter of 1 μm or more, when spread into flat layers, will form a surface, but their surface is too rough to measure the accurate contact angle by the sessile drop method. However, using a packed column of such particles, the capillary rise velocity measurements of a liquid in the column can also obtain the contact angle of the liquid with respect to the particles' surface. This wicking method can also be used to determine the contact angle of porous solid materials or fiber forms. In these cases, one uses the Washburn equation by Adamson and Roberts *et al.* [2, 61],

$$h^2 = \frac{tR\gamma_L \cos\theta}{2\eta} \tag{3-31}$$

where h is the height of the column of liquid L, that has reached due to the capillary rise action in time t, R is the average radius of the pores of the porous bed, θ is the contact angle and η is the viscosity of liquid L.

By measuring values of h at a constant time for a number of low energy liquids, which can be spread on the surface of the particles, the value of R can be obtained for the type of column at hand, packed with a given type of particle in these cases, $\cos\theta = 1$. Once R is determined, $\cos\theta$ can be examined with a number of well known apolar and polar liquids, which then permits the solution of Eq. (3-28) through γ_S^L, γ_S^+, and γ_S^- of the particles.

To use the wicking method for the determination of $\cos\theta$, it is essential to use the well-packed columns with more mono-dispersed particles. If tighter packing can locally occur, a gap will be created between the particles, leading to a strongly asymmetrical rise of the liquid in the packed column. Thus, it is very difficult to measure the precise length of travel of the liquid column. However, when only polydisperse suspensions of irregular particles are available, an extremely useful alternate method is the coating of such particles onto glass surfaces, followed by the measurement of the capillary rise of various types of liquid, as the manner of thin layer chromatography [62].

For the determination of the surface free energy by contact angles on the high-energy carbon yarns, Chwastiak [63] also used the wicking rate measurements but developed equations for that purpose that appear more appropriate than Washburn's, as shown in Eq. (3-31). As with Washburn's equation, Chwastiak's method can only accommodate values for $\theta < 90°$; at $\theta > 90°$ no capillary rise, along the porous materials can occur.

The surface tension components and parameters, as well as the viscosities of the liquids used in wicking, are shown in Table 3-3. In wicking, as in direct contact angle measurements, the preferred order in which to use the contact angles obtained with these liquids, in calculating the various surface tension components and parameters of polar solids is to calculate γ_S^L first, and then γ_S^+ and γ_S^-. For wicking, the use of glycerol is not recommended due to its very high viscosity. Wicking is particularly useful for the characterization of the non-swelling clay particles.

Instead of the Washburn method by the rate of capillary rise, one can also use the method of Bartell and Whitney [64], who described an apparatus for capillary pressure determination, defined as follows:

$$\Delta P = \frac{2\gamma_L \cos\theta}{R} \tag{3-32}$$

Table 3-3 Surface Tension Components and Parameters of a Number of Liquids Used in Direct Contact Angle Tetermination, or in Wicking, at 20°, in mJ/m², as well as their Viscosities, η at 20 °C in Poises.

Liquid	γ	γ^L	γ^{SP}	γ^+	γ^-	η
Apolar						
Dodecane	23.83	23.83	0	0	0	0.0092
Tetradecane	26.6	26.6	0	0	0	0.0218
Pentadecane	27.07	27.07	0	0	0	0.029
Cis-Decalin	32.2	32.2	0	0	0	0.0338
α-Bromonaphthalene	44.4	44.4	0	0	0	0.0489
Diiodomethane	50.8	50.8	0	0	0	0.028
Polar						
Water	72.8	21.8	51.0	25.5	25.5	0.01
Glycerol	64	34	30	3.92	57.4	14.9
Formamide	58	39	19	2.28	39.6	0.0455
Ethylene glycol	48	29	19	1.92	47.0	0.199

White [65] described a way to circumvent the need to determine *R* by a liquid using Eq. (3-32), *R* can be represented as the effective radius (R_{eff}), which is a function of the volume fraction (Φ) occupied by the particles, the specific surface area (A) per unit weight of the particles and the specific density of the particle material (ρ):

$$R_{eff} = \frac{2(1-\phi)\gamma_L \cos\theta}{\phi\rho A} \tag{3-33}$$

In practice, however, R_{eff} cannot equal R exactly, especially in the case of porous particles. The capillary pressure method cannot be used in the thin-plate configuration described above, thus, one must revert to the use of vertical tubes packed with powder. Accordingly, the thin-layer rate of the capillary rise method is by far the most convenient and usually the most accurate method.

3.3.5.3. Sessile Drop Method

The sessile drop technique, which was developed by Neumann *et al.*, has been used successfully for almost two decades. A detailed technical description of this technique has been shown in several studies [66–68]. The sessile drop technique has two different application cases; one is applied for the sessile drop case to measure the contact angle and the surface/interfacial tension, simultaneously and the other is for the pendant drop case to measure the surface/interfacial tension only. Here, the sessile drop technique for the sessile drop case is shown in Fig. 3-5. In the setup, a solid surface is leveled and placed

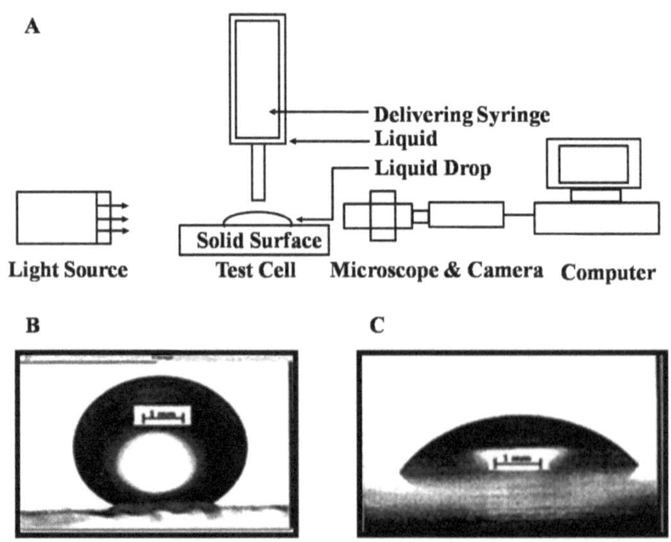

Fig. 3-5 (A) Schematic diagram of the sessile drop technique, for the sessile drop case, used for contact angle measurement. (B) Digital image of a silicone oil drop on a clean glass slide immersed in the pure water. ($\theta = 148.6°$). (C) Digital image of a silicone oil drop on the FC725 precoated glass slide immersed in the pure water ($\theta = 56.6°$).

between a light source and a microscope. A liquid drop is deposited onto the solid surface through a syringe. After the liquid drop reaches its equilibrium state and its digital image can be well obtained the video signal of its instantaneous image is transmitted to a videopix digital processor.

In general, the sessile drop technique for the sessile drop case accomplishes the contact angle measurement by finding the best fit of the theoretical drop profile to the actual values. Theoretically, the equilibrium shape of a moderately curved liquid drop can be determined by solving the Laplace equation of capillarity with the suitable boundary conditions. On the other hand, the sessile drop can automatically digitize the actual drop image, as shown in Fig. 3-5B [69].

These two contact angle data clearly show the difference of the wettability between the hydrophilic surface and the hydrophobic surface, when they are examined with a silicone oil, where both the solid surface and the oil surface are immersed in an aqueous phase. Compared to the classical methods of the contact angle measurement, the sessile drop technique is more accurate and fully automatic. As well known, the sessile drop technique for the sessile drop case is a standard method of contact angle and surface/interfacial tension measurement.

3.3.5.4. Atomic Force Microscopy Method

The corrosion of material surfaces is proven by many researches on the corrosion mechanism by the thin water films onto the surface [70, 71], whereas the distribution and morphology of the nano-sized water droplets and water wettability has not been clarified. Recently, the morphologies of the nano-sized water on various material surfaces have been observed using the atomic force microscopy (AFM) [72–74] and a wettability evaluation of the nano-sized water droplets on materials and the relationship between the nano-sized and the milli-sized water droplets on wettability has been reported.

To distinguish the forms of the particles, such as liquid or solid on the material surfaces, contact mode scanning has been used Fig. 3-6. After a non-contact mode, the contact mode was carried out near the center of the non-contact mode zone by the same cantilever and then the zone was observed with the non-contact mode again. If the particles are liquid, their locations and shapes are changed by contact with the cantilever, whereas if the particles are solid, there are no changes of their locations and shapes.

The average radius (r) and height (h) of the water droplets were measured from four cross-section profiles through the top of water droplet. However, the horizontal radius (r) in these cases will be magnified by the effect of the radius of curvature of the cantilever tip (R_l), shown schematically in Fig. 3-7. The enlarged radius (L) will appear as the water droplet radius in the observed AFM images. The enlargement ratio (L/r) can be described by Eq. (3-35) [75]. The actual radius of each water droplet (r) can be obtained by Eq. (3-34).

$$\frac{L}{r} = \sqrt{\frac{2R_1 + 2R_2 - h}{2R_2 - h}} \tag{3-34}$$

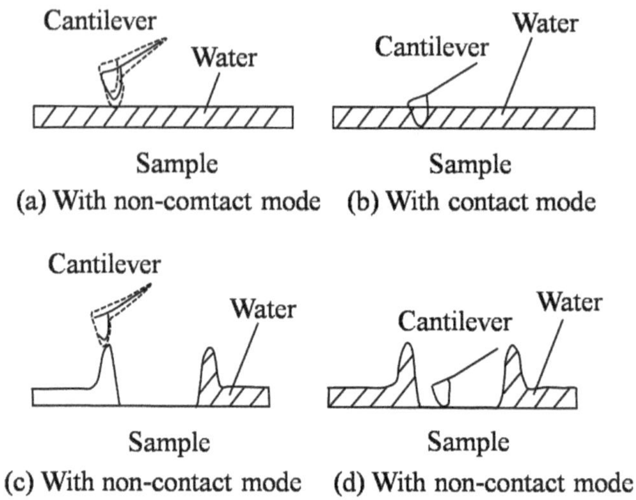

Fig. 3-6 Schematic drawings of observation and confirmation of the nano-sized water droplets by atomic force microscopy.

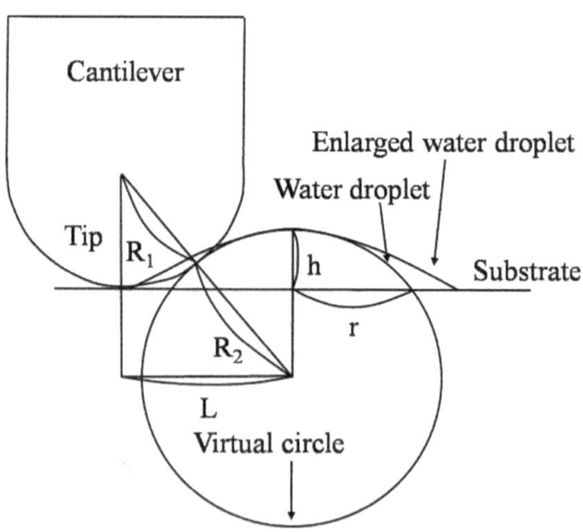

Fig. 3-7 Schematic drawing of the nano-size water droplet enlargement, by the cantilever tip of atomic force microscopy.

The direct measurement of the contact angles between nano-sized water and the substrate is difficult because the distinction of interface between water and substrate is too difficult due to the nano-size roughness of the material surfaces. On the other side, the amount of the nano-sized water onto the substrate is small

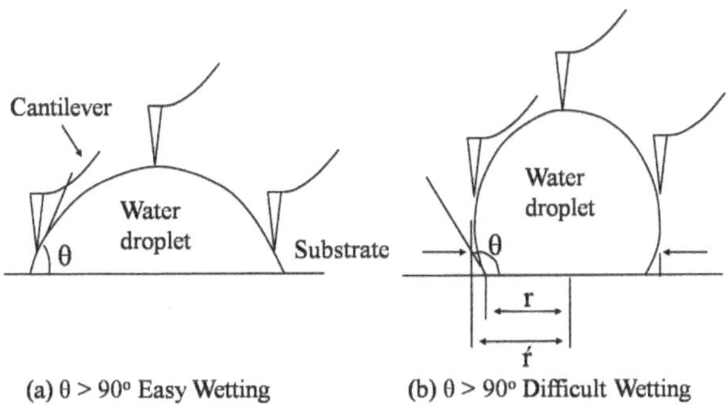

Fig. 3-8 Schematic drawing of precise measurement of the droplet shape, by the movement of cantilever of atomic force microscopy.

enough to neglect its weight, and the shape of the water droplet is thus considered as spherical form. If assumed spherical water droplets are placed on a flat substrate, then Eq. (3-35) [76] can be used to calculate the contact angle between the substrate and the water droplet. Also, the observation of many water droplets is necessary to obtain an average value of the nano-size wettabilities. However, when the contact angle is higher than 90°, the accurate measurement of the shapes of the nano-sized water droplet is very difficult because the cantilever cannot scan the lower part of nano-sized water droplet, as shown in Fig. 3-8. Thus, if the contact angle is more than 90°, Eq. (3-35) can be used because the height (h) of the droplet will be larger than the measured horizontal radius (r') of the droplet.

$$\theta = 2\tan^{-1}\left(\frac{h}{r}\right) \qquad (3\text{-}35)$$

3.3.6. Surface Tension Parameters of Liquids and Solids

3.3.6.1. Apolar Liquids

As shown in Table 3-4, if an apolar liquid needs to measure the London dispersive (γ^L) component of a solid surface using Eq. (3-36), then,

$$(1+\cos\theta) = 2\sqrt{\gamma_S^L/\gamma_L} \qquad (3\text{-}36)$$

As soon as one deals with solids with γ_S^L values greater than about 18 mJ/m^2, the choice of the alkanes, as the liquid is rather poor. However, there are a few higher-energy apolar liquids for all practical purposes (Table 3-4). Some examples are as follows:

Diiodomethane, for which $\gamma_L^L \approx \gamma_L = 50.8\ \mathrm{mJ/m^2}$
α-Bromonaphthalaene, for which $\gamma_L^L \approx \gamma_L = 44.4\ \mathrm{mJ/m^2}$

Table 3-4 Surface Tension Component and Parameters of Alkanes at 20 °C in mJ/m² [77]

Liquid	γ	γ^L	γ^{SP} (γ^+, γ^-)
Pentane	16.05	16.05	0
Hexane	18.04	18.04	0
Heptane	20.14	20.14	0
Octane	21.62	21.62	0
Nonane	22.85	22.85	0
Decane	23.83	23.83	0
Undecane	24.66	24.66	0
Dodecane	25.35	25.35	0
Tridecane	25.99	25.99	0
Tetradecane	26.56	26.56	0
Pentadecane	27.07	27.07	0
Hexadecane	27.47	27.47	0
Nonadecane	28.59	28.59	0
Eicosane	28.87	28.87	0
Cyclohexane	25.24	25.24	0

The γ_L^L values, for most of the solid surfaces can be obtained with above two liquids. It is reasonable to calculate γ_S^L values through Eq. (3-36) for polar solids, because as mentioned above, contact angle measurements with the two apolar liquids are more accurate than those examined with the more polar liquids. The various apolar and the polar liquids used in direct contact angle measurement are listed in Table 3-4.

3.3.6.2. Polar Liquids

The γ-components and polar γ-parameters of some water-miscible liquids are listed in Table 3-5. Generally, these data can't easily be obtained in all cases. For example, with ethanol and methanol, the γ^{SP} values can be determined by contact angle measurements of non-polar surface, such as Teflon, but the γ^+/γ^- ratios have to be evaluated, as contact angle measurements on polar surfaces are excluded due to the spreading of these low energy liquids on the high energy surfaces. In addition, the inclusion of polar liquids in a gel has worked relatively well for formamide and ethylene glycol however, in the gel method, the ethanol and methanol are not acceptable liquids due to their rapid evaporation. It also can't easily obtain very reliable polar data for dimethylsulfoxide by its extraordinary hygroscopicity. Therefore, the accurate values for glycerol and formamide were measured using the contact angle measurements on a variety of well-known monopolar surfaces. For ethylene glycol, the values for γ_{EG}^- and γ_{EG}^+ are derived from its known values of γ_{EG} and γ_{EG}^{SP}, and from solubility data.

Table 3-5 Surface Tension Compounds and Parameters of Water-miscible Liquid, at 20 °C in mJ/m² [56]

Liquid	γ	γ^L	γ^{SP}	γ^+	γ^-
Methanol	22.5	18.2	4.3	0.06	77
Ethanol	22.4	18.8	2.6	0.019	68
Methyl ethyl ketone	24.6	24.6	0	0	24.0
Tetrahydrofuran	27.4	27.4	0	0	15.0
Ethylene glycol	48	29	19	1.92	47.0
Glycerol	64	34	30	3.92	57.4
Formamide	58	39	19	2.28	39.6
Dimethylsulfoxide	44	36	8	0.5	32
Water	72.8	21.8	51	25.5	25.5

3.3.6.3. Synthetic Polymers

Using contact angle liquids, such as diiodomethane, α-bromonaphthalene, water, glycerol, and formamide, as mentioned above Table 3-4, contact angles have been used to measure the characteristics of polymer surfaces, as shown in Table 3-6. It is indicated that most polar polymers have a γ^L value of ≈ 40 mJ/m². Most of

Table 3-6 Surface Tension Components and Parameters of Various Synthetic Polymers, at 20 °C in mJ/m² [56]

Polymer	γ	γ^L	γ^{SP}	γ^+	γ^-
Teflon FEP	17.9	17.9	0	0	0
Polyisobutylene	25.0	25.0	0	0	0
Plypropylene	25.7	25.7	0	0	0
Corona-treated polypropylene	33.0	33.0	0	0	11.1
Polyethylene	33.0	33.0	0	0	0
Nylon 6,6	37.7	36.4	1.3	0.02	21.6
Polymethylmethacrylate	40.0	40.0	0	0	14.6
Polystyrene	42	42	0	0	1.1
Polyvinylalcohol	42	42	0	0	17–57
Polyvinylpyrrolidone	43.4	43.4	0	0	29.7
Polyvinylchloride	43.8	43.0	0.75	0.04	3.5
Polyethyleneoxide (PEG-6,000)	43.0	43.0	0	0	64
Polyoxytetramethylene glycol (MW ≈ 2,000)	44.0	41.4	2.6	0.06	27.6
Co-poly(ethylene glycol, propylene glycol) (MW ≈ 2,000)	47.5	42.0	5.5	0.13	58.8
Co-poly(ethylene glycol, propylene glycol) (MW ≈ 1,000)	47.9	40.9	7.0	0.22	55.6

these polar polymers are also mainly electron donors. Among the various polymers, the strongest electron donor of these is polyethylene oxide.

3.3.7. Solubility

The cohesive properties of a substance including organic materials or polymers can be directly expressed by its solubility in various organic solvents. In addition, the cohesive properties of a substance are quantitatively expressed through the cohesive energy. Among many parameters, on the solubility of the substance, this quantity is closely related to the internal pressure.

As early as 1916, Hildebrand [78] defined, that the order of solubility of a given solute in various solvents is determined by the internal pressures of the solvents. Later, Scatchard (1931) [79] introduced the concept of 'cohesive energy density' from Hildebrand's theories, verifying this quantity with the cohesive energy per unit volume. Finally, Hildebrand (1936) [80] represented a comprehensive treatment of this concept and proposed the square root of the cohesive energy density, as a parameter verifying the behaviors of specific solvents. In 1949, he proposed the new term, i.e., solubility parameter and the symbol, δ, again.

The solubility of a given polymer in various solvents is largely examined by their chemical structure. Generally, structural similarity between substance and solvent can provide superior solubility. In terms of the aforementioned quantities, this means that the solubility of a given polymer in a given solvent is higher if the solubility parameters of the polymer and the solvent are equal or similar. Besides, the chemical structure the physical state of the polymers is also a key factor for its solubility properties. Crystalline polymers are relatively insoluble compared to amorphous polymers and can easily be dissolved at temperatures below their crystalline melting points. In addition, the solubility is proportioned to molecular weight and it decreases on increasing the molecular weight of the polymers. These properties can also be used to fractionate polymers according to the molecular weight. The solubility parameter of the polymer is usually identified using the square root of the cohesive energy density in an amorphous state at room temperature. Most of this section will be focused on the cohesive energy and the solubility parameter, expanded the solubility parameters, and their applications to estimate the solubility parameter of various substances

3.3.7.1. Cohesive Energy

The cohesive energy (E_{coh}) of a material is the increase in the internal energy per mole of the materials, if all of its intermolecular forces are eliminated [81]. The cohesive energy density (E_{coh}'), which is defined by Eq. (3-37), is the energy required to break all the intermolecular physical links in a unit volume of the material. In polymers, these physical links mainly consist of interchain interactions of various types.

$$E_{coh} = \frac{E_{coh}}{V} \quad (3\text{-}37)$$

These definitions imply that E_{coh} is a very fundamental property of a polymer and that it should play a major role in the prediction of many other physical properties. This is indeed the case, as demonstrated in the following:

1) The most important use of E_{coh} is in calculating the solubility parameter, which quantifies the interactions between polymer and solvent, and
2) E_{coh} is also useful in correlating or predicting many other important properties of a polymer, such as its glass transition temperature, surface tension, dielectric constant, mechanical properties, and permeability to small molecules.

The cohesive energy is an important quantity for characterizing the physical state of a given polymer. It is related to other polymer properties for which the cohesive forces are important. The cohesive energy has found its most important applications, however, in the interactions between polymers and solvents. For this purpose, the solubility parameter (δ) is generally used. Therefore, the greater part of this section will be devoted to the properties and applications of the solubility parameter (see 2.5).

3.3.7.2. Solubility Parameter

The thermodynamic criteria of solubility are based on the free energy of mixing ΔG_M. Two substances are mutually soluble, if ΔG_M is negative. By definition,

$$\Delta G_M = \Delta H_M - T\Delta S_M \tag{3-38}$$

where ΔH_M and ΔS_M are the enthalpy of mixing, and the entropy of mixing, respectively.

As ΔS_M is generally positive there is a certain limiting positive value of ΔH_M below which dissolution is possible.

It has been known for several decades that there is a close connection between solubility and surface tension. Hildebrand and Scott [82] reported an empirical correlation between the solubility parameter δ and the surface tension γ:

$$\delta_\alpha V^{-2/3}\gamma^{0.44} \approx V - 2/3\sqrt{\gamma} \tag{3-39}$$

Girifalco and Good [83] showed that the equation for the interfacial tension γ_{12}, between the apolar phases 1 and 2 can be put into the form

$$\gamma_{12}^{L} = \left(\sqrt{\gamma_1^L} - \sqrt{\gamma_2^L}\right)^2 \tag{3-40}$$

where γ_{12}^L and γ_{12}^L are the London dispersive components of the surface tension of the pure phases 1 and 2 [84].

Good and Girifalco also point out that the right side of Eq. (3-40) is closely analogues to the Hildebrand-Scatchard expression for the partial molar internal energy of mixing in regular solutions:

$$\Delta U_i^{\text{mixing}} = \Phi_j^2 V_i \left(\delta_i - \delta_j\right)^2 \tag{3-41}$$

here, Φ_j is the volume fraction of *j*, and V_i the molar volume of component *i*.

In general, $(\delta_i - \delta_j)^2$ must be small for the components to be miscible. The solubility parameter of a given material can be calculated either from the cohesive energy or from the molar attraction constant (F), as $\delta = F/V$. In the derivation of Eq. (3-41), it was assumed that no specific forces are active between the structural units of the substances involved. Therefore, the equation does not hold for crystalline polymers.

In addition, if one of the substances involved contains strongly polar groups or hydrogen bridges, ΔH_M can become higher than predicted by Eq. (3-71), implying that ΔG_M becomes positive even for $\delta_1 = \delta_2$, and that dissolution does not occur. Conversely, if both substances contain polar groups or hydrogen bridges, solubility can be promoted. The δ-values of some polymers are listed in Table 3-7.

3.3.7.3. Expanded Solubility Parameters

The solubility parameter describes the enthalpy change upon the mixing of non-polar solvents well but does not give uniform results when extended to polar systems. Complete miscibility is expected to occur if the solubility parameters are similar and the degree of hydrogen bonding is similar between the components. Other investigators have decomposed the Hildebrand parameter into several terms, representing different contributions to the energy of mixing. For example, van Arkel [86] and Small [87] divided the total solubility parameter into two main components, defining a non-polar solubility (δ_λ) and a polar solubility parameter (δ_τ). The polar-nonpolar parameters are related to the Hildebrand parameter by

$$\delta^2 = \delta_\lambda^2 + \delta_\tau^2 \tag{3-42}$$

Here, the solubility parameter can be identified with the dispersive term and corresponds to the polar orientational term. This approach has also been applied to polymer solutions and complex formations.

Hansen *et al.* [88] proposed a practical extension of the Hildebrand parameter method to polar and hydrogen-bonding systems, primarily for use in polymer-liquid interactions. It was assumed that dispersion, polar, and hydrogen-bonding parameters were valid simultaneously, as related by Eq. (3-43), with the values of each component being determined empirically on the basis of many experimental observations:

$$\delta_t^2 = \delta_L^2 + \delta_P^2 + \delta_H^2 \tag{3-43}$$

here, δ_t is Hansen's total solubility parameter, δ_L the London dispersive term, δ_P the polar term, and δ_H the hydrogen-bonding term. The Hansen's total solubility parameters of some polymers are listed in Table 3-8.

Hansen's total solubility parameter should be equal to the Hildebrand parameter, although the two quantities can differ for materials with specific interactions when they are determined by different methods. The three component

Table 3-7 Experimental and Calculated Values of δ for some Polymers [85]

Polymer	δ exp. (MPa$^{1/2}$)	δ cal. (MPa$^{1/2}$)
Polyethylene	15.8	16.0
Polypropylene	16.8	17.0
Polyisobutylene	16.0	16.4
Polystyrene	17.4	19.1
Poly(vinyl chloride)	19.2	19.7
Poly(vinyl bromide)	19.4	20.3
Poly(vinylidene chloride)	20.3	20.6
Poly(tetrafluoroethylene)	12.7	11.7
Poly(chlorotrifluoroehtylene)	14.7	15.7
Poly(vinyl acetate)	19.1	19.6
Poly(vinyl propionate)	18.0	18.8
Poly(methyl acrylate)	19.9	19.9
Poly(ethyl acrylate)	18.8	19.2
Poly(propyl acrylate)	18.5	18.7
Poly(butyl acrylate)	18.0	18.3
Poly(isobutyl acrylate)	17.8	18.7
Poly(methyl methacrylate)	18.6	19.0
Poly(ethyl methacrylate)	18.2	18.6
Poly(butyl methacrylate)	17.8	17.9
Poly(isobutyl methacrylate)	16.8	18.3
Poly(tert.-butyl methacrylate)	17.0	18.0
Poly(benzyl methacrylate)	20.1	19.3
Poly(ethoxyethyl methacrylate)	18.4	18.6
Polyacrylonitrile	25.6	25.7
Poly(α-cyamomethyl acrylate)	28.7	23.8
Polybutadiene	16.6	17.5
Polyisoprene	16.2	17.4
Polychloroprene	16.8	19.2
Polyformaldehyde	20.9	20.5
Poly(tetramethylene oxide)	17.0	17.6
Poly(propylene oxide)	15.4	18.9
Polyepichlorohydrine	19.2	20.1
Poly(ethylene sulphide)	18.4	18.9
Poly(styenen sulphide)	19.0	19.6
Poly(ethylene terephthalate)	19.9	20.5

parameters were plotted on a set of three mutually perpendicular axes. The Hansen parameters result in improved agreement with the data but are still not completely accurate in predicting solution thermodynamics.

The solubility parameters provide a simple method of correlation and prediction of the cohesive and adhesive properties of materials from the knowledge of the properties of the components only. Particularly for polymers, applications include finding compatible solvents for coating resins, predicting the swelling of cured elastomers by solvents, estimating the solvent pressure under

Table 3-8 Hansen's Specified Solubility Parameters for some Polymers [85]

Polymer	δ	δ_L	δ_P	δ_H
Polyisobutylene	17.6	16.0	2.0	7.2
Polystyrene	20.1	17.6	6.1	4.1
Poly(vinyl chloride)	22.5	19.2	9.2	7.2
Poly(vinyl acetate)	23.1	19.0	10.2	8.2
Poly(methyl mehacrylate)	23.1	18.8	10.2	8.6
Poly(ethyl methacryalte)	22.1	18.8	10.8	4.3
Polybutadiene	18.8	18.0	5.1	2.5
Polyisoprene	18.0	17.4	3.1	3.1

devolatilization and reactor equipment and predicting polymer-polymer, polymer-binary-solvent, random co-polymer, and multi-component solvent equilibria, among others.

The solid and liquid additives used in polymers can be characterized by the Hansen solubility parameters. The assignment of Hansen solubility parameters to both additives and the polymers to which they are added allows predictions of behavior that can improve the performance, ensure stability, reduce costs, and maintain or improve environmental acceptability. Compatibility is found when these parameters for the polymer and the additive are close enough to each other. Mismatches lead to phase separation, or in the case of solid surfaces, to poorer adhesion than that otherwise attainable. Where the additives have distinctly different segments, such as the surface-active agents, the respective segments seek regions of similar energy, i.e., Hansen solubility parameters.

Hansen [89] noted that many plasticizers have been characterized by the Hansen solubility parameters and that their general behavior with regard to solubility and compatibility follows that of the solvents. Some differences are encountered due to generally higher molecular weights. This means a lowered compatibility in boundary situations and slower transport through polymer matrices. The general behavior can be interpreted from Fig. 3-9.

There is a circular region indicating complete compatibility with the polymer when the given Hansen solubility parameters for the plasticizer indicate a position well within the circle. Marginal compatibility is found for a plasticizer near the boundary of the circle, and incompatible situations will be found when there is a greater mismatch in the Hansen solubility parameters of the polymer and the plasticizer. There are possibilities for synergistic mixtures when two plasticizers are mixed and when the mixture has Hansen solubility parameters closer to the Hansen solubility parameters for the polymer (Table 3-9).

The fiber-reinforced composites are used in a wide range of applications where high stiffness and strength-to-weight ratios are required. The mechanical performance of composite materials depends not only on the matrix and the

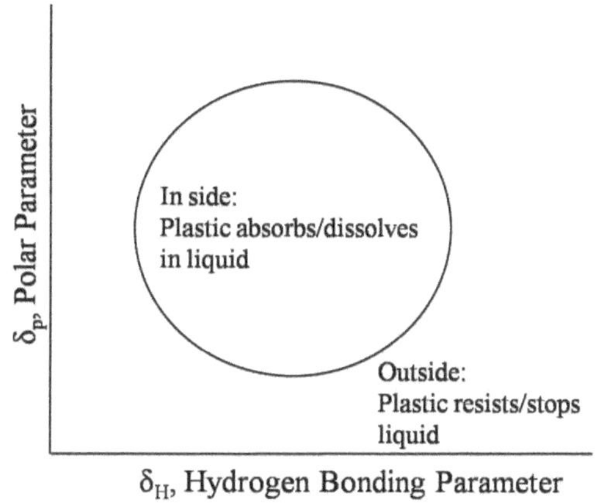

Fig. 3-9 Hansen solubility parameter diagram.

Table 3-9 Hansen Solubility Parameters for some Materials of Interest in Connection with di(2-ethylhexyl phthalate)

Materials	L ($MPa^{1/2}$)	P ($MPa^{1/2}$)	H ($MPa^{1/2}$)	V (cm^3/mole)
di(2-ethylhexyl phthalate)	16.6	7.0	3.1	377.0
2-Ethylhexyl phthalate	17.3	6.2	6.8	265.0
dioctyl adipate	16.7	2.0	5.1	400.0
di(isononyl phthalate)	16.6	6.6	2.9	432.4

reinforcing fiber properties but also to a great extent on the fiber/matrix adhesion. The physical compatibility between the fibers and the matrix can have a profound effect on the interface properties.

An experimental investigation of the fiber/matrix adhesion often involves the use of glassy carbon as a planar model surface for carbon fibers. Due to the structural similarities with carbon fiber, Hansen solubility (cohesion) parameters are widely used for predicting compatibility between two materials. Launay *et al.* [90] determined the Hansen solubility parameters of an epoxy resin, a carbon fiber, and glassy carbon in order to infer the physical compatibility of the epoxy resin with the surfaces of the carbon materials. A comparison of carbon fibers and glassy carbon Hansen solubility parameters allows a discussion of the relevancy of using glassy carbon as a model for the carbon fiber for the study of the adhesive properties of carbon fibers.

Table 3-10 presents the Hansen solubility parameters of the three materials studied here calculated from the experimental data [90]. The Hansen solubility

Table 3-10 Hansen Solubility Parameters, Radius of Interaction, R_o, and Correlation Coefficients (Data Fit) of Epoxy Resin, Glassy Carbon, Carbon Fibers and Various Carbon Materials Correlations

Materials	ΔL ($MPa^{1/2}$)	ΔP ($MPa^{1/2}$)	ΔH ($MPa^{1/2}$)	R_r ($MPa^{1/2}$)
Epoxy resin	20.0	10.0	8.0	9.0
Glassy carbon	20.0	10.0	8.0	9.0
Carbon fibers	21.3	8.7	11.5	9.3
Epikote 1001[a]	20.0	10.3	10.1	10.0
Polypropylene[b]	18.0	0.0	1.0	6.0
Petroleum coke[b]	16.4	4.0	10.0	10.7

[a]*Semisolid, low molecular weight epoxy [91].*
[b]*Refs. [91].*

parameters for epoxy resin and glassy carbon were, by coincidence, identical, with an excellent correlation coefficient of 1.0. The high data fit may be partly due to the low number of tested solvents. This helped to allow a perfect separation of the good and bad solvents according to the Hansen solubility parameter correlation. Fig. 3-10 shows a three-dimensional plot of the Hansen solubility parameters found with the glassy carbon correlation. The sphere, whose center is the Hansen solubility parameters for glassy carbon with a radius of R_r, encompasses the region of the higher surface affinities of the glassy carbon.

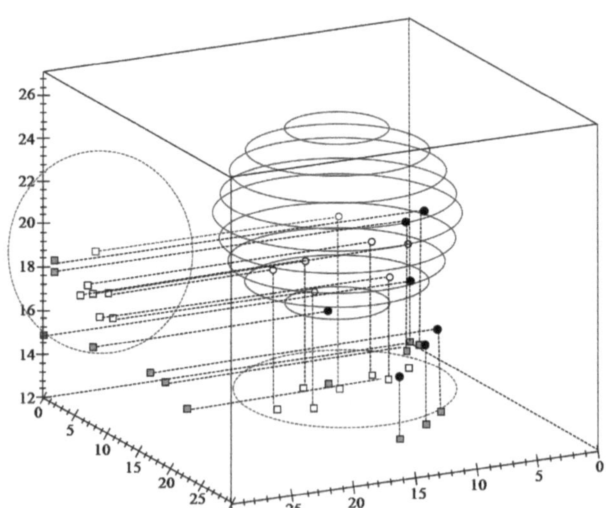

Fig. 3-10 Three-dimensional plot and projections of the glassy carbon sphere of surface affinity and of the solvents used for the correlation (○, inside the Hansen solubility parameters sphere; ●, outside the Hansen solubility parameters sphere).

The tested solvents are marked with dots. All of the good solvents are inside the Hansen solubility parameters sphere and all of the bad ones, are outside.

3.3.8. Surface Treatments

The knowledge of the surface energetics of a solid has recently allowed significant progress in many academic and scientific areas, such as adsorption, wettability, and adhesion, as the surface energetics determines the molecular interaction of the surface with its environment. To control the surface energetics surface modification techniques are widely used because they have many advantages. They are inexpensive, require only simple equipment, and they can be completed quickly. The surface treatment techniques involving wet and dry processes have been applied to various materials, such as metal, ceramics, polymers, and carbon, to improve their utilization; these treatments are necessary to promote the formation of functional groups, which can control the surface energetics. The most widely used surface modification techniques involve wet treatments, such as chemical or electrochemical treatments, and metallic coating and dry treatments, such as photoirradiation, microwave, radiofrequency, and plasmas or fluorination using ionized gas. The ultimate goal of these treatments is to change the surface chemistry and the microstructure of the material, and thus control a number of properties including the biocompatibility, roughness, reactivity, wettability, and conductivity [92, 93]. In this section, we have focused on the surface treatments of carbon materials.

3.3.8.1. Wet Treatments

3.3.8.1.1. Chemical Oxidation

The surface modification of the carbon materials has received attention because it is important in catalytic, analytical, and biotechnological applications. The oxidation of carbon materials by a wet chemical treatment is a basic method that enables the functionality of a material by surface modification. The main goal of the oxidation of a carbon surface is to obtain a more hydrophilic surface structure with a relatively large number of oxygen-containing surface groups. In general, the oxygen-containing groups behave as acids or bases which possess ion-exchange properties [94].

The oxidative methods of the surface modification of non-polar materials includes wet chemical methods, such as immersion in concentrated nitric or sulfuric acid, sodium hypochlorite, permanganate, bichromate, hydrogen peroxide, and transition metals; electrochemical oxidation in various electrolytes; and intercalation [95–99].

Among oxidation treatments, nitric acid oxidation is the most widely used method to increase the total acidity in a wet oxidation treatment. There are reports that the oxidation treatment of carbon materials in boiling nitric acid results in a significant increase in the number of acidic surface groups, such as carbonyl and carboxyl groups [100]. It is well established that the chemisorption of oxygen is strongly dependent on the crystalline nature of the carbon. That is, under the same conditions, well-ordered carbons, that contain a relatively small concentration of edge sites will chemisorb less oxygen than more amorphous carbons, such as activated carbons. Once oxygen is chemisorbed, it can only be removed as an oxide of carbon (CO and CO_2). The presence of oxygen surface complexes influences the surface behavior of carbons to a great extent. As examples, the wettability and the adsorptive behavior of a carbon, as well as its catalytic and its electrical properties are influenced by the nature and extent of the complexes [101, 102].

The porous carbon materials have been widely used in the separation, purification, and catalytic processes due to their extended specific surface area, high adsorption capacity, microporous structure, and functional surface reactivity. The adsorption of activated carbon fibers (ACFs) depends on many types of factors, such as the raw materials, activation process, nature of the pore structure, and surface functionalities [103].

In a study of the effect of an acidic treatment on surface characteristics of activated pitch-based carbon fibers and activated carbon fibers underwent an acidic treatment [104]. 1 M of boiling nitric acid was used in the oxidation of the functional groups without any damage to the fiber surfaces. The surface oxides and the isoelectric point of the activated carbon fibers studied were determined by Boehm's titration [105] and Noh-Schwarz's mass titration [106] methods.

The acid/base titration curves given for a number of activated carbon fibers are shown in Fig. 3-11. This plots pH values as a function of the mass fraction of carbon in the solution. The deviations in the asymptotic values of the curves are possibly due to the specific adsorption of background electrolyte ions, such as Na^+ and NO_3^- ions; however, the discrepancy appears to be very small and the isoelectric point of the activated carbon fibers can be estimated within an error of 0.2 pH, if the average value of the three asymptotic pH values is taken as the

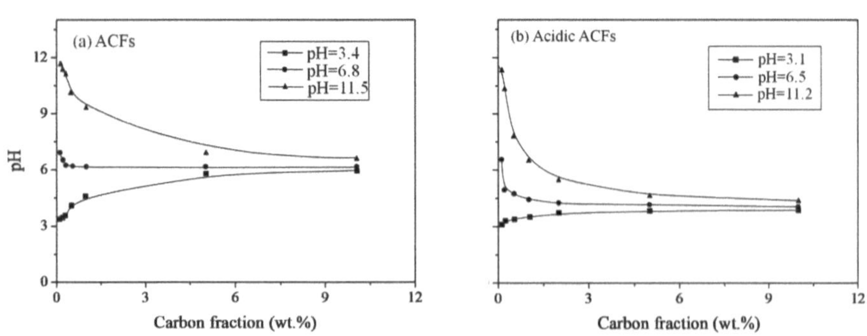

Fig. 3-11 Mass titration results of (a) activated carbon fibers (ACFs) and (b) acidic activated carbon fibers.

Table 3-11 Surface Acidic Functionalities of Activated Carbon Fibers by Boehm's Titration

	Surface acidic functionalities (meq/g)		
	Phenol	Lactone	Carboxyl
ACFs	0.58	0.28	0.02
Acidic ACFs	0.31	0.64	0.44

isoelectric point. As shown in Fig. 3-11, the asymptotic isoelectric point is about six for the activated carbon fibers and four for the acidic activated carbon fibers. The isoelectric point of the acidic activated carbon fibers shifts to a pH value much lower than that of the activated carbon fibers. This phenomenon indicates that additional acidic groups, such as carboxyl groups, may be formed through a HNO_3 treatment.

Numerous studies of the surface functionalities of carbon exist in the literature. The differences in the surface functionalities of basic activated carbon fibers and acid-treated activated carbon fibers are determined by Boehm's titration. These are listed in Table 3-11. The three bases used in the titration here are regarded as approximate probes of the acidic surface functionalities according to the scheme of $NaHCO_3$ (carboxyl), Na_2CO_3 (carboxyl and f-lactone), and NaOH (carboxyl, f-lactonic, and phenolic). As shown in Table 3-11, the phenol groups are decreased, but the lactone and carboxyl groups are significantly increased after the nitric acid treatment.

3.3.8.1.2. Electrochemical Oxidation

Several functionalization techniques have been used to introduce oxygen functional groups onto carbon fiber surfaces. However, most of these methods rely on harsh and/or corrosive conditions, such as immersion in a strong acid, while heating. Such harsh treatments can induce deformations in the original carbon fiber structures and may alter the basic properties of the carbon fibers. Thus, a proper selection of the oxidation methods and processing conditions is essential to introduce the highly effective surface functional groups onto carbon fibers, while maintaining the structural integrity [107].

One promising method used to create oxygen functional groups at carbon fiber surfaces is the mild electrochemical oxidation [108]. Electrochemical oxidation boasts of several advantages over the traditional oxidation methods, such as a HNO_3 treatment or hot-air oxidation: electrons are the only reagents and can be supplied by a direct current source; the reaction conditions can be precisely reproduced; the reaction stoichiometry can be adjusted by a constant power supply, assuming that any side reactions are negligible; and the

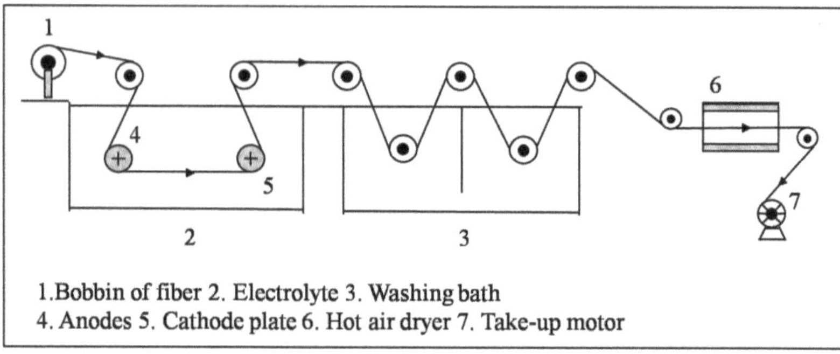

Fig. 3-12 Schematic representation of the continuous electrochemical oxidation process for carbon fibers [112].

mechanical strength and surface area of the carbon material remains unchanged, even if the surface activity is increased considerably.

For a continuous processing of the surface modification of carbon fibers, the electrochemical oxidation treatment is the most effective technique because it can introduce numerous types of active surface functional groups and can increase the surface roughness. Thus, it can substantially improve the interfacial bonding strength; moreover, the technique is relatively simple to control and can allow continuous processing of the carbon fibers. [109]. The modification process is carried out by electrochemical oxidation in aqueous solutions containing acid or alkali [110]. One irreversible oxidation or reduction peak appears under an electrochemical process, and the covalent grafting of a modifier on a carbon fiber surface takes place only once per potential cycle [111]. The scheme of the electrochemical oxidation process for carbon fibers is presented in Fig. 3-12

In a previous work, the effect of the electrochemical oxidation on the surface characteristics of activated carbons was observed [113]. The electrolytic oxidation was performed in a batch system where activated carbons in a graphite bath filled with the electrolyte were fixed on a graphite anode roller. A cathode graphite plate was also submerged in the electrolyte solution. The surface treatment conditions are listed in Table 3-12. Prior to use the residual solution

Table 3-12 Experimental Conditions of Anodic Surface Treatment for Activated Carbons

Electrolyte	NaOH (35 wt.%)
Temperature	25 °C
Oxidation time	60 sec
Current intensities	0, 0.3, 0.6, 1.5, and 3.0 A

Table 3-13 Surface and Adsorption Characteristics of the Activated Carbons Studied

Current intensity (A)	pH	Acid value (meq/g)	Base value (meq/g)
0	7.8	123	227
0.3	8.0	125	244
0.6	8.0	132	277
1.5	8.3	142	355
3.0	8.5	164	383

of the electrolyte on the treated activated carbons was removed by Soxhlet extraction by boiling with acetone at 80 °C for 2 h. The treated activated carbons were washed several times with distilled water and dried in a vacuum oven at 85 °C for 24 h.

Table 3-13 shows the experimental surface properties of the activated carbons modified by an anodic treatment as a function of the current intensity. The surface pH of the activated carbons gradually increases as the current intensity increases. The acid and base values as measured by a titration technique show that the surface of the untreated activated carbons is amphoteric with a slight excess of basic properties and an increase in the current intensity of the activated carbons leads to a greater increase of base value than the acid value. These results indicate that an anodic treatment in NaOH electrolyte plays an important role in increasing the basic (or electron donor or anion-rich) properties of the activated carbons. The base values are closely correlated with the surface pH of the activated carbons measured. It is noted that this behavior is very similar to that recently reported by Park *et al.* [114] for the anodic surface treatment of carbon fibers in NaOH electrolyte for the high-performance composite materials. His data suggested that both the acid (K_A) and base (K_D) parameters of carbon surfaces based on the linear free energy relationships (LFERs) using the acceptor (NA) and donor (ND) numbers of Gutmann [115] increase as the current density increases; moreover, $K_D > K_A$ applies for the anodized carbon fibers.

3.3.8.1.3. Modification by Coupling Agents

It is well known that the surface activity of a solid depends largely on the surface energy, active functional groups, acid–base interaction, and energetically different crystallite faces. The level of surface activity may be indirectly determined from adsorption (gas-solid molecular interaction) and wettability (liquid–liquid) analyses and can be measured directly by a mechanical interfacial adhesion technique (solid–solid) [116, 117].

Coupling agents are also used to modify surfaces. Polysiloxane of various types is widely used as a coupling agent between substrates and polymeric resins [118]. In general, a coupling agent is a bifunctional compound developed commercially to improve the filler dispersion and to prevent adsorption of curatives on the filler surface. It is composed of two functionally active end groups, i.e., a readily hydrolysable alkoxy group and an organo-functional group, and the coupling process can be accomplished, via the chemical reaction between the trialkoxy groups of silane molecules and the hydroxyl groups on the substrates to form a siloxane bond, whereas other functional groups of silane molecules, which are generally ethylene, amine, epoxy, or thiohydroxy, can remain. The usefulness of the coupling agents, for example, in improving the interfacial adhesion between fillers and matrices, is well reported in literature. It has also been applied to functionalize the surface of carbon materials [119, 120].

The surface energetics of a solid surface is a key factor in a composite system, as this parameter can determine the strength of the interaction of the solid surface with its environment. This indicates that different silane coupling agent chemical structures or energetic properties can lead to entirely different properties of the resulting composites. However, a systematic study of the effects of silane coupling agent treatments on both the surface energetics of fillers and the mechanical interfacial properties of the filler/matrix composites has yet to be comprehensively undertaken.

The silane coupling agents were used to modify montmorillonite [119]. Na^+-montmorillonite (specifically Na-MMT) was used and the silane coupling agents used were γ-amino propyl triethoxy silane (APS), γ-glycidoxy propyl methoxy silane (GPS), and γ-methacyloxy propyl trimethoxy silane (MPS). The samples were denoted as Na-MMT (non-treated MMT), as-received (MMT after the ion-exchange reaction by 2MHTL8), AP-MMT (APS-treated MMT), GP-MMT (GPS-treated MMT), and MP-MMT (MPS-treated MMT). The scheme of the reaction between the silane coupling agents and the MMT is presented in Fig. 3-13.

While investigating the surface modification effects on the surface properties of MMTs, the surface free energy of the MMTs was determined by the contact angle measurements. In Fig. 3-14, the polar components of AP-MMT and GP-MMT were increased compared to those of the as-received sample, whereas MP-MMT showed a different trend. This result can considering the epoxide ring and the amine group in GPS and APS have strong polarity, whereas the ester group in MPS has weak polarity. This polar component of the surface free energy can increase the interfacial interaction between functional groups on the MMT surfaces and epoxy resins.

The dispersive components of the MMT were slightly increased in the GP-MMT and MP-MMT but were significantly enhanced in the AP-MMT. These changes of the surface free energy can be regarded as the result of the different chemical structures or physical volumes of the silane coupling agent species studied. It is well known that the dispersive component in surface energetics

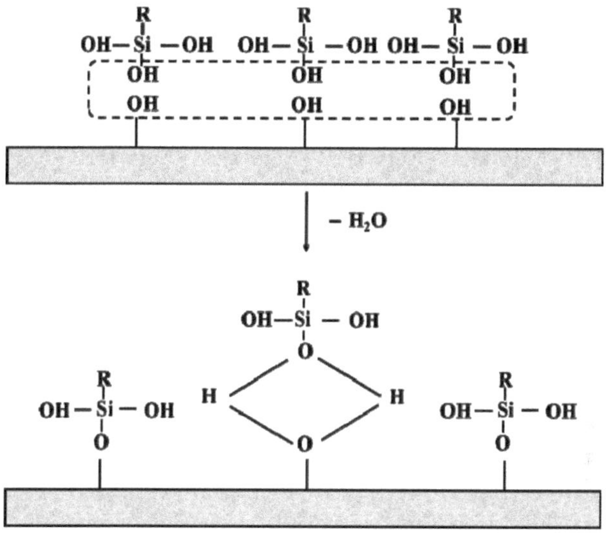

Fig. 3-13 Schematic of surface treatment of MMT by silane coupling agents.

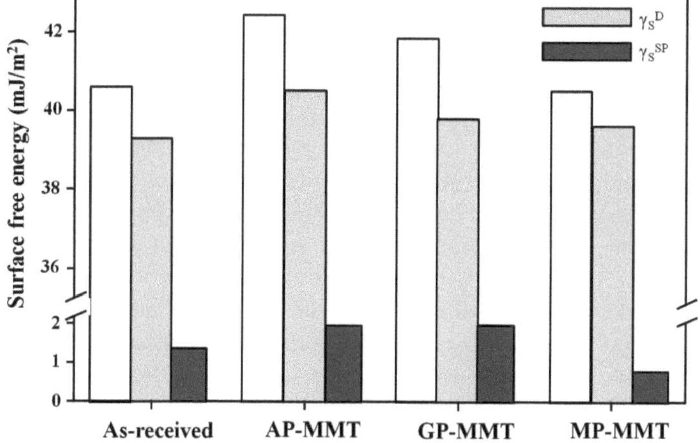

Fig. 3-14 Surface free energies of MMTs before and after silane coupling agent treatments.

can be calculated by the Lorentz–Lorenz–Debye equation, as expressed by Eqs. (3-37) to (3-44) [84, 120]:

$$\left(\frac{n^2-1}{n^2+2}\right)\frac{M}{\rho}=\frac{N_A\alpha_0}{3\varepsilon_0} \tag{3-44}$$

In this equation, n, M, ρ, and ε_0 are the refractive index [121], the relative molecular mass, the molecular density, and the permittivity of a vacuum, respectively.

Eq. (3-44) indicates, that the dispersive component (α_0) is a function of 'M/ρ', which represents the specific molecular volume. Among the three silane coupling agents, APS specifically had a larger molecular volume compared to the MPS and GPS, resulting in a higher dispersive component and higher surface free energies.

3.3.8.1.4. Metal Coating Method

In the carbon-reinforced composite systems, the improvement in the physical properties, the improvement in the interfacial adhesion between the carbon materials and the matrix is attributed to the presence of polar groups on the carbon surface which are able to interact with the active groups present in the matrix. Electroplating has been used to produce metal matrix composites reinforced with carbon materials. The carbon surfaces are metalized by electrolysis in molten salt solutions in the presence of a reductant [122]. For example, the nickel process is based on a redox reaction in which the reducing agent is oxidized and Ni^{2+} ions are reduced on the substrate surface. Once the first layer of nickel is deposited, it acts as a catalyst for the process. As a result, a linear relationship between the coating thickness and time usually occurs. The adhesion of electroplated metal/carbon composites has been shown to be excellent when the metal coating is grown from the carbon surface. In addition, the interaction between the metal-plated carbon and the matrix is strong mainly due to the higher metal surface energy that allows extensive wetting by the polymer. This indicates that metal plating onto a carbon surface is a viable means of increasing of the utilization of carbon materials through introducing improved functionality.

Currently, activated carbon fibers are widely used in separation, purification, and catalytic processes due to their extended specific surface area, high adsorption capacity, highly porous structure, and surface reactivity. Nevertheless, problems remain when the activated carbon fibers are used to purify drinking water because bacteria preferably adhere to the solid support made of carbon materials, indicating that activated carbon fibers have good biocompatibility. Bacteria may breed on activated carbon fibers during the purification process, themselves becoming a pollutant. In order to avoid this disadvantage, antibacterial, activated carbon fibers are required. Therefore, a considerable effort has been made to prepare metal-containing

Table 3-14 Compositions and Plating Conditions of Electroplating Baths

Cu plating		Ag plating	
Composition			
$CuSO_4$	10 g/L	AgCN	5 g/L
H_2SO_4	20 ml/L	Na_2CO_3	3 g/L
		NaCN	12 g/L
Conditions			
Ph	3.0	pH	6.0
Temperature	25 ± 1 °C	Temperature	25 ± 1 °C
Current density	70 A/m^2	Current density	70 A/m^2

activated carbon fibers by various surface treatment methods to solve these problems [123–125].

In a previous work, the surface characteristics of metal-plated carbon fibers were studied via their acid–base values [126]. The starting materials of commercial activated carbon fibers (around 2100 m^2/g of specific surface area) were plated for different times with Cu and Ag. The plating conditions and compositions of the plating baths are listed in Table 3-14.

To investigate the morphology of the Cu- or Ag-plated carbon fibers, the SEM imagery was used with the resulting images shown in Fig. 3-15 and it was found that grain-like metal nanoparticles existed on the activated carbon fibers.

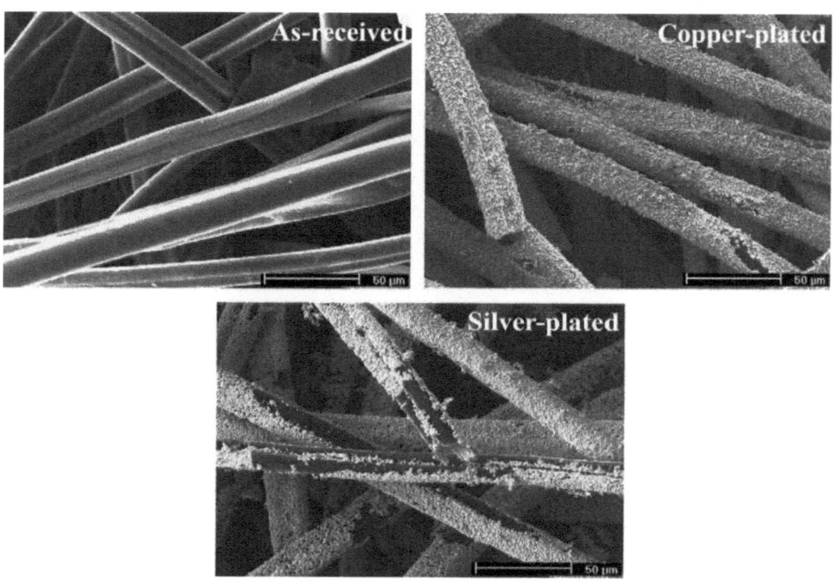

Fig. 3-15 SEM images of activated carbon fibers before and after copper and silver plating.

Table 3-15 pH and Acid–base Values of Cu- or Ag-plated Activated Carbon Fibers (Cu- or Ag-ACFs)

	pH	Acid value (meq/g)	Base value (meq/g)
Pristine	7.43	180	230
Cu/ACF-2	6.71	230	160
Cu/ACF-5	6.67	242	155
Cu/ACF-15	6.62	255	135
Ag/ACF-2	6.89	205	192
Ag/ACF-5	6.85	210	185
Ag/ACF-10	6.80	225	175

Table 3-15 shows the pH and the acid–base values of Cu- or Ag-plated activated carbon fibers as a function of the plating time. It was found that all metal-plated activated carbon fibers showed acidic features and that the acid values were increased. It is interesting to note that the Cu-plated activated carbon fibers showed acid values higher than those of Ag-plated activated carbon fibers, most likely due to the differences in the chemical activity between the two metals.

3.3.8.2. Dry Treatments

The dry chemical treatments, such as a plasma treatment, an ozone treatment, oxidation in air, or fluorination can remove impurities and introduce various functional groups such as COOH, OH, and C=O, –F on the surface of a carbon material. These techniques are simple and enable control of the quantity of the new functional groups on the carbon surfaces depending on the treatment conditions. In addition, the dry chemical treatment does not lead to the environment problems associated with wet chemical treatments. Therefore, numerous evaluations have been performed with carbon materials [127].

3.3.8.2.1. Plasma Treatments

Plasma treatments are very popularly used with a variety of materials including carbon materials. A plasma treatment on a carbon surface produces a highly active species. Depending on the nature of the gases used in the plasma zone, it can produce free radicals, ions, and meta-stable species, which cause ablation, cross-linking, or oxidation reactions [128]. There are two principal advantages to using a mild plasma treatment method for materials: (1) The reaction takes place only on the material surface without significantly changing the bulk properties and (2) It is possible to utilize any atmosphere, such as an oxidative, reductive, or inactive atmosphere. This technique has become increasingly important owing to its high efficiency and operational simplicity. The surface energetics is also of importance owing to the ability to exert careful control of the layer affected by the treatment, usually 0.02 to 50 μm in depth [129]. Generally, the plasma treatment methods are classified into three methods: atmospheric-pressure plasma treatments, corona discharge treatments, and microwave treatments.

Atmospheric-pressure Plasma Treatments The plasma treatments are generally done under low-pressure conditions to achieve stable plasma with a variety of gases doped into the reaction chamber. The use of different gases allows an increased control of the final surface, chemical and physical properties. Low-pressure plasma treatments are well known for their ability to impart a variety of surface chemistries to polymers through the use of various gases [130, 131]. The equipment used in the plasma treatment of continuous polymer webs at a low pressure is costly and requires the upkeep of the large vacuum systems [132].

Atmospheric-pressure plasma is a new technology for the plasma surface treatments under ambient atmospheric conditions rather than in a vacuum chamber. The nonthermal atmospheric-pressure plasma at temperatures ranging from 50 to 200 °C is used to enhance the hydrophilicity and adhesiveness of a surface. It can also be used to clean such surfaces. The high-energy electrons existing in plasma produce ions and reactive species that interact with surfaces to modify their characteristics. The plasma treatments are generally done under low-pressure conditions to achieve stable plasma with a variety of gases doped into the reaction chamber [133].

The nonthermal atmospheric-pressure plasma systems have structures that are simpler than those of vacuum systems; they are capable of uninterrupted treatments in a single step using a simple process. Possibly, the atmospheric-pressure plasma system is the most suitable means of improving the wettability. It is also very attractive due to its high productivity and low cost operation [134, 135]. The scheme of atmospheric-pressure plasma treatment apparatus is presented in Fig. 3-16

Corona Plasma Treatment In addition, for the generation of ozone, corona discharge is generally the most popular type for most of the industrial uses. The corona discharge treatment, one of the processes most commonly used in industry for oxidation, is widely used to improve the bondability, wettability, and printability of polymers, particularly polyolefin films because it allows continuous treatment in a single step and is a simple process to perform. The mechanism of these improvements is thought to be the introduction of oxygen- or nitrogen-containing groups. Following the chain-breaking reaction and oxidation, this treatment leads to the formation of a surface layer consisting of macromolecules of low mass and with many carbonyl groups and ethylene groups [136, 137]. The scheme of the corona discharge apparatus is presented in Fig. 3-17.

Microwave Plasma Treatment Currently, there is an increasing interest in industry to replace the low-pressure plasma surface treatment processes by the

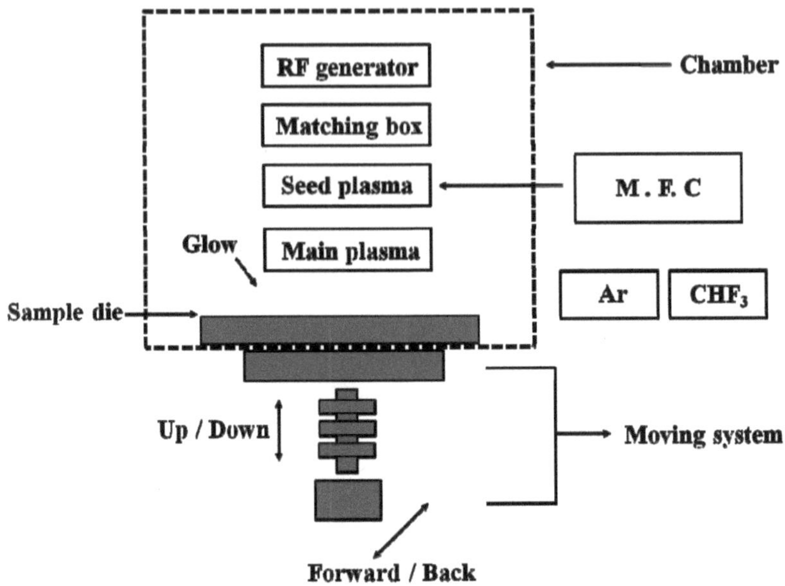

Fig. 3-16 Schematic diagram of atmospheric-pressure plasma treatment apparatus.

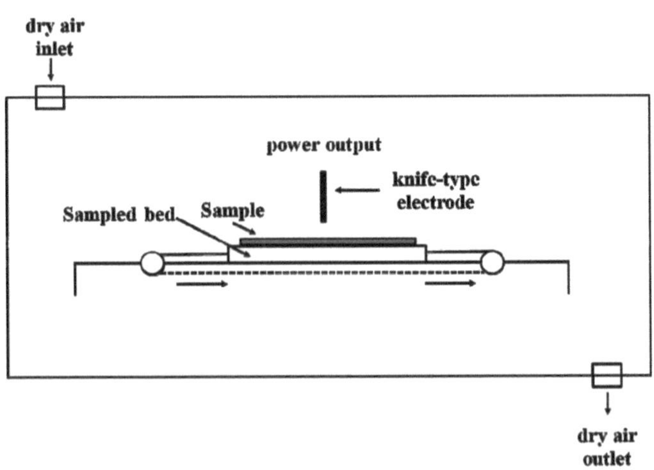

Fig. 3-17 Schematic diagram showing the corona discharge apparatus.

normal-pressure plasma processes. This is due to the expected higher reaction rates, lower investment costs, and lower operational costs of normal pressure processes. Hence, the research concerning the microwave plasma treatment process is in particular demand due to the industrial application of microwave-generated plasmas to various surface modifications. The afterglow of the microwave plasma can be applied for surface modification of a material. The surface modification process depends on the production of electrons, ions, free radicals, and other molecular species inside the plasma flame [138, 139].

The normal pressure plasmas, however, also pose problems. Due to the high pressure, the free mean path length is very limited and, therefore, the efficiency of several of these processes can be low in comparison to the low-pressure discharges. Moreover, most microwave-sustained normal-pressure discharges exhibit higher electron densities than the low-pressure plasmas and are closely related to the thermodynamic equilibrium conditions. In consequence, these plasmas have large heat transfer rates which limit the treatment of heat-sensitive materials. In general, the argon microwave plasma operated at a continuous wave of 2.45 GHz produces a high-temperature flame close to 4000 K at the center part of the plasma. Therefore, the afterglow part of the microwave plasma can be used in a surface modification process by drastically cooling the plasma flame temperature and blowing it out into the open air from a discharge tube under a high gas-flow rate. In several plasma devices, this fact facilitates stable operation conditions during the use of high gas flows. The pulsed operation modes can also help to reduce the heat transfer. However, the usage of a pulsed operation with long off-cycles is often limited because ignition for the microwave-sustained normal-pressure discharges is difficult under these conditions. Therefore, at present, the studies of microwave-sustained normal-pressure discharges have focused on reducing the heat transfer [140]. The scheme of the plasma treatment device is presented in Fig. 3-18.

The polyimides (PIs) are widely used as protective overcoats and dielectric layers for semiconductor devices owing to several valuable properties,

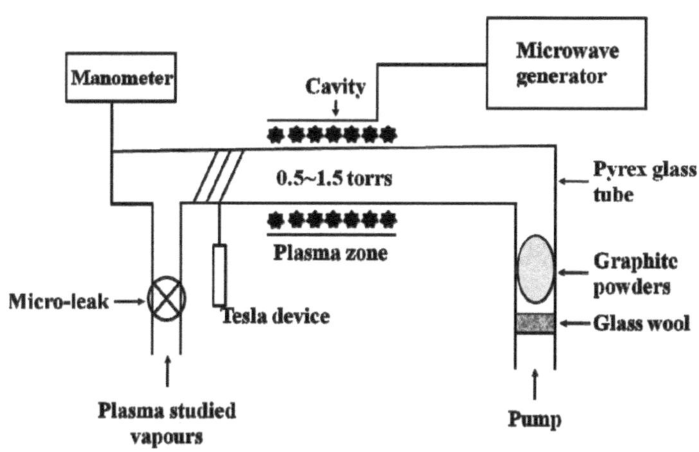

Fig. 3-18 Schematic diagram of plasma treatment device [50].

such as their low dissipation factors, excellent thermal and mechanical properties, inherent surface inertness, and low dielectric constants. These materials have been considered for use in numerous applications that require robust organic materials including composites and precursors for the high-performance aerospace materials as well as membranes for the gas separation processes [141–143].

However, in spite of the extensive use as well as the detailed characterization of polyimides (PIs) applications of these materials are limited due to their hydrophobic surfaces, which result in poor wettability and adhesion. The adhesion of a metal to a polyimide film is generally poor unless the PI surface is pretreated prior to metallization [144].

The surface properties of polyimide films treated at different treatment powers of the atmospheric-pressure plasma were investigated [133]. The gases used in the atmospheric-pressure plasma surface treatment were a mixture of Ar (99 percent) and O_2 (1 percent). The nominal treatment time was fixed at 1 min and the treatment power was varied between 0 and 300 W at a radio frequency 13.56 MHz.

Fig. 3-19 shows the surface free energy results, of the atmospheric-pressure plasma-treated polyimide films. The results show that the polar components of the surface free energy increase greatly as the atmospheric-pressure plasma treatment power increases. These results can be interpreted as showing that the atmospheric-pressure plasma treatment of polyimide film surface produces various oxygen complexes on hydrophobic surfaces. This can be attributed to an increase in the surface polarity and wettability, which improves the total surface free energies. It was also found that dispersive components of the atmospheric-pressure plasma-

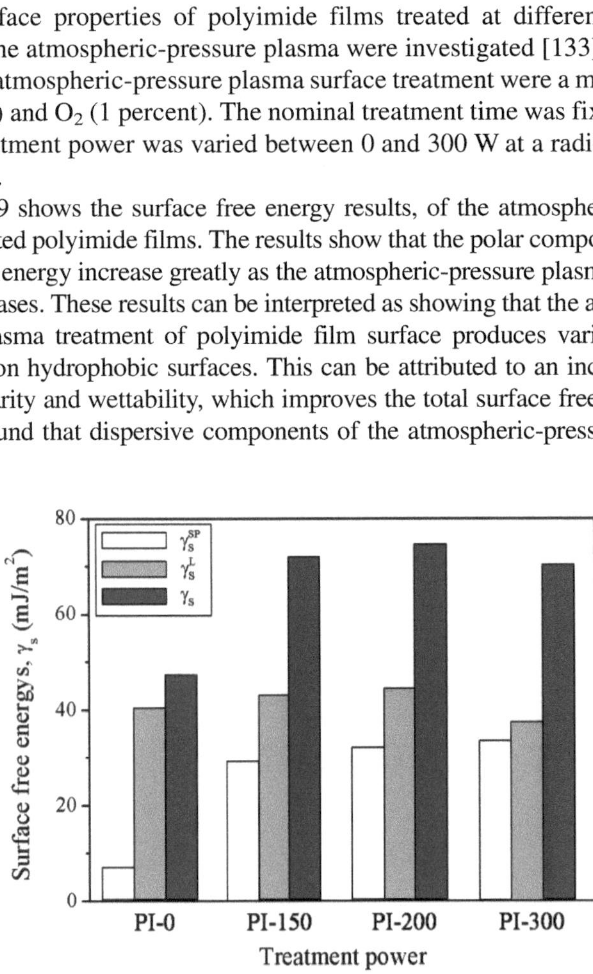

Fig. 3-19 Surface free energies of atmospheric-pressure plasma-treated polyimide films (γ_S, surface free energy; γ_S^{SP}, specific component; γ_S^L, London dispersive component).

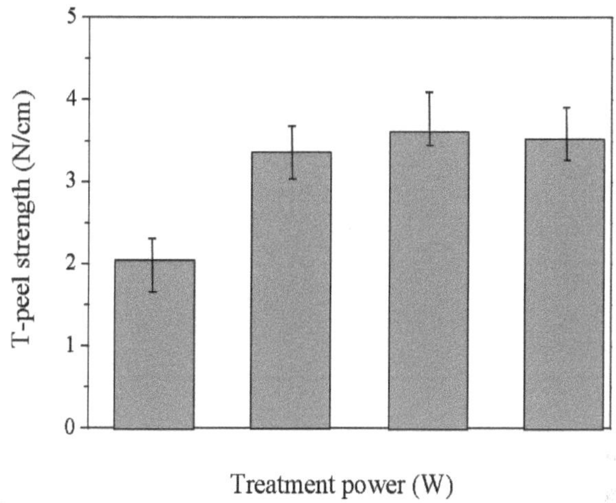

Fig. 3-20 T-peel strengths of atmospheric-pressure plasma-treated polyimide film/copper foil adhesive joints as a function of atmospheric-pressure plasma treatment power.

treated polyimide films are slightly increased for polyimide-200 and then decreased for polyimide-300. This result indicates that an atmospheric-pressure plasma treatment at a suitable treatment power can cause an increase of not only the polar component but also the dispersive component due to the scratching effects of the plasma treatment itself, as caused by the increasing specific surface area of the polyimide films, resulting in a slight increase of the hydrophobicity in the atmospheric-pressure plasma-treated polyimide films.

Fig. 3-20 represents the peel strength values of the polyimide (PI)/copper samples with different atmospheric-pressure plasma-treatment powers. These adhesion strengths are much higher in atmospheric-pressure plasma-treated polyimide/copper systems than in a pristine polyimide/copper system. The strong adhesion of the electrolessly deposited copper onto the atmospheric-pressure plasma treated polyimide film is attributed to the strong interaction between the oxygen complexes or to the spatial distribution of the cross-linked aggregation on the polyimide surfaces and the metal matrix [145].

One promising procedure for generating polar groups on fiber surfaces consists of applying a plasma treatment. Thereby, the oxygen functional groups (hydroxyl, carbonyl, carboxyl groups, etc.) can be formed on the surfaces consecutively reacting with the polymer matrix and improving the adhesion and wettability characteristics [146, 147].

The vapor-grown graphite nanofibers were modified by the plasma treatments using low-pressure plasmas with different gases (Ar gas only and/or Ar/O_2 gases), flow rates, pressures, and powers [148]. The London dispersive (γ_S^L) and polar (γ_S^{SP}) components of the surface tension (γ_S) of the graphite nanofibers after the plasma treatment were determined by measuring the contact angles of

Table 3-16 Surface Free Energies (γS), London Dispersive (γ_S^L), Specific Component (γ_S^{SP}), and Concentration of Acidic Functional Groups of Atmospheric Plasma-treated Graphite Nanofibers

Specimen	γ_S^L (mJ/m^2)	γ_S^{SP} (mJ/m^2)	γ_S (mJ/m^2)	Acidic groups (10^{-5} mol/g)
GNFs-0	38.5	5.6	44.0	4.1
GNFs-1	36.5	18.1	54.6	7.6
GNFs-2	36.1	24.7	60.8	20.5
GNFs-3	36.5	22.4	58.9	15.1
GNFs-4	33.6	13.7	47.3	8.3

the graphite nanofibers in n-hexane, water, diiodomethane, and ethylene glycol using the Wilhelmy method. These are listed in Table 3-16.

The value of the polar component of the plasma-treated graphite nanofibers is greater than that of untreated nanofibers. A plausible explanation for this finding is that a higher graphitization degree leads to lower nonreactivity. Fig. 3-21 shows the correlation between the polar components and the total oxygen concentration of the graphite nanofibers (regression coefficient R = 0.96). The increase in the oxygen concentration of the graphite nanofibers does lead to a good correlation with the results of the surface free energy. Among the most obvious results, the polar component of the investigated graphite nanofibers increases as the oxygen concentration increases, depending largely

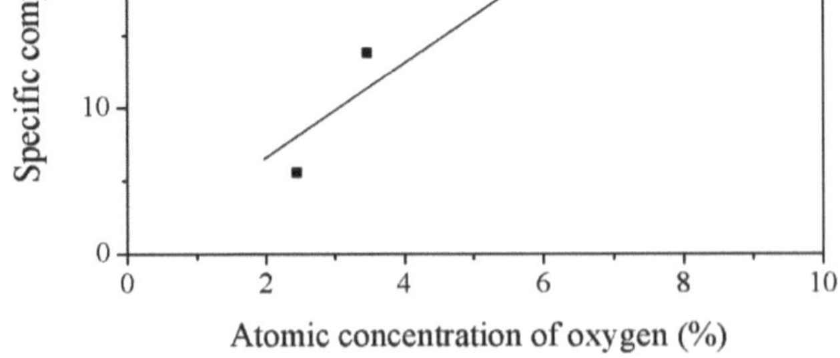

Fig. 3-21 Correlation between the polar component and the total oxygen concentration of plasma-treated graphite nanofibers.

on the increase in the surface free energy, as shown in Table 3-16. Consequently, an Ar + O_2 atmosphere plasma treatment on graphite nanofibers is an effective method to improve the stability of the structure and is also important for the removal of weak or unstable boundary layers that can form during the fabrication of a part. It is, therefore, noted that the polar component of the surface free energy is largely dominated by the surface characteristics [149]. At this time, good agreement can be obtained between the polar component of the surface free energy and the atomic concentration of the oxygen of the investigated graphite nanofibers, as shown in Fig. 3-21.

Low-density polyethylene (LDPE) is widely used in various applications due to its abundant supply, low cost, good processability, low energy demand for processing, and resistance to chemicals, and harsh environments. Currently, the packaging industry demands LDPE owing to its high specific modulus and strength; it is used directly or in the form of laminates, with the materials such as aluminum foil or paper. However, its surfaces are inert and hydrophobic and must be made more hydrophilic in nature, e.g., to improve the printability, paintability, or the degree of adhesion to adhesives or inorganic substances [150].

Therefore, it is important to employ surface treatments that modify only the surface characteristics without altering the bulk properties of materials. Considerable effort has been made to modify the surface properties and numerous researchers have reported surface modifications of polymers through plasma treatments, grafting with the use of glow discharges, and irradiation with ionizing and UV radiation [151].

To modify an additive-free low-density polyethylene, a radio frequency (RF) corona discharge method was used for the preparation of gradient surfaces [136]. The corona discharge powers were varied at 0 (LDPE 0), 50 (LDPE 50), 100 (LDPE 100), 150 (LDPE 150), and 200 W (LDPE 200) at 100 kHz.

Fig. 3-22 shows the expected mechanism through which the oxidation that occurs on a polymer surface, such as LDPE is affected by the corona treatment. The corona discharge treatment of the polymer surface can produce carbon radicalsfrom the hydrocarbon backbone with the subsequent formation of unstable hydroperoxides through the rapid binding with oxygen in air and then the decomposition of the hydroperoxides to produce various oxygen–based functionalities (hydroxyl groups, ethers, ketones, aldehydes, carboxylic acids, or carboxylic esters) in reactions with additional oxygen. The carbon radicals formed by the corona treatment may also bind among themselves to produce various types of unsaturated carbon bonds [152].

Fig. 3-23 shows the contact angle data for a LDPE film treated under a corona discharge. From the experimental results, the angles of water on the LDPE films mostly decrease with an increase in the corona discharge power. As mentioned above, it is clear that corona treatments lead to a change in the nature of the LDPE film surface, resulting in increased hydrophilic properties.

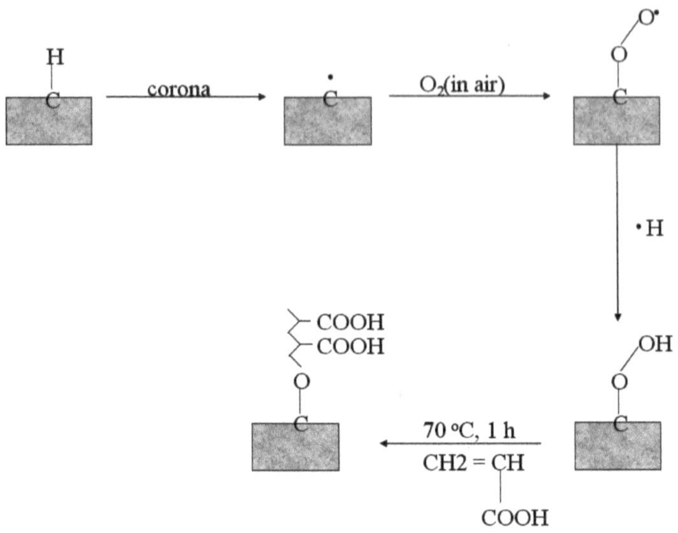

Fig. 3-22 Mechanism for the formation of –COOH group-grafted surfaces, by corona discharge treatment followed by graft copolymerization.

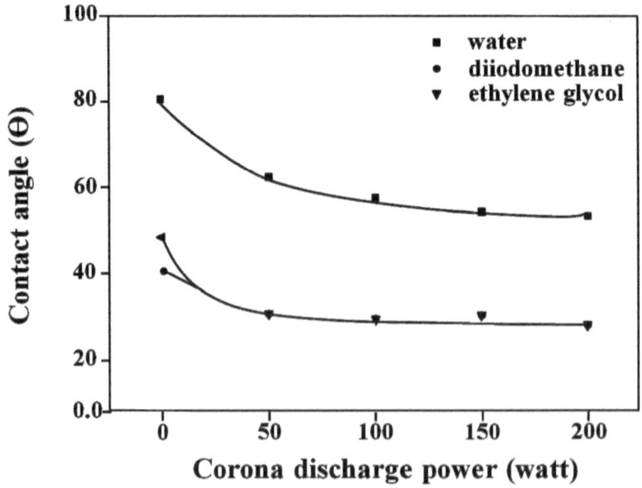

Fig. 3-23 Evolution of contact angle of corona-treated LDPE films with corona discharge powers.

Table 3-17 Surface Free Energy and their Components of LDPE Films with Corona Discharge Power

	γ_S^L (mJ/m²)	γ_S^{SP} (mJ/m²)	γ_S (mJ/m²)
LDPE 0	36.5	3.8	40.3
LDPE 50	36.8	11.5	48.3
LDPE 100	35.8	14.4	50.2
LDPE 150	34.1	16.3	50.4
LDPE 200	33.4	17.6	51.0

The surface free energy and its components in the corona-treated LDPE film with different corona discharge powers are listed in Table 3-17. As shown in the table, the surface free energy of the LDPE 0 is very low, and consists mainly of the dispersive component. These results reflect the poor wettability and non-polarity of untreated LDPE films. By analyzing the components of the surface free energy, it can be found that the specific component increases remarkably after the corona treatment, whereas the dispersive component changes only slightly. These results show that this treatment is important in improving the wettability of LDPE film because they establish hydrophilic properties in these films [153].

3.3.8.2.2. Fluorination

Recently, several studies concerning the fluorination of carbon materials have been published [154, 155]. The outstanding reactivity of fluorine and fluorinating atmospheres has been demonstrated over the past few decades by the synthesis of a wide variety of inorganic fluorine–based compounds and also by the drastic modification of their surface properties. One of the techniques to increase the surface polarity and the functional groups of carbon materials is direct fluorination with oxygen in the presence of a moderate temperature which introduces hydrophilicity onto carbon surfaces. For carbon materials, the oxygen-containing fluorination treatment offers several advantages over other treatments. One primary advantage is that the mechanical properties of the carbon materials are not significantly degraded under optimum conditions. Other important advantages are related to the desire to the development of more environment friendly processes. The physical properties and chemical reactivity of organic molecules and carbon materials can be affected dramatically by fluorination. Today's diverse commercial applications of the organo-fluorine materials are a clear evidence of the potential beneficial effects of oxyfluorination as well as the oxyfluorinated carbon materials. Actually, many of the characteristic effects of oxyfluorination could have been anticipated simply by comparing some of the fundamental atomic properties of fluorine with those of other elements. The modification of carbon materials is important for a variety of application fields, such as composite systems or electronic devices. High reactive fluorination on the surface of carbon can provide carbon materials with advanced functionality, i.e., wettability or electrical properties [156, 157].

The fluorinated carbon blacks have recently been fabricated as a new class of reinforcing or conducting fillers. The reaction of carbon blacks with elemental fluorine via proprietary processes can adjust both the surface energy and electrical conductivity of the carbon blacks due to the creation of a layer of strong carbon-fluorine (C–F) bonds at the surfaces [158].

In a previous work, carbon blacks were modified by fluorination under different conditions [159]. The changes in the surface properties of the fluorinated carbon blacks were investigated using the contact angle measurements using the sessile drop method. The testing liquids used were deionized water and diiodomethane. The scheme of general fluorination reactor os presented in Fig. 3-24

Fig. 3-25 shows the surface free energy of the fluorinated carbon blacks as a function of the fluorination pressure. The surface free energy of the fluorinated carbon blacks decreases as the fluorination pressure increases. The resulting higher fluorine content on the surfaces of the carbon blacks appears in relation to the decrease of the surface free energy. In other words, it appears that the polar part of the surface free energy increases at higher degrees of fluorination with increases in the fluorination pressure. The nature of the C–F bonding is affected mainly by the fluorine content and the fluorination temperature. The decrease in

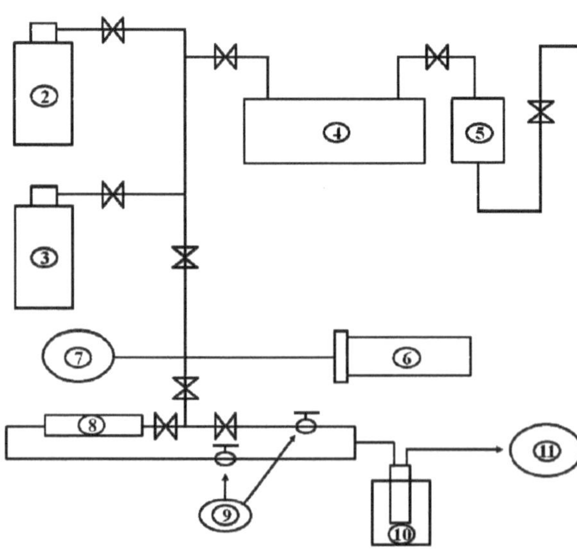

Fig. 3-24 Schematic diagram of fluorination reactor; (1) F_2 gas cylinder, (2) N_2 gas cylinder, (3) O_2 gas cylinder, (4) Buffer tank, (5) HF absorber (NaF pellet), (6) Reactor, (7) pressure gauge, (8) F_2 absorber (Al_2O_3), (9) Glass cock, (10) Liquid nitrogen, (11) Rotary vacuum pump.

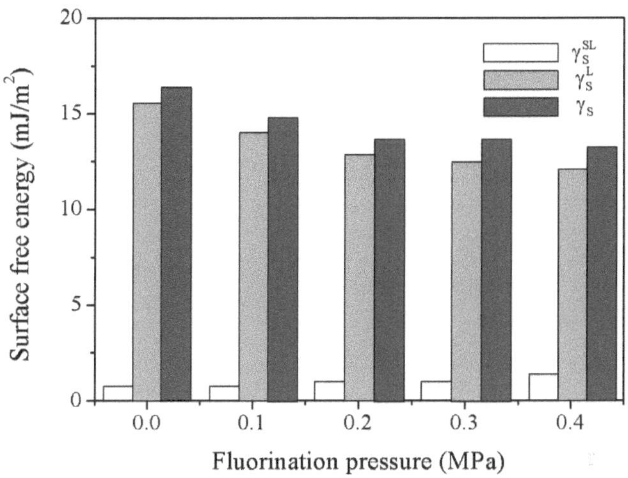

Fig. 3-25 Surface free energy of fluorinated carbon blacks as a function of fluorination pressure.

the dispersive component (γ_S^L) of the surface free energy may be caused by the destruction of the graphitic characteristic of the surface of the carbon blacks due to fluorination. The γ_S^L values are usually assumed to be fairly independent of the method of surface modification of carbon materials. It is assumed that the specific component (γ_S^{SP}) reflects the essential surface characteristics of carbon or the partial graphitized carbon framework of the particles. In contrast, we observe a distinct dependence of the values of the fluorine content on the surfaces of the carbon blacks. When a certain amount of fluorine is present on these surfaces, the amorphous part of the surface is eliminated; therefore, the amount of graphite must decrease in accordance with the results of the surface analysis.

Precise contact-angle measurements of fibrous materials present special challenges but two methods have been proposed [160]. Among the early literature, Chwastiak [161] introduced a wicking rate measurement procedure that entails enclosing the carbon fiber bundle in a glass tube so that the porosity is fixed for a given strand of carbon fibers. This procedure has been used successfully to evaluate the wettability of Thornel 50 carbon yarn for water, glycol, heptane, and epoxy resins, and was used in the present study of oxyfluorinated carbon fibers.

The carbon fibers were also treated using oxyfluorination for the modification under different conditions using the fluorine and oxygen mixtures (F_2/O_2 gases) [162]. The dynamic contact-angle measurements of carbon fibers were performed using a wicking method. The experimental setup is illustrated in Fig. 3-26. The wetting liquids used for the contact-angle measurements were *n*-hexane, deionized water, diiodomethane, and ethylene glycol.

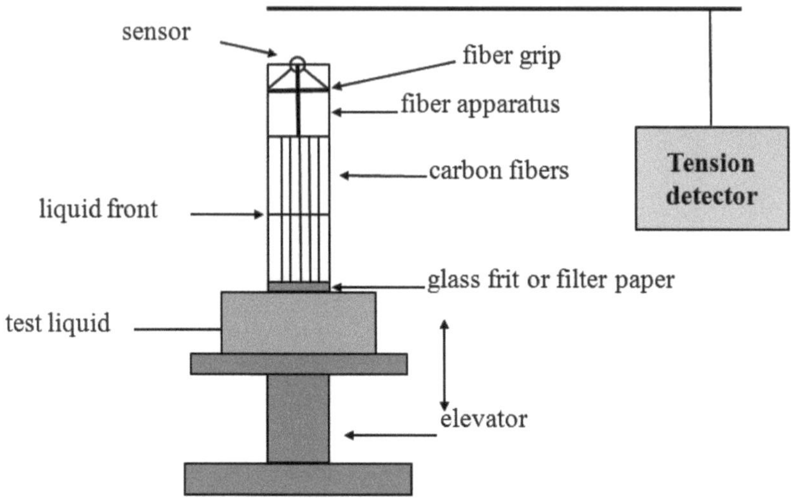

Fig. 3-26 Schematic diagram of principles of tensiometric wicking method.

It is generally known, that the specific component, γ_S^{SP}, is highly dependent on the surface functional groups. This dependency reflects the essential surface characteristics of carbon or the partial graphitized carbon framework of the fiber materials, whereas the dispersive component, γ_S^L, is largely dependent on the total electron density in the carbon fibers and thus does not vary much from one system to another. The surface free energies and their oxyfluorinated carbon fiber components are listed in Table 3-18. A decrease in the surface free energies is observed. The higher fluorine content on the carbon fiber surfaces leads to weaker surface free energies. In other words, the polar component of the surface

Table 3-18 Surface Free Energy (γ_S), London Dispersive (γ_S^L), and Specific (γ_S^{SP}) Component of Oxyfluorinated Carbon Fibers (CFO) as Functions of Oxyfluorination Temperature [unit: mJ/m2]

Specimens	γ_S^L	γ_S^{SP}	γ_S	X_P
As received	40.35	3.06	43.41	0.07
CFO-RT	30.80	8.61	39.41	0.22
CFO-100	29.70	11.68	41.38	0.28
CFO-300	28.62	9.51	38.13	0.25
CFO-400	28.45	9.10	37.55	0.24

free energy seemingly increases at higher degrees of fluorination and then decreases slightly again for lower degrees of fluorination and with higher fluorination temperatures. All of the experimental results listed in Table 3-18 are those usually observed for fluorinated carbon materials, considering that C–F bonding varies from chemical to physical bonding with increases in the fluorination temperatures (for CF_x prepared at 100 °C or higher) [163]. Consequently, the nature of C–F bonding is affected mainly by the fluorination temperature and the fluorine content; therefore, fluorination can cause a decrease in the γ_S^L value of the surface free energy. That is, when a certain amount of fluorine is present on the carbon fiber surfaces, the amount of graphite may be decreased, the result of which is in good accordance with those of the present study's surface analysis of the oxyfluorinated carbon fibers.

3.3.8.2.3. Ozone Treatment

For the surface modification of non-polar carbon materials, the oxidation can be achieved by heat treatment in air, oxygen (O_2) or ozone (O_3). Among the various sources of oxygen, an ozone treatment, one of the most common processes utilized in industry is easy to operate. It is used to introduce oxygen functional groups on carbon surfaces [164].

Ozone is a powerful oxidant (oxidation-reduction potential, E_0: +2.07 V) that has high selectivity for unsaturated bonds, whereas its reaction with the saturated bonds proceeds very slowly. It also reacts with inorganic and organic compounds directly or indirectly via the formation of hydroxyl radicals. It is known that ozone produces hydroperoxy radicals (· OOH) and superoxide radicals ($\cdot O_2^-$) as intermediates during decomposition. They typically react again with ozone and produce hydroxyl radicals (· OH) with ozonide radials ($\cdot O_2^-$) as an intermediate. The hydroxyl radicals have more powerful oxidizing power (E_0: +3.06 V) than ozone and its reactions with organics are very fast [165, 166]. Given these easy processes and strong reactivity, the ozone treatments of organic or inorganic materials have been applied in a number of industrial fields, such as bioengineering, food, and environmental engineering as well as surface modifications.

The effects of ozone treatment, on the Cr(VI) and Cu(II) adsorption behaviors of activated carbon fibers have been investigated [166]. The phenolic resin-based activated carbon fibers were used in the ozone treatments. The concentration of the ozone was measured up to 0.1 g O_3/m^3 in inert O_2 gas and was dispersed in the NaOH or H_2O_2 solutions.

Fig. 3-27 shows new phenolic groups onto the activated carbon fibers created by the deformation of the lactonic group, resulting in carbon dioxide production. As shown in Table 3-19, the number of phenolic groups was increased by the ozone treatment. Therefore, the increase in the total pore volume is mostly due to the increase of the micropore volume which is considered to have occurred

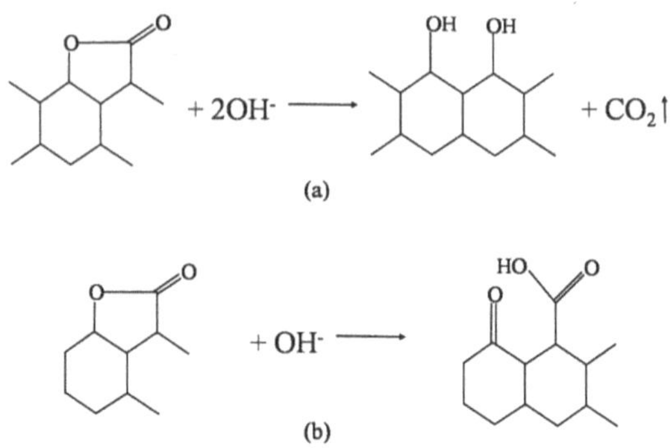

Fig. 3-27 Deformation of lactonic groups by hydroxyl radical (OH); formation of (a) phenolic groups, and (b) carboxylic groups.

because the hydroxyl radicals produced by ozone decomposition produce new micropores on the activated carbon fibers surfaces.

As shown in Table 3-19, the total acidity increased with the oxidation treatment with liquid initiators. Additionally, the content of each acidic functional group changes with the treatment. When a basic NaOH solution is used as an initiator, the hydroxyl radical is produced by the decomposition of the ozone as it reacts with the lactonic groups, leading to the formation of phenolic groups. In contrast, when an acidic H_2O_2 solution is used as an initiator, hydroxyl radicals react with the lactonic groups to produce carboxylic and phenolic groups, as in the case of a basic solution. However, it is estimated that these two functional groups are stabilized by the acidic environment, resulting in an increase of the total acidity, as shown in Table 3-19.

Table 3-19 Surface Acidity of the Activated Carbon Fibers (ACFs) Before and After Ozone Treatments

Specimens	pH_{ZPC}	Surface functional groups (meq/g) Carboxylic group	Lactonic group	Phenolic group	Total acidity (meq/g)
Non-treated ACF	3.0	0.10	0.24	0.29	0.63
Ozone-treated ACF in 1M NaOH	2.4	0.00	0.05	0.71	0.76
Ozone-treated ACF in 3M H_2O_2	2.2	0.12	0.02	0.69	0.82

3.4. ASSOCIATED PHENOMENA AND APPLICATIONS

3.4.1. Electrostatic Forces

When ions are present in a system that contains an interface, the ion density will vary near that interface. The boundary we identify as the surface defines the surface excess charge. If it is possible to separate the two bulk phases at this boundary, each of the separated phases carries an equal and opposite charge. These two charged portions of the interfacial region are known as an electrical double layer.

The purpose of this section is to introduce the basic ideas concerning electrical double layers and to develop equations for the distribution of charges and potentials, in the double layers. We also develop expressions of the potential energies and forces that result from the overlap of the double layers of different surfaces and the implication of these as regards colloid stability.

The electrostatic and electrical double-layer forces play a very important role in a number of contexts in science and engineering. The stability of a wide variety of colloids is affected by surface charges on the particles. The filtration efficiency of submicron particles can be diminished considerably by electrical double-layer forces. The coagulants are added to neutralize the electrostatic effects to promote aggregation and to enhance the ease of separation.

The electrostatic forces also play an important role in the conformation and structure of macromolecules, such as polymers, polyelectrolytes, and proteins. The self-assembly of proteins from disks to viruses is triggered by the electrostatic interactions between neighboring subunits. In the case of polyelectrolytes and charged colloids, the transport behavior (e.g., rheology) is also affected significantly by the charge effects.

3.4.1.1. Electric Double Layer

The situations in which van der Waals forces alone determine the total interaction are restricted to a few simple systems, for example, to interactions in a vacuum or to non-polar wetting films on surfaces. In more complex and more interesting systems, the long-range electrostatic forces are also involved. The interplay between these two interactions has many important consequences.

As mentioned earlier, the van der Waals force between similar particles in a medium is always attractive; hence, if only van der Waals forces are operating, we might expect all dissolved particles to stick together (coagulate) immediately and precipitate out of the solution as a mass of solid material. Our own bodies would be subject to the same fate considering that we are composed of about 75 percent water. Fortunately, this does not occur, because particles suspended in water and any liquid with a high dielectric constant are usually charged and can be prevented from coalescing due to repulsive electrostatic forces. Other repulsive forces that can prevent coalescence are solvation and steric forces. This chapter concentrates on the electrostatic forces.

The charging of a surface in a liquid can come about in two ways: i) By the ionization or dissociation of surface groups (e.g., –COOH → –COO^- + H^+ and ii) By the adsorption (binding) of ions from a solution onto a previously uncharged surface, e.g., the binding Ca^{2+} onto the zwitterionic head groups of lipid bilayer surfaces, which charges the surfaces positively. Depending on the ionic conditions, even the air-water and hydrocarbon-water interfaces can become charged in this way. The adsorption of ions from a solution can of course, also occur onto oppositely charged surface sites, e.g., the adsorption of cationic Ca^{2+} to anionic COO^-sites vacated by H^+ or $Na.^+$ This is known as an ion exchange.

Regardless of the charging mechanism, the final surface charge is balanced, by an equal but oppositely charged region of counterions, some of which are bound, usually transiently, to the surface within what is known as the Stern or Helmholtz layer, while others form an atmosphere of ions in rapid thermal motion close to the surface, known as a diffuse electric double layer Fig. 3-28.

Two similarly charged surfaces usually repel each other electrostatically in a solution, though under certain conditions they may attract at small separations. The zwitterionic surfaces, i.e., those characterized by surface dipoles but no net charge, also interact electrostatically with each other, though, here we typically find, that the force is attractive.

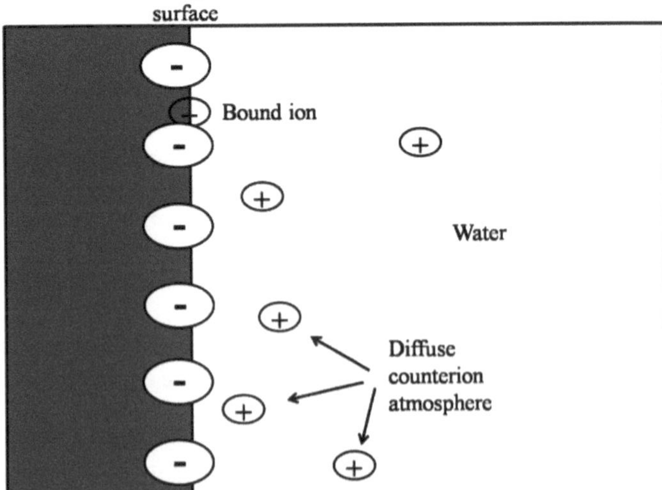

Fig. 3-28 Ions bound to a surface are not rigidly bound, but can exchange with other ions in solution; their lifetime on a surface can be as short as 10^{-9} s or as long as many hours.

3.4.1.2. Charged Surface in Water

The following sections consider the counterion distribution and the force between two similarly charged planar surfaces in a pure liquid such as water, where (apart from H_3O^+ and OH^- ions) the only ions in the solution are those that have come off the surfaces. Such systems occur when, for example, colloidal particles, clay sheets, surfactant micelles or bilayers whose surfaces contain ionizable groups interact in water and also when thick films of water build up on an ionizable surface such as glass. However, we must first consider several fundamental equations that describe the counterion distribution between two charged surfaces in a solution.

For a case when only counterions are present in a solution, the chemical potential of any ion may be described by the Poisson–Boltzmann equation:

$$\mu = \mathrm{ze}\Psi + \mathrm{kT}\log\rho \qquad (3\text{-}45)$$

here, Ψ is the electrostatic potential and ρ is the number density of ions of valency z at any point x between two surfaces Fig. 3-29. Because only differences in potential are ever physically meaningful, $\Psi_0 = 0$ can be set at the midplane ($x = 0$), where $\rho = \rho_0$ and $(\mathrm{d}\Psi/\mathrm{d}x)_0 = 0$, as well, due to symmetry.

From the equilibrium requirement that the chemical potential be uniform throughout, the expected Boltzmann distribution of counterions at any point x, is as follows:

$$\rho = \rho_0 e^{-\mathrm{ze}\Psi/kT} \qquad (3\text{-}46)$$

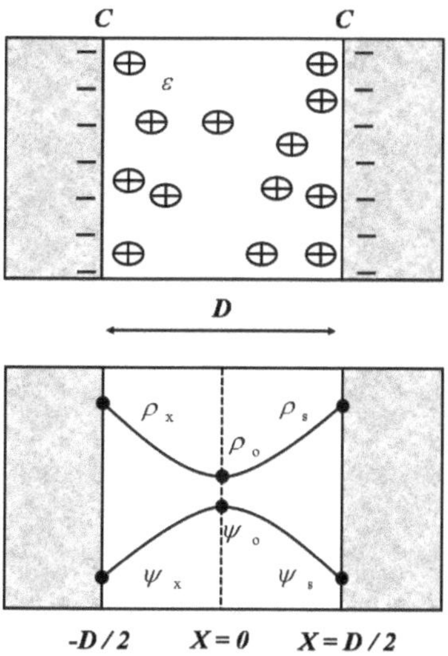

Fig. 3-29 Two negatively charged surface of surface charge density σ, separated a distance D in water.

One further important fundamental equation will be required. This is the well-known Poisson equation for the net excess charge density at x:

$$ze\rho = -\varepsilon\varepsilon(d^2\Psi/dx^2) \tag{3-47}$$

When combined with the Boltzmann distribution, Eq. (3-46), this gives the Poisson–Boltzmann equation:

$$d^2\Psi/dx^2 = -ze\rho/\varepsilon\varepsilon_0) \tag{3-48}$$

$$= -(ze\rho_0/\varepsilon\varepsilon_0)e^{-ze\Psi/kT} \tag{3-49}$$

When solved, the Poisson-Boltzmann equation gives the potential Ψ, electric field $E = \partial\Psi/\partial x$, and counterion density ρ at any point x in the gap between the two surfaces. Let us first determine these values at the surfaces themselves. These quantities are often referred to as the contact values, of Ψ_s, E_s, and ρ_s, among others.

The Poisson-Boltzmann equation is a non-linear second-order differential equation. To solve for Ψ we need two boundary conditions which determine the two integration constants. The first boundary condition follows from the symmetry requirement that the field must vanish at the midplane, i.e., that $E_0 = (D\Psi/dx)_0 = 0$. The second boundary condition follows from the requirement of overall electroneutrality, i.e., that the total charges of the counterions in the gap must be equal (and opposite) to the charge on the surfaces. If σ is the surface charge density on each surface (in C/m^2) and D is the distance between the surfaces (Fig. 3-29), the condition of electroneutrality implies the following:

$$\sigma = -\int_0^{D/2} ze\rho \, dx \tag{3-50}$$

$$= +\varepsilon\varepsilon_0 \int_0^{D/2} (d^2\Psi/dx^2)dx \tag{3-51}$$

$$= \varepsilon\varepsilon_0(d\Psi/dx)_{D/2} \tag{3-52}$$

$$= \varepsilon\varepsilon_0(d\Psi/dx)_s \tag{3-53}$$

$$= \varepsilon\varepsilon_0 E_s \tag{3-54}$$

i.e.,

$$E_S = \sigma/\varepsilon\varepsilon_0 \tag{3-55}$$

Eq. (3-55) gives an important boundary condition relating the surface charge density σ to the electric field E_s at each surface (at $x = \pm D/2$), which we may note is independent of the gap width D.

Turning now to the ionic distribution, there exists, an important general relationship between the concentrations of counterions at either surface and at the midplane. Differentiating Eq. (3-46) and then using Eq. (3-49), we obtain

$$\frac{d\rho}{dx} = -\frac{ze\rho_0}{kT} e^{\frac{-ze\Psi}{kT}} \left(\frac{d\Psi}{dx}\right) \tag{3-56}$$

$$= \frac{\varepsilon\varepsilon_0}{kT} \left(\frac{d\Psi}{dx}\right) \left(\frac{d^2\Psi}{dx^2}\right) \tag{3-57}$$

$$= \frac{\varepsilon\varepsilon_0}{2kT} \frac{d}{dx} \left(\frac{d\Psi}{dx}\right)^2 \tag{3-58}$$

hence

$$\rho_x - \rho_0 = \int_0^x d\rho \tag{3-59}$$

$$= \frac{\varepsilon\varepsilon_0}{2kT} \int_0^x d\left(\frac{d\Psi}{dx}\right)^2 \tag{3-60}$$

$$= +\frac{\varepsilon\varepsilon_0}{2kT} \left(\frac{d\Psi}{dx}\right)_x^2 \tag{3-61}$$

so that

$$\rho_x = \rho_0 + \frac{\varepsilon\varepsilon_0}{2kT} \left(\frac{d\psi}{dx}\right)_x^2 \tag{3-62}$$

which gives ρ at any point x in terms of ρ_0 at the midplane and $(d\Psi/dx)^2$ at x. In particular, at the surface, $x = D/2$, we obtain using Eq. (3-55), the contact value of ρ:

$$\rho_s = \rho_s + \sigma^2/2\varepsilon\varepsilon_0 kT \tag{3-63}$$

This result shows that the concentration of counterions at the surface depends only on the surface charge density σ and the counterion concentration at the midplane. Note that ρ_s never falls below $\sigma^2\varepsilon\varepsilon_0 kT$, even for isolated surfaces, i.e., for two surfaces far apart when $\rho_0 \to 0$.

This shows that regardless of the counter ion distribution profile ρ_x away from a surface, most of the counterions that effectively balance the surface charge are located in the first few angstroms from the surface [167]. However, for the lower surface charge densities, because $\rho_s \propto \sigma^2$, the layer of counterions extends well beyond the surface and becomes much more diffuse.

3.4.1.3. Charged Surfaces in Electrolyte

It is far more common for charged surfaces to interact across a solution containing electrolyte ions. In the oceans, there are 0.6 M ions consisting of NaCl or KCl with smaller amounts of $MgCl_2$ and $CaCl_2$, and pure water at pH 7 is strictly an electrolyte solution containing 10^{-7} M of H_3O^+ and OH^- ions. For a charged surface, exposed to a solvent containing no electrolyte ions, the equation $\left(\Psi = \left(\frac{kT}{ze}\right)\log(\cos^2 Kx)\right)$ readily shows that for an isolated surface for which $D \rightarrow \infty$, the result is $KD \rightarrow \pi$ and $\Psi_x \rightarrow \infty$. However, this situation of the solvent is removed by the addition of a little concentration of electrolyte ions.

The presence of a bulk 'reservoir' of electrolyte ions is highly affected to the electrostatic potential and the force between charged surfaces. In this chapter, this interaction including the total interaction potential will represent, when the van der Waals force is added. However, an understanding of the double-layer interaction between two surfaces is priority than that of the ionic distribution adjacent to an isolated surface in contact with an electrolyte solution. Consider an isolated surface or two surfaces far apart in an aqueous electrolyte, as shown in Fig. 3-30. For convenience, when x is 0 at the surface rather than at the midplane all of the fundamental equations are applicable to solutions containing

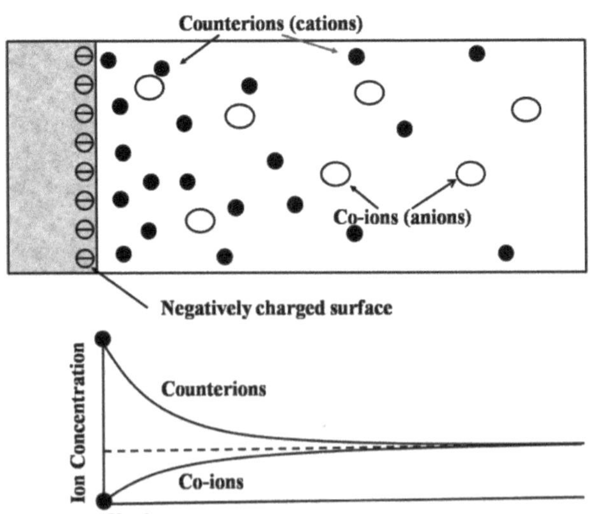

Fig. 3-30 Near a charged surface there is an accumulation of counterions and a depletion of co-ions, shown graphically below for a 1:1 electrolyte, where ρ_∞ is the electrolyte concentration in the bulk or 'reservoir' at $x = \infty$.

different types of ions i so long as this is taken into account by expressing the net charge density at any point x as $\sum_i z_i^i e\rho_{x1}$ and the total ionic concentration as $\sum \rho_{x1}$. Thus, Eq. (3-46) for the Boltzmann distribution of ions i at x now becomes

$$\rho_{xi} = \rho_{\infty i} e^{-z_i e \Psi_x / kT} \tag{3-64}$$

while at the surface, at $x = 0$, the contact values of ρ and Ψ are related by

$$\rho_{0i} = \rho_{\infty i} e^{-z_i e \Psi_0 / kT} \tag{3-65}$$

where $\rho_{\infty i}$ is the ionic concentration of ions i in the bulk (at $x = \infty$) where $\Psi_0 = 0$.

3.4.1.4. Applications

3.4.1.4.1. Electric Double Layer Capacitor

The electrochemical capacitors are unique energy-storage devices that exhibit high power density and a long cycle life. According to the energy-storage mechanism, electrochemical capacitors can be divided into two types: electrochemical double-layer and redox supercapacitors [168]. In the former, the electric double layer capacitors (EDLCs) are based on the double-layer capacitance at the solid/solution interface of the high-surface-area materials. Energy storage arises mainly from the separation of electronic and ionic charges at the interface between high-specific-area electrode materials and the electrolyte solution. Principally, it is an electrostatic phenomenon. In the latter, the fast Faradic reactions take place at the electrode materials at characteristic potentials, as in batteries, and give rise to what is known as pseudo-capacitance.

Recently, electric double layer capacitors have shown new and promising applications in the capacitive deionization of water and as pulse power sources for the digital communication devices and electric vehicles. The popularity of these devices derives from their higher energy density relative to conventional capacitors as well as their longer cycle life and the higher power density relative to batteries [169].

Generally, carbon materials are available with specific surface areas of up to 500 to 3000 m^2/g, including activated carbons, carbon blacks, glassy carbons, carbon nanorods, carbon aerogels, and nanotubes [170], which possess a variety of microstructures and electrochemical properties at a wide range of costs. The combinations of these materials provide a flexible means of optimizing the properties of electrodes for the electric double layer capacitors to balance the performance and cost. Among them, many attempts have been made to develop activated carbons for use in the electric double layer capacitors. Because the activated carbons-based electric double layer capacitors have relatively low capacitance, which could limit their practical application many researchers have worked on increasing the surface area of activated carbons in an attempt to realize high capacitance.

However, adverse effects inevitably arise, unfortunately, when the surface area of activated carbons is increased, such as series connections of resistance, decreased electric conductivity, greater difficulty in the fabrication process, and increased cost. It is known that activated carbon electrodes are somewhat polarizable. The resulting polarization adversely affects the performance of the capacitors; it reduces the accumulation of ions on the double-layer of activated carbons. Therefore, reducing the polarization of activated carbons is considered an effective approach toward attaining improved capacitance in the EDLCs.

The effect of nano-sized titanium oxide (TiO_2) on reducing the polarization of activated carbons has been investigated [171]. The energy-storage composite electrodes were prepared by mixing activated carbons modified with TiO_2. The electrochemical characteristics of the activated carbons/TiO_2 composite electrodes were determined using cyclic voltammetry. In general, the cyclic voltammogram of an ideal double-layer capacitor has a rectangular shape, while that of the pseudo-capacitance, deviates from the rectangular shape and shows reversible redox peaks associated with pseudofaradiac reactions [172]. The composite electrode was used as the working electrode, Hg/Hg_2SO_4 as the reference electrode, and platinum foil as the counter electrode. Cyclic voltammetry (CV) of the electrode materials was performed in a potential range of 0 to 0.9 V at a scan rate of 10 mV/s. The discharge capacitance (C) of the electrodes in the electric double layer capacitors was calculated from the slope of the discharge on the basis of Eq. (3-66) [172, 173],

$$C = i\left(\frac{\Delta t}{\Delta V}\right) \tag{3-66}$$

where C is the capacitance of the cell in farads, i the discharge current in amperes (A), and $\Delta V/\Delta t$ the slope in volts per second (V/s).

All of the electrochemical measurements were performed in 1 M of H_2SO_4 electrolyte. For a symmetrical system, the specific capacitance C_m in farads per gram of the activated carbon/TiO_2 composite electrodes (F/g) was calculated from the capacitance of the cell C,

$$C_m = 2C/m \tag{3-67}$$

where m is the weight (g) per electrode of the activated carbon/TiO_2 composite electrodes.

In Fig. 3-31, two rectangular-like voltammograms support the above hypothesis that only electrostatic attraction is involved in the activated carbon/TiO_2 composite electrode. The specific capacitance of the activated carbons is significantly enhanced from 100 to 155 F/g for the composite with 10 wt percent titanium in the hydrous oxide form. It was also observed that the introduction of TiO_2 into the activated carbons dramatically increases the capacity of the activated carbons despite the fact that the TiO_2 itself has very low capacitance. This phenomenon can be ascribed to the effect of the single-direction polarity of

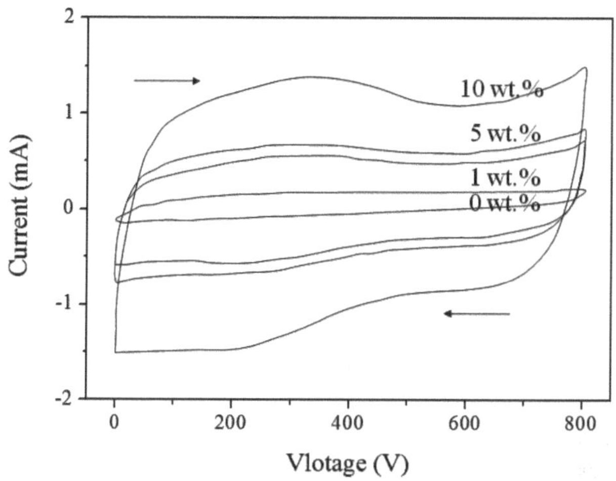

Fig. 3-31 Cyclic voltammogram of activated carbons/TiO_2 composite electrodes as a function of TiO_2 content at 1 mV/s in 1 M H_2SO_4.

TiO_2 nanoparticles. Therefore, the present method offers many advantages due to the formation of stable TiO_2 colloidal particles in the amorphous form to yield a high specific capacitance at relatively low titanium loading. It has been reported that the capacitance of electric double layer capacitors is proportional to the ion concentration and 1/thickness of the double-layer and that the ion concentration is affected by the voltage between two electrodes and the polarization of the carbon electrodes. Based on the above results, the increase of the capacitance can be attributed to the reduced polarization of the activated carbon electrodes, which is believed to be related to the concentration of the charges on the TiO_2 surface.

Fig. 3-31 also shows that the cyclic voltammogram becomes somewhat deformed during the negative sweep. This is likely due to a combination of: (i) The effect of kinetic polarization on the activated carbon/TiO_2 electrode, (ii) The contact resistance on the activated carbon/TiO_2 interface, and (iii) The diffusion barrier of ions within the pores of the activated carbons. In a capacitor, based on high-area porous materials, distributed resistance (R) unavoidably arises (mainly of the ion transfer). This is coupled in a complex manner with the distributed interfacial capacitance elements [174]. Although the addition of TiO_2 does not change the porous structure of the activated carbons, it can reduce the polarization of the activated carbons. This reduction can dramatically increase the mobility of ions in the pores of the activated carbons, which subsequently reduces the total resistance of the devices.

Electrospinning, a fiber-spinning technique that relies on electrostatic forces to produce fibers in the nanometer-to-micron diameter range has been extensively explored as a simple method of preparing nanofibers from polymer solutions or melts. The fiber morphology is controlled by the experimental design and is dependent upon the solution conductivity, solution concentration, polymer molecular weight, viscosity, and applied voltage. The main advantage of the electrospinning process is that it offers a simple means of preparing continuous fibers with unusual large surface-to-volume ratios and pore structure surfaces. As such, it is capable of producing a paper form composed of the nonwoven carbon nanofiber webs in a very simple and convenient way manner [175].

Various forms of carbonaceous materials, i.e., powders, fibers, papers or cloth (fabric or web), carbon nanotubes, carbon nanofibers, and related nanocomposites are candidates as the electrodes of electric double layer capacitors [176]. A paper-type material that is normally useful, particularly for application as an electrode as a binder addition, normally degrading the performance of a capacitor, is not necessary.

Recently, activated carbon nanofibers were studied as electrode materials [177]. A series of activated carbon nanofibers with developed mesoporosity were prepared from the electrospun polyamide/polyimide nanofibers using a carbonization/activation process to shed light on the relationship between the porosity characteristics and the electrochemical capacitance. The experimental results indicated that carbon nanofibers activated at 800 °C afforded the highest specific surface area but resulted in a low mesopore volume, as shown in Table 3-20.

The electrochemical performance of an activated carbon nanofiber electrode was assessed in a potential range of 0 to 1.0 V using cyclic voltammetry in 1 M of H_2SO_4. As shown in Fig. 3-32, the capacitive behavior of carbon nanofibers and the activated carbon nanofiber-800 shows the typical i–E responses of carbon materials. For both curves in Fig. 3-32(a), the background current density, negative at ca. 0.6 V upon both the positive and negative sweeps with a broad pair of redox peaks around 0.4 to 0.5 V is much higher than that obtained in the relatively positive potential region. This phenomenon has been

Table 3-20 Surface Characterization of Carbon Nanofibers (CNFs) and Activated Carbon Nanofibers (ACNFs) Samples

Samples	Activation yield (%)	Specific surface area[a] (m^2/g)	V_{meso}[b] (mL/g)	V_{micro}[c] (mL/g)
CNFs		240		0.12
ACNFs-700	59	1080	0.06	0.28
ACNFs-80	55	1360	0.2	0.80
ACNFs-900	50	1250	0.23	0.72

[a] *Specific surface area calculated by BET method [178].*
[b] *Mesopore volume calculated by Barret, Joyner and Halenda (BJH) method [179] based on the Kelvin equation.*
[c] *Micropore volume calculated by Horvath–Kawazoe (HK) method [180].*

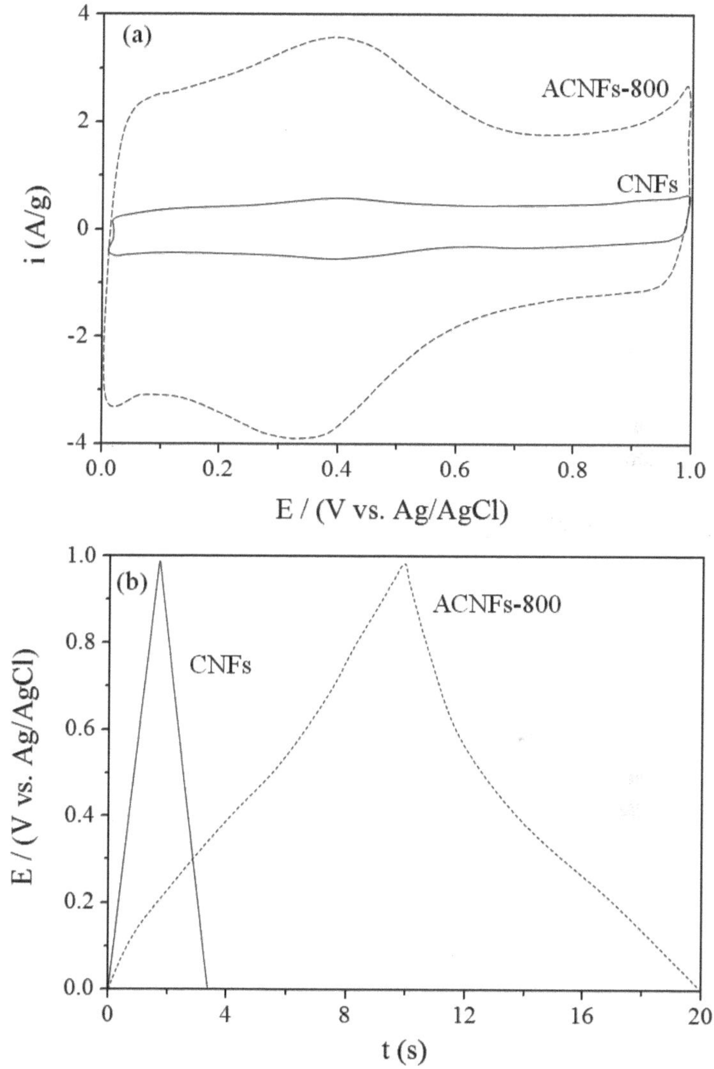

Fig. 3-32 Cyclic voltammograms (a) measured in 1M H_2SO_4 at 10 mV/s and chronopotentiograms (b) measured in 1 M H_2SO_4 at a current density of 10 A/g of carbon nanofibers, and activated carbon nanofibers-800 samples.

attributed to a combination of the electrical double-layer charge/discharge and the redox transitions of interfacial oxygen-containing functional groups, resulting from synthesis and/or purification processes [181]. As the current density of the activated carbon nanofiber-800 is much larger than that of carbon nanofiber, the specific capacitance (and electric energy) stored in the activated carbon nanofiber-800 is much larger than that of the activated carbon nanofiber. The much higher current density of the former carbon materials is attributable to its higher specific surface area and to the higher content of its surface functional groups. This is further supported by the potential-dependent choronopotentiogram of the activated carbon nanofiber-800, as shown in Fig. 3-33(b). For an ideal polarized electrode, the double-layer capacitance is only weakly dependent upon the applied potentials (i.e., $dE/dt \approx$ a constant value). However, the choronopotentiometry responses of the activated carbon nanofiber-800 do not obey this phenomenon, revealing a significant contribution of redox reactions from surface functional groups, although the charge curve is completely symmetric to its discharge counterpart, indicating excellent reversibility. As indicated in Fig. 3-32(a), the specific capacitance of carbon nanofibers and the activated carbon nanofiber-800 (measured at 10 mV/s) is equal to 30 and 150 F/g, respectively, which are very close to the values estimated from choronopotentiometry (measured at 10 A/g); 28 and 140 F/g, respectively.

The carbon nanotubes have attracted increasing interest since Iijima first discovered them in 1991 [182]. Studies of the application of carbon nanotubes as electrode materials for electric double layer capacitors have been performed and the results show that carbon nanotubes are promising candidates as electrode materials for electric double layer capacitors [183]. The carbon nanotubes have potential advantages over activated carbons due to their hollow structure, nanometer diameters, narrow size distribution, highly accessible surface area, low resistivity and high stability. These properties make them potentially quite suitable for application in the fabrication of electric double layer capacitors. However, the specific capacitance of electric double layer capacitors based on carbon nanotubes is not very high at present and the specific capacitance of the carbon nanotube for electric double layer capacitors using an organic electrolyte is only about 20 to 30 F/g. It has been found, that the carbon nanotubes oxidized with CO_2 have about two times the capacitance of carbon nanotubes without oxidization [184, 185].

In a previous study, multi-walled carbon nanotubes were used for electrode materials [186]. This was done to evaluate the specific capacitance of the multi-walled carbon nanotube-based polarizable electrodes in electric double layer capacitors, after the air-oxidation at various temperatures (400 to 600 °C). The equivalent series resistance of the capacitor unit was measured by applying an alternating current of 10 mV amplitude at a frequency of 1 kHz. Cyclic voltammetry of the air-oxidized multi-walled carbon nanotube electrodes was performed in the potential range of 0 to 0.9 V at a scan rate of 10 mV/s.

As the annealing temperature increases, more multi-walled carbon nanotubes are broken and uncapped multi-walled carbon nanotube aggregates are disassembled. Moreover, oxygen attacks the cylindrical walls of the multi-walled carbon nanotube. As shown in the TEM images (Fig. 3-33) of the multi-walled carbon nanotubes air-oxidized at 600 °C, some outer graphitic layers have been etched off, giving rough 'convex-concave' walls.

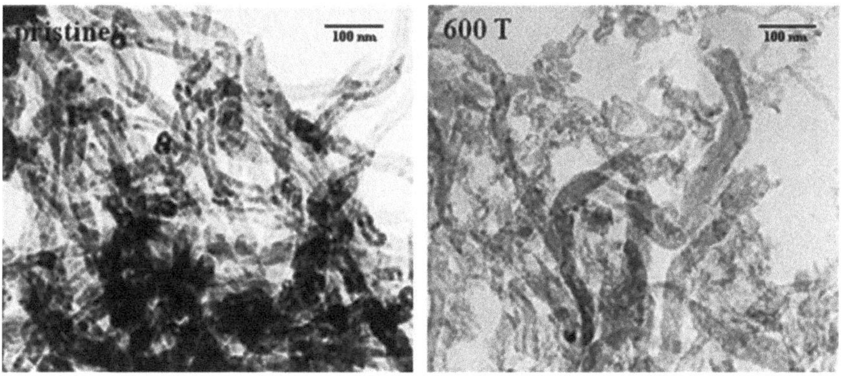

Fig. 3-33 TEM images of the pristine and multi-walled carbon nanotube, air-oxidized at 600 °C.

This results in the further enhancement of the specific area, as listed in Table 3-19. Subsequently, the specific surface area decreases slightly with further increases in the air-oxidation temperature to the maximum 600 °C. The pore volumes of the pristine multi-walled carbon nanotube and the multi-walled carbon nanotube air-oxidized at 400 °C, 500 °C and 600 °C measured by the BJH method are listed in Table 3-21. Also, the relative volume ratios of the micropores (<2 nm) of every sample calculated from the pore-size-distribution curves are listed in Table 3-21.

Fig. 3-34 shows the results of the electrochemical capacitance of the multi-walled carbon nanotube air-oxidation at different temperatures. The multi-walled carbon nanotube electrodes air-oxidized at 600 °C show the highest specific capacitance, 50 F/g, more than twice as high as that of the electrodes made of pristine multi-walled carbon nanotubes owing to the different structures and the changes of surface characteristics (increment of oxygen functional groups) of the multi-walled carbon nanotubes, as mentioned in Table 3-21.

Table 3-21 Characteristics of the Pristine and Multi-walled Carbon Nanotubes Air-oxidized Various Temperatures

Samples	Specific capacitance (F/g)	Specific surface area (m^2/g)	Pore volume (cm^3/g)	Relative volume ratio of microspores (<2 nm, %)
Pristine	25	180	0.60	12.5
400 °C	32	230	0.72	8.2
500 °C	36	236	0.73	7.1
600 °C	50	290	0.84	4.6

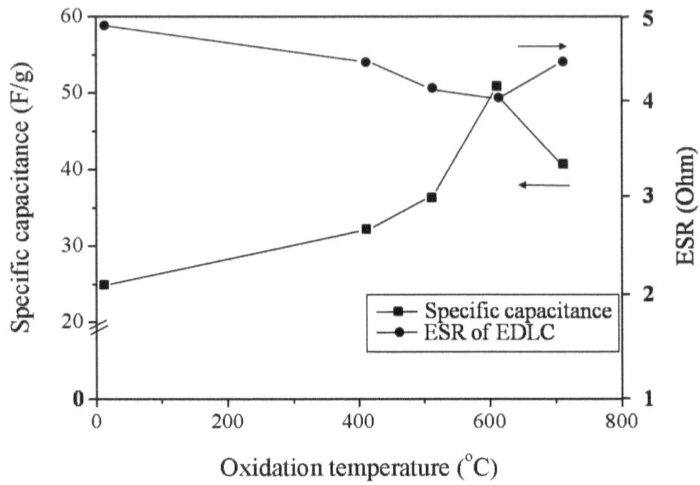

Fig. 3-34 Specific capacitances of the air-oxidized multi-walled carbon nanotubes electrodes, and equivalent series resistances of EDLCs, based on the pristine and multi-walled carbon nanotubes air-oxidized at various temperatures.

Subsequently, the specific capacitance decreases slightly with further increases in the air-oxidation temperature to the maximum 600 °C. This decrease in the capacitance is mainly due to the decreased specific surface area, reflecting damage to the novel structure of the multi-walled carbon nanotubes caused by the high temperature. The equivalent series resistances of capacitors based on the air-oxidized multi-walled carbon nanotubes are slightly reduced compared with those of capacitor units based on the pristine multi-walled carbon nanotubes. Moreover, the data of the pore volume showed that though the relative micropore volume of the oxidized multi-walled carbon nanotubes decreased with air-oxidation, the relative mesopore (2 to 50 nm) volume of the oxidized multi-walled carbon nanotubes increased which makes them viable as electrode materials for the electric double layer capacitors. It is known that the data of the BET specific surface area, pore volume, and pore diameter distribution are the crucial parameters regarding electrode materials for electric double layer capacitors, which is the main reason that the oxidized multi-walled carbon nanotubes had two times the capacitance of the pristine multi-walled carbon nanotubes.

The effects of a heat treatment on the electrochemical performance of the single-walled carbon nanotubes (SWNTs) were investigated [187]. They indicated that an increase in the textural properties after the heat treatment led to the enhanced electrochemical performance of the single-walled carbon nanotubes. In Fig. 3-35, with an increasing temperature, the specific surface area increases, whereas the average pore diameter decreases and saturates at a high temperature. The specific surface area at 1000°C is 357 m^2/g. The raw sample shows a

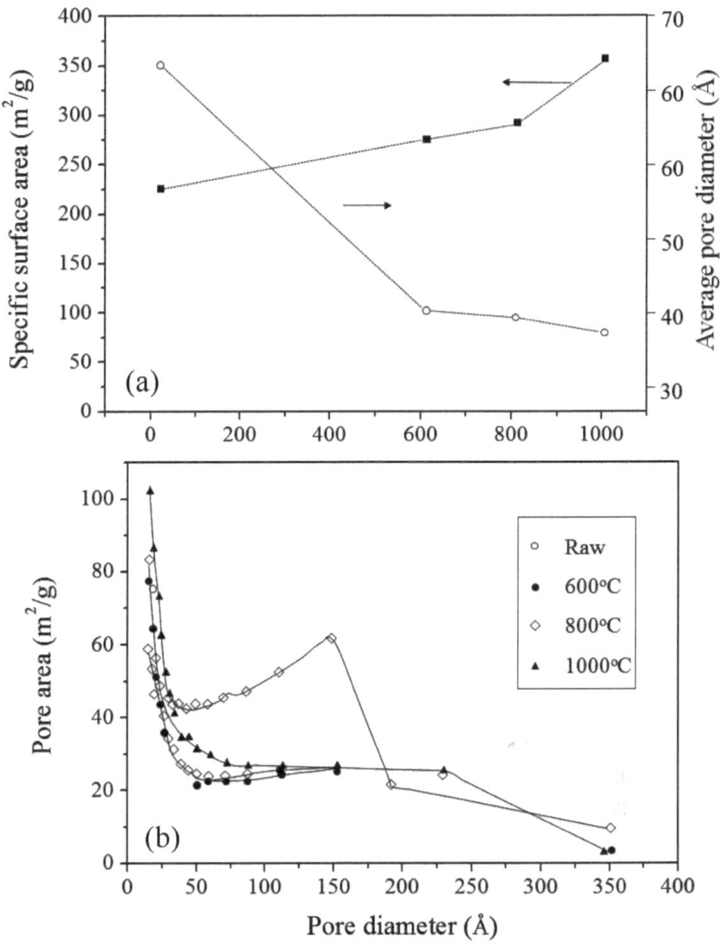

Fig. 3-35 The BET (N_2) specific surface areas and the average pore diameters of the carbon nanotubes electrode as a function of heat-treatment temperature (a) and the pore size distribution of the carbon nanotubes electrodes (b).

peak at 150 Å and has a lower distribution of smaller pore diameters near 20 Å. Upon increasing the heat-treatment temperature, the number of smaller pore diameters increases, reaching its maximum at 1000 °C, whereas the number of pore diameters, ranging from 50 to 250 Å decreases.

In general, it is known that pore sizes in the range of 30 ± 50 Å are required to maximize the capacitance in an electric double-layer capacitor. Fig. 3-36 shows the specific capacitances of electrodes heat-treated at various temperatures as a function of the charging time. The capacitances increase abruptly and reach nearly 80 percent of the maximum capacitance during the initial 10 min

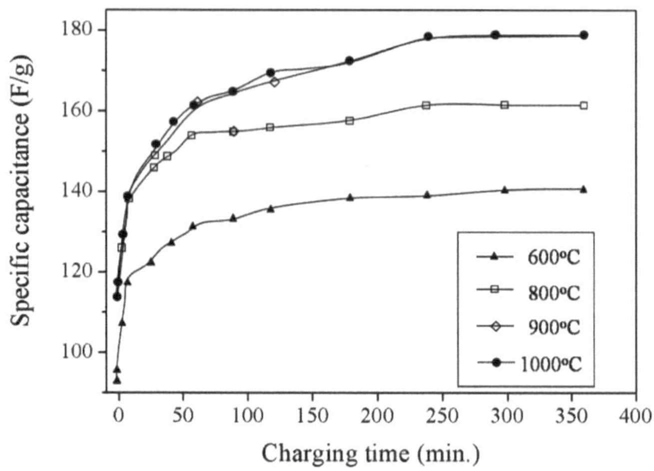

Fig. 3-36 The specific capacitances of the heat-treated electrodes at various temperatures, as a function of the charging time at a charging voltage of 0.9 V, where the capacitance was measured at a discharging current of 1 mA/cm^2.

regardless of the heat-treatment temperature. The capacitances gradually increase further and saturate to their maximum values at long charging times. A persistent increase of the capacitance over a long time is generally observed from the porous electrodes and is attributed to the existence of various forms of pores and pore diameters in the electrode. The saturated capacitance increases as the heat-treatment temperatures increase and saturates to 180 F/g at 1000 °C. This value is larger than the previously reported value of 113 F/g from the multi-walled carbon nanotubes [188]. The earlier study noted that the capacitance is directly proportional to the specific surface area with increases in the temperature. The average pore diameter approaches 30 Å at 1000 °C and, moreover, the populations of such small diameters are highly weighted at 1000 Å, as shown in Fig. 3-35.

Novoselov and Geim [189] found graphene as an entirely new class of carbon in 2004. Graphene, the basal plane of graphite, has attracted tremendous attention due to its unique physical, chemical, and mechanical properties [190]. More recently, many meaningful research results concerning the preparation of graphene as a supercapacitor electrode material have been obtained.

Du *et al.* [191] prepared two types of functionalized graphene sheets for supercapacitor electrodes. Graphene was produced by the thermal exfoliation of graphite oxide. The first type of functionalized graphene sheet was obtained by the thermal exfoliation of graphite oxide at a low temperature (250 to 400 °C) in air. The second was prepared by the carbonization of the first type of functionalized graphene sheet in a higher temperature range (700 to 900 °C) in N_2. Graphite oxide was produced from natural graphite powders according to Hummers method [192]. They investigated the effect of the specific surface area and conductivity on the electrochemical performance of both types of graphene. Cyclic voltammetry (CV) and galvanostatic charge–discharge assessments were done in a three-electrode experimental setup using 2 mol of a L^{-1} KOH aqueous solution. The cyclic voltammetry and galvanostatic charge–discharge measurements were carried out at a potential ranging from –1.0 V to –0.2 V (*vs.* a saturated calomel electrode) at room temperature.

Scanning the electron microscopy images (SEM) showed that both samples possessed nanoporous structures. The results of a N_2 adsorption-desorption analysis indicate that both of types had a high BET surface areas. Moreover, the second type of functionalized graphene sheet (574 to 737 m^2/g) has a relatively higher BET surface area compared to the first type (328 to 418 m^2/g). Fig. 3-35 shows the cyclic voltammetry curves of functionalized graphene sheets samples obtained in 2 mol/L KOH solution at a scan rate of 5 mV/s. A nearly rectangular shape (Fig. 3-37) is a characteristic of the electric double layer capacitors made from the carbon-based electrodes. Moreover, the cyclic voltammetry curve of the functionalized graphene sheets, 300 to 900 is close to the ideal rectangular shape, which indicates that the electrochemical behavior based on functionalized graphene sheets at 300 to 900, is very similar to an ideal supercapacitor. The capacitance values of the 300 to 700 functionalized graphene sheets and the 300 to 900 functionalized graphene sheets are clearly lower than those of the 250 functionalized graphene sheets the 300 functionalized graphene sheets and the 400 functionalized graphene sheets at a low scan rate of 5 mV/s.

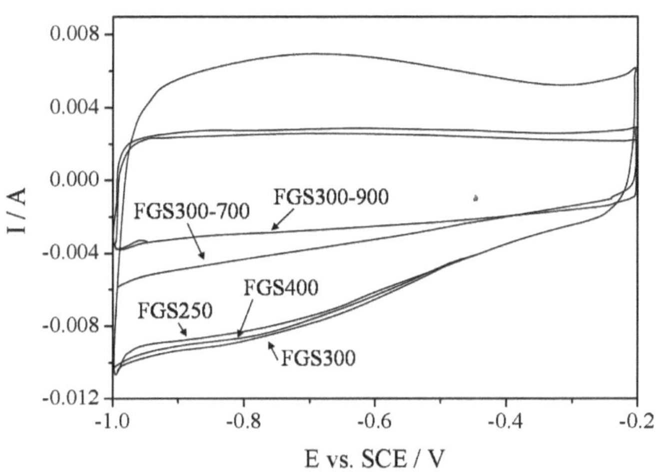

Fig. 3-37 Cyclic voltammograms of the functionalized graphene sheets samples obtained in 2 mol/L KOH at scan rates of 5 mV/s.

Table 3-22 Specific Capacitances Obtained in 2 mol/L KOH from Galvanostatic Charge–discharge Method and Capacitance Retentions from 1 A/g to 20 A/g for the Functionalized Graphene Sheets (FGS) Samples

Sample	Cm (F/g) at different current densities					Capacitance
name	1 A/g	2 A/g	4 A/g	10 A/g	20 A/g	retention (%)
FGS250	219	213	205	170	125	57
FGS300	232	224	213	171	113	49
FGS400	224	216	209	171	123	55
FGS300-700	107	104	102	89	70	65
FGS300-900	91	89	86	76	63	69

The electrochemical performances of the two samples are listed in Table 3-22. The results of the electrochemical tests were as follows: the specific capacitance values of the first type of the functionalized graphene sheet in the aqueous KOH electrolyte were about 230 F/g. The specific capacitance values of the second type of functionalized graphene sheet with a higher BET surface areas was only about 100 F/g; however, compared to the first type of functionalized graphene sheets, the second type has higher capacitance retention at a large current density owing to its good conductive behaviors.

The conducting polymers, such as polypyrrole, polyaniline, and polythiophene are very attractive due to their easy synthesis, environmental stability, high and controllable conductivity, and potential for use in applications such as lightweight battery electrodes, capacitors, energy storage devices, electromagnetic shielding devices, anticorrosion coatings, and sensors. However, the conductive polymers are usually used with other polymers or fillers due to their poor mechanical properties [193].

Recently, graphene/conducting polymer composites have also become attractive as electrode materials due to their combined effect as a low dimensional organic conductor and graphene having a high surface area and excellent conductivity. In addition, several studies have been performed regarding the electrochemical properties of the metal-doped graphene whereas there has been limited work in the area of the metal-doped graphene/conducting polymers composites [194, 195].

A graphene-reinforced polypyrrole for a supercapacitor electrode was prepared and the effect of inducing the Ag nanoparticles on the electrochemical properties of the graphene/polypyrrole electrode was discussed using a three-electrode electrochemical cell [196]. The cyclic voltammetry measurements were carried out in 1.0 M of sulfuric acid at a scan rate of 50 mV/s at a voltage range of –0.2 to 1.0 V.

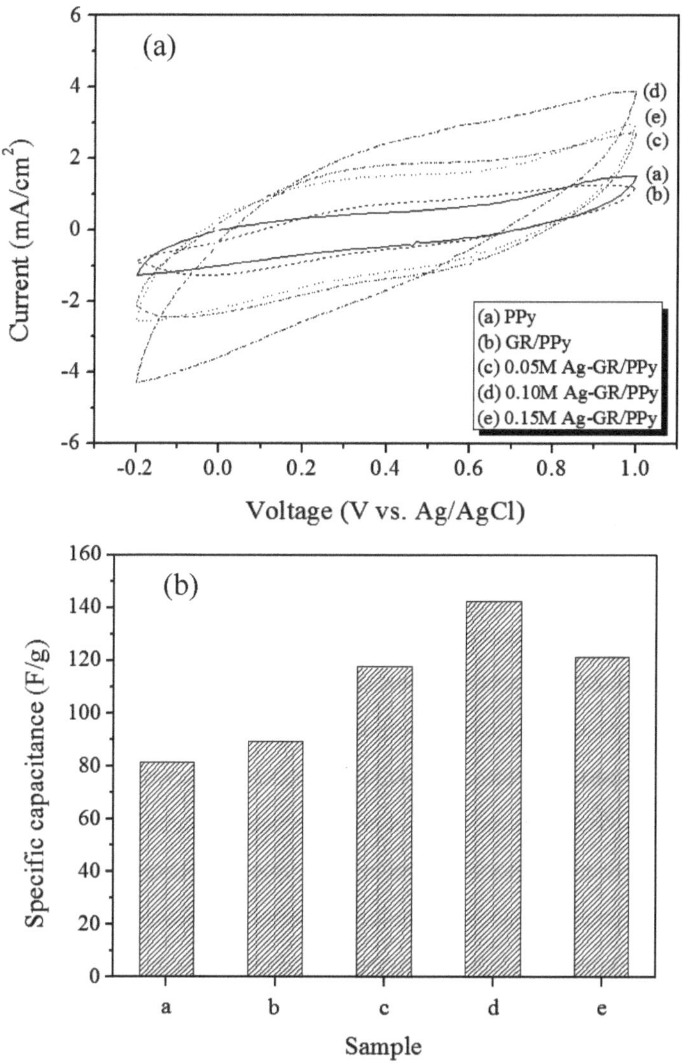

Fig. 3-38 Cyclic voltammetry (a) and specific capacitance (b) of Ag-graphene/polypyrrole nanocomposites as a function of Ag concentration (electrolyte: 1.0 M H_2SO_4, scan rate: 50 mV/s); (a) polypyrrole, (b) graphene/polypyrrole, (c) 0.05 M Ag- graphene/polypyrrole, (d) 0.10 M Ag-graphene/polypyrrole, and (e) 0.15 M Ag-graphene/polypyrrole.

Fig. 3-38(a) and Fig. 3-38(b) show the cyclic voltammograms and the specific capacitances of the polypyrrole, graphene/polypyrrole, and Ag-graphene/polypyrrole. To investigate the effect of the Ag deposition on graphene, the electrochemical properties were measured for polypyrrole nanocomposites using the pristine graphene and Ag-graphene. The results showed that graphene/

polypyrrole exhibits higher broad redox peaks and a more active surface area compared to pristine polypyrrole due to the synergistic effect between the graphene and the polypyrrole. In addition, compared to graphene/polypyrrole, the active surface area of the Ag-graphene/polypyrrole showed a remarkable increase upon an increase of the Ag concentration indicating the enhanced electrochemical properties stemming from the Ag nanoparticles deposited onto the graphene. Therefore, the Ag nanoparticles doped onto graphene serve as an efficient supporting and catalytic material to enhance the electrochemical properties of the Ag-graphene/polypyrrole. As shown in Fig. 3-38(b), the specific capacitance of the polypyrrole nanocomposites increased with the addition of graphene and Ag-graphene. The highest C_{spec} value was found in the 0.1 M Ag-graphene/polypyrrole nanocomposites. However, a further increase in the Ag concentration leads to a reduction of C_{spec} with 0.15 M Ag. This is attributable to the large particle size and the decrease in the effective surface area of the Ag nanoparticles [197].

3.4.1.4.2. Direct Methanol Fuel Cell

A direct methanol fuel cell (DMFC) is an electrochemical energy conversion device that converts the chemical energy of liquid methanol into electrical energy directly. Due to its unique advantages, such as a higher energy density, facile liquid fuel storage capability, and simpler system structure, the direct methanol fuel cell has been identified as one of the most promising power sources for the portable and clean mobile applications in the near future. Although promising, the direct methanol fuel cell technology is facing some challenging technical issues that need to be resolved before the widespread commercialization of this type of fuel cell [198].

Generally, platinum (Pt) or platinum alloy–based nanocluster impregnated carbon supports are the best electrocatalysts for the anodic and cathodic reactions in fuel cells. These electrocatalysts are produced mainly by impregnation and colloid methods [199]. Although the impregnation method is a simple procedure involving a Pt reduction process by hydrogen gas at a rather high temperature, the carbon-supported metal nanoparticle thus obtained has a large average size and a broad distribution. In comparison, the colloidal methods have the advantage of producing very small and homogeneously distributed carbon-supported metal nanoparticles; however, the methodologies are very complex.

Thus, the search for alternative routes to produce the carbon-supported metal nanoparticles is a critical theme in these research areas. The graphite nanofibers were used as support materials for platinum [200]. Pt nanoclusters supported by carbon materials were prepared by the electrochemical deposition by controlling the applied potential during the process of Pt reduction. The catalysts were characterized by stationary current voltage curves at room temperature in a three-electrode cell from 300 to 1100 mV (versus Ag/AgCl) with a scan rate

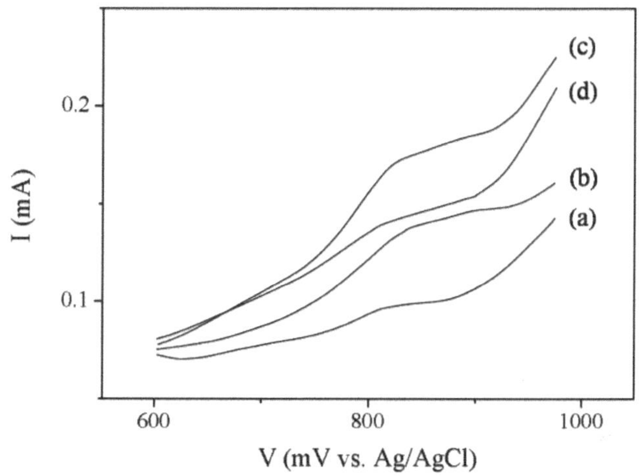

Fig. 3-39 I-V characteristic curves of Pt catalysts prepared by potential sweep method as a function of sweep times of (a) 6, (b) 12, (c) 18, and (d) 24 times (in 0.5 M H_2SO_4 + 0.5M CH_3OH).

of 20 mV/s. One molar methanol solution in 0.5 M H_2SO_4 was used as an electrolyte solution.

It was found that Pt nanoparticles can be successfully deposited on the carbon-supported electrodes by an electrochemical deposition method [200]. The average size and loading level of the Pt particles increased gradually with an increase of the potential sweep time. In terms of the Pt content, Pt obtained by the 24 sweep-times shows the highest value of 11 percent, which implies that the loading level increases gradually with the sweep time. Fig. 3-39 shows the electroactivity of the Pt catalyst supported on carbon supports according to the linear sweep voltammograms. The peak potentials and current density values are shown in Table 3-23. The anodic peaks for methanol oxidation are

Table 3-23 Peak Parameters and Specific Activities for Cyclic Voltammetry Results of Pt/Graphite Nanofibers Prepared by Potential Sweep Method as a Function of Sweep Time from 6 to 24

Sample	Peak potential (mV)	Peak current (μA)	Specific activity (mA/mg)
Pt-6 sweep times	856	85	145
Pt-12 sweep times	870	125	186
Pt-18 sweep times	872	166	204
Pt-24 sweep times	860	131	120

shown at 850 to 900 mV for each sample. The current density of the anodic peaks increased from 0.08 mA/cm^2 to 0.17 mA/cm^2, after increasing the sweep-time. However, a further increase to more than 18 sweep-times led to a decrease in the current density. The specific current densities of each sample are also shown in Table 3-23. The mass activity showed a maximum value of 204 mA/mg at 18 sweep-times. This value is similar ($\sim$ 200 mA/mg) to those in previous reports. This indicates that the electrocatalytic activity decays when the number of sweeps exceeds 18. Consequently, the electrocatalytic activity increases up to 18 sweep-times due to an increase in the Pt loading level. However, the electroactivity decays upon excessive sweep-times due to the large sizes of the Pt nanoclusters stemming from aggregation and resulting in a decrease of the interaction between the Pt nanoclusters and the electrolyte. As a result, the electroactivities of the Pt catalyst became dependent on the preparation conditions of the electrochemical deposition methods.

For direct methanol fuel cell, carbon blacks are the common support materials for electrocatalysts, along with other carbon materials, such as carbon or graphite nanofibers, and carbon nanotubes [201]. At present, the electrocatalysts generally are supported on high-surface-area carbon blacks with a high mesoporous distribution and graphite characteristics; Vulcan XC-72 carbon blacks are the most commonly used carbon support owing to their good compromise between the electronic conductivity and the Brunauer–Emmett–Teller (BET) surface area [202].

However, the effect of the surface characteristics of various carbon materials had not been fully studied to the best of our knowledge. The size and the loading efficiency of metal particles due to changes in the preparation method of the carbon-supported platinum catalysts have been investigated [203]. Furthermore, the acid/base treatment effects of carbon blacks on the nanostructure as well as the electroactivity of the carbon-supported platinum catalysts for direct methanol fuel cells have been studied.

Virgin carbon black 24 nm in size with a specific surface area of 112 m^2/g was used in our experiments. Neutral-treated carbon black base-treated carbon black and acid-treated carbon black were prepared by treating the virgin carbon black with 99.9 percent C_6H_6, 0.2 N KOH and 0.2 N H_3PO_4, respectively. Carbon-supported Pt nanoclusters were prepared by a modified polyol synthesis method, as reported previously [204]. The size and loading efficiencies of the Pt nanoclusters on the virgin carbon black, the neutral-treated carbon black, the base-treated carbon black and the acid-treated carbon black are listed in Table 3-24.

The average Pt nanoparticle sizes are summarized in Table 3-24. In the case of the Pt/virgin carbon black, the average Pt size is 3.01 nm of which the typical average range is 3 to 5 nm. In case of the Pt/base-treated carbon black, the average size decreases to the smallest size of 2.65 nm. The Pt/base-treated carbon black shows the highest value at 97 percent, whereas the Pt/acid-treated carbon black shows the lowest value at 38 percent, clearly indicating that Pt

Table 3-24 Size and Loading Efficiencies of Platinum Nanoclusters of Pt/Virgin Carbon Blacks (Pt/VCB), Pt/Neutral-treated Carbon Blacks (Pt/NCB), Pt/Base-treated Carbon Blacks (Pt/BCB), and Pt/Acid-treated Carbon Blacks (Pt/ACB)

Sample	Average size (nm)	Loading efficiency (%)	Loading efficiency (%)
Pt/VCB	3.07	90	88
Pt/NCB	3.74	93	89
Pt/BCB	2.65	97	94
Pt/ACB	–	38	36

deposition onto the carbon surface is enhanced when using base-treated carbon black. Consequently, a base treatment causes a decrease in the size of the Pt nanoclusters and produces the positive effect of enhancing the loading efficiency of Pt deposition onto the carbon surface. It is thought that the high loading efficiency of Pt is related to a probable Pt-reduction reaction with the carbon surface, thus manifesting the base characteristics. The small Pt nanocluster size of the Pt/base-treated carbon black may have originated with well-distributed base functional groups (e.g., the hydroxyl group) with which Pt reduction can occur. Upon chemical treatment of carbon blacks, we were able to obtain a smaller size and a higher loading efficiency of the Pt nanoclusters deposited onto carbon supports.

Fig. 3-40 shows the cyclic voltammetry results of the catalyst electrodes, in the electrolytes of 1.0 M methanol +0.5 M sulfuric acid. In all of the curves, at the forward scan, the current began to rise and shows an anodic peak I at a potential of about 540 mV. The curves of the Pt/base-treated carbon black and the Pt/neutral-treated carbon black began to rise at about 300 mV, which can be considered the onset potential 100 mV, lower than that of the Pt/virgin-treated carbon black. This implies that the Pt/base-treated carbon black and the Pt/neutral-treated carbon black show easier and faster electrochemical reactions due to the reduction of the overpotential during methanol oxidation. In the backward scan, another anodic peak II appeared at a potential of about 290 mV. This peak was attributed to the removal of the incompletely oxidized carbonaceous species (e.g., CO poisoning) that had formed in the forward scan [205]. For peak I, the current density followed this order: Pt/base-treated carbon black > Pt/neutral-treated carbon black > Pt/virgin carbon black > Pt/acid-treated carbon black. For peak II, the current density followed this order: Pt/neutral-treated carbon black > Pt/base-treated carbon black > Pt/virgin carbon black > Pt/acid-treated carbon black. Consequently, it is thought that the electroactivity was improved in the cases of Pt/base-treated carbon black

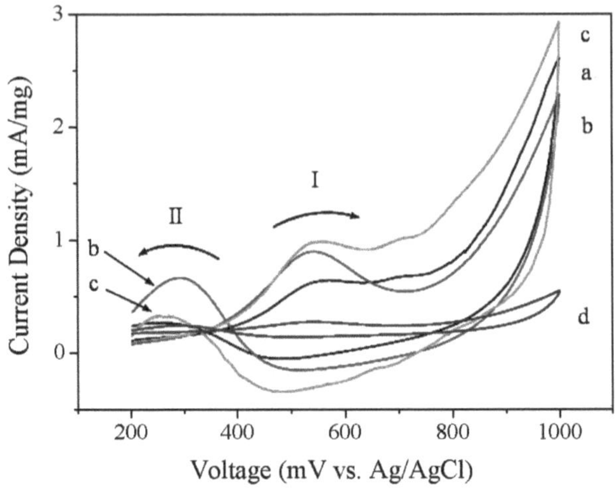

Fig. 3-40 CV graphs of: (a) Pt/virgin carbon blacks, (b) Pt/neutral-treated carbon blacks, (c) Pt/base-treated carbon blacks and (d) Pt/acid-treated carbon blacks, measured in 1.0 M CH_3OH + 0.5 M H_2SO_4.

and Pt/neutral-treated carbon black. Furthermore, Pt/base-treated carbon black showed a suppressing effect on CO poisoning, resulting from the rather small anodic peak II compared to the anodic peak I.

A heat treatment is a useful method of the activation of graphite nanofibers as the carbon supports for PtRu catalysts. Thus, the effect of a heat treatment on the efficiency of PtRu catalysts for direct methanol fuel cells was investigated [206]. Fig. 3-41 shows the transmission electron microscope (TEM) images of the PtRu/graphite nanofiber catalysts. The treatment led to the uniform dispersion of PtRu on the graphite nanofibers surfaces. In particular, the treatment at 800 °C led to uniformly deposited PtRu on the graphite nanofibers and showed the highest loading level and the smallest size. However, for the treatments at 900 °C and 1000 °C, the PtRu was slightly more aggregated and slightly less dispersed compared to the treatment at 800 °C. It is thought that the structure of the graphite nanofibers may have been decomposed by the high heat treatment temperature. These results also suggest that a heat treatment is an effective method of introducing PtRu into the catalysts.

Fig. 3-42 shows the electrocatalytic activity of the PtRu/graphite nanofiber catalysts. The catalysts deposited on the graphite nanofibers show a rather definite oxidation peak at about 640 mV *vs.* Ag/AgCl for the forward scan. The electrocatalytic activity increases as the heat treatment temperature increases to 800 °C and then decreases. The highest peak current density is obtained at the heat treatment of 800 °C. This is related to the fact that the smaller particle size and high loading level lead to an improvement of the electrocatalysts of the catalysts. Consequently, it is considered that a smaller

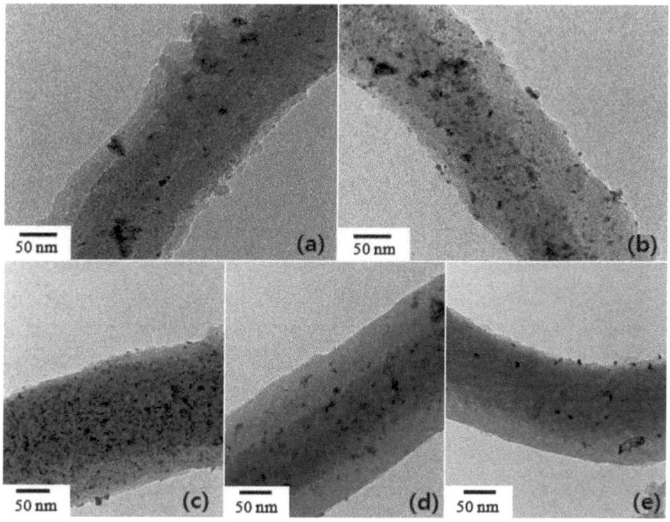

Fig. 3-41 TEM image of PtRu/graphite nanofiber catalysts: (a) as-received, (b) heat treatment at 700°C, (c) heat treatment at 800°C, (d) heat treatment at 900°C, and (e) heat treatment at 1000°C.

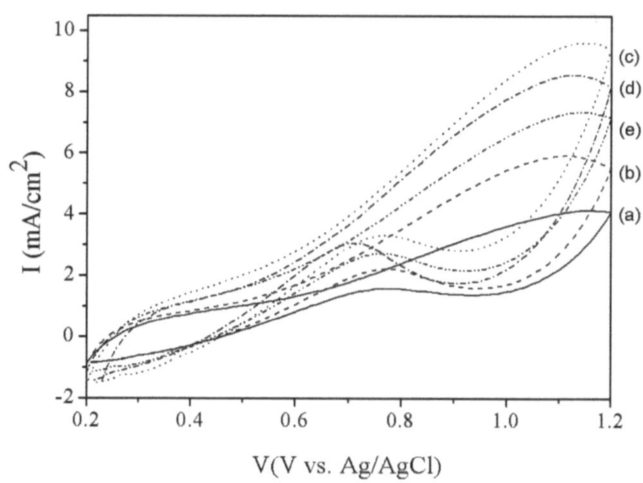

Fig. 3-42 Cyclic voltammograms of methanol oxidation obtained on PtRu/graphite nanofiber catalysts: (a) as-received, (b) heat treatment at 700°C, (c) heat treatment at 800°C, (d) heat treatment at 900°C, and (e) heat treatment at 1000°C (in 1 M CH_3OH in a 0:5 M H_2SO_4, with a scan rate of 20 mV/s at 25°C).

particle size and well-dispersed particles yield a larger available catalyst active surface, resulting in better electrocatalytic properties for methanol oxidation.

3.4.1.4.3. Lithium Battery

The lithium-ion batteries are considered the most promising energy storage technologies for mobile electronics, electric vehicles, and renewable energy systems operating on intermittent energy sources, such as wind and solar systems [207]. Usually, the lithium-ion batteries are connected in series or in parallel to deliver the user-specified electrical characteristics. The active cathode and anode materials in a typical lithium-ion battery are $LiCoO_2$ and graphite, respectively, electrically insulated by a porous polypropylene membrane separator and an intervening electrolyte of $LiPF_6$ in a mixture of organic solvents (e.g., ethylene carbonate and diethyl carbonate).

While discharging, the electrochemical potential difference between the anode and the cathode drives the Li^+ ions to move from the anode to the cathode internally through the electrolyte. The reverse process occurs during charging. The reversibility of the charge and discharge reactions at the electrodes determines the reversibility of the battery. The three most important performance indicators for lithium-ion batteries are the capacity, cycleability and rate capability, which are strongly dependent on the properties of the active electrode materials. For emerging applications, however, fundamental improvements are needed with regard to power, safety, the cycle life, and the cost [208].

A lithium-sulfur (Li–S) battery with a positive electrode made of sulfurbased compounds is one of the most promising approaches to satisfying these demands. The ambient temperature lithium–sulfur cell has attracted the attention of many research and development groups due to its inherently high specific energy [209]. This type of cell contains a sulfur cathode, a lithium metal anode and an electrolyte consisting of organic solvents and lithium salts. In terms of specific energy, lithium–sulfur batteries are the most attractive candidates among the batteries under development because lithium has a specific capacity of 3830 mAh/g while sulfur has a specific capacity of 1,675 mAh/g assuming that sulfur is fully reduced to sulfide ion (S^{2-}) during the discharge stage. Furthermore, the sulfur-based compounds are environment friendly and less costly than other materials.

Recently, the effects of electrolytes containing mixed imidazolium salts and lithium salts to improve the discharge performance and cycle-life characteristics for lithium-sulfur cells were studied [210]. Elemental sulfur coated onto an aluminum current collector was used as a positive electrode with the negative electrode being a lithium metal foil, 0.5 M of $LiSO_3CF_3$ (LiTf: lithium triflate) or 0.5 M of $LiPF_6$ in a dimethoxyethane/dioxolane (8/2, v/v) solvent was used as an electrolyte. To this electrolyte, as imidazolium salts, (1-ethyl-3-methyl-imidazolium, bis (perfluoroethyl sulfonyl) imide was introduced.

To examine the effect of the imidazolium salts, on the discharge performance of Li–S cells, the salts were introduced to 0.5 M LiTf in a dimethoxyethane/

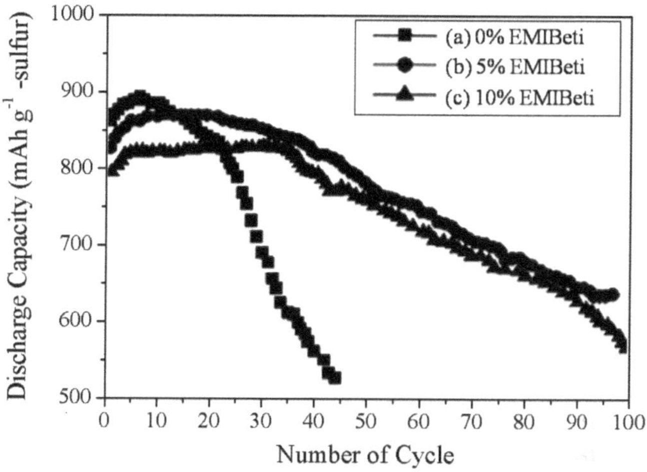

Fig. 3-43 Discharge capacity vs. cycle number for Li–S cells using organic electrolytes with salts of: (a) 0.5 M $LiSO_3CF_3$; (b) 0.5 M $LiSO_3CF_3$ + 5% (1-ethyl-3-methyl-imidazolium bis (perfluoroethyl sulfonyl) imide; (c) 0.5 M $LiSO_3CF_3$ + 10% (1-ethyl-3-methyl-imidazolium bis(perfluoroethyl sulfonyl) imide.

dioxolane (8/2, v/v) solution. The discharge capacity versus the cycle number relationship for cells containing different amounts of (1-ethyl-3-methyl-imidazolium bis (perfluoroethyl sulfonyl) imide is given in Fig. 3-43. Each test cell shows a gradual increase in the capacity up to around the 10th cycle. This probably can be explained by the fact that the electrochemical redox reaction of sulfur includes an activation step due to a gradual phase change from a solid state to the dissolved polysulfide state. On the first cycle, a test cell without (1-ethyl-3-methyl-imidazolium bis (perfluoroethyl sulfonyl) imide salt exhibited a discharge capacity of 850 mAh/g-sulfur. This value is approximately 50 percent of the theoretical capacity and is quite similar to that repeated elsewhere. During the early stages of cycling, the discharge capacities of cells containing 5 or 10 vol. percent (1-ethyl-3-methyl-imidazolium bis (perfluoroethyl sulfonyl) imide salt were slightly lower than that of a cell without (1-ethyl-3-methyl-imidazolium bis (perfluoroethyl sulfonyl) imide. On the other hand, the capacity of a cell without (1-ethyl-3-methyl-imidazolium bis (perfluoroethyl sulfonyl) imide rapidly decreased after the 30th cycle, whereas the capacities of cells with 5 or 10 percent (1-ethyl-3-methyl-imidazolium bis (perfluoroethyl sulfonyl), imide slowly decreased at approximately the same rate up to the 100th cycle. Thus, the stability of the discharge capacity is greatly improved with the addition of 5 or 10 percent (1-ethyl-3-methyl-imidazolium bis (perfluoroethyl sulfonyl) imide.

The carbonatebased electrolyte systems have been proven to show extremely good performance in the lithium ion batteries. However, in the lithium-sulfur batteries, the carbonate-based electrolyte systems lead to deterioration of the battery performance [211]. This deterioration results from the high instability of the electrochemical reaction of the polysulfide in a carbonatebased electrolyte. An electrolyte, applicable to the lithium-sulfur batteries must be chemically stable with the polysulfide so as to have high polysulfide solubility and thus be compatible with a lithium metal anode.

The organic salts with melting points below room temperature also known as ionic liquids have been the main focus of many recent scientific investigations due to their unique physical and chemical properties. They are non-volatile, non-flammable, and miscible with a number of organic solvents. Moreover, they have high thermal stability. From an electrochemical point of view, ionic liquids offering high ionic conductivity and a wide potential window are excellent electrolytes [212, 213].

The effect of an imidazolium cation on the cycle-life characteristics of Li–S cells in dimethoxyethane/dioxolane (4/1)-mixed solvents in which two types of salt systems were used, (1) the conventional lithium salts and (2) the mixed lithium/imidazolium salts, were investigated [214]. They introduced the following three types of imidazolium salts with the same anion into mixed-salt systems to evaluate the effect of the imidazolium cation on the cycle-life characteristics of the Li–S cells: 1-ethyl-3-methylimidazolium bis (trifluoromethylsulfonyl) imide, 1-butyl-3-methyl-imidazolium bis (trifluoromethylsulfonyl) imide, and 1,2-dimethyl-3-propylimidazolium bis (trifluoromethylsulfonyl) imide.

To analyze the cation influence of the imidazolium ion on the discharge performance, lithium salt/imidazolium salt and three types of imidazolium salts, 1-ethyl-3-methylimidazolium bis (trifluoromethylsulfonyl) imide 1-butyl-3-methyl-imidazolium bis (trifluoromethylsulfonyl) imide, and 1,2-dimethyl-3-propylimidazolium bis (trifluoromethylsulfony) imide with the same anion were investigated, as shown in Fig. 3-44. The test cells containing 10 vol percent 1-ethyl-3-methylimidazolium bis (trifluoromethylsulfonyl) imide and 10 vol percent 1-butyl-3-methyl-imidazolium bis (trifluoromethylsulfonyl) imide showed a better stability of discharge capacity relative to the cycle number compared to the cells using lithium salt/imidazolium salt. The test cell using 1-ethyl-3-methylimidazolium bis (trifluoromethylsulfonyl) imide showed the highest capacity at 721 mAh/g upon the 100th cycle, whereas the test cell using 1-butyl-3-methyl-imidazolium bis (trifluoromethyl sulfonyl) imide showed a slightly lower capacity of 642 mAh/g. Comparably, the test cell using lithium salt/imidazolium salt showed a capacity of 534 mAh/g and in stark contrast, the test cell using 1,2-dimethyl-3-propylimidazolium bis (trifluoromethylsulfony) imide showed an abrupt decline in its capacity at 200 mAh/g at only the 10th cycle. It was concluded that the capacity at the 100th cycle had been enhanced by the addition of 1-ethyl-3-methylimidazolium bis (trifluoromethylsulfonyl) imide and 1-butyl-3-methyl-imidazolium bis (trifluoromethylsulfonyl) imide to the electrolytes.

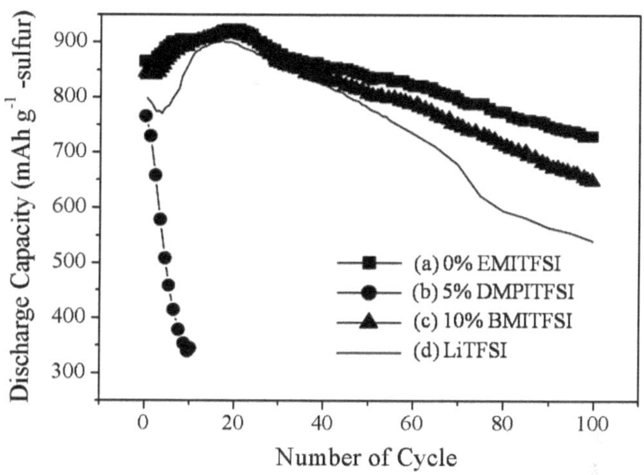

Fig. 3-44 Effect of cations of imidazolium salts on the cycle life characteristics of Li–S cells employing mixed electrolytes of both 90 vol.% of dimethoxyethane/dioxolane (4/1, v/v) -0.5 M $LiSO_3CF_3$ systems and 10 vol.% of the following imidazolium salts: (a) 1-ethyl-3-methylimidazolium bis(trifluoromethylsulfonyl)imide, (b) 1-butyl-3-methyl-imidazolium bis(trifluoromethylsulfonyl)imide, (c) 1,2-dimethyl-3-propylimidazolium bis(trifluoromethylsulfony)imide.

The improvement of the cycle characteristics upon the addition of imidazolium cation salts can be accounted for with reference to the following factors. First, it is considered that the electrochemical reaction of polysulfide may be influenced by the introduction of the large volume-size imidazolium cations. Second, the charge transfer resistance at the surface of the carbon-based electrode could be decreased by a drop in the deactivation of carbon, resulting from a side reaction of polysulfide. Thirdly, this additional cation species could assist with the formation of a stable protection layer on the Li anode surface. In order to achieve a long cycle life in Li–S cells, the stable Li anode interfacial properties are absolutely necessary.

3.4.2. Self-Assembling Systems

This section represents the interactions of small molecular aggregates, such as micelles, bilayers, and vesicles, which form in aqueous solutions by spontaneous self-assembly of amphiphilic molecules Fig. 3-45. These structures are formed from conventional colloidal particles: unlike solid particles or rigid macromolecules, which are soft and flexible, i.e., fluid-like. It is because the forces that the amphiphilic molecules hold in the micelles are not strong covalent or ionic bonds but arise from the weaker van der Waals, hydrophobic, hydrogen bonding, and electrostatic interactions. Therefore, in the solution conditions, an aqueous suspension of micelles or vesicles can be changed by the electrolyte concentration or the pH. It affects the interactions between the

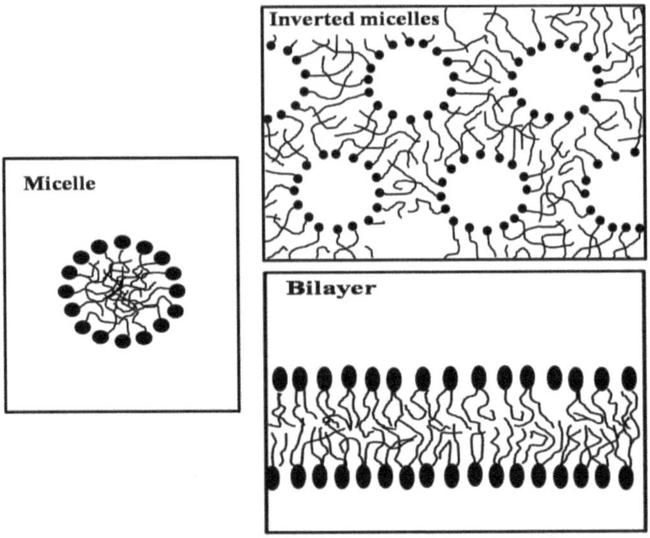

Fig. 3-45 Classification of small molecular aggregates by self-assembly [215].

aggregates and the intermolecular forces within each aggregate, resulting in the modification of the size and shape of the structures.

In this section, the thermodynamic and physical principles of self-assembly in general and of amphiphilic molecules, such as surfactants and lipids in particular will be discussed with related applications. In addition, the basic equation of self-assembly in general statistical thermodynamic terms will be concerned and then the investigation of the relevant intermolecular interactions will be considered using different amphiphiles.

3.4.2.1. Thermodynamic Equations of Self-assembly

In the literature, this subject is often confusing the most rigorous treatment has used the theory of Hall and Pethica [216] based on Hill's classic book [217] on the small-system thermodynamics. A more simplified approach was later extended to larger lipid aggregates, such as bilayers, vesicles, other micellar phases, and microemulsion droplets by many researchers [218–226].

The equilibrium thermodynamics is applied in a system of molecules that form aggregated structures in a solution, as shown in Fig. 3-46, the chemical potential of all identical molecules in different aggregates is identical and expressed as

$$\mu = \mu_1^0 + kT\ \log X_1\,(\text{monomer})\ (\text{monomers}) \tag{3-68}$$

$$= \mu_2^0 + \frac{1}{2}kT\ \log\frac{1}{2}X_2\ (\text{dimer}) \tag{3-69}$$

$$= \mu_3^0 + \frac{1}{3}kT\ \log\frac{1}{3}X_3\ (\text{trimer})\quad \ldots. \tag{3-70}$$

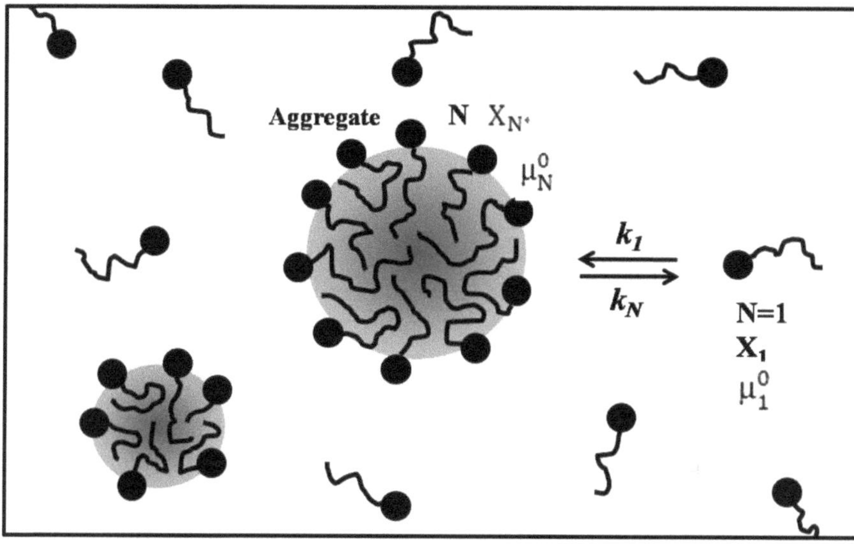

Fig. 3-46 Association of N monomers into an aggregate (e.g., a micelle). The mean lifetime of an amphiphilic molecule in a small micelle is very short, typically $10^{-5} \sim 10^{-3}$ s.

or

$$\mu_N = \mu_N^0 + \frac{kT}{N}\log\left(\frac{X_N}{n}\right) \tag{3-71}$$

$$= \text{constant}, N = 1, 2, 3, \ldots, \tag{3-72}$$

where μ_N is the mean chemical potential of a molecule in an aggregate of aggregation number N, μ_N^0 the standard part of the chemical potential in aggregates of aggregation number N, and X_N the concentration of molecules in aggregates of number N.

Eq. (3-72) can also be derived by the well-known law of mass action [227], as follows: referring to Fig. 3-46, this can be written as

$$\text{rate of association} = k_1 X_1^N \tag{3-73}$$

$$\text{rate of dissociation} = k_1(X_1/N) \tag{3-74}$$

where

$$K = k_1/k_N = exp[-N(\mu_N^0 - \mu_1^0)/kT] \tag{3-75}$$

is the ratio of the two reaction rates. These combine to give Eq. (3-72), which can also be expressed

$$X_N = N\{(X_M/M)exp[M(\mu_M^0 - \mu_N^0)]\}^{N/M} \tag{3-76}$$

and, putting M = 1,

$$X_N = N\{X_1 exp[M(\mu_1^0 - \mu_N^0)/kT]\}^N \tag{3-77}$$

where M is any arbitrary reference state of aggregates with the aggregation number M from the Eqs. (3-76) and (3-77), the conservation relationship for the total solute concentration C is

$$C = X_1 + X_2 + X_3 + \ldots . C \tag{3-78}$$

$$= \sum_{N=1}^{\infty} X_N \tag{3-79}$$

From the free energies (μ_1^0, μ_{1N}^0), the dimensionless concentrations C and X_N can be demonstrated by the volume fraction or mole fraction units. In general, C and X_N can never exceed unity. Eq. (3-75) is used for assuming ideal mixing and is restricted to dilute systems.

3.4.2.2. *Formation of Different Aggregates*

Generally, a difference in the cohesive energies between the molecules in the aggregated and the dispersed states is required for the formation of the aggregates. If the molecules, in the differently sized aggregates, have the same interaction with their surroundings, the value of μ_N^0 will remain constant in different aggregates and Eqs. (3-76) and (3-77) then become

$$X_N = NX_1^N \ (for\ \mu_1^0 = \mu_2^0 = \mu_3^0 = \cdots = \mu_N^0) \tag{3-80}$$

In given condition ($X_1 < 1$ and, $X_N \ll X_1$), most of the molecules will be in the monomer state (N = 1). If μ_N^0 increases with increasing N, Eqs. (3-76) and (3-77) show that the formation of large aggregates becomes less probable.

The condition for the formation of large stable aggregates is that $\mu_N^0 < \mu_1^0$ for some value of N, for example, when μ_N^0 gradually decreases with increasing N or μ_N^0 has a minimum value at some finite value of N. In addition, the exact functional variation of μ_N^0 with N determines many physical properties of the aggregates, such as their mean size and polydispersity. Furthermore, because this variation can be complex, it is clear that a number of structurally different populations can coexist within a single phase in the thermodynamic equilibrium with each other. The functional forms of μ_N^0 for some simple structures using Eqs. (3-75) to (3-77) will consider and investigate their physical properties.

3.4.2.2.1. One-dimensional Aggregate

As mentioned above, aggregates will form by the decrease of μ_N^0 with N. The dependence of μ_N^0 on N can be easily determined by the geometrical shape of the aggregate. Generally, a suspension of rod-like aggregates is made up of linear chains of monomer units in equilibrium with monomers in a solution. Here, αkT is the bonding energy of the monomer-monomer in the aggregate relative to the isolated monomers in the solution, as shown in Fig. 3-47. The total interaction free energy ($N\mu_N^0$) of an aggregate of monomers is writed as follows:

$$N\mu_N^0 = -(N-1)\alpha kT \tag{3-81}$$

that is

$$\mu_N^0 = -(1-1/N)\alpha kT \tag{3-82}$$

$$= \mu_\infty^0 + \alpha kT/N \tag{3-83}$$

With increasing N, the mean free energy (μ_N^0) decreases asymptotically towards μ_∞^0 the bulk energy of a molecule in an infinite aggregate. A similar expression for μ_N^0 is obtained for rod-like structure [228].

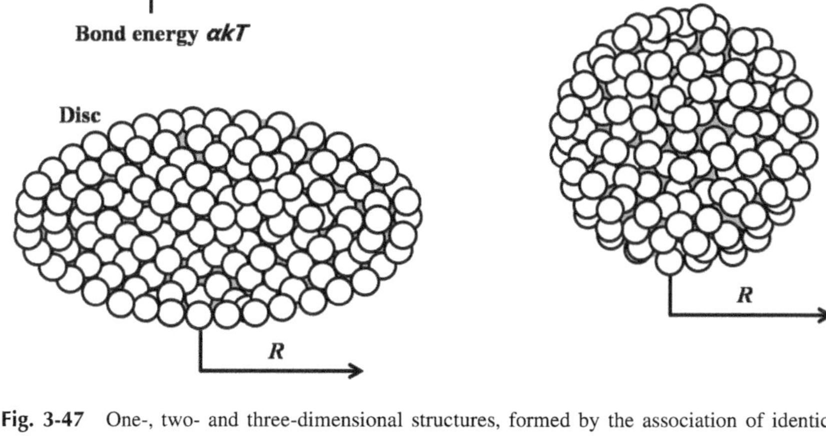

Fig. 3-47 One-, two- and three-dimensional structures, formed by the association of identical monomer units in solution.

3.4.2.2.2. Two-dimensional Aggregates

In disc-like or sheet-like aggregates, as shown in Fig. 3-47, the number N of molecules per disc is proportional to the area πR^2, whereas the number of unbounded molecules in the rim is proportional to the circumference $2\pi R$ and $N^{1/2}$. The mean free energy per molecule of the aggregate is represented as follows:

$$\mu_N^0 = \mu_\infty^0 + \alpha kT/N^{1/2} \tag{3-84}$$

where again α is some constant characteristic of the monomer-monomer and monomer-solvent interaction [229].

3.4.2.2.3. Three-dimensional Aggregates

For spherical aggregates in a solvent, as shown in Fig. 3-47, N is proportional to the volume $\frac{4}{3}\pi R^3$, whereas the number of unbounded surface molecules is proportional to $4\pi R^2$ and $N^{2/3}$. Therefore, μ_∞^0 can be expressed as follows:

$$\mu_N^0 = \mu_\infty^0 + \alpha kT/N^{1/3} \tag{3-85}$$

For example, in the association of small hydrocarbon molecules in water, if v is the volume per molecule, $N = 4\pi R^3/3v$. The free energy of the spherical aggregates is given by $N\mu_\infty^0 + 4\pi R^2\gamma$, where μ_∞^0 is the bulk energy per molecule and γ the interfacial free energy per unit area.

$$\mu_N^0 = \mu_\infty^0 + \frac{4\pi R^2\gamma}{N} \tag{3-86}$$

$$= \mu_\infty^0 + \frac{4\pi\gamma(3v/4\pi)^{2/3}}{N^{1/3}} \tag{3-87}$$

$$= \mu_\infty^0 + \frac{\alpha kT}{N^{1/3}} \tag{3-88}$$

where

$$\alpha = \frac{4\pi\gamma(3v/4\pi)^{2/3}}{kT} \tag{3-89}$$

$$= \frac{4\pi r^2\gamma}{kT} \tag{3-90}$$

r is being the effective radius of a molecule.

As mentioned above, for the simplest shapes, such as rods, sheets, and spheres, the interaction free energy of the molecules can be represented as

$$\mu_N^0 = \mu_\infty^0 + \frac{\alpha k T}{N^p} \tag{3-91}$$

where α is a positive constant that depends on the strength of the intermolecular interactions and p is a number that depends on the shape or dimensionality of the aggregates.

As shown above, Eq. (3-91) can be applied to numerous micellar structures and spherical vesicles. In general, it noted that for these types of structures, μ_N^0 decreases progressively with N. This is a necessary condition for aggregate formation [230].

3.4.2.3. Critical Micelle Concentration

In the general functional form of μ_N^0 of Eq. (3-91), the aggregates are formed above certain concentration of the monomer. The incorporation of Eq. (3-91) into the fundamental equations Eqs. (3-76) to (3-79) of self-assembly can provide very important conclusion. First, the monomer concentrations are as follows:

$$X_N = N\{X_1 \exp[(\mu_1^0 - \mu_N^0)/kT]\}^N \tag{3-92}$$

$$= N\{X_1 exp[(\alpha(1 - 1/N^p)]\}^N \approx N[X_1 e^\alpha]^N \tag{3-93}$$

For sufficiently low monomer concentrations X_1 such that $X_1 \exp[(\mu_1^0 - \mu_N^0)/kT]$ or $X_1 e^\alpha$ is much less than unity ($X_1 > X_2 > X_3 >$ for all α). Therefore, at low concentrations, most of the molecules in the solution will be the isolated monomers ($X_1 \approx C$), as shown in Fig. 3-48. However, because X_N can never exceed unity, it is clear that once X_1 approaches $\exp[(\mu_1^0 - \mu_N^0)/kT]$ or $e^{-\alpha}$ it can increase no further. The monomer concentration $(X_1)_{crit}$ at which this occurs can be regarded as the critical aggregation concentration though using more conventional term i.e., critical micelle concentration (CMC) to express the critical concentration of all self-assembled structures.

$$(X_1)_{crit} = CMC \approx exp[-(\mu_1^0 - \mu_N^0)/kT] \tag{3-94}$$

If μ_N^0 is determined by Eq. (3-91), the monomer concentration for the critical micelle concentration is

$$(X_1)_{crit} = CMC \approx e^{-\alpha} \quad \text{for all p} \tag{3-95}$$

The above two equations define the monomer concentration and further addition of monomers lead to the formation of more aggregates while leaving the monomer concentration essentially unchanged at the critical micelle concentration value, as shown in Fig. 3-48.

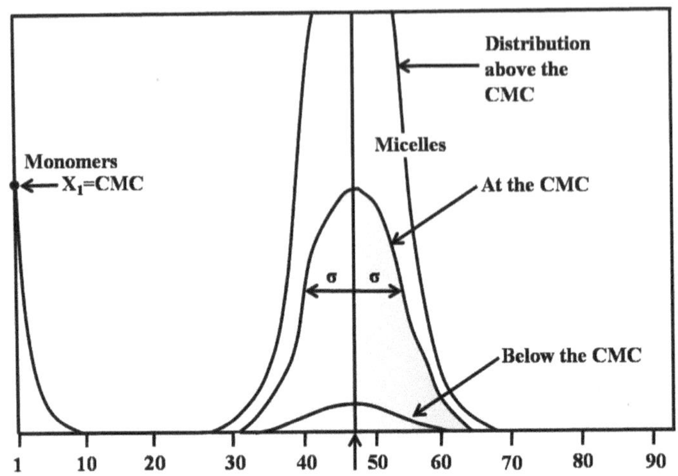

Fig. 3-48 Monomer and aggregate concentrations, as a function of total concentration.

3.4.2.4. Phase Separation Versus Micellization

The nature of the aggregates depends very much on their shape. For simple disc-like and spherical aggregates, Eq. (3-92) becomes

$$X_N = N[X_1 e^{\alpha}]^N e^{-\alpha N1/2} \quad \text{for discs } (p = 1/2) \tag{3-96}$$

$$X_N = N[X_1 e^{\alpha}]^N e^{-\alpha N2/3} \quad \text{for spheres } (p = 1/3) \tag{3-97}$$

Over the critical micelle concentration ($X_1 e^{\alpha} \approx 1$), the two equations above can be approximated by $X_N \approx Ne^{-\alpha N1/2}$and $X_N \approx Ne^{-\alpha N2/3}$, respectively. At this point, for any reasonable positive value of α which is higher than 1, these equations show that there will be very few aggregates of any appreciable size ($N > 5$).

For discs and spheres, there is a phase transition to a separate phase strictly to an aggregate of infinite size ($N \rightarrow \infty$) at the critical micelle concentration. Israelachvili *et al.* [231] showed that a transition to the large macroscopic aggregates occurs at $p < 1$ in Eq. (3-91). This applies, quite generally, to all planar or disc-like aggregates composed of identical molecules. It is for this reason that finite crystalline sheets, one-component lipid bilayers, and even biological membranes with exposed edges are rarely found floating around in a solution. Above the critical micelle concentration, infinite bilayers are formed from lipid monomers.

Moreover, for simple spherical structures, the association of oil molecules in water can be explained as follow: after adding oil to water, the molecules disperse as monomers up to the critical concentration given by Eq. (3-95) and Eq. (3-90),

$$(X_1)_{crit} \approx e^{-\alpha} \approx e^{-4\pi r^2 \gamma/kT} \tag{3-98}$$

above which they will separate out into an oil phase, indicating a large spherical aggregates.

In a system of two immiscible liquids, the solubility of a solute in a solvent (α) can be represented by the free energy of transferring a solute molecule from the solute into the solvent phase. For example, the solubility of hydrocarbons in water can be examined from the free energy of transferring the alkyl molecule from the hydrocarbon into water. The hydrophobic energy of transferring the alkyl molecule from water into hydrocarbon or into micelles can be also analyzed in a similar phenomenon.

This applies only to pure alkane chains being transferred from water into a pure hydrocarbon phase. In the case of the surfactant molecules being transferred into micelles or bilayers, the hydrophobic energy increment is significantly lower, ranging from 1.7 to 2.8 per CH_2 group. The reduced hydrophobicity of an amphiphilic chain compared to that of a pure alkane chain is believed to be due to the proximity of the hydrophilic headgroup and to the higher chain ordering of chains within the micelles, which act to reduce the energy even more. The above range of values means that typical micellar-critical micelle concentrations fall by 0.3 to 0.5 per CH_2 group added to the surfactant chain.

The important difference between the alkanes and amphiphilic molecules is not so much in their solubility or critical micelle concentration values but in the ability of the amphiphiles to assemble into structures in which μ_N^0 reaches a minimum or constant value at some finite value of N. It is for this reason that the aggregates formed are phase separation aggregates but stemming from micellization. Why and how the amphiphilic molecules do this will be investigated fully in the following chapter.

3.4.2.5. Applications

Given the range of micro/nanocapsule functions needed for active material encapsulation, a large number of processes, including dispersion, spray-drying, chilling, fluidized bed coating, granulation, extrusion, coacervation, polymerization and self-assembly are practiced by the industry. Encapsulation in the liposomes and niosomes is also popular [232].

Other less conventional encapsulation processes include the adsorption of active materials on inorganic cement particles, porous nanoparticles of silica, zeolite, and the clay of inorganic particles. The encapsulation process is simply a matter of adsorption and capillary sucking of the active materials by the pores. The hybrid capsules including both organic and inorganic components and encapsulation combining two or more of the processes mentioned above are also employed to produce the cost-effective encapsulated products [233].

Active materials may be introduced into the encapsulation process as oil in water emulsion alone, in combination with other active materials, or with a hydrophobic carrier. They are encapsulated, mostly by coacervation or spray-drying. The inorganic salts (e.g., Na_2CO_3) and crystals (e.g., SmCo and $Nd_2Fe_{14}B$) are also encapsulated for use in the antiperspirant and deodorant compositions or for specific applications, such as the treatment of sensitive tissues wounds and hair.

The aqueous solubility of the active materials is an important parameter for the manufacture of the capsule delivery systems. The high aqueous solubility means active material loss due to partitioning into water during the dispersion of active material droplets in the continuous aqueous phase. Therefore, there is a limit on the maximum concentration of polar ingredients that can be included in the delivery systems. A typical composition for producing spray-dried capsules is as follows: 10 percent active material, 40 percent modified corn starch and 50 percent water.

The reservoir capsules have a distinct protective shell enclosing a distinct single active agent core in the center, whereas in the matrix (monolithic) capsules, the active agent is dispersed throughout the particle matrix (or pores).

Generally, the formulations of the micro- or nanocapsules are covered under 'reservoir capsules', and those described as spheres (or beads) are covered under 'matrix capsules'. In practice, however, this distinction may not always be clear, and capsules often have morphologies that are an intermediate between a matrix and a reservoir, in a multicore or a multi-walled structure. In some cases, the active agent and a binder may actually be coated onto the surface of the inert core particles [234]

3.4.2.5.1. Drug-loaded Polymer Nanocapsules by Emulsion Polymerization

During the last decade, polymer carriers have become important in the field of pharmaceutical research. Success in this area requires very small and stable nanoparticles of approximately 200 nm that can incorporate hydrophobic components for a drug delivery system. To obtain such nanoparticles, the emulsion technology has attracted much interest in recent years owing to its great practical importance in terms of its drug delivery potential and interesting physical properties [235].

In the past 50 years, several particle nucleation mechanisms have been proposed for conventional emulsions, generally referred to as macroemulsions. This has led to the production of thermodynamically unstable droplets ($1 \sim 10\ \mu m$) that are easy to separate out. On the other hand, microemulsions consist of spontaneously formed droplets (~ 50 nm), that are thermodynamically stable but that generally require high surfactant and co-stabilizer ratios [236].

In the conventional emulsion in Harkins–Smith–Ewart theory, particle nucleation occurs with the capture of radicals by the monomer-swollen micelles and latex particles are then generated by propagation of the monomer [237]. However, if the droplet size is sufficiently small, the monomer droplet can be the predominant place for particle nucleation. This is due to the fact that a very large interfacial area between the monomer droplets and water is effective in comparison with micelles for the adsorption of radicals produced in the aqueous phase. This method is referred to as mini-emulsion. It is classically defined as aqueous dispersion of relatively stable oil droplets within a size range 50 to 500 nm and is composed of the monomer, water, surfactant and hydrophobic materials [238]. In addition, a mini-emulsion produced with shearing systems, i.e., a high-pressure homogenizer or an ultrasonicator, is rather stable with respect to the aggregation, sedimentation and greater shear stability, compared to conventional emulsions.

Mini-emulsion polymerization of vinyl acetate is well suited to obtain nanoparticles for the encapsulation of the hydrophobic drugs. Drug-loaded poly (vinylacetate) nanoparticles were prepared by mini-emulsion polymerization under the presence of a hydrophobe [239]. They are also discussed in terms of an application of poly (vinyl acetate) as a drug delivery carrier through drug release from the drug-loaded poly (vinylacetate) nanoparticles.

For a comparison of emulsion results depending on the polymerization method, poly (vinylacetate) latex was prepared via conventional emulsion and the hydrophobe-involved mini-emulsion using poly (vinyl acetate) as a hydrophobe. The characteristics of the prepared poly (vinylacetate) latexes are summarized in Table 3-25.

A mini-emulsion, shows superior shelf stability (1848 h), implying the complete phase separation of latexes, as compared to that in a conventional emulsion (3 h). Moreover, a mini-emulsion leads to a smaller particle size than a conventional emulsion. This results because the addition of the hydrophobe suppresses mass exchange between the different oil droplets and causes a decrease in the surface tension of the mini-emulsion (25 mJ/m^2) as compared to the conventional emulsion (28 mJ/m^2). It results in the prevention of Ostwald ripening among the monomer droplets, allowing smaller and more stable particles to form [240].

Table 3-25 Shelf Stability of Conventional Emulsion and Miniemulsion Polymerization

	Particle size before phase separation	Particle size after phase separation (μm)	Shelf life (h)
Conventional emulsion	12 μm	21.6	3
Miniemulsion	105 nm	16.7	1848

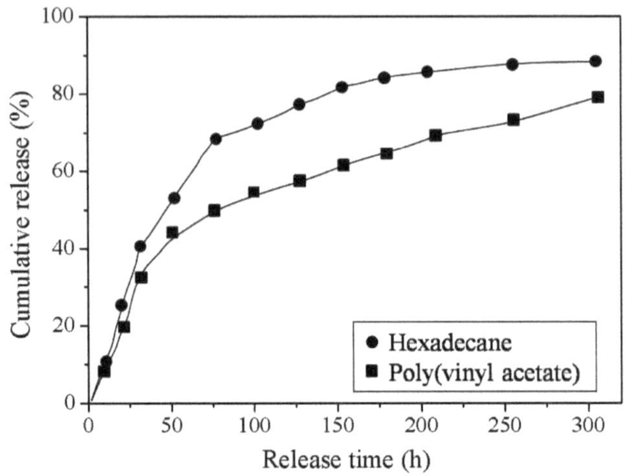

Fig. 3-49 Release profile of drug from poly (vinyl acetate) nanoparticles, using poly(vinyl acetate) and hexadecane as hydrophobe, in phosphate buffer solution (pH 6.8).

Fig. 3-49 shows an example of drug release from nanoparticles in a phosphate buffer solution containing artificial gastric juice at 37 °C. This figure shows the typical release behavior of poly (vinyl acetate) latex obtained using the different types of hydrophobes. According to the release profile of the drug from the nanoparticles, the release amount differs depending on the type hydrophobe. The cumulative release of poly (vinyl acetate) used as a hydrophobe is faster than that of hexadecane when used as a hydrophobe. This can be attributed to the fact that hexadecane at the nanoparticle surface prevents the release of a drug derived from the lower solubility of hexadecane compared to poly (vinyl acetate). The initial higher release rate, may be mainly the result of the large surface area and desorption of the drug from the nanoparticle surface. The lower release rate may be related to the degradation rate of the poly(vinyl acetate), as the drug is encapsulated into the poly,(vinyl acetate) nanoparticles during the polymerization process.

3.4.2.5.2. Drug-loaded Polymer Microcapsules by Solvent Evaporation Method

A number of microencapsulation techniques have been studied in various fields concerning the protection and controlled release of active materials from microcapsules or microspheres. A typical controlled-release device, i.e., a microcapsule is an important tool for the controlled release of chemical reactants. It can consist of active liquid or solid core materials and a protective wall. It has been widely used to create carbonless copying paper and in the preservation or targeted delivery of active materials [241].

There are several methods that can be used to make microcapsules from biodegradable polymers. Emulsion solvent evaporation and phase separation are two main processes used to prepare microcapsules. The solvent evaporation procedure is an aqueous system process, while the phase separation procedure is a non-aqueous process. It follows that the solvent evaporation method is suitable for the water-insoluble drugs, whereas the phase separation method is preferable for the water-soluble drugs. Other methods, such as spray-drying, are also used to microencapsulate some drugs [242].

Since the mid-1970s, the biodegradable polymers have been studied as matrix materials of microcapsules in anticipation that they will facilitate the production of effective drug delivery devices and find applications in controlled drug delivery. The biodegradable polymers can be decomposed into non-toxic and low-molecular-weight species after the release of the core material and then metabolized by the organism. Among the biodegradable polymers, poly (ε-caprolactone) (PCL), having a low T_g of approximately –60 °C, is very attractive due to its feasible permeability and biocompatibility. Many studies have been carried out, regarding the biodegradation of this polymer [243].

In the drug delivery systems, Eudragit RS 100 polymers, developed by RoHM Pharma GmbH, are the most widely used materials for the time-controlled delivery of drugs. Eudragit RS 100 is commonly used for the coating of tablets and the preparation of controlled-release oral pharmaceutical forms [244]. It is a poly (ethyl acrylate, methyl-methacrylate, and chlorotrimethy ammonium ethylmethacrylate) co-polymer containing an amount of quaternary ammonium groups ranging between 4 percent and 8 percent. It is insoluble at physiological pH levels but undergoes swelling in water, thus becoming permeable to water. Therefore, Eudragit RS 100 is commonly used for the enteric coating of the tablets and in the preparation of controlled-release drug forms. It is a good material for controlled oral administration of drugs [245].

In a previous work [246], the porous poly(ε-caprolactone)/Eudragit RS 100 microcapsules containing tulobuterol base as a model drug were prepared by a solvent evaporation method and the effect of the quaternary ammonium groups of Eudragit RS 100 on the release behaviors of the microcapsules was investigated. It was found that poly (ε-caprolactone) and Eudragit RS 100 microcapsules containing tulobuterol were successfully formed by encapsulation via the solvent evaporation method. Tulobuterol dispersed in the water phase was captured by the polymer matrix, and the hardened microcapsules were obtained after the solvent completely evaporated. The microcapsules thus formed have spherical forms and smooth surfaces, as determined by the SEM analysis, as shown in Fig. 3-50. However, in the case of Eudragit RS 100 microcapsules, the surface of these microcapsules shows a large number of small pores. It is suggested that the mutual repulsion of the cationic groups of the Eudragit RS 100 creates these pores during formation of the microcapsules [247].

Fig. 3-50 SEM images of (a) poly (ε-caprolactone) and (b) Eudragit RS 100 microcapsules.

To investigate the effect of Eudragit RS 100 on the release behaviors of poly (ε-caprolactone)/Eudragit RS 100 microcapsules, the dissolution of the microcapsules was carried out in a 0.5 M succinic acid solution. The release profile of the microcapsule as a function of the poly (ε-caprolactone)/Eudragit RS 100 ratio is shown in Fig. 3-51. The poly (ε-caprolactone) microcapsules show a relatively slow drug release rate compared to the poly (ε-caprolactone)/Eudragit RS 100 microcapsules. The release rate of the drugs increases remarkably with an increase in the Eudragit RS 100 content. These behaviors are likely due to the interaction between the organic acid and the quaternary ammonium group of Eudragit RS 100. The dissolution and swelling of the microcapsules under an acidic condition would increase the interaction between the organic acid and the

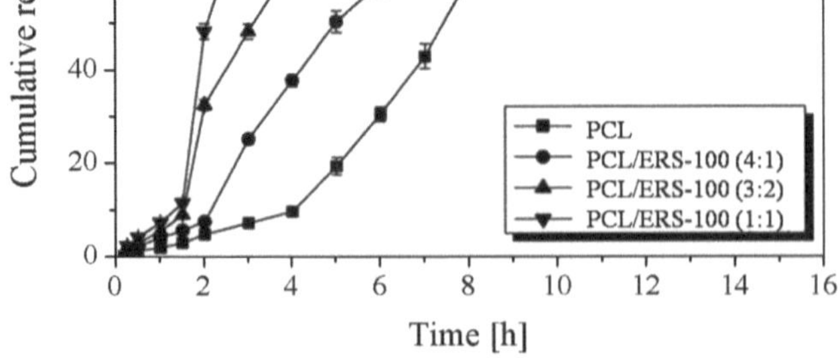

Fig. 3-51 Release behaviors of the microcapsules, as a function of poly(ε-caprolactone)/Eudragit RS 100 ratios in 0.5 M succinic acid (n = 3).

Eudragit RS 100 material, resulting in an increased diffusion rate of the drugs from the microcapsules via the increased permeability of the shell materials. Consequently, the quaternary ammonium group of Eudragit RS 100 has a significant effect on the release rate of drugs from the poly (ε-caprolactone)/ Eudragit RS 100 microcapsules [246].

Park *et al.* [248] prepared the biodegradable poly (ε-caprolactone) microcapsules containing tocopherol using an emulsion solvent evaporation method in which the microcapsules were treated by oxygen plasma to make them hydrophilic. A plasma treatment of the microcapsules including 35 wt. percent tocopherol (w/w) (poly(ε-caprolactone)/tocopherol) was carried out using the atmospheric pressure and radiofrequency for Ar/O_2 mixed gas. The microcapsules were characterized by their surface free energy or through the work of adhesion through contact angle measurements. The results are listed in Table 3-26.

The measurement of the contact angle is well known as a useful technique to investigate the characterization of the surfaces by polar liquids and non-polar liquids. It also incorporates the hydrophilic–hydrophobic properties, acid–base interaction, van der Waals force, and thermo-dynamical parameters. The London dispersive and specific components of the surface free energy of the microcapsules are determined by measuring the contact angles of two testing liquids with the known London dispersive (γ_S^L) and specific components (γ_L^{SP}). These values resulted in accordance with the method proposed by Owens, Wendt, and Kaelble. The work of adhesion (W_A) between a solid and a liquid is expressed by Eq. (3-14). The surface free energy and the work of adhesion of the plasma-treated poly(ε-caprolactone)/tocopherol are summarized in Table 3-26.

As shown in Table 3-26, the surface free energy or specific component was increased with the oxygen plasma treatment, while the London dispersive component was decreased. As expected, the work of adhesion on the water of the plasma-treated poly (ε-caprolactone)/tocopherol increased due to the plasma treatment. Moreover, the work of the adhesion of poly (ε-caprolactone)/tocopherol is proportional to the number of plasma treatments under equal conditions. These results indicate that the surface of plasma-treated poly (ε-caprolactone)/

Table 3-26 Surface Free Energy (γ_S) Factors and Work of Adhesion (W_A) of Microcapsules and Plasma-treated Microcapsules Measured at Room Temperature

	poly(ε-caprolactone)/ tocopherol	poly(ε-caprolactone)/ tocopherol-3	poly(ε-caprolactone)/ tocopherol-5
γ_S (mJ/m^2)	46.9	47.0	47.6
γ_S^L (mJ/m^2)	41.0	40.7	40.2
γ_S^{SP} (mJ/m^2)	5.9	6.4	7.5
W_A	96.4	97.9	100.3

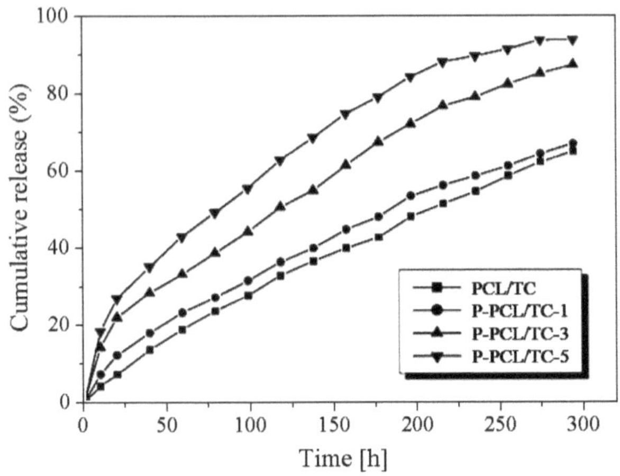

Fig. 3-52 Release behaviors of the microcapsules, prepared with different core concentrations.

tocopherol was modified with increased hydrophilic groups due to the oxygen plasma treatment as compared to the poly (ε-caprolactone)/tocopherol. It also shows that the surface characteristics of the microcapsules can be controlled with the number of plasma treatments.

Fig. 3-52 shows a release profile of the microcapsules after undergoing the plasma treatment. As a result, the release rate by the microcapsule increased with an increase in the number of plasma treatments. The release rate of the plasma-treated poly (ε-caprolactone)/tocopherol-5 was the fastest, whereas that from the poly (ε-caprolactone)/tocopherol showed the opposite trend. This result stemmed from the increase in the number of hydrophilic groups containing oxygen on the microcapsules due to the plasma treatment, resulting in the enhanced wettability between the microcapsule and the release medium. In poly (ε-caprolactone)/tocopherol, there was no sign of an initial burst, with this sample exhibiting a continuous and lag time of approximately 300 h followed by a linear release of 65 percent of the load of tocopherol, whereas the plasma-treated poly (ε-caprolactone)/tocopherol-5 showed a complete release of tocopherol before 300 h.

Among the various biodegradable polymers, the poly (ε-caprolactone) is the most promising candidate, for the biodegradable medical and pharmaceutical applications. In addition, poly (ε-caprolactone) has attracted attention in research related to controlled drug delivery due to its lack of toxicity and low cost as compared to other biodegradable polyesters. However, the application of poly (ε-caprolactone) has the drawback of a slow degradation rate in vivo due to its high crystallinity and hydrophobicity. It has been reported that the biodegradability of poly (ε-caprolactone) can be enhanced by the co-polymerizing or blending with other polymers having hydrophilicity. Among them, poly

Table 3-27 pH and Acid–base Values of SiO_2 Before and After Chemical Treatments

	pH	Acid value (meq/g)	Base value (meq/g)
SiO_2- as received	6.4	96	78
SiO_2-HCl	4.1	120	51
SiO_2-NaOH	10.3	79	134

(ethylene glycol) (PEG) as a water-soluble and non-toxic polymer can be used to increase the rate of biodegradation in the delivery system [249].

The influences of the poly (ethylene glycol) ratio on the fragrant oil release behaviors and variations of the surface free energy of the poly (ε-caprolactone)/ poly(ethylene glycol) microcapsules containing silica dioxide (SiO_2) absorbed fragrant oil are shown in Fig. 3-53 [250]. We also investigated the influence of a chemical surface treatment of SiO_2 on the release behaviors of fragrant oil from the poly (ε-caprolactone) microcapsules. SiO_2 was chemically treated with 30 wt. percent hydrochloric acid and sodium hydroxide to modify the surface at room temperature, specifically SiO_2-HCl and SiO_2-NaOH. The surface characteristics of SiO_2 are listed in Table 3-27.

The surface treatments with an adsorbent with acidic or basic chemical solutions are performed to provide a functional group that improves the adsorption capacity and selectivity of certain adsorbents in the gaseous or liquid phase [251]. Table 3-27 shows the surface properties of unmodified SiO_2 and modified SiO_2 after the acid and base chemical treatments. The acid and base values measured by a titration technique show that the surface properties of the untreated SiO_2 show slightly acidic properties. As expected, the base value of the SiO_2 surface treated with NaOH increases noticeably while the base value of the acidic-treated SiO_2 shows a decrease. These results indicate that the acid and base values are greatly influenced by the pH of the SiO_2 surfaces measured [46].

The contact angle data of the poly(ε-caprolactone)/poly(ethylene glycol) samples was measured by two testing liquids having polar and non-polar features, i.e., water and diiodomethane, and the γ_S, γ_S^L, and γ_S^{SP} values of poly (ε-caprolactone)/poly(ethylene glycol) samples obtained are given in Fig. 3-53 As an experimental result, they show that the specific component of the surface free energy, γ_S^{SP} and the hydrophilicity increase with an increase in the poly (ε-caprolactone) ratio. In contrast the London dispersive component γ_S^L slightly decreases as the poly (ethylene glycol) ratio increases. This result indicates that an increase in the poly (ethylene glycol) ratio, leads to an increase of the surface free energy, γ_S, and the hydrophilicity, which is mainly influenced by its specific components. Thus, it is expected that the fragrant oil release rate

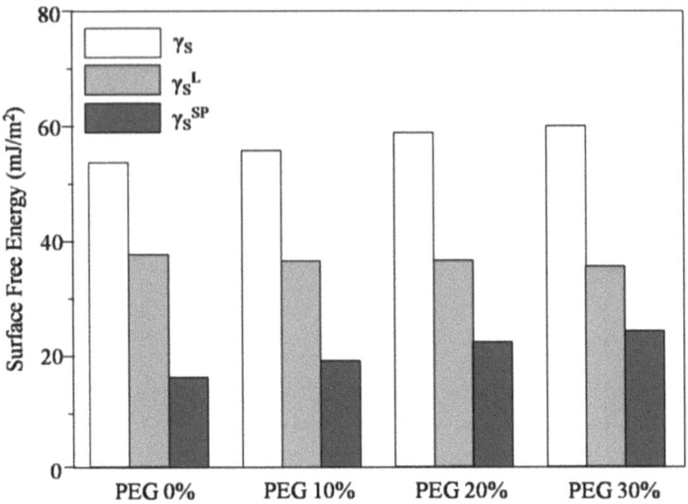

Fig. 3-53 Surface free energies of poly (ε-caprolactone)/poly(ethylene glycol) mixtures, measured by a two-liquid harmonic method.

of poly (ε-caprolactone)/poly (ethylene glycol) microcapsules is increased with increases in the poly (ethylene glycol) ratio in the phosphate buffer solution [252].

Fig. 3-54 shows the release behaviors of fragrant oil from poly (ε-caprolactone) microcapsules containing SiO_2-HCl and SiO_2-NaOH absorbed fragrant oil. As

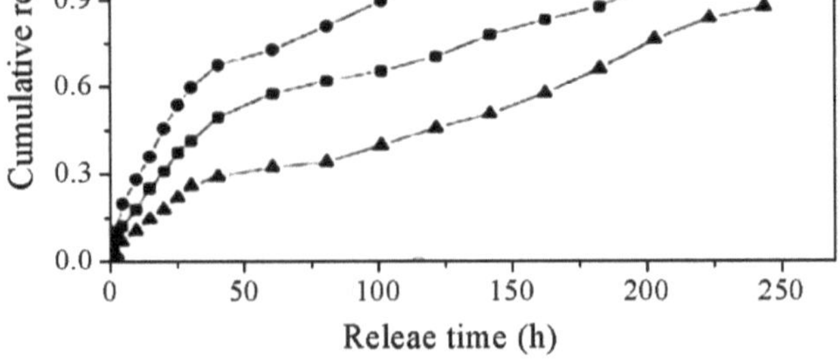

Fig. 3-54 Release behavior of poly(ε-caprolactone) microcapsules, containing surface treated or untreated SiO_2 absorbed fragrant oil.

shown in Fig. 3-54, the release rates and release amounts of the fragrant oil of poly (ε-caprolactone) microcapsules containing SiO_2-HCl are increased by 30.6 percent compared to the as-received poly(ε-caprolactone) microcapsules containing SiO_2-absorbed fragrant oil due to the increase of the specific surface area and acid values. In contrast, the release rates and release amounts of the fragrant oil from the poly (ε-caprolactone) microcapsules containing SiO_2-NaOH show a decrease of 22.7 percent as a result of the decrease in the specific surface area and the increase in the acid–base interactions between the SiO_2-NaOH and the acidic fragrant oil as the base value of SiO_2increases [253].

3.5. SUMMARY

The intermolecular properties of all forms of matter depend directly on the nature of their molecules and intermolecular forces. However, very few considerations are devoted to all the aspects of this important subject. The objective of this chapter therefore is presenting a comprehensive and unified introduction to the intermolecular forces, describing their role in determining the solid surface properties. For this background, the theoretical and applied procedures are discussed to characterize such surface features using contact angles. In addition, three kinds of intermolecular forces based on the solubility, electrostatic interaction, and self-assembling of a molecule are described with their applications to aid the easy understanding of a very broad area of phenomena in physics, chemistry, and biology.

For a long time, the solid-liquid interaction is a powerful tool capable of determining the surface features of numerous materials and applied in the fields of various industries, such as adhesive, films, prediction of solubility, emulsion systems (micro- and nanocapsulaiton techniques), energy storage systems, including the supercapacitor, direct methanol fuel cell, lithium battery, and polymer electrolyte. Practically, all the results of the solid-liquid interaction in the chemical, electrochemical and pharmaceutical industries, are freely available, and fully adaptable for the development of solid surface chemistry.

REFERENCES

[1] D.K. Chattoraj, K.S. Birdi, Adsorption and the Gibbs Surface Excess. Plenum Press, New York (1984).
[2] A.W. Adamson, Physical Chemistry of Surfaces. John-Wiley & Sons, New York (1990).
[3] S.J. Park, F.M. Brendlé, Journal of Colloid and Interface Science 188 (1997) 336.
[4] T. Young, Philosophical Transactions of the Royal Society (London) 95 (1805) 65.
[5] R. Avenyard, D.A. Haydon, An Introduction to the Principle of Surface Chemistry. Marcel Dekker, New York (1973).
[6] H.W. Fox, W.A. Zisman, Journal of Colloid and Interface Science 5 (1950) 514.
[7] F. Garbassi, M. Morra, E. Occhiello, Polymer Surfaces. John-Wiley & Sons, New York (1994).
[8] A. Kim, S.J. Park, J.R. Lee, Journal of Colloid and Interface Science 197 (1998) 119.
[9] P.L. du Nouy, Journal of Genaral Physiology 1 (1919) 512.

[10] Y. Tabata, Y. Ikada, Journal of Colloid and Interface Science 127 (1989) 132.
[11] M.K. Chaudhury, Journal of Adhesion Science and Technology 7 (1993) 669.
[12] J.G. Li, Composite Interfaces 1 (1993) 37.
[13] B. Janczuk, J.M. Bruque, M.L. Gonzalez-Martin, J. Moreno Del Pozo, Journal of Colloid and Interface Science 161 (1993) 209.
[14] M. Shigematsu, M. Morita, I. Sakada, Macromolecular Chemistry and Physics 197 (1996) 177.
[15] E. Kim, G.M. Whitesides, Journal of Physics Chemistry 101 (1997) 855.
[16] S. Wu, Polymer Interface and Adhesion. Marcel Dekker, New York (1986).
[17] P.C. Hiemenz, Principles of Colloid and Surface Chemistry. Marcel Dekker, New York (1986).
[18] F.M. Fowkes, D.C. McCarth, M.A. Mostafa, Journal of Colloid and Interface Science 78 (1980) 200.
[19] H.J. Busscher, H.M.J.C. Uyen, A.H. Weerkamp, W.J. Postma, J. Arends, FEMS Microbiology Letters 35 (1986) 303.
[20] J.B. Donnet, R.C. Bansal, Carbon Fibers. Marcel Dekker, New York (1990).
[21] J. Guastalla, J.F. Coste, Journal of Chemical Physics 72 (1975) 1007.
[22] J. Schultz, C. Cazeneuve, M.E.R. Shanahan, J.B. Donnet, Journal of Adhesion 12 (1981) 221.
[23] J. Schultz, L. Lavielle, C. Martin, Journal of Adhesion 23 (1987) 45.
[24] M.K. Chaudhury, Short-range and Long-range Forces in Colloidal and Macroscopic Systems, Ph. D. Dissertation, SUNY, Buffalo (1984).
[25] R.J. Good, Surface and Colloid Science 11 (1979) 1.
[26] S. Ohki, N. Düzgüne, K. Leonards, Biochemistry 21 (1982) 2127.
[27] C.J. van Oss, M.K. Chaudhury, R.J. Good, Molecular Mechanics of Membrane Fusion. Plenum Press, New York (1988).
[28] C.J. van Oss, R.J. Good, Journal of Protein Chemistry 7 (1988) 179.
[29] S.R. Holmes-Farley, R.H. Reamey, T.J. McCarthy, J. Dent, G.M. Whitesides, Langmuir 1 (1985) 725.
[30] L.K. Shu, In Contact Angles and Determination of the Surface Energy of Polymer Surfaces. Ph. D. Dissertation SUNY, Buffalo (1991).
[31] A.W. Neumann, R.J. Good, Surface and Colloid Science 11 (1979) 31.
[32] W.A. Zisman, F.M. Fowkes, Contact Angle Wettability and Adhesion: The Kendall Award Symposium Honoring. American Chemistry Society, Washington, D. C. (1964).
[33] F.M. Fowkes, Journal of Physical Chemistry 66 (1962) 382.
[34] F.M. Fowkes, Journal of Physical Chemistry 67 (1963) 2538.
[35] F.M. Fowkes, W.D. Harkins, Journal of American Chemical Society 62 (1940) 3377.
[36] R.J. Good, Journal of American Chemical Society 74 (1952) 5041.
[37] J. Schultz, K. Tsutsumi, J.B. Donnet, Journal of Colloid and Interface Science 59 (1977) 272.
[38] J.J. Jasper, E.V. Kring, Journal of Physical Chemistry 59 (1955) 1019.
[39] D.H. Kaelble, K.C. Uy, Journal of Adhesion 2 (1970) 50.
[40] B. Janczuk, T. Bialopiotrowicz, Journal of Colloid and Interface Science 127 (1989) 189.
[41] B. Janczuk, T. Bialopiotrowicz, W. Wojcik, Journal of Colloid and Interface Science 127 (1989) 59.
[42] J. Panzer, Journal of Colloid and Interface Science 44 (1973) 142.
[43] F.M. Fowkes, Hydrophobic Surfaces. Academic Press, New York (1969).
[44] M. Morra, E. Occhiello, F. Garbassi, Journal of Colloid and Interface Science 164 (1994) 325.
[45] E.J. Berger, Journal of Adhesion Science and Technology 4 (1990) 373.

[46] S.J. Park, In Long-range Force Contributions to Surface Dynamics in Interfacial Forces and Fields: Theory and Applications, Chap 9. Marcel Dekker, New York (1999).
[47] J.B. Donnet, M. Brendlé, T.L. Dhami, O.P. Bahl, Carbon 24 (1986) 757.
[48] S.J. Park, Ph. D. Dissertation, CNRS, France (1989).
[49] P. Commercon, J.P. Wightman, Journal of Adhesion 38 (1992) 55.
[50] J.B. Donnet, S.J. Park, M. Brendle, Carbon 30 (1992) 263.
[51] J.B. Donnet, R.Y. Qin, S.J. Park, S.K. Ryu, B.S. Rhee, Journal of Materials Science 28 (1993) 2950.
[52] J.B. Donnet, S.J. Park, W.D. Wang, Polymers of Advanced Technologies 5 (1994) 395.
[53] B. Barbier, M. Vilatte, G. Desarmot, Proceedings of Third European Conference on Composite Materials ECCM, Bordeaux (1989).
[54] B. Barbier, M.H. Auvray, J.P. Landesman, M. Villatte, G. Desarmot, Surface and Interface Analysis 16 (1990) 336.
[55] R.J. Good, Contact Angle Wetability and Adhesion. Utrecht, Netherlands (1993).
[56] C.J. van Oss, Interfacial Forces in Aqueous Media. Marcel Dekker, New York (1994).
[57] A.W. Neumann, R.J. Good, Experimental Methods in Surface and Colloid Science. Plenum, New York (1979).
[58] C.J. Budziak, A.W. Neumann, Colloids and Surfaces 43 (1990) 279.
[59] Z.M. Potoczny, E.I. Vargha-Butler, T.K. Zubovitd, A.W. Neumann, AOSTRA Journal of Research 1 (1984) 107.
[60] Z.M. Potoczny, In Determination of Surface Thermodynamic Parameters of Tar Sand Systems, Thesis, Toronto University, Onrario (1983).
[61] C.A. Ku, J.D. Henry, R. Siriwardane, L. Roberts, Journal of Colloid and Interface Science 106 (1985) 377.
[62] C.J. van Oss, R.F. Giese, Z. Li, K. Murphy, J. Norris, J. Norris, M.K. Chaudhury, R.J. Good, Journal of Adhesion Science and Technology 6 (1992) 413.
[63] S. Chwastiak, Journal of Colloid and Interface Science 42 (1973) 299.
[64] F.E. Bartell, Journal of Physical Chemistry 326 (1932) 3115.
[65] L.R. White, Journal of Colloid and Interface Science 90 (1982) 536.
[66] Y. Rotenberg, L. Boruvka, A.W. Neumann, Journal of Colloid and Interface Science 93 (1983) 169.
[67] P. Cheng, D. Li, L. Boruvka, Y. Rotenberg, A.W. Nuemann, Colloids and Surfaces 43 (1990) 151.
[68] D. Li, P. Cheng, A.W. Neumann, Advances in Colloid and Interface Science 39 (1992) 347.
[69] Y. Gu, Interactions between Small Oil Droplets and Solid Surface in Aqueous Solutions Ph. D. Dissertation, Alberta University, Alberta (1999).
[70] T. Doi, M. Yamashita, H. Nagano, Journal of the Japan Institute of Metals 62 (1998) 64.
[71] M. Yamamoto, H. Masuda, T. Kodama, Zairyo-to-Kankyo 48 (1999) 633.
[72] H. Masuda, Journal of the Japan Institute of Metals 62 (1998) 173.
[73] J. Hu, X.D. Xiao, M. Salmeron, Applied Physics Letters 67 (1995) 476.
[74] R. Wang, M. Takeda, M. Kido, Scripta Materialia 46 (2002) 83.
[75] M. Mantel, J.P. Wightman, Surface and Interface Analysis 21 (1994) 595.
[76] T. Tyou, Metals Technology 65 (1995) 1092.
[77] J.J. Jasper, Journal of physical and Chemical Reference Data 1 (1972) 841.
[78] J.H. Hildebrand, This J 28 (1916) 1452.
[79] G. Scatchard, Chemical Reviews 8 (1931) 321.
[80] J.H. Hildebrand, The Solubility of Non-Electrolytes. Reinhold, New York (1936).

[81] A.F.M. Barton, Handbook of Solubility Parameters and Other Cohesion Parameters. CRC Press, Boca Raton (1983).
[82] J.H. Hildebrand, R.L. Scott, Solubility of Nonelectrolytes. Reinhold, New York (1950).
[83] L.A. Girifalco, R.J. Good, Journal of Physical Chemistry 61 (1957) 904.
[84] A.F.M. Barton, CRC Handbook of Solubility Parameters and Other Cohesion Parameters. CRC Press, Boca Raton (1983).
[85] D.W. Van Krevelen, Properties of Polymers. Elsevier, New York (1990).
[86] A.E. van Arkel, Transanctions of the Faraday Society 42 (1946) 81.
[87] P.A. Small, Journal of Applied Chemistry 3 (1953) 71.
[88] C.M. Hansen, In Ph D. Dissertation, Danish Technical Press, Copenhagen (1967).
[89] C.M. Hansen, Progress in Organic Coatings 51 (2004) 109.
[90] H. Launay, C.M. Hansen, K. Almdal, Carbon 45 (2007) 2859.
[91] C.M. Hansen, Hansen Solubility Parameters-A user's Handbook. CRC Press, Boca Raton (1999).
[92] S.J. Park, H.J. Sohn, S.K. Hong, G.S. Shin, Journal of Colloid and Interface Science 332 (2009) 246.
[93] M.C. Granger, G.M. Swain, Journal of The Electrochemical Society 146 (1999) 4551.
[94] J.W. Shim, S.J. Park, S.K. Ryu, Carbon 39 (2001) 1635.
[95] F. Coloma, J. Narciso-Romero, A. Sepulveda-Escribano, F. Rodriguez-Reinoso, Carbon 36 (1998) 1011.
[96] S.J. Park, B.R. Jun, Journal of Colloid and Interface Science 284 (2005) 204.
[97] I.N. Ermolenko, I.P. Lyubliner, N.V. Gulko, Chemically Modified Carbon Fibers and Their Application. Weinheim, New York (1990).
[98] A. Lisovskii, R. Semiat, C. Aharoni, Carbon 35 (1997) 1639.
[99] S.J. Park, J.S. Kim, Carbon 39 (2001) 2011.
[100] Y. Otake, R.G. Henkins, Carbon 31 (1993) 109.
[101] F.H. Healey, Y.F. Yu, J.J. Chessick, Journal of Physical Chemistry 59 (1959) 399.
[102] M. Smisek, S. Cerny, Active Carbon Manufacture Properties and Applications. Elsevier, Amsterdam (1970).
[103] R.C. Bansal, J.B. Donnet, F. Stoeckli, Active Carbon. Decker, New York (1988).
[104] S.J. Park, J.S. Shin, J.W. Shim, S.K. Ryu, Journal of Colloid and Interface Science 275 (2004) 342.
[105] H.P. Boehm, Advances in Catalysis 16 (1966) 179.
[106] J.S. Noh, J.A. Schwarz, Carbon 28 (1990) 675.
[107] M.L. Toebes, J.M.P. Toebes, J.H. Bitter, A.J. van Dillen, K.P. de Jong, Carbon 42 (2004) 307.
[108] K. Kinoshita, J. Bett, Electrochimica Acta 11 (1973) 237.
[109] J. Liu, Y. Tian, Y. Chen, J. Liang, Applied Surface Science 256 (2010) 6199.
[110] K. Morita, A. Yamaguchi, N. Teramae, Journal of Electroanalytical Chemistry 563 (2004) 249.
[111] M. Delamar, R. Hitmi, J. Pinson, J.M. Savéant, Journal of the American Chemical Society 114 (1992) 5883.
[112] S.J. Park, B.J. Park, S.K. Ryu, Carbon 37 (1999) 1223.
[113] S.J. Park, K.D. Kim, Journal of Colloid and Interface Science 218 (1999) 331.
[114] S.J. Park, E. Papirer, J.B. Donnet, Journal of Chemical Physics 91 (1994) 203.
[115] V. Gutmann, The Donor-Acceptor Approach to Molecular Interactions. Plenum, New York (1978).
[116] S.J. Park, J.B. Donnet, Journal of Colloid and Interface Science 206 (1998) 29.

[117] E.P. Plueddemann, Silane Coupling Agent. Plenum, New York (1991).
[118] C. Velasco-Santos, A.L. Martınez-Hernandez, M. Lozada-Cassou, A. Alvarez-Castillo, V.M. Castano, Nanotechnology 13 (2002) 495.
[119] S.J. Park, B.J. Kim, D.I. Seo, K.Y. Rhee, Y.Y. Lyu, Materials Science and Engineering: A 526 (2009) 74.
[120] J.B. Donnet, S.J. Park, H. Balard, Chromatographia 31 (1991) 434.
[121] J.A. Riddick, W.B. Bunger, Organic Solvents. Wiley-Interscience, New York (1970).
[122] S.J. Park, B.J. Kim, Materials Science and Engineering: A 408 (2005) 269.
[123] K.E. Noll, V. Gounaris, W.S. Hou, Adsorption Technology for Air Water Pollution Control. Lewis, Michigan (1992).
[124] S.J. Park, Y.S. Jang, Journal of Colloid and Interface Science 261 (2003) 238.
[125] Y. Urabe, T. Ishikura, K. Kaneko, Journal of Colloid and Interface Science 319 (2008) 381.
[126] B.J. Kim, S.J. Park, Journal of Colloid and Interface Science 325 (2008) 297.
[127] X. Li, K. Horita, Carbon 38 (2000) 133.
[128] C.A. Frysz, D.D.L. Chung, Carbon 35 (1997) 1111.
[129] S.J. Park, M.K. Seo, K.Y. Rhee, Materials Science and Engineering: A 356 (2003) 219.
[130] M. Tatoulian, The Journal of Adhesion 15 (1995) 177.
[131] R. Foerch, J. Izawa, G. Spears, Journal of Adhesion Science and Technology 5 (1991) 549.
[132] S. Kanazawa, M. Kogoma, T. Moriwaki, S. Okazako, Journal of Physics D: Applied Physics 21 (1988) 838.
[133] S.J. Park, H.Y. Lee, Journal of Colloid and Interface Science 285 (2005) 267.
[134] L.J. Ward, W.C.E. Schofield, J.P.S. Badyal, A.J. Goodwin, P.J. Merlin, Chemistry of Material 15 (2003) 1466.
[135] Q. Zhang, C. Wang, Y. Babukutty, T. Ohyama, M. Kogoma, M. Kodama, Journal of Biomedical Materials Research 60 (2002) 502.
[136] S.J. Park, J.S. Jin, Journal of Colloid and Interface Science 236 (2001) 155.
[137] G. Xiao, Journal of Colloid and Interface Science 171 (1995) 200.
[138] J. Ehlbeck, A. Ohl, M. Maaß, U. Krohmann, T. Neumann, Surface and Coatings Technology 174–175 (2003) 493.
[139] G. Borcia, C.A. Anderson, N.M.D. Brown, Applied Surface Science 221 (2004) 203.
[140] D.H. Shin, C.U. Bang, J.H. Kim, K.H. Han, Y.C. Hong, H.S. Uhm, D.K. Park, K.H. Kim, Surface and Coatings Technology 201 (2007) 4939.
[141] K.L. Mittal, Polyimides: Synthesis, Characterization and Applications. Plenum, New York (1984).
[142] W.C. Wang, R.H. Vora, E.T. Kang, K.G. Neoh, Macromolecular Materials and Engineering 288 (2003) 152.
[143] T. Liang, Y. Makita, S. Kimura, Polymer 42 (2001) 4867.
[144] A. Weber, A. Dietz, R. Pockelmann, C.P. Klages, Journal of The Electrochemical Society 3 (1997) 1131.
[145] S.J. Park, K.S. Cho, S.K. Ryu, Carbon 41 (2003) 1437.
[146] S.J. Park, B.J. Kim, Journal of Colloid and Interface Science 275 (2004) 590.
[147] G.M. Wu, Materials Chemistry and Physics 85 (2004) 81.
[148] M.K. Seo, S.J. Park, S.K. Lee, Journal of Colloid and Interface Science 285 (2005) 306.
[149] X. Fu, W. Lu, D.D.L. Chung, Carbon 36 (1998) 1337.
[150] D.S. Bag, V.P. Kumar, S. Maiti, Journal of Applied Polymer Science 71 (1999) 1041.
[151] E. Uchida, Y. Uyama, Y. Ikada, Langmuir 9 (1993) 1121.
[152] J.H. Lee, H.W. Jung, I.K. Kang, H.B. Lee, Biomaterials 15 (1994) 705.

[153] S.J. Park, J.S. Kim, Journal of Colloid and Interface Science 244 (2001) 336.
[154] A. Tressaud, V. Gupta, C. Guimon, F. Moguet, Materials Science and Engineering: B 30 (1995) 61.
[155] R.B. Mathur, V. Gupta, O.P. Bahl, A. Tressaud, Synthetic Metals 114 (2000) 197.
[156] S.J. Park, M.K. Seo, Y.S. Lee, Carbon 41 (2003) 723.
[157] T. Nakajima, Fluorine-Carbon and Fluoride-Carbon Materials. Dekker, New York (1995).
[158] N. Watanabe, T. Nakajima, H. Touhara, Studies in Inorganic Chemistry: 8. Graphite Fluoride. Elsevier Science Publisher, New York (1988).
[159] M.K. Seo, S.J. Park, Bulletin of the Korean Chemical Society 30 (2009) 1337.
[160] M.C. Michalski, B.J.V. Saramago, Journal of Colloid and Interface Science 227 (2000) 380.
[161] S. Chwastiak, Journal of Colloid and Interface Science 42 (1973) 298.
[162] M.K. Seo, S.J. Park, Journal of Colloid and Interface Science 330 (2009) 237.
[163] A. Bismark, R. Tahhan, J. Spriner, A. Schulz, T.M. Klapotke, H. Zell, W. Michaeli, Journal of Fluorine Chemistry 84 (1997) 127.
[164] B.J. Kim, S.J. Park, Journal of Colloid and Interface Science 311 (2007) 619.
[165] S.J. Park, S.Y. Jin, Journal of Colloid and Interface Science 286 (2005) 417.
[166] K.R. Ko, S.K. Ryu, S.J. Park, Carbon 42 (2004) 1864.
[167] B. Jönsson, H. Wennerström, B. Halle, Journal of Physics Chemistry 84 (1980) 2179.
[168] H.L.F. von Helmholtz, Annals of Physics 3 (1879) 337.
[169] G. Gryglewicz, J. Machnikowski, E. Lorenc-Grabowska, G. Lota, E. Frackowiak, Electrochimica. Acta 50 (2005) 1197.
[170] H. Zhang, G.P. Cao, Y.S. Yang, Journal of Power Sources 172 (2007) 476.
[171] M.K. Seo, S.J. Park, Current Applied Physics 10 (2010) 391.
[172] E. Frackowiak, F. Beguin, Carbon 39 (2001) 937.
[173] D. Qu, Journal of Power Sources 109 (2002) 403.
[174] W.G. Pell, B.E. Conway, Journal of Power Sources 96 (2001) 57.
[175] C. Shao, H.Y. Kim, J. Gong, B. Ding, D.R. Lee, S.J. Park, Materials Letters 57 (2003) 1579.
[176] C. Kim, Y.O. Choi, W.J. Lee, K.S. Yang, Electrochimica Acta 50 (2004) 883.
[177] M.K. Seo, S.J. Park, Materials Science and Engineering: B 164 (2009) 106.
[178] S. Brunauer, P.H. Emmett, E. Teller, Journal of the American Chemical Society 60 (1938) 309.
[179] E.P. Barret, L.G. Joyner, P.P. Halenda, Journal of the American Chemical Society 73 (1951) 373.
[180] G. Horvath, K. Kawazoe, Journal of Chemical Engineering of Japan 16 (1983) 470.
[181] K. Kinoshita, Carbon: Electrochemical and Physicochemical Properties. Wiley, New York (1988).
[182] S. Iijima, Nature 354 (1991) 56.
[183] C.T. Hsieh, H.H. Teng, Carbon 40 (2002) 667.
[184] Q. Jiang, M.Z. Qu, G.M. Zhou, B.L. Zhang, Z.L. Yu, Materials Letters 57 (2002) 988.
[185] K.H. An, W.S. Kim, Y.S. Park, J.M. Moon, D.J. Bae, S.C. Lim, Y.S. Lee, Y.H. Lee, Advanced Functional Materials 11 (2001) 387.
[186] M.K. Seo, S.J. Park, Current Applied Physics 10 (2010) 241.
[187] K.H. An, W.S. Kim, Y.S. Park, Y.C. Choi, S.M. Lee, D.C. Chung, D.J. Bae, S.C. Lim, Y.H. Lee, Advanced Materials 13 (2001) 497.
[188] C. Niu, E.K. Sichel, R. Hoch, D. Moy, H. Tennent, Applied Physics Letters 70 (1997) 1480.
[189] K.S. Novoselov, A.K. Geim, S.V. Morozov, D. Jiang, Y. Zhang, S.V. Dubonos, S.V. Grigorieva, I.V. Grigorieva, A.A. Firsov, Science 306 (2004) 666.

[190] K.S. Novoselov, A.K. Geim, S.V. Morozov, D. Jiang, M.I. Katsnelson, I.V. Grigorieva, S.V. Dubonos, A.A. Firsov, Nature 438 (2005) 197.
[191] Q. Du, M. Zheng, L. Zhang, Y. Wang, J. Chen, L. Xue, W. Dai, G. Ji, J. Cao, Electrochimica. Acta 55 (2010) 3897.
[192] W.S. Hummers, R.E. Offeman, Journal of the American Chemical Society 80 (1958) 1339.
[193] K. Lee, Sh. Cho, S.H. Park, A.J. Heeger, Ch.W. Lee, S.H. Lee, Nature 41 (2006) 65.
[194] M.D. Stoller, S.J. Park, Y.W. Zhu, J. An, R.S. Ruoff, Nano Letters 8 (2008) 3498.
[195] H. Wang, Q. Hao, X. Yang, L. Lu, X. Wang, Electrochemistry Communications 11 (2009) 1158.
[196] K.S. Kim, S.J. Park, Synthetic Metals 160 (2010) 2355.
[197] S. Kim, S.J. Park, Analytica Chimica Acta 619 (2008) 43.
[198] T.S. Zhao, C. Xu, R. Chen, W.W. Yang, Progress in Energy and Combustion Science 35 (2009) 275.
[199] K. Scott, W.M. Tamma, P. Argyropoulos, Journal of Power Sources 79 (1999) 43.
[200] S. Kim, Y.J. Jung, S.J. Park, Colloids and Surfaces: A 313–314 (2008) 220.
[201] W. Li, C. Liang, J. Qui, W. Zhou, J. Zhou, G. Sun, Q. Xin, Journal of Physics Chemistry B 107 (2003) 6292.
[202] M. Carmo, A.R. dos Santos, J.G.R. Poco, M. Linardi, Journal of Power Sources 173 (2007) 860.
[203] S. Kim, S.J. Park, Electrochimica Acta 52 (2007) 3013.
[204] W. Li, C. Liang, J. Qiu, W. Zhou, H. Han, Z. Wei, G. Sun, Q. Xin, Carbon 40 (2002) 791.
[205] L. Gao, H. Huang, C. Korzeniewski, Electrochimica Acta 49 (2004) 1281.
[206] S.J. Park, J.M. Park, M.K. Seo, Journal of Colloid and Interface Science 337 (2009) 300.
[207] P.G. Bruce, B. Scrosti, J.M. Tarascon, Angewandte Chemie International Edition 47 (2008) 2930.
[208] D. Deng, M.G. Kim, J.Y. Lee, J.P. Cho, Energy & Environmental Science 2 (2009) 818.
[209] S. Kim, Y. Jung, H.S. Lim, Electrochimica Acta 50 (2004) 889.
[210] S. Kim, Y.J. Jung, S.J. Park, Journal of Power Sources 152 (2005) 272.
[211] M.Y. Chu, US Patent, 5, 814, 420 (1998).
[212] J. Fuller, R.T. Carlin, R.A. Osteryoung, Journal of The Electrochemical Society 144 (1997) 3881.
[213] T. Sato, G. Masuda, K. Takagi, Electrochimica Acta 49 (2004) 3603.
[214] S. Kim, Y.J. Jung, S.J. Park, Electrochimica Acta 52 (2007) 2116.
[215] J.N. Israelachvili, In Intermolecular and Surface Forces. Academic Press, London (1992).
[216] D.G. Hall, B.A. Pethica, Nonionic Surfactants. Marcel Dekker, New York (1967).
[217] T.L. Hill, Thermodynamics of Small Systems. Benjamin, New York (1963).
[218] R. Nagarajan, E. Ruckenstein, Journal of Colloid and Interface Science 60 (1977) 221.
[219] R. Nagarajan, E. Ruckenstein, Journal of Colloid and Interface Science 71 (1979) 580.
[220] J.N. Israelachvili, S. Marcelja, R.G. Horn, Quarterly Review of Biophysics 13 (1980) 121.
[221] H. Wennerström, B. Lindman, Physics Reports 52 (1979) 1.
[222] D.J. Mitchell, B.W. Ninham, Journal of the Chemical Society Faraday Transactions 77 (1981) 601.
[223] I. Szleifer, A. Ben-Shaul, W.M. Gelbart, Journal of Chemical Physics 83 (1985) 3612.
[224] I. Szleifer, A. Ben-Shaul, W.M. Gelbart, Journal of Chemical Physics 85 (1986) 5345.
[225] D. Blanckstein, G.M. Thurston, G.B. Benedek, Journal of Chemical Physics 85 (1986) 7268.
[226] S. Puvvada, D. Blanckstein, Journal of Chemical Physics 92 (1990) 3710.

[227] A.E. Alexander, P. Johnson, Colloid Science. Oxford University Press, London & New York (1950).
[228] H.V. Kesten, S. Vicious, Annals of Probability 36 (2008) 1838.
[229] F. Graner, J.A. Glazier, Physical Review Letters 69 (1992) 2013.
[230] D. Vollhardt, T. Gutberlet, Colloids and Surfaces: A 102 (1995) 257.
[231] J.N. Israelachvili, D.J. Mitchell, B.W. Ninham, Journal of the Chemical Society Faraday Transactions I 72 (1976) 1525.
[232] G. Blume, E.E. Teichmuller, New Evidence of the Penetration of Actives by Liposomal Carrier System Cosmetics and Toiltries Manufacture Worldwide. Aston Publishing, London (1997).
[233] L.M. Popplewell, Evaluating Encapsulation Economics, Perfumer & Flavorist (2001).
[234] S.J. Park, R. Arshady, Microcapsules for Fragrances and Cosmetics. In Microcapsule Patents and Products, Citus Books, London (2003).
[235] M.E. Page-Clisson, H. Pinto-Alphandary, M. Ourevitch, A. Andremont, P. Couvreur, Journal of Controlled Release 56 (1998) 23.
[236] H.C. Chang, Y.Y. Lin, C.S. Chern, S.Y. Lin, Langmuir 14 (1998) 6632.
[237] C.S. Chern, T.J. Chen, Y.C. Liou, Polymer 39 (1998) 3767.
[238] S. Sajjadi, F. Jahanzad, European Polymer Journal 39 (2003) 785.
[239] S.J. Park, K.S. Kim, Colloids and Surfaces: B 46 (2005) 52.
[240] L. Dai, W. Li, X. Hou, Colloids and Surfaces: A 125 (1997) 27.
[241] S.J. Park, Y.M. Lee, S.K. Hong, Colloids and Surfaces: B 47 (2006) 211.
[242] Y.Y. Yang, T.S. Chung, N.P. Ng, Biomaterials 22 (2001) 232.
[243] A.G.A. Coombes, S.C. Rizzi, M. Williamson, J.E. Barralet, S. Downes, W.A. Wallac, Biomaterials 25 (2004) 315.
[244] R. Pignatello, D. Amico, S. Chiechio, P. Giunchedi, C. Spadaro, G. Puglisi, Drug Delivery 8 (2001) 35.
[245] F. Siepmann, J. Siepmann, M. Walther, R.J. MacRae, R. Bodmeier, Journal of Controlled Release 125 (2008) 1.
[246] K.S. Kim, S.J. Park, Colloids and Surfaces: B 76 (2010) 404.
[247] S. Narisawa, M. Nagata, Y. Hirakawa, M. Kobayashi, H. Yoshino, Journal of Pharmaceutical Sciences 85 (1996) 184.
[248] S.J. Park, K.S. Kim, Colloids and Surfaces: B 43 (2005) 138.
[249] R.L. Dunn, J.P. English, J.D. Strobel, D.R. Cowsar, & T.R. Tice, In 'Preparation and Evaluation of Lactide/Glycolide Copolymers for Drug Delivery' Polymers in Medicine, Elsevier Science Publishers, Amsterdam (1988).
[250] S.J. Park, Y.J. Yang, H.B. Lee, Colloids and Surfaces: B 38 (2004) 35.
[251] R. Leboda, B. Charmas, Colloids and Surfaces: A 135 (1998) 267.
[252] S.J. Park, S.H. Kim, Journal of Colloid and Interface Science 271 (2004) 336.
[253] K.S. Kim, S.J. Park, Colloids and Surfaces: B 80 (2010) 240.

Chapter 4

Solid-Solid Interfaces

4.1. INTRODUCTION

The surface or interfacial phenomena between different phases with a planned structure and planned properties have become extremely important in recent years [1–3]. This has provided opportunities to produce interfaces with promising surface structures and properties on various solid interfaces. Whenever different solid surfaces are brought into contact adhesion can occur at the solid interface; the different solid surfaces include the glass-reinforced plastics, laminated plastics, coatings, woodchip boards, and compounded and reinforced rubbers. The properties of these materials are affected by interaction between different solids, for instance, polymers used for the filling, and reinforcing materials.

The problem of improving adhesion strength is not only for the adhesive-bonded joints, but applications of bonding by adhesives can also be commercialized in wide varieties of industries. Adhesion properties at the solid–solid interface have many advantages over the other joining methods' ability to join dissimilar or thin materials, give rise to an improved appearance and design, an excellent corrosion resistance, and so on [4–6].

The subject of the main problems in adhesion mechanisms, is at the boundary of several scientific fields, including macromolecular science, physical chemistry of surfaces and interfaces, materials science, mechanics and micromechanics of fracture, and rheology. Consequently, the study of adhesion uses various concepts, according to the special field of expertise, and therefore treatment of the phenomena observed can be considerably different. These various approaches involve many adhesion theories that are both complementary, and contradictory:

1. Mechanical interlocking
2. Electronic theory (also known as electrical double layer, or electrostatic or parallel plate capacitor theory)
3. Theory of boundary layers and interphases (also known as weak boundary layer theory)

Interface Science and Composites, Volume 18
ISSN: 1573-4285, doi: 10.1016/B978-0-12-375049-5.00004-9

4. Diffusion theory
5. Intermolecular bonding theory

Among the models of these theories, one usually distinguishes between mechanical and specific adhesion, the latter being based on the different types of bonds (electrostatic, secondary, acid–base, chemical), that can improve the adhesion between dissimilar solids. Each of these theories is available to some extent, and depends on the solids in contact and the conditions of formation of the bonded system [1].

This chapter collects the results of studies of investigation in the field of adhesion, and adhesives of composite materials. It considers some regularities of the mechanism of adhesive bonded joints, presents thermodynamic and physical verification of new principles, for controlling the adhesive strength and other important properties of composite materials, and describes application of these principles, in the process of developing adhesives for various fields of engineering and materials.

4.2. ADHESION AT SOLID-SOLID INTERFACES

4.2.1. Theories of Adhesion

The previous section showed two major phenomena; interfacial interactions and energy dissipation in adhesion.

A close contact between an adhesive and a substrate is needed for an exchange of interfacial interactions. A good spreading of the adhesive onto the substrate surface is necessary to obtain this atomic closeness. A condition of a surface that determines how fast a liquid will wet and spread on the surface, or if it will be repelled and not spread on the surface [7, 8]. The formation of numerous physical interactions, including van der Waals, London dispersion, and acid–base interactions, between the adhesive and the substrate, is the origin of good wetting. A low adhesive viscosity and sometimes, the application of pressure are needed to confirm correct spreading. The number of interfacial bonds relies on the interface area. The rugosity of a substrate increases the interfacial area and accordingly, the number of interactions, assuming, that the adhesive is able to totally wet the substrate asperity. Another effect of higher substrate rugosity is the possibility, as regards the adhesive penetration inside the hollows, of a mechanical anchoring effect. This phenomenon is the mechanical theory of adhesion, which was proposed by Mac Bain and Hopkins [9]. Substrate rugosity is generally accomplished by mechanical treatments. Chemically surface treatments can also increase the rugosity and porosity of substrates.

However, the effect of rugosity can sometimes be contrary; occasionally the polymer is not able to totally fill the substrate cavities and asperity. This partial wetting causes trapped air bubbles to form in the hollows, creating defects and the potential initiation, of a site of failure.

Good wetting is the first required criterion for the formation of interfacial interactions. Both, the number and nature of these interactions will act on the adhesion energy. An increase of adhesion energy induces the formation of strong interfacial bonds, such as acid–base or chemical interactions, which are more active than van der Waals interactions. Acid–base interactions need the presence of polar sites between the substrate and the adhesive. The formation of acid–base interactions are involved in the use of polar polymer and polar substrates. The polarity of the substrate can be increased, by surface treatments, which lead to an oxidized surface.

Sometimes, the adhesive is able to exchange intermolecular bond with a substrate generally producing a strong interfacial strength. The possibility of interfacial bonds is described in the physicochemical theory of adhesion. The formation of such chemical interactions requires the presence of reactive sites between the adhesive and the substrate. Polar groups, such as carboxylic acid, amine, hydroxyl, epoxide, or isocyanate and also the insaturation present in polymers, grafted polymers, copolymers, or monomers that directly polymerized on the substrate can exchange chemical bonds with the reactive sites of the substrate surface. Then, the substrate also has to exhibit enough surface reactivity. For example, it can be induced by the presence of an oxide layer [10]. The interdiffusion of mutual chains leads to the formation of a three-dimensional zone that is located at the interface. This phenomenon is usually called interphase. The characteristic of this interphase differs from the properties of both materials.

A zone of interphase is also be evidenced in others cases. The migration of additives or low molecular weight of a contamination layer, including the presence of chains at the material surface can cause the formation of an interphase. If the interphase is weak and has low cohesive properties a low adherence can be expected. This phenomenon is the so-called, weak boundary layer theory of adhesion, proposed by Bikerman [11].

The interphasial layers are also frequent in cross-linked adhesives. In the interphasial layers, the cross-linking degree is able to differ in the vicinity of the substrate with the bulk adhesive. However, the crystalline organization of semi-crystalline polymers can be modified in the vicinity of the substrate, which leads to the formation of a trans-crystalline layer. In the trans-crystalline layer, the thickness can sometimes be important [12]. The formation of interfacial bonds is also able to cause specific chain orientations, and conformations, with the possibility of higher or restricted mobility modification. Therefore, the presence of an interphasial zone, located at the interface of the adhesive and substrate, is frequent in assemblies. However, it is often difficult to obtain experimental evidence of interphases. These confined layers are usually inaccessible by the in situ analysis techniques.

The following steps schematically organize the different theories of adhesion.

1. The first required step is a correct wetting of the substrate by the adhesive. The consequence of the wetting is the formation of physical interfacial interactions. In some cases, a large rugosity of the substrate can be a favorable factor.

2. Formation of interfacial interactions that include London dispersion and acid–base bonds will increase the adhesion. An interdiffusion process that occurs in materials–materials, or in materials-grafted substrate systems, can also improve adhesion.
3. The formation of an interphase that is located at the substrate-adhesive interface, and exhibits specific properties is frequent.
4. The adhesive viscoelastic properties play a major role in an adherence test. These rheological properties are partly caused by dissipative phenomena that induce greater adherence energy than that of interfacial energy. The rate or temperature dependence of adherence value is a result of the viscoelastic effects.
5. The exact determination of the locus of failure can supply productive information on rupture mechanisms.

4.2.2. Contribution of Thermodynamic Adsorption to Adhesion

The term adhesion between two materials appears vague or confusing for two reasons: 1) The formation of the interface between a pair of materials, i.e., the establishment of interfacial bonds through forces at the interface that cause materials to attract one another. 2) The breaking stress or energy, required to break a formed assembly. One can easily see that both interfacial forces and mechanical properties of adherents in the vicinity of the interface, and in the bulk contribute to the total mechanical response of the assembly. Such a fundamental issue indicates a paradox that has inspired intensive research, for decades: if adhesion is only a matter of thermodynamics works needed for separation, they have relevance to the surface energies of adherents, which are typically from 10–100 mJ/m^2. Practical works of separation can be 10 to 10^6 greater. On the other hand, separation has often been viewed as affecting processes at the molecular level. The force needed for separation should be as large as $10^{-2}/10^{-10}$ or 10^8 N/m^2 to $10^{-1}/10^{-10}$ or 10^9 N/m^2 (where 10^{-10} m is a typical atomic size). Therefore, the first part of this review is the role of interfacial forces in determining the mechanical strength of bonded structures such as adhesive joints, composites, and coatings.

In other words, interfacial forces are determined by the weakness or strength of joint systems.

4.2.3. Free Energies and Work of Adhesion

It is now well demonstrated that the advancement of close intimate molecular contact at an interface is a necessary condition for good adhesion. Therefore, polymers are a good candidate for adhesives. Made of long flexible molecules, they flow under stress like viscous liquids at the region of longer time. On the other hand, they transform like soft elastic solids at the region of shorter time. The bulk properties are affected by entanglements. Without close molecular contact, the interfacial attraction will be weak. Moreover, the applied stress can be transmitted from one phase to the other, although the interface will be very low. Therefore, innumerable adhesives are applied in a rather liquid-like state. From the reason of the formation arranged by a liquid–solid contact step, a standard of good adhesion becomes the standard of good wetting. However, close molecular contact alone is not sufficient. The properties and the magnitude of the interfacial forces are also significant. Most of interfacial forces result from van der Waals and Lewis acid–base interactions, and are detailed later. The magnitudes of these forces can commonly rely on fundamental thermodynamics quantities such as surface free energies [1].

4.3. LONDON DISPERSION AND ACID–BASE INTERACTION

4.3.1. London Dispersion Force

The London dispersion force (also known as dispersion forces, London forces, induced dipole–induced dipole forces) is a type of force, acting between atoms and molecules [13]. They are part of the van der Waals forces. The London dispersion force is named after the German-American physicist Fritz London. The London dispersion force is a weak intermolecular force, arising from the quantum induced instantaneous polarization multipoles in molecules. It is can therefore, act between molecules without permanent multipole moments.

London forces are confirmed by non-polar molecules, due to the correlated movements of the electrons in interacting molecules. Electron density in a molecule becomes redistributed in the proximity of another molecule; because the electrons from different molecules start feeling and avoiding each other (see quantum mechanical theory of dispersion forces). This is frequently mentioned as the formation of instantaneous dipoles that attract each other. London forces are present between all chemical functional groups, and usually represent the main part of the total interaction force in condensed matter, even though they are largely weaker, than ionic bonds and covalent bonds.

This is the only attractive intermolecular force exhibited between neutral atoms (e.g., a noble gas). Without London forces, there could be no attractive force between noble gas atoms, and they could not exist in liquid form.

London forces become stronger as the atom or molecule in question becomes larger [14]. This is due to the increased deformation polarizability of molecules with larger more dispersed electron clouds. This trend is exemplified by the halogens (from smallest to largest: F_2, Cl_2, Br_2, I_2). Fluorine and chlorine are gases at 25 °C, bromine is a liquid, and iodine is a solid. London forces also become stronger with larger amounts of surface contact. Greater surface area means closer interaction between different molecules. (See Chapter 1.2.1.1.)

4.3.1.1. Quantum mechanical theory of dispersion force

The first explanation of the attraction between noble gas atoms was given by Fritz London in 1930 [15, 16]. He used a quantum mechanical theory based on a second-order perturbation theory. The perturbation is the Coulomb interaction

(V) between the electrons and nuclei of the two monomers (atoms or molecules) that constitute the dimer. The second-order perturbation expression of the interaction energy contains a sum over states. The states appearing in this sum are simple products of the excited electronic states of the monomers. Thus, no intermolecular antisymmetrization of the electronic states is included and the Pauli exclusion principle is only partially satisfied.

London developed the perturbation V in a Taylor series in $1/R$, where R is the distance between the nuclear centers of a mass of monomers.

This Taylor expansion is known as the multipole expansion of V, because the terms in this series can be regarded as energies of two interacting multipoles, one on each monomer. Substitution of the multipole-expanded form of V into the second-order energy yields an expression that resembles, somewhat, an expression describing the interaction between instantaneous multipoles (see the qualitative description above). Additionally an approximation, named after Albrecht Unsöld, must be introduced in order to obtain a description of London dispersion in terms of deformation polarizability.

In this manner, the following approximation is obtained for the dispersion interaction E^{L}_{AB} between two atoms A and B. Here α^A and α^B are the deformation polarizabilities, of the respective atoms. The quantities I_A and I_B are the first ionization potentials of the atoms, and R is the intermolecular distance.

$$E^{\mathrm{L}}_{AB} \approx -\frac{3\alpha^A\alpha^B I_A I_B}{4(I_A + I_B)} R^- \tag{4-1}$$

Note that this final London equation does not contain instantaneous dipoles (see molecular dipoles). The explanation of the dispersion force as the interaction between two such dipoles was invented after London gave the proper quantum mechanical theory. See the authoritative work [17], for a criticism of the instantaneous dipole model, and [18] for a modern and thorough exposition of the theory of intermolecular forces.

The London theory has much similarity to the quantum mechanical theory of light dispersion, which is why London coined the phrase ‘dispersion effect’.

4.3.2. Acid–Base Interactions

4.3.2.1. *Introduction*

When other interactions are derived from the strength of an interface, one should precisely indicate their fundamental properties. Forming of interactions can take place and lead to the formation of acid–base bonds or to covalent bonding. We will consider here the importance of acid–base interactions in the improvement of adhesion.

From the idiom of molecular orbital theory, acid–base definitions are as follows: A base is a species that employs an empty orbital in initiating a reaction. The donor orbital is commonly the highest occupied molecular orbital (HOMO), and the acceptor orbital is commonly the lowest unoccupied molecular orbital

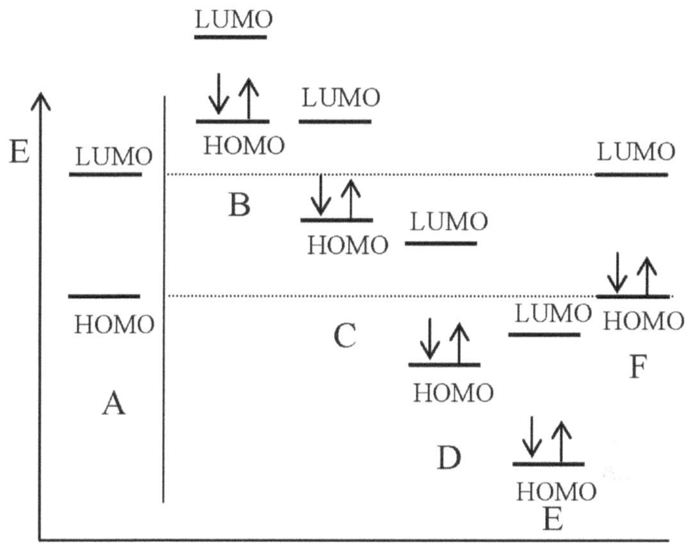

Fig. 4-1 How donor-acceptor molecular orbital interactions work.

(LUMO). To sum up how acid–base reactions work on the basis of molecular orbitals perturbation theory, Fig. 4-1 presents the relative energies of the frontier orbitals HOMO and LUMO of a hypothetical species A, and the frontier orbitals of several hypothetical reaction partners B, C, D, E, and F.

Fig. 4-1 demonstrates that with respect to B, electron transfer from B to A will be favorable, and A will act as an oxidizing agent. With respect to C, the A (LUMO)-C (HOMO) perturbation will be favorable, and A will perform as an acid. With respect to D, the A (HOMO)-D (LUMO) perturbation will be favorable and A will perform as a base. With respect to E, complete electron transfer from A to E will be favorable and A will perform as a reducing agent. For F species, the frontier orbitals are degenerated with A. Neither of these species are obviously the donor or acceptor, and species may display, both the behaviors simultaneously.

Moreover, from the general discussion of Lewis acid–base interactions, hydrogen bonding represents a particular case. It is apparent that electrostatic and charge transfer interactions are predominant, for hydrogen bonds. Generally, the principal charge transfer contribution originates from the proton acceptor–proton donor charge transfer complex, through the σ-type interactions.

4.3.2.2. Hydrogen Bonding

A hydrogen bond is the interaction of a hydrogen atom with an electronegative atom, such as nitrogen, oxygen, or fluorine from another molecule. The hydrogen must form covalent bond with another electronegative atom to create the bond. These bonds can occur between molecules or within different parts of a single molecule [19]. The hydrogen bond that was generally from 5 to 30 kJ /mol is stronger than a van der Waals interaction, but weaker than covalent or ionic bonds.

A hydrogen attached to carbon can also participate in hydrogen bonding when the carbon atom is bound to electronegative atoms, as is the case in chloroform, $CHCl_3$. The electronegative atom attracts the electron cloud, from around the hydrogen nucleus and, by decentralizing the cloud, leaves the atom with a positive partial charge. Because of the small size of hydrogen relative to other atoms and molecules, the resulting charge, though only partial, nevertheless, represents a large charge density. A hydrogen bond results, when this strong positive charge density attracts a lone pair of electrons on another heteroatom, which becomes the hydrogen-bond acceptor.

Usually, the partially covalent nature of a hydrogen bond is simple to determine on the basis of inter-atomic distances in the X-H-Y system. X-H distance is typically about 110 pm, whereas H-Y distance is from 160 to 200 pm. Hydrogen bonds can vary in strength from very weak (1–2 kJ/mol) to extremely strong (over 155 kJ/mol) [20]. The length of hydrogen bonds depends on bond strength, temperature, and pressure. The bond strength itself is dependent on temperature, pressure, bond angle, and environment.

Many materials are strengthened by hydrogen bonds in their main chains. The best known example is nylon, where hydrogen bonds occur in the repeat unit, and plays a major role in the crystallization of the material. The bonds occur between carbonyl and amine groups in the amide repeat unit. They effectively link the adjacent chains to create crystals, which help reinforce the materials. The effect is greatest in the structure of aramid fiber, where hydrogen bonds stabilize the linear chains laterally, as shown in Fig. 4-2. The chain axes are aligned along the fiber axis, making the fibers extremely stiff and strong. The hydrogen bond networks make both the natural and sensitive materials to humidity levels in the atmosphere because water molecules can diffuse into

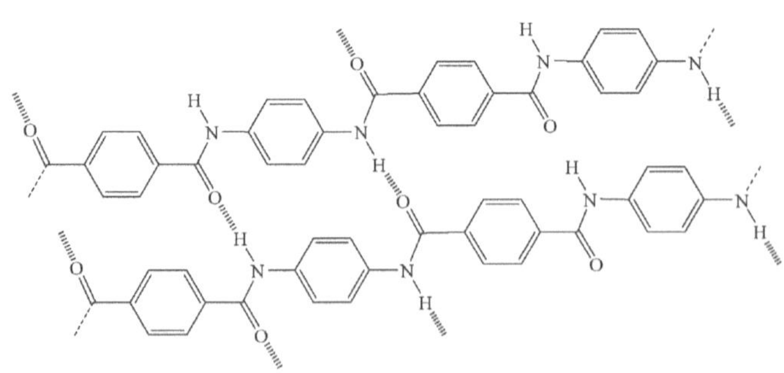

Fig. 4-2 Para-aramid structure.

the surface and disrupt the network. Some materials are more sensitive than others.

The acid–base bonds including hydrogen bonding are exothermic and their energy ranges do not exceed 55 kJ/mol [21, 22]. This compares with the energy of London dispersion forces but exceeds that of Debye interactions, and Keesom interactions. With a large and negative value, the heat of an acid–base interaction can overcome the positive or the negligibly small negative entropic term, $-T\Delta S$, so that adhesion and mixing can be substantially improved. The high energy associated with the acid–base interactions is due to intermolecular forces, acting at intermolecular distances of about 0.3 nm. The acid–base interactions are thus of the long-range type rather than the short-range type. For a series of single-fiber composites, Nardin and Schultz [22] demonstrated that the maximal work of adhesion (W) was obtained for fiber-matrix systems interacting via both dispersive and acid–base interactions.

Hydrogen bonds are involved in the species, which are combined bond distances of the A–H···B compared to the sum of the van der Waals radius, as listed in Table 4-1 [21]. The types of various hydrogen bonding between atoms A and B are located closer than the estimated distance of the sum of the van der Waals radii. One of several features is that a hydrogen atom can approach fairly close over a distance to gain access to the atom B, compared with the estimated distance of the sum of the van der Waals radii, which suggests that a hydrogen atom is penetrated in the electron cloud of atom B. (See Chapter 1.2.2.)

Table 4-1 van der Waals Bonding Ddistances *vs.* the Observed Distances for the Typical Hydrogen Honding

Bonding state	A·B[a] (Calculated value)	A·B (Observed value)	H·B (Calculated value)	H·B (Observed value)
F–H–F	270	240	260	120
O–H–O	280	270	260	170
O–H–F	280	270	260	170
O–H–N	290	280	270	190
O–H–Cl	320	310	300	220
N–H–O	290	290	260	200
N–H–F	290	280	260	190
N–H–Cl	330	330	300	240
N–H–N	300	310	270	220
N–H–S	340	340	310	240
C–H–O	300	320	260	230

[a] *Not be calculated by Bondi's method [23], but using the calculated by method of Hamilton et al. [21].*

4.3.2.3. Work of Adhesion

The work that is needed to separate reversibly the interface between the two bulk phases 1 and 2 from their equilibrium interacting distance to infinity is called the 'work of adhesion'.

$$W_{12} = \gamma_1 + \gamma_2 - \gamma_{12} \tag{4-2}$$

where γ_1 and γ_2 are the surface free energies of phases 1 and 2, respectively, and γ_{12}, the interfacial free energy, between phases 1 and 2.

When an interface comprises from two surfaces, the work of adhesion, is the decrease of Gibbs free energy per unit area. Therefore, greater the interfacial attraction greater is the work of adhesion, and the smaller the interfacial free energy, between phases 1 and 2. Thereby, γ_{12} can be related, to the character of the two phases.

In the study of low surface energy materials such as polymers, many researchers have evaluated the thermodynamic surface free energy from wetting contact angle measurements between the interface, phases 1 and 2, which are commonly presented as the solid phase (*S*) and liquid phase (*L*) so that the interfacial free energy between the liquid and the solid may be written as:

$$\gamma_{SL} = \gamma_S + \gamma_L - W_{SL} \tag{4-3}$$

The wetting equilibrium is defined from the profile of a liquid drop (*L*) in the presence of the vapor phase (*V*) on a planar, smooth, and homogeneous solid surface (*S*). The surface energies of materials and the equilibrium contact angle θ at the three-phase contact point are related by Young's equation [24]:

$$\gamma_{SV} = \gamma_{SL} + \gamma_{LV}\cos\theta \tag{4-4}$$

For a first approximation and ignoring the spreading pressure of the vapor phase of the liquid onto the solid, the solid-liquid work of adhesion is

$$W_{SL} = \gamma_S + \gamma_{LV} \tag{4-5}$$

$$\gamma_{SL} = \gamma_{LV}(1 + \cos\theta) \tag{4-6}$$

To consider the contribution of the fundamental property of the forces, across the interface it has been suggested that the surface free energies, and the work of adhesion may be expressed by the sum of two terms: The London's dispersion (L) interactions and the non-dispersion forces (SP), that include Debye, Keesom, and acid–base interactions [25, 26]. Note that three different types of forces contribute to the total long-range interaction, between the polar molecules. Generally known as 'van der Waals force' these are the dispersion force (London interaction) the induction force (Debye interactions), and the orientation force (Keesom interactions).

$$\gamma = \gamma^{\mathrm{L}} + \gamma^{\mathrm{SP}} \tag{4-7}$$

$$W_{12} = W_{12}^{\mathrm{L}} + W_{12}^{\mathrm{SP}} \tag{4-8}$$

Fowkes proposes that the London dispersive part of the work of adhesion, is measured as twice the geometric mean of the dispersive component of the surface energy, of solids 1 and 2 [27]. The total work of adhesion is reduced to

$$W_{12} = 2\sqrt{\gamma_1^L \lambda_2^L} + W_{12}^{SP} \tag{4-9}$$

In the particular case such as London's forces across the interface, the work of adhesion is determined because the non-dispersive part becomes zero. Clearly, $\gamma_1{}^L$ and $\gamma_2{}^L$ can be determined, by wettability experiments. The only existence of London interactions at the interface between two adherents leads to

$$G \propto W_{12}^L \tag{4-10}$$

where G is the adhesive strength of the assembly.

For knowledge of the strength of an acid–base interaction both the specific Gibbs free energy of interaction and the polar interaction (Debye, Keesom, and Hydrogen bonding) across the interface allow the determination of the reversible acid–base work of adhesion. In the case of adhesion between different solids, the contribution of the polar interactions to the thermodynamic work of adhesion occurs rapidly. Therefore, the acid–base of the adhesion energy is expressed as the following equation [28]:

$$W_{12}{}^{SP} \approx W_{12}{}^{AB} = \Delta G_D + \Delta G_K + \Delta G_H \tag{4-11}$$

where D is the Debye interaction, K the Keesom interaction, and H the Hydrogen bonding.

The quantitative determination of these quantities could become either a theoretical or experimental value. Theoretical determinations are mostly based on the hypothesis that the initial perturbation of the orbital of interaction species, determines the process of a reaction or an interaction. In the equation of this state, the development of perturbation theory was studied by Hudson and Klopman [29]. From the energy gap between HOMO and LUMO, acids and bases were defined by their hardness and softness [30]. The Hard Soft Acid Base principle explains some basic rules about kinetics, and equilibrium of the acid–base interactions and is explained on the basis of the density-functional theory, and the frontier orbital method [31].

The results of γ^L and γ^{SP} are investigated by water and diiodomethane, on each carbon black and thermoplastic polymer, as shown in Table 4-2 [32]. From the matrix results, the γ^{SP} on water was significantly changed in the high density polyethylene specimen. It was clearly expected that the positive temperature coefficient properties of composites made with high density polyethylene polymer would be greatly influenced.

The results of the work of adhesion of carbon black/thermoplastic polymer composites derived from Eq. (4-3) are shown in Fig. 4-3. It is interesting to note that carbon black/high density polyethylene composites produce a minimum

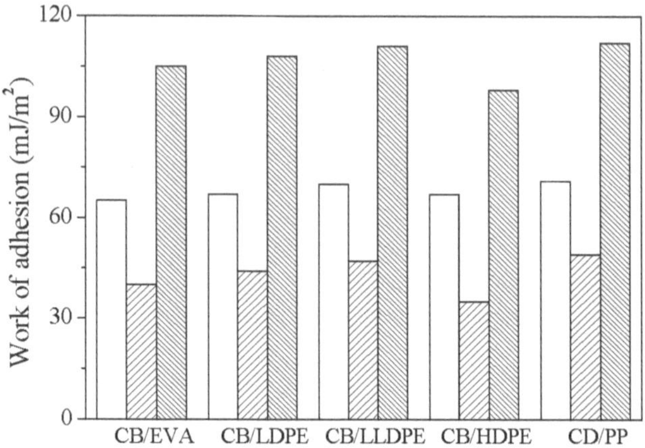

Fig. 4-3 Work of adhesion of carbon black/thermoplastic polymer composites, using a two-liquid geometric method; ((left) W_A^L, (center) W_A^{SP}, and (right) W_A).

Table 4-2 Results of γ_S^L and γ_S^{SP} on the Carbon Black and Thermoplastic Polymers Studied

Positive temperature coefficient materials	γ_S^L (mJ/m^2)	γ_S^{SP} (mJ/m^2)	γ_S (mJ/m^2)
CB (Carbon black)	36.8 ± 1.8	34.4 ± 1.7	71.2 ± 3.5
EVA (Ethylene vinyl acetate)	27.6 ± 0.5	13.1 ± 0.7	40.7 ± 1.2
LDPE (Low density polyethylene)	28.0 ± 0.3	14.3 ± 0.8	42.3 ± 1.1
LLDPE (Linear low density polyethylene)	30.2 ± 0.7	14.6 ± 0.5	44.8 ± 1.2
HDPE (High density polyethylene)	27.5 ± 0.5	10.5 ± 0.5	38.0 ± 1.0
PP (Polypropylene)	30.3 ± 0.4	15.0 ± 0.3	45.3 ± 0.7

value of work of adhesion, but carbon black/linear low density polyethylene composites produce a relatively higher work of adhesion, due to the abundant terminal groups of the linear low density polyethylene molecular chains that can be bonded with carbon black.

A good correlation between the work of adhesion and the maximum resistivity, is shown in Fig. 4-4. It can be clearly observed that the resistivity is increased as the work of adhesion between carbon blacks and thermoplastic polymers. It is then noted that the decrease in the work of adhesion induced by interactions between carbon black surfaces and polymer chains, plays an important role in improving the positive temperature coefficient behaviors of a composite in this system. As a result, the largest positive temperature coefficient

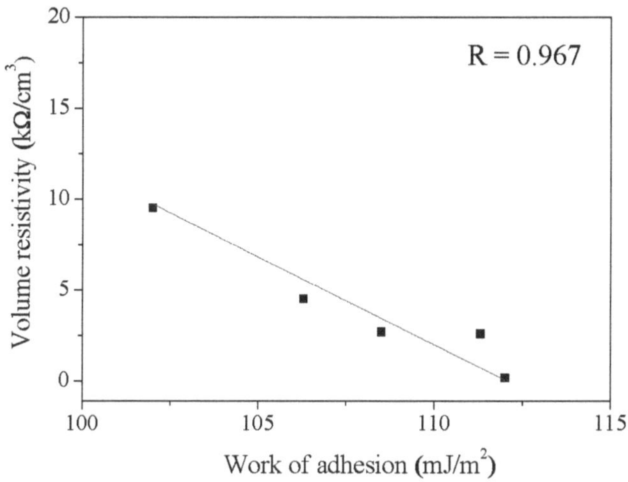

Fig. 4-4 Dependence of resistivity on the work of adhesion of carbon black/thermoplastic polymer composites.

intensity and the maximum work of adhesion occur on the carbon black/high density polyethylene composites since the resistivity of carbon black/high density polyethylene composites is higher than that of carbon black/low density polyethylene, carbon black/linear low density polyethylene, and carbon black/polypropylene composites at room temperature. However, the carbon black/linear low density polyethylene composites showed a relatively higher work of adhesion, due to the abundant terminal groups, of the linear low density polyethylene molecular chains that can be bonded with carbon black. Consequently, a good relationship between the work of adhesion and the maximum resistivity is shown in these positive temperature coefficient composites [32].

4.3.2.4. Drago's Approach

Drago and Wayland [33] suggested an equation similar to the perturbation theory, for predicting reaction enthalpies between the acid and base. Both the acid and base are defined by two independent parameters: The E value measures their ability to participate in electrostatic bonding and the C value measures their capability to participate in covalent bonding leading to

$$-\Delta H_{12}{}^{AB} = E_A \cdot E_B + C_A \cdot C_B \tag{4-12}$$

where ΔH^{AB} is the enthalpy of the acid–base adduct formation, E_A the capability of the acid to participate in electrostatic bonding, E_B the ability of the base to participate in electrostatic bonding, C_A the ability of the acid to participate in covalent bonding, and C_B the capability of the base to participate in covalent bonding.

A self-consistent set of E and C values is now available from 33 acids and 48 bases, allowing ΔH prediction for over 1584 adducts [34]. The cogency of Eq. (4-12) was clearly demonstrated for polymer adsorption on various substrates [35]. (See Chapter 1.3.7.)

4.3.2.5. Gutmann's Numbers

In addition to the previous discussion of the dual-polarity properties of Lewis acids and bases, it should be noted that many of them are amphoteric, by definition. The donor number (*DN*) was developed to correlate the behavior of a solute in various donor solvents with a given basicity or the degree of the donor electron. A relative measurement of the basicity of a solvent D is given by the enthalpy of its reaction with the reference to the acid which was chosen as $SbCl_5$ by Gutmann's scale. Subsequently, Mayer *et al.* [36] introduced an acceptor number (*AN*) as the relative P NMR shift induced by triethylphosphine and relative to the acidic strength between hexane and $SbCl_5$. In 1989, Riddle and Fowkes modified this acceptor number in order to express from these, to acceptor number (AN^*) number in correct enthalpic unit (kcal/mol) [37]. The most important assumption of Gutmann's approach is that the order of the determined base strengths remains constant for the other acids the Gibbs free energy of formation of a given adduct, being linearly related to the donor number (*DN*) of the base through Eq. (4-13) [38]:

$$-\Delta G_{12}{}^{AB} = a_{\mathrm{A}} \cdot DN_{\mathrm{B}} + b_{\mathrm{A}} \tag{4-13}$$

where ΔG is the Gibbs free energy of the acid–base adducts formation, DN_{B} the donor number of the base, and a_{A} and b_{A} the constants characteristic of the acid.

This means that a plot of the *DN* for a series of donors vs. $-\Delta G$ of their adduct formation with a given acid brings up a straight line, allowing the determination of a_{A} and b_{A}. By experimentally measuring the formation free energy of only two adducts from a given acid, one can predict through the resulting a_{A} and b_{A} values, the adduct formation free energy of this acid with any other donor for which *DN* is known. Gutmann also suggested that the free energy of an acid–base interaction could be approximated by a two-parameter equation of the form

$$-\Delta G_{12}{}^{\mathrm{AB}} = \frac{AN_{\mathrm{A}} \cdot DN_{\mathrm{B}}}{100} \tag{4-14}$$

where $\Delta G_{12}{}^{\mathrm{AB}}$ is the Gibbs free energy of the acid–base adduct formation, DN_{B} the donor number of the base, and AN_{A} the acceptor number of the acid.

The factor of 100 converts the acceptor number value from a percentage of the $SbCl_5$ value to a decimal fraction. However, it is important that of the 171 donor number values reported in the literature only 50 were determined precisely by calorimetry [39]. In a recent research [40], Gutmann's semi-empirical scale was used in terms of acceptor number and donor number values, which

Table 4-3 Characteristics of Polar Probes Used

Probe	DN	AN
chloroform	0	23.1
acetonitrile	14.1	18.9
acetone	17.0	12.5
ethylacetate	17.1	9.3
diethylether	19.2	3.9
tetrahydrofuran	20.0	8.0
nitromethane	2.7	20.5

described the ability of the polar component or the Lewis acid–based or electron acceptor-donor interaction, and these are given in Table 4-3 (See Chapter 1.3.8).

4.3.2.6. Approaches of van Oss, Good, and Chaudhury

van Oss *et al.* [41] introduced the notion of acidic and basic components of the surface energy (γ^+ and γ^-, respectively) to characterize the acid–base properties of materials and predict W^{AB}:

$$W^{AB} = 2(\gamma_1^+\gamma_2^-)^{1/2} + 2(\gamma_1^-\gamma_2^+)^{1/2} \quad (4\text{-}15)$$

The γ^+ and γ^- for a solid, can be determined by the contact angle measurements using three reference liquids of known γ_L^d, γ_L^+, and γ_L^-. The γ_L^+ and γ_L^- for test liquids were established with model surfaces and test liquids, on the arbitrary assumption, that for water $\gamma_L^- = \gamma_L^+ = 25.5\ \text{mJ/m}^2$ and

$$\gamma_L^{AB} = 2(\gamma_L^+\gamma_L^-)^{1/2} \quad (4\text{-}16)$$

Application of Eq. (4-16), to water (w) yields

$$\gamma_w^{AB} = 2(25.5 \times 25.5)^{1/2} \quad (4\text{-}17)$$

$$= \gamma^{SP}(51\ \text{mJ/m}^2) \quad (4\text{-}18)$$

where γ^{SP} is the non-dispersive or the so-called polar component of the surface energy of water.

Once γ^+ and γ^- are determined for a given material the overall acid–base ('polar') contribution to its γ can be evaluated for water in Eq. (4-16). Table 4-4 shows γ^+ and γ^- values for liquids, polymers, and other materials.

van Oss and Good [42] have shown that their approach can predict the solubility of poly (ethylene oxide) and dextran in water but the 'γ^{Polar} approach' cannot, and this leads to a positive interfacial free energy for the polymer-water systems (5.8 and 6 mJ/m^2 for poly(ethylene oxide), and dextran, respectively) and thus to the insolubility of these polymers in water.

Table 4-4 Surface Energy Components for Commonly Used Test Liquids and for Polymers and Other Materials (mJ/m^2)

	γ	γ^{L}	γ^{SP}	γ^{+}	γ^{-}
TestingLiquid					
Water	72.8	21.8	51.0	25.5	25.5
Diiodomethane	50.8	50.42	0.38	0	0
Ethylene glycol	47.7	31.0	16.7	1.92	47.0
Polymer					
Poly(ethylene oxide) 6000	43	43	0	0	64
Dextran 10000	61.2	47.4	13.8	1.0	47.4
Poly(methyl methacrylate)	39–43	39–43	0	0	9.5–22.4
Poly(vinyl chloride)	43.7	43	0.7	0.04	3.5
Polystyrene	42.0	42	0	0	1.1
Polyethylene[a]	33	33	0	0	0.1
Polyethylene[b]	57.9–62.5	42	15.9–20.5	2.1	30–50
Iodinated polyacetylene	52.6	42.8	9.8	1.04	23.15
Cellulose acetate	40.2	35	5.2	0.3	22.7
Cellulose nitrate	45	45	0	0	16
Agarose	44.1	41	3.1	0.1	24
Gelatin	38	38	0	0	19
Human serum albumin	44.3	41	3.3	0.15	18
Inorganic filler					
Red mud	55.4	39.1	16.2	2.27	28.98
Acid-treated red mud	55.9	40.6	15.3	2.37	24.77
Base-treated red mud	54.8	38.6	16.2	2.19	29.94

***Source**: Data in Ref. [1, 48, 49].*
[a]*Based on advancing angles.*
[b]*Based on receding angles.*

Although the novel approaches to interfacial inter-facial interactions just described were shown to be more valid than the 'polar interactions' approach in predicting adhesion, solubility, and miscibility phenomena [42–45], they must be viewed with caution, as the controversial publications have shown their shortcomings [45, 46]. Nevertheless the qualitative and/or quantitative determination of the acid–base properties for polymers and other materials is an important step toward the understanding and the prediction of their interfacial interactions. This is a very delicate task, because it requires the determination of the polar component of heat or the free energy changes of acid–base interactions of reference acidic and basic chemical species (of known acid–base properties) with the materials [47]. The choice of reference test acids and bases is also crucial and usually depends on the nature of the material under investigation and the experimental conditions associated with the technique used for the assessment of acid–base properties.

4.3.2.7. IR spectroscopic tools to access acid–base strength

Fowkes [50] proposed that the enthalpy of acid–base adduct formation is related to the shift of infrared frequency (Δv) of its absorption band for specific functional groups involved in the acid–base interaction, according to the following equation:

$$\Delta G_{12}{}^{AB} = k_{12}{}^{AB} \cdot \Delta v_{12}{}^{AB} \tag{4-19}$$

where $\Delta G_{12}{}^{AB}$ is the Gibbs free energy of the acid–base adduct formation, $\Delta k_{12}{}^{AB}$ the correlation constant between the IR wave number shift and enthalpy, and $\Delta v_{12}{}^{AB}$ the infrared frequency shift.

k_{AB} is a characteristic constant of the functional group determined on the basis of infrared and micro-calorimetrical from the results of adduct formation. For example, the latter is equal to –0.99 kJ/mol cm in the carbonyl group. The stretching frequency of the C=O vibration band is reduced by an amount of proportional Δv_{AB} to the Gibbs free energy of the acid–base bonding (ΔG_{AB}), according to k_{AB}. This methodology has recently been proved not only for polymer-solvent adduction but for polymer-polymer [51] and polymer-metal [52] adduction. Furthermore, the infrared shifts can be used to determine Drago's E and C constants of unknown polymers or substrates [50]. If a test of acid of known C_A and E_A gives Δv_{AB} shifts, allowing the determination of ΔG^{AB}, then Eq. (4-12), is reduced to

$$E'_b = \Delta G_{12}{}^{AB}/E_A - C'_B \cdot (C_A/E_A) \tag{4-20}$$

in which E'_B and C'_B are 'the trial' values. The actual values of E_B and C_B exist at one point of the straight line acquired from Eq. (4-20). The operation of various tests of acids shows that straight lines intersect at E_B and C_B.

4.3.2.8. Density of interacting sites

On the foundation of reflectance spectroscopy [51], the section of functional groups that develop acid–base adducts at an interface can be estimated according to

$$f^{AB} = \frac{A^{AB}}{A^{AB} + A^{F} \cdot \frac{\varepsilon^{AB}}{\varepsilon^{F}}} \tag{4-21}$$

where A represents the integrated intensity of the considered stretching mode for free, F, or acid–base, AB, species, and ε the molar absorption coefficient of Beer's law.

As an explaining example, we present the case of the adsorption of ethylene-vinyl-acetate copolymers. From various amounts of acetate comonomer, we expect to affect the growth of acid–base interactions of the acetate C=O groups and aluminols of the substrate. Fig. 4-5 showed the evolution of the fraction of interfacial acid–base interactions *vs.* the 12 nm thick vinyl-acetate copolymer, which was adsorbed on aluminum coated silicon wafers.

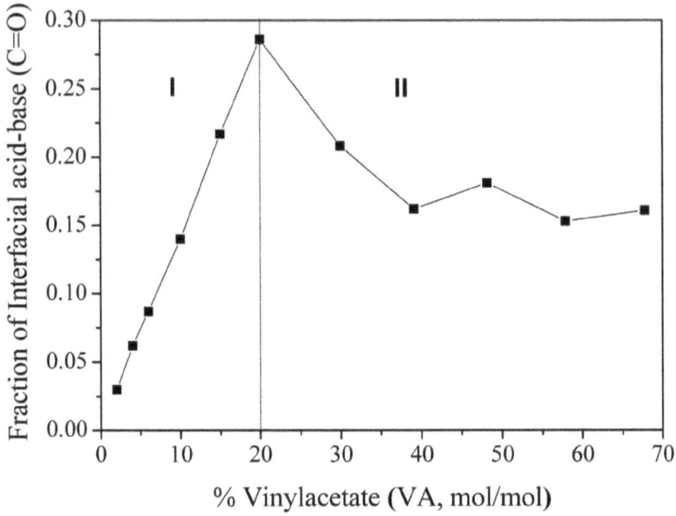

Fig. 4-5 Interfacial acid–base adducts vs. vinylacetate comonomer density.

This result shows two major regimes. The first one, with a low content of vinylacetate, consists of a sharp increase of connector density, with co-monomer density, until a maximum is obtained. Second, a constant decrease of interfacial connecting sites is studied, until the density of the co-monomer increases. The first important point, is, that the co-monomer density allows us to control the level of interfacial acid–base bonds.

The result of the foundation of interfacial polymer-metal bonds that is the failure strength of the polymer-metal assemblies can be estimated by evaluating peel forces in different media, such as air, or ethanol, in 180 peel test geometry. Peel analyzes were performed for a large range of separation rates R_s from 8.4×10^{-6} to 8.4×10^{-3} m/s. The rheological model of adhesion [53] states that the peel energy G can be expressed as

$$G^{A,L} = G_0{}^{A,L} \cdot \Phi^{A,L}(R_s, T) \tag{4-22}$$

where G_o is the intrinsic fracture energy of the interface Φ a dissipative function corresponding to the energy irreversibly expanded in visco-elastic and plastic deformations in the bulk adhesive. The function of Φ depends on both the separation rate (R_s) and temperature (T), and follows a time-temperature equivalence law [54]. The superscripts A and L refer to air or liquid.

Under specific conditions [55], the dissipative function Φ is identical in air and in liquid media. Therefore, supposing as a first assumption that the intrinsic fracture energy of the interface (G_o) is proportional, to the reversible energy of adhesion (W_o) Eq. (4-22) is obtained:

$$\frac{G^{L}}{G^{A}} = \frac{G_0{}^{L}}{G_0{}^{A}} = \frac{W_0{}^{L}}{W_0{}^{A}} \Leftrightarrow \frac{G^{A} - G^{L}}{G^{A}} \tag{4-23}$$

$$= \frac{W_0{}^{A} - W_0{}^{L}}{W_0{}^{A}} \tag{4-24}$$

Given that the dissipation factors cancel, changes in the interfacial characteristic can be immediately predicted, from peel measurements. The ratios of Eqs. (4-23) and (4-24), can be determined promptly, either experimentally (G^A–G^L/G^A) or by calculation ($W^A{}_0$–$W^L{}_o$/$W^A{}_0$) according to thermodynamic considerations.

Presenting the definition of the thermodynamic work of adhesion Eq. (4-9), we can indicate the reduction of adhesion strength due to the presence of a liquid medium as

$$\frac{\frac{G_A - G_L}{G_A} = 2\left\lfloor \left(\gamma_P^L \cdot \gamma_L^L\right)^{1/2} + \left(\gamma_S^L \cdot \gamma_L^L\right)^{1/2} - \gamma_L \right\rfloor + W_{PL}^{SP} + W_{SL}^{SP}}{2\left(\gamma_P^L \cdot \gamma_S^L\right)^{1/2} + W_{PS}^{SP}} \tag{4-25}$$

where superscripts *P*, *S*, and *L* is the polymer, the substrate, and the liquid, respectively.

Physical interactions occur only during the solid-solid and solid-liquid contact. The values of the dispersive component of the surface free energy of polymers and substrates as well as W_{PL}^{SP} and W_{SL}^{SP} are determined by the wettability measurements. A comparison between the experimental values and the calculated values allows us to demonstrate the properties of interfacial polymer-substrate bonds and to evaluate the magnitude of the interfacial acid–base work of adhesion according to

$$W_{PS}^{AB} = \frac{2\left\lfloor \left(\gamma_P^L \cdot \gamma_L^L\right)^{1/2} + \left(\gamma_S^L \cdot \gamma_L^L\right)^{1/2} - \gamma_L \right\rfloor + W_{PL}^{SP} + W_{SL}^{SP}}{\frac{G^A - G^L}{G^A}} - 2\left(\gamma_P^L \cdot \gamma_S^L\right)^{1/2} \tag{4-26}$$

Understanding the enthalpy of interaction at the interface, and the reversible acid–base work of adhesion, allows us to evaluate the number of acid–base interacting sites, at the interface as in Eq. (4-11).

Fig. 4-6 showed the evolution of the fraction of vinyl-acetate involved in acid–base adducts as well as the density of interacting sites vs. the acid–base work of adhesion for EVA copolymer-aluminum assemblies. The acid–base work of adhesion is directly proportional to the fraction of acid–base adducts

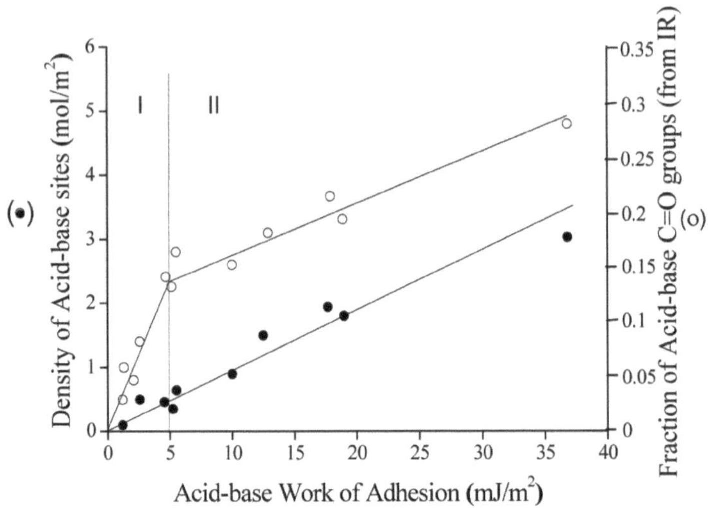

Fig. 4-6 Acid–base adducts versus work of adhesion.

determined from IR analysis. Two regimes are observed. The first one demonstrates, that a distinct increase of interfacial connecting sites contributes to low acid–base adhesion energy until a threshold value is obtained. Also, the acid–base adhesion energy rapidly increases for a low variation the density of the connecting sites. We suppose that these phenomena are largely related to the interfacial mobility reduction and thus molecular dissipation [56].

4.4. MECHANISMS OF ADHESION

Generally, the solid–solid interface or interfacial properties can correspond to the molecular interaction properties between different phases, and be distinguished by the interfacial interaction energies which represent Gibbs surface free energy or excess free energy in physical chemistry. Among the phenomena of solid–solid interface, adhesion has exhibited two major phenomena: The interfacial interactions and energy dissipation [3].

Unless different solid surfaces are clean and all of the residual adsorbents are eliminated, adhesion or bonding of one solid always occurs in different ways. The bond strength formed at the interface between the different solid surfaces is achieved by the materials in contact. If it is achieved at the metal surfaces, generally the interfacial bonding between two dissimilar metals that are brought into contact with each other is stronger than the cohesive bond strength of the weaker of the two metals. As various separating forces are applied to the bonded junction, fracture occurs in the cohesively weaker of the two materials, and this allows the cohesively weaker materials to be transferred to the cohesively stronger one. Fracture generally can occur at the interface [57].

The terms adhesion and cohesion should be distinguished. In the strict sense, cohesion represents the atomic bonding forces closely connected within a material; cohesion represents the forces that exist in the bulk of the material bonding one atom to another or one molecule to another. For instance, if one divides a crystalline material in the bulk and generates two new surfaces, the cohesive bonds are fractured. When, however, two dissimilar materials are brought into solid-state contact with an interface, the bonding of the surface of one solid to the surface of another results in the formation of adhesive bonds. This commonly represents adhesion as opposed to cohesion.

4.4.1. Mechanical Interlocking

The mechanical interlocking model was suggested by McBain and Hopkins [58]. This model considers that the adhesion strength was determined by the major factor of mechanical keying, or interlocking, the adhesive into the cavities, pores, and asperities of the solid surface. One of the most consistent examples, showing the contribution of mechanical anchoring was given many years ago by Borroff and Wake [59], who investigated adhesion between rubber and textile fabrics. These authors mainly showed that penetration of the protruding fiber edges into the rubber was the most important factor in the bonded systems. However, the theory of mechanical keying notes, that the possibility of founding excellent adhesion between smooth surfaces cannot be considered universal. In order to solve this problem, following the approach primarily proposed by Gent *et al.* [60, 61], Wake [62] suggests the effects, of both mechanical interlocking and thermodynamically interfacial interactions that could be considered as significantly multiplying factors for calculating the joint strength G:

$$G = (\text{constant}) \times (\text{mechanical keying component}) \times (\text{interfacial interactions component}) \tag{4-27}$$

Therefore, according to Eq. (4-27), the higher adhesion properties should be achieved by improving both the surface morphology and the physicochemical surface properties, of both the substrate and the adhesive. On the other hand, most of the time, the enhancement of adhesion by mechanical keying can be attributed to the increase of interfacial area due to surface roughness, if wetting conditions are sufficient, to permit the penetration of the adhesive into pores and cavities. However, the penetration of the adhesive into the cavities is not always possible owing to the existence of a back pressure stemming from entrapped air in these pores and cavities. The final depth and the rate of penetration are strongly dependent on the structure of the pores (cylindrical, conical), as shown in Fig. 4-7.

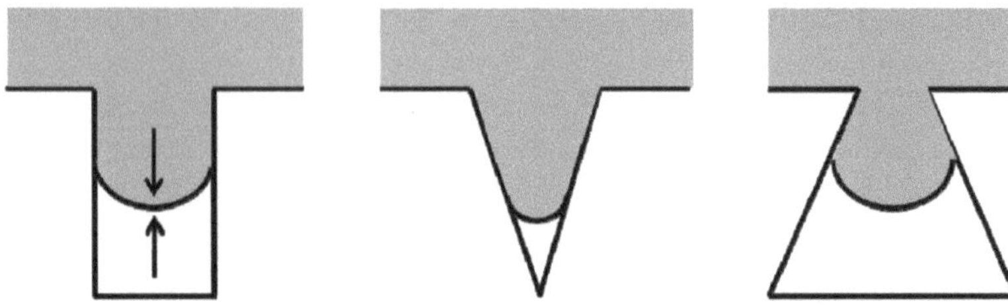

Fig. 4-7 Schematic representation of the effect of porous shape, on the penetration of the adhesive. (arrows are related, respectively, to the wetting forces in the liquid, and to the back pressure due to air entrapped in the cavity.)

As mentioned in different studies compared to the above mechanical interlocking theory [63–65], improvement of adhesion does unnecessarily result from a mechanical keying mechanism but the surface roughness can enhance the energy dissipated visco-elastically or plastically around the crack tip and in the bulk of the materials during joint failure. It is well known that this energy loss is the major consist of adhesion strength in the recent years.

4.4.2. Electronic Theory

The electronic theory of adhesion was first proposed by Deryaguin in 1948 [66–68], who suggested that the mechanism of an electron transfer between the substrate and the adhesive have different electronic band structures, which could form in order to equalize the Fermi levels [69, 70]. This phenomenon induced the formation of the electric double-layer at the interface. Furthermore, Deryaguin *et al.* proposed that the resulting electrostatic forces could contribute significantly to the adhesion strength. Therefore, the adhesive-substrate junction can be investigated as a capacitor. During the interfacial failure of this mechanism, separation of the two plates of the capacitor leads to different increasing potentials during the discharge process. It is considered that the adhesion strength resulted from the attractive electrostatic forces across the electrical double-layer. The energy of separation of the interface, G_e, is related to the discharge potential V_e calculated from Eq. (4-28):

$$G_e = \frac{h\varepsilon_d}{8\pi}\left(\frac{\partial V_e}{\partial h}\right)^2 \tag{4-28}$$

where h is the discharge distance and ε_d the dielectric constant of the medium.

According to such an approach, adhesion would vary with the pressure of the gas in which the measurement was performed. Hence, Deryaguin *et al.* [67] measured by a peel test, the peel energy at different polymer-substrate interfaces, such as poly(vinyl chloride)/glass, natural rubber/glass, or steel systems in argon, and air atmospheres. A significant variation of peel energy versus gas pressure was verified and very good agreement between the theoretical values and the experimental values of G_e was obtained, regardless of the condition of the gas used. According to Deryaguin [68], the adhesion depends on the magnitude of the potential barrier at the adhesive-substrate interface. Although this potential barrier does exist in many cases [71, 72], not shows exactly the correlation between the electronic interfacial parameters, and the experimental adhesion strength. Consequently, it could be suggested that the electrical phenomena often found during the failure processes are the conclusion rather than the cause of high bond strength.

4.4.3. Theory of Weak Boundary Layers

It is well known that alterations and modification of the adhesive and/or adherent can be found in the neighborhood of the interface leading to the formation of an interfacial zone, exhibiting properties that differ from the bulky materials [1].

This problem raised a question by Bikerman [73], who proposed that the cohesive strength of a weak boundary layer (WBL) could always be considered the main factor in determining adhesion, even when failure occurs in the interfacial layer. According to Bikerman's assumption, the experimental adhesion energy G is always equal to the cohesive energy G_c (weak boundary layer) of the weaker interfacial layer. The theory of weak boundary layer based on probability considerations showed that a fracture never propagates only along the adhesive-substrate interface for pure statistical reasons, and, that cohesive failure within the weaker material near the interface is a more favorable condition. Bikerman proposed several types of weak boundary layers using the above assumption, such as those, resulting from the presence at the interface of impurities, small molecules, or short polymer chains.

4.4.4. Diffusion Theory

The diffusion theory of adhesion is base on the adhesion strength of polymers to themselves (autohesion). Mutual diffusion (inter-diffusion) of macromolecules across the interface results in creating an inter-phase. As a mechanism, mainly stated by Voyutskii [74] indicated that the macromolecular chains or chain segments are sufficiently moved and mutually soluble. This mechanism is very important for handling many adhesion problems, such as healing and welding processes. Therefore, if the inter-diffusion phenomena are initiated the adhesion strength should be influenced by different factors, such as contact time, temperature, and the nature and molecular weight of polymers. Such dependences are experimentally observed for many polymer–polymer interfaces. Vasenin [75] developed the quantitative model derived from Fick's first law for founding the diffusion theory that correlates an amount of material w, diffusing in a given x direction across a plane of unit area to the concentration gradient $\partial c/\partial x$ and the time; t calculated from the following Eq. (4-29):

$$\partial w = -D_f \partial t \frac{\partial c}{\partial x} \tag{4-29}$$

where D_f is the diffusion coefficient.

In order to estimate, the depth of penetration of the molecules that inter-diffused into the junction region during the process of contact t_c, Vasenin [75] suggested that the variation of the diffusion coefficient with time was of the specimen of $D_d\, t_c^{-\beta}$, where D_d indicates a constant of the mobility of the polymer chains, and β is of the order value of 0.5. Therefore, it is possible to deduce the depth of penetration l_p as well as the number N_c of chains crossing the interface, as:

$$l_p \approx k(\pi D d t_c^{1/2})^{1/2} \tag{4-30}$$

$$N_c = \left(\frac{2N\rho}{M}\right)^{2/3} \tag{4-31}$$

where k is a constant, N the Avogadro's number, and ρ and M, the density and the molecular weight of the polymer, respectively.

Finally, Vasenin [75] proposed that the measured peel energy G was proportional to both the depth of penetration and the number of chains crossing the interface between the different polymers. From Eqs. (4-30) and (4-31), G indicates the equation below:

$$G \cong K\left(\frac{2N\rho}{M}\right)^{2/3} D_d^{1/2} t_c^{1/4} \tag{4-32}$$

where K is a constant that depends on the molecular characteristics of the polymers in contact.

Experimental results and theoretical predictions from Eq. (4-32) were proposed by [75] to be in excellent agreement with junctions between polyisobutylenes of different molecular weights. In particular, the influence of G on $t_c^{1/4}$ and $M^{-2/3}$ was explicitly evidenced.

An important criticism of the mechanism proposed by Vasenin is that mentioned the energy dissipated viscoelastically or plastically during peel measurements is not exhibited in Eq. (4-9). Nevertheless, the values of coefficients K and D_d are not theoretically quantified but estimated by fitting. Therefore, it can be concluded that the contribution of hysteretic losses to the peel energy is implicitly contained in these constants.

In fact, the major scientific aspect of inter-diffusion phenomena is dependent on the dynamics of polymer chains in the interfacial region. The fundamental understanding of the molecular dynamics of entangled polymers has been significantly developed by the theoretical approach proposed by de Gennes [76], extended by Doi and Edwards [77] as well as by Graessley

[78]. Their approach was derived from the idea that the chains cannot move through each other in a concentrated polymer solution, a melt, or a solid polymer. Therefore, a chain with a random coil conformation is captured in an environment of fixed obstacles. This constraint limits the ability of each chain to move inside a tube. de Gennes analyzed the motion, limited mainly to the effective one-dimensional diffusion along a given path of a polymer chain subjected to such limitation. He stated that this type of motion is wormlike and called it reptation. The reptation relaxation time τ is associated with the movement of the center of gravity of the entire chain through the polymer, which was found to vary with the molecular weight M as M^3. Moreover, the diffusion coefficient D, which indicates the diffusion of the center of the mass of the chain indicates the form $D \sim M^{-2}$.

One of the most important applications of the reptation concept is influenced by crack healing, which is primarily defined as the diffusion of macromolecules across the interface. The healing process was particularly studied by Jud and co-workers [79]. The problem of healing is closely related to the macroscopic strength measurements with a microscopic description of motion. The difference between self-diffusion phenomena in the bulk polymer and healing is that the polymer chains in the former case move over distances many times larger than their gyration radii, whereas in the latter case, healing can be considered as efficient, in terms of adhesion strength in the time that a macromolecule initially close to the interface, needs to move about halfway across this interface. This problem has been analyzed by several authors; who considered that the healing process is controlled by different factors, for example:

1. The number of bridges across the interface (de Gennes [76])
2. The crossing density of molecular contacts or bridges (Prager and Tirrell [80])
3. The center-of-mass Fickian interdiffusion distance (Jud *et al.* [79])
4. The monomer segment interpenetration distance (Kim and Wool [81]).

The resulting scaling laws for the fracture energy G versus time t during the healing process are expressed by the following equation:

$$G \cong t^{1/2} M^{-3/2} \quad \text{for (1) and (2)} \tag{4-33}$$

$$G \cong t^{1/2} M^{-1} \quad \text{for (3)} \tag{4-34}$$

$$G \cong t^{1/2} M^{-1/2} \quad \text{for (4)} \tag{4-35}$$

Even if there are some differences in the value of the exponent of the molecular weight in these expressions, every approach agrees with the dependence of G on the square root of the healing time; t, for example, a dependence has been clearly verified by experiment for poly(methyl methacrylate) polymer [73], which contradicts Vasenin's consideration.

Consequently, it is clear that diffusion phenomena greatly contribute to the adhesion strength in many situations related to the polymer-polymer junctions. Nevertheless, the inter-diffusion of macromolecular chains requires both polymers to be sufficiently soluble and the chains to have sufficient mobility. These conditions are obviously carried out for autohesion, healing, or welding of identical polymers. On the other hand, diffusion can flow a most improbable mechanism if the polymers have low solubility, are highly cross-linked or crystalline, or are placed in contact at temperatures far below their T_g. Nevertheless, with junctions between two immiscible polymers, the interface could be strongly bonded by the existence of a diblock copolymer, in which each molecule consists of a block of the first polymer bonded to a block of the second polymer, or each of the two blocks is miscible with one of polymers. Each block diffuses or 'dissolves' into the corresponding polymer when the copolymer molecules concentrate at the interface. Therefore, the improvement in joint strength can be closely related to an inter-diffusion process. When the molecular weight M of each block of the copolymer is less than the critical entanglement weight M_e, for which the entanglements of chains occur in the polymer, the adhesion strength can be represented in terms of a suction mechanism. However, when $M > M_e$, the failure of the joint generally requires the rupture of the copolymer chains.

4.4.5. Intermolecular Bonding

It is clear that the intermolecular bonds formed across the adhesive-substrate interface can greatly increase the level of adhesion between the two different materials. The terms primary and secondary are derived from the relative strength or bond energy of each type of interaction. As the typical strength of a covalent bond has a value of 100 to 1000 kJ/mol, whereas those of van der Waals' interactions, and hydrogen bonds, do not exceed 55 kJ/mol [47]. It is noted that the intermolecular bonds depend on the reactivity of both the adhesive and the substrate. The most important factor at adhesion areas involving the interfacial bonds is the use of adhesion promoter molecules (generally called coupling agents), in order to enhance the joint strength between the adhesive and substrate. This agent is able to participate in an intermolecular reaction at both ends, with the substrate on one side, and the polymer on the other side, thus performing a role in the intermolecular bridged agent at the interface. The coupling agents generally consist of silane molecules that perform the most common type of adhesion promoter [82, 83]. They are widely applied in systems involving glass or silica substrates, particularly with the polymer-based composites reinforced by glass fibers [84]. Additionally, they play an important role in the improvement in joint strength and provide a significant enhancement of the environmental resistance of the interface, which can be achieved by the participation of such coupling agents in the reaction.

The influence of intermolecular bonds, on the joint strength G, and more precisely on the intrinsic adhesion of fracture energy G_0, has been analyzed in several studies. The most appropriate and more precise work in this field was performed by Gent and Ahagon [85], who estimated the influence on the adhesion of bondings of polybutadiene to glass found at the interface by using the silane coupling agents.

In these experiments, the surface density of the interfacial bonds between the glass substrate and the cross-linked elastomer were investigated by treating the glass plates with different mixtures of vinyl- and ethyl-terminated silanes. Obviously, both the species form siloxane bonds on the glass surface. Moreover, it was assumed that the vinylsilane could react with the polybutadiene during the cross-linking treatment of this rubber, in which a functional reaction was involved. On the other hand, an intermolecular reaction between the ethyl group of the latter silane and the elastomer is improbable. Therefore, Gent and Ahagon [85] suggested that the intrinsic fracture energy G_0, increases linearity, as in Fig. 4-8, with the surface concentration of vinylsilane, in excellent agreement with their assumptions, thus verified the important influence of intermolecular bonds on adhesion strength [86]

Other experimental results [87] exhibit the inhibition of the negative temperature coefficient (NTC) by using the silane coupling agents, as illustrated in Fig. 4-9. The presence of a negative temperature coefficient effect has an adverse influence on the application of the positive temperature coefficient (PTC) switch because of the instability of conductivity. Therefore, it is

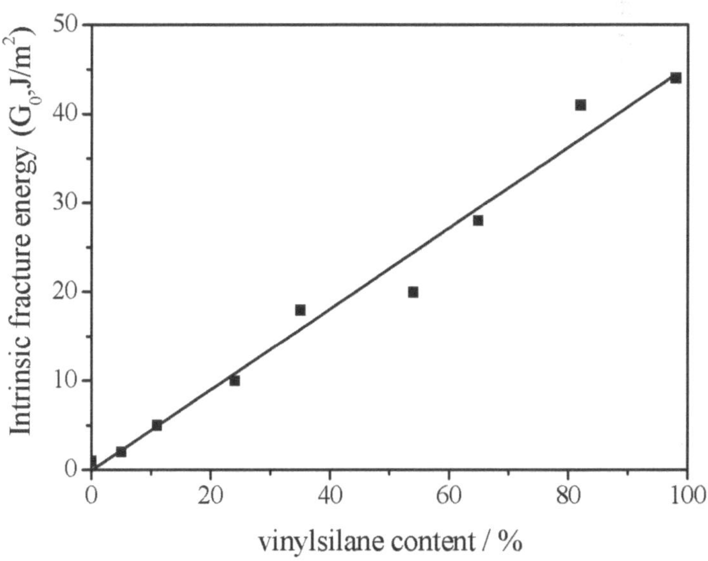

Fig. 4-8 Intrinsic fracture energy G_0 of polybutadiene-glass assemblies as a function of the concentration of vinylsilane in the mixture of vinyl- and ethylsilanes on the glass surface.

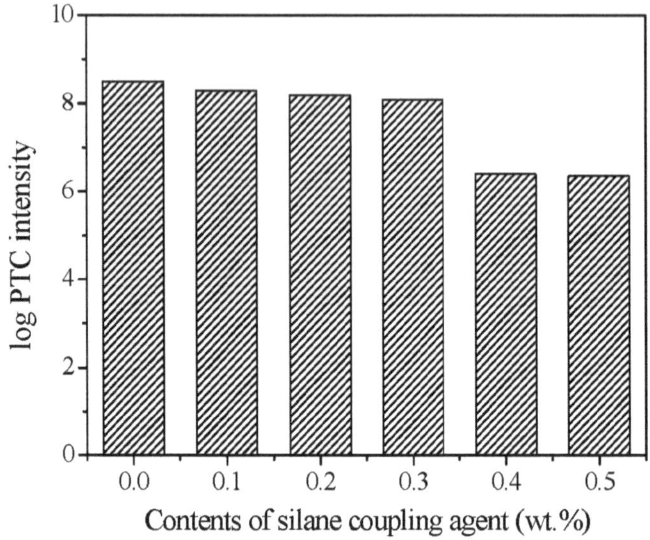

Fig. 4-9 Positive temperature coefficients (PTC) intensity of carbon black/high density polyethylene composites as a function of silane coupling agent contents.

important that a strategy is found to eliminate the negative temperature coefficient effect. The carbon black/high density polyethylene composites were prepared by carbon black as a filler of the treated silane coupling agents. From the results, it is achieved that the inhibition of negative temperature coefficient took place when added to above 0.3 wt.% of silane coupling agent. Consequently, the silane coupling agents not only enhance the joint strength between the adhesive and substrate, but also reduce the negative effect of the composites.

Analyzing the effect of the connecting chains grafted onto solid surfaces, but now considering that chain scission is the major phenomenon in controlling the adhesion strength instead of a suction process de Gennes and coworkers [88, 89] proposed that the intrinsic interfacial fracture energy G_0 is as follows:

$$G_0 \cong W + W_r N\sigma \tag{4-36}$$

$$\approx W_r N\sigma \tag{4-37}$$

where W, N, and σ is the work of adhesion, the interfacial areal density, and the number of monomers consist of the chains, or the chain segments sucked out during the crack propagation, respectively.

Furthermore, W_r is the energy required to break the network of covalent bonds formed by the connecting chains (a few J/m^2). Eqs. (4-36) and (4-37) mainly show that if W_r can be regarded as a constant, the intrinsic fracture energy increases in linearity with the surface density of interfacial bonds σ, which is in excellent agreement with the experimental value previously mentioned. Moreover, for the results gained by Gent and Ahagon [85], it is possible that the treatment of the glass surface by pure vinylsilane may represent the destruction of intermolecular bonds in Eq. (4-12), but the treatment with pure ethylsilane, may be related to the suction phenomenon. Therefore, the ratio of G_0 obtained in the former occasion to that in the latter occasion should be equalized to the ratio W_r/W. A theoretical value of 35 was investigated [89] for this ratio, and was in excellent agreement with the results illustrated in Fig. 4-6.

In some papers [90–92], Brown and co-workers suggested the enhancement of adhesion between two immiscible polymers. Since one composition of the block is poly (methyl methacrylate) and the other is polystyrene, which is totally miscible with poly (propylene oxide), it was reasonably expected that the copolymer organized at the interface, because each block dissolved in the respective homo-polymer. The molecular weight of these blocks is always higher than the critical molecular weight M_e, for which the entanglements of chains take place, in the homo-polymers. Experimentally, Brown *et al.* [91] applied partially or fully deuterated copolymers to investigate the deuterium on the fracture surface, after dissociation by secondary ion mass spectrometry and forward-recoil spectroscopy. A scission of the copolymer chains near the junction point of the two blocks was exhibited, suggesting that the diblock copolymers were well arranged at the interface, regardless of their molecular weights, with their junction accurately located at the poly(methyl methacrylate)-poly(propylene oxide) interface. Consequently, proposing that all the effectively entangled chains in the material are attracted to the fibril, the fracture energy G of the polymer is established to be directly related to the square of both the areal density σ of the entangled chains and the force f needed, to break the polymer chain:

$$G \cong \sigma^2 f^2 \frac{D}{S} \tag{4-38}$$

where D is the fibril diameter and S the stress at the craze-bulk interface, which is represented to be constant.

Brown [92] considered that diblock copolymer coupled interfaces between poly (methyl methacrylate) and poly (propylene oxide) are ideal experimental systems for testing the validity of his model. Indeed, a linear dependence of the interfacial fracture energy G on the diblock copolymer surface density σ on logarithmic scales is observed for the copolymers of different molecular weights. A slope of 1.9 ± 0.2 was found for the master straight line, in good agreement with Eq. (4-38). Nevertheless, it is worth noting that Brown's results involving chain scission at the interface and leading to a dependence of G on σ^2 contradict previous examples in which linear relationships between G and σ were established. More precise analyzes of the effect on the adhesion strength of crack velocity, molecular weight and intermolecular structure of the copolymers, etc. have been made [93–95].

4.4.6. Characterization of Adhesion

The Williams–Landel–Ferry (WLF) time-temperature equivalence predicts that the strength of the adhesive bonds depends both on the temperature and the separation rate. Gent *et al.* [85] suggested that the measured value of adherence energy (G_a) is composed of two terms, the thermodynamic work of adhesion (W_a), and a dissipation function ($f(r)$: $G_a = Waf(r)$, where r is the rate of crack propagation and $f(\mathrm{r})$ a function of r corresponding to the dissipated energy).

The dissipative phenomena have been extensively studied [96, 97]. However, the exact relation among adherence, adhesion, and polymer viscoelastic properties is still uncertain. Interfacial bonds allow stress transmission between the substrate and the adhesive. These stresses will constrain the chains directly linked to the substrate and also the neighboring chains (due to a cooperative effect, assured by entanglement or cross-linked structures), and all these constrained chains will consequently move until they rupture, thus dissipating energy. As the solicitation rate increases, the chain movements come to be more difficult and require more energy, leading to an increase of the adherence energy (G). A greater number of interfacial bonds generally create an increase in the number of constrained chains in the crack-tip region and in the energy dissipation. Moreover, stronger interfacial interactions will increase both the adhesion and energy dissipation. However, strong interfacial bonds are favorable only if the adhesive is able to sufficiently dissipate. Both, the interfacial interactions and dissipation capacities have to be taken into account to enhance the adhesive behavior of a polymer. In some cases, it can be more efficient to increase the dissipation properties by varying the molecular weight and crosslink density rather than increasing the interfacial bond number and strength.

The previous paragraphs demonstrate that the adhesive performances of a polymer depend on various parameters, involving among others the surface properties and rheological and mechanical properties. Before an adhesive is used, these properties should be characterized [98].

Most of all, it is useful to quantify the molecular parameters, such as molecular weight, degree of branching, crystallinity degree, cross-link density, stereoregularity, chemical composition, grafting density, etc. [99].

The surface characteristics can be investigated by using wettability techniques by determination of the surface energy and spectroscopic techniques and by infrared spectroscopy or x-ray photoelectron spectroscopy (XPS), for the determination of acid–base character. These methods can also be used in the formation of acid–base or polar interactions [100].

Rheological and mechanical analysis uses tensile tests to calculate the Young modulus at different speeds or temperatures and to investigate the creep and relaxation properties and dynamic mechanical measurements and rheometers for the viscoelastic behavior [101].

Substrate rugosity can be guessed by using rugosimetry and microscopy techniques i.e., optical, electronic, or atomic force microscopy [102].

Various methods can be used to characterize an inter-phasial zone. In all methods, the aim is to distinguish morphological, chemical, mechanical, and crystalline properties from the forming of bulk material properties [103, 104].

The adherence tests allow the prediction of the strength and the service lift of adhesive connections. Destructive tests that are based on the mechanical separation of the assembly and the measurement of a force or a fracture length are commonly used. The adopted test must be a representative of the loading, to be experienced in the finished product. The non-destructive tests like the ultrasonic test can be used during the lifetime of a product to evaluate the remaining life if the structure is subjected to fatigue, creep, or chemical attack.

The peel test is commonly used in the flexible substrate. Diverse experimental factors like sample dimension, adhesive thickness, or peel angle and obviously the rate and temperature are able to affect the measured force [105].

Shear tests are adapted to rigid substrates. The single-lap joint induces a non-uniform shear distribution and is unsuitable for establishing the correct stress-strain data. A different shear test which allows a more uniform shear stress, is the napkin ring, constituted by an adhesive layer inserted between two tubular rigid substrates. A fracture mechanics approach could lead to a better interpretation, of the adherence tests results [101].

4.5. ADHESIVE CONTROL

Various continuum models have been developed to indentify the contact phenomena between the solids. There has been much disagreement as to the appropriateness of these models (Derjaguin *et al.* [106–108] and Tabor [109–111]). Experimental verification can be more difficult owing to uncertainties over the influence of contaminants and asperities dominating the contact. A range of models representing the contact between both the non-deformable and deformable solids in various environments are discussed in more detail in this section. In every example, the mechanism of a sphere on a plane is considered, because this is the most fitting to the experimental techniques used to perform nanoscale adhesion.

4.5.1. Non-deformable Solid Interfaces in Various Conditions

4.5.1.1. In vacuum

For smooth, ideal, rigid solids, the Derjaguin approximation [112] relating the force law between a sphere of radius R and a flat surface to the energy per unit area $W(D)$ between two planar surfaces, separated by a distance D, gives:

$$F(D)_{\text{sphere}} = 2\pi RW(D)_{\text{plane}} \qquad (4\text{-}39)$$

This equation is included, because it is applicable to any type of force law as long as the range of interaction and the separation, are much less than the radius of the sphere. Therefore, the force to overcome the work of adhesion between a rigid sphere and a flat surface written in terms of the surface energy $\Delta\gamma$ is:

$$F_{\text{pull-off}} = 2\pi R\Delta\gamma \tag{4-40}$$

where

$$\Delta\gamma = \gamma_{\text{sphere}} + \gamma_{\text{flat}} - \gamma_{\text{interface}}(\text{Dupré}) \tag{4-41}$$

If the sphere and the flat surface are the equal material give:

$$\gamma_{\text{interface}} = 0 \tag{4-42}$$

and

$$\gamma_{\text{sphere}} = \gamma_{\text{flat}} = \gamma_s \Rightarrow F_{\text{pull-off}} = 4\pi R\gamma_s \tag{4-43}$$

This equation assumes that the only source of adhesion occurs in the solid–solid contact.

4.5.1.2. Forces due to capillary condensation

For experiments performed in air, the adhesive force acting between the two bodies may be dominated by the presence of capillary condensed water. These additional forces generated by capillary condensation are calculated for smooth, ideal, rigid solids. For a sphere and a flat surface participating in a liquid bridge the force F due to the Laplace pressure within the meniscus is as in the following equation [113]:

$$F = 4\pi R\gamma_{LV}\cos\theta \tag{4-44}$$

where γ_{LV} is the surface tension of the liquid in the condensate and θ the contact angle of this liquid on the solid.

4.5.1.3. Non-deformable solids in condensable vapor

The capillary forces just discussed act as an additional force; thus the force needed to separate a rigid sphere and a flat surface of the same material joined by a liquid bridge is given by:

$$F_{\text{pull-off}} = 4\pi R\gamma_{LV}\cos\theta + 4\pi R\gamma_{LV} \tag{4-45}$$

4.5.2. Deformable Solids

The above mentioned models considered the incompressible bodies; however, this is never the case in practice. The following section suggests models of the actual contact between the deformable solids.

4.5.2.1. Hertz

For a deforming solid, Hertzian analysis [114] gives the simplest assumption, because adhesive forces are ignored, i.e., no pull-off force and zero contact area. An applied force P, a tip radius R, and the contact diameter $2a$, are given:

$$2a = 2\left(\frac{4RP}{3E*}\right)^{1/3} \quad (4\text{-}46)$$

where

$$E* = \left(\frac{1-\nu_1^2}{E_1} + \frac{1-\nu_2^2}{E_2}\right)^{-1} \quad (4\text{-}47)$$

where E_1, ν_1 and E_2, ν_2 represent Young's modulus and the Poisson's ratio of the sphere and plane, respectively.

Because the surface forces are ignored in this model, it cannot be applied to adhesion measurements. However, it is included here, because it is used as the basis for the other models.

4.5.2.2. Johnson, Kendall, and Roberts (JKR)

The theory of Johnson, Kendall, and Roberts [115] incorporates adhesion from the change in the surface energy only where the surfaces are in contact. This is represented as represents:

$$2a = 2\left\{\left(P + 3\pi R\Delta\gamma + \sqrt{6\pi RP\Delta\gamma + (3\pi R\Delta\gamma)^2}\right)\frac{3R}{4E*}\right\} \quad (4\text{-}48)$$

where $\Delta\gamma$ is the work of adhesion, P the applied force, R the tip radius, and $E*$ as shown in Eq. (4-47).

This is dependent on the force required to divide the sphere and the flat surface after contact (known as the adhesive force) representing $F_{\text{pull-off}}$ by the following equation:

$$F_{\text{pull-off}} = \frac{3}{2}\pi R\Delta\gamma \quad (4\text{-}49)$$

Note from Eq. (4-48), that with contact radius a_0 even at zero applied load there is a finite contact area due to the adhesive forces alone, which is given by:

$$a_0 = \left(\frac{9\pi R^2\Delta\gamma}{2E*}\right)^{1/3} \quad (4\text{-}50)$$

Although this model clearly indicates some experimental observations of contact and separation [110, 111], the assumptions made in its formation are in fact unphysical. They propose that the solid does not interact at the contact region, whereas electrostatic and van der Waals' forces are non-zero at separations of several nanometers. The assumptions, proposed by JKR lead to the highest values of stress nearest of the connecting neck between the sphere and the plane.

The model is established to be most appropriate for contact between the low-elastic-moduli materials with large radii when they have the higher work of adhesion. In comparison, the following model suggests that the surface forces extend over a limited range and act in the region just outside the contact. It is found to be more appropriate for systems with small radii of curvature, lower work of adhesion, and higher modulus [116].

4.5.2.3. Derjaguin, Muller, and Toporov (DMT)

Derjaguin, Muller, and Toporov [106] suggested that under the influence of surface forces, the sphere will deform in the contact region in accordance with the Hertzian model. Since the deformation is taken as Hertzian, the surfaces do not separate until the contact area is reduced to zero. At this instant, the pull-off force, $F_{\text{pull-off}}$, is represented as

$$F_{\text{pull-off}} = 2\pi R \Delta\gamma \tag{4-51}$$

This is the same as Eqs. (4-40) and (4-41) for the value for non-deformable solids in vacuum. However, for the deformable solids, DMT theory, a_0 gives a limited contact radius that zero applied load:

$$a_0 = \left(\frac{3\pi R^2 \Delta\gamma}{2E^*}\right)^{1/3} \tag{4-52}$$

Although the DMT theory attempts to include distance-dependent surface interactions into the adhesion problem, it does not explain the influence that the surface forces have on the elastic deformation. In other words, it does not predict the neck formation proposed by JKR [115].

4.5.2.4. Maugis and Dugdale

Maugis [117] suggested that a surface force acts both inside and outside the contact perimeter. The attractive interaction is proposed to be constant up to a separation of h_0, at which point, it suddenly falls to zero. The problem of this arbitrary approximation is that it is only concerned with low values of the elastic parameter, λ,

$$\lambda = \sigma_0 \left(\frac{9R}{2\pi w E^{*2}}\right)^{1/3} \tag{4-53}$$

where $\sigma_0 h_0$ is w, the work of adhesion, R the radius, and E^* the combined modulus, as mentioned in Eq. (4-46). The value of h_0 is indicated as the maximum attractive force, and the work of adhesion corresponds to a Lennard-Jones potential [118] and the Dugdale approximation [119].

Even for lower values of λ the discrepancy in the elastic compression is lower than the atomic spacing. The net force acting on the contact consists of a Hertzian pressure related to the contact radius and an adhesive Dugdale component extending beyond the contact region up to a second radius. This radius can be found by solving two simultaneous equations.

4.5.2.5. Muller, Yushchenko, and Derjaguin (MYD)/Burgess, Hughes, and Whit (BHW)

Muller *et al.* [116] formed a more complete representation for the adhesion between a sphere and a plane by approving the solid-solid interaction to be a specified function of the local separation between the surfaces. Occasionally, the uncertainties inherent in experiments of this type suggest that a more precise model often brings up little advantage over the JKR and DMT models.

As mentioned above, there are a number of different possible models that can be applied to the contact of an elastic sphere and a flat surface. Depending on the scale of the objects, their elasticity and the load to which they are pointed, one particular model can be more suitable than the others. The evaluation of the combination of relevant parameters can be made via the two non-dimensional coordinates λ and P [120]. The first parameter, λ, can be represented as the ratio of elastic deformation, resulting from the adhesion to the effective range of the surface forces. The second parameter, P, is the load parameter, and corresponds to the ratio of the applied load to the adhesive pull-off force. An adhesion map of the model can be illustrated in Fig. 4-10 [121].

Johnson and Greenwood [121] indicated both a numerical analysis using the Lennard-Jones potential and Maugis's analysis [117] using the Dugdale's approximation [119] applied throughout a map with the Hertz, JKR, DMT and rigid zones being the effective regions where some simplification is possible. However, in practice both analyzes become less appropriate at higher values of λ (in the JKR zone).

Unfortunately, determining the appropriate field of an adhesion requires the knowledge of a number of parameters, which may not be easily achievable,

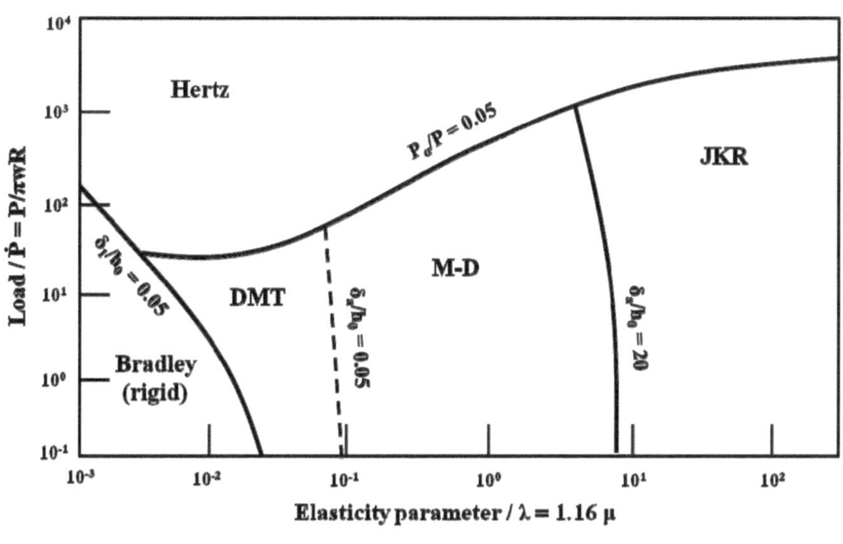

Fig. 4-10 Map of the elastic behavior of bodies.

particularly for nanoscale experiments. Such parameters include the radius of the probe or the combined modulus, which can be affected even by a factor of the contamination. In literature, the most popular models tend to be DMT and JKR. The induced problems from the incorrect use of such models are usually less than the errors encountered in the experimental determination of the suitable parameters.

4.5.2.6. Liquid bridge

When working in the presence of condensable vapors (i.e., water under ambient conditions), it is possible for an annulus of capillary condensate to form as the tip approaches the surface of the sample, as shown in Fig. 4-11. Capillary forces can then dominate the interaction between the sphere and the plane. The general formula for the adhesive force between the two due to capillary condensation is given by:

$$F = F(\Delta P) + F_{\text{solid-solid}} + F(\gamma_{LV}) \quad (4\text{-}54)$$

where $F(\Delta P)$ is the force due to the Laplace pressure ΔP within the meniscus, $F_{\text{solid-solid}}$ the force due to direct solid-solid interaction, and $F(\gamma_{LV})$ the separated force due to the liquid/vapor surface tension.

Theory gives:

$$F(\Delta P) = 4\pi R \gamma_{LV} \cos\theta \quad (4\text{-}55)$$

$$F_{\text{solid-solid}} = 4\pi R \gamma_{SL} \quad (4\text{-}56)$$

$$F(\gamma_{LV}) = 2\pi R \gamma_{LV} \sin\phi \sin(\theta + \phi) \quad (4\text{-}57)$$

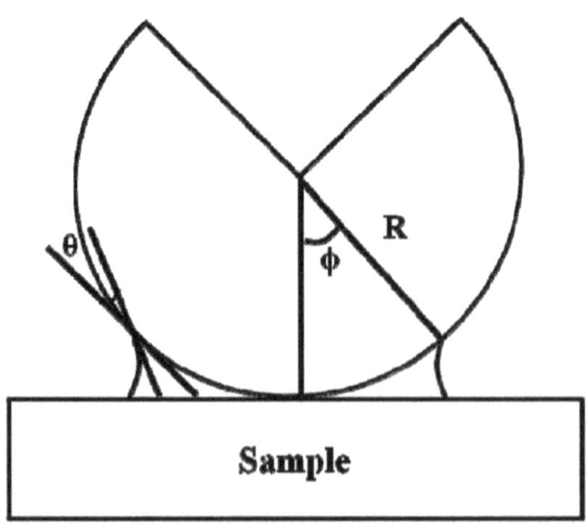

Fig. 4-11 Schematic of a liquid meniscus around the sphere.

where R is the radius of the sphere, θ the solid/liquid contact angle, γ_{SL} the solid/liquid interfacial free energy, γ_{LV} the liquid/vapor interfacial free energy, and ϕ as shown in Fig. 4-11.

With deformable solids, Eq. (4-56) may be replaced, by $F_{\text{solid-solid}} = 3\pi R\gamma_{SL}$. The previous equations depend on the liquids acting in a macroscopic, hydrodynamic manner. If the contact angle is suggested to be small and the radius of the sphere greatly exceeds the neck radius of the liquid condensate such that ϕ is small, then $\sin\phi\ \sin(\theta + \phi) < 1$. Thus the contribution of the direct surface tension can be ignored when it is compared to the Laplace pressure contribution.

When the sphere and plane are separated by a small distance D, as illustrated in Fig. 4-12, then the force owing to the Laplace pressure in the liquid bridge may be calculated so that the total surface free energy, U_{total}, of the system changes with separation [122]:

$$U_{\text{total}} = \gamma_{SV}(A_1 + A_2) - \gamma_{SL}(A_1 + A_2) - \gamma_{LV}A_3 + \gamma_{SV}A_{\text{solids}} \qquad (4\text{-}58)$$

where γ_{SV} is the solid/vapor interfacial free energy and A represents the area of each interface as shown in Fig. 4-10.

If $A_1 \approx A_2 \approx \pi R^2 \sin^2\phi$, then for small ϕ,

$$U_{\text{total}} = 2\pi R^2\phi^2(\gamma_{SV} - \gamma_{SL}) + \text{constant} + \text{smaller terms} \qquad (4\text{-}59)$$

But $\gamma_{SV} - \gamma_{SL} = -\ \gamma_L\cos\theta$ (Young equation), so

$$U_{\text{total}} \approx -2\pi R^2\phi^2\gamma_L\cos\phi + \text{constant} \qquad (4\text{-}60)$$

giving

$$F = -\frac{dU_{\text{total}}}{dD} = 4\pi R^2\phi\gamma_L\cos\theta\frac{d\phi}{dD} \qquad (4\text{-}61)$$

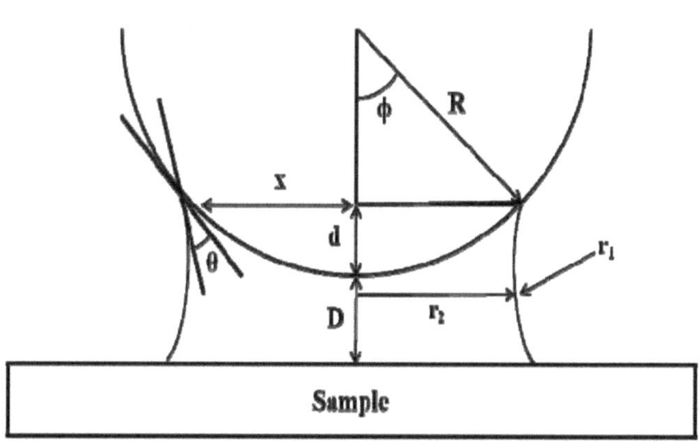

Fig. 4-12 Schematic of a liquid meniscus between the sphere and plane at finite separation.

Calculating the volume by considering a straight-sided meniscus and suggesting that the liquid volume indicated by a constant, Eq. (4-61) then gives:

$$F = \frac{4\pi R\gamma_L \cos\theta}{1 + D/d} \tag{4-62}$$

Use of the condition of constant meniscus volume is most suitable when growth and dissolution of the meniscus is comparatively slow. An alternative is considered by Kelvin equilibrium condition. The Kelvin equation relates the equilibrium meniscus curvature to the relative vapor pressure; and if the Kelvin equilibrium is maintained during the separation process then the adhesive force gives [123]:

$$F = -4\pi R\gamma_{LV}\cos\theta\left(1 - \frac{D}{2r_m\cos\theta}\right)(\text{Kelvin condition of constant } r_{\rm m}) \tag{4-63}$$

where the meniscus curvature,

$$r_m = \left(\frac{1}{r_1} + \frac{1}{r_2}\right) \tag{4-64}$$

and r_1 and r_2 are as exhibited in Fig. 4-12.

In the limitation of the small surface separations, the adhesive force and its gradient tend to have an equalized value for both, the constant-meniscus-volume and Kelvin-equilibrium conditions.

The derivations of the above equations have been based on the principles of thermodynamics and the macroscopic concepts of density, surface tension, and radius of curvature. They may therefore, cease to be appropriate as the mean radius of curvature approaches molecular dimensions.

As proposed by Fogden and White [124], including capillary condensation with the Hertz approximation suggests that the pressure outside the contact area; i.e., adhesion faced with the non-energetic problem through the tensile normal stress exerted by the condensate in an annulus around the contact circle. The resulting equations are difficult to solve analytically; their asymptotic analysis, may be summarized as follows.

For the non-adhering bodies in contact when there is capillary condensation, the previous result for rigid solids is found to apply more generally to systems of small, hard, but deformable spheres in contact in the vapor near saturation:

$$F_{\text{pull-off}} = 4\pi R\gamma_{LV} \text{ for } \theta = 0 \tag{4-65}$$

In the limitation of large, softer solids in vapor pressure closer to the value marking the onset of capillary condensation, the generalized Hertz and the original JKR theories are qualitatively identical. However, the contact area for a zero applied load will in general be different, since it is dependent upon the source of adhesion:

$$F_{\text{pull-off}} = 3\pi R\gamma_{LV} \tag{4-66}$$

and

$$a_0 = \left[9\pi\gamma_{LV}\frac{R^2}{E*}\right]^{1/3} \tag{4-67}$$

Thus $\frac{\Delta\gamma}{2}$ in the JKR approximation is replaced by γ_{LV}. It is noted that the elastic modulus affects the contact area but not the adhesion force.

4.6. ADHESIVE BEHAVIORS AT INTERFACES

4.6.1. Introduction

Glass is a uniform material, because every part of its structure is the same, down to the molecular scale. However, other materials, such as metals, polymers, ceramics, and composites, are not generally homogeneous in structure, but consist of different interfaces. The domains are readily exhibited by viewing polished surfaces with a microscope, as seen in Fig. 4-13 [125], which shows the individual crystal grains, composed of the separating interfacial boundaries, in which the grains are fixed together, through molecular adhesion. Such interfaces can generate huge effects on the properties of the material as Newton observed. However, the most significant property enhancements are established, when the interface structure and adhesion are optimized, to specific values. This is best proposed by considering the specific fillers added to polymers.

4.6.2. Particular Composites

A typical case is rubber filled with colloidal carbon particles that are mixed into the polymer. Hancock [126], one of the first scientists to study the mixing of particles into polymers, decried the use of fillers in rubber, because he viewed it

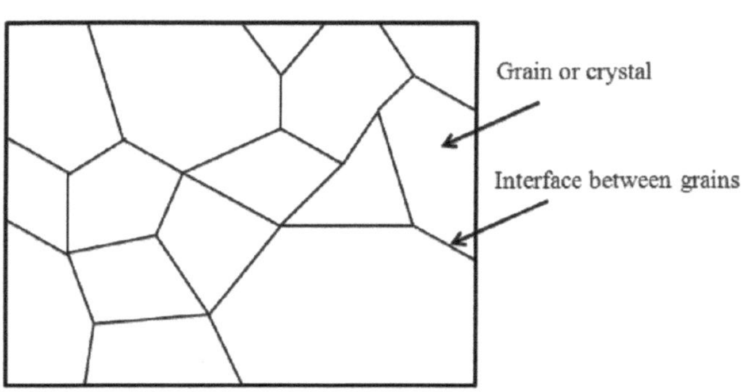

Fig. 4-13 Grain boundaries traced on alumina surface after firing.

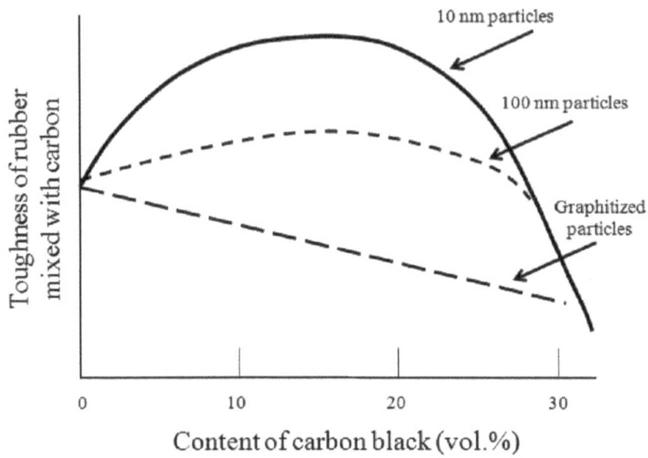

Fig. 4-14 Toughness changes in carbon black reinforced rubber as the interface is modified.

as adulteration. However, he verifies very large beneficial effects, of fine particulate sulfur, zinc oxide, and carbon, which could be added into the rubber, to improve strength and toughness. The interfaces between the particles, and the rubber were responsible for the property enhancement in the composite. This was confirmed by experiments that varied the interface systematically. For example, as an amount of carbon black is dependent on increasing the interfacial region, the toughness improved, reaching a maximum at around 20 percent of volume of the material, in Fig. 4-14 [125]. Furthermore, when the carbon surfaces were graphitized by heating to change the molecular structure of the carbon, the toughening effect was reduced, and when the particles were increased in diameter to reduce the interfacial area, the toughening was reduced [127–129].

Since those experiments of mixing the particles into natural rubber, an enormous industry has developed to mix powders into polymers for a wide range of applications. Some of these are just to recycle the former waste products, such as wood chips or slate dust. Mixing these powders with a polymerizable resin, such as polyurethane, styrene, or methyl methacrylate, produces moldable sheets with excellent properties. These techniques are applied to resin concretes, for repairing roads or buildings, kitchen work-tops, sinks, and sanitary ware. They are manufactured by grinding silica, calcium carbonate, or alumina trihydrate with a monomer and casting after a polymerizing process in a mold. Recently, in the most technical applications, specific powders and resins are used to obtain specific effects, such as wear resistance of silica filled acrylics for dental repair or hydroxyapatite filled polymers for bone implants.

Furthermore, according to the previous work by Park *et al.* [130], the oxygen functional groups on carbon black lead to enhancing the properties of polar

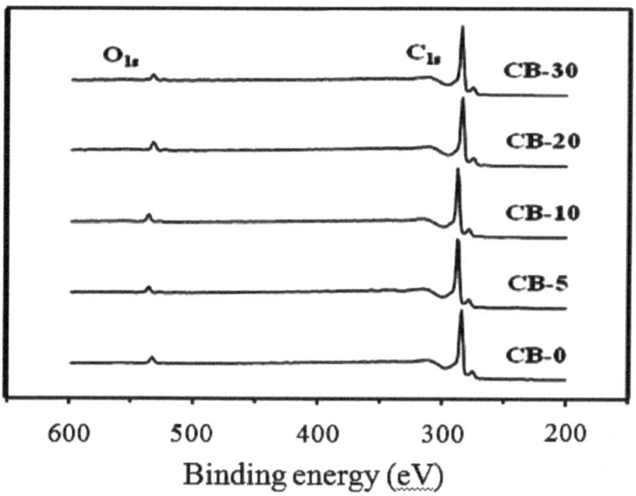

Fig. 4-15 X-ray photoelectron spectroscopy survey scan spectra of the carbon blacks.

rubber (Nitrile butadiene rubber, NBR). Fig. 4-15 shows the x-ray photoelectron spectroscopy (XPS) survey scan spectra of the carbon blacks after the oxygen plasma treatment. The x-ray photoelectron spectroscopy spectra, show carbon and oxygen (binding energy, 285 and 532 eV, respectively) peaks. The O_{1s} peak is probably due to the oxygen-containing functional groups created on the carbon black surfaces by the treatment. This indicates that the oxygen plasma treatment produces active carbon radicals, from the hydrocarbon backbone. Then, upon exposure to oxygen gas, the active radicals react, increasing the oxygen-containing functional groups on the carbon black. These functional groups can react with the nitrile butadiene rubber (NBR), which has enhanced mechanical properties.

4.6.3. Effect of Interfaces

A simple demonstration about interface adhesion is essential to composite material strength can be achieved by doping molecules at the interface, which reduce the grain boundary adhesion. For example, adding copper to molybdenum decreases the intergrain adhesion considerably, and the resulting composite is significantly brittle, falling apart at the grain boundaries. Only one layer of molecules covering the interface is adequate to generate this effect, so it influences the composite materials, and can be tremendously sensitive to small amounts of impurities of this kind. Fig. 4-16 illustrates the sensitivity effect for cubic grains 100 μm in size [125]. The volume of interface molecules relative to the total material volume is of the order $3d/L$, where d is the molecule diameter, and L is the grain size. If the contaminant molecule is 0.1 nm in

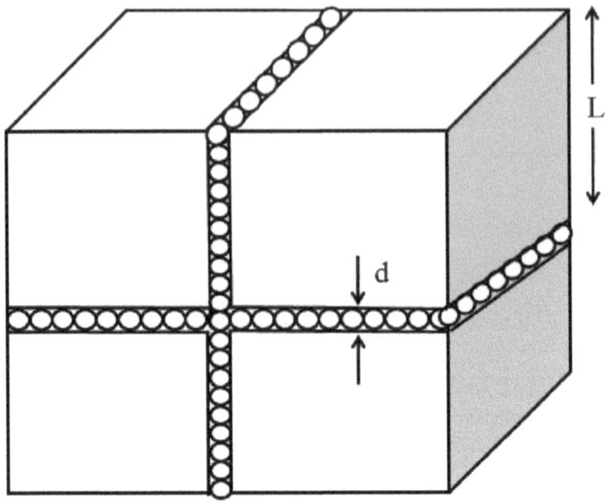

Fig. 4-16 Cubic grains with molecules controlling adhesion at the grain boundaries.

diameter, then $3d/L$ is 3 ppm. This investigation verifies that very small quantities of interface contamination could ruin the properties of a composite material.

The most dramatic effect of interfacial adhesion on material properties is found with glass fiber reinforced epoxy resins. Both the glass and the epoxy are brittle materials, which have a resistance to cracks, less than 100 J/m^2. However, when the two brittle materials are combined by mixing glass fibers into liquid epoxy, followed by polymerizing the resin to solidify it, then the crack resistance of the composite structure can increase up to 10 kJ /m^2, several orders of magnitude higher than each component separately. Such high toughness is suitable for many structural applications, for example aircraft, cars, and buildings. This principle of improving glass by interposing the polymer layers, was invented by Benedictus who found that glass flasks, which had been used to boil cellulose nitrate solutions were much more impact resistant than clean flasks. The glass laminates, produced by sticking glass sheets, together with polyvinyl butyral are now widely commercialized in armor and in car windshields.

Nowadays, many buildings are made from glass fiber reinforced plastic. Thus, it is apparent that no material should be dismissed, because it shows weak or brittle characteristics. Principally, even the most brittle materials, such as diamond, could be made tough by the introduction of suitable adhesive interfaces. This could be achieved, by adding a few parts per million of the chemical additive.

Of course, the additive would have to be able to arrange itself in the structure, as illustrated in Fig. 4-17(b), which shows how a crack disappears at interfaces. If an additive arranges itself uniformly, as in shown Fig. 4-17(a), then the crack

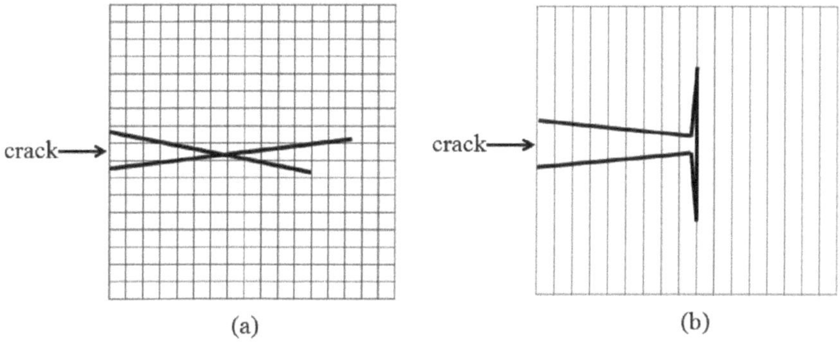

Fig. 4-17 (a) Crack penetrating along isotropic interfaces and (b) crack deflected by perpendicular interfaces.

could propagate easily, and the material would be brittle. However, if the additive produced controlled the interfaces, perpendicular to the crack path, as shown in Fig. 4-17(b), then the crack would disappear [125]. This idea that cracks can be much inhibited by interfaces, is a relatively recent one, derived from the work of Cook and Gordon [131]. Previously, it was thought, that the strength of the fibrous components was the most important parameter [132].

A good model for the effect of the interface in a composite is represented by the bimetallic strip [133]. This consists of two strips of metal with different expansion coefficients, stuck together at the interface. Each strip remains flat when heated, however, when joined together, the strips bend as the temperature changes. These new properties of the composite materials i.e., bending on heating depend on adhesion at the interface between the component strips, since no bending would occur, if the strips were not bonded. In the same way, two strips of materials may indicate little resistance to cracking, when tested separately, but when they are combined adhesively, the strips may indicate higher crack resistance. These properties of composite toughness mainly depend on the interface adhesion [134], but material properties are now the most important, owing to the elastic modulus, but not to the extension coefficient, as revealed when the passage of a crack, through an interface is investigated.

4.6.4. Crack Meeting and Interfaces

The problem of a crack meeting an interface in a composite, as shown in Fig. 4-18(a) [125] was posed by Cook and Gordon [131] who wrongly concluded that the strength of the interface was important. Zak and Williams [135], followed by Rice and Sih [136] also studied this problem using elastic stress analysis, but the problem was complicated mathematically and led to strange oscillating stresses near the crack tip. There were also substantial experimental difficulties, because cracks inside the composite materials could not easily be

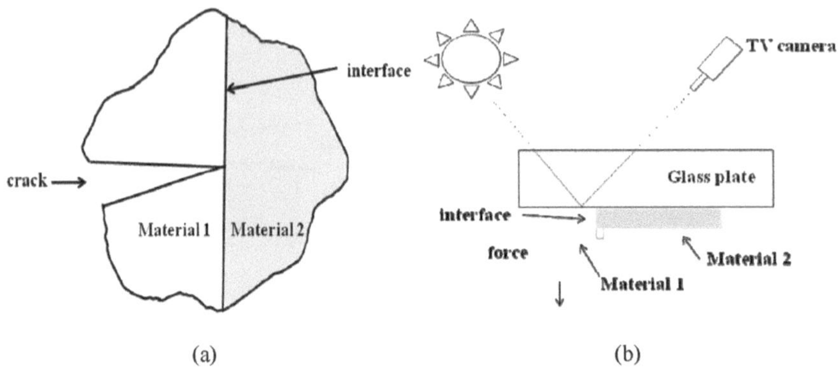

Fig. 4-18 (a) A crack meeting an interface between two materials, (b) Experimental arrangement for observing a peel crack meeting an interface between two materials.

observed though it was known from fracture surface inspection that a crack slowed down as it approached an inclusion, and then accelerated as it emerged through it.

The solution to this problem was found by adopting simple crack geometry, shown in Fig. 4-18(b) [125], in which an adhesive peel crack was viewed with a TV camera, through a glass plate [137]. Rubber peeling from glass was the experimental model. The crack speed was measured accurately, as the crack approached an interface, at which the elastic modulus of the material increased. This stiffening was achieved by moulding stiff fibers into the rubber over half its length.

The theoretical explanation of these effects can be derived, from the energy theory of fracture [137]. Consider a crack, as it just penetrates into a stiffer region of materials, shown in Fig. 4-18. When the crack is at the interface, it exhibits the shape of a long bent beam under load. However, when the crack tries to penetrate into stiffer materials, there is more elastic resistance to bending deformation, and so the shape changes. These shapes were measured microscopically on peeling samples. The theoretical shapes were calculated from the elastic beam theory. Since the stiffer material is more resistant to bending, the peeling load cannot deflect so much and consequently does less work. However, the interface fracture work remains constant so the peeling force must be raised to maintain the same fracture work and peeling speed.

Putting this theory mathematically, the beam deflection is given as

$$\delta = \frac{FL^3}{4Ebd^3} \tag{4-68}$$

where F is the force, L the beam length, E the elastic modulus, b the width, and d the film thickness, respectively.

To maintain the crack propagation at constant speed, this deflection must remain constant, i.e., F/E is constant. Therefore, at the interface, the condition for cracking follows to

$$\frac{F_2}{E_2} = \frac{F_1}{E_1} \tag{4-69}$$

and since $F_1 = bR$,

$$F_2 = \left(\frac{E_2}{E_1}\right) bR \tag{4-70}$$

Thus, while it is clear that the elastic modulus of a material, may not affect its crack resistance as indicated in $F_1 = bR$, a change in the elastic modulus at an interface, toughens the materials by a factor of (E_2/E_1).

These assumptions reveal that composites obtain their toughness primarily from the changes in elastic modulus at the interfaces. Thus, epoxy resin may have a toughness of only 100 J/m,^{2} but with glass, this can increase substantially. Because the elastic modulus of glass is 70 GPa and 3.5 GPa for resin, the ratio of moduli is thus 20, which multiplies the epoxy resin toughness from 100 J/m^2 to 2000 J/m^2. This theory also addresses the question of why ceramics are so difficult to toughen because the ratio of elastic moduli for ceramics is low. For example, the elastic modulus of zirconia is 200 GPa and that for alumina is 400 GPa. The possible condition of generating an elastic toughening effect at an interface in a zirconia/alumina composite is therefore only a factor of two [138].

4.6.5. Crack Resistance of Composites

In this section, the criteria for catastrophic propagation of a crack, and the process for investigating the slow growth of a crack under fatigue loading conditions are discussed. The purpose of this section is to discuss the basic importance to the investigation of fracture mechanics, because it provides the means for the development and measurement of the so-called fracture toughness.

Catastrophic failure is characterized by the following four criteria.

1. The Strain Energy Release Rate (G_c) of the body during crack extension.
2. The stress field just ahead of the crack tip also called the Stress Intensity Criterion (K_{Ic}).
3. A path-independent integral around the crack tip, which represents the loss of strain energy in the body. This criterion is called the J-Integral Criterion (J_c).

As mentioned above, all these criteria equalize each other for elastic behaviors. However, they represent independent methods, when energy dissipation, owing to plastic flow is involved. In other words, we discuss the effect of fatigue

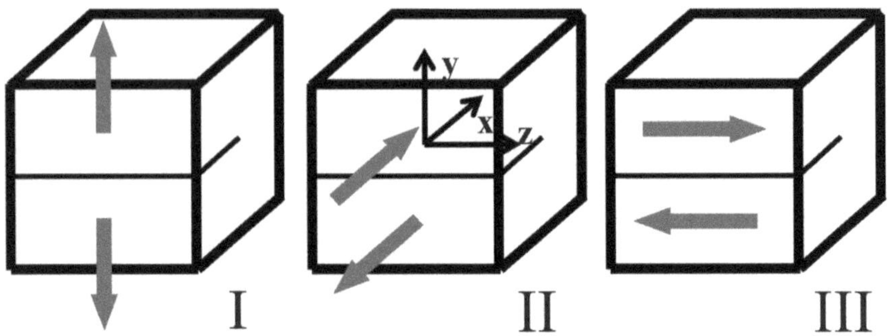

Fig. 4-19 Mode I, mode II, and mode III cracks.

loading, and the extension per load cycle of the crack. These approaches allow the number of cycles that an initially cracked body can endure to be calculated.

Stress intensity factor, K, is a parameter that amplifies the magnitude of the applied stress that includes the geometrical parameter Y (load type). These load types are categorized as Mode I, II, or III, as illustrated in Fig. 4-19. Mode I is the opening or tensile mode where the crack surfaces move directly apart. Mode II is the sliding or in-plane shear mode where the crack surfaces slide over one another in a direction perpendicular to the leading edge of the crack. Mode III is the tearing and antiplane shear mode where the crack surfaces move relative to one another and parallel to the leading edge of the crack. Mode I is the most common load type encountered in engineering design [139].

A criterion based on the stress intensity is given preferential treatment simply because it applies the equations of elasticity directly, and thus it allows numerical calculations for a cracked structure.

4.6.5.1. Fracture theory

Griffith [140] suggested a theory for the strength of brittle materials, in which it was assumed that cracks were the determining factor in the strength of such materials. In order to increase the length of a crack, energy at least equal to the surface energy of the two new surfaces produced by the growth of the crack must be available. An applied load stores elastic energy in a material. If a crack is to increase its length, the rate of decrease of elastic energy in the volume of material surrounding the crack must at least equal the rate at which surface energy is created by the growth of the crack. The tensile strength σ_b of a sheet or plate is then

$$\sigma_b = \left(\frac{2\gamma E}{\pi a}\right)^{1/2} \tag{4-71}$$

The surface energy per unit area, of crack surface is γ, E Young's modulus of the material, and a the length of the crack.

The strength of polymers and many other materials is much greater than predicted by Eq. (4-71). It is found that γ is the order of 105 to 106 mJ /m^2 for polymers, instead of a few hundred mJ /m^2 as expected for pure surface energy [141–144]. The reason for the high value of γ is that there is plastic flow and cold-drawing of the polymer during the crazing and/or shear banding that can accompany the growth of cracks in polymers; the surface energy is a very small portion of the total energy involved. Therefore, in reformulating the theory to consider the plastic flow, γ becomes the plastic work or fracture energy [145]. Chen [146] showed that an equation of the same form as Griffith's equation holds when the specimen contains multiple cracks. Williams [147, 148] proposed a modified Griffith's equation, such as that used by design engineers in solving practical problems of the strength of objects.

$$\sigma_b = \left(\frac{E\gamma}{a}\right)^{1/2} \tag{4-72}$$

The geometric constant k is generally approximately 1.0. This constant can be determined for objects of any shape, from classical mechanics with the aid of computers. However, reliable tables for the values of the fracture energy γ of many polymers will be required before equation [149] can be adopted for practical design work.

The experimental technique known as fracture mechanics is a way of measuring fracture toughness, the fracture energy γ, or its equivalent [150–152]. The fracture toughness of a material measures the ability of that material to resist the extension of a pre-existing crack by a specified type of stress. This technique of fracture mechanics is an attempt to relate the behavior of real, brittle materials that contain flaws, to various kinds of applied loads. Tensile and other experimental tests, such as a cleavage test, are made on specimens containing artificial cracks of known length. From the load at which the crack starts to propagate rapidly, the fracture surface energy can be calculated for a specimen of specified geometry by fairly complex equations. In this technique, fracture toughness is defined in terms of either of two factors: A critical stress intensity factor K_{Ic}, or a critical strain energy release rate G_{Ic}. These factors are calculated by the following equations:

$$K_{Ic}^2 = EG_c(\text{for thin sheets}) \tag{4-73}$$

or

$$K_{Ic}^2 = EG_c(1 - v^2)(\text{for thick sheets}) \tag{4-74}$$

where E is Young's modulus, and v Poisson's ratio.

The fracture surface energy γ of Griffith's equation is calculated to the critical strain energy release rate G_{Ic} by

$$G_c = 2\gamma \tag{4-75}$$

The interaction of fracture toughness to impact strength is not clearly detected. Sometimes impact strength increases as the fracture toughness increases at other times the impact strength decreases, as the fracture toughness increases.

A relationship holds for many rigid materials, between Young's modulus, and the tensile strength or yield strength [152–154]

$$\sigma_B = \frac{E}{30} \tag{4-76}$$

and

$$\frac{1}{60} \leq \frac{\sigma_y}{E} \leq \frac{1}{30} \tag{4-77}$$

These approximations hold for many polymers, under all kinds of conditions, including changes in temperature, pressure, degree of orientation, and degree of crystallinity [150].

4.6.5.2. Stress analysis of cracks

Before mentioning the K_{Ic} criterion of fracture, we can discuss the stress field, just ahead of the crack tip in a mode I fracture.

The elastic stress analysis of sharp cracks leads to infinite stresses at the crack tip. This can be exhibited by first approaching the stresses at the boundary of an elliptical opening in an infinite plate loaded with constant uniaxial stress σ_0, as shown in Fig. 4-20. A sharp crack is obtained by letting the ellipse flattern continuously, until its minor axis is zero. When $a = b$ the results coincide with that of a circle.

A more informative approach to crack tip stress analysis is to consider the admissible solutions (eigen solutions) for the so-called Stress Functions [155] in a polar coordinate system, generated at the crack tip, as shown in Fig. 4-21. The exact procedures for calculating the stresses are outside the scope of this text. Rather, we need to know a few characteristics of the stress field at the crack tip. The point P is identified by its polar coordinates, but the stress components are represented in the Cartesian system

1. For an infinite plate under a uniaxial load, as in Fig. 4-21, the solutions for stress, are of the form,

$$\sigma_{xx} = \frac{K}{\sqrt{2\pi r}} \cos\frac{\theta}{2}\left[1 - \sin\frac{\theta}{2}\sin\frac{3\theta}{2}\right] \tag{4-78}$$

$$\sigma_{YY} = \frac{K}{\sqrt{2\pi r}} \cos\frac{\theta}{2}\left[1 + \sin\frac{\theta}{2}\sin\frac{3\theta}{2}\right] + \sigma_0 \tag{4-79}$$

$$\sigma_{zz} = \nu\left(\sigma_{xx} + \sigma_{yy}\right) \tag{4-80}$$

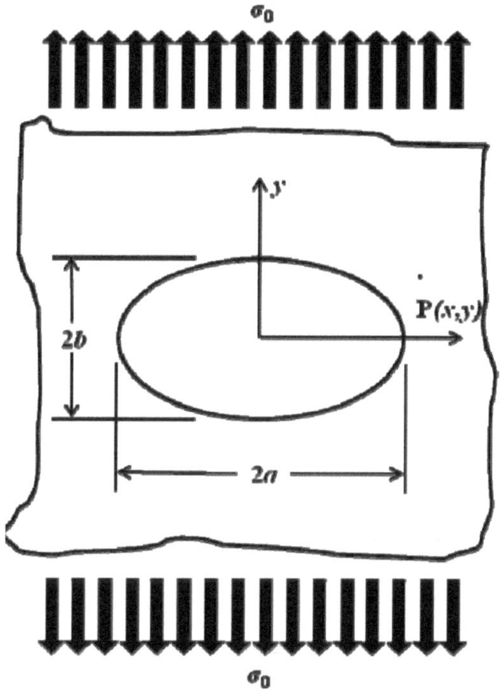

Fig. 4-20 Stresses at points along the major axis of an elliptical hole in an infinite plate.

for plane strain

$$\sigma_{zz} = 0 \tag{4-81}$$

for plane stress

$$\tau_{xy} = \frac{K}{\sqrt{2\pi r}} \cos\frac{\theta}{2} \sin\frac{\theta}{2} \cos\frac{3\theta}{2} \tag{4-82}$$

$$\tau_{yz} = \tau_{xz} \tag{4-83}$$

$$\tau_{xz} = 0 \tag{4-84}$$

where, $K = \sigma_0\sqrt{\pi\alpha}$.

2. Although in Fig. 4-21 an infinite plate is approached, the solution at a crack tip is formally the equalization for all shapes and crack orientations. This is a direct consequence of the generality with the obtained eigen solutions. Any component of stress σ_{**} is expressible in the following equation

$$\sigma_{**} = \frac{K}{\sqrt{2\pi r}} \times (\text{a function of } \theta) \tag{4-85}$$

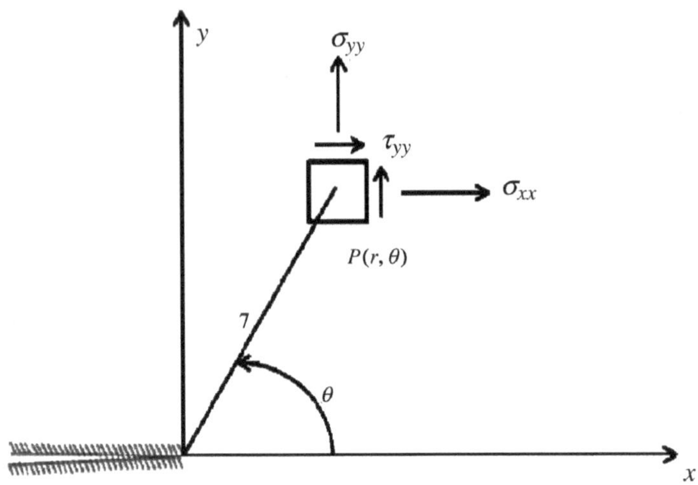

Fig. 4-21 Stress field at a crack tip.

where K is a constant that depends on the applied stress, crack size, and its position and orientation relative to the geometry and size of the component. K is always expressible in the form $K = C\sigma\sqrt{a}$ where a is the crack length, σ the applied stress, and C the geometry dependent constant.

This fact agrees with other methods of fractures analysis, and hence for the combination.

3. From the above, the parameter K can be seen to be the only variable, that is problem-specific, and not the value of the local stress. This point of view dispenses with infinite stresses. In other words, the manner of approach of the stress values to infinity is important. The parameter K describes exactly that K is called the 'stress intensity factor'. The idea of using this factor to characterize crack behavior was proposed by Irwin [156]. The calculation of K for even the simplest case, such as an infinite plate, uses the advanced mathematical theory of elasticity and is outside the scope of our discussion. However, a few pre-engineered solutions are available in previous work [157]. These solutions are generally used much in the same way as in stress analysis handbooks. A practical problem is simply modeled as closely as possible to one of the listed cases.

4.6.5.3. Stress intensity factor

The term K is applied in fracture mechanics to more accurately predict the stress state (stress intensity) close to the tip of a crack caused by a remote load or residual stresses. It is a theoretical construct applicable to a homogeneous elastic material and is useful for indicating a failure criterion for brittle materials.

The magnitude of K depends on sample geometry, the size and location of the crack, and the magnitude and modal distribution, of loads on the material.

Linear elastic theory exhibits a variation for Mode I, close to the crack tip of the form

$$\sigma_{ij} = \frac{K_I f_{ij}(\theta)}{\sqrt{(2\pi r)}} \tag{4-86}$$

This breaks down close to the tip (small r), where plastic distortion mainly occurs at critical stress intensity.

Stress intensity in any mode situation is directly proportional to the applied load on the material. If a very sharp crack can occur in a material, the minimum value of K_I can be empirically determined, which is the critical value of stress intensity required to propagate the crack. This critical value for Mode I loading in a plane strain is referred to as the critical fracture toughness (K_{Ic}) of the material. K_{Ic} has units of stress times derived from a distance. The units of K_{Ic} it is suggested that the fracture stress of the material should be achieved over some critical distance in order for K_{Ic} to be reached and crack propagation to occur. The Mode I critical stress intensity factor, K_{Ic}, is the most often used engineering design parameter in fracture mechanics and hence must be understood, if we are to design fracture tolerant materials used in bridges, buildings, aircraft, or even bells. Polishing just won't do if we detect a crack. If typically for most materials generating a crack can be exhibited, it is very close to the critical stress predicted by the stress intensity factor.

4.6.5.4. Critical strain energy release rate

The strain energy release rate (or energy release rate, G_{Ic}) is the energy dissipated, during the fracture per unit process of a newly created fracture surface area. This quantity is central to fracture mechanics, because the energy that must be supplied to a crack tip for it to grow must be balanced by the amount of energy dissipated due to the formation of new surfaces and other dissipative processes, such as plasticity.

For the purposes of calculation, the energy release rate is defined as

$$G := -\frac{\partial U}{\partial a} \tag{4-87}$$

where U, is the potential energy available for crack growth, and a, the crack area (crack length for two-dimensional problems). The units of G are J/m^2.

The energy release rate failure criterion states that a crack will grow when the available energy release rate G, is greater than or equal to a critical value G_c

$$G \geq G_c \tag{4-88}$$

The quantity G_c is the fracture energy, and is considered to be material properties that are independent of the applied loads and the geometry of the body.

In relation to fracture toughness, for two-dimensional problems (plane stress, plane strain, antiplane shear) involving cracks that move in a straight path, the Mode I stress intensity factor (K_I) is related to the energy release rate (G) by

$$G = \frac{K_I^2}{E'} \tag{4-89}$$

where E is Young's modulus, and $E = E'$ for the plane stress, and $E' = E/(1 - \nu^2)$ the plane strain.

Therefore, the energy release rate failure criterion, may also be expressed as

$$K_I \geq K_{Ic} \tag{4-90}$$

where K_{Ic} is the Mode I fracture toughness.

Gent *et al.* [158] suggested that the tearing energy (G_{IIIc}) was characterized by a trouser beam test for mechanical behaviors, as shown above Mode III, in Fig. 4-19. Following this method, Park *et al.* [159] investigated to obtain the mechanical properties of rubber filled with the carbon black information about the consequence of the filler surface modification on its reinforcing ability. For the tearing energy measurements, the crosshead speed was fixed at 100 mm/min. For the swelling studies, rectangular specimens about 100 mm long, 5 mm wide, and 1 mm thick were cut from a sheet, which were then immersed in toluene at room temperature. The length was measured at intervals until the measured length became constant, indicating an equilibrium swelling condition. Then, the linear swelling ratio, λ_s, defined as the ratio of length, L_t, at time t to the original length, L_0, was calculated.

The tearing energy, G_{IIIc} is independent of crack length, which is usually difficult to determine. The G_{IIIc} of rubber, filled with the carbon blacks was determined to obtain information about the consequence of the filler surface modification on its reinforcing ability. The tearing energy was calculated using the following equation [140, 159]:

$$G_{IIIc} = \frac{2\lambda_s^2 F}{t} \tag{4-91}$$

where F is the applied force, λ_s the linear swelling ratio in toluene at room temperature in this work, and t the width of the tear path, measured on the torn strip after tearing was completed.

In Fig. 4-22 [159], the swelling behavior of the carbon black/rubber compounds studied is characterized by measuring the length of toluene absorbed, the time taken to reach the steady state, and the degree of swelling. The acidic chemical treatment of carbon blacks leads to a considerable decrease of the swelling ratio of the composites, which can be expected to diminish the reinforcing ability or the network chain density of the composites. On the other hand, an increase of swelling ratio is seen in the case of the polar basic (BCB) or nonpolar (NCB) treatment on the carbon blacks. According to Kraus [160], the

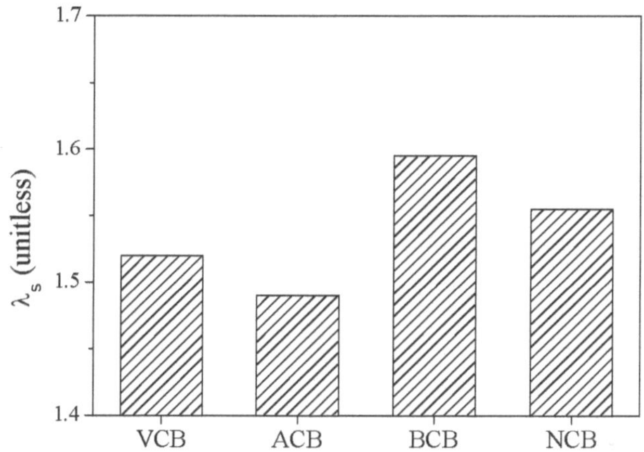

Fig. 4-22 Swelling ratios (λ_s) of different carbon black/rubber composite specimens.

degree of adhesion between the filler surface and rubber can be assessed from the swelling behavior of the composite sample in a solvent.

In Fig. 4-23 [159] the tearing energies measured in a trouser beam test are shown. These results suggested that the tearing energies of the rubber composites made of basic treated (BCB) and non-polar-treated (NCB) carbon blacks, are higher than those of rubber composites made of acid treated carbon black, or even as-received carbon black (VCB). This seems to be a consequence of the

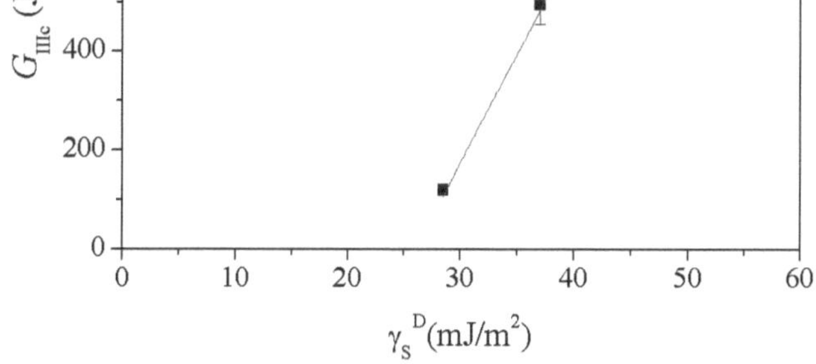

Fig. 4-23 Dependence of the tearing energy of composites on the London dispersive component of surface free energy of the carbon blacks studied.

interaction between acidic carbon black and basic rubber matrix in the rubber compounding composites [161]. According to Park *et al.* [159], the London dispersive component of the surface free energy, as investigated by carbon blacks leads to a good correlation with the G_{IIIc} of rubber compound composites, with a constant regression coefficient (R = 0.997), as seen in Fig. 4-23. Therefore, it can conclude that an increase of London dispersive component, γ_S^D, of the surface free energy of carbon blacks improves the degree of adhesion at interfaces between carbon black and hydrocarbon-based rubber matrix.

4.6.5.5. J-integral

The J-integral represents a way to calculate the strain energy release rate, or work (energy) per unit fracture surface area, in a material [162]. The theoretical concept of J-integral was developed in 1967 by Cherepanov [163] and in 1968 by Rice [164] independently. They showed that an energetic contour path integral (called J) was independent of the path around a crack. Later, experimental methods were developed, which allowed the measurement of critical fracture properties using laboratory-scale specimens for materials in which sample sizes are too small, and for which the assumptions of linear elastic fracture mechanics (LEFM) do not hold, and to infer a critical value of fracture energy J_{Ic}. The quantity J_{Ic} defines the point, at which large-scale plastic yielding during propagation, takes place under mode one loading [162, 165]. The J-integral is equal to the strain energy release rate for a crack in a body subjected to monotonic loading [166]. This is true, under quasi-static conditions, both for the linear elastic materials and for materials that experience small-scale yielding at the crack tip.

The two-dimensional J-integral was originally defined as [164] (see Fig. 4-24 for an illustration)

$$J := \int_{\Gamma} \left(W dx_2 - t \cdot \frac{\partial u}{\partial x_1} ds \right) \tag{4-92}$$

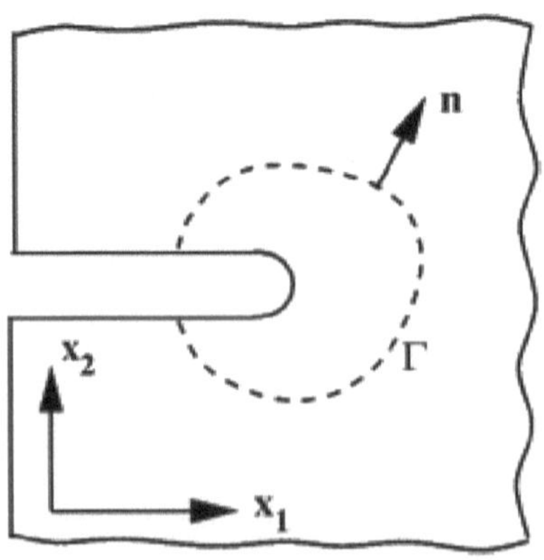

Fig. 4-24 Line J-integral around a notch in two dimensions.

where $W(x_1, x_2)$ is the strain energy density, x_1 and x_2 the coordinate directions, $t = n \cdot \sigma$ the traction vector, n the normal to the curve Γ, σ the Cauchy stress tensor, and u the displacement vector.

The strain energy density is given by

$$W = \int_0^{\varepsilon} \sigma : d\varepsilon \tag{4-93}$$

$$\varepsilon = \frac{1}{2}\left[\nabla \mathrm{u} + (\nabla \mathrm{u})^T\right]] \tag{4-94}$$

The J-integral around a crack tip is frequently expressed in a more general form (and in index notation) as

$$J_i = \lim_{\varepsilon \to 0} \int_{\Gamma_\varepsilon} \left(W n_i - n_j \sigma_{jk} \frac{u_k}{x_i} \right) d\Gamma \tag{4-95}$$

where J_i is the component of the J-integral for the crack opening in the x_i direction and ε the small region around the crack tip.

Green's theorem [167] can show that this integral is zero, when the boundary Γ is closed and encloses a region that contains no singularities and is simply connected. If the faces of the crack do not have any traction on them, then the J-integral is also path independent.

Rice [166] also showed that the value of the J-integral represents the energy release rate for the planar crack growth. The J-integral was developed because of the difficulties involved in computing the stress close to a crack in a nonlinear elastic or elastic-plastic material. Rice showed that if monotonic loading was assumed (without any plastic unloading) then the J-integral could be used to compute the energy release rate of plastic materials too. For isotropic, perfectly brittle, linear elastic materials, the J-integral can be directly related to fracture toughness.

For plane strain, under Mode I loading conditions, this relation is

$$J_{Ic} = G_{Ic} = K_{Ic}^2 \left(\frac{1 - \nu^2}{E} \right) \tag{4-96}$$

where G_{Ic} is the critical strain energy release rate, K_{Ic} the fracture toughness in Mode I loading, ν Poisson's ratio, and E Young's modulus of the materials.

For Mode II loading, the relation between the J-integral and the mode II fracture toughness (K_{IIc}) is

$$J_{IIc} = G_{IIc} = K_{IIc}^2 \left[\frac{4(1 - \nu^2)}{3E} \right] \tag{4-97}$$

For Mode III loading, the relation is

$$J_{IIIc} = G_{IIIc} = K_{IIIc}^2 \left(\frac{1 + \nu}{E} \right) \tag{4-98}$$

This relation is unambiguous. In relating crack opening displacement to G_{Ic}, there is some question as to whether or not the value of the yield strength should be elevated due to constraints. Eqs. (4-96), (4-97), and (4-98) in a sense allow a direct extension of linear elastic fracture mechanics into the elastic-plastic and general yielding range [168].

4.6.5.6. Experimental data and applications

If the load applied to either materials or composites containing a crack is too high, the crack may suddenly grow and cause the member to fail by fracturing in a brittle manner, that is, with little plastic deformation. From the theory of fracture mechanics, a quantity called the stress intensity factor, K, can be defined as characterizing the severity of the crack situation as affected by crack size, stress, and geometry. In approaching K, the material is assumed to behave in a linear-elastic manner, according to Hooke's Law, so that the approach being used is called linear-elastic fracture mechanics (LEFM).

A given material can withstand a crack without brittle fracture occurring as long as this K is below a critical value K_c, which is a property of materials, generally called the fracture toughness. The values of K_c vary widely for different materials and are influenced by temperature and loading rate and secondarily by the thickness of the member. Some representative values of K_{Ic} which is the worst case K_c considering thickness effects, are illustrated in Tables 4-5 and 4-6.

For instance, consider a crack in the center of a wide plate of stressed materials, as illustrated in Fig. 4-25. In this occasion, K influenced by the remotely applied stress S and the crack length a, is calculated from the following equation.

$$K = S\sqrt{\pi a}(a \ll b) \tag{4-99}$$

where this equation is accurate only if it is small compared to the half-width b of the member.

For a given crack length and a material with fracture toughness K_c, the critical value of a remote stress necessary to cause fracture is

$$S_c = \frac{K_c}{\sqrt{\pi a}} \tag{4-100}$$

hence, longer cracks have a more severe effect on strength than do shorter ones, as might be expected.

Experimental data exhibiting the effect of different crack lengths on strength are shown in Fig. 4-25. The curve given by Eq. (4-99) is shown, in which the particular value of K_c corresponds to the case at hand namely 2014-T6

Table 4-5 Fracture Toughness and Corresponding Tensile Properties for Representative Metals at Room Temperature

Material	Toughness K_{Ic}	Yield σ_0	Ultimate σ_u	Elongation $100\varepsilon_f$	Reduction of Area %RA
	$MPa \cdot m^{1/2}$ ($ksi \cdot in^{1/2}$)	MPa (ksi)	MPa (ksi)	%	%
(a) Steels					
AISI 1144	66(60)	540(78)	840(122)	5	7
ASTM A470-8 (Cr-Mo-V)	60(55)	620(90)	780(113)	17	45
ASTM A517-F	187(170)	760(110)	830(121)	20	66
AISI 4130	110(100)	1090(158)	1150(167)	14	49
18-Ni maraging air melted	123(112)	1310(190)	1350(196)	12	54
18-Ni maraging vacuum melted	176(160)	1290(187)	1345(195)	15	66
300-M 650 °C temp.	152(138)	1070(156)	1190(172)	18	56
300-M 650 °C temp.	65(59)	1740(252)	2010(291)	12	48
(b) Aluminum and Titanium Alloys (L-T Orientation)					
2014-T651	24(22)	415(60)	485(70)	13	-
2024-T351	34(31)	325(47)	470(68)	20	-
2219-T851	36(33)	350(51)	455(66)	10	-
7075-T651	29(26)	505(73)	570(83)	11	-
7475-T7351	52(47)	435(63)	505(73)	14	-
Ti-6Al-4V annealed	66(60)	925(134)	1000(145)	16	34

Source: *Data in Ref. [169–173].*

aluminum (1.5 mm thick) at –195 °C. Fracture strengths from several tests are shown. Most of the experimental data agree with the curve indicating a degree of success here for the linear-elastic fracture mechanics. On the other hand, the data fall below the line where the stress approaches the yield strength σ_0. This phenomenon occurs, as in Eq. (4-100), which assumes a linear-elastic behavior, and applies only if the plastic zone is small compared to the crack length and other geometric dimensions, and this fails for the highest stresses representing failure at small crack lengths.

In previous research [159], for an epoxy resin/polyurethane blend system, the fracture toughness can be measured by a three-point bending test for the critical stress intensity factor (K_{Ic}) according to ASTM E 399. For the single-edge notched (SEN) beam fracture toughness test the value of K_{Ic} is calculated

Table 4-6 Fracture Toughness of some Polymers and Ceramics at Room Temperature

Materials	K_{Ic} MPa · $m^{1/2}$	K_{Ic} (ksi · $in^{1/2}$)
Polymers		
Acrylonitrile butadiene styrene	3.0	(2.7)
Acrylic	1.8	(1.6)
Epoxy	0.6	(0.55)
Polycarbonate	2.2	(2.0)
Poly(ethylene terephthalate)	5.0	(4.6)
Polyester	0.6	(0.55)
Polystyrene	1.15	(1.05)
Poly(vinyl chloride)	2.4	(2.2)
Rubber modified Poly(vinyl chloride)	3.35	(3.05)
Ceramics		
Soda-lime glass	0.76	(0.69)
Magnesia, MgO	2.9	(2.6)
Alumina, Al_2O_3	4.0	(3.6)
Al_2O_3, 15% ZrO_2	10	(9.1)
Silicon carbide, SiC	3.7	(3.4)
Silicon nitride, Si_3N_4	5.6	(5.1)
Dolomitic limestone	1.30	(1.18)
Westerly granite	0.89	(0.81)
Concrete	1.19	(1.08)

Source: *Data in Ref. [174–180].*

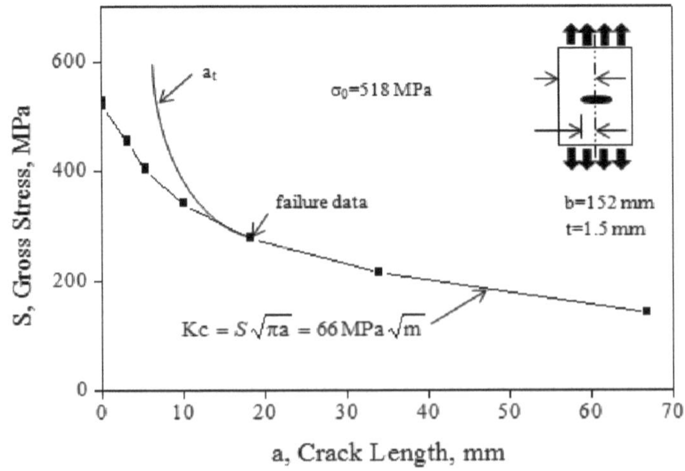

Fig. 4-25 Failure data for cracked plates of 2014-T6 aluminum tested at –195 °C.

as follows:

$$K_{Ic} = \frac{PS}{BW^{3/2}} f(a/W) \tag{4-101}$$

and

$$f(a/W) = \frac{3(a/W)^{1/2}\left[1.99 - (a/W)(1-a/W)(2.15 - 3.93a/W + 2.7a^2/W^2)\right]}{2(1+2a/W)(1-a/W)^{3/2}} \tag{4-102}$$

where P is the rupture force, S the span between the supports, and W and B the specimen width and thickness, respectively.

Fig. 4-26 shows the evolution of K_{Ic} in flexure of the epoxy resin/polyurethane blend system with the content of polyurethane. This result indicates, that the maximum K_{Ic} value is about 3.8 MPa = m at 40 phr of polyurethane. As mentioned earlier, this result can be explained by means of hydrogen bonding between the hydroxyl group in the epoxy resin and the isocyanate group in the polyurethane [159].

The carbon–carbon composites reported by Park *et al.* [181] show the results for the fracture toughness (K_{Ic}) of the composites according to the $MoSi_2$ content, as shown in Fig. 4-27 [181]. Excellent relationships are shown between the $MoSi_2$ content and the resulting fracture toughness of the composites, that is, the value of K_{Ic} increases with an increasing $MoSi_2$ content, which corresponds to the work of fracture for the degree of adhesion at interfaces. The maximum strength value of K_{Ic} is obtained at a $MoSi_2$ content of 12 wt%. It was suggested

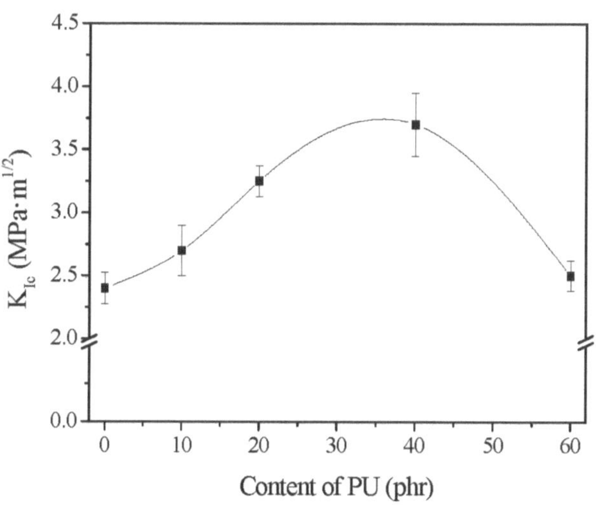

Fig. 4-26 Evolution of K_{Ic} with the content of polyurethane (PU).

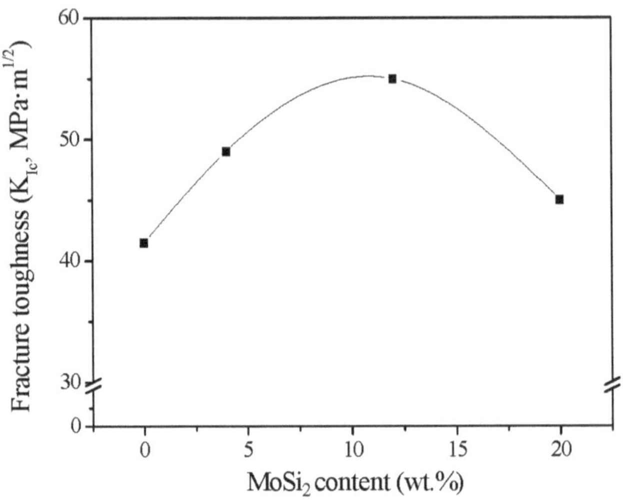

Fig. 4-27 Fracture toughness (K_{Ic}) of carbon/carbon composites, as a function of the filler content of the prepreg resin.

that additional energy is needed to extend the interfacial crack under this condition, which is attributed to the increasing interfacial adhesion between the fibers and the matrix. In addition, this is in good agreement with the result for work of fracture of the composites.

Table 4-6 lists the results of fracture toughness (K_{Ic}) for the composites as a function of the oxyfluorination condition according to the previous literature [173]. As seen in Table 4-6, there is a correlation between the characteristics of the oxyfluorinated multi walled carbon nanotube (MWNT) surfaces and the fracture properties of the composites. The K_{Ic} values increase with an increasing polarity and oxygen functionality of the multi walled carbon nanotube surfaces. The maximum value of K_{IC} was obtained at an oxyfluorination temperature of 100 °C. Therefore, we suggest that the additional energy needed to extend the interfacial crack at this condition may be attributed to increased interfacial adhesion between the reinforcement and the matrix.

The impact strength is the important mechanical property for evaluating the toughness of composite materials. Table 4-7 [182] lists the impact strengths of the oxyfluorinated multi walled carbon nanotubes-reinforced composites. The impact strength of these composites increases with increasing fluorine/oxygen content. This result reveals that the fracture behavior by impact loading is often interrupted by the presence of hydrogen bonding between the multi walled carbon nanotubes and the matrix resins in the composites. The hydrogen bonding has a good effect on hindering the formation of crack growth paths, resulting in increasing the impact strength of the oxyfluorinated multi walled carbon nanotube-reinforced composites.

Table 4-7 K_{Ic} **and Impact Strengths of Oxyfluorinated Multi Walled Carbon Nanotubes (MWNTs)-reinforced Epoxy Matrix Composites as a Function of Temperature**

Specimens	Fracture toughness (K_{Ic}, MPa · $m^{1/2}$)	Impact strength (kPa)
Neat-MWNTs	2.46	21.34
RT-MWNTs	2.63	22.23
100-MWNTs	3.05	24.04
200-MWNTs	2.80	22.16
300-MWNTs	2.50	21.93

4.6.6. Delamination at Interfaces

The crack stopping effect stated above predicts that the interfacial adhesion is excellent. An alternative effect is suggested if the adhesion at the interface is reduced. Then the crack can deflect along the interface, inhibiting catastrophic failure by the Cook–Gordon [131] mechanism, as illustrated in Fig. 4-28 [125].

Imagine a block of material that has been cut in half and rejoined at the interface. In this case there is no change in the elastic modulus at the interface, so the crack is not interrupted by an interface mismatch. If the adhesion at the interface is excellent, then the crack can move straight through as Fig. 4-28(b) shows, but if the interface has poor adhesion, then the crack deflects along the interface and the catastrophic failure is inhibited. The principle is that of a rope; one strand of the rope can break without much reducing the strength of the rope

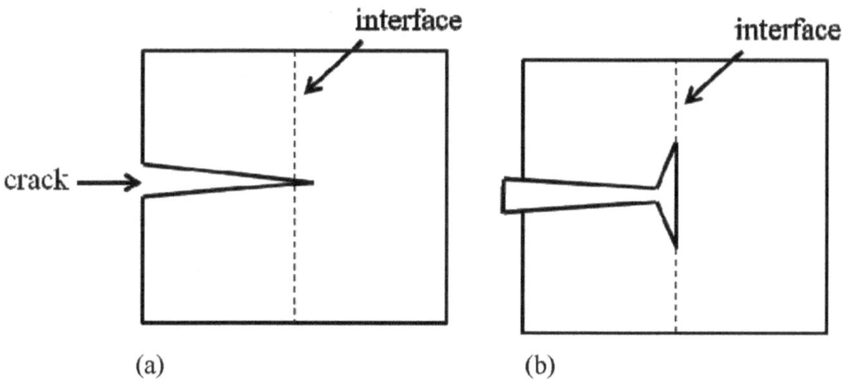

Fig. 4-28 Block of material with an adhesive interface; (a) good crack penetration of adhesion, and (b) poor crack deflection of adhesion.

as a whole but if the rope strands become bonded together with adequate adhesion, then eventually the crack will move straight through and the whole rope fails. The problem is to achieve an interface adhesion that just inhibits crack penetration of the interface, because this is the condition for maximum performance in bending or compression, while retaining the toughness derived from the low interface adhesion.

The model proposed by Cook and Gordon [131] derived from using numerical stress analysis concluded that the interface strength was five times less than the material strength, at the critical deflection point. A more perfect theory [180] was obtained by considering the crack propagation criteria in two directions, at an interface a short distance c inside a block of solid material. The crack approaching the interface is suggested by the Griffith equation which gives the stress σ required to draw the crack through the interface:

$$\sigma = \left[ER_c/(1-\nu^2)/\pi c\right]^{1/2} \tag{4-103}$$

where the fracture toughness R_c is the cohesive toughness of the material, E the elastic modulus, and the block is thick so that the (1-ν_2) term for plane strain conditions is contained.

In contrast, the stress required to initiate the crack along the interface follows the lap-join equation:

$$\sigma = \sqrt{\frac{4ER_a}{c}} \tag{4-104}$$

where R_a is the interface adhesion energy.

Dividing these equations gives the condition for the crack to just deflect:

$$\frac{R_c}{R_a} = 4\pi(1-\nu^2) \tag{4-105}$$

Showing, that the ratio of cohesive to adhesive fracture energy play an important role in maintaining at a value, at around 10, depending on the geometry and Poisson's ratio.

This equation was experimentally verified using ethylene propylene rubber as the model material. When cross-linked with dicumyl peroxide, this rubber was brittle as well as transparent, allowing cracks to be drawn easily through the materials. The advantage of using such rubber is that the cracks can be drawn quite slowly, so that the deflection phenomenon can be seen very easily without high speed cameras [183].

Interfacial adhesion could also be varied simply in this material by bringing two smooth surfaces of rubber together then changing the cross-linking reaction, to give different values of the interface toughness R_a. The main block of rubber was moulded against polished steel to give a smooth surface after the cross-linking process. A thin sheet of rubber was also cross-linked during a press, using a glass mould to give a smooth surface after it adhered to the main block.

The thin sheet was brought into good molecular contact with the block, as it revealed the removal of bubbles at the interface, and the final composite was cross-linked to give the full rubber reaction. This experiment obtained medium adhesion at the interface. By varying the times of the initial and final cross-linking, while maintaining the longer reaction time, the interface adhesion could be modified over a wide range, while maintaining the same bulk properties of the rubber materials. Then a razor slit was cut into the rubber surface, a tensile stress was applied, and the crack was observed travelling through the interface, as in Fig. 4-29 [125]. The behavior of the crack when it reached the controlled interface was stated as a function of the interfacial adhesion.

When a crack was driven towards the interface at a speed of 34 μm /s, the crack deflection was observed at low adhesion, but not at high adhesion. The

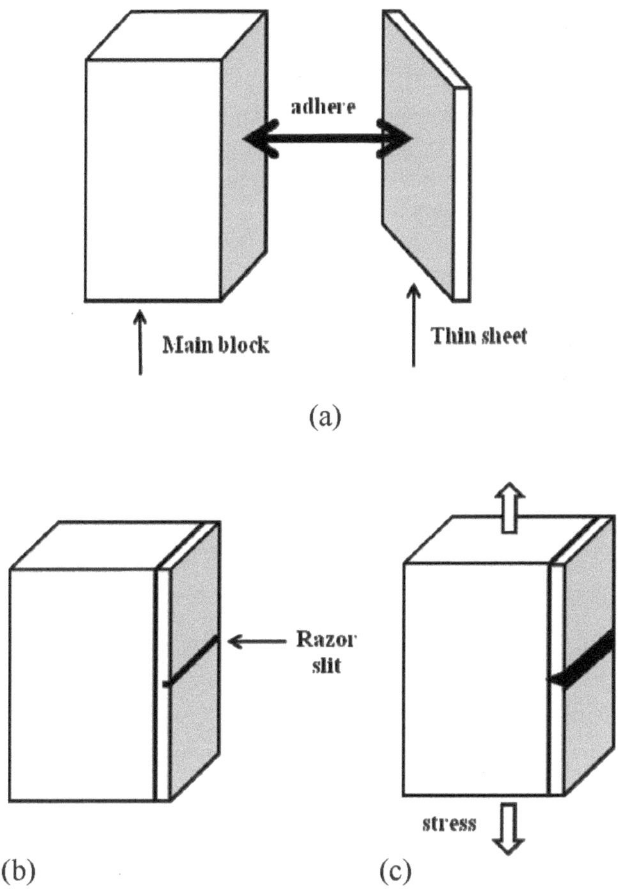

Fig. 4-29 Marking the test specimens by (a) adhering a thin strip onto a block (b) making a cut in the strip, and (c) stretching the sample to drive the notch to the interface.

ratio of R_c/R_a at the changeover transition was 9.1, in a low adhesion interface and to increase the crack speed. For this rubber system, the interface toughness increased more rapidly with crack speed than the cohesive toughness of the rubber. Thus the crack crossed through the interface, at high speed whereas it was deflected at lower speed. By this method the transition ratio was 11, somewhat higher than the theoretical prediction of 9.1.

The energy balance theory of fracture also predicted that crack deflection should be somewhat easier in the centre of the sample than on the outer surface. The theoretical condition giving the ratio of the cohesive to interface fracture energies for deflection at the centre was $R_c/R_a = 4.7$. The value obtained experimentally was 5.6 [137], showing reasonable agreement with the theory. The effect of a modulus mismatch on the crack deflection was also tested. It was demonstrated that for a modulus mismatch of a factor 10, the deflection criterion became easier, with $R_c/R_a = 2.5$.

4.6.7. Bending and Compression

The composite structures are significantly tough when investigated in tension. However, when stressed in other ways the composite may be more brittle than expected. The tensile properties are excellent, but the rope is loose when bent or compressed. Furthermore, wood is difficult to chop across the grain but cleaves easily along the grain. The problem is to describe the composite properties under different loading circumstances.

A simple laminate is loaded in four different ways such as tension, bending, compression, and peeling, as shown in Fig. 4-30 [125]. In each case a crack

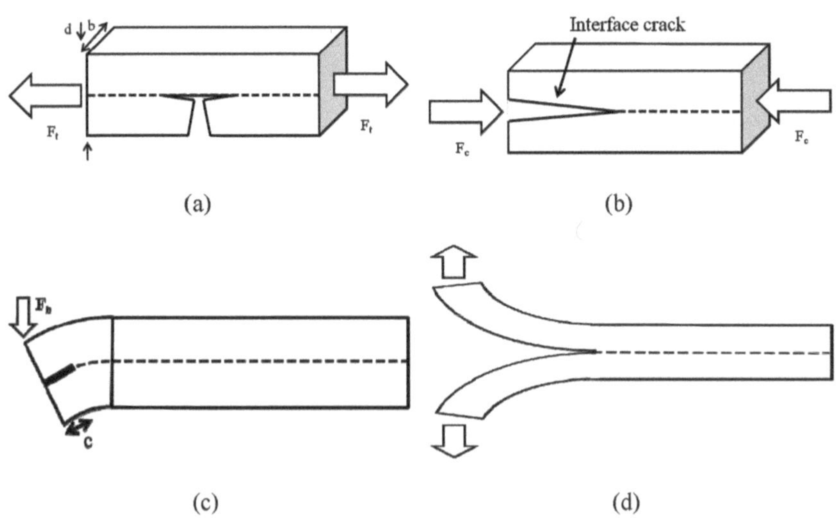

Fig. 4-30 Four methods for cracking a laminate: (a) tension, (b) compression, (c) bending, and (d) T peel.

travels along the interface, but the force required turns out to be completely different. This was first analyzed by Outwater and Murphy [183] who proposed the mechanism by which cracks move through a body, and applied an energy conservation theory, to the analyzed results. A very simple display of such interfacial debonding in laminates was obtained by measuring the delamination in the four geometries, using an interface produced by pressing Perspex sheets together at 110 °C under a load of 10 tonnes [125]. This interface was planar, visible through the material so that cracking could be followed easily, and had a fracture energy of around 30 J/m^2 influenced by the crack speed. By applying forces in tension, bending, compression, and peeling, the difference in composite failure could be estimated for the equal interface [184–186].

From the energy balance theory of fracture, four equations were produced to explain the failure force F in the different geometries, given the same interfacial fracture energy R.

The simplest equation for peeling was given by:

$$F_{\mathrm{p}} = Rb/2 \tag{4-106}$$

where b is the sample length of the perpendicular direction to the tension, as shown in Fig. 4-30(a).

For tension, the force for debonding was

$$F_{\mathrm{t}} = b(2REd)^{1/2} \tag{4-107}$$

where E was the elastic modulus and d the sample thickness, respectively.

Compression failure was slightly easier than the tension given by:

$$F_{\mathrm{c}} = b(2REd/3)^{1/2} \tag{4-108}$$

whereas bending failure was easiest of all, especially as the crack length c increased

$$F_{\mathrm{b}} = b(REd^3/18c^2)^{1/2} \tag{4-109}$$

These theories were defined to compare with the experimental results in Fig. 4-31 [125]. Good agreement was achieved. The observation was that composites must give different performance characteristics depending on the loading method. Tension gave the best values followed by compression, with bending and peeling leading to weak behavior. They do not show this behavior, though it is clear from the equations above that the problem of bending can be weakened by reducing the crack length at the interface, in other words by reducing the flaw size between the laminae.

A study on the influence of wet chemical treatments on the evolution of polymer surface roughness is important in order to evaluate the adhesion strength of plated metals on top of the polymers. By the chemical treatment of the surface, its adhesion can be increased. The surface properties of the polymer are important for the adhesion of the metal to the polymer. Hence,

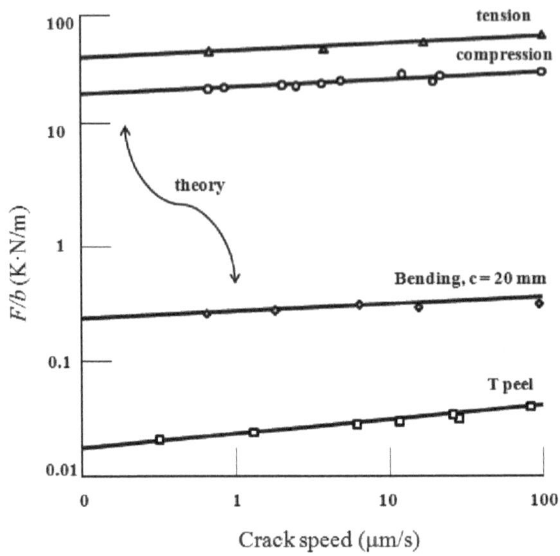

Fig. 4-31 Failure forces in the four different loading geometries: tesnsion, compression, bending, and T peelure.

there has recently been intense research, on improving the adhesion of plated aluminum onto polymer surfaces. Therefore, the polyimide films were initially modified with KOH solution as a function of KOH treatment time and then reacted with HCl solutions to polyamic acid surfaces [187]. They investigated the effect of the mechanical properties of the polyimide films and the influences of surface properties on the mechanical properties of polyimide films were also studied by a peel test with a metal layer. The T-peel test (ASTM 1876-72) was performed at a peel rate of 254 mm/min using Lloyd LR5K.

In Fig. 4-32(a) [187], the characteristic absorption bands of the pure polyimide appear near 1780 (C=O in phase), 1720 (C=O out of phase), 1513 (C=C in C_6H_6), 1370 (C=O in imide), 1100 ($(OC)_2NC$ in imide), and 720 cm^{-1} (imide ring deformation). After treatment in KOH solution in Fig. 4-32(b), the intensities of the imide structures at 1780 and 1720 cm^{-1} peak were reduced. The shifting of the peak at 1370 cm^{-1} indicates that the imide structures of the pure polyimide were hydrolyzed into the amide structures in potassium polyamates and polyamic acids after they were washed and acidified. The characteristic absorption bands of the polyamic acid, appear near 3400-2400 (O-H in carboxylic), 3500-3100 (NH in amide), 1650 (C=O in amide), 1535 (C=O in amide), and 1513 cm^{-1} (C=C in C_6H_5) and the 1595 cm^{-1} peak disappears. The hydrolysis reactions can be further confirmed by the appearance of the peaks for carboxylate ions, in potassium polyamates and carboxyl acids in polyamic acids. The peak at 1418 cm^{-1} is due to the vibrations of –OH in –COOH and

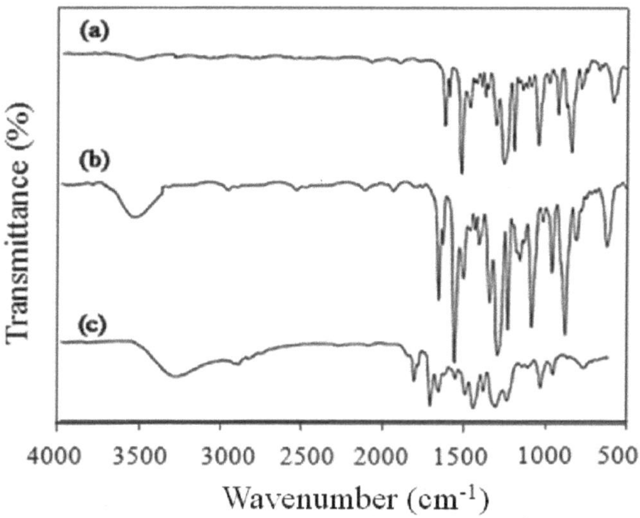

Fig. 4-32 FT-IR spectra of polyimide films; (a) pure polyimide, (b) 10 wt.% KOH treatment, and (c) 10 wt.% HCl treatment.

C=O in –COO–. In both Figs. 4-32(b) and (c), the intensities of the peaks of the imide structures at 1780, 1720, and 1370 cm^{-1} were reduced but not completely eliminated, suggesting that only the uppermost layers of the polyimide were hydrolyzed.

Fig. 4-33 [187] represents the peel strength values of polyimide/aluminum samples as a function of KOH-treatment times. These adhesion strengths are

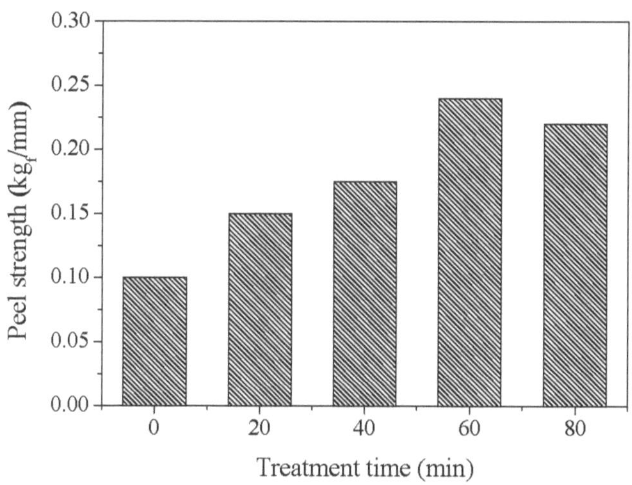

Fig. 4-33 Peel strength of polyimide films as a function of treatment time.

much higher in KOH-treated polyimide/aluminum systems than that of a pristine system. This data indicates that the peel strength increased from about 0.1 kg_f/mm for the pure polyimide to about 0.24 kg_f/mm for the KOH-treated polyimide. The increased formation of the functional groups identified as oxygen bonding, can contribute to the enhanced chemical interactions between the polyimide and aluminum surfaces, resulting in an increase in the performed peel strengths.

4.6.8. Adhesion of Fibers in Composites

Broutman [125] demonstrated the detachment of a single fiber from a matrix in three types of test, as shown in Fig. 4-34. In Fig. 4-34(a), a resin block was stretched, and shear stresses enhanced at the embedded fiber ends, causing failure. As shown in Fig. 4-34(b), Poisson's ratio tensions, developed across the neck of the compression sample to pull the resin from the fiber by interface tension. As shown in Fig. 4-34(c), it was essentially a debonding and friction test as a fiber was pulled through a block of resin. Such tests have an interesting pedigree because they have traditionally been used on well-known macrocomposites such as reinforced concrete [188].

The interfacial strength has largely been discredited as a parameter for describing such failures. The energy method of fracture mechanics is more satisfying because it shows that failure is really a function of two separate variables; crack length and fracture energy, with other elastic, geometric, and kinetic factors also needing consideration. The correct fiber debonding equation

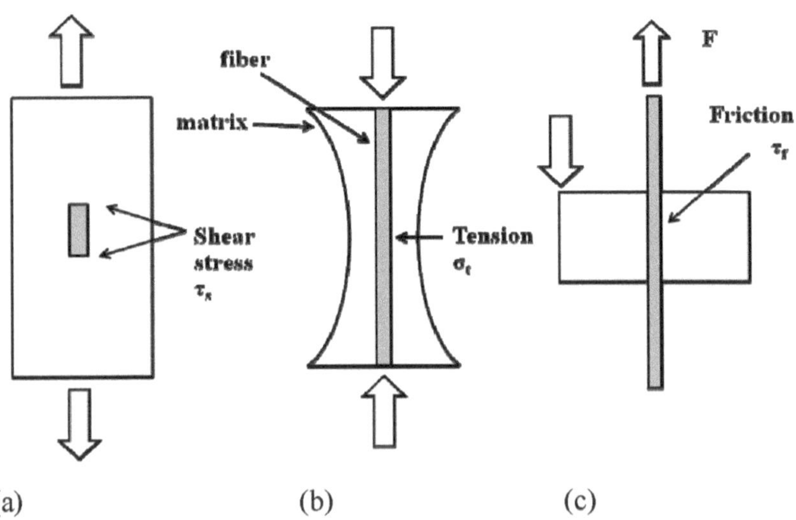

Fig. 4-34 Three test devised by Broutman for testing the fiber/matrix interface: (a) shear test, (b) tension test, and (c) friction test.

based on the energy method was predicted theoretically by Gurney and Hunt [189] in 1967, as mentioned above: $\sigma^2 = 8ER/d$. A failure criterion dependent on a critical stress or 'interface strength' is altogether too simplistic. In the same way, the idea that a fiber should maintain its higher strength in a composite matrix is bizarre. Obviously, a matrix crack can damage the fiber and cause it to weaken considerably. Reductions by two factors influence the fracture mechanics as suggested in strength tests on fibers embedded in a cracked matrix [190].

A schematic description of the test shown in Fig. 4-34(a), now known as the single fiber fragmentation test, was provided by Nairn and his colleagues [191]. The single fiber embedded in a large amount of matrix is stretched, causing fragmentation of the fiber as well as interface debonding, resin fracture, and other complex effects. When T50 carbon fibers were measured in an epoxy matrix, the crack density developed as the strain increased. A debonding of the fibers also occurred. When the fiber fracture energy was 10 J/m^2 and the interface fracture energy was 30.5 J/m^2 the theoretical plots suggested by the energy theory of fracture assumed the data. Such models can now be verified by direct measurement at the interface using Raman spectroscopy [192]. The debond growth can also be used to determine the interface bonding energy. For example, a value of 200 J/m^2 was obtained by measuring the debond growth between E glass fibers and the polymer matrix [193].

Recently, a generally researched test method which exhibited good mechanical properties and long durability of the composites largely depended on the fiber-matrix interfacial adhesion of the composites, since load stress transfers from one matrix to the others via the fiber. For example, the inter-laminar shear strength (ILSS) is improved when the constitutive elements of the composites are fabricated in modifications that increase the interfacial surface area and surface functionality, as shown in Fig. 4-35 [122]

In previous work [194], the degree of adhesion at the interfaces between the fiber and the matrix made with and without fillers was measured by the short-beam flexural test for the inter-laminar shear strength of the interfacial

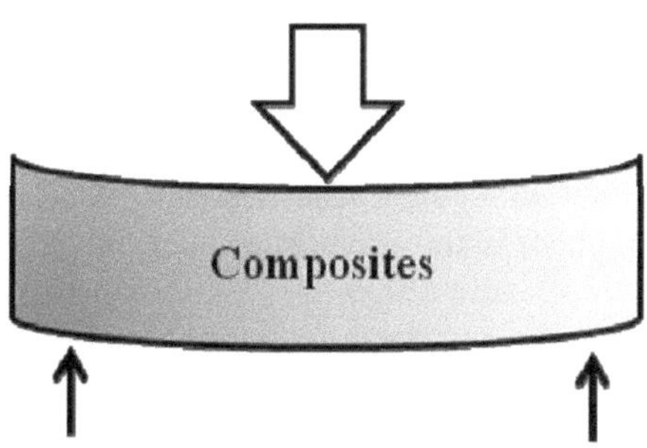

Fig. 4-35 Schematic diagram of interlaminar shear strength (ILSS) test.

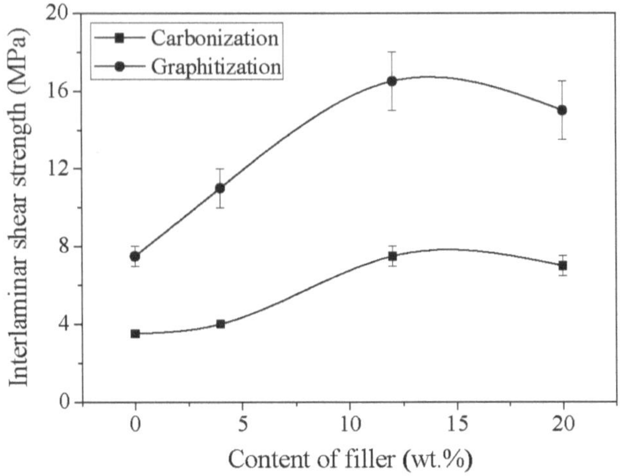

Fig. 4-36 Interlaminar shear strength of carbon-carbon composites measured at 25 °C.

mechanical behavior ($L/d = 6$; cross-head speed, 0.5 mm/min) using the following equation:

$$ILSS = \frac{3}{4} \cdot \frac{F}{bd} \tag{4-110}$$

where F (N) is the failure load at the maximum moment, b (m) the width of the specimen, and d (m) the thickness of the specimen.

Fig. 4-36 and Fig. 4-37 show the inter-laminar shear strength results measured at 25 °C and 1000 °C as a function of the filler contents, respectively. These results demonstrate that inter-laminar shear strength at both temperatures gradually increases as the filler content increases to 12 wt.%. This is due to the presence of filler in the open and closed pores of the composites, resulting in increases in the interfacial area of the constitutive elements in the composites. Meanwhile, a slight marginal decrease in the inter-laminar shear strength is observed in the carbonized composites. It seems that an excess of filler in the carbonized composites results in the formation of another interphase between the fiber and the matrix leading to the creation of an aggregated dispersion phase. However, the stabilization of the inter-laminar shear strength in graphitized composites made with 20 wt.% filler seems to result in a reduction of the aggregated dispersion phase in the composites after the graphitization occurred above their melting temperature (2030 °C) [195]. (See Chap. 9.2.2.)

More direct measures of interface debonding energy are provided by the pull-out tests, as illustrated in Fig. 4-38 [125]. When a tensile force is applied to a fiber to extract it from its composite matrix, an interface crack eventually starts to run along the fiber. It is obvious from a simple fracture mechanics argument

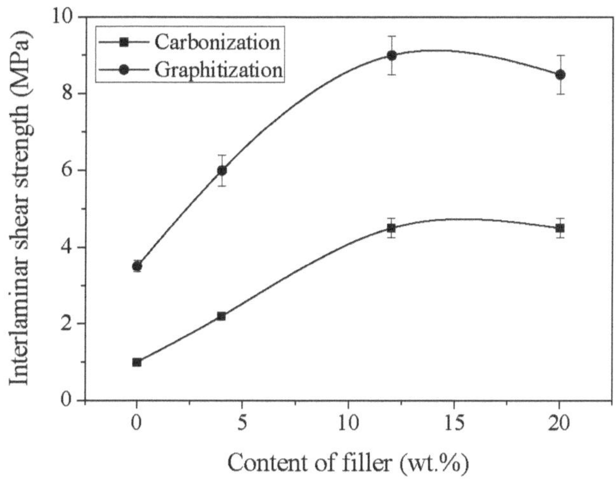

Fig. 4-37 Interlaminar shear strength of carbon-carbon composites, measured at 1000 °C.

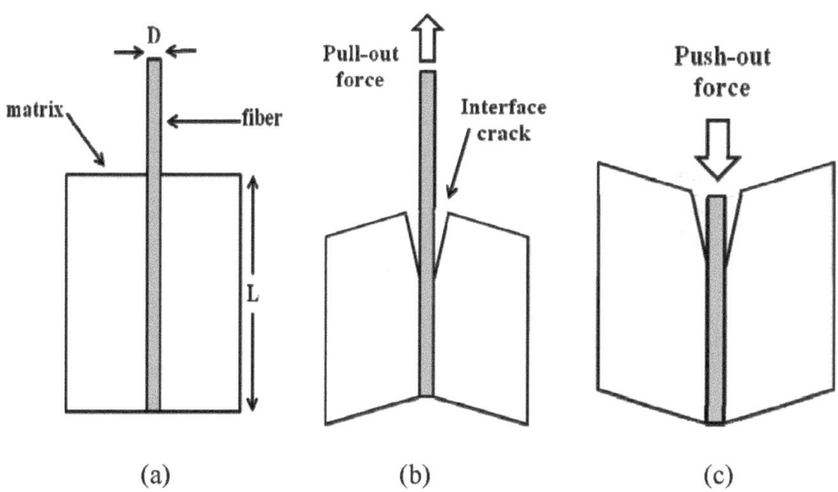

Fig. 4-38 (a) Single fiber embedded in a block of resin, (b) pull-out force causes an interfacial crack, and (c) as does push-out.

[196] that the stress on the fiber to propagate the crack, proposing a very compliant matrix, must be followed by an expression of the equation

$$\text{stress} = F/\pi DL \tag{4-111}$$

$$= [8GR/D \ln(1/V_f)]^{1/2} \tag{4-112}$$

where F is in a state of simple extension, under the action of pull-out force, D the fiber diameter, L the bonded length, G the shear modulus of the matrix, R the interface fracture energy, and V_f the volume fraction of fibers in the composite.

Thus the stress on the fiber at the point of debonding, predicts a rise for finer fibers and for higher volume fractions, as expected from the size effect in the Griffith theory of fracture. By the same line of reasoning, the strength of a fiber, damaged by a deep matrix crack must be of the form [190]

$$\text{Strength} = [8ER/\pi D(1 - v^2)]^{1/2} \tag{4-113}$$

where E is Young's modulus of the fiber and matrix (assumed equal), R the fiber fracture energy, v Poisson's ratio and D the fiber diameter. This equation shows that a smaller diameter fiber must seem to be stronger within the cracked composite materials. (See Chap. 9.3.)

In fine fibers with very smooth surfaces, healing can also occur after the debonding process, so that interfacial dislocations are produced. For fat fibers with rough surfaces, molecular contact cannot be reformed easily. Finally, frictional pull-out is then found with damage at the interface. Due to the problems inherent in conventional pull-out methods and with other interfacial testing methods, a new pull-out test was developed by Miller *et al.* [197, 198] and used by others [199, 200], which provides a more convenient method for the measurement of interfacial shear strengths of fiber/resin interfaces. Because this method uses very small amounts of resin, it is commonly referred to as the microbond pull-out technique or test. From this suggestion, Drzal *et al.* [201] proposed a fiber holder and straining device, mounted horizontally, and positioned under an optical microscope, to collect the data. One end of the fiber specimen is fixed with the adhesive to a metal tab which is connected to a load cell (the microdroplets are sheared off the fiber at a rate of about 0.1 mm/min using a moveable stage). To grip the droplet, an adjustable micrometer equipped with flat, rectangular cross-section blades is used. The blades of the micrometer are first positioned on one side of the droplet, then the blades are brought into contact with the fiber and then opened slightly to let the fiber, but not the droplet, move between them. The moveable stage is used to translate the fiber and droplet laterally in the horizontal plane. As the blades continue to move, they make full contact with the droplet, and an axial force detected by the load cell is exerted on the droplet. The axial force on the droplet is then transferred to the fiber through a shearing force at the fiber/matrix interface. When the shearing force exceeds the interfacial bond strength, detachment occurs, and the droplet is displaced horizontally along the axis of the fiber. The maximum in the force curve is taken as the point at which the droplet has debonded from the fiber.

Fig. 4-39 represents the micro-bond data taken using different curing schemes and plotted in the calculations from Eqs. (4-111) and (4-113) [201]. in Fig. 4-39, note that the micro-bond interfacial shear strength (ISS) calculated for all cases is low compared to the corresponding fragmentation test result of

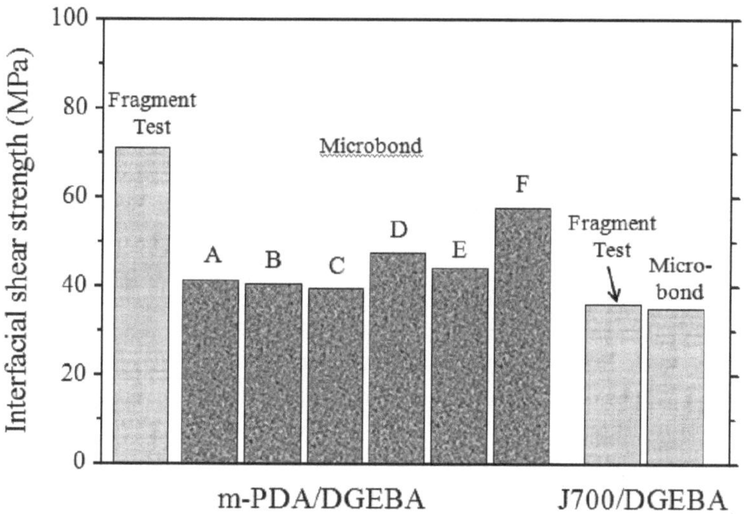

Fig. 4-39 A comparison of the microbond and the fragmentation test, for determining the interfacial shear strength of the di-glycidyl ether of bisphenol A (DGEBA)/meta-phenylene diamine (m-PDA) system and the di-glycidyl ether of bisphenol A/J-700 system.

about 65-70 MPa (fragmentation tests conducted on specimens subjected to all curing schemes showed an interfacial shear strength, between 65 and 70 MPa). Fig. 4-39 gives representative data with (curing schemes A, B, and C) and without (curing schemes D, E, and F) a 25 °C cure portion. The lines reveal the fact that with the 25 °C cure the micro-bond interfacial shear strength (the slope) has increased (for the micro-bond interface shear strength values, in Fig. 4-39, however, individual 'best-fit' lines through the origin for each set of data were used to calculated the slope and the micro-bond interface shear strength). Fig. 4-39 indicates that the interfacial shear strength, calculated for these curing schemes are slightly higher and thus closer to the interface shear strength values measured using the fragmentation test. The cure step at 25 °C, allows some reaction to occur between amine and epoxy which retards the diffusion process as indicated by the higher interfacial shear strength results. This result suggests that allowing the droplets to cure at room temperature before being exposed to a high temperature environment effectively causes the system to gel so that the loss of the curing agent at high temperatures is reduced.

4.7. SUMMARY

Adhesion is a very important field beyond the reach of any single model or theory, involving several mechanisms governed simultaneously by physical and mechanical properties. A number of interfacial phenomena involved in adhesion are influenced by the various materials to be bonded. The correlations among adhesion, London dispersive and/or acid–base interactions, review with the emphasis on inter-molecules and complex materials, in the field of polymers, fillers, fibers, and pigments.

These interactions and phenomena are characterized by mechanical properties, surface properties testing techniques. These techniques are now fairly well established and routinely used, to interrogate the dispersive or acid–base characteristics of solid-solid interfaces, which are developing processes in recent times. For researchers, the exact knowledge of the relative contribution of interfacial interaction in adhesion measurement is still an exciting challenge, whose final objective is to propose new control method in the formulation of adhesive systems with controlled adhesive properties.

Finally, these interactions and phenomena at solid-solid interfaces have received a great deal of attention in studies in the field of physical chemistry of surfaces and fracture mechanics, resulting in contributing the challenge and complete understanding of the adhesion properties in various materials and composites.

REFERENCES

[1] K.L. Mittal, A. Pizzi, Adhesion Promotion Techiniques: Technological Applications, 1st edit. Marcel Dekker, New York (1999).
[2] L. Gong, A.D. Friend, R.P. Wool, Macromolecules 31 (1998) 3706.
[3] S.J. Park, M.K. Seo, Polymer (Korea) 29 (2005) 221.
[4] K.L. Mittal, Polymer Engineering & Science 17 (1977) 467.
[5] R.A. Veselovsky, V.N. Kestelman, in Adhesion of Polymers: The Process of Adhesive-bonded Joint Formation, 1st edit. McGraw-Hill, New York (2002).
[6] F.L. Jin, M. Han, S.J. Park, Polymer International 55 (2006) 1265.
[7] L.H. Sharpe, H. Schonhorn, Chemical & Engineering News 15 (1963) 67.
[8] W.A. Zisman, Industrial & Engineering Chemistry 55 (1963) 18.
[9] J.W. Mac Bain, D.G. Hopkins, Journal of Physics Chemistry 29 (1925) 188.
[10] S. Bistac, M.F. Vallat, J. Schultz, Applied Spectroscopy 12 (1997) 1823.
[11] J.J. Bikerman, The Science of Adhesive Joints, 2nd edit. Academic Press, New York (1968).
[12] S. Bistac, D. Cheret, M.F. Vallat, J. Schultz, Journal of Applied Polymer Science 65 (1997) 347.
[13] F. London, Transactions of the Faraday Society 33 (1937) 8.
[14] S.J. Park, van der Waals Interaction at Surfaces, in Encyclopedia of Surface and Colloid Science, 2nd edit, Taylor & Francis, New York (2006).
[15] R. Eisenschitz, F. London, Zeitschrift für Physik 60 (1930) 491.
[16] F. London, Zeitschrift für Physik C 11 (1930) 222.
[17] J.O. Hirschfelder, C.F. Curtiss, R.B. Bird, Molecular Theory of Gases and Liquids. Wiley, New York (1954).
[18] A.J. Stone, The Theory of Intermolecular Forces. Clarendon Press, Oxford (1996).
[19] A.D. McNaught, A. Wilkinson, Compendium of Chemical Terminology, 2nd edit. Blackwell Scientific Publications, Oxford (1997).
[20] J. Emsley, Chemical Society Reviews 9 (1980) 91.
[21] W.C. Hamilton, J.A. Ibers, Hydrogen Bonding in Solids. Benjamin, New York (1968) P. 16.

[22] M. Nardin, J. Schultz, Langmuir 12 (1996) 4238.
[23] A. Bondi, Journal of Physics Chemistry 68 (1964) 441.
[24] T. Young, Philosophical Transactions of the Royal Society (London) 95 (1805) 65.
[25] P. Debye, Physik Z. 13 (1912) 295.
[26] W.H. Keesom, Physik Z 22 (1921) 129.
[27] F.M. Fowkes, Industrial & Engineering Chemistry 56 (1964) 40.
[28] S.J. Park, M. Brendle, Journal of Colloid and Interface Science 188 (1997) 336.
[29] R.F. Hudson, G. Klopman, Tetrahedron Letters 12 (1967) 1103.
[30] M. Berkowitz, R.G. Parr, Journal of Chemical Physics 88 (1988) 2554.
[31] R.G. Pearson, Journal of the American Chemical Society 85 (1963) 3533.
[32] S.J. Park, H.C. Kim, H.Y. Kim, Journal of Colloid and Interface Science 255 (2002) 145.
[33] R.S. Drago, B. Wayland, Journal of the American Chemical Society 87 (1965) 3571.
[34] R.S. Drago, Structure Bonding 15 (1973) 73.
[35] F.M. Fowkes, Journal of Adhesion Science and Technology 1 (1987) 7.
[36] U. Mayer, V. Gutmann, W. Greger, Monatsh Chemical 106 (1975) 1235.
[37] F.L. Riddle, F.M. Fowkes, Journal of the American Chemical Society 112 (1990) 3259.
[38] S.J. Park, E. Papirer, J.B. Donnet, Journal of Chemical Physics 91 (1994) 203.
[39] Y. Marcus, Journal of Solution Chemistry 13 (1984) 599.
[40] R. Schmid, V.N. Sapunov, Non-Formal Kinetics, 1st edit. Verlag-Chemie, Weinheim (1982).
[41] C.J. van Oss, R.J. Good, M.K. Chaudhury, Langmuir 4 (1988) 884.
[42] C.J. van Oss, R.J. Good, Langmuir 8 (1992) 2877.
[43] F.M. Fowkes, M.A. Mostafa, Industrial and Engineering Chemistry Product Research and Development 17 (1978) 3.
[44] N.P. Cheremisinoff, P.N. Cheremisinoff, Handbook of Advanced Materials Testing. Marcel Dekker, New York (1995).
[45] F.M. Fowkes, Journal of Adhesion Science and Technology 4 (1990) 669.
[46] M.D. Vrbanac, J.C. Berg, Journal of Adhesion Science and Technology 4 (1990) 255.
[47] S.J. Park, Long-range Force Contributions to Surface Dynamics. In Interfacial Forces and Fields: Theory and Applications,Marcel Dekker, New York (1990).
[48] S.J. Park, D.I. Seo, C.W. Nah, Journal of Colloid and Interface Science 251 (2002) 225.
[49] S.J. Park, B.R. Jun, Journal of Colloid and Interface Science 284 (2005) 204.
[50] F.M. Fowkes, D.O. Tischler, J.A. Wolfe, L.A. lannigan, C.M. Ademu-John, M.J. Halliwell, Journal of Polymer Science – Polymer Chemistry 22 (1984) 547.
[51] M. Brogly, M. Nardin, J. Schultz, Polymer 39 (1998) 2185.
[52] M. Brogly, S. Bistac, J. Schultz, Macromolecules 31 (1998) 3967.
[53] A.N. Gent, J. Schultz, Journal of Adhesion 3 (1972) 281.
[54] M.L. Williams, R.F. Landel, J.D. Ferry, Journal of the American Chemical Society 77 (1955) 3701.
[55] A. Carré, J. Schultz, Journal of Adhesion 18 (1984) 171.
[56] T. Hubbard, Encyclopedia of Surface and Colloid Science, 1st edit. Marcel Dekker, New York (2002).
[57] D.H. Buckley, Surface Effects in Adhesion, Friction, Wear and Lubrication, 1st edit. Elsevier, New York (1981).
[58] J.W. McBain, D.G. Hopkins, Journal of Physics Chemistry 29 (1925) 188.
[59] E.M. Borroff, W.C. Wake, Trans Institute of Rubber Industries 25 (1949) 190.
[60] A. N. Gent, & J. Schultz, in Proceedings Symposium on Recent Advances in Adhesion, 162nd ACS Meeting 31 (1971) 113.

[61] A.N. Gent, J. Schultz, The Journal of Adhesion 3 (1972) 281.
[62] W.C. Wake, Adhesion and the Formulation of Adhesives, 2nd edit. Applied Science, London (1982).
[63] P.J. Hine, S. El Muddarris, D.E. Packham, The Journal of Adhesion 17 (1984) 207.
[64] H.E. Bair, S. Matsuoka, R.G. Vadimsky, T.T. Wang, The Journal of Adhesion 3 (1971) 89.
[65] T.T. Wang, H.N. Vazirani, The Journal of Adhesion 4 (1972) 353.
[66] B.V. Deryaguin , N.A. Krotova, V.V Karassev, Y.M. Kirillova & I.N. Aleinikova, Proceedings of 2nd International Congress on Surface Activity, 3, International Congress of Surface Activity, London (1957).
[67] B.V. Deryaguin, V.P. Smilga, Adhesion: Fundamentals and Practice, 1st edit. McLaren & Son, London (1969).
[68] B.V. Deryaguin, N.A. Krotova, V.P. Smilga, Adhesion of Solids, 1st edit. Consultants Bureau, New York (1978).
[69] E. Fermi, Rend. Accad. Lincei 3 (1926) 145.
[70] P.A.M. Dirac, Proceedings of the Royal Society (London) A 112 (1926) 661.
[71] H. Krupp, W. Schnabel, The Journal of Adhesion 5 (1973) 269.
[72] L. Lavielle, J.L. Prévot, J. Schultz, Angewandte Makromolekulare Chemistry 169 (1989) 159.
[73] J.J. Bikerman, The Science of Adhesive Joints, 2nd edit. Academic Press, New York (1961).
[74] S.S. Voyutskii, Autohesion and Adhesion of High Polymers, 1st edit. Wiley-Inter-science, New York (1963).
[75] R.M. Vasenin, Adhesion: Fundamentals and Practice, 1st edit. Mclaren & Son, London (1969).
[76] P.G. de Gennes, C.R. Academy of Science Paris, Series B 291 (1980) 219.
[77] M. Doi, S.F. Edwards, Journal of the Chemical Society Faraday Transactions 2 (1978) 1789.
[78] W.W. Graessley, Advances in Polymer Science 47 (1982) 67.
[79] K. Jud, H.H. Kausch, J.G. Williams, Journal of Materials Science 16 (1981) 204.
[80] S. Prager, M. Tirrell, Journal of Chemical Physics 75 (1981) 5194.
[81] Y.M. Kim, R.P. Wool, Macromolecules 16 (1983) 1115.
[82] E.P. Plueddemann, Silane Coupling Agents, 2nd edit. Plenum Press, New York (1982).
[83] K.L. Mittal, Silanes and Other Coulping Agents. VSP, Utrecht (1992).
[84] S.J. Park, J.S. Jin, Journal of Polymer Science B 41 (2003) 55.
[85] A.N. Gent, A. Ahagon, Journal of Polymer Science Part B: Polymer Physics 13 (1975) 1285.
[86] P. Delescluse, J. Schultz, M.E.R. Shanahan, K.W. Allen, Adhesion 8, 1st edit. Elsevier Applied Science, London (1984).
[87] S.J. Park, S.W. Song, M.K. Seo, E.K. Kim, J.S. Shin, Journal of the Korean Industrial and Engineering Chemistry 14 (2003) 76.
[88] E. Raphaël, P.G. de Gennes, Journal of Physics Chemistry 96 (1992) 4002.
[89] F. Brochard-Wyart, P.G. de Gennes, The Journal of Adhesion 57 (1996) 21.
[90] H.R. Brown, Macromolecules 22 (1989) 2859.
[91] H.R. Brown, V.R. Deline, P.F. Green, Nature 341 (1989) 221.
[92] H.R. Brown, Macromolecules 24 (1991) 2752.
[93] C. Creton, E.J. Kramer, C.Y. Hui, H.R. Brown, Macromolecules 25 (1992) 3075.
[94] W.F. Reichert, H.R. Brown, Polymer 34 (1993) 2289.
[95] E. Boucher, J.P. Folkers, H. Hervet, L. Léger, Macromolecules 29 (1996) 774.
[96] D. Maugis, M. Barquins, Journal of Physics D: Applied Physics 11 (1978) 1989.
[97] S. Bistac, Journal of Colloid and Interface Science 219 (1999) 210.
[98] A.V. Shenoy, Rheology of Filled Polymer Systems. Kluwer Academic Pub, Dordrecht (1999).
[99] S.J. Park, K. Li, F.L. Jin, Materials Chemistry and Physics 108 (2008) 214.

[100] S.J. Park, B.J. Kim, Materials Science and Engineering: A 408 (2005) 269.
[101] V. Shah, Handbook of Plastics Testing and Failure Analysis, 3rd edit. Wiley-Interscience, California (2007).
[102] C. Yang, Role of Surface Roughness in Tribology: from atomic to macroscopic scale, 7th edit. Forschungszentrum Jülich, Berlin (2008).
[103] L.I. Minkova, M. Velcheva, M. Paci, P. Magagnini, F.P. La Mantia, D. Sek, Journal of Applied Polymer Science 73 (1999) 2069.
[104] A.J. Kinloch, M.S.G. Little, J.F. Watts, Acta Materialia 48 (2000) 4543.
[105] K.S. Kim, N. Aravas, International Journal of Solids and Structures 24 (1988) 417.
[106] B.V. Derjaguin, V.M. Muller, Y.P. Toporov, Journal of Colloid and Interface Science 53 (1975) 314.
[107] B.V. Derjaguin, V.M. Muller, Y.P. Toporov, Journal of Colloid and Interface Science 67 (1978) 378.
[108] B.V. Derjaguin, V.M. Muller, Y.P. Toporov, Journal of Colloid and Interface Science 73 (1980) 293.
[109] D. Tabor, Journal of Colloid and Interface Science 58 (1977) 2.
[110] D. Tabor, Journal of Colloid and Interface Science 67 (1978) 380.
[111] D. Tabor, Journal of Colloid and Interface Science 73 (1980) 293.
[112] B.V. Derjaguin, Zhurnal Fizicheskoi Khimii 6 (1935) 1306.
[113] J.S. McFarlane, & D. Tabor, Proceedings of the Royal Society A202 (1950) 224.
[114] H. Hertz, J. Reine, Angew. Math 92 (1881) 156.
[115] K.L. Johnson, K. Kendall, & A.D. Roberts, Proceedings of the Royal Society London A 324 (1971) 301.
[116] V.M. Muller, V.S. Yushchenko, B.V. Derjaguin, Journal of Colloid and Interface Science 77 (1980) 91.
[117] D. Maugis, Journal of Colloid and Interface Science 150 (1992) 243.
[118] J.E. Lennard-Jones, B.M. Dent, Transactions of the Faraday Society 24 (1928) 92.
[119] D.S. Dugdale, Journal of the Mechanics and Physics of Solids 8 (1960) 100.
[120] L.R. Fisher, J.N. Israelachvili, Colloids and Surfaces 3 (1981) 303.
[121] K.L. Johnson, J.A. Greenwood, Journal of Colloid and Interface Science 192 (1997) 326.
[122] J.N. Israelachvili, Intermolecular and Surface Forces, 2nd edit. Academic Press, London (1992).
[123] H.K. Christenson, Journal of Colloid and Interface Science 104 (1985) 234.
[124] A. Fogden, L.R. White, Journal of Colloid and Interface Science 138 (1990) 414.
[125] K. Kendall, Molecular Adhesion and its Applications: The Sticky Universe, 1st edit. Kluwer Academic/Plenum Publisher, New York (2001).
[126] T. Hancock, Personal Narrative: American Chemical Society Centennial Volume 1857 (1939) 206.
[127] K. Kendall, British Polymer Journal 10 (1978) 35.
[128] K. Kendall, F.R. Sherliker, British Polymer Journal 12 (1980) 85.
[129] G. Kraus, Reinforcement of Elastomers, 1st edit. Wiley-Interscience, New York (1965).
[130] S.J. Park, K.S. Cho, S.K. Ryu, Carbon 41 (2003) 1437.
[131] J. Cook, & J. E. Gordon, Proceedings of the Royal Society (London) A 282 (1964) 508.
[132] A. Kelly, Strong solids, 2nd edit. Clarendon Press, Oxford (1973).
[133] D. Hull, An introduction of Composite Materials, 2nd edit. Cambridge University Press, Cambridge (1981).
[134] M.K. Seo, S.J. Park, Bulletin of the Korean Chemical Society 30 (2009) 124.

[135] A.R. Zack, M.L. Williams, ASME Journal of Applied Mechanics 34 (1963) 967.
[136] J.R. Rice, G.C. Sih, ASME Journal of Applied Mechanics 32 (1965) 418.
[137] K. Kendall, Proceedings of the Royal Society. (London) A 341 (1975) 409.
[138] R.A.J. Sambell, D.H. Bowen, D.C. Phillips, Journal of Materials Science 7 (1972) 663.
[139] H. Tada, P.C. Paris, G.R. Irwin, The Stress Analysis of Cracks Handbook, 3rd edit. ASME Press, New York (2000).
[140] A.A. Griffith, Philosophical Transactions of the Royal Society (London) A 221 (1920) 163.
[141] J.D. Ferry, R.F. Landel, M.L. Williams, Journal of Applied Physics 26 (1955) 359.
[142] J.J. Benbow, F.C. Roesler, Proceedings of the Royal Society (London) B 70 (1957) 201.
[143] J.P. Berry, Journal of Polymer Science 50 (1961) 313.
[144] B. Rosen, Fracture Processes in Polymeric Solids, 1st edit. Wiley-Interscience, New York (1964).
[145] E. Orowan, Report of Progress in Physics 12 (1948) 185.
[146] P.E. Chen, Polymer Engineering & Science 11 (1971) 51.
[147] M.L. Williams, International Journal of Fracture Mechanics 1 (1965) 292.
[148] M.L. Williams, F.N. Kelley, in Proceeding of 5th International Congress on Rheology, Tokyo University Press, Tokyo (1970).
[149] G.R. Irwin, Journal of Applied Mechanics 61 (1939) A49.
[150] S. Mostovoy, E.J. Ripling, Journal of Applied Polymer Science 10 (1966) 1351.
[151] E.J. Ripling, S. Mostovoy, H.T. Corten, The Journal of Adhesion 3 (1971) 107.
[152] A.H. Cottrell, Dislocations and Plastic Flow in Crystals, 9th edit. Oxford University Press, Oxford (1956).
[153] R. Buchdahl, Journal of Polymer Science 28 (1958) 239.
[154] N. Brown, Materials Science and Engineering 8 (1971) 65.
[155] J.G. Williams, Fracture Mechanics of Polymers, 1st edit. Ellis Horwood Ltd, Chichester (1984).
[156] G.R. Irwin, Lastic zone near a crack and fracture toughness, Proceeding of 7th Sagamore Ordnance Mater. Res. Conf., New York (1960).
[157] T. Hiroshi, P.C. Paris, G.R. Irwin, The Stress Analysis of Cracks Handbook, 3rd edit. Del Research Corporation, Pennsylvania (1973).
[158] A.N. Gent, R.H. Tobias, Journal of Polymer Science Polymer Physics 20 (1982) 2051.
[159] S.J. Park, J.S. Kim, Journal of Adhesion Science and Technology 15 (2001) 1443.
[160] G. Kraus, Journal of Applied Polymer Science 7 (1963) 861.
[161] C.J. van Oss, Interfacial Forces in Aqueous Media. Marcel Dekker, New York (1994).
[162] V.S. John, V. van Krystyn, in 3.302 Mechanical Behavior of Materials, MIT, Cambridge (2006).
[163] G.P. Cherepanov, Journal of Applied Mathematics and Mechanics 31 (1967) 503.
[164] J.R. Rice, Journal of Applied Mechanics 35 (1968) 379.
[165] M.A. Meyers, K.K. Chawla, Mechanical Behavior of Materials, 2nd edit. Cambridge Univrersity Press, Cambridge (2009).
[166] J.R. Rice, Journal of Applied Mechanics 35 (1968) 379.
[167] G. Green, An Essay on the Application of Mathematical Analysis to the Theories of Electricity and Magnetism, Green's Mathematical Papers, London (1828).
[168] J.A. Begley, J.D. Landes, in Fracture Toughness, ASTM International, Philadelphia (1972).
[169] H.E. Boyer, T.L. Gall, in Metals Handbook: Desk Edition, ASTM International, Materials Park, Ohio, 6 (1985) 34.

[170] H.E. Boyer, T.L. Gall, in Metals Handbook: Desk Edition, ASTM International, Materials Park, Ohio (1985) 9.8.
[171] MILHDBK, Military Handbook: Metallic Materials and Elements for Aerospace Vehicle Structures, MIL-HDBK-5G, U. S. Dept. of Defense, MIL-HDBK-5 Coordination Activity, Wright-Patterson AFB, Ohio, 2, Chap. 3. 10 (1994).
[172] MILHDBK, Military Handbook: Metallic Materials and Elements for Aerospace Vehicle Structures, MIL-HDBK-5G, U. S. Dept. of Defense, MIL-HDBK-5 Coordination Activity, Wright-Patterson AFB, Ohio, 2, Chap. 5.3 (1994).
[173] R.O. Ritchie, Journal of Engineering Materials and Technology, Transactions of the American Society of Mechanical Engineers 99 (1977) 195.
[174] C.A. Dostal, in Engineered Materials Handbook Vol. 2 Engineering Plastics, 1st edit, ASTM International, Ohio (1988).
[175] M.G. Karfakis, M. Akram, Rock Fracture Toughness in Zero Point of Charge Environment, Proceeding of the 8th Annual Workshop, Generic Mineral Technology Center, Mine Systems Design and Ground Control, Virginia (1990).
[176] M.G. Karfakis, M. Akram, International Journal of Rock Mechanics and Mining Science & Geomechanics Abstracts 30 (1993) 1253.
[177] A. Kelly, N.H. Macmillan, Strong Solids, 3rd edit. Clarendon Press, Oxford (1986) 376.
[178] S.P. Shah, S.E. Swartz, C. Ouyang, Fracture Mechanics of Concrete: Applications of Fracture Mechanics to Concrete. Wiley-Interscience, New York (1995) 176.
[179] J.G. Williams, in Fracture Mechanics of Polymers and Adhesives: Fracture of Non-Metallic Materials. Kluwer Academic Pubs, Dordrecht (1987) 243.
[180] K. Kendall, Proceedings of the Royal Society London A 344 (1975) 287.
[181] S.J. Park, M.K. Seo, D.R. Lee, Carbon 41 (2003) 2991.
[182] S.J. Park, H.J. Jeong, C.W. Nah, Materials Science and Engineering: A 385 (2004) 13.
[183] J.O. Outwater, M.C. Murphy, Modern Plastics 47 (1970) 160.
[184] K. Kendall, Journal of Materials Science 11 (1976) 638.
[185] K. Kendall, Journal of Materials Science 11 (1976) 1263.
[186] K. Kendall, Journal of Materials Science 11 (1976) 1267.
[187] S.J. Park, E.J. Lee, S.H. Kwon, Bulletin of the Korean Chemical Society 28 (2007) 188.
[188] G.V. Shiriajeva, G.D. Andreevskaya, Plasticheskie Massy 4 (1962) 43.
[189] C. Gurney, J. Hunt, Proceedings of the Royal Society London A 299 (1967) 508.
[190] N.McN. Alford, J.D. Birchall, K. Kendall, 330 (1987) 51.
[191] J.A. Nairn, Y.C. Liu, Computer Interface 4 (1997) 241.
[192] N.J. Wadsworth, I. Spilling, British Journal of Applied Physics 1 (1968) 1049.
[193] Y. Huang, R.J. Young, Computer Science and Technology 52 (1994) 505.
[194] S.J. Park, M.S. Cho, Carbon 38 (2000) 1053.
[195] S.J. Park, M.S. Cho, J.R. Lee, P.K. Pak, Carbon 37 (1999) 1685.
[196] K. Kendall, Journal of Materials Science 10 (1975) 1011.
[197] B. Miller, P. Muri, L. Rebenfeld, Computer Science and Technology 28 (1987) 17.
[198] U. Gaur, B. Miller, Computer Science and Technology 34 (1989) 35.
[199] K.P. Mcalea, G.J. Besio, Polymer Composites 9 (1988) 285.
[200] L.S. Penn, G.C. Tesoro, H.X. Zhou, Polymer Composites 9 (1988) 184.
[201] V. Rao, P. Herrera-Franco, A.D. Ozzello, L.T. Drzal, The Journal of Adhesion 34 (1991) 65.

Chapter 5

Interface Applications in Nanomaterials

5.1. INTRODUCTION

Interfacial science and technology deal with the behaviors of fine-particle dispersions, fibers and thin films, and other systems strongly influenced by the properties of their interfaces. On the other hand, the field is defined by a dimension, not by a type of materials, and it therefore includes inorganic, organics, and biomaterials. Many composite materials can create the fibers, rods, and spherical foam with highly ordered molecular orientation in reactions at liquid/liquid interfaces. These multifunctional composite materials can have potential utility as novel bio-, medical-, environmental-, and nano-materials, and hence will experience a bright future as enabling agents in science [1–3].

The number of publications dealing with various aspects of the applications of multifunctional materials and structures related to interfacial science and technology has increased markedly in recent years. Fig. 5-1 shows how the number of English language refereed journal articles in multifunctional materials and structures has steadily increased since 2000, based on data collected from the Engineering Village© web-based information service.

Multifunctional materials are necessarily composite materials, and the strong growth in the use of composites has been greatly influenced by multifunctional design requirements. There are increasing reports in the literature that significant improvements of multiple structural functions can be achieved with new hybrid multiscale composites that incorporate nanoscale reinforcements as well as conventional micron scale fibers or particle reinforcements. For example, while fiber-dominated properties (i.e., longitudinal tensile strength and elastic modulus) of conventional unidirectional polymer composites with micron size fiber reinforcements are excellent, the corresponding matrix-dominated transverse tensile strength and longitudinal compressive strength properties are often poor. Fig. 5-2 shows a description of length scales related to common materials relevant to microscale and nanoscale systems.

Interface Science and Composites, Volume 18
ISSN: 1573-4285, doi: 10.1016/B978-0-12-375049-5.00005-0

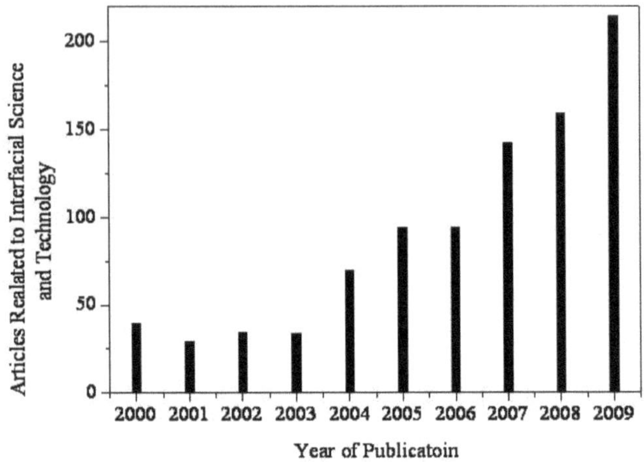

Fig. 5-1 Recent English language referred journal publications related to multifunctional materials and structures.

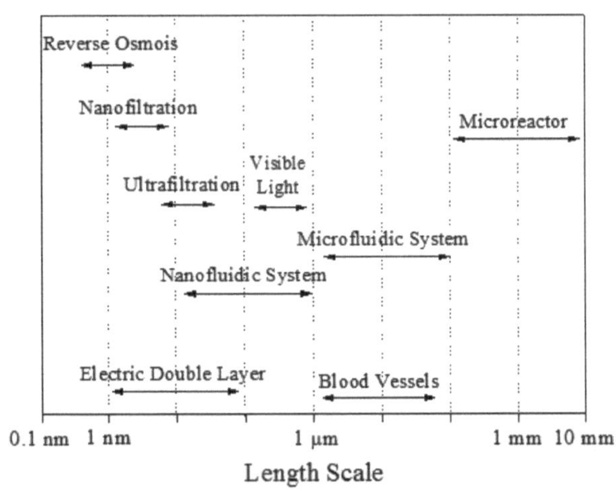

Fig. 5-2 A description of length scales as they relate to common materials, physical objects, phenomena, and techniques relevant to microscale and nanoscale systems.

5.2. ENERGY STORAGE AND CONVERSION DEVICES

5.2.1. DYE-SENSITIZED SOLAR CELLS

Dye-sensitized solar cells [4] have attracted great interest because of their potential application as a cost effective alternative to p–n junction solar cells [5]. In conventional solar cells, the semiconductor has the tasks of light absorption and charge-carrier transport, whereas in dye-sensitized solar cells, the two functions are separately controlled. A photosensitizing dye, anchored to the surface of a wide band gap semiconductor, absorbs light. The photogenerated excitons rapidly split, with electrons transferred to the conduction band of the semiconductor leaving oxidized dye molecules behind. The electrons, through the external circuit, travel to the counter electrode to reduce the iodide ion of the electrolyte. The iodide ions transport the electrons back to the dye (reduction process). So far, ruthenium complexes and porous TiO_2 have mostly been used as sensitizers and the semiconductor, respectively, in dye-sensitized solar cells [6].

In the early 1990s, a new class of dye-sensitized solar cells was reported and showed surprisingly high efficiency of over 4 percent given that very inexpensive TiO_2 was used as the bulk of the photovoltaic cells [7]. The performance of dye-sensitized solar cells depends on several key components of the cells. First, the dye molecules have strong optical absorbance in the visible light range. The excited dye molecules transfer an electron from the metal to the p* orbital of the carboxylated bipyridyl ligand attached to the metal-oxide surface (anatase TiO_2), and then release the electrons to the oxide within 100 fs.

ZnO has been the subject of a great deal of research interest in dye-sensitized solar cells due to some of its fascinating properties. Compared to other semiconductors, ZnO has unique properties such as higher binding energy (60 meV), wide band gap (3.37 eV), high breakdown strength, good cohesion, and good exciton stability. Moreover, ZnO is one of the hardest materials in the family of II–VI semiconductors. Electron mobility in ZnO is higher than that in TiO_2, making the former suitable for dye-sensitized solar cells. Recently, it has become possible to form vertical nanowires of ZnO [8].

Pal *et al.* [9] introduced a new concept by replacing nanocrystalline films with oriented, long, high-density ZnO nanowires prepared from solution seeded synthesis; Fig. 5-3 (a) shows a schematic diagram of this concept. As shown in Fig. 5-3 (b), the high surface area is favorable for trapping the dye molecules, and the electron transport in oriented nanowires should be orders of magnitude faster than the percolation in polycrystalline films.

In dye-sensitized solar cells, after absorbing photon energy from the illuminated white light, the dye molecules become excited (D*) and inject electrons into the ZnO nanowires (Fig. 5-4). The electron transfer occurs due to the favorable energy difference between the lowest unoccupied molecular orbital of the dye and the conduction band of ZnO. The photogenerated electrons percolate rapidly through the ZnO nanowire and are collected by the conducting glass support. The highest occupied molecular orbital of the dye is energetically lower than the redox potential, E_{Redox}, of the iodine/triiodide couple. The energy difference provides the driving force for hole injection into the electrolyte. Recombination of the charge carrier is also minimized in such devices since transport of only one type of carrier (electrons, in general) is energetically possible from the dye to the semiconductor. Vertical growth of the nanowires was achieved via a simpler chemical route [9].

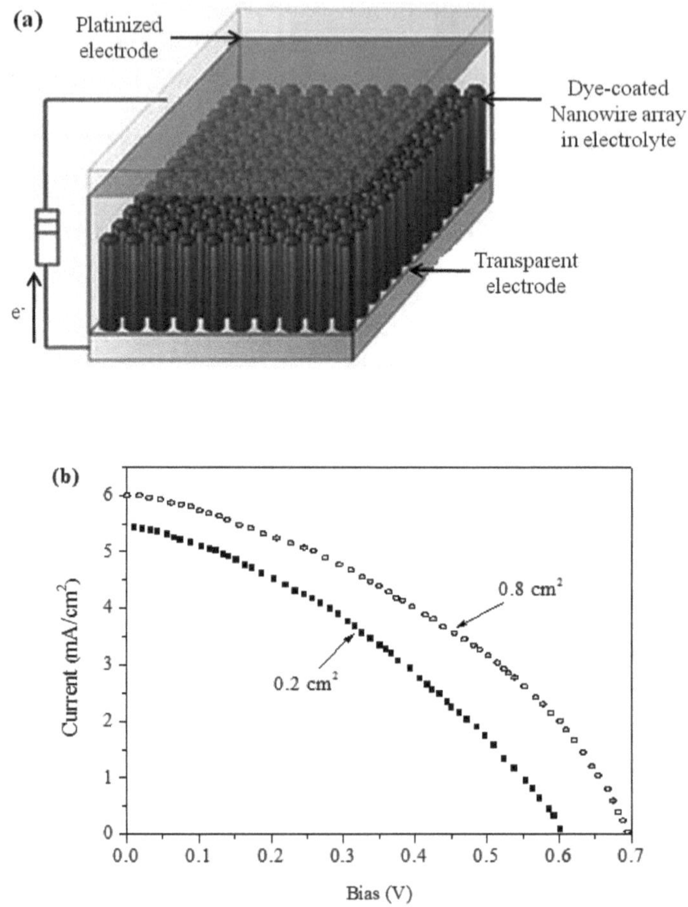

Fig. 5-3 Dye-sensitized solar cells based on ZnO nanowires: (a) schematic diagram of the cell and (b) current density as a function of bias.

5.2.2. Lithium-Ion Batteries

At the turn of the millennium, the consciousness of the need to use energy more efficiently spread, at least in the 'developed' countries. The efficient use of energy also includes efficient storage of electricity, meaning that one needs to have high energy density batteries that have low energy losses during storage, electrical charging, and discharging; finally, batteries should have long lifetimes and minimum production and disposal costs [10].

In the last decade, a number of different battery systems were found to meet the requirements for electric vehicles (EV), among which are the nickel-metal hydride, the sodium-sulfur, and the lithium-ion systems [11]. In the United States, the lithium-ion battery is considered to be most promising and thus its development is considerably supported.

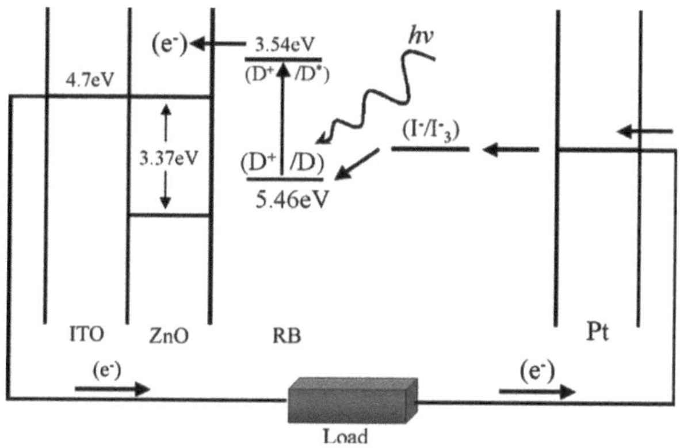

Fig. 5-4 Schematic band diagram showing the working principle of the dye-sensitized solar cells.

There are two major arguments for using a lithium-ion battery: lithium is the lightest element that can safely be handled in electrochemical processes and it exhibits the highest oxidation potential of any element; its properties make lithium ideally suited for high energy density batteries. However, lithium-metal electrodes in contact with liquid electrolytes cause a variety of problems, which in the worst case can lead to fires and explosions. Therefore, the most promising approach is that of the 'rocking chair' cell, in which the lithium metal is replaced by a lithium-ion source. During charging and discharging of such batteries, lithium ions are rocked between lithium-carbon and lithium-metal oxide intercalation compounds, which act as the electrode couple. A typical example is displayed in Fig. 5-5, in which Li_xC_6 is used as the anode and $Li_xMn_2O_4$ (a spinel) is used as the cathode.

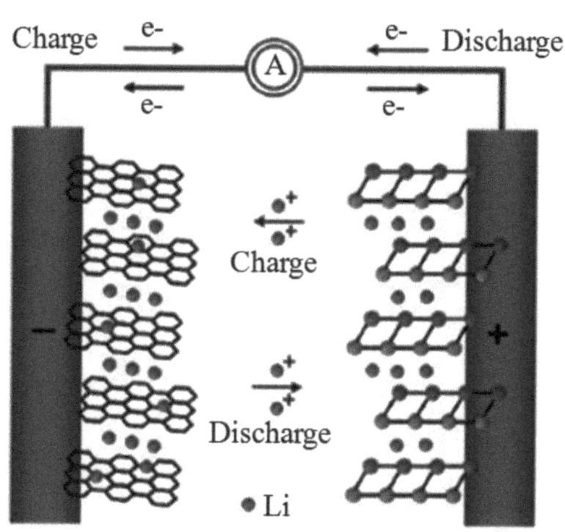

Fig. 5-5 Schematic illustration of a lithium rocking chair battery with graphite and spinel as intercalaction electrodes and its electrode reactions.

Anode reaction:

$$Li_xC_6 \Leftrightarrow xLi^+ + xe^- + C_6 \quad (5\text{-}1)$$

Cathode reaction:

$$\begin{array}{c} xLi^+ + xe^- + Li_{1-x}Mn_2O_4 \Leftrightarrow LiMn_2O_4 \\ \text{Charging} \leftrightarrow \text{Discharging} \end{array} \quad (5\text{-}2)$$

Due to the numerous applications of lithium batteries in the automotive and aerospace industries and in portable devices, significant interest exists from both developers and manufacturers to improve the chemistry of such batteries [12].

Solid and solid-like polymer electrolytes offer unique advantages, such as satisfactory mechanical properties, ease of fabrication as thin films and formation of proper electrode-electrolyte interfaces. Polymeric electrolytes have found wide application in electric, medical, electroluminescence, and photoelectric devices. Since Wright *et al.* [13] discovered in 1973 that polyethylene, based on polyethylene oxide $(OCH_2CH_2)_n$ with an alkaline metal salt, offered ionic conductivity, polyethylene has received widespread attention. Moreover, it is possible to substitute liquid electrolytes in transitional lithium-ion batteries, thereby achieving an electrolyte system for solid-like lithium polymer batteries; therefore, the promising next-generation anodes could be used in lithium polymer cells as well.

Competitor electrolytes to polyethylene, ones that can be used to realize solid lithium batteries, are organic salts with melting points below room temperature, the so-called room-temperature molten salts, which have been the main focus of many recent scientific investigations due to their unique physical and chemical properties [14]. They are nonvolatile, nonflammable, miscible with a number of organic solvents, and have high thermal stability. From an electrochemical point of view, they offer high ionic conductivity and a wide potential window. For these reasons, these salts have attracted much attention for their potential application to electrochemical capacitors or nonaqueous lithium ion batteries.

The most common and extensively studied polyethylene compound is based on the polyethylene oxide-lithium salt complex. This polyethylene oxide-lithium salt complex shows good electrochemical and mechanical performance, and many studies of polyethylene oxide combined with various lithium salts, such as $LiClO_4$ and $LiBF_4$, has been reported. However, polyethylene oxide-lithium salt complexes have a conductivity of only 10^{-4}–10^{-5} S/cm, which is low for some applications. This unsatisfactory conductivity can be ascribed in part to the high degree of polyethylene oxide crystallinity at room temperature, as well as to the strong interactions between lithium ions and polymers [15]. One efficient way to decrease the crystallinity of a polyethylene compound based on the polymer matrix, polyethylene oxide, is the addition of nanoscale inorganic fillers such as TiO_2, SiO_2, and Al_2O_3, or organic plasticizers such as ethylene carbonate or, dimethylformamide [16].

The recent interest in polyethylene carbonate containing layer-structured nanoparticles, such as montmorillonite (MMT; $(Na, Ca,)_{0.33}(Al, Mg)_2(Si_4O_{10})(OH)_2{\cdot}nH_2O)$) and Mobil crystalline material-41 has risen from the fact that such polyethylene carbonates exhibit a dramatic increase in tensile strength, heat resistance, and solvent resistance, as well as a decrease in gas permeability. All these properties are desirable in electrolytes for rechargeable batteries [17–19] and are attributed to the addition of nanoparticles.

5.2.3. Supercapacitors

Electrochemical capacitors or supercapacitors have high power density, long charge-discharge cycle life, and high energy efficiency. A combined system of supercapacitors and rechargeable batteries or fuel cells is considered to be a potential power source for electric vehicles. In such a system, supercapacitors provide the necessary high power for acceleration and allow for recuperation of brake energy [20]. According to the energy storage mechanism, electrochemical capacitors can be classified as carbon-based electrochemical double layer capacitors (EDLCs) [21, 22] or conducting polymers and metal oxide based pseudo-capacitors. In electrochemical double layer capacitors, energy storage arises from the accumulation of electronic and ionic charges at the interface between the electrode materials and the electrolyte solution. The capacitance of the latter is due to the reversible faradic reaction of the electro-active species of the electrode, such as the surface functional groups and transition metal oxides. It is clear that the electrode is the key element in the development of supercapacitors. Fig. 5-6 shows the Helmholtz model of an electric double layer system.

The development of advanced composite materials based on metal oxide-carbon nanotubes has been studied recently under the anticipation that these materials will provide improved capacitive behaviors due to their enhanced stability, high conductivity, and pseudocapacitive property [23]. One of the

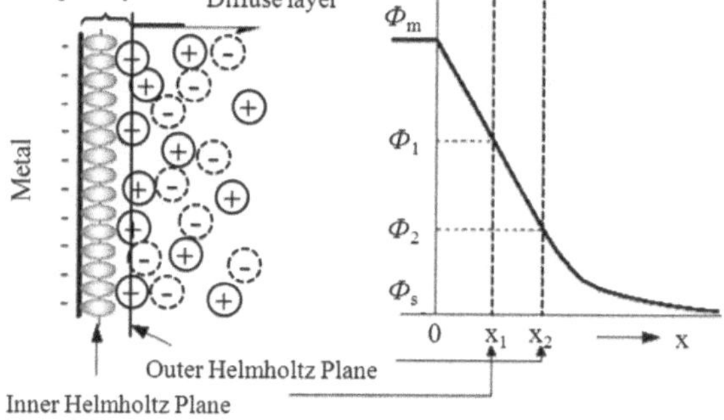

Fig. 5-6 Helmholts model of electric double layer system.

most extensively studied metal oxides for use in electrochemical capacitors is hydrous ruthenium oxide (RuO_2), which has a pseudo-capacitance of 720 F/g [24].

Park *et al.* [25] presented the first report on the synthesis and capacitive behaviors of Fe_3O_4/multiwalled carbon nanotube composites. Nano-sized Fe_3O_4 particles with an average diameter of about 14 nm were attached to the purified surfaces of multiwalled carbon nanotubes, as shown in Fig. 5-7. From Fig. 5-8 (a), it can be seen that Fe_3O_4 shows a pseudo-capacitive behavior originating from the redox reactions with sulfite ions and iron oxide in 1 M Na_2SO_3 electrolyte. Charge-discharge measurement (Fig. 5-8 (b)) at 0.2 A/g yielded a specific capacitance of 58 and 165 F/g for pure multiwalled carbon nanotubes and Fe_3O_4/multiwalled carbon nanotube composites, respectively. These results indicate that the Fe_3O_4/multiwalled carbon nanotube composites have potential applicability as electrode materials of supercapacitors.

One of the most critical requirements is the optimization of the microstructures of the carbon materials. For example, nanostructured porous carbon

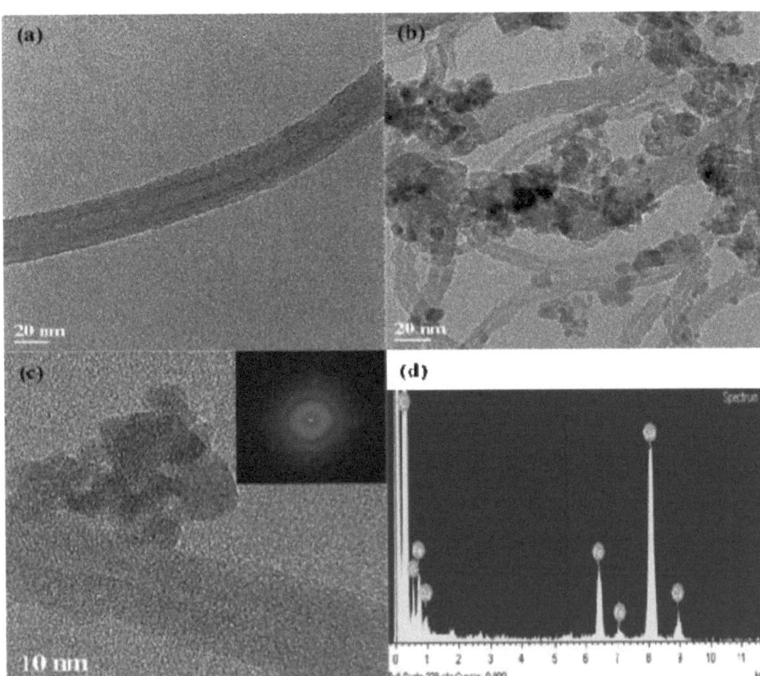

Fig. 5-7 Microstructure of multi walled carbon nanotube and Fe_3O_4/multi-walled carbon nanotube composites: (a) TEM image of normal multi-walled carbon nanotube; (b) TEM image of Fe_3O_4/multi-walled carbon nanotube composites; (c) and (d) show the high-resolution TEM image and electron dispersive X-ray spectrum of the Fe_3O_4/multi-walled carbon nanotube composites, respectively.

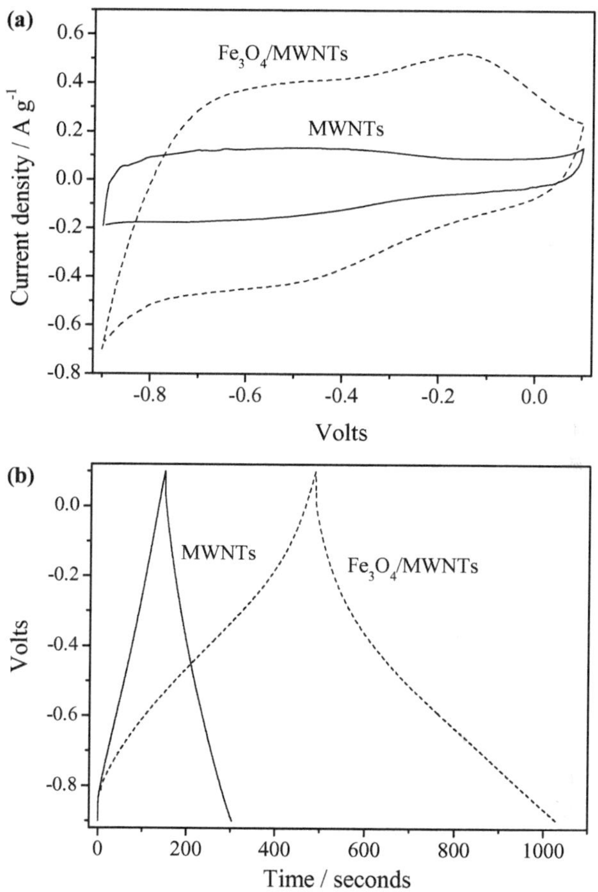

Fig. 5-8 (a) Cyclic voltammetric curves of multi-walled carbon nanotubes (MWNTs) and Fe_3O_4/multi-walled carbon nanotube composites electrodes in 1 M Na_2SO_3 solution. (b) Chronopotentiograms of multi-walled carbon nanotube s and Fe_3O_4/multi-walled carbon nanotube (Fe_3O_4/MWNTs) composite electrodes at 0.2 A/g.

electrodes with carefully controlled surface chemistry and tuned microporous and mesoporous structures have recently been fabricated by means of sol-gel processing using resorcinol and formaldehyde as precursors, followed by aging, solvent exchange after gelation, removal of the solvent by freeze-drying, and finally pyrolyzing to remove hydrogen and oxygen from the carbon gel at around 1050 °C in nitrogen [26]. The resultant porous carbon is referred to as a carbon cryogel, which is similar to the carbon aerogel that is fabricated using supercritical drying. The microporous structure, electrochemical properties, and energy-storage performance are all controlled by the fabrication conditions. The specific capacitance critically depends on the microstructure of the carbon, including the surface area and pore size distribution [27].

Chemical modification of carbon cryogels is used as an efficient approach to alter both the porous structure and the surface chemistry, which alterations result in much improved electrochemical properties. Preliminary experiments involved the transportation of ammonia-borane (NH_3BH_3), dissolved in anhydrous THF, to the pores of resorcinol-formaldehyde hydrogels during post-gelation solvent exchange. After being soaked in the NH_3BH_3 solution, the modified hydrogels were subjected to the same freeze-drying and pyrolysis processes as were the unmodified hydrogels [28]. The resultant modified carbon cryogels are referred to as boron and nitrogen-doped, indicating that the samples were modified with ammonia borane as precursor although they are in fact co-doped with boron and nitrogen, whereas the unmodified carbon cryogels are referred to s carbon cryogel samples. Figs. 5-9 (a), (b) compare the typical SEM images of arbon cryogels and boron and nitrogen-doped carbon cryogel samples, respectively. The SEM images reveal the highly nanoporous structure in both samples.

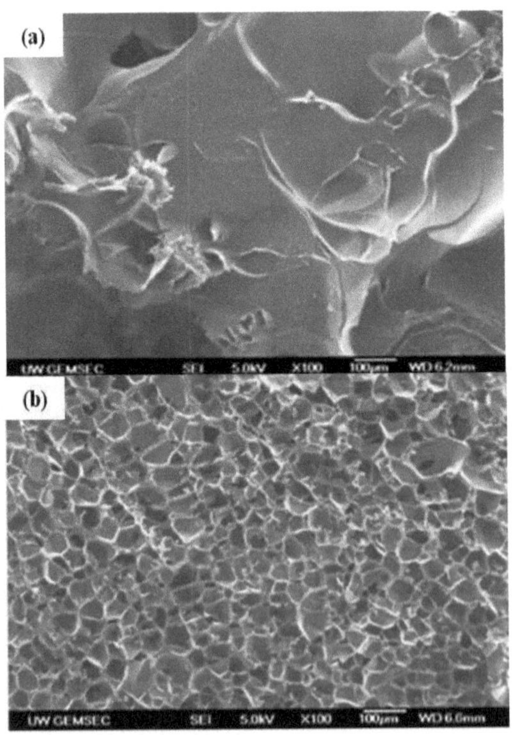

Fig. 5-9 SEM images of (a) carbon cryogels and (b) boron and nitrogen-doped carbon cryogels.

After pyrolysis, the pores in the carbon cryogel sample shift from a 5 nm to 16 nm radius and have a broad distribution of pore sizes. Conversely, in the boron and nitrogen-doped carbon cryogel sample has a more uniform pore size distribution with a dominant concentration of pores of radius 6 nm appears (Fig. 5-10 (a)). These structural changes suggest that the boron–nitrogen co-doping processes alter the carbon structure and prevent changes, at least in the mesopore region. The pseudocapacitive behavior of the boron and nitrogen-doped carbon cryogel sample is presented in Fig. 5-10 (b).

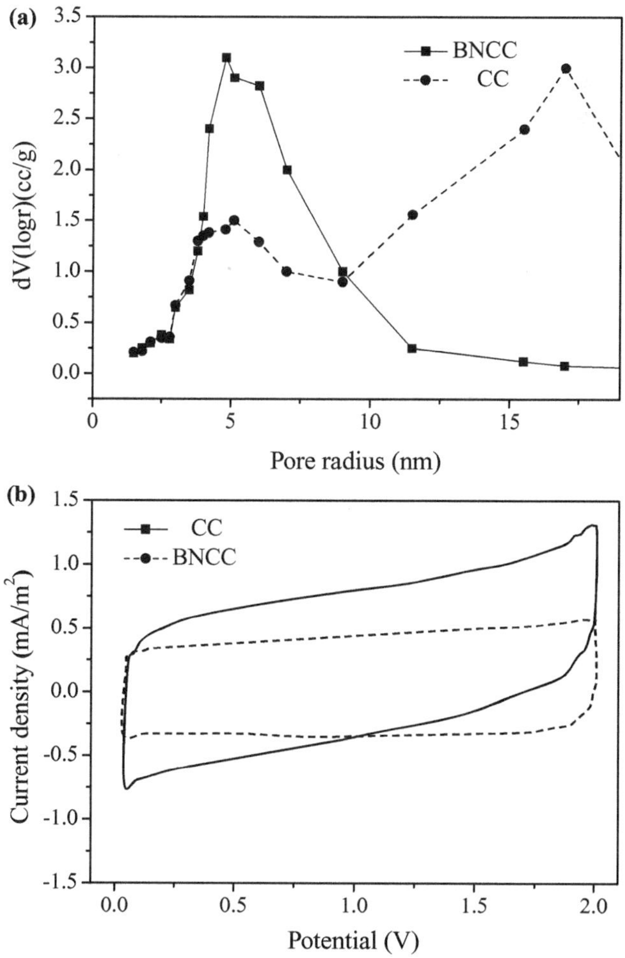

Fig. 5-10 (a) Comparison of pore size distribution of carbon cryogels (CC) and boron and nitrogen-doped carbon cryogels (BNCC) samples. (b) Cyclic voltammetric curves for carbon cryogels (CC) and boron and nitrogen-doped carbon cryogels (BNCC).

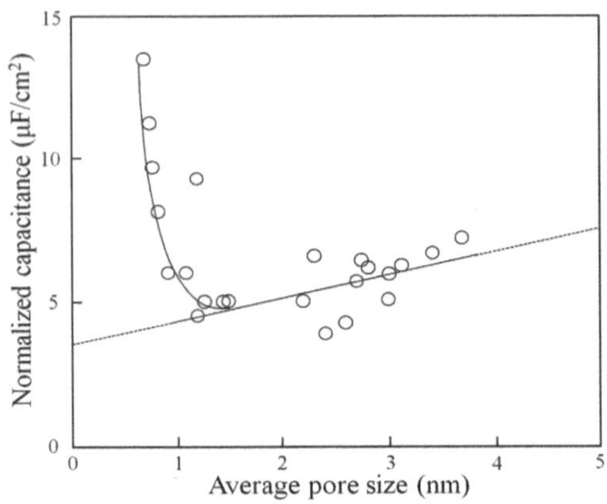

Fig. 5-11 Effect of pore sizes on charge storage in carbide-derived carbon materials.

Interconnected mesoscale porosity plays an important role in ensuring that charged ions can freely access all the surfaces (2–50 nm). Many groups have thus investigated surfactant templated mesoporous carbon with controllable pore sizes. However, recently, a new study reported the effect of pore size on charge-storage properties and provided new information on the relative roles of mesoscale and microscale porosity [29]. The charge storage in carbide-derived carbon by high temperature chloration is reported in Fig. 5-11. This material displays good control of the pore sizes down to less than 1 nm. Three regions are observed: in region I, where the mesopore dominates, the capacitance increases with the pore sizes owing to better pore accessibility and less overlapping of the double layer structure. However, as the pore size becomes smaller in region II, the capacitance begins to increase. In region III, the capacitance increases sharply with decreasing pore sizes. The effect of the ultra small pore on the capacitance is attributed to the distortion of the double layers in the small pores and to the decrease in the double layer thickness [29].

5.3. ENVIRONMENTAL TECHNOLOGIES

5.3.1. NO_x and SO_x Removals

The growth in environmentalism has seen the introduction of the concept of 'environmental quality', which is typically applied to the air we breathe and the water we consume. In terms of NO_x/SO_x, if the air contains more than 0.1 parts per million (ppm) NO_2 or SO_2, persons with respiratory complaints may experience breathing difficulties; if it contains more than 2.5 ppm NO_2 or 5 ppm SO_2, healthy persons can also be affected [30]. Policymakers have acknowledged the potential dangers posed by excessive NO_x/SO_x release, and the Kyoto Protocol set out measures to reduce such emissions by the year 2008 to levels below those recorded in 1990 [31].

NO_x/SO_x release is explicitly linked to consumption of fossil fuel, i.e., coal, oil, and natural gas. Even allowing for steady improvements in energy efficiency, future generations will use massive quantities of energy. If current trends prevail and energy demand is met by burning fossil fuels, the environmental implications are grave. Energy technologies drawing on renewable energy serve to minimize the negative environmental impacts associated with the fossil fuel cycle. Such technologies, which are either reasonably well established or in the formative stages, convert sunlight, wind, flowing water, and the heat of the earth and oceans, certain plants, and other resources, into useful energy. The use of renewable energy can still impact the environment, but the effect is far smaller than that of present energy sources, which are dependent on the deployment of nonrenewable resources. Because vehicle exhaust contains appreciable levels of toxic emissions, much can be done to alleviate the environmental burden through economics of motor fuel consumption and engine/combustion modifications. Fuel cell developments suggest that such devices will make a valuable contribution to future power generation [32].

Fuel cells that operate on pure hydrogen as fuel produce only water as byproduct, thus eliminating all emissions associated with standard methods of electricity production. Hydrogen production/storage remains something of an obstacle in fuel cell commercialization. Fuel cells have yet to make a serious impression on the energy market, and mass market zero-emission automobiles are far from realization.

The most abundant nitrogen oxide in the environment is nitrous oxide (N_2O), which is relatively unreactive and not regarded as a primary pollutant. Nitric oxide (NO) and nitrogen dioxide (NO_2) comprise the predominant atmospheric burden and are denoted by the collective term NO_x. NO_x is produced mainly in high-temperature combustion processes involving atmospheric nitrogen (or as a fossil fuel/biomass component) and oxygen and is associated with power stations, refineries, transport, agriculture, and domestic applications. In addition to contributing, as a heat-trapping pollutant, to the greenhouse effect, NO_x directly impacts the environment in three ways: depletion of the ozone layer, production of acid rain, and general air pollution. Of the two oxides of sulfur, SO_2 and SO_3 (collectively SO_x), the former is far more abundant in the atmosphere [33]. Sulfur dioxide reacts on the surface of a variety of airborne solid particles, is soluble in water, and can be oxidized within airborne water droplets. Natural sources of sulfur dioxide include releases from volcanoes, oceans, biological decay, and forest fires. The most important man-made SO_2 sources are fossil fuel combustion, smelting, manufacture of sulfuric acid, conversion of wood pulp to paper, incineration of refuse, and production of elemental sulfur. Coal and oil burning are the predominant sources of atmospheric SO_x, which can contribute to respiratory illness, alterations in pulmonary defenses, and aggravation of existing cardiovascular disease. In the atmosphere, SO_x mixes with water vapor, producing sulfuric acid, which can be transported over hundreds of kilometers and deposited as acid rain. Sulfur dioxide and the sulfuric acid that it generates have four established adverse effects: (1) toxicity to humans, (2) acidification of lakes and surface waters, (3) damage to trees and crops, and (4) damage to buildings.

Control of NO_x/SO_x emissions can follow two strategies: a direct curtailment of NO_x/SO_x formation (primary measures), and a secondary, downstream treatment (end-of-pipe solutions). Effective emissions reduction requires controls on both stationary and mobile sources. One viable approach in reducing NO_x production focuses on fuel de-nitrogenation, in which the nitrogen component is removed from liquid fuels by intimate mixing with hydrogen at elevated temperatures to produce ammonia and cleaner fuel. This technology can reduce the nitrogen contained in both naturally occurring and synthetic fuels.

5.3.1.1. Pollution Problems

Key combustion-generated air pollutants include sulfur oxides (principally SO_2), particulate matter, carbon monoxide, unburned hydrocarbons, and nitrogen oxides (NO_x) as shown in Fig. 5-12. Table 5-1 shows typical concentrations range of emissions in the flue gases from efficient electric power generation plants as a function of the fuel utilized [34].

Natural gas, for example, is a relatively clean fuel with respect to NO_x formation when compared with heavy fuel or coal. Natural gas does not contain any fuel nitrogen, i.e., nitrogen atoms bound the hydrocarbon molecules; therefore, the only NO_x production route is via oxidation of the molecular nitrogen contained in the combustion air.

Sulfur occurs in fuels as inorganic minerals (primarily pyrite, FeS_2), organic structures, sulfate salts, and elemental sulfur. Sulfur contents range from parts per million (ppm) in pipeline natural gas, to a few tenths of a percent in diesel and light fuel oils, to 0.5 percent to 5 percent in heavy fuel oils and coals. Sulfur compounds are pyrolized during the volatilization phase of oil and coal combustion and react in the gas phase to form predominantly SO_2 and some SO_3.

Fig. 5-12 Examples of atmosphere pollutions.

Table 5-1 Typical concentrations and ranges of emission in fuel gases from power generation plants

	Natural gas	Fuel oil	Coal
NO_x (ppm)	25 ~ 160	100 ~ 600	150 ~ 1000
SO_x (ppm)	< 0.5 ~ 20	200 ~ 2000	200 ~ 2000
CO_2 (%)	5 ~ 12	12 ~ 14	10 ~ 15
O_2 (%)	3 ~ 18	2 ~ 5	3 ~ 5
H_2O (%)	8 ~ 19	9 ~ 12	7 ~ 10
N_2 (%)	Balance	Balance	Balance

Conversion of fuel sulfur to these oxides is generally high (85–90 percent) and is relatively independent of the combustion conditions. From 1 percent to 4 percent of the SO_2 is further oxidized to SO_3; the latter is highly reactive and extremely hygroscopic. It combines with water to form sulfuric acid aerosol. Combustion-related particulate emissions may consist of one or more of the following types, depending on the fuel: mineral matter derived from ash constituents of liquid and solid fuels that can vaporize and condense as submicron-size aerosols; sulfate particles formed in the gas phase that can condense; unburned carbon that includes unburned char (the remaining solid particle); coke; and soot (occurs as fine particles (0.02–0.2 Am), often agglomerated into filaments or chains that can be several millimeters long). Carbon monoxide is a key intermediate in the oxidation of all hydrocarbons. In a well-adjusted combustion system, essentially all the CO is oxidized to CO_2 and final emission of CO is very low indeed (a few parts per million). However, in systems that have low temperature or that are in poor adjustment, CO emissions can be significant. Additionally, various unburned hydrocarbon species may be emitted from hydrocarbon flames. In general, there are two classes of unburned hydrocarbons: (1) small molecules that yield the intermediate products of combustion (for example, formaldehyde) and (2) larger molecules that are formed by pyrosynthesis in hot, fuel-rich zones within flames, e.g., benzene, toluene, xylene, and various polycyclic aromatic hydrocarbons.

5.3.1.2. Emission Regulation

Starting in the early 1970s, policy has gradually evolved in various countries, notably Japan, the United States, and Germany, to establish a program of reducing emission levels of pollutants in flue gases from stationary sources. Table 5-2 gives an overview of different targets [35].

The last protocol, the Gothenburg Protocol, is unique in the sense that it establishes reductions of four pollutants to abate three effects (acidification, eutrophication, and the effects of tropospheric ozone on human health and

Table 5-2 Air emission reduction targets for the European Commission

Policy/Pollutant	Base year	Target year	Reduction (%)
United Nations Economic Commission for Europe Convention on Long-Range Transboundary Air Pollution			
Sulfur dioxide[a]	1980	2000	62
Sulfur dioxide[b]	1990	2010	75
Nitrogen oxides[c]	1987	1994	Stabilization
Nitrogen oxides	1990	2010	50
Ammonia	1990	2010	12
5th Environmental Plan			
Sulfur dioxide	1985	2000	35
Nitrogen dioxide	1990	2000	30
Directive on National Emission Ceiling (NECD)[c]			
Sulfur dioxide	1990	2010	78
Nitrogen oxides	1990	2010	55
Ammonia	1990	2010	21

[a] *Target from 1994 second sulfur protocol. The different emission ceiling for each member state corresponds to a 62% emission reduction for the U.S.*
[b] *Targets from multipollutant Gothemburg Protocol.*
[c] *Targets from first NO_x protocol.*

vegetation). The protocol has been signed by 29 European countries together with Canada and the United States. The countries themselves estimate the critical loads. Apart from the protocols, the member states have to fulfill several directives. In May 1999, the European Commission (EC) presented a proposal for a directive on national emission ceilings (NECD) for the same pollutants as those covered in the convention on long-range transboundary air pollution (CLRTAP) and, for the first time, for ammonia.

Updated information related to pollutant emissions can be obtained at the European Environment Agency, which interprets emission data, coordinates the development of the spatial distribution of emissions, and provides information on policies and scenarios [36]; information can also be found in the European Pollutant Emission Register, which is the first Europe-wide register of emissions into air and water from large- and medium-sized industrial facilities in Europe. The first European Pollutant Emission Register report includes data for the year 2001 from about 10,000 industrial facilities in the European Union and Norway.

5.3.1.3. NO_x and SO_x Storage and Reduction

This concept has emerged whereby the problem of destruction of NO_x is undertaken as a two-stage operation, the combination of two air/fuel ratios instead of a single fixed ratio. During a fuel-lean stage, NO_x is trapped on a selective sorbent in the form of nitrate (NO_3^-), as proposed. Because there is plenty of oxygen

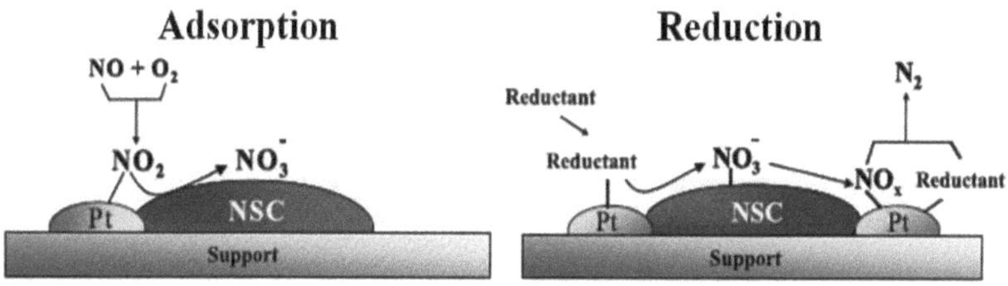

Fig. 5-13 Adsorption and reduction of NO_x by NO_x storage and reduction concept.

present at this stage, HC, H_2, and CO are readily oxidized to water and carbon dioxide. Then, when the engine is switched to fuel-rich (normal air/fuel ratio) operation, the resulting exhaust becomes comparatively oxygen-deficient and HC, H_2, and CO remain unoxidized. Hence, the three components react with the NO_3^- and change into harmless nitrogen, water, and carbon dioxide (Fig. 5-13) [37].

The first catalyst of this type was developed by Toyota [38], for which the composition was based on that of a three-way catalyst. The preparation was by impregnation of noble metals upon alumina (essentially Pt) of several alkali and alkali earth metals (Na^+, K^+, Ba^{+2}) and of rare earth oxides (mainly La_2O_3) [39]. The basicity of the storage component determines the amount of NO_x adsorbed. Nevertheless, the performance of platinum, especially the hydrocarbon oxidation, decreases with support basicity. The particle size of platinum and BaO and the distance between them are other factors that govern the NO_x storage: small particles in contact with each other adsorb more than large particles [40].

5.3.1.4. Carbonaceous Materials

NO is adsorbed on some specific carbonaceous materials with the highest known capacities per weight of adsorbent, as shown in Fig. 5-14. These materials include carbon fibers and carbon nanotubes. Whilst carbon fibers consist of

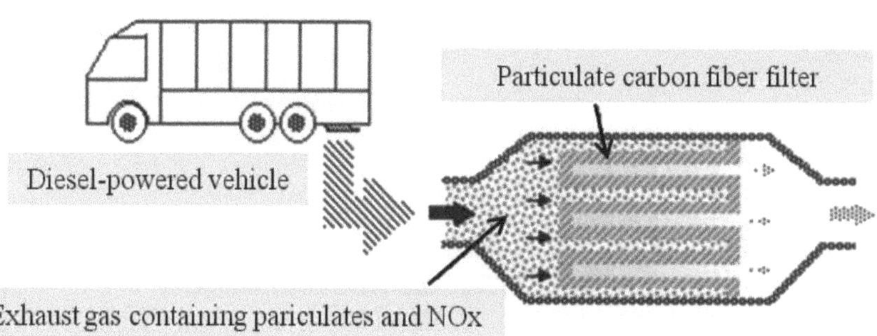

Fig. 5-14 Exhaust gas conversion for diesel-powered vehicle.

graphite sheets arranged in various orientations along the fiber axis with many edges (disposed like a fishbone), both single-walled- (SWNT) and multiwalled- (MWNT) carbon nanotubes show a concentric wall structure consisting of ordered graphite platelets. The morphology and the size of the above materials, especially because they present huge length vs. diameter ratios, can play a significant role in catalytic applications.

For active carbon fibers, modified by iron or copper oxides on their surface, the NO_x adsorption ranges up to 320 mg at 303 K. Even higher values are reported for NO_2. Maximal NO_x desorption occurs between 393 and 443 K. NO is fully recovered by heating above 473 K. Sorption is thought to proceed via an enhanced physisorption mechanism comprising dimerization of NO and a kind of liquefaction in the micropores (supercritical micropore filling). On unmodified active carbon fiber, NO is well sorbed, but in far lower amounts than it would be with a metal oxide coating; further, NO is released at lower temperatures. The mechanism of micropore filling of supercritical NO is not clear at the present stage. However, the presence of highly dispersed Fe species is obviously essential for this phenomenon to take place. From the temperature dependence of the isotherm, this type of NO adsorption seems to possess both chemisorption and physical adsorption characteristics. Actually, gaseous NO molecules produce strong chemisorptions on Fe_2O_3 highly dispersed around the entrance of the slit-shaped micropores of activated carbon fibers. After the adsorption step, almost all the absorbed NO molecules migrate to fill the micropores as $(NO)_2$, as presented in Fig. 5-15. This intermolecular interaction plays a key role in the micropore filling.

Surprisingly, the modified active carbon fiber sorbents have a high selectivity for NO_x: in a mixture with 300 ppm NO, the presence of O_2, SO_2, H_2O, and CO_2 as co-gases did not prevent an almost complete removal of NO. However, these materials present several problems, such as their low oxidative stability: the first traces of CO_2, by oxidation of the sorbent, are released at 383–433 K [41]. An uptake amount of 78 mg of NO g^{-1} was observed on carbon nanotubes at room temperature in the presence of oxygen (5 percent). This adsorption is also reversible and multiwalled carbon nanotubes could be regenerated by

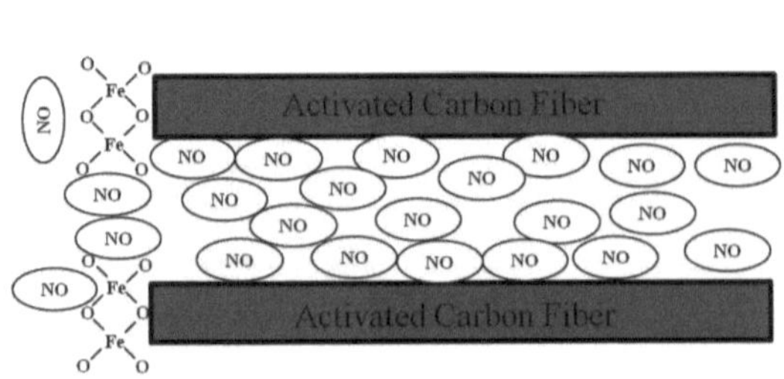

Fig. 5-15 Schematic model of the NO dimmer in the slit-shaped active carbon fibers micropores.

heating to 573 K. By comparison, SO_2 can also be adsorbed on the carbon nanotubes, but the adsorption rate and amount are less than those for NO [42]. Although this carbon material has an attractive performance, there are still two possible drawbacks that have not yet been cleared up: the oxidative stability (and hence, regenerability) and the capture of SO_2.

NO_x removal using activated carbon fibers (ACFs) with a transition metal (Ni, Fe, Cu, Ag, etc.) has been studied in many works. Recently, several methods have been developed for the prevention and control of air pollution. Many researchers have studied air pollution removal using activated carbons (ACs) and carbon fibers activated by metal electroplating, impregnation, sputtering, or controlling the pore structures of the activated carbon fibers themselves [43–47].

The roles of metals introduced by an electroplating method in the NO_x reduction over the metal/carbon surfaces have been reported [48]. Fig. 5-16 shows the NO reduction efficiency of the samples of activated carbon fibers, copper plated carbon fibers activated for 5 s (Cu-5), and copper plated carbon fibers activated for 10 s (Cu-10). The as-received sample has low activity, whereas the copper/activated carbon fiber samples have high reduction activity. NO reduction efficiency curves maintain almost a fixed concentration after 170 min of reaction time and increase with the copper content.

In the case of the copper/activated carbon fibers composites, the copper metal plays a role as a reductant in the NO reduction by forming CuO and some CO_2 production, as revealed by the x-ray diffraction (Fig. 5-17.) results, which can be presented as the following Eq. [48]:

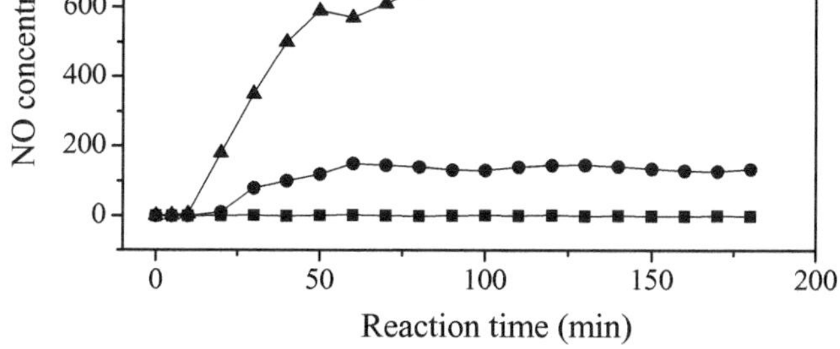

Fig. 5-16 NO removal efficiency of the copper-plated activated carbon fibers as a function of the reaction time.

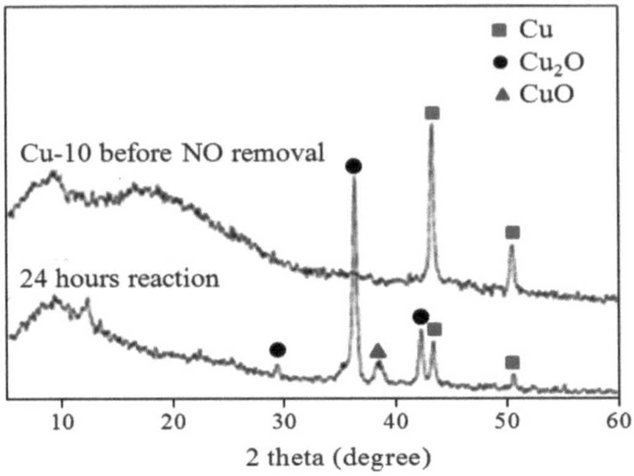

Fig. 5-17 X-ray diffraction profiles of the activated carbon fibers before and after NO removal reaction for 24 h.

$$\mathrm{NO} + \mathrm{ACF} - \mathrm{Cu} \rightarrow \mathrm{ACF} - \mathrm{Cu_2O} + \frac{1}{2}\mathrm{N_2} \tag{5-3}$$

$$\mathrm{NO} + \mathrm{ACF} - \mathrm{Cu_2O} \rightarrow \mathrm{ACF} - \mathrm{CuO} + \frac{1}{2}\mathrm{N_2} \tag{5-4}$$

These results do not mean that the activated carbon fibers of the copper/activated carbon fibers do not participate in the reduction mechanism. The activated carbon fibers also do, as was revealed in Fig. 5-18. However, the copper metals on the Cu/activated carbon fibers play the major role in this experiment. From the results, it should be recognized that the copper metal on the activated carbon fiber surfaces plays the role of reductant through the formation of copper oxide (CuO), resulting in the reduction of NO in this system.

5.3.2. Water Purification

A major global technological challenge for the next century lies in developing new technologies for water purification [49]. Membrane processes are an integral part of developing freshwater supplies through filtration and desalination. Historically, membranes have pore or nanocapillary diameters varying from 1 nm to 1 mm [50]. Membrane fouling has been a challenge due to biofouling, scaling, and colloid-driven fouling. Many different strategies have been employed to tackle this problem by varying membrane chemical composition, surface charge, and morphology. Other strategies include growth of hydrophilic polymers such as poly (ethylene oxide) or poly (ethylene glycol) brushes to reduce biofouling of membranes.

For example, some researchers have focused on developing different surface modification and selective layer copolymerization strategies to mitigate the effects of membrane fouling; these new strategies cause a reduction in the output flux of the membrane [51]. Fig. 5-18 shows images and data from such recent efforts. There has also been a recent effort trying to enhance the flux of clean water through membranes and understanding the fundamentals of transport phenomena through confined nanoscale spaces; this has included the development of nanofluidic membrane sensors.

Recently, activated carbon fibers have been widely used in separation, purification, and catalytic processes due to their extended specific surface area, high adsorption capacity, highly porous structure, and surface reactivity [52, 53]. Nevertheless, problems still remain when the activated carbon fibers are used to purify drinking water, because bacteria preferably adhere to the solid support made of carbon materials, indicating that activated carbon fibers have good biocompatibility. Bacteria may breed on activated carbon fibers during the purification process, themselves becoming a pollutant. In order to avoid this disadvantage, antibacterial activated carbon fibers are required [52]. Therefore, many efforts have been made to prepare metal-containing activated carbon fibers by various surface treatment methods to solve these problems [53].

A previous work [54] reported on the antibacterial behavior of transition metals decorated activated carbon fibers. An electroplating method was used to prepare metal/activated carbon fiber. As shown in Fig. 5-19, the grain-like metal nanoparticles (Cu and Ag) existed on the activated carbon fibers. Fig. 5-20 shows the changes of metal residues with time for the samples. As expected, all samples showed a time-dependent curve of this erosion rate. It was found that

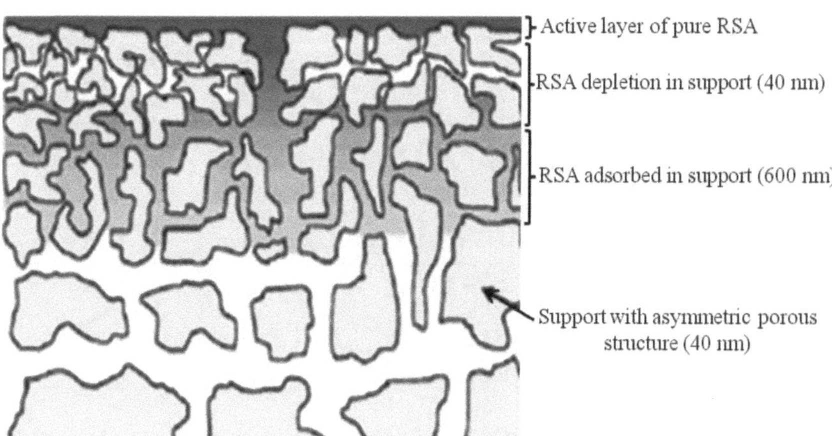

Fig. 5-18 Images showing two strategies based on formation of rigid-star amphiphiles (RSA) or co-polymerization of polyacrylonitrile-polyethylene oxide for reducing membrane fouling and developing high permeability membranes for water purification.

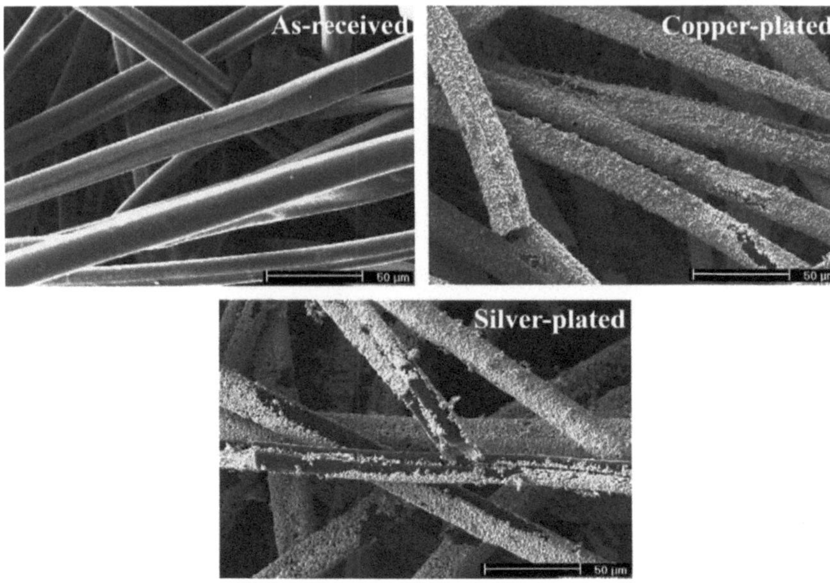

Fig. 5-19 SEM images of activated carbon fibers before and after copper and silver plating on activated carbon fibers.

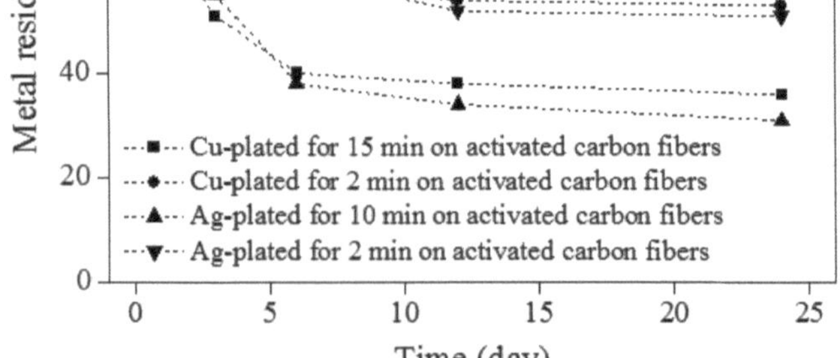

Fig. 5-20 Water erosion resistance of the metal-plated activated carbon fibers.

erosion curves decreased rapidly in the initial time and then flattened. It was also observed that samples that had high metal content showed higher decrease rates and lower specific metal residues than samples with low metal content. It can be assumed that copper plated for 15 min-activated carbon fiber or silver plated for

10 min-activated carbon fiber have rougher surfaces and are more susceptible to erosion than those of copper plated for 2 min-activated carbon fiber or silver plated for 2 min-activated carbon fiber due to the excessively plated metals, resulting in their showing low washing resistance.

It was thus concluded that the electroplating of carbon surfaces with Cu or Ag can be a good method for adding antibacterial activity to biocompatible activated carbon fibers in order to use them as filters for water purification and avoid secondary pollution by bacteria [54].

Application of carbon nanoparticles and nanotubes for removal of metal ions (Cd^{2+}, Cu^{2+}, Ni^{2+}, Pb^{2+}, Zn^{2+}, etc.) from water have been described recently [55]. In order to improve the separation of carriers of metal ions from treated water, the metal ions can be bound to polymeric molecules and carbon nanoparticles forming nanocarbon conjugated polymer nanocomposites (NCPC) in water that are able to precipitate rapidly. Fig. 5-21 shows the nanocarbon conjugated polymer nanocomposites.

Simultaneously the positively charged complex metal-poly(ethyleneimine) combines with carboxyl groups of nanocarbon conjugated polymer nanocomposites as follows: (i) by hydrogen bond between carboxyl group of nanocarbon and amine group of poly(ethyleneimine)which have not formed coordination bonds with metal, (ii) by electrostatic interaction between negatively charged carboxyl ion of nanocarbon and positively charged metal-poly(ethyleneimine) complex. The ability to remove metal ions Zn(II), Cd(II), Cu(II), Hg(II), Ni(II),

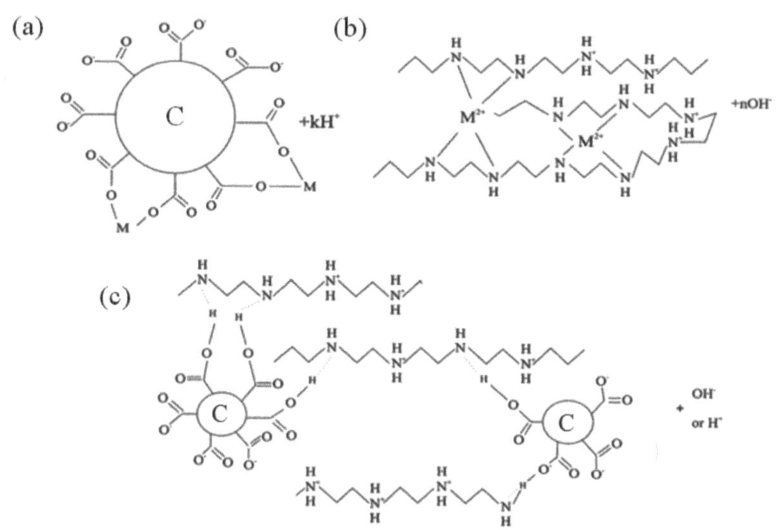

Fig. 5-21 The structures of (a) carbon nanoparticles with an attactched metal ion, (b) polymeric complex poly(ethyleneimine)-metal ions and (c) the formation of nanocarbon conjugated polymer nanocomposites.

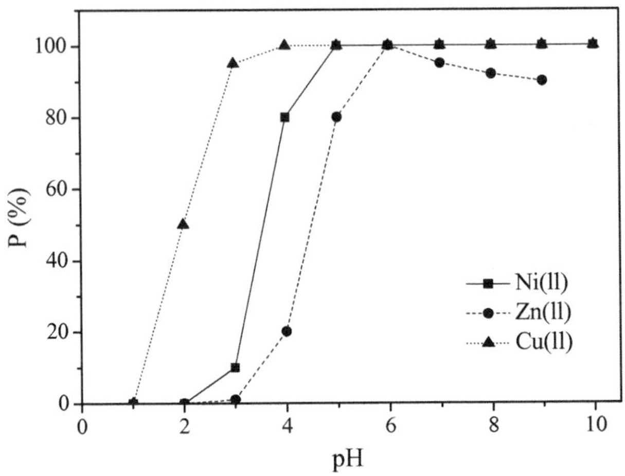

Fig. 5-22 The dependence of the removal ratio (P) for Cu(II), Zn(II), and Ni(II) on pH of solutions.

Cr(VI) was investigated [56]. The mean size of nanocarbon composites was 20 nm. Poly(ethyleneimine) and metal ion concentrations were 150 mg/L and 1 mg/L, respectively. Figs. 22 and 23 represent the dependence of the removal ratio P against pH of solutions containing metal ions with concentration of 10 mg/L.

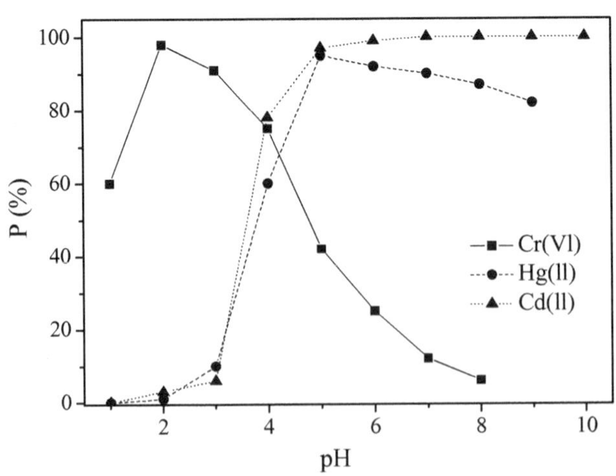

Fig. 5-23 The dependence of the removal ratio (P) for Cd(II), Hg(II), and Cr(II) on pH of solutions.

5.4. GAS STORAGE

5.4.1. Introduction

The use of adsorption phenomenon can be helpful in the design of modern chemical technologies that are friendlier for the environment. Physisorption under supercritical conditions has been actively studied [57]. Recently, high-pressure adsorption of methane [58] and hydrogen [59, 60] has become of general interest in connection with secondary energy storage. Adsorption of nitrogen monoxide at an ambient temperature is concerned in processes that deal with the protection of urban environments.

However, the mechanism of supercritical gas adsorption is not fully established yet because of the lack of accurate experimental results and an established theory. High pressure adsorption of a supercritical gas has an essential problem: how can the surface excess mass or absolute adsorbed amount for an exact thermo-dynamical analysis be determined? Surface excess mass isotherms under supercritical conditions have shown the following problem from experiments, computer simulations, and theoretical equations: the surface excess mass reaches a maximum with increasing pressure and then decreases until it reaches a negative value with a further increase in pressure [61].

The surface excess mass is useful in industrial applications. However, the surface excess mass is not necessarily enough for exact thermodynamic treatment. Agarwal *et al.* [58] showed that a molecular interpretation of high-pressure adsorption data can be done by using absolute adsorption quantities. It is desirable to elucidate the relationship between the surface excess and the absolute adsorption. That is, a structural understanding of the interface between the supercritical gas and the physisorbed phase is necessary. Fig. 5-24 shows the relationship between the surface excess mass and the absolute adsorbed amounts.

The surface excess mass, n_{ex} (region A), and the absolute adsorbed amount, n_{ab} (regions A and B); n_{ex} and n_{ab} are given by

$$n_{ab} = \int_{V_{ad}} \rho_{ad}(r)dr \tag{5-5}$$

$$n_{ex} = n_{ab} - \rho_{Bulk} V_{ad} \tag{5-6}$$

where ρ_{ad} and ρ_{Bulk} are densities of adsorbed and bulk gas phases, respectively, and V_{ad} the volume ad of the adsorbed phase.

In the case of the vapor adsorption measurement (e.g., nitrogen adsorption at 77 K), we can approximate the surface excess mass from the absolute adsorbed amount, because the bulk density is low and the volume of the adsorbed phase V_{ad} is negligible in comparison with the gas phase volume.

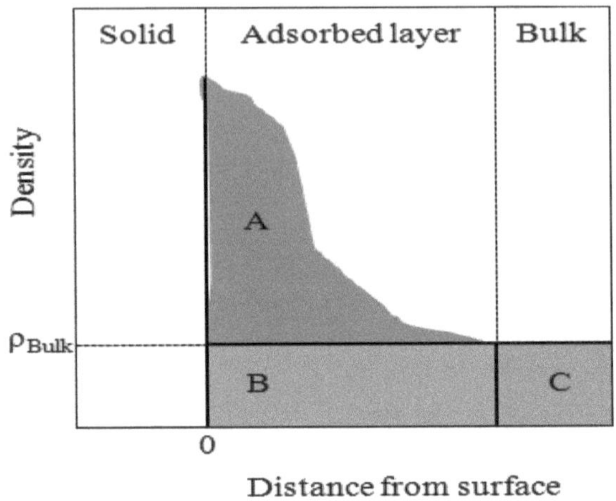

Fig. 5-24 Relationship between the surface excess mass and absolute adsorbed amount. Region A expresses the surface excess mass, the sum of regions A and B corresponds to the absolute adsorbed amount.

On the other hand, one of the examples is the gas separation and purification processes commonly used in industry. A very popular cryogenic method as shown in Fig. 5-25 proposed by von Linde in 1895 can be replaced by modern technologies based on the idea of Skarstrom [62].

By passing the gas mixture through the adsorption columns and suitably manipulating the pressures and directions of the gas flow, the mixture can be

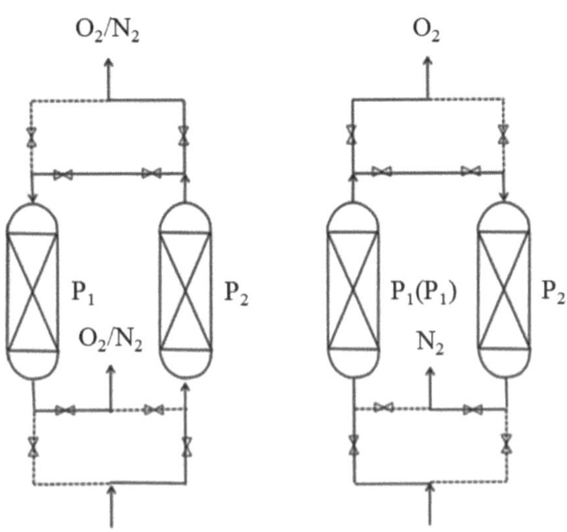

Fig. 5-25 Scheme of pressure swing adsorption process.

separated. This technique called Pressure Swing Adsorption (PSA) has been developed recently. In comparison to the Linde method, PSA is less energy consuming, involves applying 'soft' technological regimes, and allows for the purity of separated gases to be similar to that obtained by the cryogenic method. PSA systems are controlled by computers so they are fully automated. Industrial gas separation by adsorption processes is usually based on the equilibrium properties of mixture components. Separation can also be performed by utilizing the kinetics of adsorption/desorption of mixture components. For example, if large differences between adsorption rates of components exist, one of them is quickly removed from a mixture. In the present paper we focus our attention on the theoretical description of the time dependence of partial surface coverage of gas mixture components. There are only few papers devoted to this subject to be found in literature [63].

The obtained results can be used in the computer software controlling PSA apparatus. As the conditions of the gas separation process change quickly, it is necessary to determine very fast how to change the physical conditions of the process. The accuracy and quickness of this determination have major influence on the purity of the separated gases. As a theoretical tool we use the Statistical Rate Theory (SRT), proposed by Ward *et al.* [64, 65]. SRT has been applied successfully to represent the rates of various interfacial transports, like the rate of gas adsorption at the liquid-gas interface, hydrogen adsorption by metals, electron exchange between ionic isotopes in solution, permeation of ionic channels in biological membranes, rate of liquid evaporation and kinetics of adsorption on energetically heterogeneous surfaces.

Recently, Azizian *et al.* [66] used the SRT approach to study the kinetics of competitive adsorption at the solid/solution interface. This is still one more reason that we took interest in the problem of theoretical description of the kinetics of mixed-gas adsorption. Similarly to the adsorption equilibrium, the experimental measurement of the adsorption rate of gas mixture components is difficult and time consuming in comparison to analogous measurements for pure gases. We have recently made theoretical predictions of mixed-gas adsorption equilibrium. Generally, to theoretically predict the mixed-gas adsorption equilibrium, only information about single-gas isotherms is necessary. We tried to adopt such a rule for the case of theoretical description of kinetics of mixed-gas adsorption.

5.4.2. Hydrogen

H_2, the cleanest fuel, on combustion produces energy and water vapor [67]. The energy density of H_2 (38 kWh/kg), being higher than gasoline (14 kWh/kg), makes H_2 very attractive as a transportation fuel. The main problem associated with H_2 technology is the availability of safe and practically possible H_2 storage devices, which can easily load and unload H_2 to provide a continuous and consistent supply to the fuel cell.

Several materials have been assessed as candidate materials for hydrogen storage, including ceramics [68], metal hydrides [69], polymers, and carbons [70–74]. Metal hydrides store hydrogen well but need high desorption temperature. Ceramics and polymers are theoretically effective at storing hydrogen reversibly, but there have been no reports confirming the storage capability of these materials at room temperature.

5.4.2.1. Metal Hydrides

Metal hydrides, such as MgH_2, $NaAlH_4$, $LiAlH_4$, LiH, $LaNi_5H_6$, and $TiFeH_2$, with varying degrees of efficiency, can be used as storage media for hydrogen, often reversibly. Some are easy-to-fuel liquids at ambient temperature and pressure; others are solids that can be turned into pellets. These materials have good energy density by volume, although their energy density by weight is often worse than that of the leading hydrocarbon fuels. Most metal hydrides bind with hydrogen very strongly. As a result, high temperatures of around 120–200 °C are required to release their hydrogen content, as shown in Fig. 5-26. This energy cost can be reduced by using alloys that consist of a strong hydride former and a weak one such as in $LiNH_2$, $NaBH_4$, and $LiBH_4$ [67].

These are able to form weaker bonds, thereby requiring less input to release stored hydrogen. However, if the interaction is too weak, the pressure needed for rehydriding is high, thereby eliminating any energy savings. The target for onboard hydrogen fuel systems is roughly <100 °C for release and <700 bar for recharge (20–60 kJ/mol H_2).

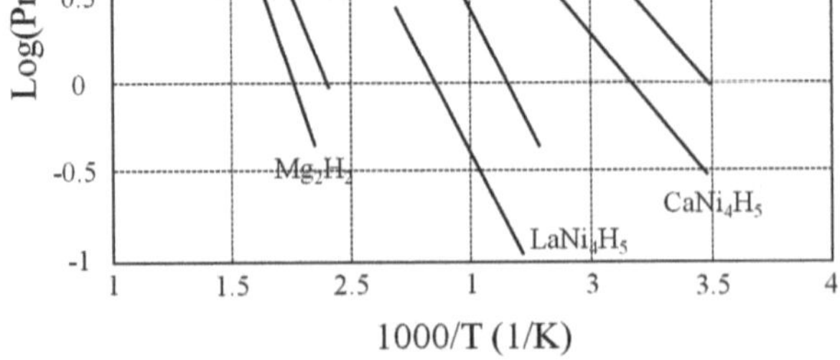

Fig. 5-26 Various metal hydrides for hydrogen storages.

5.4.2.2. Carbohydrates

Carbohydrates (Polymeric $C_6H_{10}O_5$) release H_2 in a bioreformer mediated by the enzyme cocktail—cell-free synthetic pathway biotransformation. Carbohydrate provides high hydrogen storage densities as a liquid with mild pressurization and cryogenic constraints: it can also be stored as a solid power. Carbohydrate is the most abundant renewable bioresource in the world. In May 2007, biochemical engineers from Virginia Tech University and biologists and chemists from the Oak Ridge National Laboratory announced a method of producing high-yield pure hydrogen from starch and water. In 2009, they demonstrated a production of nearly 12 moles of hydrogen per glucose unit from cellulosic materials and water. Thanks to the complete conversion and modest reaction conditions, they propose to use carbohydrate as a high energy density hydrogen carrier, with a density of 14.8 wt.%.

5.4.2.3. Metal-organic Frameworks

Metal-organic frameworks represent another class of synthetic porous materials that store hydrogen. In 2006, chemists at UCLA and the University of Michigan have achieved hydrogen storage concentrations of up to 7.5% weight in Metal-organic frameworks-74 [75]. However, the storage was achieved at the low temperature of 77 K [76]. In 2009, researchers of the University of Nottingham reached 10 wt.% at 77 bar (1, 117 psi) and 77 K with Metal-organic frameworks.

5.4.2.4. Carbon Materials

Carbon materials have high surface area, cavities, and micropores where H_2 can be adsorbed. Since the kinetic diameter of H_2 (0.4059 nm) is slightly higher than that of graphite (0.3355 nm), no H_2 intercalation in graphite is reported. Activated carbon has also been tested for H_2 storage but was found to be ineffective due to the fact that a very small percentage of the surface interacts with H_2 molecules at ambient temperature and pressure.

A previous work [73] reported on novel high specific surface area activated carbon prepared from polymeric precursors for hydrogen storage; this report also investigated the effects of specific surface area and micropore fraction on hydrogen uptake. Activated carbon was prepared by KOH activation method as a function of KOH:polymer ratio in the range of 1:1–4:1 and activated for 2 h at temperature ranging from 500 to 1100 °C. As shown in Fig. 5-27, well-developed porous structures formed after activation. This result indicates that K^+ ions, successfully and homogeneously captured by SO^{-3} ions on the resin surfaces, functioned as a powerful activation agent at high-activation temperature.

Fig. 5-28 (a) and (b) shows the pore-size distributions and hydrogen storage behaviors of the activated carbons as a function of activation temperature at 298 K and 100 bar, respectively. Activated carbons prepared at 700 °C showed

Fig. 5-27 SEM images of polymeric resin based activated carbons prepared at (a) 700 °C and (b) 900 °C.

the best results, and the trend of hydrogen uptake capacity at 298 K was the same as that at 77 K. This indicates that very small pores in carbon structures play the principal role in determining hydrogen uptake capacity, at 298 K. They claimed that 0.6–0.7 nm of pore diameter was the optimal pore size for H_2 storage due to the dynamic molecular size of hydrogen ($d = 0.456$ nm); they also asserted that the optimal carbon materials for hydrogen storage possess the following attributes: (1) high-specific surface area, (2) small pore size, and (3) high-micropore fraction [73].

Carbon nanotubes are perennial candidates for H_2 storage [77]. Carbon nanotubes are produced by arc-discharge, laser ablation, and chemical vapor deposition methods. There are two variations of carbon nanotubes. Single-walled carbon nanotubes (SWCNTs) are composed of individual graphene sheets rolled into seamless hollow cylinders (diameter = 1–20 nm). Multiwalled carbon nanotubes (MWCNTs), by contrast, are made up of a few tens of cylindrical graphitic sheets. Carbon nanotubes [78] provide large external surface area and internal hollow cavities for gas storage by chemisorptions or physiosorption.

H_2 can be adsorbed on the outer wall of carbon nanotubes by H—C bonding, or in inside cavities as H—H bonds. However, adsorption inside the cavity is not stable [79]. In the case of multiwalled carbon nanotubes, H_2 can be adsorbed between layers of carbon nanotubes; however, the presence of H_2 between tubes increases the radius of the carbon nanotubes and makes them unstable. In the case of bundled carbon nanotubes, H_2 can also rest in places between carbon nanotube bundles [80]. Research results for H_2 storage in carbon nanotubes are quite diverse. H_2 adsorption [81] (7 wt.%) at 0.67 bar and 327 °C surpasses the minimum requirement of 6.5 wt.% recommended by the US Department of Energy (DOE). Amount of H_2 adsorption or desorption can be increased up to 4.2 wt.% from 3.3 wt.% at ambient temperature and pressure.

The hydrogen storage of activated carbon nanotubes and metal-doped carbon nanotubes were reported early [82, 83]. The optimum condition for higher

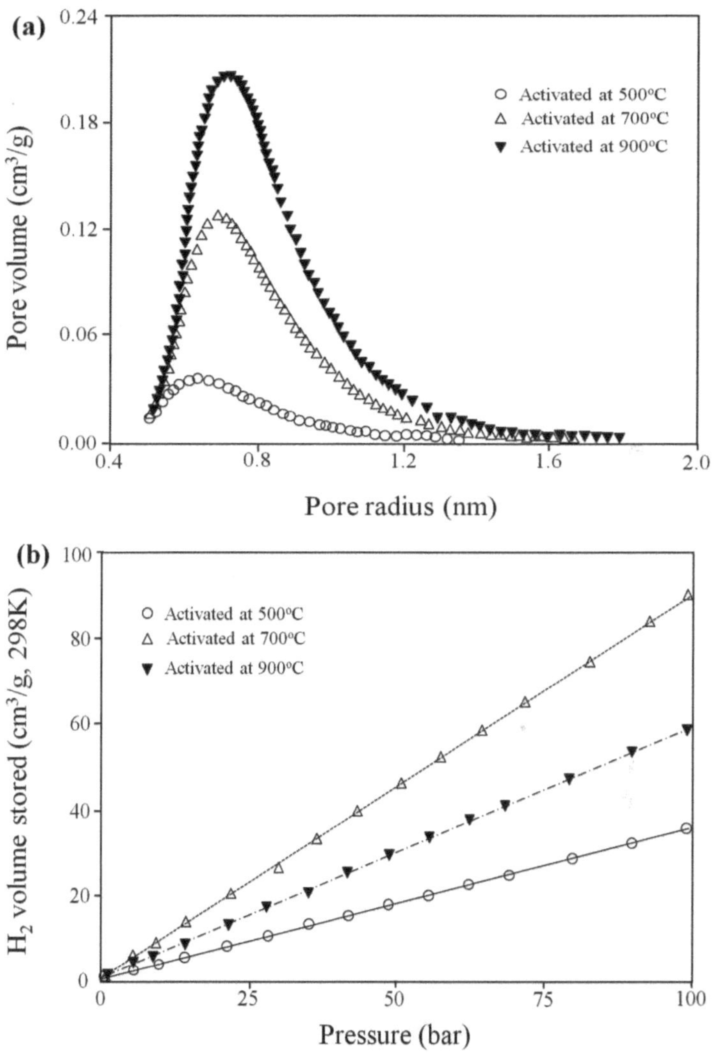

Fig. 5-28 (a) Micropore size distribution and (b) H_2 storage behaviors at 298 K of activated carbons as a function of activation temperature.

hydrogen storage capacities of the multiwalled carbon nanotubes existed after they were treated with CO_2 activation at 900 °C, at which point a high hydrogen storage capacity of 0.78 wt.% was obtained at 298 K and 100 bar. Therefore, a high hydrogen storage capacity can be expected in activated multiwalled carbon nanotubes with a large surface area, pore volume, and advantageous structure characteristics (defects on the external surface).

Generally, gas adsorption depends strongly on the micropores of the carbon nanomaterials. The actual order of hydrogen storage capacities is very similar to that of the micropore volume. Therefore, the hydrogen storage capacities of activated multiwalled carbon nanotubes under common operating conditions (298 K/100 bar) are attributed to micropores within the hydrogen molecules' friendly-sites of the defective surface. The main drawback for the commercialization of carbon nanotube based H_2 storage technology is the high cost of carbon nanotubes (50–600 \$/g) depending upon purity.

Recently, hydrogen storage on carbon nanotubes using a transition metal has drawn the interest of many researchers. It is known that the presence of platinum (Pt) on carbon materials can enhance the hydrogen storage capacity due to a hydrogen spillover effect. Platinum nanoparticles play an important role in the dissociation of hydrogen molecules, which process subsequently gives rise to the adsorption of hydrogen molecules by van der Waals forces [84]. Fig. 5-29 shows the TEM and high resolution TEM images of platinum/single-walled carbon nanotubes. The TEM images of platinum/single-walled carbon nanotubes show a more or less uniform distribution of noble metal particles of size of about 3–5 nm on the single-walled carbon nanotubes bundles. The high-resolution TEM image of platinum/single-walled carbon nanotubes clearly shows lattice planes of Pt particles, indicating the crystalline nature of the platinum nanocrystalline particles.

The hydrogen adsorption isotherms of platinum dispersed single-walled carbon nanotubes, along with as-grown and purified single-walled carbon nanotubes at 298 and 125 K, are shown in Fig. 5-30. The chemical adsorption process is improved by the presence of platinum catalysts in which the atomic hydrogen occupies sites that would be prevented from adsorption in the other case owing to the increase in hydrogen uptake. The dispersion of nanocrystalline platinum particles over single-walled carbon nanotubes creates defect sites that enhance the hydrogen storage by adsorption of atomic hydrogen at these sites.

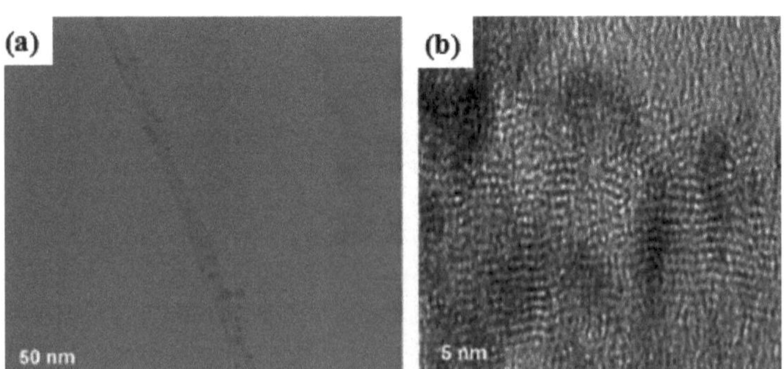

Fig. 5-29 (a) TEM and (b) high resolution TEM images of platinum/single-walled carbon nanotubes.

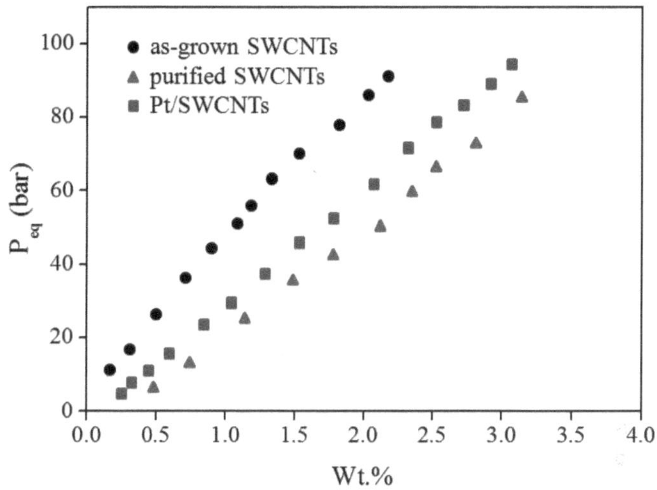

Fig. 5-30 Hydrogen adsorption isotherms of as-grown, purified, and platinum dispersed single-walled carbon nanotubes (SWCNTs).

Ramaprabhu *et al.* [84] suggested platinum/single-walled carbon nanotubes as a potential hydrogen storage material.

Another promising carbon material for storing H_2 are carbon nanofibers [85], which consist of stacked graphite layers (diameter: 5–100 mm, length: 5–100 mm), and are capable of adsorbing H_2 up to 50 wt.% at ambient temperature and high pressure. These carbon nanofibers are produced by interaction of metal-catalyst nanoparticles with hydrocarbon vapor at high temperature.

Furthermore, activated carbon fibers and metal-doped activated carbon fibers for hydrogen storage have been investigated by many researchers. Fig. 5-31 shows SEM images of the as-received activated carbon fibers and of two of the copper-plated activated carbon fibers. It was found that activated carbon fibers were partially covered with copper particles. The sizes of the observed copper particles fell in the range of 200–300 nm, indicating that some activated carbon fiber pores can be blocked in the presence of copper particles, leading to a specifically decreased surface area [74].

Fig. 5-32 shows the hydrogen adsorption and desorption data of copper-plated activated carbon fiver sample at 298 K. It was found that a desorption curve was almost same with a adsorption curve, indicating that hydrogen molecules adsorbed were mostly desorbed at 298 K, and the main mechanism of hydrogen storage was a physical adsorption.

Fullerene, another potential carbon material for H_2 storage, is the third allotrope of carbon containing only carbon atoms; unlike diamond and graphite, the atoms are arranged in a soccer ball shape. The most stable fullerenes are C_{60} and C_{70}, composed of 12 pentagons, and 20 and 25 hexagons, respectively. C_{60} can form compounds like $C_{60}H_{24}$, $C_{60}H_{36}$, and $C_{60}H_{48}$. Further addition

Fig. 5-31 SEM images of the as-received and of two of the copper-plated activated carbon fibers as a function of plating time. (a) activated carbon fibers, (b) copper-plated for 10 min activated carbon fibers, and (c) copper-plated for 30 min activated carbon fibers.

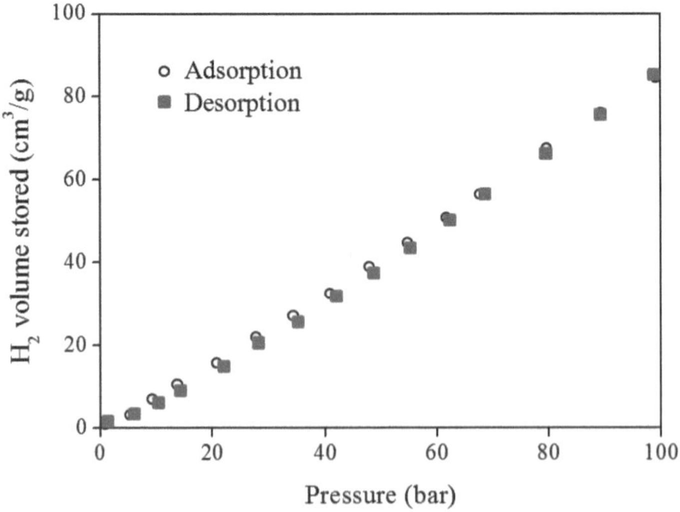

Fig. 5-32 Hydrogen adsorption and desorption data of copper-plated activated carbon fibers.

of H_2 atoms to fullerene ruptures the fullerene cage. $C_{60}H_{48}$ represents 6.3 wt.% of H_2 adsorption. Fullerene can achieve H_2 adsorption up to 6 wt.% at 180 °C and ~ 25 bar [86]. But, desorption of H_2 requires high temperatures of at least 225 °C due to very high strength of the C−H bond. If the desorption temperature could somehow be lowered, this material could be a good one for H_2 storage.

5.4.2.5. Mechanism

Fig. 5-33 shows the newly proposed mechanism for hydrogen storage. Park *et al.* [87] showed that hydrogen molecules are chemically reacted to metal particles on carbon surfaces and then spill over into the carbon structures under high pressure. However, it is virtually impossible to break chemical bonds when the gas pressure is 10 MPa. Moreover, it was found that the hydrogen adsorption was fully reversible, indicating that there was no chemical reaction or that very small reactions occurred. For this reason, a dipole-induced model is proposed.

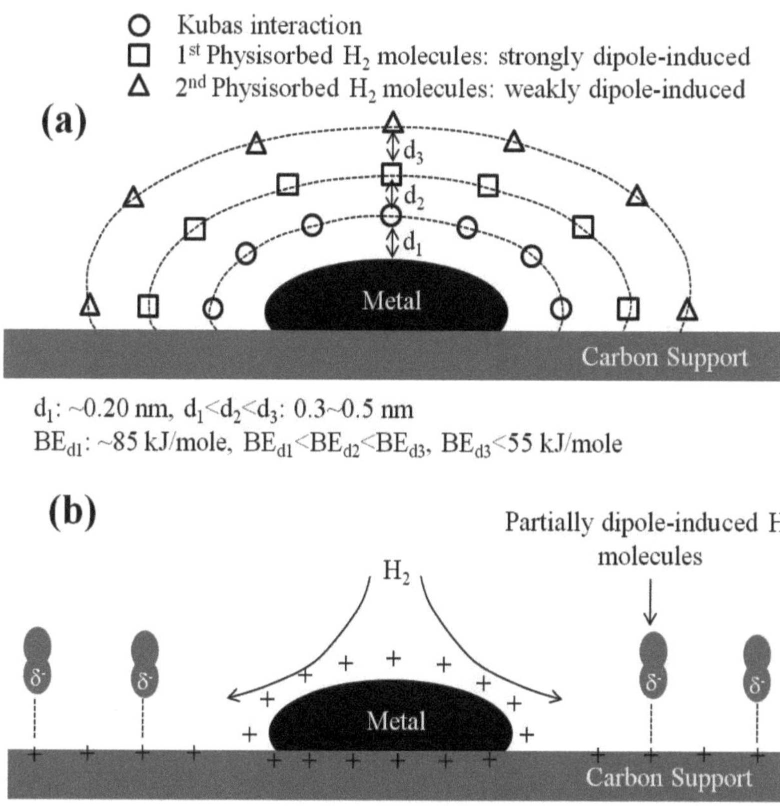

Fig. 5-33 A schematic of hydrogen storage mechanism on metal/carbon surfaces: (a) the accumulation of hydrogen molecules on metal/carbon surfaces and (b) the dispersion of hydrogen molecules from metals to carbon surfaces by dipole-induced effects.

The first hydrogen molecules introduced can be reacted with metal particles due to the high oxidation power of metals on graphite surfaces due to the Kubas reaction [88]. However, the amount of this initial reaction is very small. The binding energy and bond distance in this case are 85 kJ/mol and 0.2 nm, respectively, because the binding energy and bond distance of metal hydride are normally close to 85 kJ/mol and 0.2 nm.

The second group of hydrogen molecules introduced cannot chemically react due to a site limitation caused by the initially introduced hydrogen molecules; however, they are physically adsorbed by dipole-induced effects. The hydrogen molecules are basically nonpolar, but the strong interaction of the metal particles leads to the dipole inducing effects of the hydrogen molecules. The third group and any remaining hydrogen molecules are adsorbed by the same mechanism of the high gas pressure, but this equilibrium can be broken when the interaction of oxidized graphite surfaces and dipole-induced hydrogen molecules is stronger than that of metal-hydrogen interactions. This mechanism can be applied only under higher pressure, and graphite supports are charged strong electron acceptors [87].

5.4.3. Carbon Dioxide Adsorption

Global warming is widely attributed to an increase in atmospheric level of greenhouse gases. Carbon dioxide (CO_2) is considered to be the most important greenhouse gas with the largest impact on climate change [89]. The gradual increase in the atmospheric concentration of CO_2 due to the burning of fossil fuels is becoming a serious environmental problem [90]. CO_2 capture and sequestration from point source emissions has been recognized as a potential way to stabilize CO_2 in the atmosphere [91].

Various processes, such as liquid solvent absorption, cryogenic techniques, membrane separation, solid sorbents, and pressure (and/or temperature) swing adsorption have been proposed for the separation and recovery of CO_2 emitted by power plants. To date, most commercial CO_2 capture plants use mine-based processes and wet scrubbing systems [92], but these processes are energy intensive due to the large amount of water needed in these systems. Other problems associated with these processes are amine degradation by oxidation leading to corrosion of process equipment [93].

The goal in reducing CO_2 emissions on an industrial scale is the development of cost-effective techniques for the separation and capture of CO_2 [94]. Adsorption is considered to be one potential option because of the low energy requirement, cost advantage, and ease of applicability over a relatively wide range of temperatures and pressures.

Zeolites have shown promising results for separating CO_2 from gas mixtures. However, the presence of water inhibits the CO_2 adsorption capacity of zeolites [95]. Activated carbons are being proposed as suitable candidates for CO_2 capture: they do not require any moisture removal, present a high CO_2

adsorption capacity at ambient pressure and, moreover, are easy to regenerate. It has been recognized that the surface chemistry of activated carbon can strongly affect the adsorption capacity. Due to the acidic role of CO_2 (weak Lewis acid), it is expected that the introduction of Lewis bases onto the activated carbon surfaces can favor the CO_2 capture performance of these materials [96].

The presence of copper oxide on carbon materials can increase the adsorption capacity of the acidic adsorbate, indicating that copper oxides serve as an electron donor [97]. Fig. 5-34 shows the X-ray diffraction pattern of copper-doped activated carbons and Cu_2O-doped activated carbons. The metallic copper peaks were clearly found at $2\theta = 43°$ and $50°$ and decreased as the oxidation time increased.

Copper un-doped activated carbon showed an adsorption capacity of 13 mg/g while the neat copper-doped activated carbons showed a capacity of 10 mg/g. It was indicated that metallic copper did not enhance the adsorption or storage of carbon dioxide of the copper-doped activated carbons at room temperature and 1.0 atm. Normally, transition metals have acid features (electron acceptor). This means that acidic carbon dioxides can have repulsive forces with transition metals, resulting in a decrease in the adsorption capacity. In contrast, metal oxides such as copper oxide, calcium oxide, or magnesium oxide have base features, implying that they are electron acceptor friendly [97].

As shown in Fig. 5-35, Cu_2O-doped activated carbons showed a higher adsorption capacity compared to that of the as-received and neat copper-doped activated carbons samples. The adsorption capacity increased as the oxidation time increased. However, the adsorption capacity of Cu_2O-doped for 10 min activated carbons did not increase significantly compared to that of Cu_2O-doped for 5 min activated carbons. This result indicates that the presence of copper

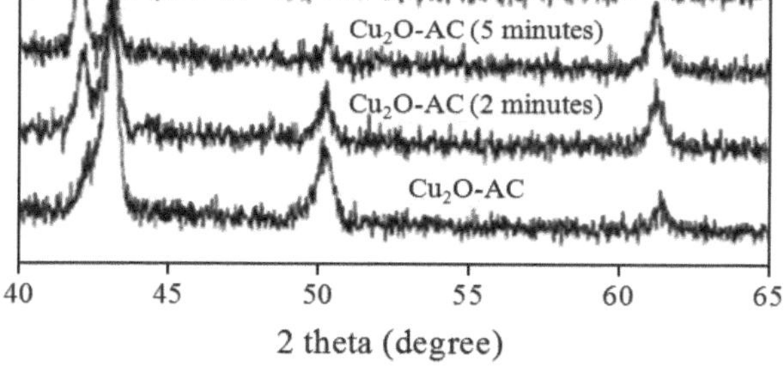

Fig. 5-34 XRD spectra of Cu_2O-doped activated carbons as a function of oxidation time.

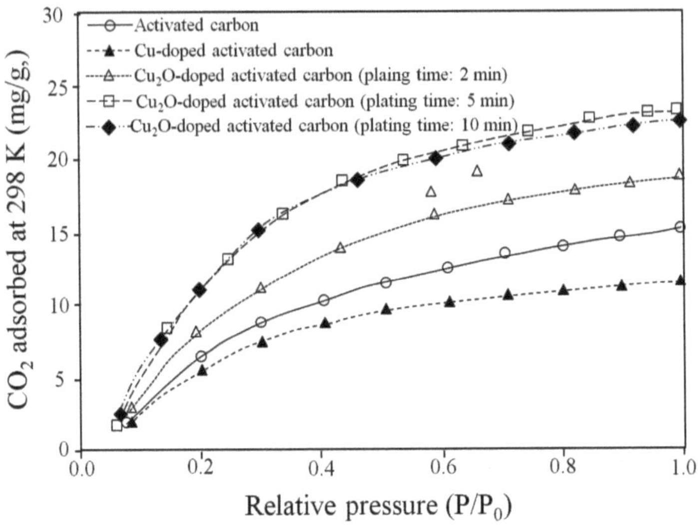

Fig. 5-35 Carbon dioxide adsorption behaviors of activated carbons, copper-doped, and copper oxide-doped activated carbons.

oxides can enhance the carbon dioxide adsorption; however, an excessive amount of copper oxide does not accelerate the adsorption ability of Cu_2O-doped activated carbons [97].

The functionalization of porous silica surfaces has received much attention for carbon dioxide adsorption. In recent years, Zelenak *et al.* studied the effects of pore size and amine ligand basicity of amine-functionalized mesoporous silica [98, 99]. Mesoporous silica samples MCM-41, SBA-12, and SBA-15, with different pore sizes, were functionalized with 3-aminopropyltriethoxysilane (APTES). The sorption capacities of these materials were found to be in the order SBA-15 > SBA-12 > MCM-41; however, SBA-12 showed the highest rate of carbon dioxide adsorption in these materials. This result was attributed to the pore size and pore structure. Moreover, SBA-12 was grafted with aminopropyltriethoxysilan, 3-(methylamino)propyltriethoxysilane (MAPS), and 3-(phenylamino)propyltriethoxysilane (PAPS) [99]. SBA-12 grafted with APTES showed the highest CO_2 sorption capacity of all of these materials.

Araki *et al.* [100] studied the adsorption capacity of CO_2 for aminopropyl-functionalized mesoporous silica microspheres (AF-MSM). Fig. 5-36 shows the effect of the aminopropyltriethoxysilan content on CO_2 adsorption capacities and amine efficiency. The adsorption capacities increased exponentially with aminopropyltriethoxysilan content. This result supports the exponential relationship between amine content and aminopropyltriethoxysilan content. Furthermore, the adsorption capacities increased with increasing basicity of the amino ligand. On the other hand, the amine efficiency gradually decreased with increasing aminopropyltriethoxysilan content. Araki *et al.* [100] reported

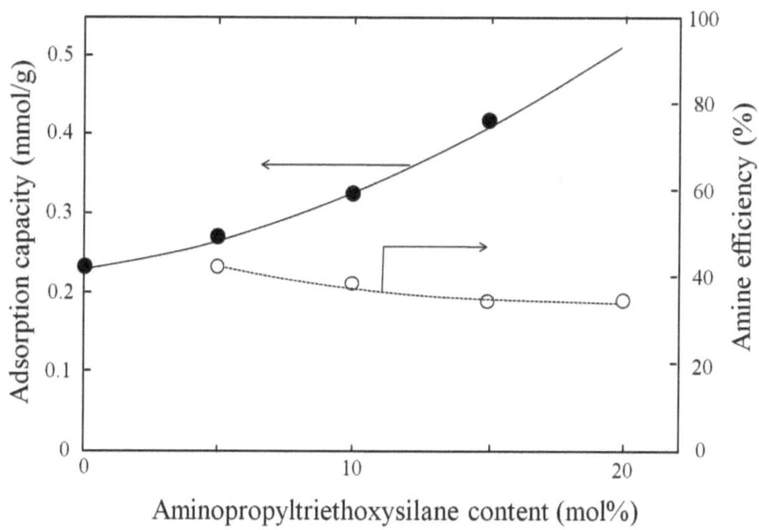

Fig. 5-36 Relationship between CO_2 adsorption capacity and aminopropyltriethoxysilan content in silica source.

on the amine efficiency is largely concerned with the surface density of the amine. The amine efficiency of AF-MSM, with low aminopropyltriethoxysilan content, is high; this result might be caused by the high surface density of amine in the low aminopropyltriethoxysilan content.

5.5. BIO TECHNOLOGIES

5.5.1. Delivery Systems for Food and Drug Products

Consumers in the industrialized world are becoming increasingly aware of the relationship between diet and health. Thus, the demand for a balanced diet and functional food products that address specific health benefits is growing steadily. Healthy food products, as compared to their standard counterparts, can be characterized by several attributes: containing (i) low to moderate sodium, sugar and trans-fat content, (ii) significantly reduced energy density, (iii) an increasing amount of whole grain and dietary fiber, (iv) high quantity of milk and vegetable proteins, or (v) bioactive ingredients, i.e., nutrients that have health sustaining properties [101].

However, many of the nutritionally attractive micronutrients used for fortification cannot just be added to a product, since they are either only hardly soluble in aqueous systems, show a limited stability against chemical or physical degradation, or reveal an uncontrolled release or bioavailability. Moreover, the stability, bioavailability or bioefficacy of active substances strongly depend on the food matrix and the chosen (micro) encapsulation or delivery system [102].

Spray-drying (micro) encapsulation has been used in the food industry since the late 1950s to provide, especially for flavor oils, some protection against degradation/oxidation, and to convert liquids to powders. Microencapsulation can also provide a physical barrier between different active ingredients in solid products [103].

One of the principal goals of microencapsulation nowadays is to protect the active ingredients from both chemical (e.g., oxidation) or physical (e.g., precipitation and crystallization) degradation induced through exposure to oxygen, light, moisture, temperature, or ionic strength changes, or to allow controlled or sustained release of active ingredients under desired conditions, i.e., during eating or digestion. Due to the low diffusion coefficient of oxygen in the glassy capsule material and due to the relatively large particle size (usually larger than 200 μm), sensitive oils, such as flavors and essential oils, can be stabilized for up to several years [104].

The microcapsule may even have multiple walls as shown in Fig. 5-37. The choice of the wall material is very important for encapsulation efficiency and microcapsule stability. The criteria for selecting suitable wall materials are mainly based on wall physico-chemical properties such as solubility, molecular weight, glass/melting transition, crystallinity, diffusibility, film forming, and emulsifying properties or costs. Most utilized wall materials are biopolymers of various sources [105].

The successful development of bioactive containing delivery systems for liquid products depends on several factors: (i) the ability to disperse active ingredients into an aqueous phase, in case the actives are water insoluble, (ii) the stability of the 'capsule' structure, preventing effects like creaming or sedimentation, (iii) minimizing the impact on the textural, rheological, or

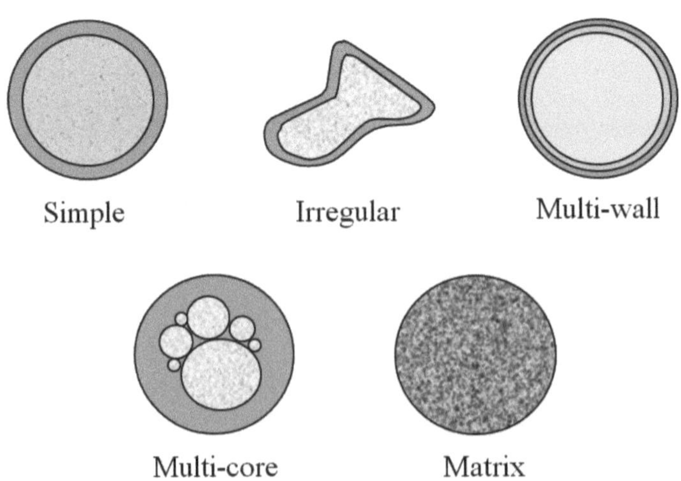

Fig. 5-37 Morphologies of different types of microcapsules.

optical properties of the final food product, (iv) protection of the encapsulated active molecules against degradation during processing and storage, and (v) controlled release during consumption, either in the mouth or during digestion in the GI tract. Most challenging in practical applications seems to be the sufficient stabilization of oxidation sensitive active molecules, such as vitamins or polyphenols. Moreover, masking possible off-taste effects or increasing (or controlling) the bioavailability or bioefficacy of the active ingredients during digestion are also quite demanding tasks to achieve in practical situations [106].

In general, a wide selection of delivery systems is available for use in food systems. Ultimately, one would like to relate the characteristics of the delivery systems to the functional attributes of the final product, such as sensory, physico-chemical, and biological/nutritional impact [107]. Table 5-3 summarizes the various types of systems that can be used for the delivery of active ingredients in aqueous liquid products. Although solid microcapsules represent the large

Table 5-3 Description of various kinds of delivery systems for liquid products

Delivery systems	Size and structure	Delivery systems	Size and structure
Powder particles	10 µm ~ 1 mm 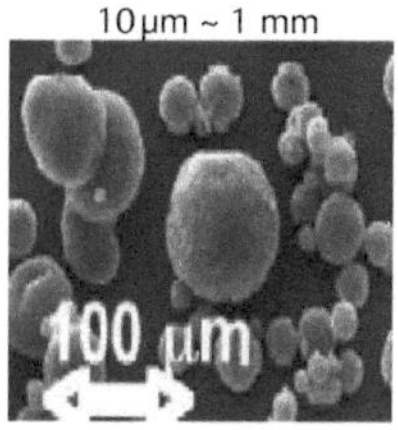	Liposomes, vesicles	20 nm ~100 µm 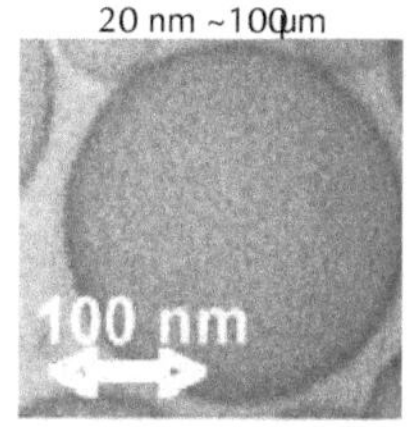
Emulsions	100 nm ~10 µm 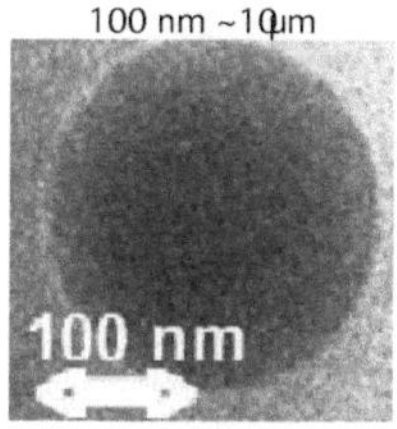	Micro-emulsion	5 ~ 100 nm 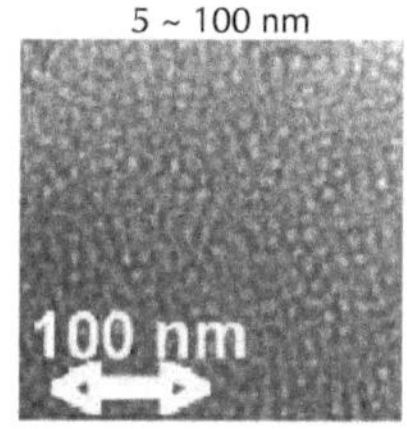
Molecular complexes	10 ~ 600 nm 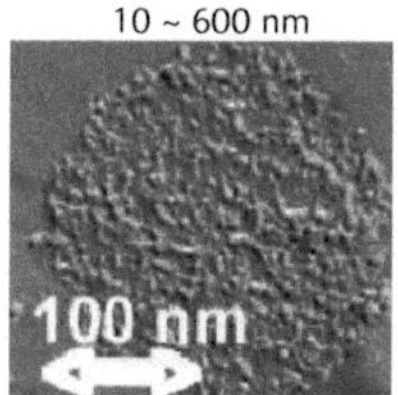	Dispersed reversed surfactant systems	100 nm ~ 1µm 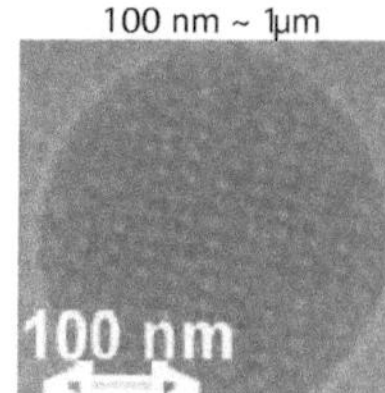

majorities of delivery systems used in food, since the available shell materials are very powerful in, for instance, protecting oxidation sensitive active ingredients from contact with oxygen or heavy metals in powders, these shell materials are much less attractive for delivery in aqueous systems, as mentioned above [108].

5.5.1.1. Oil-in-water Emulsion

Lipophilic active ingredients can be delivered in different forms. The best described and researched 'encapsulation system' for lipidic materials in aqueous products are oil-in-water emulsions. Emulsions, such as milk, yogurt drinks, dressings, sauces, or mayonnaise, are ubiquitous in food. Their oil droplets can easily be used for the delivery of lipophilic active ingredients. For example, the delivery of the antioxidant vitamin E (tocopherol) is very important in food. Vitamin E is the major and most potent lipid-soluble antioxidant in vivo [109]. It functions as the major radical scavenging antioxidant in lipoproteins and efficiently interrupts the chain propagation of lipid oxidation, thus protecting poly-unsaturated fatty acids and low density lipoproteins from oxidation. Vitamin E or its derivatives are frequently added to the oil phase of o/w emulsion products for fortification reasons or in order to stabilize unsaturated oils against oxidation [110].

McClements *et al.* [111] demonstrated the stabilization of oil droplets first with an anionic surfactant and then by adding a positively charged polymer, as depicted in Fig. 5-38. It was observed that this kind of emulsion more efficiently protects fatty acids and essential oils (citral and limonene) from oxidation that do ordinary emulsions stabilized by a single surfactant or amphiphilic layer. The observed effect against oxidation of the oil droplets in this multilayer emulsion

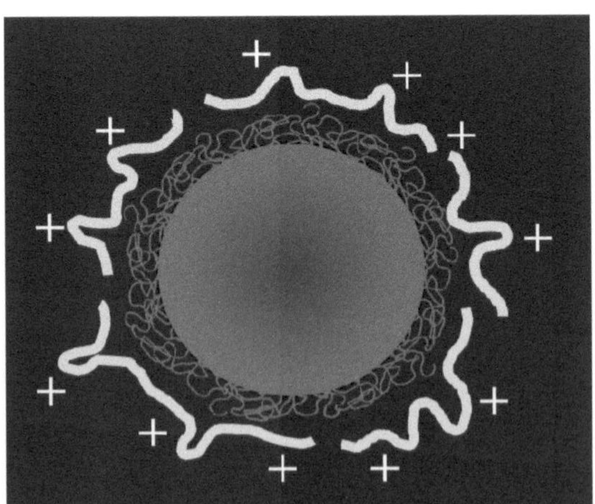

Fig. 5-38 Schematic of a multilayered emulsion droplet stabilized by an emulsifier (e.g., lecithin) and a polymer (e.g., chitosan).

system was attributed to the net positively charged interface. A positive charge around oil droplets hinders the contact with transition metals, like iron or zinc, and, as a consequence, prevents them from acting as pro-oxidants of the oil droplets.

The relatively large thickness of the interface may also have a positive influence in terms of barrier function. In conclusion, an efficient control of the water–oil droplet interface reduces oxidation of sensitive oil droplets, like poly-unsaturated fatty acids (PUFA), when compared to normally stabilized oil droplets [112]

5.5.1.2. Solid-lipid Nanoparticles

Colloidal dispersions of solid-lipids such as triglycerides, waxes, or fatty acids (solid-lipid nanoparticles) are under intensive investigation as carrier systems for different routes of drug administration [113]; for instance, for the intravenous or dermal delivery of lipophilic substances. Methods for the preparation of such dispersions include high-pressure homogenization of the melted matrix lipid in a hot aqueous phase and different precipitation techniques such as precipitation from warm microemulsions [114].

Bunjes *et al.* [115], using a combination of X-ray diffraction, Cryo-transmission electron microscopy (plunging technique), and freeze-fracture electron microscopy, as shown in Fig. 5-39, showed the different crystal structures and morphologies that particles can have. While the α structure gives a shape close to a sphere, the β structure adopts a needle or platelike shape with much fewer defects. This suggests that only the α form is suitable as a delivery system since the structure is relatively defective while the crystals of the β' and β structures are much more perfect in their crystal structure, ejecting the 'encapsulated' active elements to the outside of the particle. In addition, the plate-like crystals of the β form have a tendency to grow much more, leading to destabilization and to gel formation. Therefore, it is of prime importance to control the particle crystalline structure, i.e., preferably maintaining it as the α lattice.

Therefore, solid-liquid nanoparticles (SLNs) seem to have a large potential for the protection of active components, but the lack of physical stability (e.g., transformation of the α crystal into β or β', resulting in particle growth and ejection of loaded active molecules), is at present the major issue when trying to apply these particles in delivery systems for food. In addition, active elements may be exposed to high temperatures during the preparation of the lipid carrier materials, leading to chemical degradation. Finally, saturated lipids are needed to obtain these kinds of delivery system. Such lipids are not the preferred ones in terms of nutrition and health [116]

5.5.1.3. Molecular Complexes

Another strategy to deliver active ingredients in aqueous foods is by physically complexing them with other molecules, hoping that in this way a better

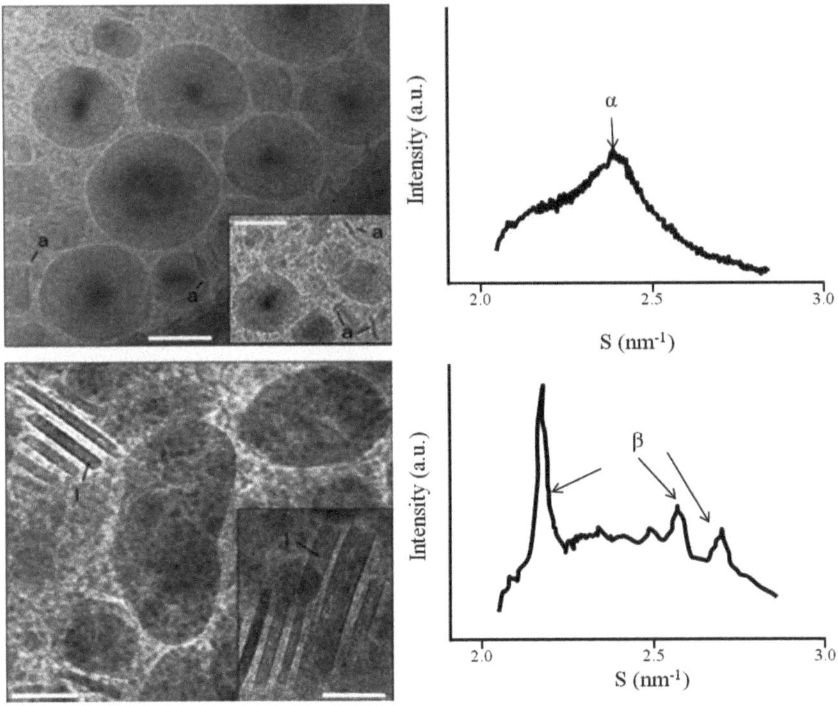

Fig. 5-39 Characteristics of tristearin dispersions stabilized in dependence on thermal history particles obtained by slow cooling from the melt and particles.

solubilization and/or an increase in the chemical stability of the complexed bioactive can be achieved. In this context the idea of a molecular complex refers to the physical association between a host and a guest (active ingredient) molecule. The most studied host molecules are the cyclodextrins. However, molecular association with other polysaccharides (e.g., amylose) or proteins or their aggregates can also be achieved [106].

Cyclodextrins are cyclic (or taurus shape) oligosaccharides having a hydrophilic outer surface insuring good dissolution of the complex in an aqueous environment (Fig. 5-40 (a). Cyclodextrins contain a lipophilic cavity enabling to host relatively small lipophilic or amphiphilic constituents (Fig. 5-40 (b)), such as fatty acids, vegetable and essential oils, nucleic acids, vitamins and hormones [117]. There are three main types of cyclodextrins: (i) α, corresponding to six glucopyranose units linked by α-(1,4) bonds, (ii) β, corresponding to seven units (U), and (iii) γ, corresponding to eight units. The dimensions of the internal cavity are crucial for the 'encapsulation' of guest molecules. Only molecules that fit physically into the cavity can be incorporated. The cavity diameter varies between 0.5 and 0.8 nm, which makes the cavity relatively small, allowing the

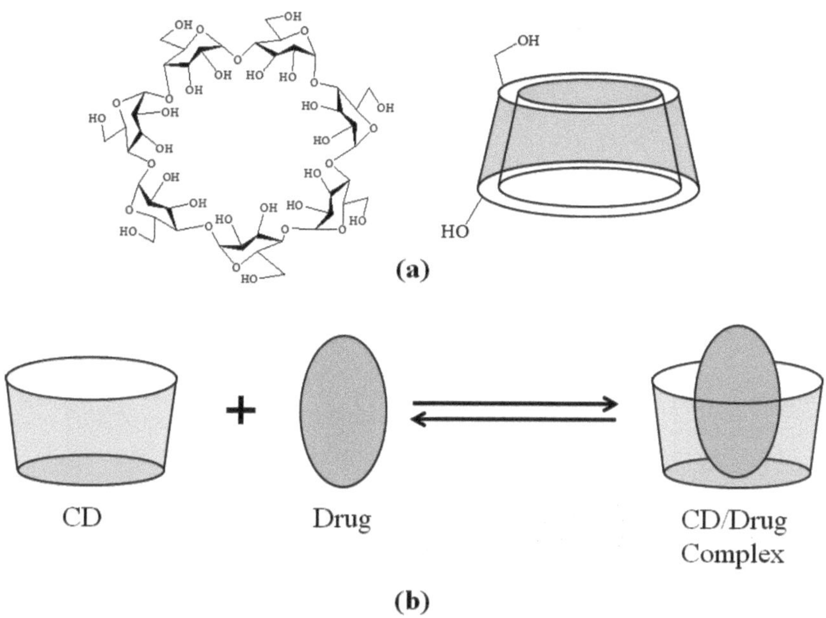

Fig. 5-40 Schematic representation of the structure of the cyclodextrin (β form in the drawing) molecule and the mechanism of drug (nutrient or aroma) complexation. (a) Cyclodextrin (CD) structure view from different angles. (b) Complexation with a drug molecule.

solubilization of only relatively small molecules. Long guest molecules may not be protected in an optimal way due to the limited height of the taurus [117].

Proteins and peptides are amphiphilic molecules. They are also relatively soluble in water and can bind lipophilic or amphiphilic active ingredients. Semo *et al.* [118] used casein micelles to solubilize vitamin D2. Sodium caseinate, $CaCl_2$ and K_2HPO_4 were used to encapsulate the vitamin and reconstitute the casein micelle solution. It was found that the casein micelle can provide a partial protection against UV-lightinduced degradation compared to the serum media (of the casein micelle dispersion), which was used as a control [118].

As shown in Fig. 5-41, the various samples tested may be ranked in order of increasing protective effect as follows: docosahexaenoic acid (DHA) in water (pH 4.5) < docosahexaenoic acid in water (pH 7.0) < β-lactoglubin + docosahexaenoic acid at pH = 7.0 < β-lactoglubin + docosahexaenoic acid + pectin at pH = 4.5. The partial protection by β-lactoglubin can be explained by its mild antioxidant activity [119]. However, the complex comprising both β-lactoglubin and pectin conferred the highest protection against oxidation to docosahexaenoic acid, resulting in only about 5–10 percent loss during 100 h, compared to about 80 percent loss when the unprotected docosahexaenoic acid was monitored. It was indicated that the formation of the β-lactoglubin – docosahexaenoic acid/pectin complexes can be an effective way to retard

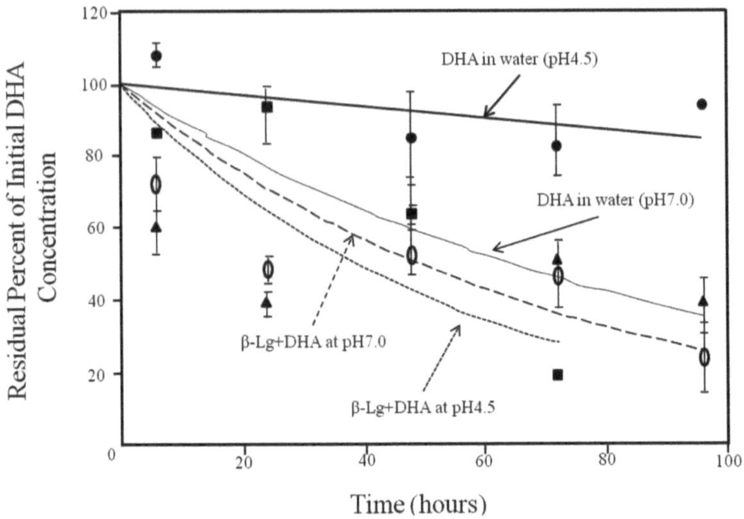

Fig. 5-41 Docosahexaenoic acid (DHA) stability during a stress test at 40°C and low ionic strength.

docosahexaenoic acid degradation, due to the fact that docosahexaenoic acid is immobilized by the protein and shielded by the protein–pectin complex and hence its reactivity is reduced, and so is the accessibility of oxidizing agents to docosahexaenoic acid.

5.5.1.4. Self-assembly Delivery Systems

Self-assembly structures, such as micelles, microemulsions, and liquid crystalline phases, are formed by the spontaneous association of surfactants in aqueous (or oil) phases. Surfactants are used in many food applications, such as in bread and cake production for the improvement of shelf-life (prevention of starch retrogradation due to formation of monoglyceride–amylose complexes) and flavor retention [120]. Another important application of surfactants deals with the control of emulsion or foam formation and stabilization. All these applications are based on the potential of amphiphiles to adsorb at interfaces or to crystallize and co-crystallize with other molecules.

One of the most useful (and simple) concepts for a semiquantitative description of the relation between surfactant molecular shape and self-assembly phase formation was given by Israelachvili *et al.* [121], who defined the so-called dimensionless surfactant packing parameter P.

$$P = V/al \qquad (5\text{-}7)$$

where V is the molecular volume of the hydrophobic moiety, l the molecular length of the hydrocarbon chain, and a the effective (or hydrated) cross-sectional area of the polar head-group.

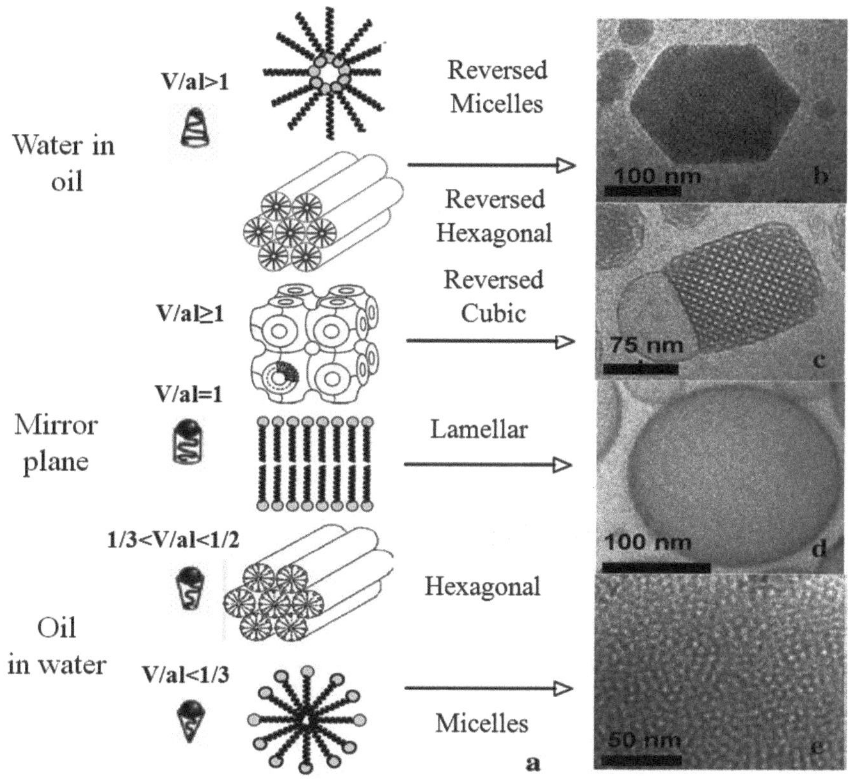

Fig. 5-42 The different existing surfactant self-assembly structures as function of their packing parameter (*P*).

Depending on *P*, different self-assembly structures can be formed as shown in Fig. 5-42. In Fig. 5-42, going from the top to bottom, we see an increase in the packing parameter [122]. If the packing parameter (*P*) is small ($P \ll 1$), structures like normal micelles, and hexagonal or cubic phases, are formed. If the packing parameter (*P*) is close to 1, a lamellar liquid crystalline (L_α) phase is formed, which when dispersed into water gives rise to vesicles or liposomes formation. If the packing parameter (*P*) is large ($P \gg 1$), reversed self-assembly structures, such as reversed micelles, reversed hexagonal structures, or reversed cubic structures, are formed.

Liposomes, often also denoted as vesicles, are formed when the surfactant molecules have a packing parameter (*P*) close to 1. Contrary to microemulsions their formation is often not completely spontaneous. When mixed with water the surfactant spontaneously forms a lamellar phase, which then needs to be dispersed to form vesicles. Liposomes can contain (i) one bilayer forming unilamellar vesicles (ULV), (ii) several concentric bilayers forming multi lamellar vesicles, or (iii) non-concentric bilayers forming multi vesicular vesicles

(MVV). The size of these structures can be rather small (in the range of 20 nm) or rather large (exceeding 1 μm) [123].

In a recent study, Folmer *et al.* [124] investigated the uptake of tocopherol using a Caco-2 cell model. The main idea was to find out whether differences in structure formation and/or hydrophilicity of the utilized tocopherol derivative would induce different tocopherol uptake behavior. Two α tocopherol succinate derivatives, in which two different hydrophilic ethyleneglycol chains were attached to the tocopherol succinate, were used. The two samples are tocopherol hexaethylene glycol succinate, which forms vesicles in water, and tocopherol dodecaethylene glycol succinate, which forms elongated micelles in water (Fig. 43). Tocopherol hexaethylene glycol succinate or tocopherol dodecaethylene glycol succinate were either solubilized into mixed micelles using bile (taurocholate) acids, monooleins, lysophosphatidylcholine

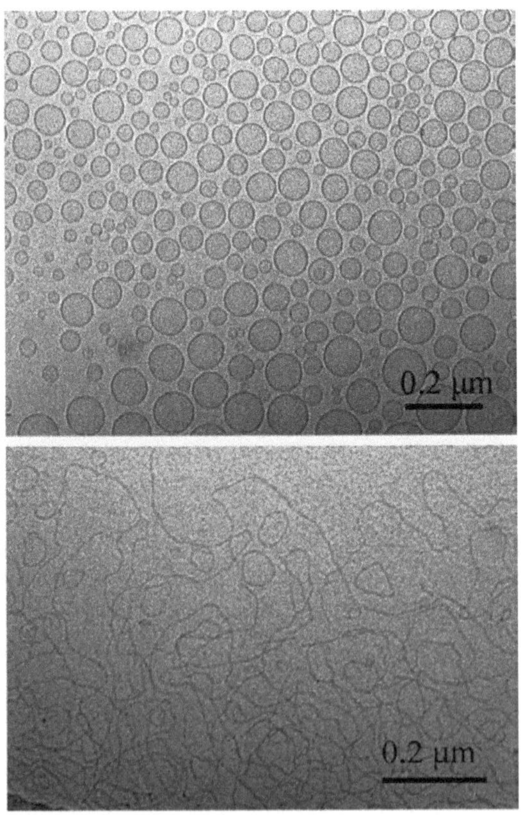

Fig. 5-43 Top: Cryo-TEM image of vesicles formed when tocopherol hexaethylene glycol succinate is mixed with water. Bottom: Cryo-Tem image of the rod-like micelles formed when tocopherol dodecaethylene glycol succinate is mixed with water.

and oleic acid or used as such in water. When, the derivatives were solubilized into mixed micelles, tocopherol hexaethylene glycol succinate showed a slightly but significantly more efficient esterase transformation into tocopherol and tocopherol succinate and better subsequent uptake of these constituants into the cells than did tocopherol dodecaethylene glycol succinate. This must be related to the different efficiency of the hydrolytic reaction of the esterase. The situation was completely changed when tocopherol hexaethylene glycol succinate and tocopherol dodecaethylene glycol succinate were just mixed into water (and not incorporated into mixed micelles). In this case, the total hydrolysis and uptake into the cells was about twofold lower for tocopherol hexaethylene glycol succinate than for tocopherol dodecaethylene glycol succinate.

Obviously, elongated micelle structures are much more efficient than vesicles in terms of tocopherol bioavailability. This result demonstrates that the type of the formed self-assembly structure can also play a crucial role in determining the bioactivity of active ingredients such as tocopherol.

Recently, pH sensitive hydrogels have drawn considerable attention because of their potential applications in controlled release systems. Typical polymers have been studied as drug carriers [125].

Park *et al.* [126] reported the pH sensitivite and insoluble poly(acrylic acid) hydrogels for drug delivery systems. pH sensitive Pluronic F127 (PF127)/poly (acrylic acid) hydrogels containing tulobuterol-loaded silica were prepared with Pluronic F127 content. Fig. 5-44 shows the pure poly(acrylic acid) and Pluronic F127 containing hydrogels.

The release behaviors of the hydrogels with different drug contents are shown in Fig. 5-45. The hydrogels containing 1 wt.% drugs show more sustained release behaviors compared to the hydrogels containing 4 wt.% drugs. The hydrogels containing 4 wt.% drugs exhibit fast initial release rate,

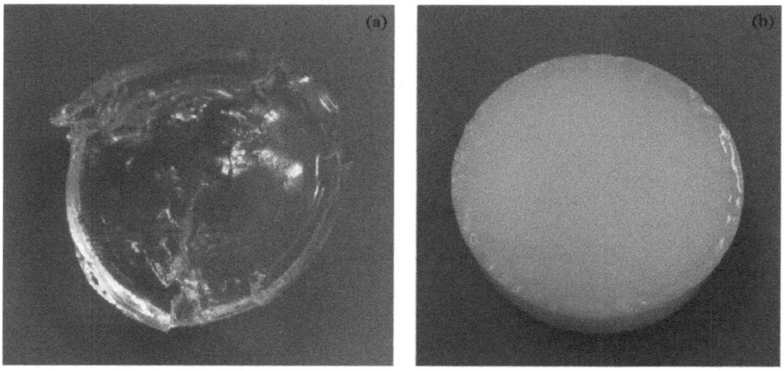

Fig. 5-44 The pure poly(acrylic acid)and Pluronic F127/poly(acrylic acid) images after swelling/deswelling of 10 times.

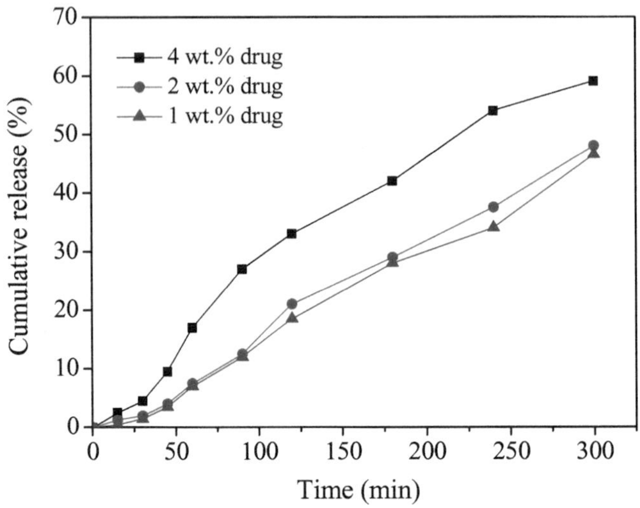

Fig. 5-45 Release behaviors 20 wt.% Pluronic F127/poly(acrylic acid) hydrogels with different drugs contents.

indicating that the non-adsorbed drugs are released with the swelling of the hydrogels and then adsorbed drugs are subsequently released [126].

Porous silica has large surface area and porosity, and can thus adsorb various substances, including drugs. In addition, the physical interaction between the silica and the adsorbed guest molecules can delay the drug release from silica [127]. Therefore, the mechanism of drug release from the hydrogel containing silica involves two routes: (1) diffusion of drugs from silica, and (2) drug release from the hydrogel matrix, as shown in Fig. 5-46. Consequently, the

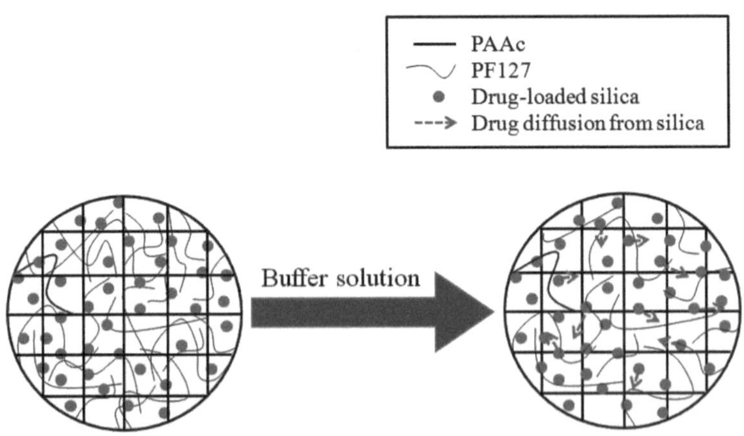

Fig. 5-46 Scheme of drug release from Pluronic F127(PF127)/poly(acrylic acid) (PAAc) hydrogel containing tulobuterol-loaded silica.

incorporation of porous materials such as silica in biocompatible polymers can provide sustained drug release behavior.

Most delivery systems are able to solubilize lipophilic nutrients into aqueous products. Concerning the appearance of the final food product, only nanoemulsions, microemulsions, and some complexes are able to keep the product transparent, if desired especially for some beverage formulations. Most of the described systems suffer either (i) from some regulatory issues concerning the utilized carrier material, or (ii) from an insufficient loading capacity restricting their use for the delivery of nutrients in only small concentrations. Most challenging, however, is to achieve also an appropriate chemical stability of the delivered active ingredients during the shelf-life of the liquid product. Realizing a sufficient chemical stability of the delivered nutrients and aromas in the fortified end product is probably the most desired functionality that the delivery system must accomplish. In contrast to the delivery in powdered products, it seems that for the delivery of active ingredients in liquid products, there is no generic solution available that can solve most of the delivery problems we are facing. Every solution is unique and depends very much (i) on the final product matrix and packaging, (ii) on the active ingredient(s) to be delivered, and (iii) on the chosen delivery system and its interfacial properties.

5.5.2. Cosmetics

Women all over the world are doing a variety of things to combat aging and maintain beauty. They want their faces to look as they did when they were younger, and are trying all kinds of new technologies to find the look of youth and beauty. Cosmetics are substances used to enhance the appearance or odor of the human body [128].

Cosmetics include skin-care products, fingernail and toe nail polish, eye and facial makeup, and permanent wave solution. These products and procedures have effects that range from the temporary, such as cosmetic colored contact lenses, to the major, such as cosmetic surgery. For decades, the popularity of cosmetics has increased rapidly. The number of novel cosmetic products that have been brought to market continues to expand exponentially [129]. Especially in the United States, cosmetics are used by girls at an increasingly young age. Due to the fast-decreasing age of make-up users, many companies have catered to this expanding market by introducing more flavored cosmetics. The social consequences of younger and younger beautification have garnered much attention over the last few years. Cosmetic chemists are always looking for interesting and exotic ingredients that improve skin's appearance and health [130].

5.5.2.1. Anti-aging

Especially, nano cosmetics are so named because they utilize extremely tiny (or nano) particles to penetrate the skin far more deeply than traditional antiaging treatments are able to do. Skin ageing results from multiple biological processes occurring both in the dermis and the epidermis. In the epidermis, the barrier function of the skin is altered, leading to an increase of trans epidermal water loss (TEWL). This natural phenomenon results from the diffusion of water from blood circulation toward the dermis and the different cell layers of the epidermis. The water finally evaporates at the surface. The skin is therefore less moisturized, losing its flexibility, softness, and firmness. In the dermis, the amount of molecules responsible for supporting the skin structure (collagen, elastin, and hyaluronic acid) decreases. Their synthesis processes decrease while their degradation processes increase [131]. Traditionally, antiaging skin care products have been designed to hold moisture in the skin by creating a barrier between the skin and the outside world. Nanoparticles in the new generation of cosmetics do not work this way. They are designed to penetrate the upper layers of the skin and stimulate new skin cell production that gives skin a new, plump, and youthful appearance. Nanoparticles make it possible to get a multitude of chemicals into the deeper layers of the skin because the chemicals can be covered by the particles. Many of these chemicals would cause irritation in other forms and may be stimulating the inner working by irritating from the inside. This is the nanocosmetic secret to antiaging and antiwrinkle skin care products. Basically, if one uses these products, one will definitely see results but will not know what effects may be in the long run.

Water-soluble phytoconstituents (mainly polyphenolics) can be converted into a lipid-compatible molecular complex known as PHYTOSOME®. A PHYTOSOME® is generally more bioavailable than a simple herbal extract due to its enhanced capacity to cross the lipid-rich biomembranes and reach circulation [132]. A PHYTOSOME® is an amphiphilic substance with a definite melting point; it is generally soluble in nonpolar solvents (in which its hydrophilic moiety is not soluble), and is moderately soluble in fats. The low solubility in aqueous media makes the formation of stable emulsions and creams possible (Fig. 5-47), improving the biopharmaceutical properties of both highly lipid insoluble and poorly water-soluble phytoconstituents.

PHYTOSOME® as a delivery system offers interesting applications and opens new opportunities for the use of active ingredients in the cosmetic field. The G. biloba PHYTOSOME® was investigated for the treatment of skin aging connected to superficial capillary blood flow circulation. Extracts from G. biloba are used orally to improve peripheral circulation, and their phospholipid complexes were found to improve skin microcirculation after topical application. Activation of microcirculation ameliorated skin aging associated with dystrophic alteration of the epidermis and dermis, and the regressive abiotrophic pannicular disease of lower limbs and breast associated with venous stasis and/or chronic venous insufficiency. The role of PHYTOSOME® in functional cosmetics on aging skin was also reported [133].

Fig. 5-47 Electron-microscopic examination of lyophilized microdispersion of Ginselect® PHYTOSOME® 0.1% in water.

Retinoids are natural and synthetic vitamin A derivatives. Biologically active forms of retinoids are capable of modulating gene expression by binding to nuclear receptors to activate transcription of specific DNA sequences. Because of their ability to modulate the genes involved in cellular differentiation and proliferation, retinoids are thought of as good candidates for both treating and preventing the photoaging process [H.J. Kim, B. Kim, H. Kim, S.J. Um, J.D. Lee, H.C. Ryoo, and H.G. Jung, Bioorganic Medical Chemistry, 16, 6387 (2008).

Fig. 5-48 shows the level of cell viability following exposure to various retinoid derivatives such as retinoic acid, retinol, and retinyl retinoate. A viability of 100 percent was determined by treating cells with medium only. As can be seen from the results presented in Fig. 5-48, retinyl retinoate exhibited a lower intracellular toxicity than retinol and retinoic acid. In addition, the cytotoxicity of retinyl retinoate was lower than that of retinol. Taken together with the photosensitivity and thermal sensitivity data, our results suggest that hybrid retinyl retinoate has tremendous potential as a powerful antiwrinkle agent.

5.5.2.2. UV Protection

Nanoparticles are used in products that delivers ingredients evenly to the skin surface and are not made to penetrate to deeper layers of the skin; one example of this concept is the use of sunscreen for maximum ultra-violet protection purposes [134]. The technology used in newer sunscreens is superior to that of the older, chemically based sunscreens as it is less likely to react

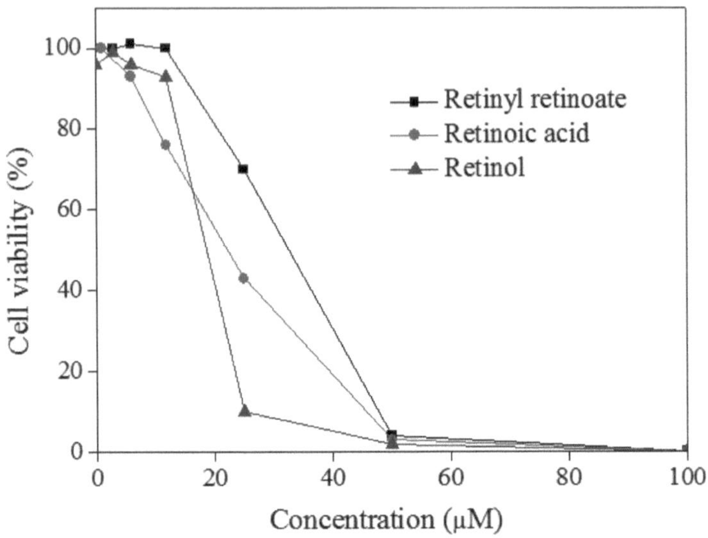

Fig. 5-48 Cytotoxic effect of retinol derivatives. Normal human fibroblast cells cultured with Modified Eagle Medium were used for the cell toxicity test. After the cells were treated with each different concentrations (0.4, 0.8, 1.6, 3.1, 6.3, 12.5, 25, 50, and 100 lM) for 1 day, the cell viability values of the derivatives were determined.

to sensitive skin types and provides both longer protection and higher levels of protection. The skin can repair many of the mutations caused by ultraviolet radiation; however, if the damage is too great, the affected cells may die, as shown in Fig. 5-49. Alternatively, the damage to the skin is often not

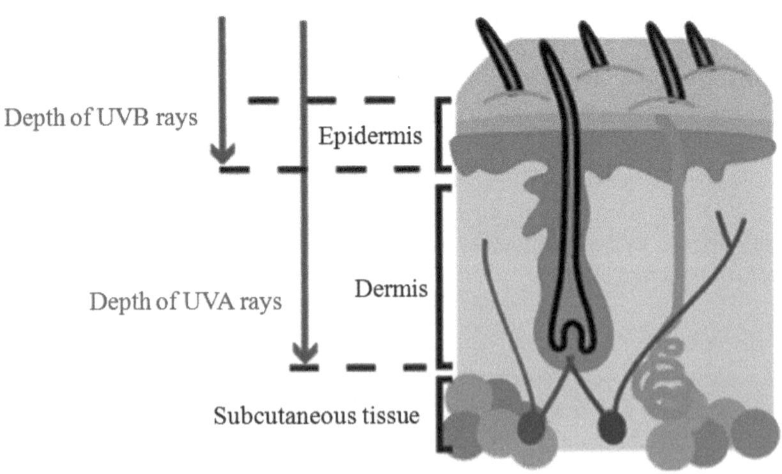

Fig. 5-49 A scheme of ultra-vilolet radiation penetrates the skin.

repaired perfectly and mutations occur. This results in premature aging, the formation of actinic keratoses or pre-cancers, and skin cancer. Ultra-violet radiation penetrates the epidermal or outer layer of the skin. Such radiation damages DNA in this layer and causes other changes in skin cells. This ultimately may result in the signs of photoaging. Over time, precancers and skin cancers may develop [135].

Ultra-violet A (UVA) radiation, while also damaging the epidermis, penetrates deeper into the skin to the level of the dermis. Ultra-violet B (UVB) not only harms epidermal cells, but also damages collagen and elastin, which make up the structure of the dermis and keep the skin resilient. Blood vessels can also be harmed. Ultra-violet C (UVC) radiation, reflected by the ozone layer, does not reach the Earth.

Ultrafine particles of titanium dioxide (TiO_2) are very attractive as an ultra-violet protection ingredient in cosmetic products. The UV-scattering behavior of TiO_2 suspensions in a silicone oil were studied by Nasu *et al.* [136].

The polyether groups of branch-type dispersants tend to form weak bond with polyether groups in other chains and finally construct the bridges between particles through multichain association. Considering that the rearrangements of chains are not allowed on the particle surfaces, the bridges are regarded as solid bonds. For alternatively repeated type ((AB)n-type) dispersants, the inter-chain association may be very weak, because the polyether group are alternately repeated in one long chain. The differences in adsorption conformation on the particles are schematically shown in Fig. 5-50.

Fig. 5-51 shows the transmittance plotted against the wavelength for suspensions diluted to 5 wt.%. All suspensions show high transmittance at long wavelengths. Good transparency is achieved in the visible range and this feature is the essential requirement in cosmetic products. For ultra-violet protection by ultra-fine TiO_2 particles, there exist optimum average diameters in the range of 50–120 nm [137].

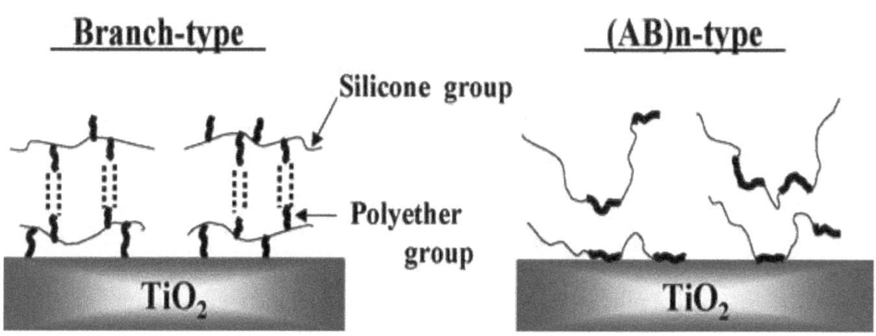

Fig. 5-50 Conformation models of adsorbed dispersants.

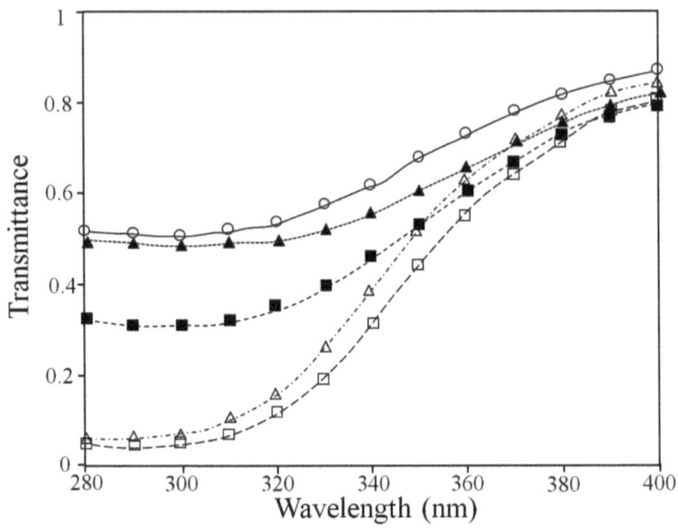

Fig. 5-51 Wavelength dependence of transmittance measured right after the preparation (open symbols) and after the storage at 50 °C during 2 weeks (filled symbols) for diluted suspensions containing different dispersants dispersants; branch-type with hydrophilic–lipophilic balance = 1 (○); branch-type with hydrophilic–lipophilic balance = 4 (Δ, ▲); alternative-type((AB)n-type) with hydrophilic–lipophilic balance = 3 (□, ■).

5.5.3. Adhesion for Biological Cells

There are three questions related to the adhesion of biological cells: (1) why cells move when they are adhered to a substrate, (2) why cells gather in multicell organisms, and (3) how to stop unwanted cells sticking, growing, and fouling on substrates such as teeth. The question of cell movement was studied by Palecek *et al.* [138], who used $CHOB_2$ cells on silane coated glass, with fibrinogen or fibronectin on the surface to promote adhesion in a particular way. At low adhesion and at high adhesion the cells moved more slowly, with an optimum adhesion of 20 nN for maximum speed, as shown in Fig. 5-52 (a). Migration speed was measured by video camera with image analysis to give the position of the cell centroid over a period. Adhesion was measured in a fluid flow device at 50 percent cell removal in 5 min, assuming a hemispherical model. The same cell speed relation to adhesion was observed when the cell integrin (fibronectin receptor) was modified by genetic engineering, or when the binding affinity of the receptor and adhesion molecule was modified. This means that several possibilities may emerge for moderating cell movement, for example in cancer treatment [139].

Adhesion is normally defined as the difference between simple rinsing of the surfaces and vigorous scraping, scrubbing, or sonication. Plaque is revealed by staining with a dye such as erythrosine or by scraping with a probe. After

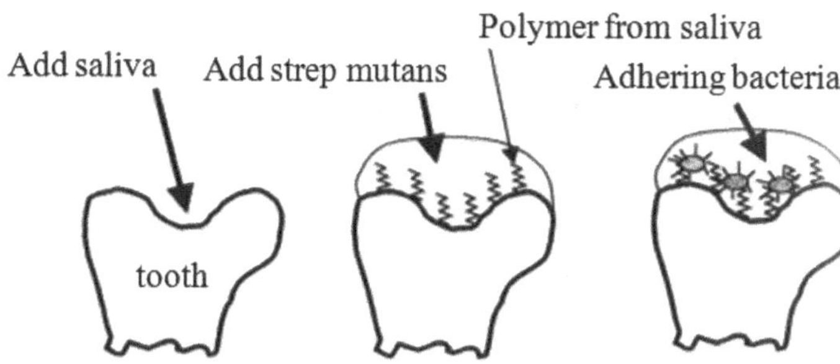

Fig. 5-52 Mechanism of adhesion of strep. Mutans to hydroxyapatite, involving the saliva polymer and the buildup of cell layers.

removal, the cells can be cultured and observed by microscopy [140]. Antiadhesive agents have been sought to prevent the sticking of the bacteria to the teeth but generally these agents have not been successful, and bacteriocides are normally used. The problem is the mechanism of bacterial adhesion, as shown in Fig. 5-52. It is thought that a film of polymer from saliva first attaches to the hydroxyapatite. Then, bacteria attach to this coating. The bacteria produce a complex layer of cell sheath that glues the cells to each other. The composition of saliva is therefore significant. Roger *et al.* [141] compared salivas with and without lysozyme.

Self-assembling polypeptides offer great potential as scaffolds for building bio-inspired materials featuring nanotopography. Polypeptides comprising repeating [142] or amphiphilic [143] sequences of amino acids have successfully been used to create a variety of fiber-like nanostructures with the ability to support cells and promote cell differentiation. The properties of fibrils may be further extended by functionalization, resulting in ligands displayed on the fibril surface on a nano-length scale. Proposed applications for functionalized fibrils include: scaffolds with enzyme activity, nanowires for the electronics industry, diagnostic devices, and templates for assembly of inorganic materials.

Gras *et al.* [144] reported on functionalized amyloid fibrils for cell adhesion. Transthyrethrin-(105-115)-peptide(TTR1)/arginine-glycine-aspartic acid (RGD) peptide was found to have strong cell binding properties and to dramatically reduce cell adhesion to fibronectin coated surfaces in a concentration dependent manner, as shown in Fig. 5-53.

The supramolecular design of artificial extracellular matrices is the key to our understanding of cell–matrix interactions, and is of great importance for the future of regenerative medicine.

The most common target has been the use of a matrix to activate receptor-mediated biological adhesion mechanisms since they are critical to cell survival and function. Biological cell adhesion to the extracellular matrix commonly occurs via binding of integrin receptors to specific epitopes present on the

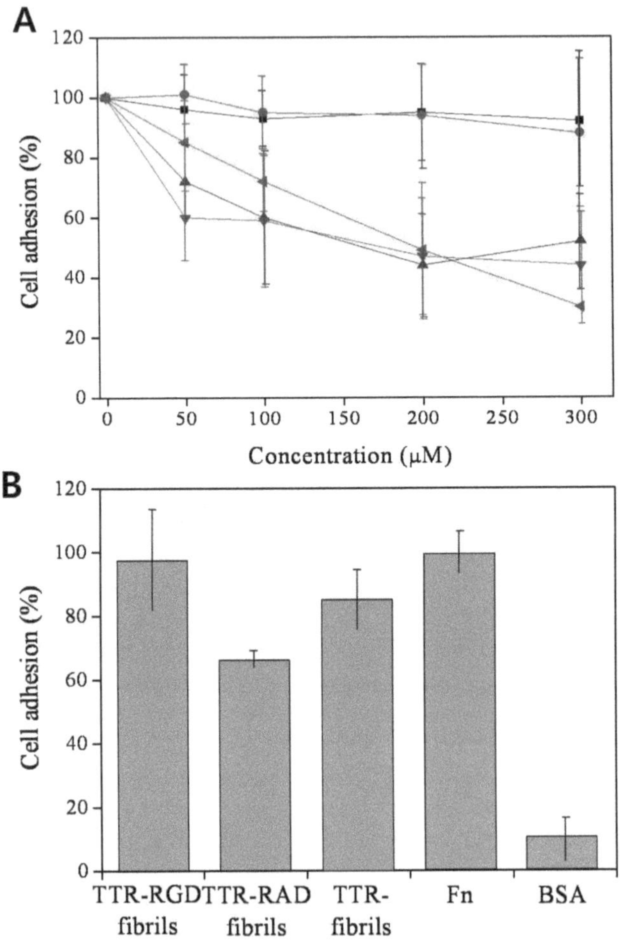

Fig. 5-53 Bioactivity of transthyrethrin-(105-115)-peptide/arginine-glycine-aspartic acid peptide and cell adhesion on transthyrethrin-(105-115)-peptide -based fibrils. A: Inhibition of 3T3 cell adhesion to fibronectin coated surfaces by soluble peptides: transthyrethrin-(105-115)-peptide/ arginine-glycine-aspartic acid (TTR-RGD) (▼), transthyrethrin-(105-115)-peptide – arginine-alanine-aspartic acid (TTR-RAD) (□), transthyrethrin-(105-115)-peptide (TTR)(■), arginine-arginine-glycine-asparagine-seiric acid (GRGDS) (●) and arginine-glycine-asparagine-seiric acid (RGDS) (○). B: 3T3 cell adhesion to a dried layer of TTR1-based fibrils or positive and negative control surfaces fibronectin and bovine serum albumin (BSA), respectively.

surface of extracellular matrix proteins such as fibronectin, leading to the formation of focal adhesions and related contacts [145].

Most previous work on artificial extracellular matrices has focused on the design of polymeric matrices with attributes such as biodegradable backbones, the ability to form gels, or functionalization with bioactive peptides, most notably arginine-glycine-aspartic acid (RGD) [146].

Fibroblast attachment and spreading was found to be dependent on the type of peptide amphiphile molecule used for to form the supramolecular nanofibers (Fig. 5-54). Enhanced cell attachment and spreading was observed on nanofibers formed by branched peptide amphiphiles (PAs) compared to linear ones, and, as one might expect, branched peptide amphiphile nanofibers presenting a single arginine-glycine-asparagine-serine acid (RGDS) epitope with D-configuration exhibit less cell attachment than the corresponding nanofibers with L-configuration [147]. Interestingly, in spite of the similar concentration of epitopes presented by the various supramolecular nanofibers (effectively van der Waals packing of epitopes), we observed significant differences in the number of adherent and spread cells. Nanofibers formed from branched peptide amphiphile (PA) molecules led to enhanced cell attachment and spreading as compared to linear peptide amphiphile (PA) nanofibers, and supramolecular fibers formed by the peptide amphiphile (PA) monomer containing a cyclic arginine-glycine-aspartic acid (RGD) peptide in branched architecture showed significantly higher cell spreading than a fibronectin control. All PAs showed enhanced cell adhesion compared to a PDL negative control. Among nanofibers formed from branched peptide amphiphiles (PAs), cell adhesion was dependent both on ligand density and ligand affinity. Varying the number of linear arginine-glycine-asparagine-serine acid (RGDS) epitopes per molecule from one to two

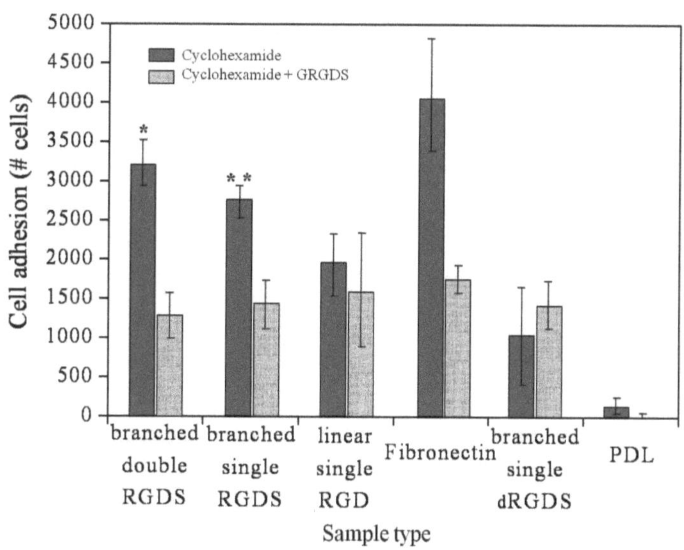

Fig. 5-54 Cell attachment in presence of cyclohexamide or cyclohexamide and soluble arginine-glycine-asparagine-serine acid (RGDS).

resulted in greater cell adhesion to the substrate; however, a branched peptide amphiphile containing a single high-affinity cyclic arginine-glycine-asparagine-serine acid (RGDS) epitope led to to enhanced cell adhesion relative to the branched peptide amphiphile containing two linear epitopes [148].

5.6. CARBON NANOTUBES-BASED COMPOSITE MATERIALS

5.6.1. Role of Reinforcement

The structural characteristics of host polymer materials can be improved by including dispersed materials in various ways [149]. The dispersed phase may be continuous (fiber, tape) or discontinuous (particulate, flake, whisker). We restrict the discussion to reinforcing phases used in fiber (or whisker) form since this article deals with mechanical reinforcement effects rather than with considerations of cost reduction, electrical properties, or other nonmechanical aspects. In polymer matrix composites, reinforcing fibers are used mainly to improve stiffness and strength. Taking stiffness as an example, the low modulus value (1–5 GPa typically) of most polymer matrices is readily and dramatically increased by the presence of glass or carbon fibers (typical modulus: 72 GPa and 250–750 GPa, respectively). Advanced fiber composite materials take advantages of the highly directional properties of fibers to generate superior structural properties. The main fibers in use for composite applications include the various types of glass and carbon/graphite, boron, silicon carbide, alumina, the aramids, and polyethylene. Fibers at the experimental stage include polybenzoxazole (PBO) and polybenzothiazole (PBT).

The diameters of most of these fibers vary from a few to about one hundred microns, whereas their length ranges between a few millimeters up to fully continuous. Whiskers have diameters in the nanometer range, and lengths down to the submicron range (as do carbon nanotubes, which are briefly considered at the end of the article). Whiskers are thin, elongated single crystals with properties much superior to those of the bulk material from which they are made [150].

Typical whiskers include aluminum oxide (Al_2O_3), silicon carbide (SiC), carbon, silicon nitride (Si_3N_4), nickel aluminide ($NiAl_3$), iron, nickel, and copper. Single fibers, which are difficult to handle individually, are usually gathered into bundles such as yarns, tows, or strands of a few hundred up to 100,000 or more single fibers. Such spools are the starting elements in the fabrication of a structure. They are combined with polymers using various techniques, to form intermediate flat sheets ('prepregs,' random mats, woven and knitted fabrics, braids, and more recent noncrimp fabrics) or direct final shapes (using filament winding, for example). The overall fiber architecture is controlled by the selection of appropriate reinforcement formats and processing routes. Table 5-4 shows the maximum achievable fiber volume fraction using different reinforcement formats, and compares the orientation factor which is 1 for a uniaxial array, 0.3 for a quasi-isotropic lay-up, and 0.12 for a random three-

Table 5-4 Connection among the reinforcement format, the manufacturing process, the fiber orientation factor, and the maximum fiber volume fraction

Reinforcement format	Applicable process	Orientation factor	Maximum fiber volume fraction
UD tow	Filament winding; pultrusion	1.00	0.80
UD prepreg	Autoclave; resin film infusion	1.00	0.65
Multiaxial prepreg	Autoclave; resin film infusion	0.31	0.60
2-D noncrimp fabric	Resin film infusion; resin transfer molding	0.30	0.55
Woven 2-D fabric	Resin transfer molding; wet lay-up	0.27	0.50
Orthogonal 3-D fabric	Resin transfer molding	0.30	0.40
Random planar	Sheet molding compound; resin transfer molding; wet lay-up	0.30	0.30
Random 3-D short fiber	Bulk molding compound; injection molding	0.12	0.20

dimensional (3-D) lay-up. No less important is the type of fiber surface treatment applied to enhance handling, processing, fiber-matrix adhesion, and to reduce fiber breakage or fiber-to-fiber adhesion. Many properties of the composite structures, including impact resistance and damage tolerance, are strongly dependent on the quality of the fiber-matrix interfaces [151].

The properties of the major fibers used in composites are given in Table 5-5. The specific strength against Young's modulus is plotted in Fig. 5-55. It is possible to use simple models to select materials that a priori should be suitable as reinforcements. The theoretical tensile strength of a solid is given by [152]:

$$\sigma_{\max} = (E \cdot \gamma / a_0)^{1/2} \tag{5-8}$$

where E and γ are the Young's modulus and the surface energy of the solid, respectively, and a_0 is the atomic spacing.

Carbon nanotubes are giant linear fullerenes (closed, convex cage molecules containing only hexagonal and pentagonal faces) having a tubular geometry, which were discovered in 1991 [153]. Nanotubes may arise as concentric cylinders and form multiwalled carbon nanotubes, or as single-shell cylinders or single-walled nanotubes, which are closed at their ends. Estimates for the Young's modulus of a carbon nanotube are consistent with the exceptionally high values already measured for a graphene sheet, about 1.2 TPa. Single-wall carbon nanotubes are also expected to be very strong and to resist fracture under extension and compression. Unlike carbon fibers, single-wall carbon nanotubes

Table 5-5 Properties of representative fiber and whisker materials

Materials	Fiber Diameter (nm)	Tensile modulus (GPa)	Tensile strength (GPa)	Density (g/cm^3)	Thermal expansion coefficient (10^{-6}/K)
E-glass	10–20	72	3.5	2.54	5.0
S-glass	12	86	4.6	2.50	5.6
High-modulus PAN-based carbon	6–8	290–590	2.5–3.9	1.70–1.94	−1.0 to −1.2
High-strength PAN-based carbon	5–8	230–295	3.5–7.1	1.76–1.82	−0.4 to −1.0
High-modulus Pitch-based carbon	10	520–830	2.1–2.2	2.03–2.18	−1.4
Boron	140	385–400	2.8	2.63	4.9
Silicon carbide	142	400	3.3	3.08	4.86
Silicon carbide (Nicalon)	15	200	2.8	2.60	3.1
Alumina (Al_2O_3) FP	20	379	1.9	3.95	8.3
Nextel (Al_2O_3) 610	10–12	380	3.1	3.9	8.0
Saffil (Al_2O_3-SiO_2)	3	100–300	1.0	3.3–3.5	–
PBT (heat-treated)	–	331	4.2	1.58	–
Kevlar 149	12	180	3.6	1.47	−2.0
Kevlar 49	12	130	3.6	1.45	−2.0
Kevlar 29	12	60	3.8	1.44	−2.0
UHMWPE (Dyneema SK60)	15	87	2.7	0.97	−12.0
UHMWPE (Spectra 1000)	30	172	3.0	0.97	–
Aluminium oxide whiskers	2.8–9.8	427	20.7	3.97	–
Berylium oxide whiskers	9.8–29	427	20.7	2.86	–
Silicon carbide whiskers	0.75–2.8	482	20.7	3.22	–

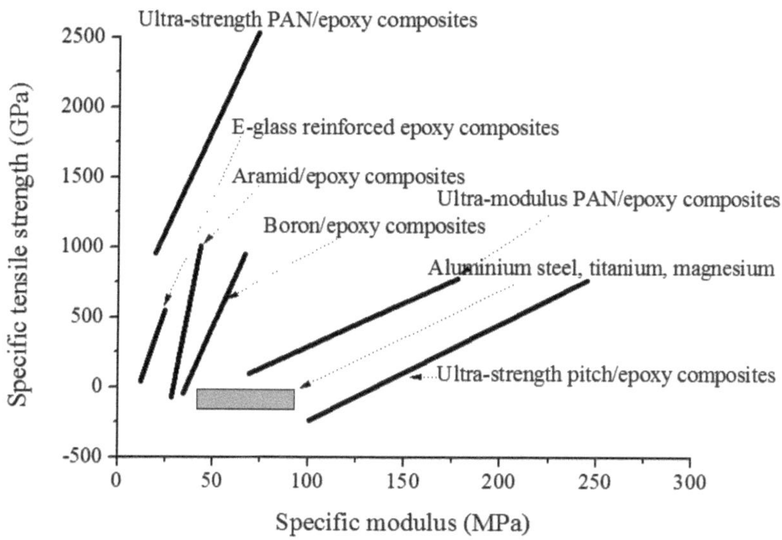

Fig. 5-55 Specific strength against specific Young's modulus for the major fibers used in polymer composites.

are remarkably flexible (by many orders of magnitude compared to conventional high performance fibers, because of their small diameter; see the definition of flexibility above), but should regain their original shape when the stresses distorting them are removed. The excellent mechanical properties of carbon nanotubes would make them ideal for composites applications, which would have significant consequences in structural applications.

According to the rolling angle of the graphene sheet, carbon nanotubes have three chiralities: armchair, zigzag, and chiral (Fig. 5-56). Many research efforts

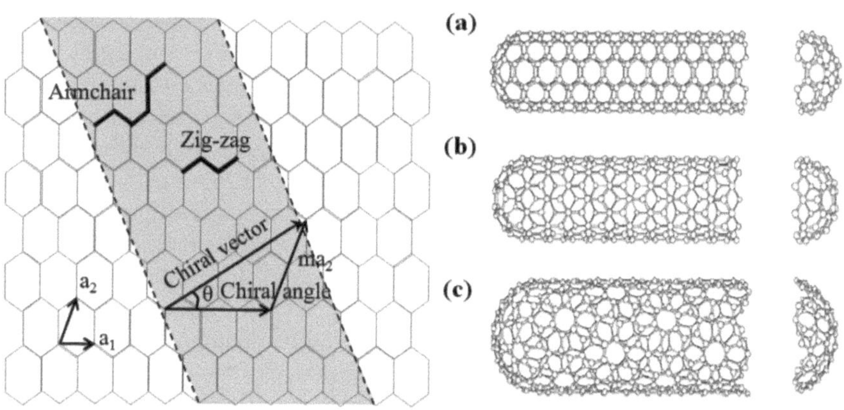

Fig. 5-56 Schematic diagram showing how a hexagonal sheet of graphene is rolled to form a carbon nanotube with different chiralities (A: armchair; B: zigzag; C: chiral).

Table 5-6 Dimension and corresponding number of particles in composites for different fillers

Filler	Description			
	Average dimension of filler	Density (g/cm^3)	N[a]	S[b]
Al_2O_3 particle	100 μm in diameter (d)	4.0	1.9	$S = \pi d^2$
Carbon fiber	5 μm in diameter (d) × 200 μm in length (l)	2.25	2.55	$S = \pi dl + \pi d^2/2$
Graphite nanoplatelet	45 μm in length (squre, l), 7.5 nm in thickness (t)	2.2	$6.58 * 10^4$	$S = 4l^2 + 2lt$
Carbon nanotube	12 nm in diameter (d) × 20 μm in length (l)	1.8	$4.42 * 10^4$	$S = \pi dl + \pi d^2/2$

[a]*N: number of particles in* 1.0 mm^3 *with* 0.1V.% *filler contents.*
[b]*S: surface area of individual particles.*

have been directed toward producing carbon nanotube/polymer composites for functional and structural applications [154]. However, even after a decade of research, the full potential of employing carbon nanotubes as reinforcements has been severely limited because of the difficulties associated with dispersion of entangled carbon nanotube during processing and poor interfacial interaction between carbon nanotubes and polymer matrix.

Table 5-6 compares the dimensions of commonly used fillers, including Al_2O_3 particles, carbon fibers, graphite nanoplatelets (GNPs) and carbon nanotubes and the number of particles corresponding to a uniform filler volume fraction of 0.1 percent in a composite of 1.0 mm^3 cube. Because there are large differences in dimension and geometry of these four different reinforcements, the number of fillers contained for a given filler volume fraction will also greatly vary. For example, there are only two pieces of Al_2O_3 particles and the number increases to 200 when carbon fibers are added in the composites with the same filler volume fraction, whereas there are some 442 million pieces of carbon nanotubes. This observation sheds light on why the dispersion of carbon nanotubes in a polymer matrix is more difficult than for other fillers.

In addition to the size effect of fillers, the physical nature of particles also plays an important role in dispersing them into the polymer matrix. The challenge is this: how to incorporate individual carbon nanotubes, or at least relatively thin carbon nanotube bundles or disentangled carbon nanotubes, inside a polymer matrix. In other words, dispersion of carbon nanotubes is not only a geometrical problem, dealing with the length and size of the carbon nanotube, but also relates to a method of how to separate individual carbon nanotubes from carbon nanotube agglomerates and stabilize them in a polymer matrix to avoid secondary agglomeration.

The performance of carbon nanotube/polymer nanocomposites depends on the dispersion of carbon nanotubes in the matrix and on the interfacial interactions between the carbon nanotubes and the polymer. However, the carbon atoms on carbon nanotube walls are chemically stable because of the aromatic nature of the bond. As a result, reinforcing carbon nanotubes are inert and can interact with the surrounding matrix mainly through van der Waals interactions, which means that they are unable to provide an efficient load transfer across the carbon nanotube/matrix interface. Therefore, significant efforts have been directed toward developing methods to modify the surface properties of carbon nanotubes. There are several comprehensive papers that review the chemistry of functionalized carbon nanotubes and the reaction mechanisms between the carbon nanotubes and functional groups [155]. The major principles of these methods, along with the corresponding advantages and disadvantages, are summarized in Table 5-7.

While many studies have hitherto been initiated to modify the carbon nanotube surface characteristics based on the below techniques, there are many processes and material variables that have not been fully optimized.

Surface functionalization is based on the covalent linkage of functional entities onto the carbon scaffold of carbon nanotubes. This can be performed at the termini of the tubes or at their sidewalls. Direct functionalization is associated with a change of hybridization from sp^2 to sp^3 and with a simultaneous loss of the p-conjugation system on the graphene layer (Fig. 5-57). This process can be made by a reaction with some molecules of a high chemical reactivity, such as fluorine. Defect functionalization is another method for surface functionalization of carbon nanotubes. This process takes advantage of the surface transformation of defect sites on carbon nanotubes (Fig. 5-57).

Table 5-7 Advantages and disadvantages of various carbon nanotube (CNT) functionalization methods

Method	Principle	Possible damage	Easy to use	Interaction with polymer matrix[a]	Agglomeration of carbon nanotubes
Polymer Wrapping	Van der Waals force, π–π stacking	X	√	V	X
Surfactant adsorption	Physical adsorption	X	√	W	X
Endohedral method	Capillary effect	X	X	W	√

[a]*S: Strong W: Weak, V: Variable according to the miscibility between matrix and polymer on carbon nanotube.*

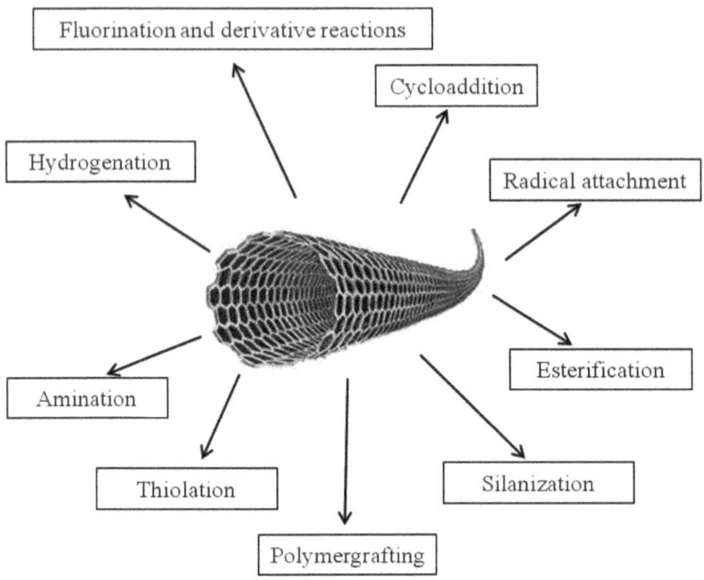

Fig. 5-57 Strategies for surface functionalization of carbon nanotubes.

Defect sites can be the open ends and/or holes in the sidewalls, pentagonal or heptagonal irregularities in the hexagon graphene framework. Carbon nanotubes functionalized in this way are soluble in many organic solvents because the hydrophobic nature of carbon nanotubes is changed to a hydrophilic one due to the attachment of polar groups. Surface functionalized carbon nanotubes can produce strong interfacial interaction with many polymers, allowing carbon nanotube-based nanocomposites to possess high mechanical and functional properties.

Epoxy-based nanocomposites comprise the majority of reports using carbon nanotube/polymer composites [156] and Fig. 5-58 illustrates a flowchart of typical processes used to produce carbon nanotube/epoxy nanocomposites.

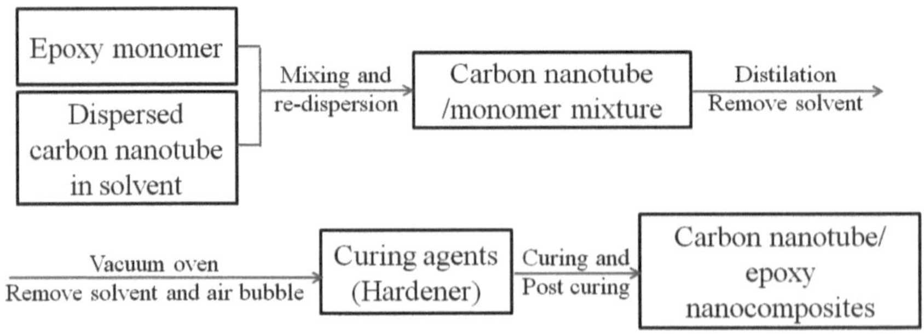

Fig. 5-58 Schematics of carbon nanotube/epoxy nanocomposite fabrication.

Carbon nanotubes exhibit excellent mechanical properties with Young's modulus as high as 1.2 TPa and tensile strength of 50–200 GPa. The combination of these exceptional mechanical properties and the low density, high aspect ratio and high surface area make carbon nanotubes an ideal candidate for reinforcement in composite materials. The carbon nanotube reinforced nanocomposites can be considered to be a kind of particulate composite with the filler dimensions on a nanometer scale and a high aspect ratio. Unlike the microscopic particulate composites, the mechanical properties of nanocomposites depend strongly on the dispersion state of nanofillers, apart from the properties of fillers and matrix materials.

A good dispersion not only makes more filler surface area available for bonding with the polymer matrix, but also prevents the aggregated filler from acting as a stress concentrator, both of which are detrimental to mechanical performance of composites [157]. However, some challenges, such as carbon nanotube content in composites, length and entanglement of carbon nanotubes, and viscosity of matrix have to be overcome to obtain uniform carbon nanotube dispersion in nanocomposites. There have been many reports showing that there is a critical carbon nanotube content in the matrix below which the strengthening effect for carbon nanotube/polymer composites increases with increasing carbon nanotube contents [158, 159]. Above this critical carbon nanotube content, however, the mechanical properties of carbon nanotube/polymer composites decrease, and, in some cases, they decrease below those of the neat matrix material. These observations can be attributed to (i) the difficulties associated with uniform dispersion of carbon nanotubes at high carbon nanotube contents; (ii) and lack of polymerization reactions that are adversely affected by the high carbon nanotube content. The latter effect becomes more pronounced when functionalized carbon nanotubes are employed to produce carbon nanotube/polymer nanocomposites.

As for fiber-reinforced polymer composites, the external load applied to carbon nanotube/polymer composites has to be transferred to the carbon nanotubes, allowing them to take the major share of the load [160]. The efficiency of the load transfer depends on the interfacial bond between the filler and the matrix materials. A strong interfacial adhesion corresponds to high mechanical properties of composites through enhanced load transfer from the matrix to carbon nanotubes. Surface functionalizations of carbon nanotubes have proven to enhance the interfacial adhesion.

Table 5-8 summarizes recent studies on the effects of carbon nanotube functionalization on the mechanical properties of carbon nanotube nanocomposites made from thermosets and thermoplastics. These results clearly indicate that the functionalization of carbon nanotubes can enhance the modulus, strength, and fracture resistance of the nanocomposites [161].

The thermal conductivity of materials is dominated by atomic vibrations or phonons, and the conduction by electrons is generally negligible for insulating materials [162]. For carbon nanotube/polymer nanocomposites, the thermal

Table 5-8 Effect of carbon nanotube functionalization on the mechanical properties of carbon nanotube(CNT)/polymer nanocomposites

Matrix		Dispersion technique	Functionalization technique	CNT content (wt %)	Enhancement on mechanical properties[a]		
					Modulus(%)	Strength (%)	Toughness (%)
Thermo-plastic[b]	PA	Twin-screw extruder	Diamine treatment	1.0	6.1	−5.3	–
	PB	Shear mixing	Polymer grafting	1.5	18	−27	−40
	PE	Shear mixing	Maleic anhydride and amine treatment	1.5	22	−17	−69
	PI	Ultrasonication	Acid treatment	7.0	39	19	–
	PP	Ultrasonication and stir	Undecyl radicals attachment	1.5	55	−10	–
	PS	Probe ultrasonication	Butyl attachment	0.25	-8.3	2.1	–
	PVA	Ultrasonication and stir	Polymer grafting	2.5	35	−4.8	–
	PMMA	Ultrasonication	Polymer grafting	0.10	57	−2.7	–
Thermo-set[c]	EP	Calendering	Amino treatment	0.10	2.1	−2.2	17
		Probe ultrasonication	Surfactant treatment	0.25	5.6	6.8	35
		Ultrasonication	Organic silane	0.25	8.7	3.6	−22
	PU	Ball mill	Polymer grafting	0.7	48	27	–
		Ultrasonication	Acid treatment	10	340	51	–
	VR	Ball mill	Acid treatment	25	444	175	–
		Ball mill	Organic silane	1	35	−7.8	–

[a] *Data in brackets indicate the enhancement on mechanical properties by employing functionalized CNTs.*
[b] *PA (Nylon): polyamide; PB: polybutylene; PE: polyethylene; PI: polyimide; PP: polypropylene; PS: polystyrene; PVA: poly (vinyl acetate); PMMA: poly(methyl methacrylate).*
[c] *EP: epoxy; PU: polyurethane; VR: vulcanized rubber.*

conductivity depends on several factors, including the content, aspect ratio, and dispersion of carbon nanotubes and their interfacial interactions with polymer matrix. The excellent thermal properties of carbon nanotubes, such as high thermal conductivity and good thermal stability, lead to the expectation that carbon nanotubes could make useful functional fillers to rectify the thermal properties of polymers. Contrary to this expectation, the enhancements of thermal properties due to the incorporation of carbon nanotubes in polymer composites have not been remarkable, with the exception of a few isolated cases.

Many efforts have been devoted to employing carbon nanotubes as thermal conducting filler in polymer composites and some enhancements have indeed been observed. For example, Biercuk *et al.* [163] used single-walled carbon nanotubes to augment the thermal transport properties of epoxy, showing 70 and 125 percent increases in thermal conductivity at 40 K and room temperature, respectively, with 1 wt.% single-walled carbon nanotubes. Evseeva *et al.* [164] found that the introduction of 0.1–1.0 wt.% multiwalled carbon nanotubes enhanced the thermal conductivity of epoxy resin by about 40 percent. However, comparison of the data obtained from the above reports revealed that the thermal conductivities of carbon nanotube/polymer composites showed only a marginal improvement compared with the electrical conductivities with the same carbon nanotube content. Taking into account the different transport mechanisms between the thermal and electrical conductivities of composites, Moniruzzaman *et al.* [165] proposed that the phonons, major carriers for thermal conduction, were much more likely to travel through the matrix than through the carbon nanotube networks because the difference in thermal conductivity between the carbon nanotubes and the polymer, i.e., about 104 W/(m·K) with $K_{CNT} \approx 10^3$ and $K_{polymer} \approx 10^{-1}$ W/(m·K), is much smaller than that of the electrical conductivity, i.e., in the range of 10^{15}–10^{19} with $\sigma_{CNT} \approx 10^2$–$10^6$ and $\sigma_{polymer} < 10^{-13}$ S/cm.

Rheology is the study of the flow behavior of materials under conditions in which they flow rather than experience elastic or plastic deformation. It is also concerned with establishing predictions for mechanical behavior (on the continuum mechanical scale) based on the micro- or nanostructure of the materials [166].

The study of rheological responses of carbon nanotube/polymer nanocomposites has both practical importance related to composite processing and scientific importance as a probe of the composite dynamics and microstructure. The rheological properties of carbon nanotube/polymer nanocomposites depend on factors such as the characteristics of the filler loading, aspect ratio and dispersion, polymer molecular weight, and the interfacial interaction between the polymer and filler [167, 168].

The variations of viscosity and storage modulus of composites as a function of frequency are two common techniques used to characterize the rheological properties of carbon nanotube/polymer nanocomposites. Fig. 5-59 shows the complex viscosity (Fig. 5-59 (a)) and storage modulus (Fig. 5-59 (b)) of carbon

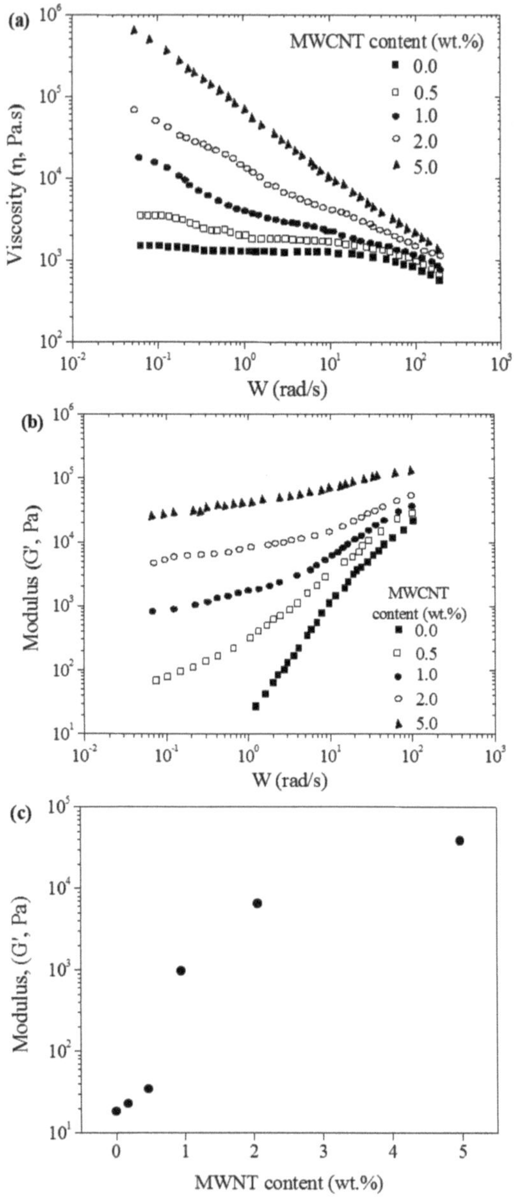

Fig. 5-59 Rheological response ((a): complex viscosity; (b): storage modulus) of carbon nanotube/polycarbonate nanocomposites as a function of frequency at 230 °C and corresponding rheological percolation threshold (c) at frequency of 1 rad/s.

nanotube/polycarbonate nanocomposites with different carbon nanotube contents [169]. At low frequencies, the fully relaxed polymer chains exhibit the typical Newtonian viscosity plateau. With the addition of carbon nanotubes, the low-frequency complex viscosity significantly increases, whereas the whole trend of viscosity drops with an increase of frequency at a given carbon nanotube content (Fig. 5-59 (a)), indicating that the relaxation of polymer chains in the nanocomposites is effectively restrained by the presence of carbon nanotubes. The storage modulus of nanocomposites gradually increases with increasing frequency and carbon nanotube content (Fig. 5-59 (b)), indicating a transition from viscous liquid to solid-like behavior. Similar to the electrical percolation behavior, the rheology of carbon nanotube/polymer nanocomposites also shows a transition from a rheological state (where the viscosity or storage modulus changes significantly with increasing filler content) to a solid-like behavior (where the viscosity or storage modulus is insensitive or has only a slight variation with increasing filler content). The concentration at which this transition occurs is known as the rheological percolation threshold, and is shown in Fig. 5-59 (c).

Mechanical reinforcement of polymers by multiwalled carbon nanotubes can be realized only by solving two main problems: (i) dispersion of multiwalled carbon nanotubes and (ii) interfacial adhesion between the nanotubes and the matrix. The first experimental work focusing on interfacial interaction in multiwalled carbon nanotube/epoxy nanocomposites was that of Cooper *et al.* [170]. They investigated the detachment of multiwalled carbon nanotubes from an epoxy matrix. In a special pull-out test of individual multiwalled carbon nanotubes, the interfacial shear strength values were found to be in the 35–376 MPa range.

The realization of a multiwalled carbon nanotube-reinforced epoxy requires, besides homogenous dispersion, strong interfacial interaction between the multiwalled carbon nanotubes and the polymer. The effectiveness of multiwalled carbon nanotubes as reinforcing elements in tough epoxy matrices is hindered by weak interfacial interactions. As such, stress-induced deformation of composites can lead to failure of the multiwalled carbon nanotubes/epoxy interface, and, finally, to pull-out. Further enhancement of multiwalled carbon nanotubes with composite materials can possibly be achieved by surface functionalization, through physical bonds to the polymeric matrix. These bonds will enable stress transfer between the polymer and the multiwalled carbon nanotubes, leading to improved interfacial interactions, as qualitatively determined previously.

The nature of the interfacial zone accords with the microstructural characteristics of the reinforcing multiwalled carbon nanotubes, in which any of the three mechanisms, physical interaction, physical-chemical interaction or mechanical interlock, may be dominant, and in which the van der Waals forces of attraction are the primary binding forces at the interface [171]. Also having a significant influence on the fracture toughness of the interface is the surface roughness of the reinforcement.

Fig. 5-60 High-resolution TEM image of dispersion of bromo-functionalized multi-walled carbon nanotubes in epoxy matrix resins.

Fig. 5-60 shows the observed dispersion in a representative section of a sample containing bromo-functionalized multiwalled carbon nanotubes [172].

Fig. 5-61 (a) and (b) show the dynamic measurements of the storage modulus and loss factor as a function of temperature [172]. The addition of both multi-walled carbon nanotube types (bromo-functionalized and non-functionalized) to the epoxy resin does not show any influence of the multiwalled carbon nanotube content on the storage modulus in the glassy region or in the rubbery region. The addition of the multiwalled carbon nanotubes improves the elastic properties of the epoxy system at elevated temperatures. Particularly, at temperatures close to or higher than the glass transition temperature (T_g) of the nanocomposites, indicated by a peak in the loss modulus, the presence of more multiwalled carbon nanotubes is able to impart greater stiffness to the polymer matrix [173]. This interfacial interaction reduces the mobility of the epoxy matrix around the multiwalled carbon nanotubes and leads to the observed increase in thermal stability. A strong increase of the storage modulus, especially below the glass transition temperature (T_g), cannot have been expected for the multiwalled carbon nanotube contents used in this study. The load transfer is only performed between the outermost layers of the multiwalled carbon nanotubes and the epoxy matrix, resulting in a much lower effective content of reinforcing multiwalled carbon nanotubes.

5.6.2. Electromagnetic Interference Shielding Properties

Electrically conducting composites with a volume conductivity higher than 10^{-10} S/cm, are considered to be an important group of relatively inexpensive

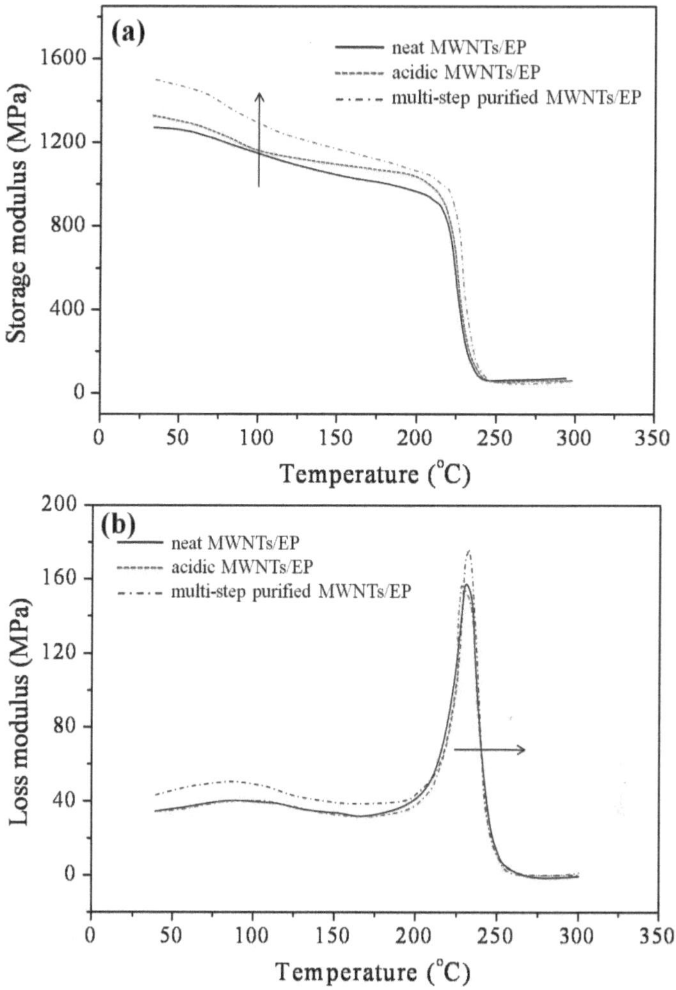

Fig. 5-61 Storage (a) and loss modulus (b) of multi-step purified 0.5 wt.% multi-walled carbon nanotubes(MWNTs)/epoxy(EP) composites.

materials for many engineering applications (Fig. 5-62) [174, 175], such as electrically conducting adhesives, antistatic coatings and films, electromagnetic interference shielding materials for electronic devices, thermal interface materials, etc.

At the very outset, in EMI shielding, the effects of both reflection and absorption of the incident EM radiation have to be considered [176]. The intensity of the reflected radiation (*R*) is related to the difference of the

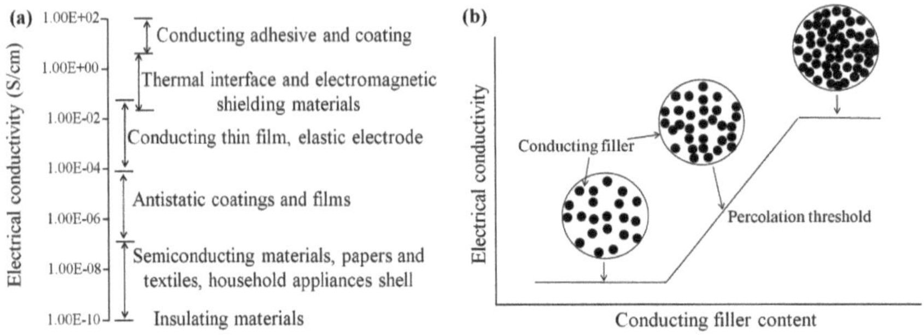

Fig. 5-62 Typical applications of conducting composites (a) and a schematic of percolation phenomenon and conducting network in conducting composites (b).

characteristic impedances (Z) of the incident medium (i.e., air of impedance Z_a) and the materials (Z_m) through

$$R = \left(\frac{Z_a - Z_m}{Z_a + Z_m}\right)^2 \tag{5-9}$$

The absorption (A) is quantified through the skin depth (δ), the extent to which the radiation, of frequency f, penetrates the material through

$$\delta = \frac{1}{\sqrt{\pi f \mu \sigma}} \tag{5-10}$$

where μ is the magnetic permeability and σ the electrical conductivity.

Consequently, while metallic materials (with low Z_m and high σ) are obviously efficient shielding materials, issues such as cost and bulk preclude their wide-scale application. Such issues have encouraged the use of polymers in which the intrinsic poor conductivity could be compensated for through the use of filler materials that form a conducting network.

Electromagnetic interference remains a technical challenge in the proper functioning of numerous electrical and electronic devices; electromagnetic interference tends to degrade the interception of signals, thus affecting the performance of such equipment adversely. So, it is quite important to shield these electrical and electronic devices from the electromagnetic interference frequency spectrum to ensure proper functioning. Conventionally, metal sheeting is used for this purpose, but the seams commonly encountered in metal sheets tend to cause leakage of the radiation and diminish the effectiveness of the shield. Polymer-matrix composites containing conductive fillers are quite attractive as electromagnetic interference shielding materials because the seams are significantly reduced or are completely eliminated. Besides this fact, polymers carry the usual advantage of ease of processibility and corrosion resistance over the metals. Polymer composites containing conductive polymers such as polyaniline or polypyrole as electromagnetic interference shields have been reported in the literature [177], but their mechanical strength is often very poor. The use of carbon materials as fillers in polymer composites for electromagnetic interference shielding has been advocated [178], as such materials can enhance the mechanical properties of these composites considerably.

Carbon nanotubes are a promising electromagnetic interference shielding candidate [179] due to their light weight, high conductivity, and exceptionally good mechanical properties. In the case of highly conductive single-walled carbon nanotube film, 10 nm thickness gives more than 20 dB at 10 GHz and over 50 dB can be easily achieved at over 10 μm thickness [180]. In the polymer composite, the high aspect ratio of carbon nanotube filler has an advantage compared to carbon nanofiber and carbon black in terms of conductivity due to its lower percolation limit. The remaining metal catalyst in the multiwalled carbon nanotube (MWCNT) and higher carbon nanotube aspect ratio lead to higher conductivity, and this leads to better shielding performance. For light-weight electromagnetic interference shielding performance and corrosion resistance, purified carbon nanotubes with less metal impurity are desired. Longer carbon nanotubes are beneficial in terms of conductivity because of reduced intertube contact [181].

In polymer composites, shielding mechanisms are more complicated than those for homogeneous conductive barriers because of the huge surface area available for reflection and multiple-reflection. The first reflection of an electromagnetic wave from a conductive material surface should be distinguished from the multiple-reflection mechanism, which is the re-reflection of the waves already reflected, as illustrated in Fig. 5-63. In the literature, based on an improperly analyzed power balance of composite plates of 1 mm in thickness

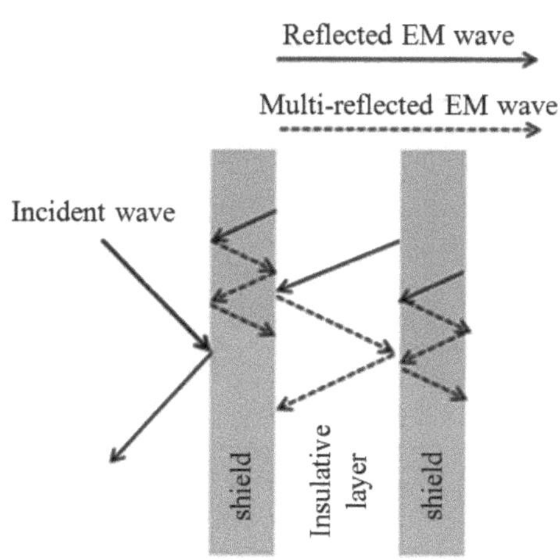

Fig. 5-63 Schematic showing the difference between reflection and multiple-reflection mechanisms.

[182], it has been wrongly concluded that reflection is the primary shielding mechanism in polymer composites. In addition, based on an inaccurate distinction between reflection and multiple-reflection mechanisms [183], multiple-reflection was wrongly considered a mechanism that increases the overall electromagnetic interference shielding effectivness.

To maximize electrical conductivity and mechanical properties, higher loading of carbon nanotubes in electromagnetic interference shielding composites is preferred. However, most carbon nanotube composites with polymer matrices have a maximum loading at around 40 wt.% [184] and 10 wt.% in the epoxy [185]. Higher loading will decrease the mechanical properties due to agglomeration.

Figs. 5-64–5-66 show the incident, reflected, transmitted, and absorbed power of different multiwalled carbon nanotube/polypropylene composites as a function of multiwalled carbon nanotube concentration and multiwalled carbon nanotube/polypropylene plate thickness. The reported power values are the average values in the X-band frequency range. In Figs. 5-64–5-66, it can be seen that the amount of power blocked by reflection increased with an increase in the multiwalled carbon nanotube content, i.e., it correlated with the increase in composite conductivity, whereas the amount of power blocked by absorption, for 0.34 mm and 1 mm thick plates, initially increased with the increase in multiwalled carbon nanotube content and then decreased again. For all plates made of 5 vol.% multiwalled carbon nanotube/polypropylene composite, it is apparent that the amount of energy blocked by reflection is higher than that blocked by absorption. Many researchers have improperly

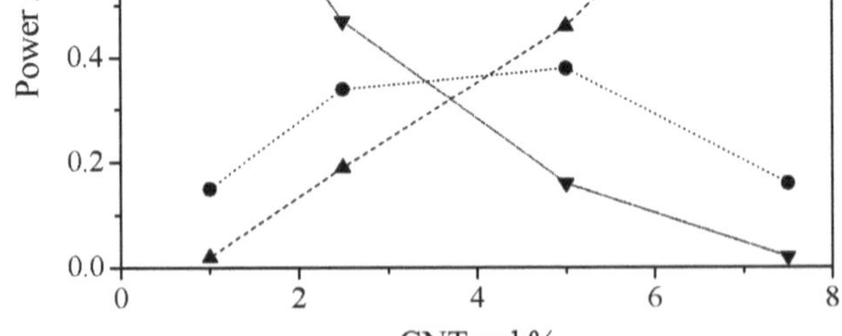

Fig. 5-64 Power balance in the X-band frequency range for 0.34 mm plates made of multi-walled carbon nanotube/polypropylene composites as a function of carbon nanotube (CNT) content.

concluded from similar observations that reflection is the dominant shielding mechanism. The lower amount of power blocked by absorption is due to the lower power transmitted into the sample as a result of the better reflection. The contribution of absorption to the overall shielding should be based on the ability of the material to attenuate the power that has not been reflected. Electromagnetic interference shielding effectiveness is a relative quantity and has nothing to do with the absolute power values [186].

Three mechanisms are involved in electromagnetic interference shielding: reflection, absorption, and multiple-reflection. For a multiwalled carbon nanotube/polymer composite sample, as this is at 340 μm, absorption is the primary shielding mechanism, followed by shielding by reflection. The theoretical analysis results show that multiple reflection within the multiwalled carbon nanotube internal surfaces might have a negative influence on the overall electromagnetic interference shielding effectiveness because the multiwalled carbon nanotube diameter is orders of magnitude smaller than the skin depth. Multiple-reflection between external surfaces of multiwalled carbon nanotubes also decreases the overall shielding, but, this influence is smaller than that between the internal surfaces.

Many multiwalled carbon nanotube-polymer composites have been developed as electromagnetic interference shielding materials due to easy fabrication processes, excellent mechanical properties and good conductivity. Multiwalled carbon nanotubes were homogeneously dispersed in pure acrylic emulsion by ultrasonication to prepare multiwalled carbon nanotube/polyacrylate

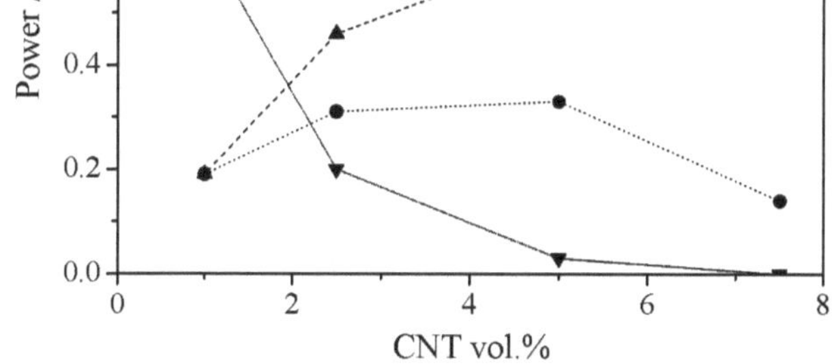

Fig. 5-65 Power balance in the X-band frequency range for 1 mm plates made of multi-walled carbon nanotube/polypropylene composites as a function of carbon nanotube (CNT) content.

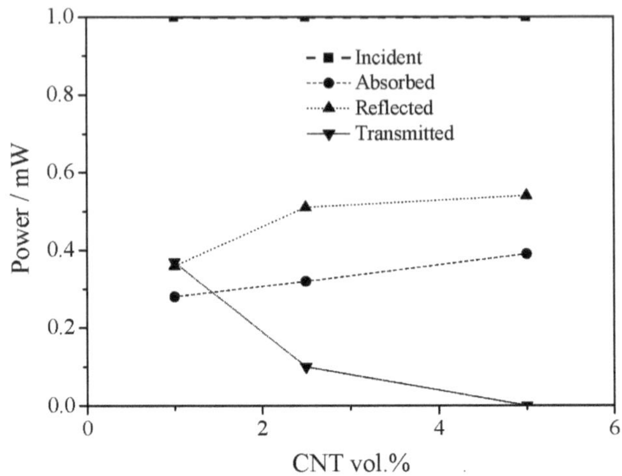

Fig. 5-66 Power balance in the X-band frequency range for 2.8 mm plates made of multi-walled carbon nanotube/polypropylene composites as a function of carbon nanotube (CNT) content.

composites applied on building interior walls for electromagnetic interference shielding applications [187].

Fig. 5-67 shows the variation of shielding effectiveness with multiwalled carbon nanotube loadings at fixed frequencies of 500 MHz and 10 GHz. It is evident that the shielding effectiveness increases progressively with the increase

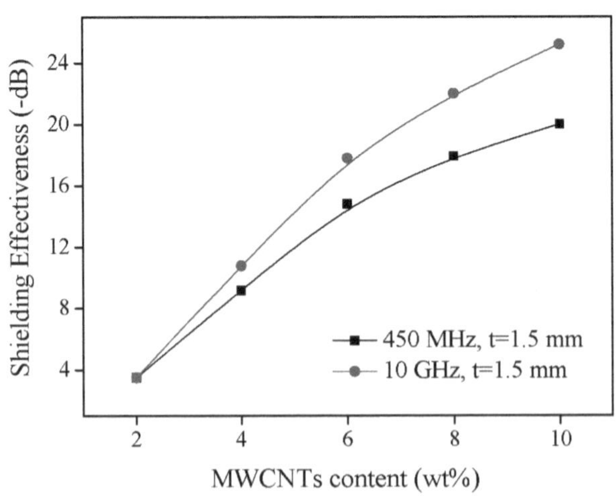

Fig. 5-67 Electromagnetic interference shielding effectiveness of multi-walled carbon nanotube/polyacrylate composites with multi-walled carbon nanotube (MWCNT) contents at 450 MHz and 10 GHz.

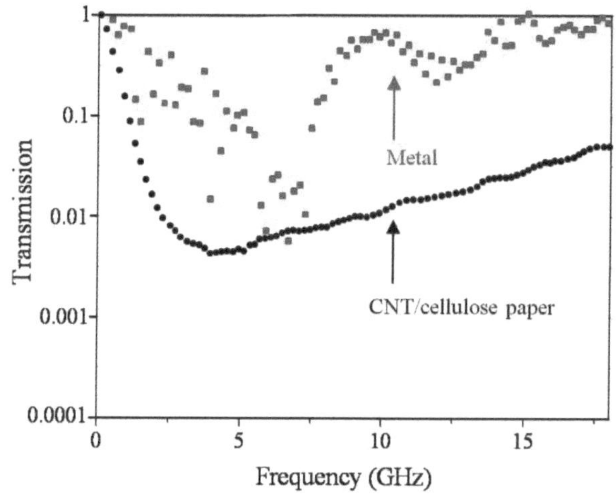

Fig. 5-68 The electromagnetic interference shielding efficiencies of the carbon nanotube (CNT)/ cellulose paper sample in lower frequency range.

of filler loading, but the rate is much faster for 10 GHz compared to that for 450 MHz. The higher electromagnetic interference shielding effectiveness for multiwalled carbon nanotube/polyacrylate composite films can mainly be ascribed to the smaller skin depth, which results in high multiple reflection. The high performance of shielding effectiveness is mainly due to the high conductivity and large number density of multiwalled carbon nanotubes, which helps a lot in forming good conductive networks in the insulating polyacrylate matrix. The conductive network formed due to the dispersion of multiwalled carbon nanotubes behaves like a conductive mesh, which intercepts electromagnetic radiation [188]. The small mesh size generally increases the shielding effectiveness.

Fugetsu *et al.* [189] reported that carbon nanotube/cellulose possesses potential for electromagnetic interference shielding effects. As seen in Fig. 5-68, the electromagnetic interference shielding effectiveness of the carbon nanotube/cellulose paper was comparable or even better than that of the metal-printed circuit-board. Carbon nanotube networks play an important role in determining the electromagnetic interference shielding effectiveness [190]. Carbon nanotubes are electrically conductive with a volume resistance of $5.3 \times 10^{-1}\ \Omega$ cm. The composite paper is capable of shielding electromagnetic interference over the tested range of 15–40 GHz, particularly in range of 30–40 GHz, with absorption as the essential shielding mechanism.

5.6.3. Optical Properties

The optical properties of carbon nanotube/polymer nanocomposites consist of optical nonlinearity and limitation, photoluminescence, light emission, and photonic properties. The nonlinear (NL) property of a material describes the behavior of light in nonlinear media in which the dielectric polarization responds nonlinearly to the electric field of the light. Nonlinearity is typically observed at very high light intensities, such as those provided by pulsed lasers. It has been demonstrated that suspensions of both single-walled carbon nanotubes and multiwalled carbon nanotubes have nonlinear optical properties [191]; thus they can be used to rectify the optical properties of polymer composites to fabricate nonlinear optical waveguide devices.

Hammel *et al.* [192] pioneered this research field: the wrapping of carbon nanotubes with poly(phenylacetylenes) (PPA) showed that carbon nanotubes had a strong photostabilization effect and that the poly(phenylacetylenes) chains were protected from photodegradation under harsh laser irradiation with incident fluence. The carbon nanotube/poly(phenylacetylenes) solutions effectively limited intense optical pulses, with the saturation fluence tunable by varying the carbon nanotube content. The nonlinear properties of carbon nanotube/polymer nanocomposites are governed not only by the dispersion state of carbon nanotubes, but also by the polymer structure. Wang *et al.* [193] studied the nonlinear properties of single-walled carbon nanotubes in *N*-methyl-2-pyrrolidone (NMP), *N,N*-dimethylformamide (DMF) and *N,N*-dimethylacetamide (DMA), and found that the nonlinear extinction coefficients of single-walled carbon nanotubes markedly increased with increasing single-walled carbon nanotube concentration. Carbon nanotubes dissolved in *N*-methyl-2-pyrrolidone exhibited a much better dispersion, although their optical limiting properties were inferior to those of *N,N*-dimethylformamide and *N,N*-dimethylacetamide. The average bundle size of carbon nanotubes and the physical properties of the solvent dominated the nonlinear extinction and optical limiting properties of carbon nanotube dispersions. Recent studies also revealed similar results: a good dispersion of carbon nanotubes did not guarantee a high performance of the nonlinear and optical limiting properties of carbon nanotube/polymer nanocomposites [194]. A possible reason for this observation is the complicated optical limiting mechanisms of composite materials. Besides the contribution by nonlinear scattering, other factors, such as nonlinear absorption and refraction and electronic absorption, also contributed to the optical limiting properties [195]. In addition to the carbon nanotube dispersion, the backbones of the polymers have a considerable effect on the nonlinear properties of carbon nanotube/polymer nanocomposites: e.g., functionalization of carbon nanotubes using conjugated polymers that have high affinity with carbon nanotubes resulted in significant improvement in nonlinear absorption and scattering due to the enhanced p–p interactions between the carbon nanotubes and the polymer matrix.

The black color of carbon nanotubes, combined with their high surface area and high electron affinity, enable carbon nanotubes to be an ideal candidate to improve the photoluminescence of carbon nanotube/polymer nanocomposites. Photoluminescence is a process in which a substance absorbs photons (i.e., electromagnetic radiation) and re-radiates photons. In carbon nanotube/polymer nanocomposites, excitons formed on the polymer by the absorption of light, which can be dissociated by electron transfer from the photo-excited polymer to carbon nanotubes. This process quenches the photoluminescence of polymers [196]. For example, Singh *et al.* [197] studied carbon nanotube/poly(3-hexylthiophene) nanocomposites, and showed that the intensity of photoluminescence emission from the composite decreased with increasing carbon nanotube content; they also demonstrated that the absorption coefficient of the composites was insensitive to carbon nanotubes of up to 5 wt.%. Carbon nanotubes functionalized using phthalocyanines also showed a considerable quenching effect of the fluorescence intensity in the photoluminescence spectrum [198].

Other potentially optical applications of carbon nanotubes are polymer based light-emitting and photonic devices. The advantages of using organic light-emitting diodes (OLEDs) based on conjugated polymers include low cost, low operating voltage, excellent processability, and flexibility. However, their low quantum efficiency and stability have limited their application and development [199]. Choi *et al.* [200] observed that device qualities, such as external quantum efficiency, were improved by 2–3 times with the addition of up to 0.2 wt.% single-walled carbon nanotubes in organic light-emitting diodes. They ascribed this observation to the facile hole injection and the polymer-carbon nanotube interactions, which stiffened the polymer chains. Subsequently, the better hole transport in the metallic single-walled carbon nanotube-polymer medium induced more efficient single exciton formation at or near the interface region. Single-layer light-emitting diodes fabricated using multiwalled carbon nanotube doped poly (p-phenylene benzobisoxazole) decreased the threshold voltage by 2 V. Up to 0.1 wt.% of multiwalled carbon nanotubes, the diode emission current increased by about two orders of magnitude over those of light-emitting diodes without multiwalled carbon nanotubes. However, a further increase of carbon nanotube content caused a successive decrease in light emission intensity due to the poor dispersion of carbon nanotubes.

Another interesting photonic application of carbon nanotube/polymer nanocomposites is the photomechanical actuator. The photomechanical effect is the change in the shape of a material when it is exposed to light. Ijima *et al.* [201] demonstrated that single-walled carbon nanotube can bend under light. This movement was sensitive to the intensity of the light, and was attributed to the localized electrostatic effects due to uneven distributions of photogenerated charges in carbon nanotubes. Actuators constructed using carbon nanotubes dispersed in several polymer matrices, such as Nafion, a sulfonated tetrafluoroethylene based fluoropolymer-copolymer [202], and polydimethylsiloxane (PDMS) [203], have been reported. Fig. 5-69 (a) shows a schematic for the characterization of the photomechanical actuation behavior of carbon nanotube/polymer nanocomposites.

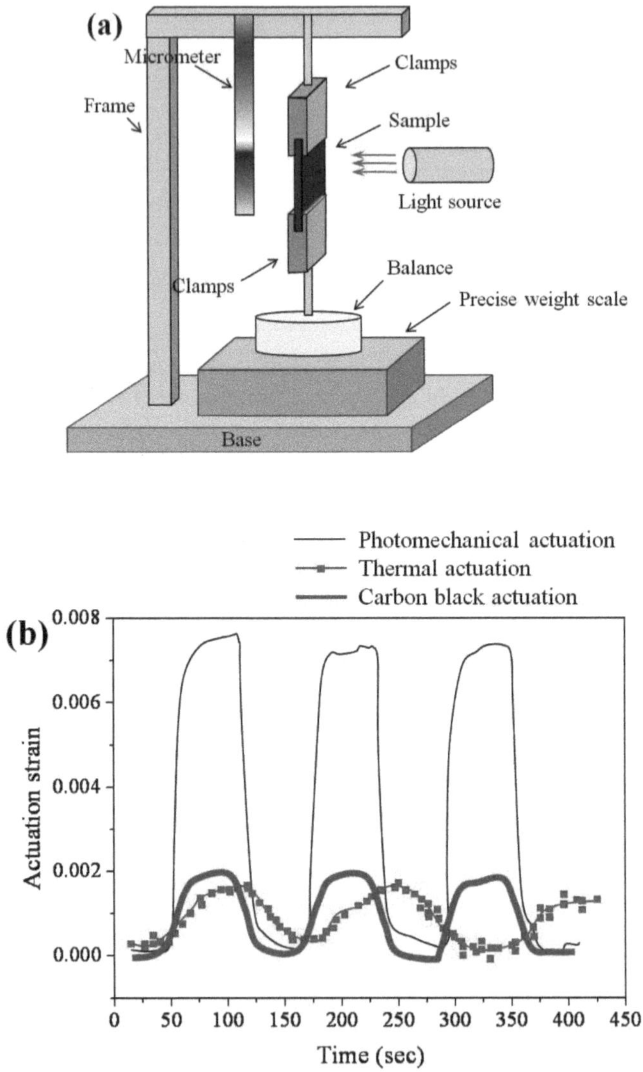

Fig. 5-69 (a) Schematic of experimental setup for characterizing the photomechanical actuation and (b) comparison of photomechanical actuation in single-walled carbon nanotube/polydimethylsiloxane multilayeractuators (solid line) with thermal actuation of the same structure (j) and the photomechanical actuation of a carbon black/polydimethylsiloxane multilayer actuator (N).

The introduction of carbon nanotubes in polydimethylsiloxane resulted in a composite actuator exhibiting better performance than those containing carbon black; this result was due to the high electrical conductivity of the nanotubes (Fig. 5-69 (b)). Further studies revealed that the photomechanical response of carbon nanotube/polymer nanocomposites was an intrinsic behavior of carbon nanotubes, depending on the carbon nanotube alignment in the polymer matrix [203]. The incorporation of aligned carbon nanotube thin films in an acrylic elastomer allowed stable expansive actuation of 0.3 percent at a light illumination below 120 mW with an actuation stroke large enough to manipulate small objects [204]. Clearly, polymer/carbon nanotube nanocomposites represent a new class of multifunctional materials that can be employed in photonic devices with improved performance.

Field emission displays (FEDs) are a very promising substitute for conventional liquid crystal displays (LCDs). While having the best available image quality, similar to that of cathode ray tube (CRT) displays, FEDs also offer a superior viewing angle (160°) and faster response time as compared to liquid crystal displays. It was realized in the late 1990s that materials other than carbon cannot be used as stable field emitters (FEs). Among different forms of carbon, carbon nanotubes are the most promising material for use in field emission displays [205].

A free-standing carbon nanotube sheet is highly transparent (70 percent transmission for polarization parallel to nanotube orientation and 90 percent for perpendicular polarization at 450 nm) and has the ability to polarize transmitted light (Fig. 5-70).

Multiwalled carbon nanotube sheets exhibit hysteretic behavior; the area of the hysteresis loop increases with increasing inter-electrode distance. Such behavior is not usually observed for stable field emission cathodes, unless there is degradation (Fig. 5-71). However, multiwalled carbon nanotube sheets

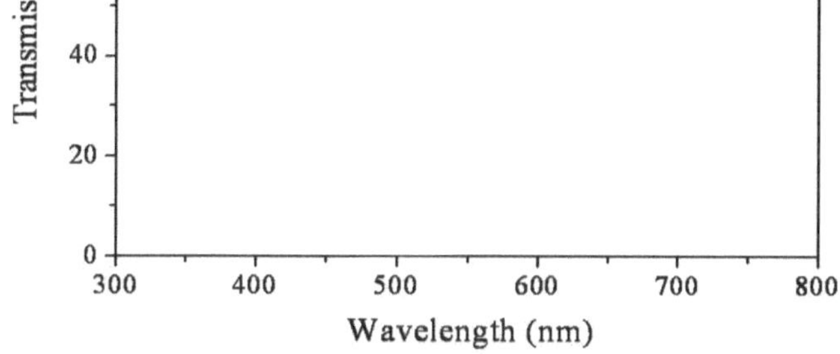

Fig. 5-70 Optical transparency of carbon nanotube sheet for two different polarizations, parallel (ll) and perpendicular (⊥) to nanotube orientation.

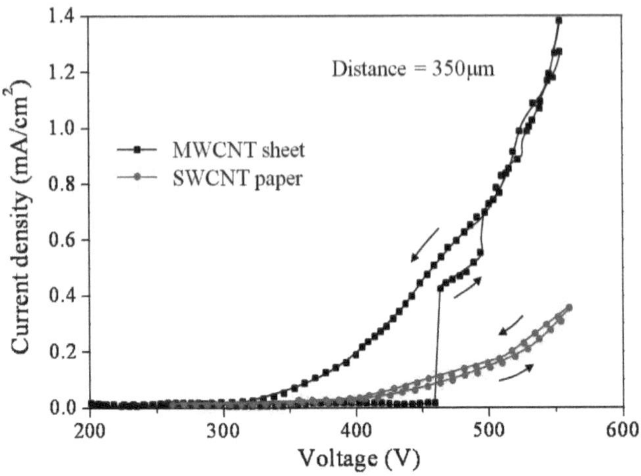

Fig. 5-71 Hysteretic and step-like field emission from multi-walled carbon nanotube (MWCNT) (solid squares) as compared to smooth field emission from single-walled carbon nanotube (SWCNT) sheets (open circles) for 350 lm inter-electrode distance.

hysteresis loops are reproducible and appear because of reversible structural changes. When applied voltage increases and the electric field between cathode and anode reaches a certain 'activation' value, the free ends of some nanotubes get detached from the carbon nanotube bundles and rise. Because the threshold field for a nanotube tip is a few times higher than for its side [206], a steep increase in the current is observed at this point. A step-like increase in the current at larger distances shows that in the beginning only a part of the nanotubes rise and at increasing voltages other tubes do (Fig. 5-71).

The first generation of flexible electrodes was based on well-established techniques used to make transparent electrodes on glass but with much lower peak processing temperatures. On glass, transparent conducting oxides (TCOs), such as indium tin oxide (ITO), tin oxide, and zinc oxide (ZnO), are sputtered at elevated temperatures (300–600 °C). ITO combines excellent optical transparency (>90 %) and low electrical resistivity ($\sim 10^{-4}\ \Omega$ cm) [207].

The visible light transmittance of indium tin oxide (ITO), poly(3,4-ethylenedioxythiophene) (PEDOT), and carbon nanotube (CNT) films is shown in Fig. 5-72. Carbon nanotube film displays high transparency across the complete visible light spectrum. In comparison, indium tin oxide has a maximum transparency in the range of 500–550 nm, at the expense of significantly lower transparency at other wavelengths. For the same level of conductivity, current carbon nanotube films show slightly lower transparency at 550 nm compared to the peak transmittance of indium tin oxide. However, carbon nanotube films exhibit significantly higher transparency across the whole visible light

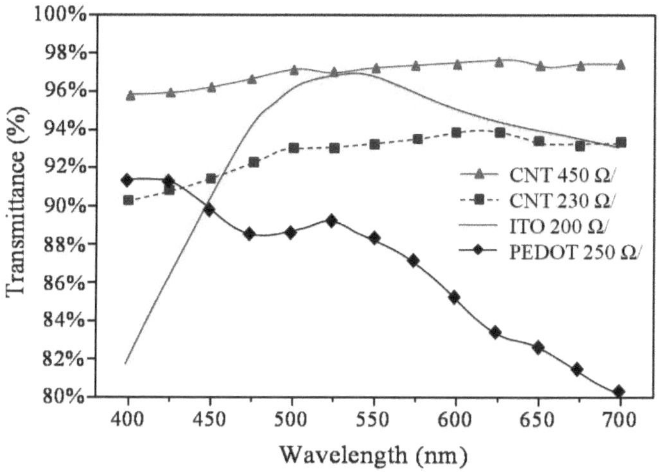

Fig. 5-72 Optical performance: visible light transmittance of indium tin oxide (ITO), Poly(3,4-ethylenedioxythiophene) (PEDOT) and carbon nanotube (CNT) films.

spectrum. Carbon nanotube films are much more transparent than poly(3,4-ethylenedioxythiophene) [208].

5.7. THE VERSATILE PROPERTIES OF GRAPHENE

Potential applications of graphene were discussed in Ref. [209] and, during the last 2 years, significant progress has been made along many of the lines started there. The major difference between now and then is the advent of mass production technologies for graphene. This has dramatically changed the whole landscape by making the subject of applications less speculative and allowing the development of new concepts unimaginable earlier (Fig. 5-73).

(1) **Transistors**. Graphene can be used to make excellent transistors. Graphene is so thin that we can easily control whether or not it conducts by applying an electric field. We would like to be able to do this with metals, but we cannot make metal films thin enough to affect their conducting state in this way. Electrons in graphene also travel ballistically over sub-micron distances. As a result, graphene-based transistors can run at higher frequencies and more efficiently that the silicon transistors we use now. Fig. 5-74 shows a graphene-based transistor device.
(2) **Gas Sensors**. Gas molecules that land on graphene affect its electronic properties in a measurable way; in fact, we have measured the effect of a single molecule associating with a graphene. This means that we can create gas sensors that are sensitive to a single atom or molecule.

Fig. 5-73 A graphene-based Hall bar device with gold contacts lying on an Si/SiO_2 substrate. The conducting state of the device can be affected by applying an electric field to the Si.

(3) **Support Membranes for Transmission Electron Microscopy**. Graphene is effectively the thinnest material that we can make out of atoms. Suprisingly, it is also very strong, thanks to a lack of crystal boundaries to break along and very strong bonds between carbon atoms (Carbon nanotubes are made from rolled up graphene, and it has been suggested that cabling made from nanotubes would be strong enough to create an elevator into space). As a result we can use it to hold any micro- and nanoscopic objects we wish to look at in an electron microscope (e.g., DNA, nanoparticles) similar to how we use glass slides in optical microscopes. Graphene is the perfect material for this job as it is made only of carbon; it is very thin, so it will not interfere with the pictures taken as much as other materials would; also, graphene has a very simple crystal structure, so it can easily be eliminated from diffraction patterns.

(4) **Inert Coatings**. Graphene is resistant to attacks by many powerful acids and alkalis such as hydrofluoric acid and ammonia, so one day it could be used to give objects an atomically thin protective coating that would provide protection against these agents.

Most of the buzz is currently around graphene's long-term prospects in computer electronics. Immediate, but often mundane, applications are the least discussed and remain unnoticed even within parts of the graphene community. An extreme example of the former is an idea about graphene becoming the base electronic material 'beyond the Si age.'

Although this possibility cannot be ruled out, it is so far beyond the horizon that it cannot be assessed accurately either. At the very least, graphene-based integrated circuits require the conducting channel to be completely closed in the off state. Several schemes have been proposed to deal with graphene's gapless spectrum and, recently, nanoribbon transistors with large on-off current ratios at room temperature were demonstrated [210, 211] (Fig. 5-74 (A)). Nevertheless, the prospect of 'graphenium inside' remains as distant as ever. This is not because of graphene shortfalls but rather because experimental tools to define

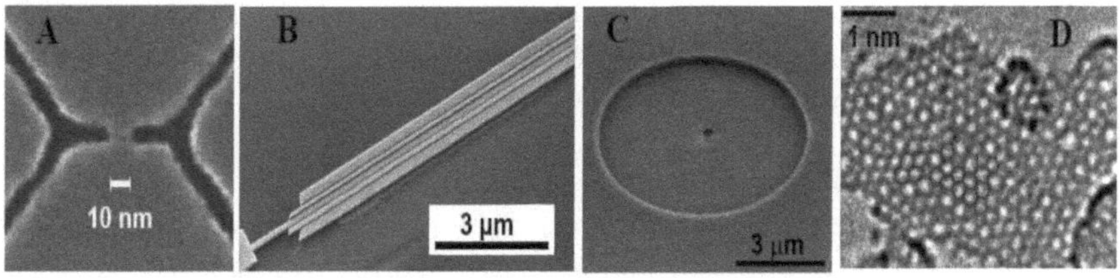

Fig. 5-74 (A) Graphene nanoribbons of sub-10-nm scale exhibit the transistor action with large on-off ratios [210, 211]. (B) All the fundamentals are in place to make graphene-based high electron mobility transistorss [212]. (C) Graphene-based normal electron mobility transistors. Shown is a drum resonator made from a 10-nmthick film of reduced graphene oxide, which covers a recess in a Si wafer [213]. (D) Ready to use: graphene membranes provide an ideal support for TEM [214].

structures with atomic precision are lacking. More efforts in this direction are needed but the progress is expected to be painstakingly slow and depend on technological developments outside the research area.

An example to the contrary is the use of graphene in transmission electron microscopy (TEM). It is a tiny niche application but it is real. Single-crystal, one atom thick and low atomic mass membranes provide the best imaginable support for atomic resolution TEM. With micron-sized crystallites now available in solution [215] for their cheap and easy deposition on standard grids and with films transferrable from metals [216, 217] onto such grids, graphene membranes are destined to become a routine TEM accessory (Fig. 5-74 (D)).

The space between graphene dreams and immediate reality is packed with applications. The one that is neither grand nor mundane is individual ultra-high frequency analog transistors (Fig. 5-74 (B)). This area is currently dominated by GaAs-based devices known as high electron mobility transistors, which are widely used in communication technologies. Graphene offers a possibility to extend high electron mobility transistors's operational range into THz frequencies. The fundamentals allowing this are well known: graphene exhibits room-T ballistic transport such that the charge transit between source and drain contacts takes only 0.1 ps for a typical channel length of 100 nm. Also, gate electrodes can be placed as close as several nm above graphene, which allows shorter channels and even quicker transit. Although graphene's gapless spectrum leads to low on-off ratios of 10 to 100, they are considered sufficient for the analog electronics. The progress toward graphene high electron mobility transistors is hindered by experimental difficulties in accessing the microwave range. Only recently, the first frequency tests of graphene transistors were reported [212]. Long channels and low mobility in these experiments limited the cut-off frequencies to less than 30 GHz [212], well below the operational range of GaAs-based high electron mobility transistors. However, the observed scaling of the operational frequency as a function of the channel length and μ indicates that the

THz range is accessible [212]. With graphene wafers in sight, these efforts are going to intensify, and high electron mobility transistors and other ultrahigh-frequency devices such as switches and rectifiers have a realistic chance to reach the market.

Graphene powders, consisting of ultrathin two-dimensional (2D) graphene sheets, have a flexible porous structure mainly with mesopores and macropores, in addition to their high electrical conductivity, superior mechanical properties, good electrochemical stability, and high surface area [218–226]. Therefore, graphene is a good candidate for the capacitive electrode since it can provide a large accessible surface area for fast transport of hydrate ions to achieve high double-layer capacitance in aqueous electrolytes [227, 228]. In addition, the superior electrical conductivity of graphene makes nanostructured MnO_2/graphene composites promising for the use as Faradic electrode in asymmetric electrochemical capacitors. More importantly, the presence of nanostructured MnO_2 is able to efficiently prevent the aggregation of graphene sheets caused by van der Waals interactions, consequently leading to an increase in the available electrochemical active surface area and a suitable porous structure for energy storage. Therefore, asymmetric electrochemical capacitors constructed from MnO_2/graphene composites as positive electrode and graphene as negative electrode are expected to exhibit high voltage, high energy, and power densities in neutral aqueous electrolytes.

Recently, Cheng *et al.* [225] report the preparation of MnO_2 nanowire/graphene composites by solution-phase assembly of graphene sheets and

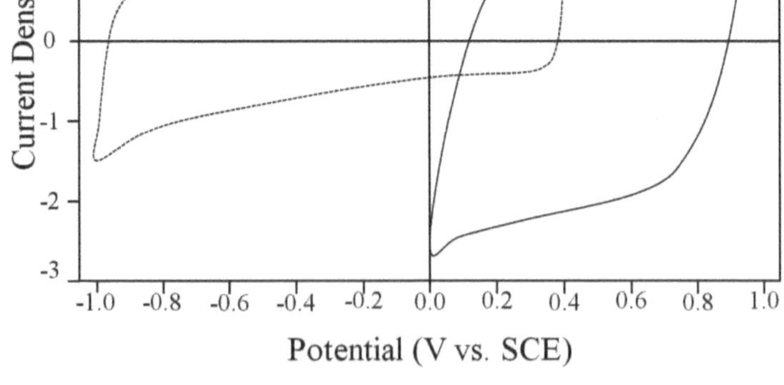

Fig. 5-75 CV curves of graphene and mangan-graphene composite (MGC) electrodes performed in a three-electrode cell in a 1 M Na_2SO_4 solution at a scan rate of 10 mV/s.

MnO_2 nanowires and the assembly of asymmetric electrochemical capacitors based on nanowire/graphene composites as positive electrode and graphene as negative electrode (see Fig. 5-75). The cyclic voltammetric curve of graphene electrode exhibits a nearly ideal rectangular shape, and no peaks of oxidation and reduction are observed, indicating a typical characteristic of electric double layer capacitor behavior (see Fig. 5-76) [221]. The cyclic voltammetry shape of

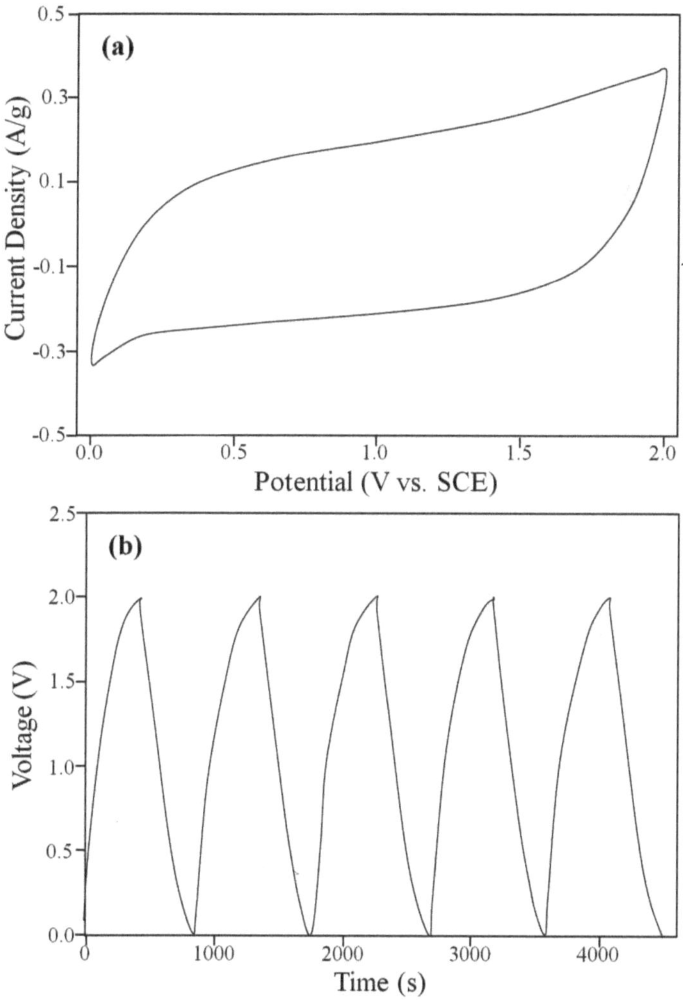

Fig. 5-76 (a) CV curve of the mangan-graphene composite (MGC) asymmetric electrochemical capacitors with a voltage of 2 V measured at a scan rate of 10 mV/s. (b) Galvanostatic charge-discharge voltage profile of the mangan-graphene composite (MGC) asymmetric electrochemical capacitors at a current density of 200 mA/g.

the nanowire/graphene composites electrode is very different from that of the graphene electric double layer capacitor and exhibits nearly a mirror-image current response on voltage reversal, indicating a good reversibility. The cyclic voltammetry and galvanostatic charge-discharge measurements were also used to evaluate the stable electrochemical windows of the asymmetric cell. Fig. 5-76 (a) shows a typical cyclic voltammetry of the asymmetric electrochemical capacitor with a voltage of 2.0 V at a scan rate of 10 mV/s. It is interesting to note that the asymmetric electrochemical capacitors show ideal capacitor behavior with a nearly rectangular cyclic voltammetry shape in a neutral aqueous electrolyte. From the charge-discharge curve of the asymmetric electrochemical capacitors (Fig. 5-76 (b)), it can be seen that the potentials of charge-discharge lines are nearly proportional to the charge or discharge time in the Na_2SO_4 electrolyte, indicating a rapid current-voltage response.

5.8. SUMMARY

Many scientific advances in surface science have affected a myriad of applications, including emerging and rapidly expanding fields such as microsystems and nanosystems. Surface modification in such small-scale systems can be a powerful tool due to the high specific surface area to volume ratio in these devices providing a direct method for manipulating transport and reaction phenomena within confined spaces by systematically varying surface properties. A few of these applications with relevance to problems of public health and welfare directly affecting modern societal needs such as energy storage and generation, environmental techniques, biomedical, cosmetics, electromagnetic interference shielding, transparent electrode, sensing, and actuation have been discussed in this chapter.

Interface science to develop in the future are very broad because it is not confined to any type of material and building up of hierarchical structures and functions is a challenge depending also on the types of materials. For example, graphene which is thinnest materials in the universe are especially promising because the high electrical properties and the interplay of their weak interactions enable a variety of hierarchical structures.

REFERENCES

[1] H. Möhwald, Colloid Polymer Science 288 (2010) 123.
[2] C.B. Duke, Proceedings of the National Academic Science 100 (2003) 3858.
[3] D. Menzel, Science 295 (2002) 58.
[4] M. Grätzel, Nature 414 (2001) 338.
[5] W.U. Huynh, J.J. Dittmer, A.P. Alivisatos, Science 295 (2002) 2425.
[6] B. O'Regan, M. Grätzel, Nature 353 (1991) 737.
[7] M. Grätze, Nature 403 (2000) 363.
[8] L.E. Greene, M. Law, D.H. Tan, M. Montano, J. Goldberger, G. Somorjai, P. Yang, Nano Letters 5 (2005) 1231.

[9] B. Pradhan, S.K. Batabyal, A.J. Pal, Solar Energy Materials and Solar Cells 91 (2007) 769.
[10] S.J. Park, M.K. Seo, S. Kim, Next-Generation Electrolytes for Li Batteries, in High Energy Density Lithium Batteries: Materials, Engineering Applications John Wiley & Sons, London (2010).
[11] J. Britz, W.H. Meyer, G. Wegner, Macromolecules 40 (2007) 7558.
[12] F.B. Dias, L. Plomp, J.B.J. Veldhuis, Power Sources 88 (2000) 169.
[13] D.E. Fenton, J.M. Parker, P.V. Wright, Polymer 14 (1973) 589.
[14] M. Yoshizawa, W. Ogihara, H. Ohno, Electrochemical Solid-State Letters 4 (2001) E25.
[15] C. Berhier, W. Gorecki, M. Minier, M.B. Armand, J.M. Chabagno, P. Riquaud, Solid State Ionics 11 (1983) 91.
[16] S. Kim, E.J. Hwang, Y. Jung, M. Han, S.J. Park, Colloids and Surfaces A 313-314 (2008) 216.
[17] S. Kim, S.J. Park, Solid State Ionics 178 (2007) 973.
[18] S. Kim, S.J. Park, Electrochimica Acta 52 (2007) 3477.
[19] S. Kim, E.J. Hwang, S.J. Park, Current Applied Physics 8 (2008) 729.
[20] B.E. Conway, Electrochemical Supercapacitors. Kluwer Academic/Plenum Publishers, New York (1999).
[21] M.K. Seo, S.J. Park, Current Applied Physics 10 (2010) 241.
[22] M.K. Seo, S.J. Park, Materials Science and Engineering B 164 (2009) 106.
[23] K.W. Nam, C.W. Lee, X.Q. Yang, B.W. Cho, W.S. Yoon, K.B. Kim, J Power Sources 188 (2009) 323.
[24] C.C. Hu, K.H. Chang, M.C. Lin, Y.T. Wu, Nano Letters 6 (2006) 2690.
[25] Y.H. Kim, S.J. Park, Current Applied Physics in press (2010).
[26] A. Feaver, G.Z. Cao, Carbon 44 (2006) 590.
[27] B.B. Garcia, A. Feaver, G.F. Zhang, R. Champion, G.Z. Cao, T.T. Fisher, K.P. Nagle, G.T. Seidler, Journal of Applied Physics 104 (2008) 1.
[28] S. Sepehri, B.B. García, Q. Zhang, G. Cao, Carbon 47 (2009) 1436.
[29] J. Chmiola, G. Yushin, Y. Gogotsi, C. Portet, P. Simon, P.L. Taberna, Science 313 (2006) 1760.
[30] G.C. Koltsakis, A.M. Stamatelos, Progress in Energy Combustion Science 23 (1997) 1.
[31] J.T. Houghton, Climate Change 1995: The Science of Climate Change. Cambridge University Press, Cambridge (1996).
[32] S.J. Park, J.M. Park, M.K. Seo, Current Applied Physics 10 (2010) 1009.
[33] A.C. Stern, R.W. Boubel, D.B. Turner, D.L. Fox, Fundamental of Air Pollution, 2nd edit. Academic Press, Orlando (1984).
[34] B. Ramachandran, R.G. Hermann, S. Choi, H.G. Stenger, C.E. Lyman, J.W. Sale, Catalysis Today 55 (2000) 281.
[35] J.W. Erisman, P. Grennfelt, M. Sutton, Environmental International 29 (2003) 311.
[36] V. Vestreng, E. Stbren, Analysis of the UNECE/EMEP Emission Data. MSC-W Status Report 2000. Norwegian Meteorological Institute, Oslo (2000).
[37] S. Matsumoto, Catalysis Today 29 (1996) 43.
[38] K. Kato, H. Nohira, K. Nakanishi, S. Igushi, T. Kihara, H. Nuraki, European Patent, 0 573 672 A1 (1993).
[39] S. Matsumoto, Catalysis Today 4 (2000) 102.
[40] N. Takahasi, H. Shinjoh, T. Ijima, T. Suzuki, T. Yamakazi, K. Yokota, Catalysis Today 27 (1996) 63.
[41] K. Kaneko, Langmuir 3 (1987) 357.
[42] R.Q. Long, R.T. Yang, Industrial Engineering Chemistry Research 40 (2001) 4288.
[43] S.J. Park, B.J. Kim, Journal of Colloid and Interface Science 282 (2005) 124.

[44] I. Hong, H. Jiang, Y.D. Park, J.Y. Kim, B.H. Ha, Chemical Physics Letters 366 (2002) 572.
[45] Z. Zhu, Z. Liu, S. Liu, H. Niu, Applied Catalysis B Environmental 30 (2001) 267.
[46] S.J. Park, J.S. Shin, Journal of Colloid and Interface Science 264 (2003) 39.
[47] J.I. Miyamoto, H. Kanoh, K. Kaneko, Carbon 43 (2005) 855.
[48] S.J. Park, B.J. Kim, Journal of Colloid and Interface Science 292 (2005) 493.
[49] M.A. Shannon, P.W. Bohn, M. Elimelech, J.G. Georgiadis, M.J. Marinas, A.M. Mayes, Nature 452 (2008) 301.
[50] S. Prakash, A. Piruska, E.N. Gatimu, P.W. Bohn, J.V. Sweedler, M.A. Shannon, IEEE Sensors Journal 8 (2008) 441.
[51] A. Asatekin, S. Kang, M. Elimelech, A.M. Mayes, Journal of Membrane Science 298 (2007) 136.
[52] K.E. Noll, V. Gounaris, W.S. Hou, Adsorption Technology for Air Water Pollution Control. Lewis, Michigan (1992).
[53] Y. Urabe, T. Ishikura, K. Kaneko, Journal of Colloid and Interface Science 319 (2008) 381.
[54] B.J. Kim, S.J. Park, Journal of Colloid and Interface Science 325 (2008) 297.
[55] G.P. Rao, C. Lu, F. Su, Separation Purification Technolology 58 (2007) 224.
[56] R.A. Khaydarov, R.R. Khaydarov, O. Gapurova, Water Resources 44 (2010) 1927.
[57] R.K. Agarwal, J.A. Schwarz, Carbon 26 (1988) 873.
[58] S.Y. Zhand, O. Talu, D.T. Hayhurst, Journal of Physical Chemistry 95 (1991) 1722.
[59] K.A.G. Amankwah, J.S. Noh, J.A. Schwarz, International Journal of Hydrogen Energy 14 (1989) 437.
[60] S. Hynek, W. Fuller, J. Bentley, International Journal of Hydrogen Energy 22 (1997) 601.
[61] P. Malbrunot, D. Vidal, J. Vermesse, R. Chahine, T.K. Bose, Langmuir 8 (1992) 577.
[62] C.W. Skarstrom, N. Ann, Academic Science 72 (1959) 751.
[63] T. Panczyk, Physical Chemistry Chemical Physics 8 (2006) 3782.
[64] C.A. Ward, Journal of Chemical Physics 67 (1977) 229.
[65] C.A. Ward, R.D. Findlay, M. Rizk, Journal of Chemical Physics 76 (1982) 5599.
[66] S. Azizian, H. Bashiri, H. Iloukhani, Journal of Physical Chemistry C 112 (2008) 10251.
[67] R.F. Service, Science 305 (2004) 958.
[68] J. Dong, X. Wang, H. Xu, Q. Zhao, J. Li, International Journal of Hydrogen Energy 32 (2007) 4998.
[69] G.L. Soloveichik, M. Andrus, Y. Gao, J.C. Zhao, S. Kniajanski, International Journal of Hydrogen Energy 34 (2009) 2144.
[70] A.C. Dillon, K.M. Jones, T.A. Bwkkedahl, C.H. Kiang, D.S. Bethune, M.J. Heben, Nature 386 (1997) 377.
[71] R.T. Yang, Carbon 38 (2000) 623.
[72] B.J. Kim, Y.S. Lee, S.J. Park, International Journal of Hydrogen Energy 33 (2008) 4112.
[73] B.J. Kim, Y.S. Lee, S.J. Park, International Journal of Hydrogen Energy 33 (2008) 2254.
[74] S.J. Park, B.J. Kim, Y.S. Lee, M.S. Cho, International Journal of Hydrogen Energy 33 (2008) 1706.
[75] Y. Liu, H. Kabbour, C.M. Brown, D.A. Neumann, C.C. Ahn, Langmuir 24 (2008) 4772.
[76] G. Antek, W.F. Adam, J. Matzger, O.M. Yaghi, Journal of American Chemical Society 128 (2006) 1304.
[77] H. Tagaki, H. Hitory, H. Soneda, N. Yoshizawa, Y. Yamade, Materials Science and Engineering B 108 (2004) 143.
[78] P.A. Gordon, R.B. Saeger, Industrial and Engineering Chemistry Research 38 (1999) 4647.
[79] H. Liu, Y.Y. Fan, M. Liu, H.T. Cong, H.M. Cheng, M.S. Dresselhaus, Science 286 (1999) 1127.
[80] H. Cheng, C. Liu, Q. Yang, Carbon 39 (2001) 1147.

[81] J. Alleman, A. Dillon, T. Gennett, K. Jones, P.Parilla, Carbon Nanotubes Materials for Hydrogen Storage. Proceeding of 2000 DOE/NREL Hydrogen Programme Review, USA (2000).
[82] S.Y. Lee, S.J. Park, International Journal of Hydrogen Energy 35 (2010) 6757.
[83] S.J. Park, S.Y. Lee, International Journal of Hydrogen Energy in press (2010).
[84] A.L.M. Reddy, S. Ramaprabhu, International Journal of Hydrogen Energy 33 (2008) 1028.
[85] A. Chambers, C. Park, R. Terry, R.T.K. Baker, N.M. Rodriguez, Journal of Physical Chemistry B 102 (1998) 4253.
[86] F. Chen, W. Li, R. Loutfy, R. Murphy, J. Wang, Hydrogen Storage in Fullerenes and in an Organic Hydride. Proceeding of US DOE Hydrogen Programme Review, USA (1998).
[87] S.J. Park, S.Y. Lee, International Journal of Hydrogen Energy 35 (2010) 13048.
[88] G.J. Kubas, R.R. Ryan, B.I. Swanson, P.J. Vergamini, J.J. Wasserman, Journal of American Chemical Society 106 (1984) 451.
[89] M.G. Plaza, C. Pevida, B. Arias, J. Fermoso, A. Arenillas, F. Rubiera, J.J. Pis, Journal of Thermal Analysis and Calorimetry 92 (2008) 601.
[90] A. Arenillas, K.M. Smith, T.C. Drage, C.E. Snape, Fuel 84 (2005) 2204.
[91] X. Xu, C. Song, B.G. Miller, A.W. Scaroni, Fuel Processing Technology 86 (2005) 1457.
[92] P. Tontiwachwuthikul, A. Meisen, C.J. Lim, Journal of Chemical and Engineering Data 36 (1991) 130.
[93] M.L. Gray, Y. Soong, K.J. Champagne, J. Baltrus, R.W. Stevens Jr, P. Toochinda, S.S.C. Chuang, Separation and Purification Technolology 35 (2004) 31.
[94] M.G. Plaza, C. Pevida, B. Arias, M.D. Casal, C.F. Martín, J. Fermoso, F. Rubiera, J.J. Pis, Journal of Environmental Engineering 135 (2009) 426.
[95] E. Díaz, E. Muñoz, A. Vega, S. Ordóñez, Journal of Industrial Engineering Chemistry 47 (2008) 412.
[96] M.G. Plaza, C. Pevida, B. Arias, J. Fermoso, F. Rubiera, J.J. Pis, Energy Procedia 1 (2009) 1107.
[97] B.J. Kim, K.S. Cho, S.J. Park, Journal of Colloid and Interface Science 342 (2010) 575.
[98] V. Zeleňá, M. Badaničováa, D. Halamovaa, J. Čejkab, A. Zukalb, N. Murafac, G. Goerigkd, Chemical Engineering Journal 144 (2008) 336.
[99] V. Zelenak, D. Halamowa, L. Gaberova, E. Bloch, P. Lewellyn, Microporous and Mesoporous Materials 116 (2008) 358.
[100] S. Araki, H. Doi, Y. Sano, S. Tanaka, Y. Miyake, Journal of Colloid and Interface Science 339 (2009) 382.
[101] S. Palzer, Trends in Food Science 20 (2009) 194.
[102] E. Acosta, Current Opinion in Colloid and Interface Science 14 (2009) 3.
[103] R. Wegmuller, M.B. Zimmermann, V.G. Buhr, E.J. Windhab, R.F. Hurrell, Journal of Food Science 71 (2006) S181.
[104] J. Ubbink, A. Schoonman, Flavor Delivery Systems. John Wiley & Sons, London (2003).
[105] A. Gharsallaoui, G. Roudauta, O. Chambina, A. Voilleya, R. Saurela, Food Research International 40 (2007) 1107.
[106] L. Sagalowicz, M.E. Leser, Current Opinion of Colloid and Interface Science 15 (2010) 61.
[107] D.J. McClements, E.A. Decker, Y. Park, J. Weiss, Critical Reviews in Food Science and Nutrition 49 (2009) 577.
[108] U. Lesmes, D.J. McClements, Trends in Food Science and Technolology 20 (2009) 448.
[109] P.P. Hoppe, G. Krennrich, European Journal of Nutrition 39 (2000) 183.
[110] Y. Serfert, S. Drusch, K. Schwarz, Food Chemistry 113 (2009) 1106.

[111] S. Ogawa, E.A. Decker, D.J. McClements, Journal of Agricultural Food and Chemistry 51 (2003) 2806.
[112] L.A. Shaw, D.J. McClements, E.A. Decker, Journal of Agricultural Food and Chemistry 55 (2007) 3112.
[113] W. Mehnert, K. Mäder, Advanced Drug Delivery Reviews 47 (2001) 165.
[114] H. Bunjes, B. Siekmann, Microencapsulation, 2nd edit. Marcel Dekker, New York (2006).
[115] H. Bunjes, F. Steiniger, W. Richter, Langmuir 23 (2007) 4005.
[116] E.B. Souto, R.H. Müller, International Journal of Cosmetic Science 30 (2008) 157.
[117] M.E. Davis, M.E. Brewster, Nature Reviews Drug Discovery 3 (2004) 1023.
[118] E. Semo, E. Kesselman, D. Danino, Y.D. Livney, Food Hydrocolloids 21 (2006) 936.
[119] H.C. Liu, W.L. Chen, S.J.T. Mao, Journal of Dairy Science 90 (2007) 547.
[120] N.J. Krog, Food emulsions. Marcel Dekker, New York (1997).
[121] J.N. Israelachvili, S. Marcelja, R.G. Horn, Quarterly Reviews of Biophysics 13 (1980) 121.
[122] B. Jönsson, B. Lindman, K. Holmberg, B. Kronberg, Surfactants and Polymers in Aqueous Solution. John Wiley & Sons, London (1998).
[123] E. Soussan, S. Cassel, M. Blanzat, I. Rico-Lattes, Angewandte Chemie International Edition 48 (2009) 274.
[124] B.M. Folmer, D. Baron, E. Hughes, L. Miguet, B. Sanchez, O. Heudi, Biochemical Pharmacology 78 (2009) 1464.
[125] M. Hamidi, A. Azadi, P. Rafiei, Advanced Drug Delivery Reviews 60 (2008) 1638.
[126] K.S. Kim, S.J. Park, Colloids Surface B 80 (2010) 240.
[127] T. Takahashi, M. Yamaguchi, Macrocycle Chemistry 10 (1991) 286.
[128] S.K. Sahoo, S. Parveen, J.J. Panda, Biology and Medicine 3 (2007) 20.
[129] G. Cevc, U. Vierl, Journal of Controlled Release 141 (2010) 277.
[130] S.J. Park, R. Arshady, Microcapsules for Fragrances and Cosmetics. InMicrocapsule Patents and ProductsCitus Books, London (2003).
[131] K. Bogunia-Kubik, M. Sugisaka, Biosystems 65 (2002) 123.
[132] R. Rossi, F. Basilico, G. Rossoni, A. Riva, P. Morazzoni, P.L. Mauri, Journal Pharmaceutical and Biomedical Analysis 50 (2009) 224.
[133] E. Bombardelli, A. Cristoni, P. Morazzoni, Fitoterapia 65 (1994) 387.
[134] S.P. Wanichwecharungruang, Nanotechnology 4 (2007) 257.
[135] H. Hidaka, H. Kobayashi, T. Koike, T. Sato, N. Serpone, Journal of Oleo Science 55 (2006) 249.
[136] A. Nasu, Y. Otsubo, Journal of Colloid and Interface Science 296 (2006) 558.
[137] P. Stamatakis, B.R. Palmer, G.C. Salzman, C.F. Bohren, T.B. Allen, Journal of Coatings Technology 62 (1990) 95.
[138] S.P. Palecek, J.C. Loftus, M.H. Ginsberg, D.A. Lauffenberger, A.F. Horwitz, Nature 385 (1997) 537.
[139] J.A. Varner, D.A. Cheresh, Current Opinion in Cell Biology 8 (1996) 724.
[140] M. Heginbothom, T.C. Fitzgerald, W.G. Wade, Journal of Clinical Pathology 43 (1990) 253.
[141] V. Roger, J. Tenovno, M. Lenander Lumikari, E. Sonderling, P. Vilja, Caries Research 28 (1994) 421.
[142] E. Genové, C. Shen, S. Zhang, C.E. Semino, Biomaterials 16 (2005) 3341.
[143] J.D. Hartgerink, E. Beniash, S.I. Stupp, Science 294 (2001) 1684.
[144] S.L. Gras, A.K. Tickler, A.M. Squires, G.L. Devlin, M.A. Horton, C.M. Dobson, C.E. MacPhee, Biomaterials 29 (2008) 1553.
[145] N. Erez, A. Bershadsky, B. Geiger, European Journal of Cell Biology 84 (2005) 235.

[146] A.K. Dillow, S.E. Ochsenhirt, J.B. McCarthy, G.B. Fields, M. Tirrell, Biomaterials 22 (2001) 493.
[147] M.D. Pierschbacher, K.M. Yamada, M.P. Sheetz, Journal of Cellular Biochemistry 28 (1985) 115.
[148] H. Storrie, O. Mustafa, N. Suha, A. Amara, T. Volberg, M. Rao, B. Geiger, S.I. Stupp, Biomaterials 28 (2007) 4608.
[149] A. Afaghi-Khatibi, G. Lawcock, L. Ye, Y. Mai, Computer Methods in Applied Mechanics and Engineering 185 (2000) 173.
[150] E.M. Castrodeza, F.L. Bastian, J.E.P. Ipina, Thin-Walled Structure 41 (2003) 1089.
[151] U.K. Vaidya, S. Pillay, S. Bartus, C.A. Ulven, D.T. Grow, B. Mathew, Materials Science and Engineering A 428 (2006) 59.
[152] A.H. Conttrell, The Mechanical Properties of Matter. John Wiley & Sons, New York (1964).
[153] S. Iijima, Nature 354 (1991) 56.
[154] M.S. Dresselhaus, G. Dresselhaus, R. Saito, Carbon 33 (1995) 883.
[155] J.N. Coleman, U. Khan, Y.K. Gunko, Advanced Materials 18 (2006) 689.
[156] D. Tasis, N. Tagmatarchis, A. Bianco, M. Prato, Chemical Reviews 106 (2006) 1105.
[157] P.C. Ma, M.Y. Liu, J.K. Kim, B.Z. Tang, ACS Applied Materials Interfaces 1 (2009) 1090.
[158] L.Q. Liu, H.D. Wagner, Composites Science and Technology 65 (2005) 1861.
[159] P.C. Ma, B.Z. Tang, J.K. Kim, Carbon 46 (2008) 1497.
[160] T.V. Kosmidou, A.S. Vatalis, C.G. Delides, Express Polymer Letters 264 (2008) 364.
[161] P.C. Ma, J.K. Kim, B.Z. Tang, Composites Science and Technology 67 (2007) 2965.
[162] S.R. Phillpot, A.J.H. McGaughey, Materials Today 8 (2005) 18.
[163] M.J. Biercuk, M.C. Llaguno, M. Radosavljevic, J.K. Hyun, A.T. Johnson, J.E. Fischer, Applied Physics Letters 80 (2002) 1767.
[164] L.E. Evseeva, S.A. Tanaeva, Mechanics of Composite Materials 44 (2008) 87.
[165] M. Moniruzzaman, K.I. Winey, Macromolecules 39 (2006) 194.
[166] P. Prentice, Rheology and its Role in Plastic Processing Rapra Review Report. Rapra Technology Ltd., Shawbury (1995).
[167] Y.Y. Huang, S.V. Ahir, E.M. Terentjev, Physical Reviews B 73 (2006) 125.
[168] Z. Fan, S.G. Advani, Journal of Rheology 51 (2007) 585.
[169] S. Abbasi, P.J. Carreau, A. Derdouri, M. Moan, Rheologica Acta 48 (2009) 943.
[170] C.A. Cooper, S.R. Cohen, A.H. Barber, H.D. Wagner, Applied Physics Letters 81 (2002) 3873.
[171] G.B. Pye, J.J. Beaudoin, Cement and Concrete Research 22 (1992) 551.
[172] M.K. Seo, S.J. Park, Bulletin of Korean Chemical Society 30 (2009) 124.
[173] M.K. Seo, S.J. Park, Macromolecular Materials and Engineering 289 (2004) 368.
[174] N. Li, Y. Huang, F. Du, X.B. He, X. Lin, H.J. Gao, Nano Letters 6 (2006) 1141.
[175] S.Y. Yang, K. Lozano, A. Lomeli, H.D. Foltz, R. Jones, Computer Part A 36 (2005) 691.
[176] H.M. Kim, Applied Physics Letters 84 (2004) 589.
[177] Z. Wang, Z.Y. Liang, B. Wang, C. Zhang, L. Kramer, Computer Part A 35 (2004) 1225.
[178] M.H. Al-Saleh, U. Sundararaj, Carbon 47 (2009) 1738.
[179] J.E. Riggs, D.B. Walker, D.L. Carroll, Y.P. Sun, Journal of Physical Chemistry B 104 (2000) 7071.
[180] B.Z. Tang, H.Y. Xu, Macromolecules 32 (1999) 2569.
[181] J. Wang, W.J. Blau, Journal of Physical Chemistry C 112 (2008) 2298.
[182] L. Zhang, S.D. Allen, C. Woelfle, F. Zhang, Journal of Physical Chemistry C 113 (2009) 13979.
[183] Y. Chen, Y. Lin, Y. Liu, Journal of Nanoscience and Nanotechnology 7 (2007) 1268.

[184] S.J. Henley, R.A. Hatton, G.Y. Chen, C. Gao, H. Zeng, H.W. Kroto, Small 3 (2007) 1927.
[185] I. Singh, P.K. Bhatnagar, P.C. Mathur, Carbon 46 (2008) 1141.
[186] N. He, Y. Chen, J. Bai, J. Wang, W.J. Blau, J. Zhu, Journal of Physical Chemistry C 113 (2009) 13029.
[187] Y. Li, C. Chen, S. Zhang, Y. Ni, J. Huang, Applied Surface Science 254 (2008) 5766.
[188] Y. Yang, M.C. Gupta, K.L. Dudley, R.W. Lawrence, Journal of Nanoscience and Nanotechnology 5 (2005) 927.
[189] B. Fugetsu, E. Sano, M. Sunada, Y. Sambongi, T. Shibuya, X. Wang, T. Hiraki, Carbon 46 (2008) 1256.
[190] H.M. Kim, K. Kim, C.Y. Lee, J. Joo, Applied Physics Letters 84 (2004) 589.
[191] N.C. Das, S. Yamazaki, M. Hikosaka, T.K. Chaki, D. Khastgir, A. Chakraborty, Polymer International 54 (2005) 256.
[192] E. Hammel, X. Tang, M. Trampert, T. Schmitt, K. Mauthner, A. Eder, P. Potschke, Carbon 42 (2004) 1153.
[193] J. Wang, W.J. Blau, Journal of Physical Chemistry C 112 (2008) 2298.
[194] L. Zhang, S.D. Allen, C. Woelfle, F. Zhang, Journal of Physical Chemistry C 113 (2009) 13979.
[195] Y. Chen, Y. Lin, Y. Liu, Journal of Nanoscience and Nanotechnology 7 (2007) 1268.
[196] S.J. Henley, R.A. Hatton, G.Y. Chen, C. Gao, H. Zeng, H.W. Kroto, Small 3 (2007) 1927.
[197] I. Singh, P.K. Bhatnagar, P.C. Mathur, Carbon 46 (2008) 1141.
[198] N. He, Y. Chen, J. Bai, J. Wang, W.J. Blau, J. Zhu, Journal of Physical Chemistry C 113 (2009) 13029.
[199] M. Baibarac, P. Gomez-Romero, Journal of Nanoscience and Nanotechnology 6 (2006) 1.
[200] J.Y. Kim, M. Kim, H. Kim, J. Joo, J.H. Choi, Optical Materials 21 (2003) 147.
[201] Y. Zhang, S. Ijima, Physical Reviews Letters 82 (1999) 3472.
[202] B.J. Landi, R.P. Raffaelle, M.J. Heben, J.L. Alleman, W. VanDerveer, T. Gennett, Nano Letters 2 (2002) 1329.
[203] S.V. Ahir, E.M. Terentjev, Nature Materials 4 (2005) 491.
[204] S. Lu, B. Panchapakesan, Nanotechnology 16 (2005) 2548.
[205] D. Talin, K. Dean, J. Jaskie, Solid State Electronics 45 (2001) 963.
[206] A.A. Kuznetzov, S.B. Lee, M. Zhang, R.H. Baughman, A.A. Zakhidov, Carbon 48 (2010) 41.
[207] D.S. Ginley, C. Bright, Materials Research Bulletin 25 (2000) 15.
[208] K.A. Sierros, D.S. Hecht, D.A. Banerjee, N.J. Morris, L. Hu, Thin Solid Films 518 (2010) 6977.
[209] A.K. Geim, K.S. Novoselov, Nature Materials 6 (2007) 183.
[210] L.A. Ponomarenko, F. Schedin, M.I. Katsnelson, R. Yang, E.W. Hill, K.S. Novoselov, A.K. Geim, Science 320 (2008) 356.
[211] X. Li, X. Wang, L. Zhang, S. Lee, H. Dai, Science 319 (2008) 1229.
[212] J.T. Robinson, M. Zalalutdinov, J.W. Baldwin, E.S. Snow, Z. Wei, P. Sheehan, B.H. Houston, Nano Letters 8 (2008) 3441.
[213] T.J. Booth, P. Blake, R.R. Nair, D. Jiang, E.W. Hill, U. Bangert, A. Bleloch, M. Gass, K.S. Novoselov, M.I. Katsnelson, A.K. Geim, Nano Letters 8 (2008) 2442.
[214] Y. Hernandez, V. Nicolosi, M. Lotya, F.M. Blighe, Z. Sun, S. De, I.T. Mcgovern, B. Holland, M. Byrne, Y.K. GuńKo, J.J. Boland, P. Niraj, G. Duesberg, S. Krishnamurthy, R. Goodhue, J. Hutchinson, V. Scardi, A.C. Ferrari, J.N. Coleman, Nanotecnology 3 (2008) 563.

[215] A. Reina, X. Jia, J. Ho, D. Nezich, H. Son, V. Bulovic, M.S. Dresselhaus, J. Kong, Nano Letters 9 (2009) 30.
[216] K.S. Kim, Y. Zhao, H. Jang, S.Y. Lee, J.M. Kim, K.S. Kim, J.H. Ahn, P. Kim, J.Y. Choi, B.H. Hong, Nature 457 (2009) 706.
[217] Y.M. Lin, K.A. Jenkins, A. Valdes-Garcia, J.P. Small, D.B. Farmer, P. Avouris, Nano Letters 9 (2009) 422.
[218] M.M. Gudarzi, F. Sharif, Journal of Colloid and Interface Science 349 (2010) 63.
[219] V.K. Singh, M.K. Patra, M. Manoth, G.S. Gowd, S.R. Vadera, N. Kumar, New Carbon Materials 24 (2009) 147.
[220] A.K. Geim, Science 324 (2009) 1530.
[221] M.D. Stoller, S. Park, Y. Zhu, J. An, R.S. Ruoff, Nano Letters 8 (2008) 3498.
[222] Y. Wang, Z.Q. Shi, Y. Huang, Y.F. Ma, C.Y. Wang, M.M. Chen, Y.S. Chen, Journal of Physical Chemistry C 113 (2009) 13103.
[223] Z.S. Wu, W.C. Ren, L.B. Gao, J.P. Zhao, Z.P. Chen, B.L. Liu, D.M. Tang, B. Yu, C.B. Jiang, H. M. Cheng, ACS Nano 3 (2009) 411.
[224] D.W. Wang, F. Li, J. Zhao, W. Ren, Z.G. Chen, J. Tan, Z.S. Wu, I. Gentle, G.Q. Lu, H.M. Cheng, ACS Nano 3 (2009) 1745.
[225] D.W. Wang, F. Li, Z.S. Wu, W. Ren, H.M. Cheng, Electrochemistry Communications 11 (2009) 1729.
[226] L.H. Tang, Y. Wang, Y.M. Li, H.B. Feng, J. Lu, J.H. Li, Advanced Functional Materials 19 (2009) 2782.
[227] W. Lv, D.M. Tang, Y.B. He, C.H. You, Z.Q. Shi, X.C. Chen, C.M. Chen, P.X. Hou, C. Liu, Q.H. Yang, ACS Nano 3 (2009) 3730.
[228] L.L. Zhang, R. Zhou, X.S. Zhao, Journal of Material Chemistry 20 (2010) 5983.

Chapter 6

Element and Processing

6.1. INTRODUCTION

Composite materials are as much an engineering concept as they are specific materials. They represent an idea, which to the best of my knowledge was first recognized in engineering in the 'composite' construction of the latter Clipper ships in the middle of the nineteenth century where wooden planking was set on iron frames. The modern term composites arose in the aircraft industry in the middle part of the twentieth century. Their advent can be viewed either as a means of using plastics in aircraft construction or as a means of utilizing the stiffness and the strength of fibers, which were first made in the late 1950s and early 1960s. The first of these views emphasized the desirable properties of plastics, which were becoming available in the late 1930s. The modern synthetic plastics were first discovered in the early 1900s. The desirable properties were the ease of formabilityvery low density compared with metals, excellent surface finish, and lack of corrosion. They lacked stiffness and so some stiffening agent was necessary. The leading idea was to use plastics for aircraft construction.

Composite materials are formed by combining two or more materials that have quite different properties. One material (the matrix or binder) surrounds and binds together a cluster of fibers or fragments of a much stronger material (the reinforcement) [1–5]. The different materials work together to give the composite unique properties but within the composites we can easily tell the different materials apart—they do not dissolve or blend into each other. Composites exist in nature. A piece of wood is a composite with long fibers of cellulose (a very complex form of starch) held together by a much weaker substance called lignin. Cellulose is also found in cotton and linen but it is the binding power of the lignin that makes a piece of timber much stronger than a bundle of cotton fibers. Likewise, humans have been using composite materials for thousands of years.

Over recent decades many new composites have been developed, some with very valuable properties. By carefully choosing the reinforcement, the matrix, and the manufacturing process that brings them together, engineers can tailor the

Interface Science and Composites, Volume 18
ISSN: 1573-4285, doi: 10.1016/B978-0-12-375049-5.00006-2

properties to meet specific requirements. They can, for example, make the composite sheet very strong in one direction by aligning the fibers that way but weaker in another direction where strength is not so important. They can also select properties, such as resistance to heat, chemicals, and weathering by choosing appropriate matrix materials.

For the matrix, many modern composites use thermosetting polymers or thermoplastics. The plastics are polymers that hold the reinforcement together and help to determine the physical properties of the end product. Ceramics, carbon, and metals are also used as the matrix for some highly specialized purposes. For example, ceramics are used when the material is going to be exposed to high temperatures (heat exchangers) and carbon is used for the products that are exposed to friction and wear (bearings and gears).

Making an object from a composite material usually involves some form of mold. The reinforcing material is first placed in the mold and then the semi-liquid matrix material is sprayed or pumped in to form the object. Pressure may be applied to force out any air bubbles and the mold is then heated to make the matrix set solid. The molding process is often done by hand but automatic processing by machines is becoming more common. One of the new methods is called pultrusion (a term derived from the words 'pull' and 'extrusion'). This process is ideal for manufacturing products that are straight and have a constant cross section, such as bridge beams.

The greatest advantage of the composite materials is strength and stiffness combined with lightness. By choosing an appropriate combination of reinforcement and matrix materials the manufacturers can produce properties that exactly fit the requirements for a particular structure and for a particular purpose. In this chapter, we briefly cover basic ideas concerning the two essential parameters and the fabrication process methods of the composite materials.

6.2. REINFORCEMENTS

It is an important factor that the strength of the composite materials follows rule-of-mixtures if continuous filament is the reinforcing phase and the matrix-to-reinforcement phase interfacial reactions are controlled to provide good bonding without degradation of the reinforcements. The reinforced fibers require a matrix in order to be useful to the engineer. The mechanical properties of the fibrous composites depend upon the properties of the (usually) two components of the thermoelastic fibrous composites. Modern methods of describing and characterizing the distribution and orientation of fibers must also be included, and consideration of the nonmechanical properties of the composites. More research is advocated to reduce the cost of continuous filaments and of fabrication techniques. Both the ceramic and metal fibers are considered.

We will see in the next chapter that the mechanical properties of fibers of a given type and the dimensions are also determined by the structure. For example the properties of the organic fiber, aramid, are dependent on the orientation of the molecular structure whereas for carbon fibers the perfection and alignment of the graphite crystals are paramount. There are of course many properties other than strength that we have to take into account when selecting the reinforcements. In the case of fibers the flexibility is important as it determines whether the fibers may be easily woven or not and influences the choice of method for composite manufacture. The flexibility of a fiber depends mainly on the Young's modulus E_f, and diameter D of the fiber:

$$\text{Flexibility, } \alpha \frac{1}{E_t D^4} \tag{6-1}$$

Thus large diameter fibers with a high Young's modulus are not flexible.

Finally, the fact that the reinforcements are bonded to the matrix means that any loads applied to the composites are carried by both the constituents. As in most cases the reinforcements are the stiffer and stronger constituents; it is the principal load bearer. The matrix is said to have transferred the load to the reinforcements.

6.2.1. Carbon Fibers

6.2.1.1. Introduction

Carbon fibers are a new breed of high-strength materials consisting of extremely thin fibers about 0.005–0.010 mm in diameter. Carbon fibers have been described as the fibers containing at least 90% carbon obtained by the controlled pyrolysis of appropriate fibers. The carbon atoms are bonded together in microscopic crystals that are more or less aligned parallel to the long axis of the fiber. The crystal alignment makes the fiber very strong for its size. Several thousand carbon fibers are twisted together to form a yarn, which may be used by itself or woven into a fabric. Carbon fiber has many different weave patterns and can be combined with a plastic resin and wound or molded to form composite materials, such as carbon fibers reinforced plastic, to provide high strength-to-weight ratio materials [6–9]. The density of the carbon fibers is also considerably lower than that of steel making it ideal for applications requiring low weight. The properties of carbon fibers such as high tensile strength, low weight, and low thermal expansion make it very popular in aerospace, civil engineering, military, and motorsports along with other competition sports. However, it is relatively expensive when compared to similar materials such as fiberglass or plastic. Carbon fibers are very strong when stretched or bent but weak when compressed or exposed to high shock (e.g., a carbon fiber bar is extremely difficult to bend but will crack easily if hit with a hammer).

Based on modulus, strength, and final heat treatment temperature, carbon fibers can be classified into the following three categories:

(1) Based on properties of carbon fibers, they can be grouped into:
  1. Ultra-high-modulus, type UHM (modulus >450 GPa)
  2. High-modulus, type HM (modulus between 350–450 GPa)
  3. Intermediate-modulus, type IM (modulus between 200–350 GPa)
  4. Low modulus and high-tensile, type HT (modulus <100 Gpa, tensile strength >3.0 GPa)
  5. Super high-tensile, type SHT (tensile strength >4.5 GPa)

(2) Based on precursor materials, carbon fibers are classified into:
  1. Polyacrylonitrile (PAN)-based carbon fibers
  2. Pitch-based carbon fibers
  3. Mesophase pitch-based carbon fibers
  4. Isotropic pitch-based carbon fibers
  5. Rayon-based carbon fibers
  6. Gas-phase-grown carbon fibers

(3) Based on final heat treatment temperature, carbon fibers are classified into:
  1. Type-I, high-heat-treatment carbon fibers (HTT), where final heat treatment temperature should be above 2000 °C and can be associated with high-modulus type fiber.
  2. Type-II, intermediate-heat-treatment carbon fibers (IHT), where final heat treatment temperature should be around or above 1500 °C and can be associated with high-strength type fiber.
  3. Type-III, low-heat-treatment carbon fibers, where final heat treatment temperature not greater than 1000 °C. These are low modulus and low strength materials.

6.2.1.2. Structures

Carbon belongs to the materials with the highest temperature resistance, if kept in nonoxidizing atmosphere. Under these conditions carbon does not sublimate until temperatures of 3730 °C are reached. The materials can be used in technical applications up to 2800 °C.

Carbon has two well known crystalline forms (diamond and graphite) but it also exists in the quasi-crystalline and glassy states. As far as fiber technology is concerned, graphite is the most important structural form of carbon. The graphic structure consist of hexagonal layers in which the bonding is covalent and strong (~525 kJ/mol); these layers, which are called the basal planes are stacked in an ABAB—sequence as already shown in Fig. 1-2 with inter-layer bonds being weak (<10 kJ/mol) van der Waals bonds. It should be no surprise to the reader that a consequence of the marked difference in the intra-layer and inter-layer bonding is that the properties of graphite are very anisotropic. The theoretical elastic modulus of graphite is approximately 1000 GPa in the basal plane and only 35 GPa in the c-direction perpendicular to these planes. As we will see alignment of the basal plane parallel to the fiber axis give stiff fibers which, because of the relatively low density of around 2 Mg/m^3, have extremely high values for specific stiffness.

In carbon fibers, a layered structure is present but unlike true graphite the graphitic layers are not neatly stacked but have a more random arrangement [10, 11]. The layers are also not planar but rather undulated so this structure is called turbostratic [12] (Fig. 6-1). It is therefore appropriate to refer to the structure of PAN-based carbon fibers as turbostratic graphite. In fibers derived from mesophase pitch precursors, structures are formed closer to the true graphite structure.

Also, in order to achieve carbon fibers with high tensile strengths and moduli, the carbon planes have to be oriented toward the fiber axis and optimized in their structure. This can be done by temperature treatments in an inert atmosphere and under tension.

Perpendicular to the fiber axis there is usually no orientation of the carbon planes. This type of structure is present in all carbon fibers, but nevertheless, there are differences in terms of degree of orientation, arrangement of the layers in radial direction, interaction in between the planes, and the presence of different structural flaws such as microvoids and other imperfections [13–15]. This leads to fibers with different mechanical properties from high tensile strength to very high modulus.

The elastic modulus of the fibers depends mainly on the degree of orientation of the planes along the fiber axis whereas the tensile strength is limited by the number of structural flaws. During temperature treatments up to 1500 °C, the tensile strength tends to be optimized and treatments at even higher temperatures (up to 2800 °C) lead to fibers with high moduli.

For the use of carbon fibers in composite materials, the fibers have to be protected against oxidation. Otherwise they will deteriorate by oxidative

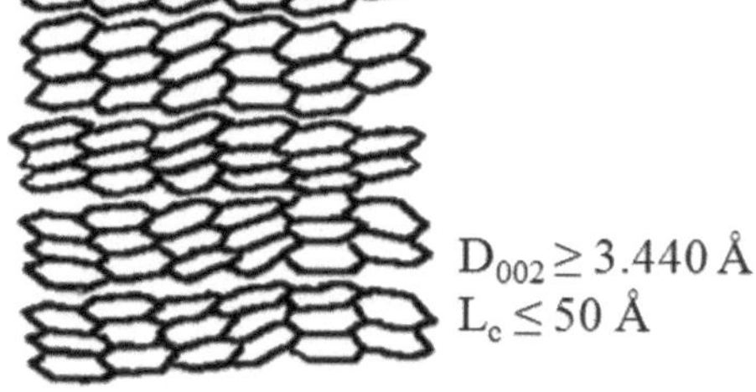

Fig. 6-1 Carbon turbostratic structure.

Table 6-1 Structural Parameters of Various Carbon Fibers [6]

Fiber	Z (°)	d (002)	L_c (nm)	L_a (0)	L_a (90)	Three dimensional order	SEM Morphology
T-300	35.1	0.342	1.5	2.2	4.1	NO	Particles
T-40	30.2	0.343	1.8	3.0	4.6	NO	Particles
T-50	16.4	0.3423	5.3	6.5	17.5	Maybe	Some sheet-like
GY-70	9.6	0.3396	14.1	12.0	40.0	YES	Sheet-like
AS-4	36.8	0.342	1.8	3.0	4.0	NO	Particles
M40J	21.4	0.3427	3.6	5.5	12.5	NO	
M60J	9.9	0.3411	7.8	7.0	28.0	YES	Particles
IM8	–	0.3431	1.9	3.1	5.1	NO	

Z: Full-width at half-maximum of the (002) azimuthal scan, d(002): d-spacing of crystallites, L_c: Crystallite size perpendicular to graphitic basal plane, L_a(0) and L_a(90): Crystallite sizes along graphitic plane perpendicular and parallel to the fiber axis.

degradation at temperatures above 450 °C. The fibers can be protected against oxidation by the surrounding matrix itself but in most cases the composites have to be protected additionally by a so called environmental barrier coating.

In general, the structural parameters, such as degree of preferred orientation along the fiber axis and crystallite sizes, increase, and the *d*-spacing (002) decreases with increasing the heat treatment temperature. The fiber density increases as the heat treatment temperature rises. Such structural parameters for the types of carbon fibers are shown in Table 6-1[16, 17]. The densities of commercial carbon fibers are shown in Table 6-2. According to Oberlin and Guigon, whatever the fiber examined L_{c002} reaches 30 Å, and L_{aTEM} (L_a by TEM) 150–200 Å at a heat treatment temperature of 1700 Å. The latter

Table 6-2 Structural Parameters of Various Carbon Fibers (Table 6-1 continued) [7]

Fiber	Z(°)	D_d	D_c	L_a	L_a (90)	L_c
C1	20.1	55.8	33.3	4.6	7.4	5.9
C2	36.1	66.0	86.7	3.0	5.3	1.7
X	21.2	50.2	40.0	4.8	7.2	4.7
Y	22.0	49.9	40.0	4.8	7.0	4.3
Z	30.8	57.5	80.0	3.4	5.9	2.6
T1000	31.5	62.2	86.7	2.9	5.2	1.7

D_d: Intercrystallite disorder, D_c: Intracrystallite disorder, L_a: Crystallite length along the a-axis, L_c: Crystallite width perpendicular to the a-axis.

increases from 200 to 250 Å at 2000 °C to 250–300 Å at 2200 °C, and then reaches about 400 Å at 2800 °C, whereas L_{c002} remains near to 35 Å [18, 19].

6.2.1.3. Production Processes

There are three successive stages in the conversion of the polyacrylonitrile precursor into high-performance carbon fibers as shown in Fig. 6-2. (a) Oxidative stabilization: the polyacrylonitrile precursor is first stretched and simultaneously oxidized in a temperature range of 200–300 °C in air, which breaks many of the hydrogen bonds and oxidizes the materials. This treatment converts the thermoplastic polyacrylonitrile to a non-plastic cyclic or ladder compound [18–23]. (b) Carbonization: after oxidation, the fibers are carbonized at about 1000 °C without tension in an inert atmosphere (normally nitrogen) for a few hours. During this process, the noncarbon elements are removed as volatiles to give carbon fibers with a yield of about 50 percent of the mass of the original polyacrylonitrile [24–27]. (c) Graphitization: Depending on the type of fibers required, the fibers are treated at temperatures between 1500 and 3000 °C, which improves the ordering and orientation of the crystallites in the direction of the fiber axis and induces graphitization of the materials changing the molecular bond structure [28, 29]. When heated in the correct conditions, these chains bond side-to-side (ladder polymers) forming narrow graphene sheets which eventually merge to form a single columnar filament. The result is usually 93–95% carbon.

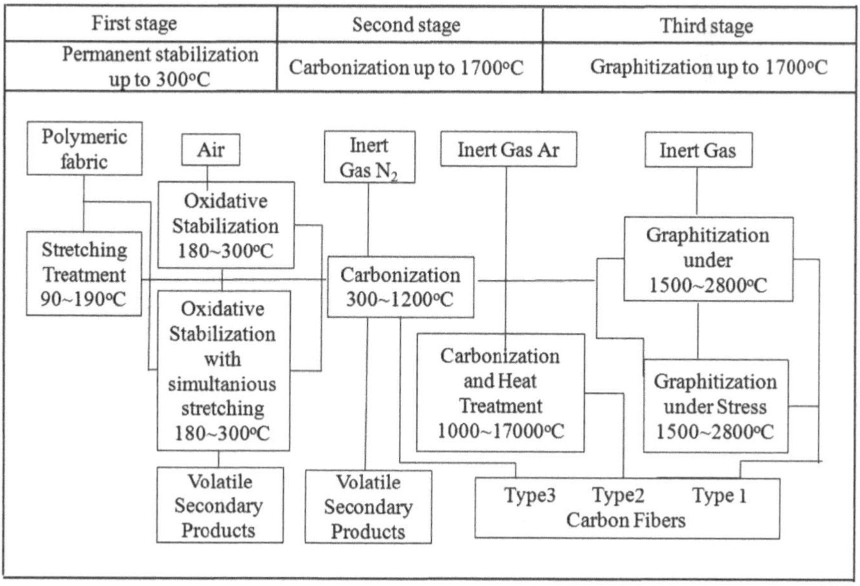

Fig. 6-2 Schematic representation of carbon fibers preparation from polyacrylonitrile (PAN) fibers.

The starting materials for this process are the PAN fibers produced especially for carbon fiber manufacturing. The composition of the polyacrylonitrile is different from the others used for textile applications (different comonomers are used for the polymerization of acrylonitrile). The polyacrylonitrile precursor gives a higher yield of up to 50% on conversion to carbon.

As for all commercial fibers the details of the production are confidential but it is known that a three stage process is involved in the conversion and the resulting structure is sensitive to the rates of heating. The first stage involves stretching and oxidation. The fibers are initially stretched 500–1300% to improve molecular alignment and then heated in air while still under tension to 200–250 °C. This results in intramolecular rearrangement to give a ladder-like polymer; chemical cross-links between the oriented chains in three dimensions are formed so that the chains can no longer bend or distort on heating (Fig. 6-3 (b)), which can withstand further treatments at higher temperatures [23]. This is followed by oxidation whereby some two-thirds of the oxygen sites are filled slowly with the release of H_2O (Fig. 6-3 (b)); this is sometimes called oxy-PAN. In the next stage, the fibers are treated at 900–1500 °C in an inert atmosphere (nitrogen), also under tension, which leads to the carbonization of the fibers. In this process, the noncarbon elements are removed as volatile products yielding carbon fibers with about 50 percent of the mass of the original polyacrylonitrile precursor. During this carbonization the carbon plains are optimized in their structure (reduction of the number of structural flaws) leading to fibers with high tensile strength. The

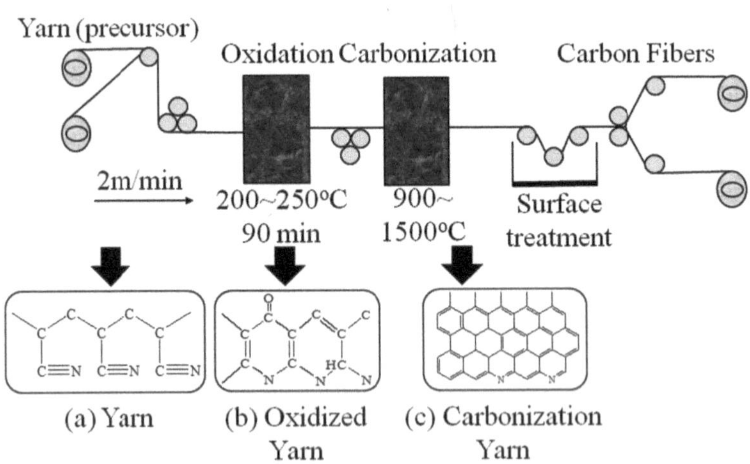

Fig. 6-3 Production processes for polyacrylonitrile (PAN)-based carbon fibers and the changes in structure during the production of carbon fibers from polyacrylonitrile precursor. (a) polyacrylonitrile; (b) ladder-like polymer or oxidation of ladder-like polymer to oxy-polyacrylonitrile; (c) carbonization of oxy-polyacrylonitrile to carbon ring structure.

idealized formation of the carbon ring structure is shown in Fig. 6-3 (c); it can be seen that there is an operational hazard at this stage as toxic volatiles such as hydrogen cyanide are given off.

In the additional high temperature steps (graphitization) in a nitrogen or argon atmosphere, the fiber structure can be further improved to produce high moduli. Temperatures up to 2800 °C are applied to arrange the carbon plains and to increase their orientation toward the fiber axis. However, X-ray diffraction patterns show that in this state no true graphitic structure is formed. Typically fibers, which have been treated only at 1000 °C exhibit tensile strength values of about 2000 MPa and tensile moduli of about 170 GPa. After temperature treatments up to 1500 °C, tensile strengths of about 3500 MPa and tensile moduli of about 275 GPa are reached. At 2500 °C, fibers with reduced tensile strengths of about 2800 MPa are formed (presumably by the evolution of prolate pores during the formation of more compact crystalline structures) but these fibers have very high tensile moduli of 480 GPa. Even higher values for tensile moduli (up to 600 GPa) can be reached if the fibers are treated at high temperature under tension. The polyacrylonitrile process is by far the most important production process for carbon fibers. A three-dimensional representation of the arrangement of the basal planes in polyacrylonitrile-based carbon fibers is given in Fig. 6-4.

The value of the modulus increases with increasing temperature of graphitization as shown in Fig. 6-5. In fact, the properties can be varied over a wide range by suitable heat treatment conditions but it is difficult to combine high strength with high modulus. Refinements in fiber process technology over the

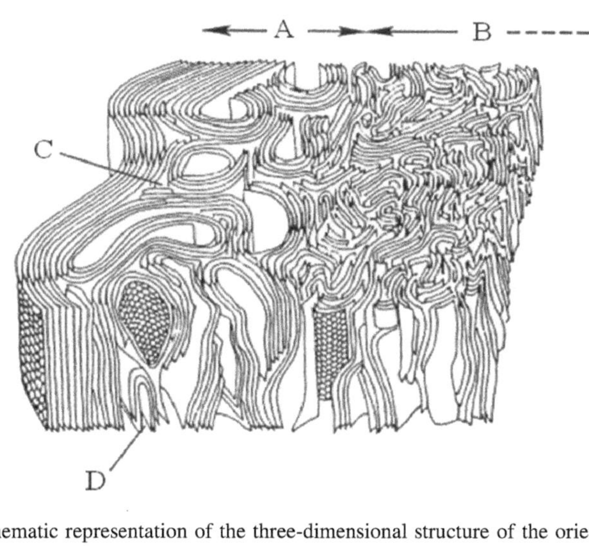

Fig. 6-4 Schematic representation of the three-dimensional structure of the orientation of basal planes of the graphite in a carbon fiber with different substructures: A = surface region; B = bulk region; C = hairpin defect; and D = disclination [24].

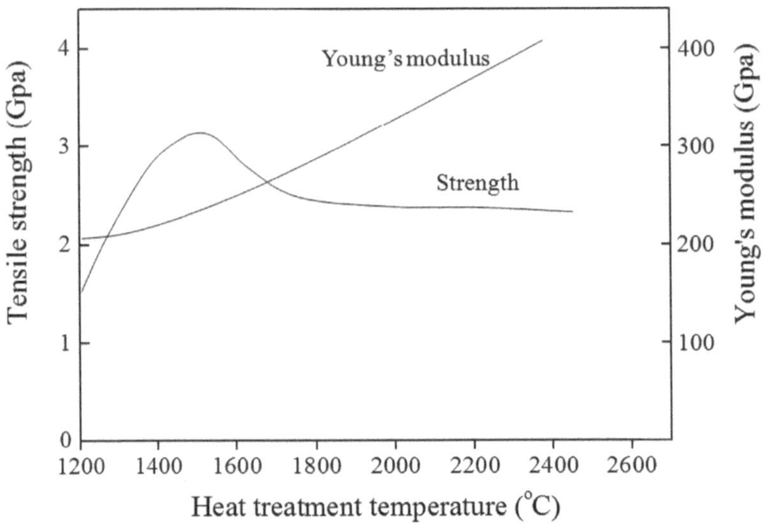

Fig. 6-5 Effect of heat treatment temperature on the Young's modulus and strength of carbon fibers produced from a polyacrylonitrile precursor.

past 20 years have led to considerable improvements in strength and in strain to fracture for polyacrylonitrile-based fibers. These can now be supplied in three basic forms: high strength (HS), high modulus (HM), and ultra high modulus (UHM) as listed in Table 6-3. The most recent developments of the high strength fibers have led to what are known as high strain fibers, which have strain values approaching 2% before fracture.

Fig. 6-6 shows the stress-strain curves for a range of carbon fibers. The strain is elastic, up to failure, and a lot of stored energy given by the area under the stress–strain curve is released when the fibers break in a brittle manner. These fibers are surface-treated during manufacture to promote adhesion with the polymer matrix whether thermosetting or thermoplastic. The surface is roughened by chemical oxidation, which causes etching and then coated with an appropriate size to aid the bonding to the specified matrix.

Table 6-3 Typical Properties of Three Grades of Carbon Fibers

Typical Properties	High Strength	High Modulus	Ultra-high Modulus
Density (g/cm^3)	1.8	1.9	2.0 ~ 2.1
Young's Modulus (GPa)	230	370	520 ~ 620
Tensile Strength (GPa)	2.48	1.79	1.03 ~ 1.31
Tensile Elongation (%)	1.1	0.5	0.2

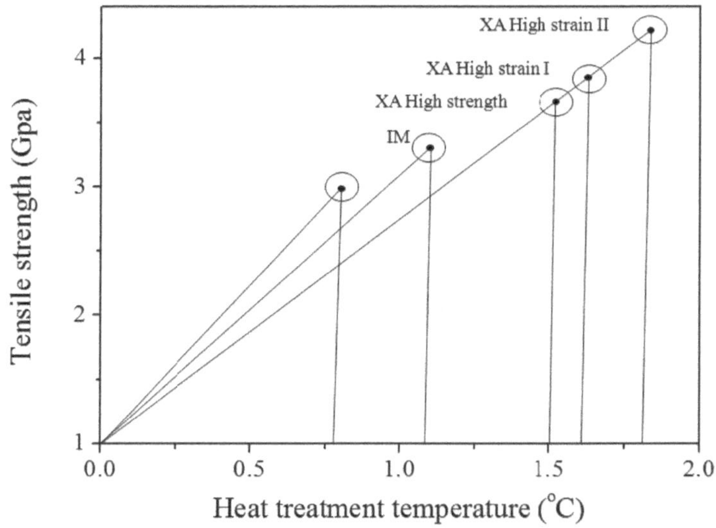

Fig. 6-6 Stress-strain curves for carbon fibers produced from a polyacrylonitrile precursor.

6.2.1.3.1. Carbon fibers from pitch precursors

Pitches derived from coal tar, petroleum residues, or poly (vinyl chloride) can be used as relatively cheap precursors for carbon fibers. Pitches are thermoplastic materials and can therefore be extruded directly via the melt spinning processes to precursor fibers. These can be subsequently transferred to carbon fibers in processes similar to the one described for the polyacrylonitrile route. This includes a stabilization step between 250 and 300 °C and then carbonization and graphitization at temperatures between 1000 and 2500 °C [30].

In untreated pitches, condensed aromatic structures are present, which are isotropic and randomly distributed. This leads to fibers with low orientation of the carbon plains along the fiber axis and moderate mechanical properties. In order to achieve enhanced fiber properties, the pitches are thermally treated at 400–450 °C by which liquid crystalline, anisotropic structures are formed, the so called mesophases.

Precursor fibers produced from the mesophase pitch exhibit high orientation values of the carbon plains in the direction of the fiber axis and can therefore be transformed to carbon fibers with very good mechanical properties. The yield of carbon from pitches can be above 75 wt.%. Tensile strength values of 3500 Mpa and tensile moduli of about 400 GPa can be reached for the carbon fibers based on pitch. Also, this fiber type can be brought to even higher moduli of about 600 Gpa by an additional high temperature treatment.

6.2.1.3.2. Carbon fibers from regenerated cellulose

For this production route, cellulose rayon fibers are used as a carbon source. Rayon fibers are produced by dissolving and spinning of cellulosic materials, such as wood pulp or cotton. Since cellulose is a polymeric carbohydrate, water evolves during the temperature treatment leaving a residue of carbon. The production process is also conducted in steps including a lower temperature treatment (below 400 °C), subsequent carbonization (up to 1500 °C), and graphitization (at about 2500 °C). The yield of carbon fibers from rayon is low, ranging between 10 and 30 wt.%.

The mechanical properties of the rayon derived carbon fibers achieved in standard processing (only carbonization) are not as good as for polyacrylonitrile derived fibers. Typically values of 700 MPa are achieved for the tensile strength and 70 GPa for the modulus. Unlike in the polyacrylonitrile process, the fibers cannot be held under higher tension during the low temperature treatment. But in the following high temperature treatments, tension can be applied to these fibers, by which good mechanical properties can be reached, that is, tensile strengths up to 2800 MPa and moduli up to 550 GPa. Since the whole process is more complex for the rayon derived fibers, polyacrylonitrile derived types are more commonly used in industry.

Finally, it should be mentioned that although we have concentrated on the effect of graphitization and the orientation on properties, there are other structural features, such as voids, microcracks, surface flaws, and impurities, which have to be controlled in order to obtain fibers of a consistent high quality [31, 32].

6.2.1.4. Surface treatment

In the late 1960s and early 1970s, several different types of surface treatment were used. (a) Wet methods: immersion of fibers in oxidizing agents, such as nitric and chromic acid, and electrochemical oxidation. Electrochemical oxidation is the most widely used industrial method for treating carbon fibers but the details of such processes are mostly proprietary. (b) Dry methods: oxidation in air or oxygen. Plasma treatment is one of the convenient and effective methods to produce functional group on carbon fiber surfaces [33–38].

The physical and chemical properties of carbon fibers are usually modified to achieve good adhesion between the reinforcement and the matrix materials. To enlarge the surface polarity of nonpolar carbon fibers, various surface treatment techniques are applied, such as oxidation in acid solutions [39, 40], dry oxidation in oxygen, anodic oxidation, plasma treatments, mild fluorination [41–47], metallic coating [48–50], and so on. Interfacial adhesion between the fibers and the matrices cannot be achieved without intimate contact, that is, unless the fiber surface contacts the resin on an intermolecular equilibrium distance level [51–53].

A large interfacial area of intimate contact by the matrix resin is a prerequisite whenever the interfacial bond is due primarily to van der Waal's physical adsorption force. In addition, complete wetting of the fibers by the liquid polymeric matrix is advantageous when the surface free energy of the fibers is well above that of the liquid polymer in view of the London dispersive-specific polar components of surface free energy.

Untreated commercial carbon fibers have a surface tension close to 40 mJ/m^2 and an extremely hydrophobic nature due to the extremely high temperature of the manufacturing process leading to carbonization (700–1500 °C) or to graphitization (above 2500 °C). On the other hand, most polymeric matrix resins, such as phenolic resin are slightly hydrophilic with a surface tension in the range of 35–45 mJ/m^2[51–54].

It is known that surface treatments improve fiber resin bonding but the mechanism by which the adhesion is improved is still controversial. There appear to be three major mechanisms contributing to the strength of the fiber/resin bond. (a) The adsorption (either physical or chemical) of the resin molecules onto surface complexes; acidic complexes have usually been considered to be of most importance. (b) Removal of contaminants providing a superior surface for adhesion. (c) A mechanical keying effect when the resin penetrates pits and channels on the roughened surface brought about by oxidation.

Fig. 6-7 shows the functional groups that are thought to be present on oxidized carbon surfaces. It was hoped that by introducing one or more of these

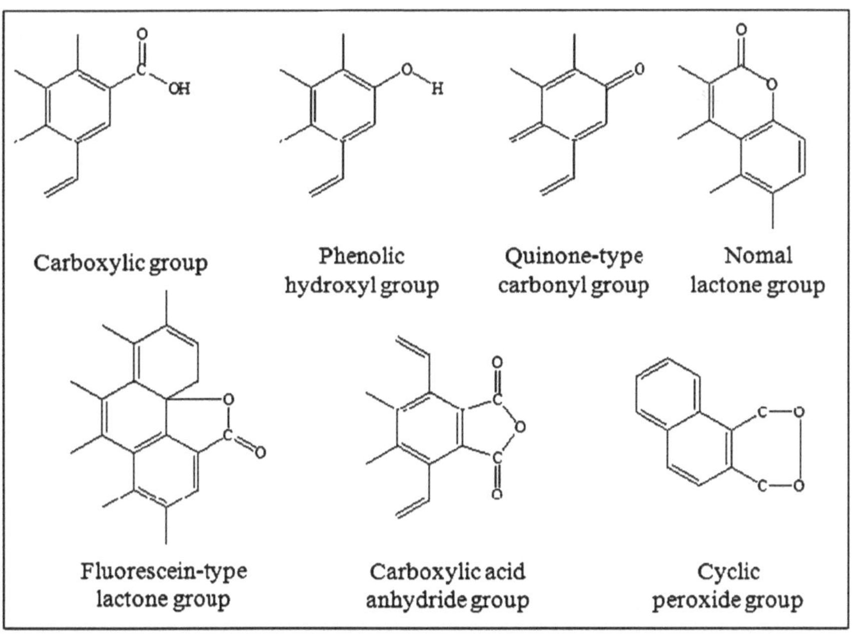

Fig. 6-7 The possible functional groups presented on oxidized carbon surfaces.

functional groups they would chemically interact with the epoxy resin promoting bonding between fiber and resin. In most cases an increase in the surface functionality is accompanied by an increase in surface area and so the relative importance of these two factors is difficult to assess. In 1968, Herrick [55] reported that the treatment of Thornel 25 fibers in nitric acid increases their surface area and also the number of surface oxygen complexes. Similar results have been reported by Donnet [56], Fitzer and Weis [57], and Rand and Robinson [58]. The interfacial shear strength of composites made from these treated fibers also increased.

Some fibers after nitric acid treatment were heated in hydrogen to reduce the surface oxygen complexes. The interfacial shear strength of the corresponding composites decreased significantly suggesting that the surface functionality was of most importance in enhancing fiber/resin bonding. Scola and Brooks [59] on the other hand found no decrease in interfacial shear strength for composites produced from treated Hitco HMG50 fibers (see Table 6-4) that had been heated in hydrogen prior to incorporation into the composites.

It has been found by many workers [60] that heating the fibers removes acidic surface oxidation almost completely. The treatment of AG (see Table 6-4) carbon fibers in Hummers reagent [56] produces a graphitic oxide layer. This lamellar graphitic oxide can be destroyed by heating. The formation of this lamellar coat greatly improves the interfacial shear strength of composites but the interfacial shear strength made from treated fibers which had been heated decreased to a similar value to that of the untreated fiber/resin composites. For AC (see Table 6-4) fibers heating did not produce such a marked effect. Although heating caused a decrease in the interfacial shear strength, the original value for the untreated fiber/resin composites was not regained.

Table 6-4 Description of Carbon Fibers

Product name	Manufacturer	Precursor type	Properties
Thornel 25	Union carbide	Rayon	High strength
HMG-50	Hitco	Rayon	High modulus
AG	Le Carbone Lerraine	Polyacrylonitrile	High modulus
AC	Le Carbone Lerraine	Polyacrylonitrile	High strength
Type I	Not reported	Polyacrylonitrile	High modulus
Type II	Not reported	Polyacrylonitrile	High strength
HMU	Hercules Inc.	Polyacrylonitrile	High modulus
HMS	Hercules Inc.	Polyacrylonitrile	High modulus
T300	Amoco	Polyacrylonitrile	High strength
P100	Amoco	Pitch	High modulus

Fitzer and Weis [57] studied the effect of electrochemically treating fibers in nitric acid on the shear strength of these composites. He used titration methods to determine the surface oxide concentration and blocking reagents such as diazomethane. Over-oxidation of the fibers was found to influence composite fracture behavior strongly causing brittle fracture in the short beam shear test (not a good indicator of the interfacial shear strength). With this in mind only short oxidation times were used. He concluded that the amount of the fiber resin bonding was solely dependent upon the amount of acidic surface functionality. Surprisingly, Fitzer found that composites made from type I and type II fibers gave the same interfacial shear strength values but type I had a surface 'oxide' concentration one order of magnitude less than type II fibers. This surely indicates that surface treatments promote bonding via some other mechanism as well as the one he suggested.

Brelant [61] has a completely different view on the subject suggesting that the size and distribution of voids at the fiber resin interface. The limiting strength is said to be the stress acquired to propagate cracks through the voids along the fiber/matrix boundary. The increase in shear strength noted in the experiments by Fitzer *et al.* [62] for the short treatment times, Brelant proposed, could be attributed to the removal of contaminants or the reduction of protuberances which facilitate void formation. He suggested that fiber resin bonding is primarily physical in nature. This was later confirmed by Drzal and coworkers, [63] who published evidence to show that the effect of commercial treatments was to remove weakly bound crystallites from the untreated fiber surface and hence provide a superior surface to which the resin could adhere. Koztowski and Sherwood [64] found that these crystallites were removed by different mechanisms depending on the pH of the electrolyte used. In basic solutions small fragments of fiber were detected in the working electrode solution whereas in acidic solutions they were oxidized to carbon dioxide. They then went on to study the effect of electrochemical oxidations on the interfacial shear strength of the resulting composites using treatment levels which were comparable to those of the industrially treated fibers.

X-ray photoelectron spectroscopy is a very powerful technique for characterizing functional groups on carbon fiber surfaces. Oxidized Torayca carbon fiber T300 has markedly larger amounts of carbonyl and hydroxyl groups and a moderately larger amount of carboxyl groups, than the unoxidized T300 fiber. The concentration of these groups was determined by X-ray photoelectron spectroscopy after the functional groups had been modified by chemical reactions with fluorine compounds. The relative amounts of oxygen, estimated from the amounts of three functional groups determined by the surface modification reaction, however, were about 20 percent of the observed values from a direct measurement of O_{1s} spectra for both the unoxidized and oxidized carbon fibers. Ishitani and coworker [65] concluded that hydroxyl, carbonyl, and carboxyl groups are major functional groups formed by the surface oxidation treatment whereas the amounts of ester and ether groups are negligible. The chemical

Table 6-5 Surface Chemical Composition of Surface-oxidized Torayca T800H Fibers

Sample	O/C × 10^3					N/C × 10^3		
	≡COH	–COOH	≡COC≡–COOC≡=CO	≡Si–O–	Total	$-NH_2$	=NH=N–	Total
Control	–	2	30	14	46	1	24	25
Weakly oxidized	1	5	75	–	81	3	28	31
Strongly oxidized	3	31	244	–	278	5	29	34

modification reagents may react only with the functional groups in the outermost layer (5–10 Å) in the carbon fiber surface because of the poor permeability of reagents in to carbon fiber. If there are more functional groups in a greater depth within the detection depth of X-ray photoelectron spectroscopy (30–50 Å), the direct O_{1s} measurement will five a higher relative intensity [66].

Nakayama *et al.* [67] determined the concentration of functional groups on the surface of the weakly and strongly oxidized Torayca T800 fibers by the curve-fitting the C_{1s} spectrum using an asymmetric peak shape or a symmetric peak shape. Amounts of groups such as hydroxyl, carboxyl, and amine were estimated by gas chemical modification using fluorine compounds (Table 6-5). Consequently, the real surface of the oxidized carbon fiber was considered to be composed of both the graphite-like and aliphatic structures with functional groups.

Estimation of functional groups on carbon fiber surfaces using chemical modification were also applied to Hercules AS-4 and Celenese Celion 6000. Fluorine-containing compounds and a mercuric fluorine-containing compound have been used as modification reagents. The results obtained are shown in Table 6-6 [68]. The X-ray photoelectron spectroscopic analysis utilizing derivatization reactions with fluorine-containing compounds of functional groups on carbon fibers was also performed by Chen *et al.* [69]. The results obtained are shown in Table 6-6. The carbon fibers used were Hercules IM7, Amoco T40, Amoco T650-42, and BASF Celion G40-800. These carbon fibers are classified as intermediate-modulus grade having a tensile modulus of about 280 GPa. All the fibers were surface treated and contained no sizing.

In this case, it is to be noted that though the −COOH group is clearly acidic, this is not always true for the hydroxyl group, R−OH. The −OH group can be acidic if the R group is a phenyl or basic, if R is aliphatic. Similarly, the −C=O is

Table 6-6 Concentration of Functional Groups per Hundred Surface Atoms Based on Functional Group Derivatization and Fluorine or Mercury X-ray Photoelectron Spectroscopic Analysis [68]

Fiber	C	O	N	OH	−C=O	COOH	OH+C = O+COOH
IM7	81	14	5.2	0.7	1.1	1.8	3.6
T650-42	85	11	2.0	0.5	1.4	1.6	3.5
G40-800	82	15	2.4	0.7	1.4	1.2	3.2
T40	95	2.7	1.9	0.2	1.8	0.6	2.6
AS-4	85	11	4.1	0.1	0	0.4	0.5
Celion 6000	84	14	1.6	0.2	0.2	0.30	0.7

[c]Hercules, [d]Amoco, [e]BASE. C, O, and N: total carbon, oxygen, and nitrogen determined by X-ray photoelectron spectroscopy prior to derivatization.

generally slightly basic, but if there is an alpha hydrogen, such as in CH–C=O, this hydrogen atom shows acidic characteristics. Therefore, the assessment of carbon-surface acidity or basicity can only be made tentatively from X-ray photoelectron spectroscopy data particularly since X-ray photoelectron spectroscopy cannot detect hydrogen atoms [69]. In such a case, surface acid-base free energy data become useful for deducing whether either one or both of them contribute to the basic nature or to the acidic nature. It has also been pointed out that carbon fibers from various manufacturers have significant differences in their surface acidity [70].

6.2.1.5. Commercial products

There are a number of companies producing carbon fibers and each has a number of carbon fiber products with different fiber properties and yarn counts. Therefore, no complete market survey of carbon fiber products is given here but some examples of fiber properties and prices are displayed. A classification of carbon fibers in terms of their mechanical properties can be conducted by using the IUPAC (International Union of Pure and Applied Chemistry) guidelines (Table 6-7).

Carbon fibers are offered in a wide range of tensile strengths and moduli. Also a wide range of filament numbers in rovings are produced ranging from 1,000 (1 K) to 400,000 (400 K) filament bundles. Depending on the fiber properties, the fibers can be priced between 20 €/kg up to 1500 €/kg. Table 6-8 shows a survey of carbon fibers from Toray, Toho Tenax, and SGL Carbon. Some fiber types are sold in different qualities, that is, a regular type and also a type that is qualified for aerospace applications. Since there are higher quality standards in aerospace applications, the fibers for these applications are 15–25% more expensive than those displayed in the table due to the additional quality controls during and after production.

The atomic structure of the carbon fiber is similar to that of graphite consisting of sheets of carbon atoms (graphene sheets) arranged in a regular hexagonal pattern. The difference lies in the way these sheets interlocked. Graphite

Table 6-7 Classification of Carbon Fibers

Types	Properties
UHM (ultra high modulus)	Elastic modulus > 450 GPa
HM (high modulus)	Elastic modulus 350 ~ 450 GPa
IM (intermediate modulus)	Elastic modulus 200 ~ 350 GPa
HT (high tensile, low modulus)	Tensile strength >3 GPa, Elastic modulus <100 GPa
SHT (super high tensile)	Tensile strength >4.5 GPa

Table 6-8 Comparison of Carbon Fibers from Different Producers

Mfr	Fiber type	Filament count	Filament diameter (m)	Surface area (m^2/g)	Tensile Strength (MPa)	Tensile modulus (GPa)	Tensile strain (%)	Density (g/m^3)
Amoco	T-300	1k, 2k	7.0	0.45	3650	231	1.4	1.76
[Thormel]	T-40	12k	5.1	0.5	5650	290	1.8	1.81
	T650/42	6k, 12k	5.1	0.5	4620	290	1.6	1.78
	T-50	3k, 6k	6.5	0.45	2900	390	0.7	1.81
Hexcel	AS4	3k, 12k			3930	221	1.7	1.79
	IM4	12k			4138	276	1.5	1.73
	IM7	6k, 12k			5379	276	1.8	1.77
	UHM	3k, 12k			3447	441	0.8	1.87
Mitsubishi	TR30	3k			3530	235	1.5	1.79
Rayon	TR50	12k			4900	235	2.1	1.80
[Pyrofil]	MR50k	12k			5490	294	1.8	1.80
	SR50	12k			4220	490	0.9	1.88
Sigrafil	C30		6.8		3000	230	1.4	1.78
	C35		7.0		3200	210	1.4	1.8
Tenax	HTA	1k, 24k	7.0		3950	238	1.5	1.77
	UTS	12k	7.0		4800	240	2.0	1.8
	IMS	6k, 24k	5.0		5500	290	1.9	1.8
	UMS	12k	4.7		4500	435	1.1	1.81
Toho	HTA	3k, 12k	7.0		3920	235	1.7	1.77
Rayon	ST4	12k	7.0		4810	240	2.0	1.78
[Besfight]	IM600	12k, 24k	5.0		5790	285	2.0	1.80
	HM35	12k	6.7		3240	345	0.9	1.79
	TM40	12k	6.2		3430	390	0.9	1.85

(continued)

Table 6-8 *(continued)*

Mfr	Fiber type	Filament count	Filament diameter (m)	Surface area (m^2/g)	Tensile Strength (MPa)	Tensile modulus (GPa)	Tensile strain (%)	Density (g/m^3)
	UM68	12k	4.1		3330	650	0.5	1.97
Toray	T300*	1-12k	7.0		3530	230	1.5	1.76
[Torayca]	T300J*	3k, 12k	7.0		4210	230	1.8	1.78
*available	T700S	12k	7.0		4900	230	2.1	1.80
From	T800H*	6k	5.		5490	294	1.9	1.81
Soficar	T1000G	12k	5.0		6370	294	2.2	1.80
	M40J*	6k	5.0		4410	377	1.2	1.77
	M50J	6k	5.0		4120	475	0.8	1.88
	M60J	3k, 6k	5.0		3820	588	0.7	1.94
	X665	6k	5.0		3430	637	0.5	1.98
	M40*	6k, 12k	7.4		2740	392	0.7	1.81
Zoltek	Panex33	48k, 320k			3600	228		1.78
[Panex]	Panex30				1552	221		1.75

is a crystalline material in which the sheets are stacked parallel to one another in regular fashion. The intermolecular forces between the sheets are relatively weak van der Waals forces giving graphite its soft and brittle characteristics. Depending upon the precursor to make the fiber, the carbon fiber may be turbostratic or graphitic or have a hybrid structure with both graphitic and turbostratic parts present. In turbostratic carbon fiber the sheets of carbon atoms are haphazardly folded or crumpled, together. Carbon fibers derived from polyacrylonitrile are turbostratic whereas carbon fibers derived from mesophase pitch are graphitic after heat treatment at temperatures exceeding 2200 °C. Turbostratic carbon fibers tend to have high tensile strength whereas heat-treated mesophase-pitch-derived carbon fibers have high Young's modulus and high thermal conductivity. The mechanical and physical properties of commercial carbon-based fibers are listed in Table 6-9.

Carbon fiber filament yarns are used in several processing techniques: the direct uses are for pre-pregging, filament winding, pultrusion, weaving, braiding, etc. Carbon fiber yarn is rated by the linear density (weight per unit length, i.e., 1 g/1000 m = 1 tex) or by number of filaments per yarn count in thousands. For example, 200 tex for 3,000 filaments of carbon fiber is three times as strong as 1,000 carbon fibers but is also three times as heavy. This thread can then be used to weave a carbon fiber filament fabric or cloth. The appearance of this fabric generally depends on the linear density of the yarn and the weave chosen. Some commonly used types of weave are twill, satin and plain.

6.2.2. Glass Fibers

Glass fibers are the materials made from extremely fine fiber of glass which is a non-crystalline material with a short-range network structure (Fig. 6-8). As such it has no distinctive microstructure and the mechanical properties which are determined mainly by composition and surface finish are isotropic. Glass fibers are usually used as a reinforcing agent for many polymer products; the resulting composite materials properly known as the fiber-reinforced polymer or glass-reinforced plastic is called 'fiberglass' in popular usage. There are many groups of glasses, for example silica, oxynitride, phosphate, and halide glasses, but from the point of view of composite technology only the silica glasses are currently of importance.

Glass fibers are formed when thin strands of silica-based or other formulation glass is extruded into many fibers with small diameters suitable for textile processing. The technique of heating and drawing glass into fine fibers has been known for millennia; however, the use of these fibers for textile applications is more recent. Until this time all glass fibers had been manufactured as staple (a term used to describe naturally formed clusters or locks of wool fibers). The first commercial production of glass fibers was in 1936. In 1938, Owens-Illinois Glass Company and Corning Glass Works joined to form the Owens-Corning Fiberglas Corporation. When the two companies joined to

Table 6-9 Properties of Carbon-based Fibers

Properties	Polyacrylonitrile			Pitch Type-P	Rayon
	Intermediate modulus	High modulus	Ultra high modulus		
Diameter (μm)	8–9	7-10	7 ~ 10	10-11	6.5
Density (g/cm³)	1.78–1.82	1.67 ~ 1.90	1.86	2.02	1.53 ~ 1.66
Tensile modulus (GPa)	228–276	331 ~ 400	517	345	41 ~ 393
Tensile strength (MPa)	2410–2930	2070 ~ 2900	1720	1720	620 ~ 2200
Elongation (%)	1.0	0.5	0.3 ~ 0.4	0.4 ~ 0.9	1.5 ~ 2.5
Coeff. of thermal expansion ($\times 10^{-6}$/°C)	–0.1 ~ –0.5(–) 7 ~ 12 (⊥)	–0.5 ~ –1.2(–) 7 ~ 12 (⊥)	–1.0 (–) 7 ~ 12 (⊥)	–0.9 ~ –1.6(–) 7.8 (⊥)	—
Thermal conductivity (W/m/°C)	20	70–105	140	—	38
Specific heat (J/kg/K)	950	925	–	–	–

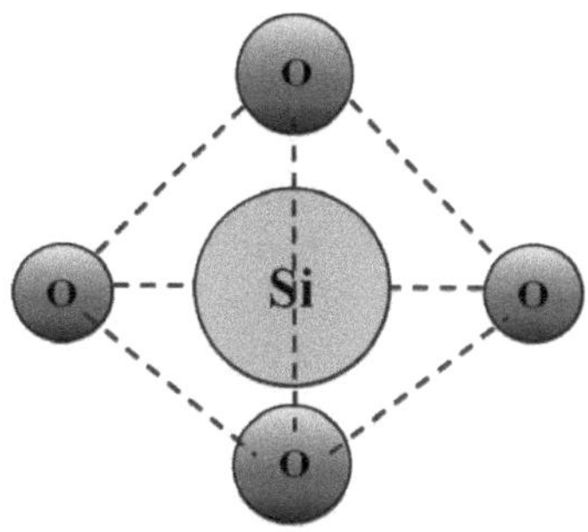

Fig. 6-8 Molecular structure of glass fibers.

produce and promote glass fibers, they introduced continuous filament glass fibers [71]. Owens-Corning is still the major glass fiber producer in the market today [72].

The types of glass fibers most commonly used are mainly E-glass (alumino-borosilicate glass with less than 1 wt.% alkali oxides mainly used for glass-reinforced plastics), but also A-glass (alkali-lime glass with little or no boron oxide), E-CR-glass (alumino-lime silicate with less than 1 wt.% alkali oxides, has high acid resistance), C-glass (alkali-lime glass with high boron oxide content used, for example for glass staple fibers), D-glass (borosilicate glass with high dielectric constant), R-glass (alumino silicate glass without MgO and CaO, with high mechanical requirements), and S-glass (alumino silicate glass without CaO but with high MgO content with high tensile strength) [73]. The compositions of some typical glasses which are used in the manufacture of fibers are given in Table 6-10.

Table 6-10 Typical Composition and Properties of Glass Fibers

	E-glass (%)	S-glass (%)	AR-glass (%)
SiO_2	54	65	64
TiO_2			3
ZrO_2			13
Al_2O_3	14	25	1
B_2O_3	9		
MgO	5	10	
CaO	18		5
Na_2O			14
Modulus (GPa)	70	80	75
Strength (MPa)	2200	2600	1700
Density (Mg/m^3)	2.54	2.49	2.70

Although pure silica is a perfectly viable glass and glass fiber, it must be worked with at very high temperatures which is a drawback unless it's specific chemical properties are needed. It is usual to introduce impurities into the glass in the form of other materials to lower its working temperature. These materials also impart various other properties to the glass that may be beneficial in different applications. The first type of glass used for fiber was soda lime glass or A glass. It was not very resistant to alkali. A new type E-glass was formed; this is an alumino-borosilicate glass that is alkali free (<2%) [74, 75]. This was the first glass formulation used for continuous filament formation. E-glass still makes up most of the fiberglass production in the world. Its particular the components may differ slightly in percentage but must fall within a specific range. The letter E is used because it was originally for electrical applications. S-glass known as R-glass in Europe is based on this SiO_2-Al_2O_3-MgO system; this fiber has higher stiffness and strength (hence the designation S) than E-glass (Table 6-10). It also retains its improved properties to higher temperatures. However, it is more difficult to draw into fibers due to its limited working range and is therefore more expensive. S-glass fibers are still used for some specialist applications but they have been largely superseded by the fibers such as carbon and aramid with superior mechanical properties. C-glass was developed to resist attack from chemicals mostly acids that destroy E-glass [75]. T-glass is a North American variant of C-glass. A-glass is an industry term for cullet glass, often bottles, made into fiber. AR-glass is alkali-resistant glass. Most glass fibers have limited solubility in water but are very dependent on pH. Chloride ions will also attack and dissolve E-glass surfaces.

Since E-glass does not really melt, but soften, the softening point is defined as 'the temperature at which a 0.55–0.77 mm diameter fiber 235 mm long, elongates under its own weight at 1 mm/min when suspended vertically and heated at the rate of 5 °C per minute' [76]. The strain point is reached when the glass has a viscosity of 1014.5 poise. The annealing point, which is the temperature where the internal stresses are reduced to an acceptable commercial limit in 15 min is marked by a viscosity of 1,013 poise [76].

The basis of textile-grade glass fibers is silica, SiO_2. In its pure form it exists as a polymer, $(SiO_2)_n$. It has no true melting point but softens at 2,000 °C (3,630 °F) where it starts to degrade. At 1,713 °C (3,115 °F), most of the molecules can move about freely. If the glass is then cooled quickly, they will be unable to form an ordered structure [74]. In the polymer, it forms SiO_4 groups that are configured as a tetrahedron with the silicon atom at the center and four oxygen atoms at the corners. These atoms then form a network bonded at the corners by sharing the oxygen atoms.

The vitreous and crystalline states of silica (glass and quartz) have similar energy levels on a molecular basis also implying that the glassy form is extremely stable. In order to induce crystallization, it must be heated to temperatures above 1,200 °C (2,190 °F) for long periods of time.

Glass fibers are useful because of their high ratio of surface area to weight. However, the increased surface area makes them much more susceptible to chemical attack. By trapping air within them, blocks of glass fiber make good thermal insulation with a thermal conductivity of the order of 0.05 W/(m·K). The strength of glass is usually tested and reported for 'virgin' or pristine fibers—those that have just been manufactured. The freshest, thinnest fibers are the strongest because the thinner fibers are more ductile. The more the surface is scratched, the less the resulting tenacity [75]. Because glass has an amorphous structure, its properties are the same along the fiber and across the fiber [74]. Humidity is an important factor in the tensile strength. Moisture is easily adsorbed and can worsen microscopic cracks and surface defects and lessen tenacity.

In contrast to the carbon fiber, glass can undergo more elongation before it breaks [74]. There is a correlation between the bending diameter of the filament and the filament diameter. The viscosity of the molten glass is very important for manufacturing success. During drawing (pulling of the glass to reduce fiber circumference), the viscosity should be relatively low. If it is too high, the fiber will break during drawing. However, if it is too low, the glass will form droplets rather than drawing out into fiber.

The use of glass fibers as reinforcement in the thermoplastic polymers likes polyamides, polyesters or polypropylene yields materials with very attractive properties [77, 78]. Glass fiber containing thermoplastic compounds show high strength and stiffness while maintaining a high level of impact strength. Combined with their low specific weight, ease of processing and tailor-made properties, these materials find widespread use in automotive applications and household/electronic appliances.

6.2.3. Aramid Fibers

Aramid fibers are a class of heat-resistant and strong synthetic fibers. They are used in aerospace and military applications, for ballistic rated body armor fabric, in bicycle tyres, and as an asbestos substitute [79]. The name is a shortened form of 'aromatic polyamide'. They are the fibers in which the chain molecules are highly oriented along the fiber axis so the strength of the chemical bond can be exploited. Aramids are generally prepared by the reaction between an amine group and a carboxylic acid halide group. Simple AB homopolymers may look like:

$$nNH_2\text{-Ar-COCl} \rightarrow \text{-(NH-Ar-CO)}_{n-} + nHCl$$

The most well-known aramids (Nomex, Kevlar, Twaron, X-fiper, and New Star) are AABB polymers. Nomex, X-Fiper, New Star, and Teijinconex contain predominantly the meta-linkage and are poly-metaphenylene isophtalamides (MPIA). The molecules are arranged in planar sheets (Fig. 6-9) with strong covalent bonds in the molecular chains parallel to the fiber axis but weak hydrogen interchain bonding in the transverse direction. The planar sheets

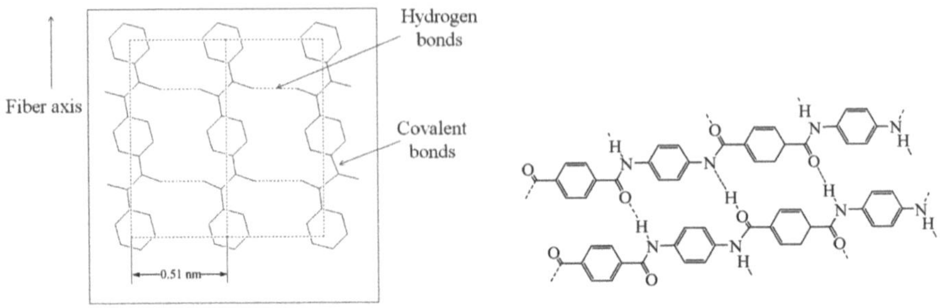

Fig. 6-9 The planar arrangement of the aramid molecules showing the weak hydrogen and strong covalent bonds.

are in turn pleated and stacked to give the three-dimensional orientated structure with radial symmetry of Kevlar 49.

Aramid fibers have good high temperature properties for polymeric materials [80, 81]. They have a glass transition temperature of about 360 °C burn with difficulty and do not melt like nylon. A loss in performance due to carbonization occurs around 425 °C but they can be used at elevated temperatures for sustained periods and even at 300 °C for a limited time. Aramid fibers do have a tendency to degrade in sunlight. Their dimensional stability is good as the coefficient of thermal expansion is low (approximately $-4 \times 10^{-6}\ K^{-1}$). Other properties which are important in certain applications are as follows: good resistance to abrasion, good resistance to organic solvents, nonconductive, very low coefficient of thermal expansion, no melting point, degradation starts from 500 °C, low flammability, good fabric integrity at elevated temperatures, sensitive to acids and salts, sensitive to ultraviolet radiation, high fracture toughness (impact resistance), high cut resistance, and textile processibility.

Para-aramid fibers such as Kevlar and Twaron provide outstanding strength-to-weight properties: high Young's modulus, high tenacity, low creep, and low elongation at break (~3.5%). The disadvantages of Kevlar are the ability to absorb moisture, difficulties in cutting, and low compressive strength.
There are several modifications of Kevlar developed for various applications:

1. Kevlar 29: high strength (520,000 psi/3600 MPa), low density (90 lb/ft^3/1440 kg/m^3) fibers used for manufacturing bullet-proof vests, composite armor reinforcement, helmets, ropes, cables, and asbestos replacing parts.
2. Kevlar 49: high modulus (19000 ksi/131 GPa), high strength (550000 psi/3800 MPa), low density (90 lb/ft^3/1440 kg/m^3) fibers used in aerospace, automotive and marine applications.
3. Kevlar 149: ultra high modulus (27000 ksi/186 GPa), high strength (490000 psi/3400 MPa), low density (92 lb/ft^3/1470 kg/m^3) highly crystalline fibers used as reinforcing dispersed phase for composite aircraft components.

Also, aramid fibers may protect carbon fibers and improve their properties: hybrid fabric (Aramid fibers + Carbon fibers) combines very high tensile strength with high impact and abrasion resistance.

6.2.4. Ultra-high-molecular-weight Polyethylene

Ultra-high-molecular-weight polyethylene is a subset of the thermoplastic polyethylene. It has extremely long chains with molecular weight numbering in the millions usually between 2 and 6 million. The longer chain serves to transfer load more effectively to the polymer backbone by strengthening intermolecular interactions. This results in a very tough material with the highest impact strength of any thermoplastic presently made. It is highly resistant to corrosive chemicals with exception of oxidizing acids. It has extremely low moisture absorption, has a very low coefficient of friction, is self-lubricating, and is highly resistant to abrasion (15 times more resistant to abrasion than carbon steel). Its coefficient of friction is significantly lower than that of nylon and acetal and is comparable to that of Teflon but ultra-high-molecular-weight polyethylene has better abrasion resistance than Teflon. It is odorless, tasteless, and non-toxic.

Ultra-high-molecular-weight polyethylene is a type of polyolefin structure and despite relatively weak van der Waals bonds between its molecules derives ample strength from the length of each individual molecule. It is made up of extremely long chains of polyethylene which all align in the same direction. Each chain is bonded to the others with so many van der Waals bonds that the whole can support great tensile loads. When formed to fibers, the polymer chains can attain a parallel orientation greater than 95 percent and a level of crystallinity of up to 85 percent.

Dyneema (Toyobo Co.) and Spectra (Honeywell Co.) are gel spun through a spinneret to form oriented-strand synthetic fibers of ultra-high-molecular-weight polyethylene, which are one of the world's strongest and lightest fibers and have yield strengths as high as 2.4 GPa and density as low as 0.97 kg/L. High-strength steels have comparable yield strengths and low-carbon steels have yield strengths much lower (around 0.5 GPa). Since steel has a density approximately equal to 7.8 kg/L, this gives the strength-to-weight ratios for these materials in a range from 10 to 100 times higher than steel. The strength-to-weight ratios for Dyneema are about 40 percent higher than for aramid fibers.

With outstanding toughness and extraordinary visco-elastic properties, ultra-high-molecular-weight polyethylene fiber can withstand high-load strain-rate velocities. Light enough to float, it also exhibits high resistance to chemicals, water, and ultraviolet light. It has excellent vibration damping, flex fatigue and internal fiber-friction characteristics and the low dielectric constant make it virtually transparent to radar.

Ultra-high-molecular-weight polyethylene fiber is used in numerous high-performance applications including police and military ballistic-resistant vests, helmets, and armored vehicles as well as sailcloth, fishing lines, marine cordage, lifting slings, and cut-resistant gloves and apparel [82, 83]. Ultra-high-molecular-weight polyethylene fiber is also used in armor, in particular, personal armor and on occasion as vehicle armor.

6.2.5. Ceramic Fibers

Ceramic fibers are mostly used as refractory fibers in uses over 1000 °C and are characterized by a polycrystalline structure rather than amorphous. Refractory ceramic fibers are mostly used for thermal insulation at high temperatures and to make special composites. They are very expensive fibers because only a small quantity is produced and they are used in particular fields such as aerospace [84–86].

Materials used are boron, silica, carbide, alumina, and zirconia. They show as filaments or very thin fibers (wiskers); bigger monofilaments or multifilaments can be used in textile transformation (yarn, ropes, parts of composites). Their characteristics are bound to thin structure of the materials, according to the kind of fiber. Finally, there are more sophisticated kinds of fibers, with a basis of borate, carbide, silicon nitrure, and borate nitrure.

Fibers act according to their shape and thickness. The thinner they are, the most flexible they are and are used for light and soft refractories (the thicker fibers give problems of skin irritation due to the penetration of these very thin fibers). When fibers have a diameter over 1 mm, they can be considered as transformable fibers according to the textile system. Wiskers are absolutely the most resistant (modulus >600 GPa, tenacity >20,000 MPa). Refractory fibers are low chemical reactivity. Their best characteristics are their oxidation resistance at high temperatures and their inactivity toward corrosion in composites with metal matrix.

6.2.6. Boron Fibers

Boron fiber is produced by chemical vapor deposition from boron trichloride on to a heated substrate [87, 88]. The temperatures involved are high and so the substrate has to have a high melting point; the most commonly used substrate is a heated tungsten wire of about 10 μm diameter. Continuous lengths of up to 3000 m have been made. The boron trichloride is mixed with hydrogen gas and decomposes according to:

$$2BCl_3\ (\text{gas}) + H_2\ (\text{gas}) = 2B\ (\text{solid}) + 6HCl\ (\text{gas})$$

to give an approximately 50 μm thick coating of boron with an extremely small grain size of only 2–3 nm. The boron exists in two crystalline forms namely rhombohedral and tetragonal. The former is thought to be predominant in the fibers but the details of the structure and properties depend critically on the deposition temperature and rate. Interaction of the boron with the tungsten core can occur to yield tungsten borides.

Boron is a metal known for its exceptional resistance and form: even if it is very light as a filament, it has a triple resistance and a double form in comparison to steel. Among reinforcement fibers, borate is the only one that is exceptionally resistant to traction, compression, and flexion together with high form and low density.
The physical-mechanical characteristics of boron fibers are:

1. Density (g/cm^3): 2.58
2. Tenacity (CN/dtex): 13.6
3. Elongation at break (%): 0.8
4. Melting point (°C): 2,000

Besides the exceptional resistance to traction and temperature, boron fiber produced by AVCO shows good resistance to acids and alkali, to organic solvents, to ultra-violet rays, and to micro-organism. It is a very particular fiber gotten with sophisticated technologies and a complex production, so its cost is high. On the other hand, boron fibers are almost only used in composites that suffer very high temperatures.

6.2.7. Metal Fibers

The properties of the metal fibers create strong mechanical interlocking of the fiber structure allowing a very low density (high flow) media with excellent isotropic strength and longevity. Fibers are available in stainless steel and many nonferrous alloys in sizes from 1 μ to 50 μ. Metal fibers, ferrous, and nonferrous are easy to produce particularly by wire drawing processes. They are less sensitive to surface damage, are extremely strong, have excellent resistance to high temperature, are easier to handle, and are inherently more ductile than ceramic, polycrystalline, or multiphase materials. The disadvantages of metal fibers are their added weight compared with ceramic or multiphase materials and their tendency to alloy with metal matrices. The incorporation of fine wires in a refractory ceramic resulted in an improvement in the impact resistance and thermal shock properties. They are also used to reinforce polymer matrices such as epoxy or polyethylene reinforced with steel wire, steel wire reinforced pneumatic tyres, and nickel reinforced electromagnetic shielding enclosures [89, 90]. The properties of fibers and conventional bulk materials are listed in Table 6-11.

6.2.8. Particulates (Fillers)

Particulates (Fillers) are another type of reinforcement in composite materials [91–93]. Fillers are used in matrix resins for a variety of reasons: cost reduction, improved processing, density control, optical effects, thermal conductivity, control of thermal expansion, electrical properties, magnetic properties, flame

Table 6-11 Properties of Fibers and Conventional Bulk Materials

Type of materials		Diameter (m)	Density ($g \cdot cm^{-3}$)	Tensile modulus (GPa)	Tensile strength (MPa)	Melting point (°C)	%Elongation at break	Relative cost
Fibers	E-glass	7	2.54	70	3.45	1540+	4.8	Low
	S-glass	15	2.50	86	4.50	1540+	5.7	Moderate
	Graphite, high modulus	7.5	1.9	400	1.8	>3500	1.5	High
	Graphite, high strength	7.5	1.7	240	2.6	>3500	0.8	High
	Boron	130	2.6	400	3.5	2300	-	High
	Kevlar 29	12	1.45	80	2.8	500(D)	3.5	Moderate
	Kevlar 49	12	1.45	130	2.8	500(D)	2.5	Moderate
Bulk materials	Steel		7.8	208	0.34 -2.1	1480	5-25	<Low
	Aluminum alloys		2.7	69	0.14 -0.62	600	8-16	Low

retardancy and improved mechanical properties such as hardness and tear resistance. Each filler type has different properties and these in turn are influenced by the particle size, shape, and surface chemistry. Filler characteristics are discussed from costs to particle morphology. Particle specific surface area and packing are important aspects. Filler loading is also a critical factor to decide of properties of composite materials. The principal filler types are outlined. These include carbon black, natural mineral fillers and synthetic mineral fillers. Carbon blacks are very important fillers especially in the rubber industry. Filler surface modification is an important topic. Most particulate fillers are inorganic and polar which can give rise to poor compatibility with the hydrocarbon polymers and processing problems among the other effects.

6.2.9. Reinforcement Forms

Regardless of the materials reinforcements are available in forms to serve a wide range of processes and end-product requirements. Materials supplied as reinforcement include roving, milled fiber, chopped strands, continuous, chopped, or thermo-formable mat. Reinforcement materials can be designed with unique fiber architectures and be preformed (shaped) depending on the product requirements and manufacturing process.

6.2.9.1. Multi-end and Single-end rovings

Rovings are utilized primarily in thermoset compounds but can be utilized in thermoplastics. Multi-end rovings consist of many individual strands or bundles of filaments which are then chopped and randomly deposited into the matrix resin. Processes such as sheet molding compound, preform, and spray-up use the multi-end roving. Multi-end rovings can also be used in some filament winding and pultrusion applications. The single-end roving consists of many individual filaments wound into a single strand. The product is generally used in processes that utilize a unidirectional reinforcement such as filament winding or pultrusion.

6.2.9.2. Mats

Reinforcing mats are usually described by weight-per-unit-of-area. For instance, a 2 ounce chopped strand mat will weigh 2 ounces per square yard. The type and amount of binder that is used to hold the mat together dictate differences between mat products. In some processes such as hand lay-up, it is necessary for the binder to dissolve. In other processes, particularly in compression molding the binder must withstand the hydraulic forces and the dissolving action of the matrix resin during molding. Therefore, two general categories of mats are produced and are known as soluble and insoluble.

6.2.9.3. *Woven, stitched, and braided fabrics*

There are many types of fabrics that can be used to reinforce resins in a composite. The multidirectional reinforcements are produced by weaving, knitting, stitched, or braiding continuous fibers into a fabric from twisted and plied yarn. Fabrics refer to all flat-sheet, roll goods, whether or not they are strictly fabrics. Fabrics can be manufactured utilizing almost any reinforcing fiber. The most common fabrics are constructed with fiberglass, carbon, or aramid. Fabrics are available in several weave constructions and thickness (from 0.0010 to 0.40 inches). Fabrics offer oriented strengths and high reinforcement loadings often found in high performance applications.

Fabrics are typically supplied on rolls of 25 to 300 yards in length and 1 to 120 inches in width. The fabric must be inherently stable enough to be handled, cut, and transported to the mold but pliable enough to conform to the mold shape and contours. Properly designed, the fabric will allow for quick wet out and wet through of the resin and will stay in place once the resin is applied. Fabrics like rovings and chopped strands come with specific sizings or binder systems that promote adhesion to the resin system.

Fabrics allow for the precise placement of the reinforcement. This cannot be done with milled fibers or chopped strands and is only possible with continuous strands using relatively expensive fiber placement equipment. Due to the continuous nature of the fibers in most fabrics, the strength to weight ratio is much higher than that for the cut or chopped fiber versions. Stitched fabrics allow for customized fiber orientations within the fabric structure. This can be of great advantage when designing for shear or torsional stability.

Woven fabrics are fabricated on looms in a variety of weights, weaves, and widths. In a plain weave, each fill yarn or roving is alternately crossed over and under each warp fiber allowing the fabric to be more drapeable and conform to curve surfaces. Woven fabrics are manufactured where half of the strands of the fiber are laid at right angles to the other half (0–90°) (Fig. 6-10). Woven fabrics are commonly used in boat manufacturing.

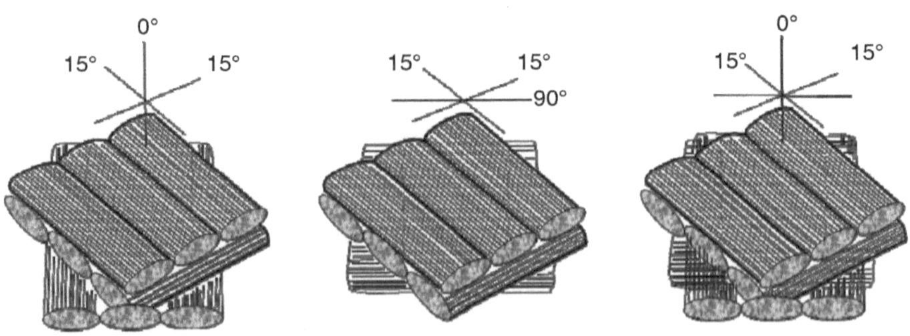

Fig. 6-10 Diagram of stitched triaxial and quadraxial fabrics.

Stitched fabrics also known as non-woven, non-crimped, stitched, or knitted fabrics have optimized strength properties because of the fiber architecture. Woven fabric is where two sets of interlaced continuous fibers are oriented in a 0° and 90° pattern where the fibers are crimped and not straight. Stitched fabrics are produced by assembling successive layers of aligned fibers. Typically the available fiber orientations include the 0° directions (warp), 90° direction (weft or fill), and ±45° direction (bias). The assembly of each layer is then sewn together. This type of construction allows for load sharing between the fibers so that a higher modulus both tensile and flexural is typically observed. The fiber architecture construction allows for optimum resin flow when composites are manufactured. These fabrics have been traditionally used in boat hulls for 50 years. Other applications include light poles, wind turbine blades, trucks, buses, and underground tanks. These fabrics are currently used in bridge decks and column repair systems. Multiple orientations provide a quasi-isotropic reinforcement.

Braided fabrics are engineered with a system of two or more yarns intertwined in such a way that all of the yarns are interlocked for optimum load distribution. Biaxial braids provide reinforcement in the bias direction only with fiber angles ranging from ±15 to ±95 °C (Fig. 6-11). Triaxial braids provide reinforcement in the bias direction with fiber angles ranging from ±10 to ±80 °C and axial (0 °C) direction.

6.2.9.4. Unidirectional

Unidirectional reinforcements include tapes, tows, unidirectional tow sheets and rovings (which are collections of fibers or strands). Fibers in this form are all aligned parallel in one direction and uncrimped providing the highest mechanical properties (Fig. 6-12). Composites using unidirectional tapes or sheets have high strength in the direction of the fiber. Unidirectional sheets are thin and multiple layers are required for most structural applications.

In some composite designs it may be necessary to provide a corrosion or weather barrier to the surface of a product. A surface veil is a fabric made from

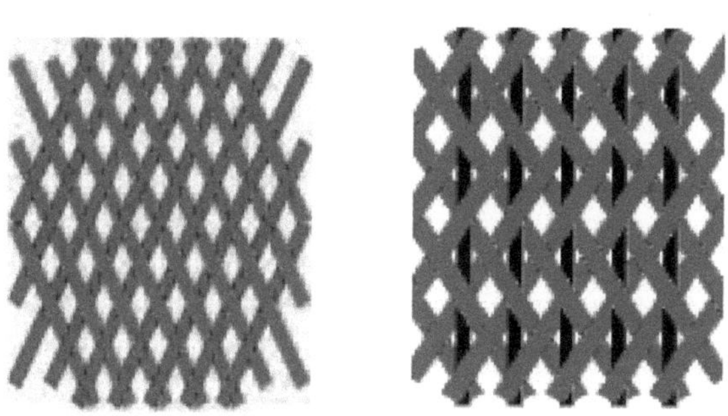

Fig. 6-11 Biaxial and triaxial braided fabrics.

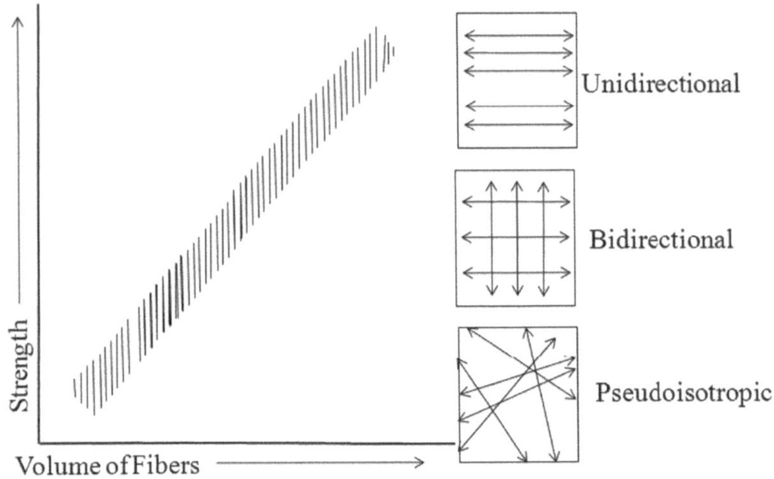

Fig. 6-12 Strength relation to fiber orientation.

nylon or polyester that acts as a very thin sponge that can absorb resin to 90 percent of its volume. This helps to provide an extra layer of protective resin on the surface of the product. Surface veils are used to improve the surface appearance and insure the presence of a corrosion resistance barrier for typical composites products such as pipes, tanks and other chemical process equipment. Other benefits include increased resistance to abrasion, ultraviolet (UV), and other weathering forces. Veils may be used in conjunction with gel coats to provide reinforcement to the resin.

6.2.9.5. Prepreg

Prepregs are a ready-made material made of a reinforcement form and polymer matrix. Passing reinforcing fibers or forms such as fabrics through a resin bath is used to make a prepreg. The resin is saturated (impregnated) into the fiber and then heated to advance the curing reaction to different curing stages. Thermoset or thermoplastic prepregs are available and can be either stored in a refrigerator or at room temperature depending on the constituent materials. Prepregs can be manually or mechanically applied at various directions based on the design requirements.

6.3. MATRICES

6.3.1. Polymer Matrices

The primary functions of the resin are to transfer stress between the reinforcing fibers act as a glue to hold the fibers together and protect the fibers from mechanical and environmental damage. For fibers to carry maximum load the matrix must have a lower modulus and against elongation than the reinforcement. The matrix determines the service operating temperature of composites as well as processing the parameters for part manufacturing.

Matrix resins are divided into two major groups known as thermoset and thermoplastic. The thermoset resins are usually liquids or low melting point solids in their initial form. When used to produce finished goods these thermosetting resins are 'cured' by the use of a catalyst, heat, or a combination of the two. Once cured, solid thermoset resins cannot be converted back to their original liquid form. Unlike thermoplastic resins cured thermosets will not melt and flow but will soften when heated (and lose hardness) and once formed they cannot be reshaped. Thermoplastic resins on the other hand become soft when heated and may be shaped or molded while in a heated semi-fluid state and become rigid when cooled. Maximum continuous-use temperatures of the various types of thermoset and thermoplastic resins are shown in Table 6-12.

6.3.1.1. Thermoset resins

Thermosets are crosslinked polymers that are cured or set using heat or heat and pressure which once cured cannot be remelted or reformed. During curing they form three-dimensional molecular chain called crosslinking as shown in

Table 6-12 Maximum Continuous-use Temperature for Various Thermosets and Thermoplastics

Materials	Maximum continuous-use temperatures (°C)
Thermosets	
Vinylester	60 ~ 150
Polyester	60 ~ 150
Phenolics	70 ~ 150
Epoxy	80 ~ 215
Cyanate esters	150 ~ 250
Bismaleimide	230 ~ 320
Thermoplastics	
Polyethylene	50 ~ 80
Polypropylene	50 ~ 75
Acetal	70 ~ 95
Nylon	75 ~ 100
Polyester	70 ~ 120
Poly(phenylene sulfide)	120 ~ 220
Poly(ether ether ketone)	120 ~ 250
Teflon	200 ~ 260

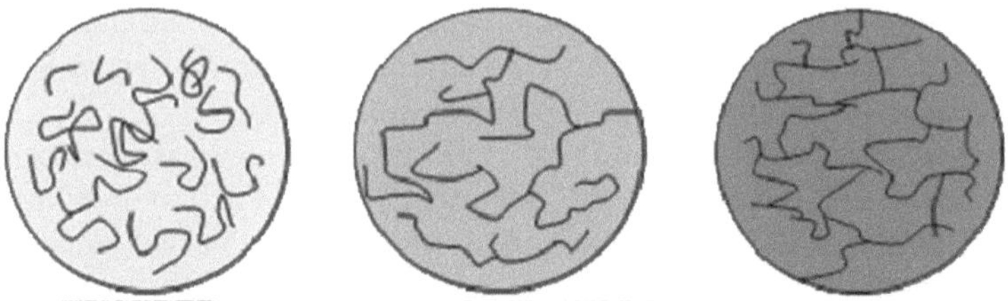

Fig. 6-13 Crosslinking of thermoset molecules during curing.

Fig. 6-13. The higher the number of crosslinkings the more rigid and thermally stable the materials will be. Cured thermoset resins may soften when heated but do not melt or flow. They generally have a higher resistance to heat than the thermoplastics. Because thermosets are brittle in nature they are used with filler materials such as powders or fibers to provide improved strength and/or stiffness [94–97]. Fibers can be either chopped or wound and commonly include glass, glass fibers, or cloth. Some products contain solid lubricant fillers such as graphite or molybdenum disulfide. Others contain aramid fibers, metal powders, or inorganic fillers with ceramics and silicates.

Thermoset resins provide easy processability and better fiber impregnation because the liquid resin is used at room temperature for various processes such as filament winding, pultrusion, resin transfer molding, and reaction injection molding. Thermosets offer greater thermal and dimensional stability, better rigidity, and higher electrical, chemical, and solvent resistance. The most common resins used in thermoset composites are epoxy, polyester, vinylester, phenolics, cyanate esters, bismaleimides, polyurethane, etc. Some of basic properties of thermoset resins are listed in Table 6-13.

Table 6-13 Typical Properties of Thermoset Resins

Resins	Density (g/cm^3)	Tensile modulus (GPa)	Tensile strength (MPa)
Epoxy	1.2 ~ 1.4	2.5 ~ 5.0	50 ~ 110
Phenolics	1.2 ~ 1.4	2.7 ~ 4.1	35 ~ 60
Polyesters	1.1 ~ 1.4	1.6 ~ 4.1	35 ~ 90
Vinylesters	1.04 ~ 1.1	3.2 ~ 3.6	68 ~ 80

6.3.1.1.1. Polyesters

Polyester resins are the most widely used resin systems particularly in the marine industry. By far the majority of dinghies, yachts and workboats built in composites make use of this resin system. Polyesters are low-cost resin systems and offer excellent corrosion resistance. The operating service temperatures for polyesters are lower than for epoxies. Polyesters are widely used for pultrusion, filament winding, sheet molding compounds, and resin transfer molding operations. A range of raw materials and processing techniques are available to achieve the desired properties in the formulated or processed polyester resin. Polyesters are versatile because of their capacity to be modified or tailored during the building of the polymer chains. They have been found to have almost unlimited usefulness in all segments of the composites industry. The principal advantage of these resins is a balance of properties (including mechanical, chemical, and electrical) dimensional stability, cost, and ease of handling or processing.

Polyesters can be a thermosetting resin or a thermoplastic resin. Thermoset polyesters are produced by the condensation polymerization of dicarboxylic acids and difunctional alcohols (glycols). In addition, unsaturated polyesters contain unsaturated materials such as maleic anhydride or fumaric acid as part of the dicarboxylic acid component. For the curing or cross-linking process a reactive monomer such as styrene is added in the 30–50 wt.% range.

There are two principle types of polyester resin used as standard laminating systems in the composite industry. Orthophthalic polyester resin is the standard economic resin used by many people. Isophthalic polyester resin is now becoming the preferred material in industries such as the marine industry where its superior water resistance is desirable. Fig. 6-14 shows the idealized chemical structure of Q typical polyester. Note the positions of the ester groups (CO–O–C) and the reactive sites (C*=C*) within the molecular chain.

Most polyester resins are viscous, pale colored liquids consisting of a solution of polyester in a monomer which is usually styrene. The addition of styrene in amounts of up to 50 percent helps to make the resin easier to handle by

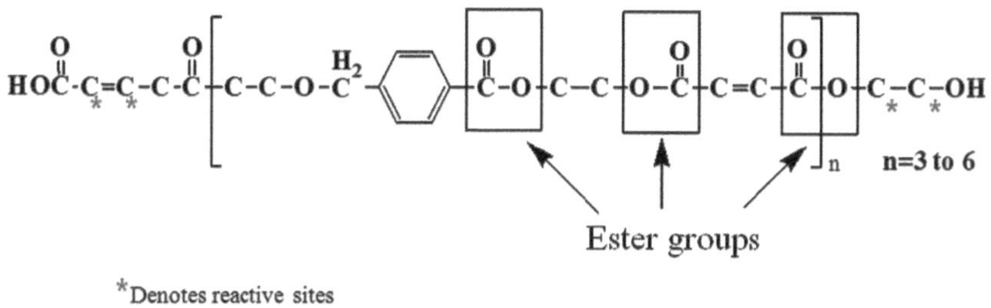

Fig. 6-14 Idealized chemical structure of typical isophthalic polyester.

reducing its viscosity. The styrene also performs the vital function of enabling the resin to cure from a liquid to a solid by 'cross-linking' the molecular chains of the polyester without the evolution of any by-products. These resins can therefore be molded without the use of pressure and are called 'contact' or 'low pressure' resins. Polyester resins have a limited storage life as they will set or 'gel' on their own over a long period of time. Often small quantities of inhibitors are added during the resin manufacture to slow this gelling action. For use in molding a polyester resin requires the addition of several ancillary products.

These products are generally

1. Catalyst
2. Accelerator
3. Additives: Thixotropic
Pigment
Filler
Chemical/fire resistance

A manufacturer may supply the resin in its basic form or with any of the above additives already included. Resins can be formulated to the molder's requirements ready simply for the addition of the catalyst prior to molding. As has been mentioned given enough time an unsaturated polyester resin will set by itself. This rate of polymerization is too slow for practical purposes and therefore catalysts and accelerators are used to achieve the polymerization of the resin within a practical time period. Catalysts are added to the resin system shortly before use to initiate the polymerization reaction. The catalyst does not take part in the chemical reaction but simply activates the process. An accelerator is added to the catalyzed resin to enable the reaction to proceed at workshop temperature and/or at a greater rate. Since accelerators have little influence on the resin in the absence of a catalyst they are sometimes added to the resin by the polyester manufacturer to create a 'pre-accelerated' resin. The molecular chains of the polyester can be represented as follows (Fig. 6-15 (a)) where 'B' indicates the reactive sites in the molecule.

With the addition of styrene 'S' and in the presence of a catalyst the styrene cross links the polymer chains at each of the reactive sites to form a highly

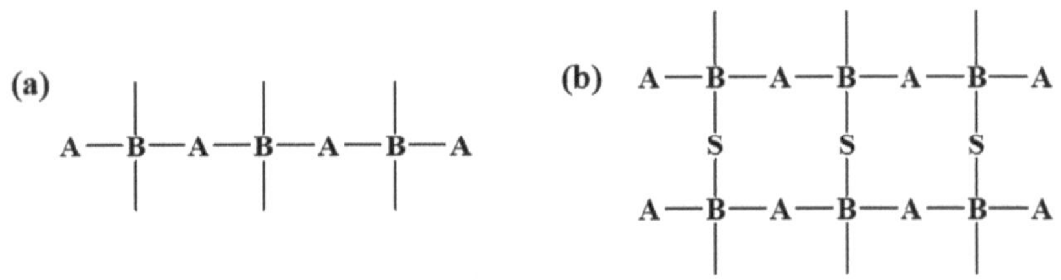

Fig. 6-15 Schematic representation of polyester resin: (a) uncured and (b) cured.

complex three-dimensional network as represented in Fig. 6-15 (b). The polyester resin is then said to be 'cured'. It is now a chemically resistant (and usually) hard solid. The cross-linking or curing process is called 'polymerization'. It is a non-reversible chemical reaction. The 'side-by-side' nature of this cross-linking of the molecular chains tends to mean that the polyester laminates suffer from brittleness when shock loadings are applied.

6.3.1.1.2. Vinyl ester

Vinyl esters were developed to combine the advantages of epoxy resins with the better handling/faster cure which are typical for unsaturated polyester resins. These resins are produced by reacting epoxy resin with acrylic or methacrylic acid. This provides an unsaturated site much like that produced in the polyester resins when maleic anhydride is used. The resulting materials are dissolved in styrene to yield a liquid that is similar to polyester resin. Vinyl esters are also cured with the conventional organic peroxides used with polyester resins. Vinyl esters offer mechanical toughness and excellent corrosion resistance. These enhanced properties are obtained without complex processing handling or special shop fabricating practices that are typical with the epoxy resins. Vinylesters are widely used for pultrusion, filament winding, sheet molding compounds, and resin transfer molding processes. They offer good chemical and corrosion resistance and are used for fiber-reinforced polymer pipes and tanks in the chemical industry. They are cheaper than epoxies and are used in the automotive and other high-volume applications where the cost is critical in making material selection.

The vinyl ester molecule also features fewer ester groups. These ester groups are susceptible to water degradation by hydrolysis which means that vinyl esters exhibit better resistance to water and many other chemicals than their polyester counterparts and are frequently found in applications such as pipelines and chemical storage tanks.

Fig. 6-16 below shows the idealized chemical structure of a typical vinyl ester. Note the positions of the ester groups and the reactive sites (C*=C*) within the molecular chain.

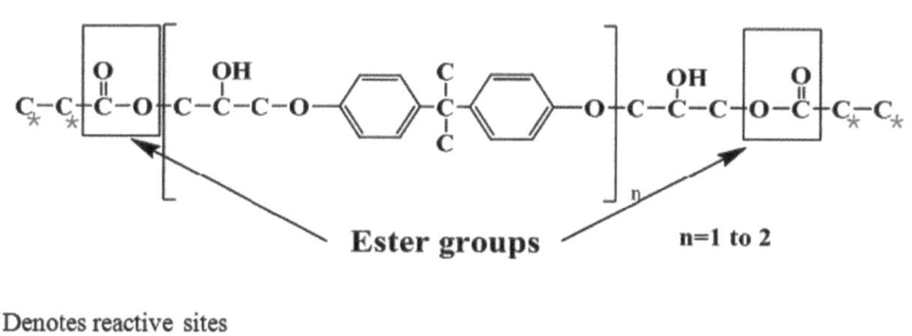

Fig. 6-16 Idealized chemical structure of a typical epoxy based vinyl ester.

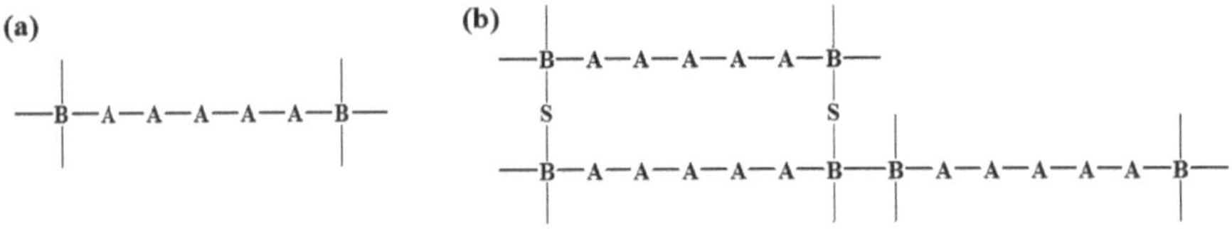

Fig. 6-17 Schematic representation of vinyl ester resin: (a) uncured and (b) cured.

The molecular chains of vinyl ester represented in Fig. 6-17 can be compared to the schematic representation of polyester shown previously where the difference in the location of the reactive sites can be clearly seen:

With the reduced number of ester groups in a vinyl ester, when compared to polyester, the resin is less prone to damage by hydrolysis. The material is, therefore, sometimes used as a barrier or 'skin' coat for a polyester laminate that is to be immersed in water such as in a boat hull. The cured molecular structure of the vinyl ester (Fig. 6-17) also means that it tends to be tougher than a polyester, although to really achieve these properties, the resin usually needs to have an elevated temperature post cure.

6.3.1.1.3. Epoxy

The large family of epoxy resins represents some of the highest performance resins of those available at this time. Epoxies generally out-perform most other resin types in terms of mechanical properties and resistance to environmental degradation which leads to their almost exclusive use in aircraft components.

The term 'epoxy' refers to a chemical group consisting of an oxygen atom bonded to two carbon atoms that are already bonded in some way. The simplest epoxy is a three-member ring structure known by the term 'alpha-epoxy' or '1,2-epoxy'. The idealized chemical structure is shown in Fig. 6-18 below and is the most easily identified characteristic of any more complex epoxy molecule.

Epoxy resins are formed from a long chain molecular structure similar to vinyl ester with reactive sites at either end. In the epoxy resin, however, epoxy groups instead of the ester groups form these reactive sites. The absence of ester groups means that the epoxy resin has particularly good water resistance. The epoxy molecule also contains two ring groups at its centre which are able to absorb both mechanical and thermal stresses better than linear groups and

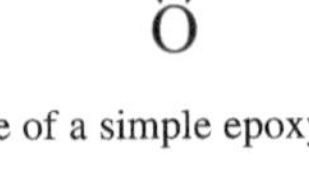

Fig. 6-18 Idealized chemical structure of a simple epoxy (ethylene oxide).

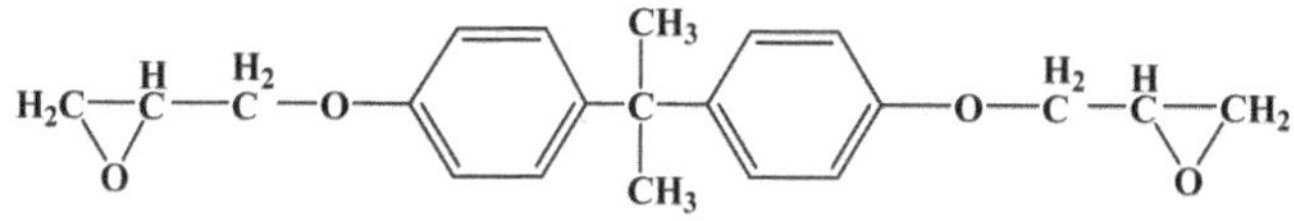

Fig. 6-19 Idealized chemical structure of a typical epoxy (diglycidyl ether of bisphenol-A) resins.

therefore give the epoxy resin very good stiffness, toughness, and heat resistant properties. Fig. 6-19 shows the idealized chemical structure of a typical epoxy. Note the absence of the ester groups within the molecular chain.

Epoxies differ from polyester resins; in that a 'hardener' rather than a catalyst cures them. The hardener often an amine is used to cure the epoxy by an 'addition reaction' where both materials take place in the chemical reaction. The chemistry of this reaction means that there are usually two epoxy sites binding to each amine site. This forms a complex three-dimensional molecular structure which is illustrated in Fig. 6-20.

Since the amine molecules 'co-react' with the epoxy molecules in a fixed ratio, it is essential that the correct mix ratio is obtained between resin and hardener to ensure that a complete reaction takes place. If amine and epoxy are not mixed in the correct ratios, unreacted resin or hardener will remain within the matrix which will affect the final properties after cure. To assist with the accurate mixing of the resin and the hardener, manufacturers usually formulate the components to give a simple mix ratio that is easily achieved by measuring out by weight or volume.

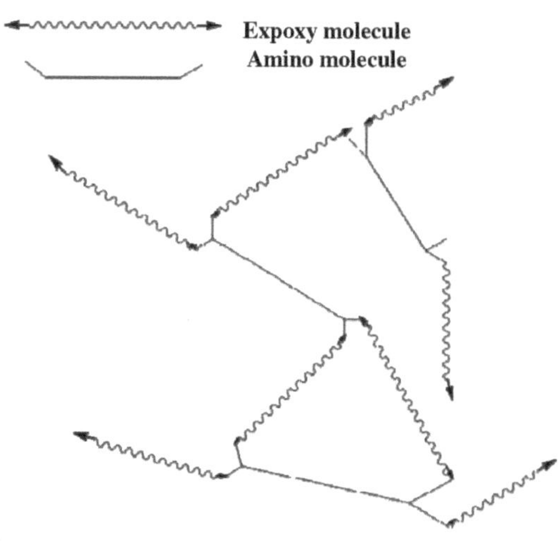

Fig. 6-20 Schematic representation of epoxy resins (cured 3-d structure).

Epoxy resins have a well-established record in a wide range of composite parts structures and concrete repair. The structure of the resin can be engineered to yield a number of different products with varying levels of performance. A major benefit of epoxy resins over unsaturated polyester resins is their lower shrinkage. Epoxy resins can also be formulated with different materials or blended with other epoxy resins to achieve specific performance features. Generally, the epoxies are cured by addition of an anhydride or an amine hardener as a 2-part system. Different hardeners as well as quantity of a hardener produce a different cure profile and give different properties to the finished composite. Cure rates can be controlled to match process requirements through the proper selection of hardeners and/or catalyst systems.

Epoxy resins are used with a number of fibrous reinforcing materials including glass, carbon, and aramid for high performance aerospace applications. Fiber reinforced epoxies are used in the sport and leisure industry and they have also replaced the reciprocating metal components particularly in the weaving industry. Epoxy resins are compatible with most composite manufacturing processes particularly vacuum-bag molding, autoclave molding, pressure-bag molding, compression molding, filament winding, and hand lay-up.

Epoxy-based composites provide good performance at room and elevated temperatures. Epoxy resins can operate well up to temperatures of 200 to 250 °F and there are epoxy resins that can perform well up to 400 °F. For high-temperature and high-performance epoxy resins the cost increases but they offer good chemical and corrosion resistance.

6.3.1.1.4. Phenolics

Phenolics are a class of resins commonly formed by the reaction of phenol (carbolic acid) and formaldehyde and catalyzed by an acid or base. Urea, resorcinol, or melamine can be used instead of phenol to obtain different properties. Their cure characteristics are different than other thermosetting resins such as epoxies due to the fact that water is generated during the cure reaction. Phenolics are used for aircraft interiors, stowbins, and galley walls as well as other commercial markets that require low-cost, flame-resistant, and low-smoke products. Pigmented applications are limited to red, brown, or black.

Phenolic composites have many desirable performance qualities including high temperature resistance, creep resistance, excellent thermal insulation and sound damping properties, corrosion resistance, and excellent fire/smoke/smoke toxicity properties. Phenolics are used for various composite manufacturing processes such as filament winding, resin transfer molding, injection molding, and compression molding. Phenolics provide easy processability, tight tolerances, reduced machining, and high strength. Because of their high temperature resistance, phenolics are used in exhaust components, missile parts, manifold spacers, commutators, and disc brakes.

6.3.1.1.5. Cyanate esters

Cyanate esters offer excellent strength and toughness better electrical properties and lower moisture absorption compared to other resins. If they are formulated correctly their high-temperature properties are similar to bismaleimide and polyimide resins. They are used for a variety of applications including spacecrafts, aircrafts, missiles, antennae, radomes, microelectronics, and microwave products. Cyanate esters are formed via the reaction of bisphenol esters and cyanic acid that cyclotrimerize to produce triazine rings during a second cure. Cyanate esters are more easily cured than epoxies. The toughness of cyanate esters can be increased by adding thermoplastics or spherical rubber particles.

6.3.1.1.6. Bismaleimide

Bismaleimide is produced by a vinyl-type polymerization of a pre-polymer terminated with two maleimide groups. Bismaleimide has an intermediate temperature capability (between epoxy and polyimide). Monomers are usually synthesized from maleic anhydride and an aromatic diamine; the bismaleamic acid formed is cyclodehydrated to a bismaleimide resin. The double bond of the maleimide is very reactive and can undergo chain extension reactions. These addition-type polyimides or maleimide-based polyimides are used in high performance structural composites requiring higher temperature use and increased toughness. Bismaleimides are hardly used in their pure, solid form. In most applications they come in conjunction with the reactive co-monomers (e.g., vinyl and allyl compounds, allyl phenols, aromatic amines, etc.). A great advantage of these mixtures over the pure powder is better processability. With the help of the co-monomers bismaleimide powders can be turned into paste-like liquids and cast to the desired shape. However, the viscosity of such systems is often very high so that solvents are added in order to make processing easier or even to make processing possible at all. A typical field of application of such bismaleimide systems is their use as matrix resins for high-performance fiber-reinforced composite materials in the aviation and space industries. Very often these matrix resins require high curing temperatures (more than 200 °C) and long curing times.

6.3.1.1.7. Polyurethane

Polyurethane is a family of polymers with widely ranging properties and uses all based on the exothermic reaction of organic polyisocyanates with polyols (an alcohol containing more than one hydroxyl group). A few basic constituents of different molecular weights and functionalities are used to produce the whole spectrum of polyurethane materials. The versatility of polyurethane chemistry enables the polyurethane chemist to engineer polyurethane resin to achieve the desired properties. Polyurethanes as an elastomer have superior toughness and abrasion, in such applications as solid tyres, wheels, bumper components, or insulation. There are many formulations of polyurethane foam to optimize the density for insulation, structural sandwich panels, and architectural components. Polyurethanes are often used to bond composite structures together. Polyurethane is widely used for structural reaction injection molding processes and reinforced reaction injection molding processes in which isocyanate and polyol are generally mixed in a ration of 1:1 in a reaction chamber and then rapidly injected into a closed mold containing short or long fiber reinforcements. The automotive industry is a big market for these processes. Polyurethane is currently used for automotive applications such as bumper beams, hoods, body panels, etc. Polyurethane is also used for wear and impact resistance coatings.

6.3.1.2. Thermoplastic resins

Thermoplastic resins are materials that soften when heated and harden again when cooled which are in general ductile and tougher than thermoset resins and are sued for a wide variety of non-structural applications without fillers and reinforcements [98–101]. Thermoplastics can be melted by heating and solidified by cooling which render them capable of repeated reshaping and reforming. Thermoplastic molecules do not crosslink and therefore they are flexible and reformable. Thermoplastics can be either amorphous or semicrystalline as shown in Fig. 6-21.

After the fast developments of new types of fibers with extremely high properties that took place in the last decades of the last century, the commercial availability of thermoplastic matrix composites represents an important recent innovation in the field of composites. The success of such a type of composites is due to different reasons interesting from the side of fabrication technologies from the point of view of eco-sustainability of the entire process and from the side of final performances.

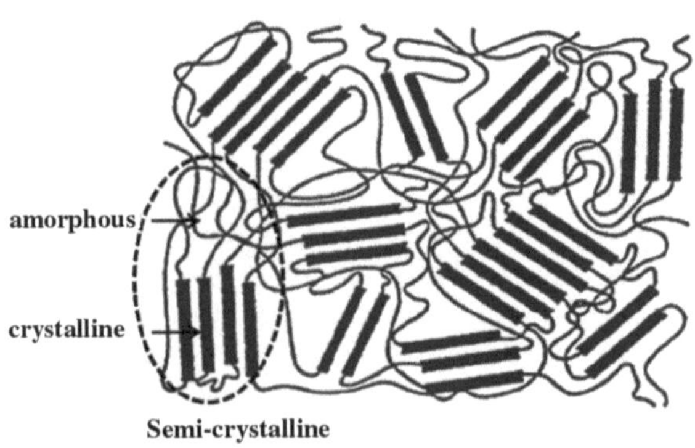

Fig. 6-21 Molecular arrangements in thermoplastic polymers.

In particular, the use of thermoplastic matrices avoids the presence of dangerous vapors in the working shops as in the case of thermosetting matrices and this represents at the same time a strong environmental improvement and an economical challenge for the fabrication, the latter being further strongly improved if one considers the necessity of expensive hardware and procedures needed in the use of preparing thermosetting composites. Thermoplastic composites imply the reversibility of thermal actions on the materials during fabrication of the final element. In other words it is possible to pre-fabricate semi-finished items and later take them to final shape and dimensions with no time limit for that. Several thermoplastic polymers can be considered as possible matrices for composites and many innovations are expected in the very close future.

Presently the most interesting for actual applications are polyethylene, polypropylene, polyester, nylon, polycarbonate, polyetherimide, poly (phenylene sulfide), and poly (ether ether ketone) produced in different places in the world. Using these matrices, there are also various semi-finished product producers whose products can be used to produce final elements. Some typical properties of common thermoplastic matrices are Table 6-14.

6.3.1.2.1. Polyethylene

PE is the most commonly used plastic in the world which is a semi-crystalline polymer with a linear molecular structure of repeating $-CH_2-CH_2$- units. In general, polyethylene is considered a strong, lightweight thermoplastic with very good chemical resistance. Further, the polymer elongates before breaking which enhances the toughness of porous parts made from polyethylene. Polyethylene is classified into several different categories based mostly on its density and branching. The mechanical properties of polyethylene depend

Table 6-14 Typical Properties of Thermoplastic Resins

Resins	Density (g/cm^3)	Tensile modulus (GPa)	Tensile strength (MPa)
Polypropylene	0.90	1–1.4	25–38
Polyethylene	0.9–1.0	0.7–1.4	20–35
Polycarbonate	1.06–1.2	2.2–2.4	45–70
Nylon	1.1	1.3–3.5	55–90
Polyester	1.3–1.4	2.1–2.8	55–60
Poly(ether ether ketone)	1.3–1.35	3.5–4.4	100
Poly(phenylene sulfide)	1.3–1.4	3.4	80
Polyetherimide	1.27	3	105

significantly on variables such as the extent and type of branching, the crystal structure, and the molecular weight. Products made from standard polyethylene grades typically have pore size diameters ranging from 7 to 150 μm. Polyethylene is resistant to concentrated acids, alkalies, and many organic solvents. Depending on the crystallinity and molecular weight, a melting point and glass transition may or may not be observable. The temperature at which these occur varies strongly with the type of polyethylene. For common commercial grades of medium- and high-density polyethylene, the melting point is typically in the range 120–130 °C. The melting point for average, commercial, low-density polyethylene is typically 105–115 °C.

6.3.1.2.2. Polypropylene

Polypropylene is in many respects similar to high-density polyethylene. However, instead of forming a long polymer chain made up of repeating $-CH_2$-components, the polypropylene molecule has a series of CH_3 groups that hang off the main carbon backbone. Isotactic polypropylenes are composed with the CH_3 groups oriented on one side of the carbon backbone. This orientation creates a greater degree of cystallinity making polypropylene stiffer and more resistant to creep (the tendency to flow under stress) compared to polyethylene. Generally, polypropylene has demonstrated certain advantages in improved strength, stiffness, and higher temperature capability over polyethylene. Polypropylene has been very successfully applied to the forming of fibers due to its good specific strength which is why it is the single largest use of polypropylene. Available in molded shapes, sheets, and tubes, the nominal pore size for polypropylene ranges from 80 to 150 μm. Polypropylene is highly resistant to most acids and alkalies and is resistant to most organic solvents below 80 °C. When not stressed, polypropylene may be used at temperatures ranging from 93 to 149 °C. Further, polypropylene can normally be subjected to steam sterilization. Polypropylene is an inexpensive, ductile, low strength material with reasonable outdoor performance. The material surface is soft wax-like and scratches easily. Stiffness and strength are often improved using reinforcement of glass, chalk, or talc. The color is opaque and white but it can be dyed in many colors.

6.3.1.2.3. Nylon

Nylon was the first commercially successful synthetic polymer. There are two common methods of making nylon for fiber applications. In one approach, the molecules with an acid (COOH) group on each end are reacted with molecules containing amine (NH_2) groups on each end. The resulting nylon is named on the basis of the number of carbon atoms separating the two acid groups and the two amines. These are formed into monomers of intermediate molecular weight, which are then reacted to form long polymer chains. Nylon fibers are used in many applications including fabrics, bridal veils, carpets, musical strings, and ropes. Solid nylon is used for mechanical parts such as machine screws, gears, and other low- to medium-stress components previously cast in metal. Engineering-grade nylon is processed by extrusion, casting, and injection molding. Nylon can be used as the matrix in composite materials with reinforcing fibers like glass or carbon fiber and has a higher density than pure nylon. Such thermoplastic composites (25 percent glass fiber) are frequently used in car components next to the engine such as intake manifolds where the good heat resistance of such materials makes them feasible competitors to metals.

6.3.1.2.4. Polyester

Polyesters, as thermoplastics may change shape after the application of heat. While being combustible at high temperatures, polyesters tend to shrink away from flames and self-extinguish upon ignition. Polyester fibers have high tenacity and E-modulus as well as low water absorption and minimal shrinkage in comparison with other industrial fibers.

6.3.1.2.5. Polyphenylene sulfide

Polyphenylene sulfide is an engineering plastic, a high-performance thermoplastic with a maximum crystallinity of 65 percent. Poly (phenylene sulfide) can be molded, extruded, or machined to high tolerances. In its pure solid form, it is an opaque white to light tan. Maximum service temperature is 218 °C. Poly (phenylene sulfide) has not been found to dissolve in any solvent at temperatures below about 200 °C. Poly (phenylene sulfide) is one of the most important high temperature polymers because it exhibits a number of desirable properties. These properties include resistance to heat, acids, and alkalies and to mildew, to bleaches, aging, sunlight, and abrasion. It absorbs only small amounts of solvents and resists dyeing. Poly (phenylene sulfide)-based composites are used for applications where great strength and chemical resistance are required at elevated temperature.

6.3.1.2.6. Poly(ether ether ketone)

Poly (ether ether ketone) is a semicrystalline, high-temperature (up to 500 °F) engineering thermoplastic that is excellent for applications where thermal, chemical, and combustion properties (UL flammability rating of V-0) are critical to performance. At the same time, poly (ether ether ketone) emits little smoke or toxic gas when exposed to flame. This material is tough, strong, and rigid and has superior creep resistance. It also resists radiation and a wide range of solvents. With its resistance to hydrolysis, poly(ether ether ketone) can withstand boiling water and superheated steam used with autoclave and sterilization equipment at temperatures higher than 482 °F. Poly(ether ether ketone) resins are available in unreinforced, 30 percent glass-fiber reinforced, 30 percent carbon-fiber reinforced, and HPV-reinforced grades (bearing grade). Typical applications include the automotive, marine, nuclear, oil-well, electronics, medical, and aerospace industries.

6.3.1.2.7. Polyetherimide

Polyetherimide is considered an advanced thermoplastic which has both ether links and imide groups in its polymer chain. Polyetherimide is an amorphous high performance polymer which is characterized by excellent thermal properties, good chemical resistance, inherent flame retardancy, and exceptional dimensional stability. Polyetherimide also exhibits high tensile strength, without the use of reinforcement, very low smoke emission, and excellent hydrolytic stability. Because of polyetherimide's high stability, its range of processing is wider than many other thermoplastics.

6.3.1.2.8. Polycarbonate

Polycarbonate is a versatile, tough plastic used for a variety of applications from bulletproof windows to compact disks (CDs). The main advantage of polycarbonate over other types of plastic is unbeatable strength combined with light weight. While acrylic is 17 percent stronger than glass, polycarbonate is nearly unbreakable. Bulletproof windows and enclosures as seen inside the banks or at drive-through are often made of polycarbonate. Add to this the advantage that polycarbonate is just one-third the weight of acrylic or one-sixth as heavy as glass and the only drawback is that it is more expensive than either acrylic or glass. Polycarbonate is also used in the electronics industry. Apple's original iMac featured polycarbonate mixed with clear colors for a transparent computer case. Many cell phones, pagers, and laptops also use clear or opaque polycarbonate in their casings.

Other uses for polycarbonate include greenhouse enclosures, automobile headlights, outdoor fixtures, and medical industry applications though the list is virtually endless. Somewhat less toxic than polyvinyl chloride, to produce, polycarbonate nevertheless requires toxic chemicals in its production phase. It is, however, recyclable and environmentally preferable to polyvinyl chloride in applications for which either material can be used.

6.3.2. Metal Matrices

The development of metal matrix composites has been concentrated on three metals: aluminum, magnesium, and titanium [102–105]. Metals are normally alloyed with other elements to improve their physical and mechanical properties and a wide range of alloy compositions is available. Final properties are strongly influenced by thermal and mechanical treatments which determine the microstructure. Some typical properties of common metal matrices are Table 6-15. The metals used for composites are usually ductile and essentially isotropic. Unlike polymers, the increases in stiffness

Table 6-15 Typical Properties of Metal and Ceramic Matrices

Matrix	Density (g/cm^3)	Tensile modulus (GPa)	Tensile strength (MPa)
Al	2.70	70	200 ~ 600
Mg	1.80	45	100 ~ 300
Ti	4.5	110	300 ~ 1000
Borosilicate glass	2.3	64	100
SiC	3.4	400	400
Al_2O_3	3.8	380	500

achieved by incorporation of the reinforcement are often relatively small. However, important improvements are often achieved in properties such as wear characteristics, creep performance, and resistance to thermal distortion. All three metals are very reactive with a strong affinity for oxygen. This has implications for the production of composites particularly in regard to chemical reactions at the interface between the matrix and the reinforcement which has proved especially troublesome for titanium.

6.3.2.1. Aluminum (Al)

Aluminum is a hard and strong, silvery-white metal which is used in an enormous variety of products due to its particular properties. It has low density, is non-toxic, has a high thermal conductivity, has excellent corrosion resistance, and can be easily cast, machined, and formed. It is also non-magnetic and non-sparking. It is the second most malleable metal and the sixth most ductile. It is, therefore, extensively used for kitchen utensils, outside building decoration, and in any area where a strong, light, and easily constructed material is needed. The electrical conductivity of aluminum is about 60 percent that of copper per unit area of cross-section but it is nevertheless used in electrical transmission lines because of its low density. Pure aluminum has a low tensile strength but when combined with thermo-mechanical processing, aluminum alloys display a marked improvement in mechanical properties especially when tempered. Aluminum alloys form vital components of aircraft and rockets as a result of their high strength-to-weight ratio. Aluminum readily forms alloys with many elements such as copper, zinc, magnesium, manganese, and silicon. Today, almost all bulk metal materials that are referred to loosely as 'aluminum' are actually alloys.

6.3.2.2. Magnesium (Mg)

Magnesium is a silvery-white, lustrous, and relatively soft metal which tarnishes slightly in air. Magnesium is the third most commonly used structural metal following steel and aluminum. The main applications of magnesium are in order: component of aluminum alloys, in die-casting, to remove sulfur in the production of iron and steel, and the production of titanium in the Kroll process. It improves the mechanical, fabrication, and welding characteristics of aluminum when used as an alloying agent. Magnesium in its purest form can be compared with aluminum and is strong and light, so it is used in several high volume part manufacturing applications including automotive and truck components.

6.3.2.3. Titanium (Ti)

Titanium is a hard, lustrous, and silvery metal which is as strong as steel but much less dense. It is therefore important as an alloying agent with many metals including aluminum, molybdenum, and iron. Due to their high tensile strength to density ratio, high corrosion resistance, fatigue resistance, high crack resistance, and ability to withstand moderately high temperatures without creeping, titanium alloys are used in aircraft, armor plating, naval ships, spacecraft, and missiles. Titanium also has potential use in desalination plants which convert sea water to fresh water. The metal has excellent resistance to sea water and so is used to protect the hulls of ships and other structures exposed to sea water. However, the largest use of titanium is in the form of titanium (IV) oxide which is extensively used in both house paint and artists' paint. This paint is also a good reflector of the infrared radiation and so is used in solar observatories where heat causes poor visibility.

6.3.3. Ceramic Matrices

Ceramic matrix is the matrix made of a ceramic material embedded between the fibers. Typical fibrous materials include carbon, silicon carbide, aluminum oxide and mullite. Aluminum oxide, zirconium oxide, and silicon carbide are chiefly used as matrix components [106–110]. Ceramic matrix composite terminology is based on the principle fiber type/matrix type (e.g., C/silicon carbide for carbon fibers reinforced silicon carbide). This designation frequently precedes the production methods, which are usually realized in three steps. Ceramic matrix composites have had a stormy development, wherever the combination of breakage resistance and strength of conventional technical ceramics such as aluminum oxide, silicon carbide, or zirconium oxide prove inadequate. Because even small production errors or scratches on the surface of conventional ceramics can lead to cracks forming, their use in many applications had always been impossible. Only the embedding of fibers proved capable of increasing the resistance to cracks and ductility, breakage strength, and thermal shock resistance sometimes drastically. Typical applications are brake disks in the automotive industry, emergency rail systems for the trans-rapid, emergency running systems for rapidly rotating machines, or friction pads for clutches. The main advantages of adding ceramic fibers to ceramics (or generating interfaces in some other way) are to improve their toughness. Ceramics are very brittle and even a small increase in toughness may be advantageous. Usually, the toughness increases cones from repeated crack deflection at interfaces and the nature of the interface assumes overriding importance.

6.3.3.1. *Borosilicate glass*

Borosilicate glass is a type of glass with the main glass-forming constituents' silica and boron oxide. Borosilicate glasses are known for having very low coefficients of thermal expansion ($\sim 5 \times 10^{-6}$/ °C at 20 °C) making them resistant to thermal shock more so than any other common glass.

6.3.3.2. *Silicon carbide (SiC)*

Silicon carbide is composed of tetrahedra of carbon and silicon atoms with strong bonds in the crystal lattice. This produces a very hard and strong material. Silicon carbide is not attacked by any acids or alkalis or molten salts up to 800 °C. In air, SiC forms a protective silicon oxide coating at 1200 °C and is able to use up to 1600 °C. The high thermal conductivity coupled with low thermal expansion and high strength gives this material exceptional thermal shock resistant qualities. Silicon carbide ceramics with little or no grain boundary impurities maintain their strength to very high temperatures approaching 1600 °C with no strength loss. Chemical purity, resistance to chemical attack at temperature, and strength retention at high temperatures has made this material very popular as wafer tray supports and paddles in semiconductor furnaces. The electrical conduction of the materials has led to its use in resistance heating elements for electric furnaces and as a key component in thermistors (temperature variable resistors) and varistors (voltage variable resistors).

Today, the materials have been developed into a high quality technical grade ceramic with very good mechanical properties. It is used in abrasives, refractories, ceramics, and numerous high-performance applications. The material can also be made an electrical conductor and has applications in resistance heating, flame igniters, and electronic components.

6.3.3.3. *Aluminum oxide (Al_2O_3)*

Aluminum oxide is an amphoteric oxide of aluminum with the chemical formula Al_2O_3. It is also commonly referred to as alumina. Aluminum oxide is an electrical insulator but has a relatively high thermal conductivity (30 Wm^{-1} K^{-1}) for a ceramic material. Annual world production of alumina is approximately 45 million tons and over 90 percent of which is used in the manufacture of aluminum metal. The major uses of specialty aluminum oxides are in refractories, ceramics, and polishing and abrasive applications. Large tonnages are also used in the manufacture of zeolites, coating titanium pigments, and as a fire retardant/smoke suppressant.

6.4. FABRICATION PROCESS OF COMPOSITES

In many cases, composite processing utilizes the same technique as matrix processing which includes injection molding, compression molding, and extrusion. There are other techniques which are unique only to composite processing. These include filament winding, pultrusion, and hand lay-up. In spite of the fact that some techniques are used commonly with polymer processing, the operational conditions can be very different; thus, it is important not to directly transfer knowledge without careful consideration. In this section, a brief explanation of various processing methods will be given and those manufacturing processes typically used to make products found in construction/civil infrastructure market are covered.

Unique to the composites industry is the ability to create a product from many different manufacturing processes. There are a wide variety of processes available to the composites manufacturer to produce cost efficient products. Each of the fabrication processes has characteristics that define the type of products to be produced. This is advantageous because this expertise allows the manufacturer to provide the best solution for the customer. In order to select the most efficient manufacturing process, the manufacturing team considers several factors such as user needs performance requirements, size of the product, surface complexity, appearance, production rate, total production volume, economic targets/limitations, labor, materials, tooling/assembly, and equipment.

6.4.1. Hand Lay-up Molding

Hand lay-up molding is the method of laying down fabrics made of reinforcement and painting with the matrix resin layer by layer until the desired thickness is obtained. This is the most time and labor consuming, composite processing method but majority of aerospace composite products are made by this method in combination with the autoclave method. Due to the hand assembly involved in the lay-up procedure, one can align long fibers with controlled orientational quality. Another advantage of this method is the ability to accommodate irregular-shaped products. Such advantages are utilized in low performance composites including fiber-glass boat and bath tub manufacturing (Fig. 6-22).

6.4.1.1. Laminate materials

There are an infinite number of laminate types that can be developed [111–119]. These materials can be categorized into three basic areas: core materials, high strength and stiffness skins, and outer protective layers. Core materials typically serve the function of connecting and spacing of the skins to develop stiffness and strength in a sandwich arrangement. The key property of core materials is shear strength to ensure shear conductivity between the skins, thus the ability to sustain loads and bending. Core materials are normally wood,

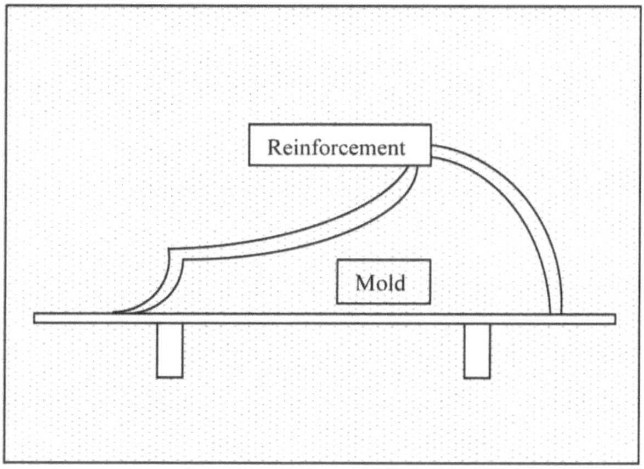

Fig. 6-22 Schematic diagram of hand lay-up process.

honeycomb, and structural foams. The outer structural layer or skins traditionally are either metal or composite either in combination with a core material or a multitude of high strength and stiffness layers. Composite materials offer the widest range of high strength skins with the ability to change the fiber type (fiberglass, carbon, and aramid) in addition to the fiber volume and orientation. Composites are well suited for large deflection applications where high strain capability and fatigue is required. Typically, they are more corrosion and environmentally resistant than metals. Composite materials in a lamina form are applied in the form of precured prepregs or 'B stage' and wet layup configurations. The final group of laminae is made up of thermoplastic and thermoset materials which act as a covering to the laminate structure. In some applications a plastic film or coating is incorporated into the laminate structure to protect the structure from impact and environmental effects.

Hand lay-up is the oldest and simplest method used for producing reinforced plastic laminates. Capital investment for the hand lay-up processes is relatively low. The most expensive piece of equipment typically is a spray gun for resin and gel coat application. Some fabricators pour or brush the resin into the molds so that a spray gun is not required for this step. There is virtually no limit to the size of the part that can be made. The molds can be made of wood, sheet metal, plaster, and FRP composites.

In a particular hand lay-up process (otherwise known as wet lay-up) high solubility resin is sprayed, poured, or brushed into a mold. The reinforcement is then wet-out with resin. The reinforcement is placed in the mold. Depending upon the thickness or density of the reinforcement, it may receive additional resin to improve wet out and allow better drapeability into the mold surface. The reinforcement is then rolled, brushed, or applied using a squeegee to remove entrapped air and to compact it against the mold surface.

Chopped strand mat is the lowest cost form of reinforcement used in wet lay-up. It also provides equal reinforcing strength in all directions due to the random orientation of the fibers that form the mat. Woven roving is especially suitable for thick laminates requiring greater strength. The woven fabric and braid can also provide a low cost reinforcement. Once the reinforcement is thoroughly wet out with the resin, it can be easily formed into complex shapes.

6.4.1.2. Surface preparation and bonding

A key component to a successful lamination is the bonding process of the layers. There are three basic components [120–122] which make up the bonding process. First is the surface preparation of the laminae which improves the substrate's ability to accept and adhere to an adhesive. Surface preparation varies depending on material type. Composites use sanding and grinding, surface texturing, or solvent cleaning. The second component is the adhesive itself including epoxies, urethanes, phenolics, polyesters, solvents, acrylics, and others. Each adhesive has its attributes depending on the substrate type in use requirements and process constraints. As a general rule a maximum bond is achieved for a given substrate type when the material itself fails during an ultimate strength test. The maximum lap shear strength of an adhesive is achieved when the adhesive exhibits a cohesive failure in the bond line. The third component of lamination is the process by which the materials are bonded together. This involves a host of parameters primarily time, heat pressure, mixture, moisture, and catalysts (initiators). It is important that the three basic components of bonding are properly employed to achieve a successful lamination.

6.4.1.3. Laminate construction

There are three types of laminated construction [123–126]. These include sandwich lamination consisting of at least two high stiffness and strength outer layers connected by a core, all laminated construction consisting of relatively high stiffness and strength layers and a third type consisting of a structural member that is reinforced on the tensile or compression or both sides of a flexural beam.

Sandwich lamination constructions are found in many applications from satellite structures to snow skis. Although both applications may utilize a sandwich approach, satellite applications generally require stiffness, strength, and extreme lightweight while the snow ski requires the laminated beam or composite structure to withstand large deflections and dynamic performance requirements. In addition, the ski structure is integrated with thermoplastic surfaces, and metal components. Typically, metal and composite materials are applied to these sandwich structures. Even complex shapes can be achieved by using composite prepregs and wet lay-ups.

All laminate constructions utilize relatively high strength/stiffness materials. An array of laminate configurations is possible. In aerospace applications, multiple plys are orientated in various directions to provide customized structural strength and stiffness. These same principles, although less complex, are used in automotive, industrial, and recreational products in the form of structural members, springs, archery limbs, and bicycles.

6.4.1.4. Multiply construction

The third type of laminated construction is utilized most often in construction and infrastructure applications [127–129]. To date, reinforcements have been integrated into Glue-Laminated wood beams (Glu-Lams) and infrastructure components for fabricating columns, beams, and walls. In many of these applications, reinforcement most typically composites is used via lamination to the tensile side of a beam or as a wrap on a column. Pre-cured prepreg or wet layup composite materials have all been utilized in these applications. These types of reinforcements improve the strength of laminated wood beams and/or reduce the use of E-Rated lumber with less costly new growth wood. Lamination in bridges and buildings provide a lower cost simplified method to revitalize existing structures, increase load-carrying capability, and increase resistance to seismic events.

In the automotive industry, composites are being combined with metals for performance, weight reduction, and cost advantages. The construction industry is applying composites to wood laminated beams and I-Joists.

The possibilities and advantages of laminated materials are significant and provide solutions to product requirements generally not achievable by using a single material. The advantages in structural performance, reduced weight, and oriented structural properties are just some of the advantages of this approach. In many cases, the result is a simplified and less costly solution to many engineering structural problems.

6.4.2. Spray-up Molding

Spray-up molding is much less labor intensive than the hand lay-up method by utilizing a spray gun and a fiber cutter. However, only short fiber reinforced composites can be made. A continuous fiber is fed into the cutter and chopped. The chopped fiber is sprayed upon a mold with the stream of resin mist and catalyst delivered through separate nozzles. The sprayed mixture of fiber and resin soon cures on the mold at room temperature and the product is produced. Because of the spraying operation, large and complex-shaped objects can be easily made.

6.4.3. Compression Molding, Transfer Molding and Resin Transfer Molding

Compression molding uses a press to compress either dough of resin and fiber mixture or the layers placed by a hand lay-up method or mechanical means typically at an elevated cure temperature [130–134]. With the compressive force, the void content is lower than the ordinary atmospheric pressure processing method.

A matched die mold allows shaping of the composite precursor into reproducible shapes. Although a compression molding machine is used, it is still a labor intensive method as the dough or layed-up materials must be weighed and hand-fed into the mold.

Transfer molding is the improved version of compression molding from the material metering point of view as the fiber/resin mixture is transfered from the reservoir into the mold cavity by the press. However, a long-fiber reinforced composite cannot be made. This method is nearly identical to a plunger-type injection molding operation based on the material flow. The term 'transfer molding' is used for a compression press operation while plunger type injection molding is obviously carried out in an injection molding machine.

Resin transfer molding (Fig. 6-23) as it is commonly referred to be a 'Closed Mold Process' in which the reinforcement materials are placed between two matching mold surfaces—one being male and one being female. The matching mold set is then closed and clamped and a low-viscosity thermoset resin is injected under moderate pressures (50–100 psi typical) into the mold cavity through a port or series of ports within the mold. The resin is injected to fill all voids within the mold set and thus penetrates and wets out all the surfaces of the reinforcing materials. The reinforcements may include a variety of fiber types in various forms such as continuous fibers, mat, or woven type construction as well as a hybrid of more than one fiber type. Vacuum is sometimes used to enhance

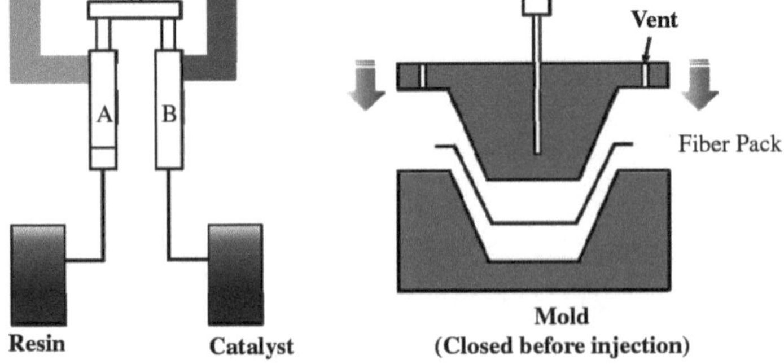

Fig. 6-23 Schematic diagram of resin transfer molding process.

the resin flow and reduce void formation. The part is typically cured with heat. In some applications, the exothermic reaction of the resin may be sufficient for proper cure.

Resin transfer molding as a process is multi-compatible with a variety of resin systems including polyester, vinyl ester, epoxy, phenolic, and modified acrylic and hybrid resins such as polyester and urethane [135–137]. Typically it requires a resin viscosity of 200 to 600 centerpoise to penetrate all surfaces of the mold cavity.

(1) Advantages of the resin transfer molding process include:

1. As a closed mold process, emissions are lower than open mold processes such as spray up or hand lay up.
2. The mold surface can produce a high quality finish (like those on an automobile).
3. This process can produce parts faster as much as 5-20 times faster than open molding techniques.
4. Resin transfer molding produces tighter dimensional tolerances to within ± .005 inch.
5. Complex mold shapes can be achieved. Cabling and other fittings can be incorporated into the mold designs.

(2) Disadvantages of the resin transfer molding process are:

1. High production volumes required to offset high tooling costs compared to the open molding techniques.
2. Reinforcement materials are limited due to the flow and resin saturation of the fibers.
3. Size of the part is limited by the mold.

In the traditional resin transfer molding process, a matched set of molds or 'closed mold' is used. The fiber reinforcements are usually preformed off line to enhance the production cycle time of the molds to perform at a respectable production rate. Resin is injected at high pressures and the process is sometimes assisted with vacuum. However, vacuum assisted resin transfer molding is different for many reasons. First, the fabrication of parts can be accomplished on a single open mold. Second, the process uses the injection of resin in combination with a vacuum and captured under a bag to thoroughly impregnate the fiber reinforcement. In the late 1980s, Bill Seemann invented and patented a variation to the vacuum assisted resin transfer molding process called SCRIMP^{TM} which is Seemann composite resin infusion molding process (Fig. 6-24). This process has been used in many new and large applications ranging from turbine blades and boats to rail cars and bridge decks. Unique to this process is the manufacturing method that allows the efficient processing of vacuum assisted resin transfer molding to produce large structural shapes that are virtually void-free. This process has been used to make both thin and very thick laminates. In addition, complex shapes with unique fiber architectures allow the fabrication of large parts that have a high structural performance.

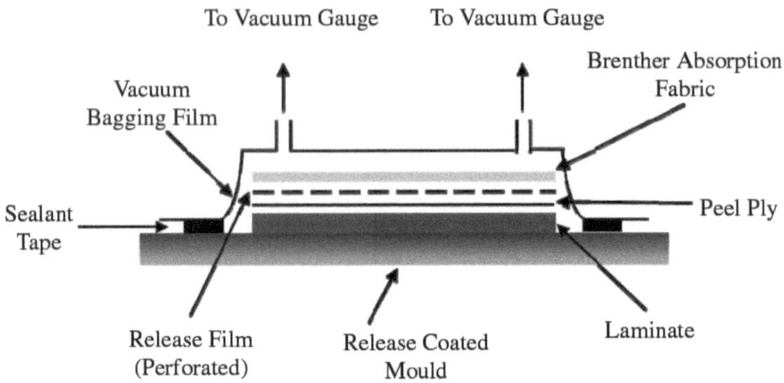

Fig. 6-24 Schematic diagram of vacuum assisted resin transfer molding process.

Parts using vacuum assisted resin transfer molding are made by placing dry fiber reinforcing fabrics into a mold applying a vacuum bag to the open surface and pulling a vacuum while at the same time infusing a resin to saturate the fibers until the part is fully cured. This process allows for easy visual monitoring of the resin to insure complete coverage to produce good parts without defects.

6.4.4. Injection Molding

Injection molding is probably the most extensively used method for processing short-fiber reinforced thermoplastics [138–142]. Injection molding is accomplished by large machines called injection molding machines as shown in Fig. 6-25. Resin is fed to the machine through the hopper. Colorants are

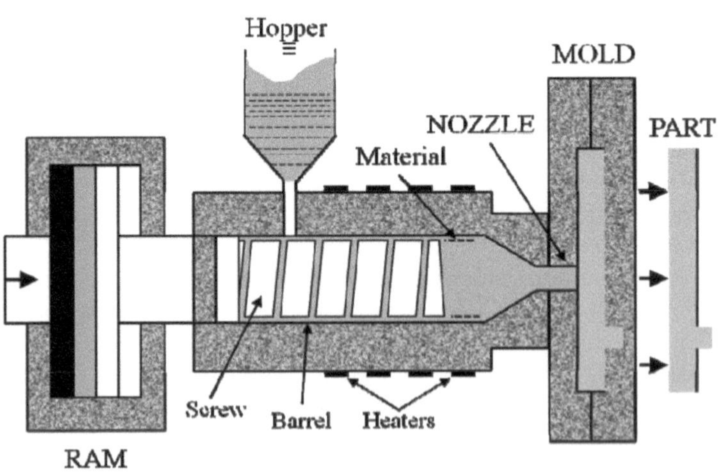

Fig. 6-25 Schematic diagram of injection molding process.

usually fed to the machine directly after the hopper. The resins enter the injection barrel by gravity through the feed throat. Upon entrance into the barrel, the resin is heated to the appropriate melting temperature.

The fiber/resin mixture whether it is pre-blended or fed as a physical mixture is fed into the hopper and transfered into the heated barrel. The material softens by the heat transfer from the barrel wall. At the same time, the screw rotates to apply high-shear process to further heat the material and fill the barrel. The molten material is collected in front of the screw by the rotation of the screw and then injected with a high pressure into the mold cavity through the runner and the gate. The mold is cooled below the solidification temperature of the resin in case of thermoplastic composites. The level of automation of this method is the highest among many processing methods. Due to the intensive mixing with high-shear and passage through a narrow gate extensive fiber damage occurs; therefore, injection molding for composite materials is suitable only with short fiber reinforced or particulate-filled polymers.

There is a critical length of fibers below which the fiber length does not degrade. The critical length is determined by the rheological properties of the composite mold, fiber properties, and instrument factors [143–145]. Less fiber damage occurs when a plunger-type injection molding machine is used rather than a screw-type injection molding machine. Because the plunger-type injection molding machine does not achieve a high degree of mixing in the machine, the raw materials must be thoroughly mixed prior to feeding.

6.4.5. Reaction Injection Molding

Reaction injection molding is widely employed in processing of various thermoset polymers. It usually contains two or more liquids accumulated in a high pressure mixing unit and then injected into a mold where they are polymerized or cured to form a desired part. Hence, it presents a typical example of complex non-isothermal and compressible flows of liquid with reaction at high pressure and high temperature. Being mostly motivated by practical applications many researchers worked in the field and made quite a few attempts to describe and experimentally or theoretically residual stresses remaining in the injection-molded parts.

Reaction injection molding is one of the newest processing methods (Fig. 6-26). Instead of using already polymerized materials as matrices, highly reactive monomeric or oligomeric ingredients are placed in two tanks which are then quickly mixed by impingement and injected into the mold cavity. As soon as the two materials are mixed, chemical reaction begins to form a polymeric matrix which completes typically within 5–30 s. Thus, the major portion of the reaction injection molding machine is a high pressure pump and a metering system. Again with high intensive shear, only short

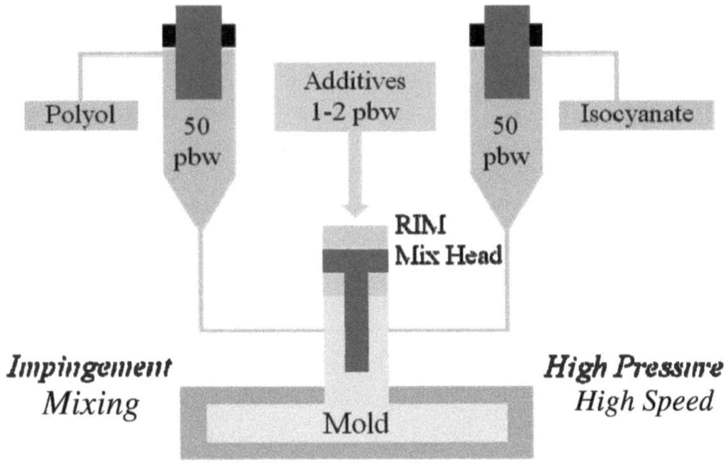

Fig. 6-26 Schematic diagram of reaction injection molding process.

fibers and fillers can be used as reinforcements. However, reaction injection molding utilizes low viscosity chemicals and this allows the preplacement of continuous fiber-woven fabrics in the mold in the same manner as resin transfer molding. Distinction is made between these two methods based on the preparation of the resin precursor. When the resin formulation is already made, the method is called resin transfer molding while if the resin is prepared in situ by an impingement or static mixer, the method is termed as reaction injection molding.

6.4.6. Pultrusion

Pultrusion is a continuous molding process that combines fiber reinforcements and thermosetting resin (Fig. 6-27). The pultrusion process is used in the fabrication of composite parts that have a constant cross-section profile [146–151]. Typical examples include various rods and bar section, ladder side rails, tool handles, and electrical cable tray components and now bridge beams and decks. Most pultruded laminates are formed using rovings aligned down the major axis of the part. Various continuous strand mats, fabrics (braided, woven and knitted), and texturized or bulked rovings are used to obtain strength in the cross axis or transverse direction.

The process is normally continuous and highly automated. Reinforcement materials such as roving, mat, or fabric, positioned in a specific location using preforming shapers or guides to form the profile. The reinforcements are drawn through a resin bath or wet-out where the material is thoroughly coated or impregnated with a liquid thermosetting resin. The resin-saturated reinforcements enter a heated metal pultrusion die. The dimensions and shape of the die

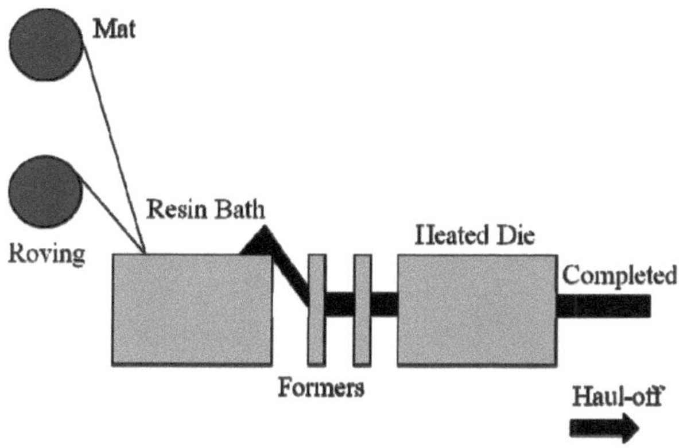

Fig. 6-27 Schematic diagram of pultrusion process.

will define the finished part being fabricated. Inside the metal die, heat is transferred initiated by precise temperature control to the reinforcements and liquid resin. The heat energy activates the curing or polymerization of the thermoset resin changing it from a liquid to a solid. The solid laminate emerges from the pultrusion die to the exact shape of the die cavity. The laminate solidifies when cooled and it is continuously pulled through the pultrusion machine and cut to the desired length. The process is driven by a system of caterpillar or tandem pullers located between the die exit and the cut-off mechanism.

The initial capital investment for pultrusion is generally higher than open mold or hand layup processes. The primary expense for the pultrusion manufacturers is the material handling guides and die fabrication costs. The net result is a low-cost process for high volume. The process provides flexibility in the design of the pultruded profiles. Currently profiles up to 72 inches wide and 21 inches high are possible. Pultrusion can manufacture both simple and complex profiles eliminating the need for the extensive post-production assembly of components. Since the process is continuous, length variations are limited to shipping capabilities. This process allows for optimized fiber architectures with uniform color eliminating the need for many painting requirements.

6.4.7. Filament Winding

Filament winding is also a unique processing method for the polymer composite processing with a continuous reinforcing fiber (Fig. 6-28) [152–155]. Resin-wet rovings are wound with a certain pattern around a mandrel. The wound mandrel is then placed into an oven and cured to a solid composite. Due to the controlled tension, squeezing action and controlled winding pattern, the fiber content can be very high to produce the composites with one of

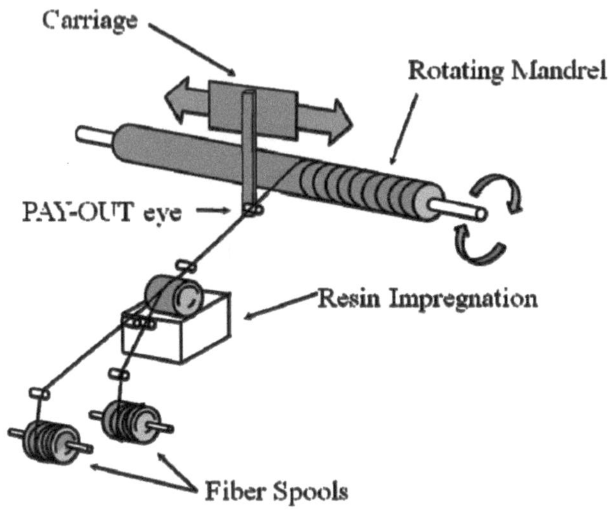

Fig. 6-28 Schematic diagram of filament winding process.

the highest mechanical properties. The winding process is time consuming and is the cause of low productivity. However, due to its very high mechanical properties with automated operation, it is actively evaluated by aerospace industries.

6.5. APPLICATIONS OF COMPOSITES

6.5.1. Sports

Sports are a test platform for carbon fiber production. Carbon fiber production enterprises to provide new materials for the sporting goods market as well as the commercial mode of operation. With the production and application of technological advances, carbon fiber frequent appearances in the sporting goods (see in Fig. 6-29). Over the years, carbon fiber golf clubs in the use has been among the first in the hockey stick and the bicycle stand of the applications are increasing year by year. In the racket, fishing rod, snowboarding, windsurfing masts, marine hull, backpack, tent pole, softball and baseball bat products can also see the carbon fiber presence.

6.5.2. Aircraft

The Boeing 777 whose maiden flight was 10 years ago is around 20 percent composites by weight with the composite materials being used for the wing's fixed leading edge, the trailing-edge panels, the flaps and flaperons, the spoilers,

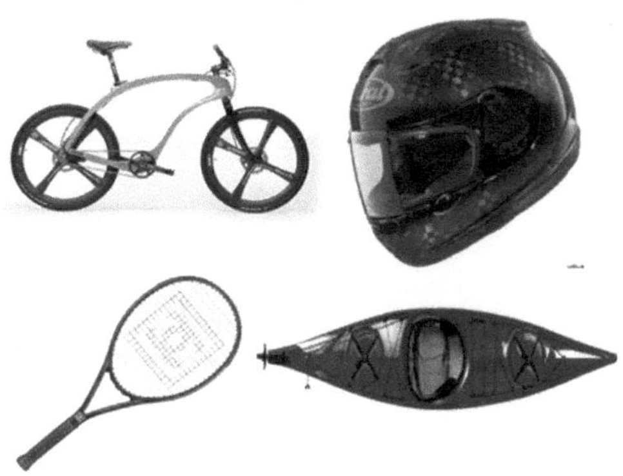

Fig. 6-29 The fibers-reinforced composites for sports.

and the outboard aileron. They are also used for the floor beams, the wing-to-body fairing, and the landing-gear doors. Using composite materials for the empennage saves approximately 1,500 lb in weight. The Boeing 787 and A380 will average extensive use of composite materials (estimates are as high as 50 percent) in the quest for very high efficiency and performance with reduced weight (see in Fig. 6-30). The composites are also used in the major structural elements of modern helicopters which are approximately 50 percent composites by weight. The formability of composites has been used to particular advantage in helicopter manufacture to reduce the numbers of component parts and therefore cost.

6.5.3. Auto-mobile Parts

The automotive companies in the today's modern world are forced to look for new ways and innovations in manufacturing a car/truck due to fierce competition. The cars today should have all the comforts needed by the customers at

Fig. 6-30 The fibers-reinforced composites for aircraft.

Fig. 6-31 The fibers-reinforced composites used in manufacturing of automotive parts.

low cost. This has led to the use of composite materials in the construction of the body, interiors, chassis, hoods, electrical components, etc. The composite materials have the desired properties to suit the requirements. Hence there is more scope for the composites today and also will be in future in the automobile industry. The recent innovations in the sphere of advanced composites in the automotive manufacturing industry have led to the use of the carbon fibers-reinforced plastic (CFRP) with various thermoset or thermoplastic composites (see in Fig. 6-31).

6.5.4. Infrastructures

In recent decades, material development in response to a call for more durable infrastructures has led to many exciting advancements. Fiber reinforced composite designs with very unique properties are now being explored in many infrastructural applications (see in Fig. 6-32). Even concrete and steel are being steadily improved to have better properties and durability.

Fig. 6-32 The fibers-reinforced composites for infrastructural parts.

6.6. SUMMARY

The role of the reinforcement in composite materials is fundamentally one of the increasing the mechanical properties of the neat resin system. All of the different fibers used in composites have different properties and so affect the properties of the composites in different ways. Although it is undoubtedly true that the high strength of composites is largely due to the fiber reinforcement, the importance of matrix materials cannot be underestimated as it provides support for the fibers and assists the fibers in carrying the loads. It also provides stability to the composite materials. Resin matrix system acts as a binding agent in a structural component in which the fibers are embedded.

Also, it may be seen that the fiber-reinforced composites may be manufactured by one of the many standard fabrication methods. The selection of the best method for a given application requires the consideration of numerous factors. The selection process must determine the appropriate constituent materials, the actual part design, the required tooling, the fabrication process, and the necessary inspection and testing requirements to assure part quality.

REFERENCES

[1] M. Grayson, Encyclopedia of Composites. John Wiley & Sons, New York (1996).

[2] D. Gay, S.V. Hoa, S.W. Tsai, Composite Materials: Design and Applications. CRC Press, Boca Raton (2003).

[3] S.W. Tsai, H.T. Hahn, Introduction to Composite Materials. Technomic Publishing Co., Lancaster (1980).

[4] D. Hull, T.W. Clyne, An Introduction to Composite Materials. Cambridge University Press, New York (1996).

[5] K.K. Chawla, Composite Materials: Science and Engineering, 2nd edit. Springer-Verlag, New York (1998).

[6] J.B. Donnet, R.C. Bansal, Carbon Fibers, 2nd edit. Marcel Dekker, New York (1990).

[7] E. Fitzer, Carbon Fibers and Their Composites. Springer, Berlin (1985).

[8] S.J. Park, Carbon Materials. Daeyoungsa, Seoul (2006).

[9] M.C.R. Paiva, C.A. Bernardo, M. Nardin, Carbon 38 (2000) 1323.

[10] M.S. Dresselhaus, G. Dresselhaus, K. Sugihara, I.L. Spain, H.A. Goldberg, Graphite Fibers and Filaments. Springer, Berlin (1988).

[11] N. Dilsiz, N.K. Erinc, E. Bayramli, G. Akovali, Carbon 33 (1995) 853.

[12] Z. Wu, C.U. Pittman Jr, S.D. Gardner, Carbon 33 (1995) 597.

[13] N. Oya, D.J. Johnson, Carbon 39 (2001) 635.

[14] L. Li, L. Liu, H. Wan, Q. Xu, G. Cheng, J. Wen, Applied Surface Science 255 (2009) 8030.

[15] H. Rennhofer, D. Loidl, S. Puchegger, H. Peterlik, Carbon 8 (2010) 964.

[16] S. Kumar, D.P. Anderson, A.S. Crasto, Journal of Materials Science 28 (1993) 423.

[17] M.G. Dobb, H. Guo, D.J. Johnsonand, C.R. Park, Carbon 33 (1995) 1553.

[18] M. Guigon, A. Oberlin, Composites Science and Technology 27 (1986) 1.

[19] M. Guigon, A. Oberlin, Fibre Science and Technology 20 (1984) 177.

[20] D.J. Johnson, Structure and properties of carbon fibers. In Carbon Fibers Filaments and Composites. Kluwer Academic, The Netherlands (1990).

[21] U. Zielke, K.J. Hüttinger, W.P. Hoffman, Carbon 34 (1996) 983.
[22] C.U. Pittmann Jr, G.R. He, B. Wu, S.D. Gardner, Carbon 35 (1997) 317.
[23] H.H. Sarraf, L. Škarpová, P. Louda, Journal of Achievements in Materials and Manufacturing Engineering 25 (2007) 24.
[24] A.R. Bunsell, Journal of Applied Polymer Science. Applied Polymer Symposium 47 (1991) 87.
[25] F.T. Wallenberger, Advanced Inorganic Fibers: Processes, Structures, Properties Applications. Kluwer Academic Publishers, Dordrecht (2000).
[26] W. Ruland, Advanced Materials 2 (1990) 528.
[27] N. Dilsiz, N.K. Erinc, E. Bayramli, G. Akovali, Carbon 33 (1995) 853.
[28] M.C.R. Paiva, C.A. Bernardo, M. Nardin, Carbon 38 (2000) 1323.
[29] M. Toyoda, H. Katoh, M. Inagaki, Carbon 39 (2001) 2231.
[30] A.H. Wazir, L. Kakakhel, New Carbon Materials 24 (2009) 83.
[31] A. Tressaud, V. Gupta, L. Piraux, L. Lozano, E. Marquestaut, S. Flandrois, A. Marchand, O.P. Bahl, Carbon 32 (1994) 1485.
[32] J.B. Donnet, M. Brendlé, T.L. Dhami, O.P. Bahl, Carbon 24 (1986) 757.
[33] Courtaulds Ltd., British Patent, 144341 (1969).
[34] J.D. Ray, S. Steinsiger, and B.A. Cass, US Patent, 16251 (1970).
[35] Morganite Research Development Ltd., British Patent, 493571 (1969).
[36] J.T. Paul, Appl. No. 25208/74, Patent Specification 1433712.
[37] M.K. Seo, S.J. Park, Journal of Colloid and Interface Science 330 (2009) 237.
[38] B. Lindsay, M.L. Abel, J.F. Watts, Carbon 45 (2007) 2433.
[39] S. Wang, A. Garton, Journal of Applied Polymer Science 45 (1992) 1743.
[40] Z. Wu, C.U. Jr. Pittman, S.D. Gardner, Carbon 33 (1995) 597.
[41] A. Fukunaga, S. Ueda, Composites Science and Technology 60 (2000) 249.
[42] M.A. Montes-Moran, F.W.J. Vanhattum, J.P. Nunes, A. Martinez-Alonso, J.M.D. Tascon, C.A. Bernardo, Carbon 43 (2005) 1795.
[43] S.J. Park, M.H. Kim, Journal of Materials Science 35 (2000) 1901.
[44] S.R.J. Oliver, N. Bowden, G.M. Whitesides, Journal of Colloid and Interface Science 224 (2000) 425.
[45] Y. Chang, H.J. Ohara, Journal of Fluorine Chemistry 57 (1992) 169.
[46] J. Liu, Y. Tian, Y. Chen, J. Liang, L. Zhang, H. Fong, Materials Chemistry and Physics 122 (2010) 548.
[47] K.Z. Li, C. Wang, H.J. Li, X.T. Li, H.B. Ouyang, J. Wei, Composites Science and Technology 68 (2008) 1105.
[48] R. Beydon, G. Bernhart, Y. Segui, Surface and Coatings Technology 126 (2000) 39.
[49] E. Hage, S.F. Costa, L.A. Pessan, Journal of Adhesion Science and Technology 11 (1997) 1491.
[50] H. Li, H. Liang, F. He, Y. Huang, Y. Wan, Surface and Coatings Technology 203 (2009) 1317.
[51] F. Hoecker, J.K. Kocsis, Journal of Applied Polymer Science 59 (1996) 139.
[52] S.J. Park, Interfacial Forces and Fields: Theory and Applications, Chap. 9. Marcel Dekker, New York (1999).
[53] S.J. Park, J.S. Jin, J.R. Lee, Journal of Adhesion Science and Technology 14 (2000) 1677.
[54] S.J. Park, J.R. lee, Journal of Materials Science 33 (1998) 647.
[55] J.W. Herrick, In Proceeding 23rd Annual Technical Conference, Reinforced Plastics Division, Sect. 16A, (1968).
[56] J.B. Donnet, Carbon 20 (1982) 267.

[57] E. Fitzer, R. Weis, Carbon 25 (1987) 455.
[58] B. Rand, R. Robinson, Carbon 15 (1977) 2547.
[59] D.A. Scola and S.C. Brooks, 25th Society of Plastics Industry Meeting, DC (1970).
[60] A. Proctor, P.M.A. Sherwood, Journal of Electron Spectroscopy and Related Phenomena 27 (1982) 39.
[61] S. Brelant, Carbon 19 (1981) 329.
[62] E. Fitzer, K.H. Geigl, W. Huttner, R. Weiss, Carbon 14 (1976) 389.
[63] L.T. Drzal, M.J. Rich, P.F. Lloyd, The Journal of Adhesion 16 (1982) 1.
[64] C. Kozlowski, P.M.A. Sherwood, Journal of the Chemical Society, Faraday Transactions 1 81 (1985) 2745.
[65] T. Takahagi, A. Ishitani, Carbon 22 (1984) 43.
[66] T. Takahagi, A. Ishitani, Carbon 26 (1988) 389.
[67] Y. Nakayama, F. Soeda, A. Ishitani, Carbon 28 (1990) 21.
[68] T.A. Devilbiss and J.P. Wightman, In Proceedings of the First International Conference on Composite Interfaces, Cleveland, Ohio, p. 307 (1986).
[69] C.T. Chen, N.V. Smith, F. Sette, Physical Review B 43 (1991) 6785.
[70] B.W. Chun, C.R. Davis, Q.A. He, R.R. Gustafson, Carbon 30 (1992) 177.
[71] K.L. Loewenstein, The Manufacturing Technology of Continuous Glass Fibers. Elsevier Scientific, New York (1973) p. 2.
[72] A Market Assessment and Impact Analysis of the Owens Corning Acquisition of Saint-Gobain's Reinforcement and Composites Business (2007).
[73] E. Fitzer, R. Kleinholz, H. Tiesler, M.H. Stacey, R.D. Bruyne, I. Lefever, A. Foley, W. Frohs, T. Hauke, M. Heine, H. Jager, S. Sitter, Fibers, 5. Synthetic Inorganic. Wiley-VCH, Weinheim (2008).
[74] V.B. Gupta, V.K. Kothari, Manufactured Fiber Technology. Chapman & Hall, London (1997) p. 544.
[75] M.B. Volf, Technical Approach to Glass. Elsevier, New York (1990).
[76] R.E. Krieger, Handbook of Fiberglass and Advanced Plastic Composites. Huntingdon, New York (1970).
[77] R. Sakin, I. Ay, Materials & Design 29 (2008) 1170.
[78] A. Bernasconi, P. Davoli, A. Basile, A. Filippi, International Journal of Fatigue 29 (2007) 199.
[79] L. Liu, Y.D. Huang, Z.Q. Zhang, Z.X. Jiang, L.N. Wu, Applied Surface Science 254 (2008) 2594.
[80] P.J. de Lange, P.G. Akker, E. Mader, S.L. Gao, W. Prasithphol, R.J. Young, Composites Science and Technology 67 (2007) 2027.
[81] B. Larin, C.A. Avila-Orta, R.H. Somani, B.S. Hsiao, G. Marom, Polymer 49 (2008) 295.
[82] M. Grujicic, G. Arakere, T. He, W.C. Bell, B.A. Cheeseman, C.F. Yen, B. Scott, Materials Science and Engineering: A 498 (2008) 231.
[83] S.R. Bakshi, J.E. Tercero, A. Agarwal, Computer Part A 38 (2007) 2493.
[84] D. Koch, K. Tushtev, G. Grathwohl, Composites Science and Technology 68 (2008) 1165.
[85] N. Yamamoto, A.J. Hart, E.J. Garcia, S.S. Wicks, H.M. Duong, A.H. Slocum, B.L. Wardle, Carbon 47 (2009) 551.
[86] A. Rossoll, C. Otto, B. Moser, L. Weber, A. Wanner, A. Mortensen, Scripta Materialia 59 (2008) 842.
[87] P.Y. Chang, P.C. Yeh, J.M. Yang, Materials Science and Engineering: A 496 (2008) 273.
[88] J.D. DeFouw, D.C. Dunand, Acta Materialia 56 (2008) 5751.
[89] M.R. Bambach, Computer Part B 41 (2010) 550.
[90] J.W. Yi, S.B. Lee, J.B. Kim, S.K. Lee, Surface and Coatings Technology 204 (2010) 1419.

[91] A. Pirondi, L. Collini, International Journal of Fatigue 31 (2009) 111.
[92] S.C. Tjong, Y.W. Mai, Composites Science and Technology 68 (2008) 583.
[93] S. Garoushi, L.V.J. Lassila, A. Tezvergil, P.K. Vallittu, Dental Materials 23 (2007) 17.
[94] Q. Guo, J. Liu, L. Chen, K. Wang, Polymer 49 (2008) 1737.
[95] V. Kaushik, J. Raghavan, Computer Part A 41 (2010) 1210.
[96] R.C. Wetherhold, M. Corjon, P.K. Das, Composites Science and Technology 67 (2007) 2428.
[97] J. Palmer, O.R. Ghita, L. Savage, K.E. Evans, Computer Part A 40 (2009) 490.
[98] C. Steggall-Murphy, P. Simacek, S.G. Advani, S. Yarlagadda, S. Walsh, Computer Part A 41 (2010) 93.
[99] S.G. Pantelakis, C.V. Katsiropoulos, G.N. Labeas, H. Sibois, Computer Part A 40 (2009) 595.
[100] J.A. Ramos, M. Blanco, I.I. Zalakain, I. Mondragon, Journal of Colloid and Interface Science 336 (2009) 431.
[101] M. Rico, J. Lopez, C. Ramirez, J. Diez, B. Montero, Polymer 50 (2009) 569.
[102] H. Ye, X.Y. Liu, H. Hong, Journal of Materials Processing Technology 200 (2008) 12.
[103] S. Tzamtzis, N.S. Barekar, N.H. Babu, J. Patel, B.K. Dhindaw, Z. Fan, Computer Part A 40 (2009) 144.
[104] R. Ekici, M.K. Apalak, M. Yildirim, F. Nair, Materials & Design 31 (2010) 2818.
[105] I. Sabirov, O. Kolednik, Materials Science and Engineering: A 527 (2010) 3100.
[106] T. Baxevanis, N. Charalambakis, Mechanics of Materials 42 (2010) 570.
[107] S.R. Choi, R.W. Kowalik, D.J. Alexander, N.P. Bansal, Composites Science and Technology 69 (2009) 890.
[108] A. Sommers, Q. Wang, X. Han, C. T'Joen, Y. Park, A. Jacobi, Applied Thermal Engineering 30 (2010) 1277.
[109] G. Fantozzi, P. Reynaud, Materials Science and Engineering: A 521 (2009) 18.
[110] S. Pompidou, J. Lamon, Composites Science and Technology 67 (2007) 2052.
[111] F. Rastellini, S. Oller, O. Salomon, E. Onate, Computers & Structures 86 (2008) 879.
[112] A.E. Mahi, M. Assarar, Y. Sefrani, J.M. Berthelot, Computer Part B 39 (2008) 1069.
[113] Y.X. Zhang, H.S. Zhang, Computers & Structures 92 (2010) 2159.
[114] J. Xu, B.V. Sankar, Computer Part A 39 (2008) 1625.
[115] S.X. Wang, L.Z. Wu, L. Ma, Materials & Design 31 (2010) 118.
[116] L.J. Deka, S.D. Bartus, U.K. Vaidya, Composites Science and Technology 69 (2009) 725.
[117] P. Feraboli, T. Cleveland, P. Stickler, J. Halpin, Computer Part A 41 (2010) 557.
[118] D.N. Betts, A.I.T. Salo, C.R. Bowen, H.A. Kim, Computers & Structures 92 (2010) 1694.
[119] A. Faraz, D. Biermann, K. Weinert, International Journal of Machine Tools and Manufacture 49 (2009) 1185.
[120] H. He, J. Wang, K. Li, J. Wang, J. Gu, Materials & Design 31 (2010) 4631.
[121] R. Jamaati, M.R. Toroghinejad, Materials Science and Engineering: A 527 (2010) 4146.
[122] G. Liu, X. Yang, Y. Wang, Polymer 48 (2007) 4385.
[123] T.J.C. Liu, H.C. Wu, Materials & Design 31 (2010) 1971.
[124] F. Rastellini, S. Oller, O. Salomon, E. Onate, Computers & Structures 86 (2008) 879.
[125] P. Roca, G. Araiza, Construction and Building Materials 24 (2010) 1372.
[126] H. Huang, A.M. Waas, Composites Science and Technology 69 (2009) 2338.
[127] J.J. Nie, Y.D. Xu, L. Zhang, L. Cheng, J.Q. Ma, Journal of Materials Processing Technology 209 (2009) 572.
[128] P.J. Liotier, V. Alain, D. Christine, Computer Part A 41 (2010) 653.

[129] T.C. Truong, D.S. Ivanov, D.V. Klimshin, S.V. Lomov, I. Verpoest, Computer Part A 39 (2008) 1380.
[130] P. Dumont, L. Orgeas, D. Favier, P. Pizette, C. Venet, Computer Part A 38 (2007) 353.
[131] P. Bhat, J. Merotte, P. Simacek, S.G. Advani, Computer Part A 40 (2009) 431.
[132] J. Merotte, P. Simacek, S.G. Advani, Computer Part A 41 (2010) 881.
[133] M.S. Kim, W.I. Lee, W.S. Han, A. Vautrin, C.H. Park, Computer Part A 40 (2009) 1192.
[134] S.R. Dhakate, R.B. Mathur, B.K. Kakati, T.L. Dhami, International Journal of Hydrogen Energy 32 (2007) 4543.
[135] P.A. Sreekumar, K. Joseph, G. Unnikrishnan, S. Thomas, Composites Science and Technology 67 (2007) 453.
[136] M. Haider, P. Hubert, L. Lessard, Composites Science and Technology 67 (2007) 3176.
[137] P.A. Sreekumar, S.P. Thomas, J. Saiter, K. Joseph, G. Unnikrishnan, S. Thomas, Computer Part A 40 (2009) 1777.
[138] H. Ye, X.Y. Liu, H. Hong, Journal of Materials Processing Technology 200 (2008) 12.
[139] M.A. Khan, J. Ganster, H.P. Fink, Computer Part A 40 (2009) 846.
[140] S.J. Liu, M.J. Lin, Y.C. Wu, Composites Science and Technology 67 (2007) 1415.
[141] Z.Y. Liu, D. Kent, G.B. Schaffer, Materials Science and Engineering: A 513 (2009) 352.
[142] P.W. Ho, Q.F. Li, J.Y.H. Fuh, Materials Science and Engineering: A 485 (2008) 657.
[143] S.X. Zhang, Z.Y. Ong, T. Li, Q.F. Li, S.F. Pook, Materials & Design 31 (2010) 2897.
[144] G. Morales, M.I. Barrena, J.M. Gomez de Salazar, C. Merino, D. Rodriguez, Composite Structures 92 (2010) 1416.
[145] P.A. Santos, M.A.S. Spinace, K.K.G. Fermoselli, Computer Part A 38 (2007) 2404.
[146] M.S. Yun, W.I. Lee, Composites Science and Technology 68 (2008) 202.
[147] A. Peled, B. Mobasher, Z. Cohen, Cement and Concrete Composites 31 (2009) 647.
[148] H.Y. Kim, K.T. Park, J.W. Jeong, Y.H. Lee, Y.K. Hwang, D.H. Kim, Computers & Structures 91 (2009) 20.
[149] W.F. Ragheb, Thin-Walled Structure 48 (2010) 709.
[150] H.M. Akil, L.W. Cheng, Z.A.M. Ishak, A.A. Bakar, M.A.A. Rahman, Composites Science and Technology 69 (2009) 1942.
[151] R. Stewart, Reinforced Plastics 53 (2009) 34.
[152] G. Li, C. Zhang, Y. Wang, P. Li, Y. Yu, X. Jia, H. Liu, X. Yang, Z. Xue, S.K. Ryu, Composites Science and Technology 68 (2008) 3208.
[153] W. Chen, Y. Yu, P. Li, C. Wang, T. Zhou, X. Yang, Composites Science and Technology 67 (2007) 2261.
[154] A. Zuraida, A.A. Khalid, A.F. Ismail, Materials & Design 28 (2007) 71.
[155] M.T. Arellano, L. Crouzeix, B. Douchin, F. Collombet, H.H. Moreno, J.G. Velazquez, Composite Structures 92 (2010) 2457.

Chapter 7

Types of Composites

7.1. INTRODUCTION

Composites are materials made up of two or more materials that are combined in a way that allows the materials to stay distinct and identifiable. The purpose of composites is to allow the new materials to have strengths from both materials, often times covering the original materials weaknesses. Composizztes are different from alloys because they are combined in such a way that it is impossible to tell one particle, element, or substance from the other. Composites are usually classified by the type of reinforcements they use. These reinforcements are embedded into a matrix that holds it together. The reinforcements are used to strengthen the composites. For example, in a mud brick, the matrix is the mud and the reinforcement is the straw. Common composite types include random-fiber or short-fiber reinforcement, continuous-fiber or long-fiber reinforcement, particulate reinforcement, flake reinforcement, and filler reinforcement.

We know that composites have two (or more) chemically distinct phases on a microscopic scale, separated by a distinct interface, and it is important to be able to specify these constituents. The constituent that is continuous and is often but not always present in the greater quantity in the composites is termed the matrix. The normal view is that it is the properties of the matrix that are improved on incorporating another constituent to produce composites. Composites may have a ceramic, metallic, and polymeric matrix.

Depending on the composition and the processing of the raw materials as well as on forming and the type of sintering, the properties of the materials can be closely tailored to the desired application. The properties of these three classes of materials differ considerably as demonstrated in Fig. 7-1. The direction of the arrow indicates the tendency in comparison with typical values. Any assessment must depend on the particular application. A broad arrow indicates a particularly strong manifestation of the feature. The characteristic properties of metals, ceramics, and plastics can be explained by the different nature of their atomic bonds. As a generalization the polymers have low strengths and Young's modulus, ceramics are strong, stiff and brittle, and metals have intermediate strengths and modulus together with good ductility, i.e., they are not brittle. The second constituent is referred to as the reinforcing phase, or reinforcements, as it enhances or reinforcements is harder, stronger and stiller than the matrix,

Interface Science and Composites, Volume 18
ISSN: 1573-4285, doi: 10.1016/B978-0-12-375049-5.00001-3

	Ceramic	Metal	Polymer
Hardness	⇧	⇩	⇩
Elastic modulus	⇧	⇧	⇩
High temperature strength	⇧	⇩	⇩
Thermal expansion	⇩	⇧	⇧
Ductility	⇩	⇧	⇧
Corrosion resistance	⇧	⇩	⇩
Resistance to wear	⇧	⇩	⇩
Electrical conductivity	⇧⇩	⇧	⇩
Density	⇩	⇧	⇩
Thermal conductivity	⇧⇩	⇧	⇩
⇧ Tendency to high values		⇩ Tendency to low values	

Fig. 7-1 A comparison of the properties of ceramics, metals, and polymers.

although there are some exceptions; for example, ductile metal reinforcements in a ceramic matrix and rubber-like reinforcements in a brittle polymer matrix. At least one of the dimensions of the reinforcements is small, say less than 500 μm, and sometimes only of the order of a micron. The geometry of the effectiveness phase is one of the major parameters in determining the effectiveness of the reinforcements; in other works, the mechanical properties of composites are a function of the shape and dimensions of the reinforcements. We usually describe the reinforcements as being either fibrous or particulate. Fig. 7-2 represents a commonly employed classification scheme for composite materials which utilizes this designation for the reinforcements (Fig. 7-2, Block A).

Particulate reinforcements have dimensions that are approximately equal in all directions. The shape of the reinforcing particles may be spherical, cubic, platelet or any regular or irregular geometry. The composites illustrated in the micrograph of Fig. 7-3 have angular particles of about 10 μm as the reinforcements. The arrangement of the particulate reinforcements may be random or with a preferred orientation, and this characteristic is also used as a part of the classification scheme (Block B). In the majority of particulate reinforced composites the orientation of the particles is considered, for practical purposes, to be random (Fig. 7-3).

A fibrous reinforcement is characterized by its length being much greater than its cross-sectional dimension. However, the ratio of the length to the cross-

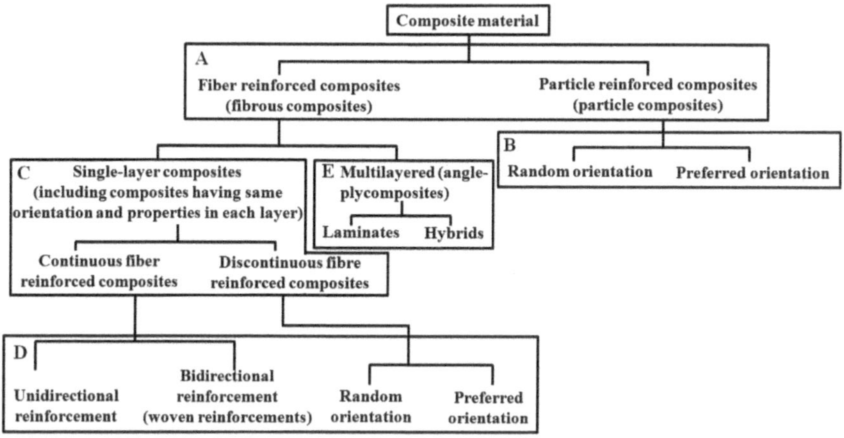

Fig. 7-2 Classification of composite materials.

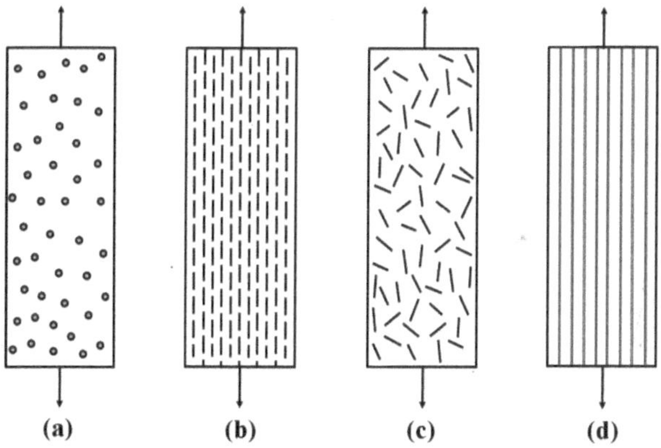

Fig. 7-3 Examples of composites: (a) particulate, random; (b) discontinuous fibers, unidirectional; (c) discontinuous fibers, random; (d) continuous fibers, unidirectional.

sectional dimension, known as the aspect ratio, can vary considerably. In single-layer composites long fibers with high aspect ratios give what are called continuous fiber reinforced composites, whereas discontinuous fiber composites are fabricated using short fibers of low aspect ratio (Block C). The orientation of the discontinuous fibers may be random or preferred (Fig. 7-3). The frequently encountered preferred orientation in the case of a continuous fiber composite (Fig. 7-3) is termed unidirectional and the corresponding random situation can be approximated to by bidirectional women reinforcement (Fig. 7-2, Block D).

With an improvement of high-temperature performance or tribological properties is often of interest. While various industrial applications have been developed or are being explored for metal matrix composites, their commercial usage is still quite limited when compared to that of polymer matrix composites. Finally, ceramic matrix composites are also being studied. The objective here is usually to impart toughness to the matrix by the introduction of other constituents, since the stiffness and strength are unlikely to be much affected. Such materials are still, for the most part, in the early stages of development, partly because they are rather difficult to manufacture.

In considering the formulation of composite materials for a particular type of application, it is important to consider the properties exhibited by the potential constituents. The properties of particular interest are the stiffness (Young's modulus), strength, and toughness. Density is of great significance in many situations, since the mass of the component may be of critical importance. Thermal properties, such as expansivity and conductivity, must also be taken into account. In particular, because the composite materials are subjected to temperature changes (during manufacture and/or in service), a mismatch between the thermal expansivities of the constituents leads to an internal residual stress. These can have a strong effect on the mechanical behaviors. Some representative property data are listed in Table 7-1 for various types of matrix

Table 7-1 Overview of Properties Exhibited by Different Classes of Materials

Type of materials	Density ($g \cdot cm^{-3}$)	Young's modulus (GPa)	Tensile strength (MPa)	Fracture toughness ($MPa \cdot \sqrt{m}$)	Thermal conductivity ($W \cdot m^{-1} \cdot K^{-1}$)	Thermal expansivity ($10^{-6} \cdot K^{-1}$)
Epoxy	1.25	3.5	50	0.5	0.3	60
Nylon	1.1	2.5	80	4	0.2	80
Polyurethane	1.2	0.01	20	0.1	0.2	200
Mild steel	7.8	208	400	140	60	17
Concrete	2.4	40	20	0.2	2	12
Alumina	3.9	380	500	4	25	8
Wood (load//grain)	0.6	16	80	6	0.5	3
(Spruce) (load⊥grain)	0.6	1	2	0.5	0.3	10
General polymer matrix composites (in-plane)	1.8	20	300	40	8	20
Adv. polymer matrix composites (load//fibers	1.6	200	1500	40	200	0
(APC-2) (load⊥fibers)		3	50	5	40	30
Metal matrix composites (Al-20% SiC_p)	2.8	90	500	15	140	18

and reinforcement, as well as for some typical engineering materials and a few representative composites. Inspection of these data shows that some attractive property combinations can be obtained with composites.

7.2. POLYMER MATRIX COMPOSITES

7.2.1. Introduction

Polymer matrix composites consist of thermoset or thermoplastic matrix resins reinforced by fibers that are much stronger and stiffer than the matrix. Polymer matrix composites are attractive because they are lighter, stronger, and stiffer than the unreinforced polymers or conventional metals, with the additional advantage that their properties and form can be tailored to meet the needs of a specific application. High-performance fiber reinforcements are of the highest interest for military and aerospace composite applications; these include carbon fibers and such organic fibers as aramids, liquid crystalline polymers, and ultrahigh-molecular-weight polyethylene [1]. The properties of composite structures depend not only on the fiber reinforcements, but also on the polymer matrix, the characteristics of the interface between the fiber and the matrix, and the manufacturing process used to form the finished structure.

7.2.2. High Performance Fiber Technology

High-performance fibers are those that are engineered for specific uses that require exceptional strength, stiffness, heat resistance, or chemical resistance. There exist a wide variety of fibers with widely ranging properties; Fig. 7-4 compares some of the different optimized categories. These fibers have generally higher tenacity and higher modulus than typical fibers. Tenacity denotes the

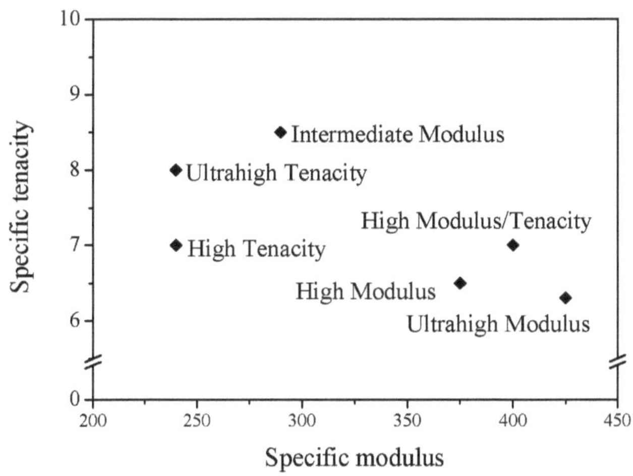

Fig. 7-4 Relative properties for the different categories of high-performance fibers.

relative strength of a textile fiber, expressed in grams of breaking force per denier unit. Denier denotes the system of measuring the weight of a continuous-filament fiber. Modulus denotes the value of the specific stress (load divided by area) to the strain (such as elongation) of a fiber. It is a measure of the relative flexibility and resilience of a material (i.e., rubber has a low modulus and steel a high one). In the larger fiber market, such high-performance fibers are generally niche products, but some are produced in large quantities.

Glass is the oldest high-performance fiber, one that has been manufactured since the 1930s. Today's glass fibers can be found in such end uses as insulation, fire-resistant fabrics, and reinforcement for fiberglass composites such as bathtub enclosures and boats. In addition, continuous filaments of optical-quality glass have revolutionized the communications industry in recent years.

Carbon fibers are one of the most important high-performance fibers for military and aerospace applications. Carbon fibers are engineered for strength and stiffness, but variations differ in electrical conductivity, thermal, and chemical properties. The primary factors governing the physical properties are the degree of carbonization (or the carbon content, usually greater than 92 percent by weight), the orientation of the layered carbon planes, and the degree of crystallization.

Commercial carbon fibers are made by extrusion of some organic precursor materials into filaments, followed by a carbonization process to convert the filaments into carbon. Different precursors and carbonization processes are used depending on the desired product properties. Precursor fibers can be specially purified rayon, pitch, or acrylics. The precursor fiber may also be converted into fabric form, which is then carbonized to produce the end product.

High-performance organic fibers have also become very important in recent years. Aramids are among the best known of the high-performance, synthetic, organic fibers. Closely related to the nylons, aramids are polyamides derived from aromatic acids and amines. Because of the stability of the aromatic rings and the added strength of the amide linkages, aramids exhibit higher tensile strength and thermal resistance than the aliphatic polyamides (nylons). The para-aramids (*p*-aramids, based on terephthalic acid and *p*-phenylenediamine, or *p*-aminobenzoic acid) exhibit higher strength and thermal resistance than those with linkages in the Meta positions of the benzene ring. The greater degree of conjugation and more linear geometry of the para linkages, combined with the greater chain orientation derived from this linearity, are primarily responsible for the increased strength. The high impact resistance of the para-aramids makes them popular for 'bullet-proof' body armor. For many less demanding applications, the aramids may be blended with other fibers.

7.2.2.1. High-performance carbon fibers

In the 1950s, it was recognized that the materials that combine light weight, high strength, and high stiffness are needed to produce ultralightweight structures [2].

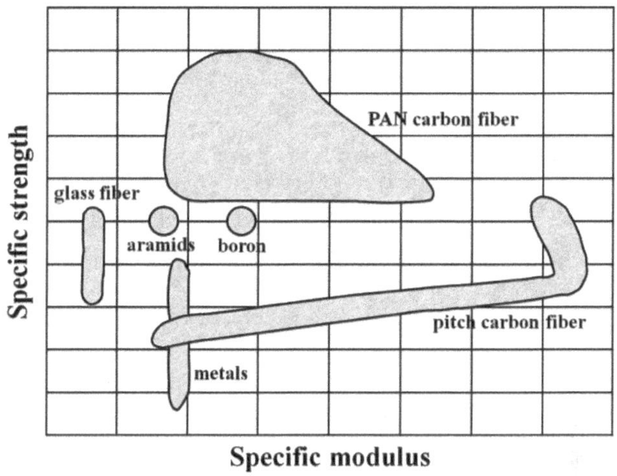

Fig. 7-5 Generalized specific modulus vs. specific strength for several high-performance fibers (carbon, aramid) compared with lower-performing materials (glass, metals).

Carbon fibers were developed to fill this need. Carbon fibers are typically produced by high-temperature processing of one of the three types of precursor fibers: polyacrylonitrile, pitch, or rayon. Depending upon the type of precursor and the processing method, the finished carbon fiber has somewhat different properties. Fig. 7-5 compares some of the properties of polyacrylonitrile-based and pitch-based carbon fibers. Typically, polyacrylonitrile-based carbon fibers have a higher specific strength and lower specific modulus than fibers made from pitch or rayon; pitch-based fibers have lower electrical resistivity and higher thermal conductivity. In general, polyacrylonitrile-based carbon fibers are less expensive than the pitch-based or rayon-based carbon fibers.

Carbon fibers generally are delivered in tows or bundles of fibers. These are found in two forms based on the starting precursor: special acrylic fiber or textile tow. Textile tow as referred to here is a modified form of an acrylic fiber used in the textile and carpet industry. The world capacity is around 5 billion pounds. It is made as large-tow materials, with 120–320 k (k = 1,000) filaments that may be split into smaller tows of around 36 k. In most cases, a crimp is applied to allow easier handling of the tow. The textile tow is most often piddled into a box containing 300–1,000 pounds for shipping. The crimp is needed to keep the piled fiber together in tows.

The special acrylic fiber tow, conversely, is manufactured solely for conversion to carbon fiber. It can be a compound similar to a textile tow, but it is made on a special spinning line. The filament count ranges from 3to 24 k or larger if required by a specific manufacturing process, such as pultrusion. This fiber is normally provided on a spool without crimp or twist. The conversion process for the two materials is the same except for the feed end and the take-up end of the line. It is more difficult to maintain a uniform tension on the textile tow precursor, which results in the large tow producing a slightly lower tensile strength and greater variability when compared to special acrylic fiber materials.

(1) PAN-based carbon fibers

More than 90 percent of all commercial carbon fibers are produced by the thermal conversion of polyacrylonitrile precursor fibers. Polyacrylonitrile-based carbon fibers are produced in three distinctly separate process steps: polymerization and wet spinning, stabilization, and carbonization. The precursor used to form polyacrylonitrile-based carbon fibers is a binary copolymer of acrylonitrile with 6–9 percent of an acid comonomer [3]. Although polyacrylonitrile fibers can be produced by either wet- or dry-spinning processes, wet spinning is used to produce nearly all commercial precursor fibers. The solution used in a wet-spinning process normally consists of 10–30 percent by weight of polyacrylonitrile (or polyacrylonitrile copolymer) dissolved in a polar solvent. This solution is first filtered and then extruded through a thin plate containing thousands of capillaries termed a spinneret. As the polyacrylonitrile solution is forced through the spinneret capillaries, the shear field tends to orient the solidifying polymeric structure parallel to the direction of flow.

As the solution exits the spinneret, it enters a coagulation bath where the copolymer precipitates into filament form. The filaments are made up of a finer structure, termed 'fibrils,' which run parallel to the filament axis. Various processing parameters, such as coagulation bath temperature, solvent concentration, and degree of stretch, can influence the fibrillar structure and its orientation within the as-spun polyacrylonitrile fibers. In other words, wet spinning yields a precursor fiber in which the polyacrylonitrile molecules are organized into fibrils that, in turn, are generally oriented parallel to the fiber axis. The bundle of filaments (referred to as a tow) is then washed to remove excess solvent and stretched to enhance molecular orientation. This step is essential for producing a final carbon fiber with adequate strength and modulus. Like most polymeric fiber processes, stretching does not greatly increase the crystallinity, which is a maximum of 50 percent, or the molecular order within the polyacrylonitrile fiber; rather, it enhances the axial orientation. This fibrillar network appears to be the precursor of the graphene network that develops during the final heat treatment [4].

The polyacrylonitrile precursor fibers are converted to carbon fibers in two separate process steps. In the first step, the fibers are heated in air to cross-link the structure. This renders the fibers infusible and inhibits relaxation of the structure during the final heat treatment step. In this step, termed carbonization, a standard-modulus fiber is heated in an inert environment to temperatures approaching 1400 °C; higher-modulus fibers may be processed up to 2600 °C. This step drives off most of the non-carbon elements, creating the carbon fibers. Since the polyacrylonitrile precursor is approximately 67 percent carbon, it is not surprising that the overall process conversion efficiency for the polyacrylonitrile-based fibers (pounds of carbon fibers per pound of precursor fibers) ranges from 50–60 percent.

Polyacrylonitrile-based precursor fibers continue to be produced using fundamental technology developed in the 1960s. Since then, fiber properties have improved significantly. Tensile strengths have more than doubled, from 2.8 GPa (400 kpsi) to more than 5.5 GPa (800 kpsi), and fibers with 290 GPa (42 Mpsi) have been introduced. Until the 1980s, a special acrylic fiber precursor was used for nearly all polyacrylonitrile-based carbon fibers. This precursor fiber is produced in tow sizes of 3,000, 6,000, and 12,000 filaments and is normally supplied on spools without applying a crimp or twist. However, the emergence of non-defense, non-aerospace markets increased the demand for fibers with acceptable properties at a lower price. This led, in part, to the introduction in the early 1990s of large polyacrylonitrile staple tows as precursor material that contains up to 320,000 individual filaments. These large staple tows, originally developed for the textile industry, must be crimped in order to be handled effectively, but this allows the manufacturer to increase the throughput of fibers, resulting in a lower-cost product. However, the defects inserted by crimping, as well as the difficulty in maintaining uniform tension, small oxygen concentration gradients, and a uniform temperature profile, result in carbon fibers from the large textile tows having slightly lower mean strength and greater strength and stiffness variability compared to special acrylic fiber-based materials.

(2) Pitch-based carbon fibers

Like polyacrylonitrile-based carbon fibers, the peculiarities of pitch-based fibers are the direct result of the precursor and the process used to convert it to fiber form. In this case, the precursor is mesophase pitch, a liquid crystalline material consisting of large polynuclear aromatic hydrocarbons derived from petroleum or coal.

The pitch precursor fibers are converted to carbon fibers in two separate process steps. Like polyacrylonitrile precursor fibers, the fibers are initially heated in air to cross-link the structure. This renders the fiber infusible and inhibits relaxation of the structure during the final heat treatment step. However, the pitch fibers (unlike the polyacrylonitrile fibers) are extremely weak at this stage, which makes it difficult, if not impossible, to pull it through the oxidation oven (the technique used in polyacrylonitrile-based carbon fiber process). This problem has led to alternate oven designs and advanced fiber-handling techniques. In the final heat treatment step termed carbonization the fibers are heated in an inert atmosphere to temperatures as high as 3000 °C. This drives off most of the non-carbon elements, creating highly oriented carbon fibers, with truly graphitic crystallinity. Since the mesophase precursor is approximately 90 percent carbon, the overall process conversion efficiency for the pitch-based fibers (pounds of carbon fiber per pound, of precursor fiber) can be much greater than the polyacrylonitrile-based carbon fibers. However, the brittle nature of pitch fibers leads to increased breakage during processing, often lowering the actual process conversion.

The first commercial mesophase precursors were produced by Union Carbide using a thermal polymerization process [5]. The precursors were prepared by thermally polymerizing (using an inert gas sparge) a highly aromatic isotropic pitch feed at temperatures of 400–410 °C for as long as 40 h. The shear forces during extrusion and subsequent drawing produce a filament with high molecular orientation in the direction of the fiber axis. This orientation is maintained during oxidation and high-temperature carbonization. Carbon fibers can be produced in this way with a variety of strength and flexibility characteristics.

The Mitsubishi Gas Chemical Company commercialized a process that uses a strong Lewis acid catalyst (hydrogen fluorene-boron fluorene, HF-BF_3) to catalyze a pure chemical feed, such as naphthalene or methyl-naphthalene, to a 100 percent mesophase product. The use of HF-BF_3 has been shown to greatly reduce the molecular weight distribution of the mesophase product compared to that produced by thermal polymerization [6].

ConocoPhillips also developed a process for producing a mesophase carbon fiber precursor [7]. The extraction step removed the smaller disordering molecules and concentrated the higher-molecular-weight materials. The higher-molecular-weight fraction was then converted to 100 percent mesophase by multiple, brief, heat treatment steps. The final product consisted of a high-molecular-weight mesophase containing a small fraction of the solvent that effectively served as a plasticizer.

The mesophase products resulting from these three processes differ considerably; for example, each process yields a product with a different molecular weight distribution (the averages range from 800–1,200 Da) and a different concentration of aliphatic side chains on the individual mesophase molecules. Consequently, their viscosity characteristics differ and their rate of stabilization differs as well. Nevertheless, all of these mesophase products reach a viscosity of approximately 200 Pa s, well below their degradation temperature, allowing them to be melt-spun into fiber form. Also, although somewhat irregular, all of the individual mesophase molecules are disk-like in shape.

In the melt-spinning process, the solid feed is melted and forced through a spinneret. As the molten precursor exits these holes, it is simultaneously quenched by the surrounding atmosphere and drawn down by the take-up device, forming solid fibers. At the first glance, this would appear to be a relatively simple process; however, the results are extremely sensitive to small changes in process conditions. Under typical process conditions, the tensile stress on the mesophase fibers is about 20 percent of that required to break the fibers [8]. (In comparison, during melt spinning the tensile stress developed within a nylon fiber is less than 1 percent of the breaking strength of the filament.) This greater tensile stress in the mesophase fibers is a direct result of two peculiarities of mesophase: its highly temperature-dependent viscosity and the brittle nature of the as-spun mesophase fibers.

Because their viscosity is highly temperature dependent, mesophase pitch fibers draw down and cool very quickly during fiber formation. In fact, at typical melt-spinning conditions, mesophase fibers are already 100 °C below their glass transition temperature by the time they are 2 cm from the spinneret. As a result, they can break easily during spinning and are extremely difficult to handle before they are carbonized. Although the rheology of mesophase makes the control of the melt-spinning process more difficult, its liquid crystalline nature gives these precursors, structural advantages compared to polymeric precursors such as polyacrylonitrile.

Since the first development of mesophase precursors, researchers have recognized that melt-spun mesophase fibers can develop remarkably high elastic moduli. This is a direct result of the transverse microstructure and axial molecular orientation created as the liquid crystalline precursor flows through a capillary and then is extended during fiber formation. Unless relaxation occurs during thermosetting, the transverse microstructure and axial orientation are merely perfected during carbonization. The transverse microstructural texture of the early commercial mesophase carbon fibers was either radial or flat layer. Thus, in the transverse direction the graphene layer planes fan out from the center of the fiber, and in the axial direction the layer planes tend to align parallel to the fiber axis.

The mesophase pitch-based carbon fibers with radial and flat-layer transverse textures readily develop three-dimensional crystallinity. Although this structure makes these pitch-based carbon fibers more flaw sensitive than the polyacrylonitrile-based fibers with their more random, filament structure, it gives them superior lattice-dependent properties (modulus and thermal conductivity).

Fig. 7-6 shows that prior to 1990, the polyacrylonitrile-based carbon fibers were characteristically high strength but low modulus, whereas pitch-based carbon fibers were high modulus but low strength. Since then, the manufacturers have discovered that they can disrupt flow during extrusion and create mesophase precursor fibers with random transverse textures. This disrupted texture mimics the filament texture of polyacrylonitrile-based fibers, by reducing flaw sensitivity. This, in addition to developing purer mesophase precursors, has allowed the introduction of new varieties of pitch-based carbon fibers with improved tensile strengths.

Similarly, manufacturers such as Cytec have begun using improved mesophase precursors, and linearizing the transverse texture during extrusion to create new varieties of pitch-based carbon fibers with thermal

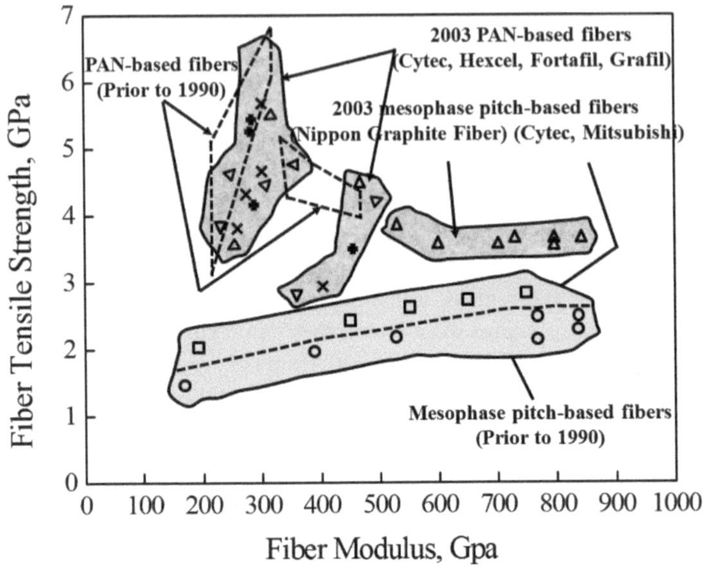

Fig. 7-6 Improvement of the mechanical properties of commercial polyacrylonitrile-based and mesophase pitch-based carbon fibers from the period prior to 1990–2003.

conductivities that are at least three times that of copper [9]. Most agree that these highly ordered textures can be produced only by using a liquid crystalline precursor. Thus, this new product may represent the natural niche for mesophase pitch-based carbon fibers.

(3) Rayon-based carbon fibers

Rayon was the original precursor for carbon fibers. Like polyacrylonitrile, rayon precursor fibers are solution spun, then stabilized and carbonized to form carbon fibers. High tension is often applied during the final step (termed stress graphitization) to increase the modulus and strength of rayon-based carbon fiber. The solution spinning process gives rayon-based carbon fibers a crenulated cross-section, a peculiarity that appears to improve the performance of carbon–carbon composites. Stress graphitization and the low carbon content of the precursor account for the high cost of this type of carbon fibers.

Rayon-based carbon fibers are used predominantly for ablative applications such as re-entry vehicle nosetips, heat shields, and solid-rocket motor nozzles and exit cones. These applications require high thermal resistance, but relatively low strength. The tensile strengths of these fibers vary from 345 to 690 MPa (50–100 kpsi) and the tensile moduli range from 20 to 55 GPa (3–8 Mpsi). Because of their relatively low tensile strength and high cost, the rayon-based carbon fibers have been largely supplanted by PAN-based carbon fibers [10].

(4) Microstructure-dependent properties of carbon fibers

Fig. 7-6 also shows the trade-off between tensile strength and modulus for each class of fibers and illustrates the improvements in properties of commercial polyacrylonitrile-based and pitch-based carbon fibers achieved during the 1990s. The reason for the significant differences in modulus and strength lies in the structure of the fibers. X-ray studies prove that polyacrylonitrile-based carbon fibers have no long-range, three-dimensional order [11]. Instead, the polyacrylonitrile-based carbon fibers contain extensively folded and interlinked turbostratic layers of carbon with interlayer spacings considerably larger than that of graphite. As a result, polyacrylonitrile-based carbon fibers have a low degree of graphitization. The turbostratic layers in polyacrylonitrile-based carbon fibers appear to follow the original fibrillar structure of the polyacrylonitrile precursor fibers. Although the turbostratic layers within these filaments tend to be oriented parallel to the fiber axis, they are not highly aligned. In contrast, pitch-based fibers are composed of highly oriented graphene sheets parallel to the fiber axis. Transverse to the fiber axis the sheets can be arranged radially, randomly, or in a flat-layer structure. Pitch-based fibers have a larger crystallite size as well. Thus, pitch-based fibers have a higher modulus because of the high degree of orientation in the fibers compared to polyacrylonitrile-based fibers. Pitch-based fibers have lower strength because the high degree of graphitization also leads to larger crystallites. Carbon fibers fail due to the failure of internal defects in the form of these crystallites. The smaller a crystallite is, the less likely its failure is to cause catastrophic failure in the fibers.

Recent work indicates that the intercrystalline and intracrystalline disorder characteristics of polyacrylonitrile fibers are also responsible for the superior compressive strength of this type of carbon fiber. Thus, the same fibrillar morphology that disrupts the transverse structure of polyacrylonitrile-based carbon fibers and gives them a higher tensile strength than the pitch-based fibers also gives polyacrylonitrile-based fibers superior compressive properties. In other words, polymeric precursor fibers, such as polyacrylonitril fibers, offer natural advantages when compression properties are critical. Developing superior compressive properties in pitch-based carbon fibers could be much more difficult.

As shown in Fig. 7-6, the tensile strengths of pitch-based carbon fibers improved significantly during the 1990s, whereas the strengths of polyacrylonitrile-based fibers, although higher than those of pitch-based fibers, remained relatively constant for two reasons. The obvious reason is that the lower strength of the early pitch-based fibers offered more opportunity for improvement. However, the major reason is that pitch-based carbon fiber producers have developed better mesophase pitch precursors and learned how to control the molecular structure of the precursor during fiber formation. The results have been improved properties and potentially lower production costs.

Nevertheless, the natural application of mesophase pitch-based fibers is not in high-strength applications, but in those requiring high thermal conductivity. This is because linearizing the pitch fiber's transverse structure to produce high thermal conductivity (approximately 100 times greater than polyacrylonitrile-based fibers) is easier than disrupting the structure enough to significantly increase the tensile strength. Furthermore, pitch-based carbon fibers are generally capable of achieving higher moduli compared to polyacrylonitrile-based fibers, making them attractive for stiffness-critical applications such as certain spacecraft components. Thus, for the foreseeable future, it is likely that the PAN-based fibers will dominate applications where strength is critical, and pitch-based fibers will dominate applications where heat transfer or stiffness is important.

(5) Surface treatment of carbon fibers

In the late 1960s and early 1970s several different types of surface treatment were used. (a) Wet methods: immersion of fibers in oxidizing agents such as nitric and chromic acid and electrochemical oxidation. Electrochemical oxidation is the most widely used industrial method for treating carbon fibers [12–15] but the details of such processes are mostly proprietary. (b) Dry methods: oxidation in air or oxygen. Plasma treatment is one of the convenient and effective methods to produce functional groups on carbon fiber surfaces.

The physical and chemical properties of carbon fibers are usually modified to achieve good adhesion between the reinforcement and the matrix materials. To enlarge the surface polarity of non-polar carbon fibers, various surface treatment techniques are applied, such as oxidation in acid solutions [16, 17], dry oxidation in oxygen, anodic oxidation, plasma treatments, mild fluorination [18–22], metallic coating [23–25], and so on. Interfacial adhesion between the fibers and the matrices cannot be achieved without intimate contact that is, unless the fiber surface contacts the resin on an intermolecular equilibrium distance level [26–29].

A large interfacial area of intimate contact by the matrix resin is a prerequisite whenever the interfacial bond is due primarily to van der Waals physical adsorption force. In addition, complete wetting of the fibers by the liquid polymeric matrix is advantageous when the surface free energy of the fibers is well above that of the liquid polymer in view of the London dispersive-specific polar components of surface free energy. The untreated commercial carbon fibers have a surface tension close to 40 mJ/m^2 and an extremely hydrophobic nature due to the extremely high temperature of the manufacturing process leading to carbonization (700–1500 ± °C), or to graphitization (above 2500 ± °C). On the other hand, most polymeric matrix resins, such as phenolic resin, are slightly hydrophilic with a surface tension in the range of 35–45 mJ/m^2[26–29].

(6) Properties of carbon fibers

The tensile stress–strain curve is linear to the point of rupture. Although there are many carbon fibers available in the open market, they can be arbitrarily divided into three grades, as also shown in Table 6.3. They have lower thermal expansion coefficients than both the glass and aramid fibers. The carbon fibers are anisotropic materials, and their transverse moduli are an order of magnitude less than its longitudinal modulus. The materials have a very high fatigue and creep resistance. Since its tensile strength decreases with increasing modulus, its strain at rupture will be much lower. Because of the material brittleness at higher modulus, it becomes critical in joint and connection details, which can have high stress concentrations. As a result of this phenomenon, carbon composite laminates are more effective with adhesive bonding that eliminates mechanical fasteners.

It is known that surface treatments improve fiber resin bonding but the mechanism by which the adhesion is improved is still controversial. There appear to be three major mechanisms contributing to the strength of the fiber/resin bond.

(a) The adsorption (either physical or chemical) of the resin molecules onto surface complexes; acidic complexes have usually been considered to be of most importance.

(b) Removal of contaminants providing a superior surface for adhesion.

(c) A mechanical keying effect when the resin penetrates in pits and channels on the roughened surface brought about by oxidation.

X-ray photoelectron spectroscopy is a very useful apparatus in the determination of chemical composition and functional groups of the fiber surfaces [30–33]. As shown in Table 7-2, the amounts of both surface oxygen and nitrogen groups increased with an increase in the current density in electrolytic solution. The large variation of the oxygen group is due to the varying acidity of the surface functional groups. The O_{1s}/C_{1s} ratios as a function of the applied electric

Table 7-2 **Variation of Surface Chemical Compositions of the Carbon Fibers with Current Density, Observed by X-ray Photoelectron Spectroscopy Measurement [unit: at%]**

	X-ray photoelectron spectroscopy			
Current density ($A{\cdot}m^{-2}$)	C_{1s}	O_{1s}	N_{1s}	O_{1s}/C_{1s}
0	91.8	6.9	1.3	7.5
0.2	80.7	12.2	7.1	15.1
0.4	79.2	13.1	7.7	16.5
0.8	76.9	14.0	9.1	18.2
1.6	78.0	14.2	7.8	18.2

Table 7-3 Variation of Surface Functional Groups of the Carbon Fibers with Current Density, Observed from X-ray Photoelectron Spectroscopy Measurement [unit: at%]

	Functional groups			
Current density ($A{\cdot}m^{-2}$)	-C-C- (284.7 eV)	-C-OH- (286.1 eV)	-C=O- (287.8 eV)	-OH-C=O- (289.2 eV)
0	63.3	25.4	4.2	2.7
0.2	54.1	28.2	9.0	6.7
0.4	53.0	28.3	9.1	6.8
0.8	50.1	28.9	10.0	9.6
1.6	51.2	29.1	10.2	6.6

potential are also indicated in Table 7-2, and the ratios increase on increasing the electric current density up to about 0.4 A/m^2. When the O_{1s}/C_{1s} ratio is considered, it is clear that during electrochemical oxidation, the hydroxyl (−COOH) and carbonyl (C=O) groups increased on the fiber surfaces, as seen in Table 7-3.

It is noted that the double bonds in the carbonyl and carboxyl functional groups on the fiber surfaces can largely enhance the physical attraction at interfaces between the fibers and epoxy resins, and simultaneously the amide-type functional groups. Therefore, the increase in amounts of oxygen-containing functional groups on the fibers plays an important role in improving the degree of adhesion at interfaces (hereby, Keesom's attraction of van der Waals force, hydrogen bonding, and the other small polar effects) between the fibers and matrix. As seen in Tables 7-2 and 7-3, it is also noted that larger increases of oxygen than nitrogen groups of the fibers treated are related to enhancing the bulk degree of physical attraction, as acidic functionality.

X-ray photoelectron spectroscopic spectra of untreated and nickel-plated carbon fibers are shown in Fig. 7-7. As anticipated, the untreated carbon fibers (Fig. 7-7(a)) show a C_{1s} peak and a substantial O_{1s} peak at 284.6 and 532.8 eV, respectively [34]. The O_{1s} peak is probably due to the intrinsic surface carbonyl or carboxyl groups. Otherwise, for the nickel-plated carbon fibers (Figs. 7-7(b–e)), carbon, oxygen, and nickel (BE = 857.6 eV) peaks are observed in X-ray photoelectron spectroscopic spectra [34]. The O_{1s} peak of nickel-plated carbon fibers is probably due to the NiO, C=O−OH, and O−C−O groups on the fiber surfaces.

7.2.2.2. *High-performance organic fibers*

The precise definition of high-performance organic fibers is controversial. High-performance functionalities may include high strength, high stiffness, high toughness, damage tolerance, durability, dimensional stability, and flame resistance.

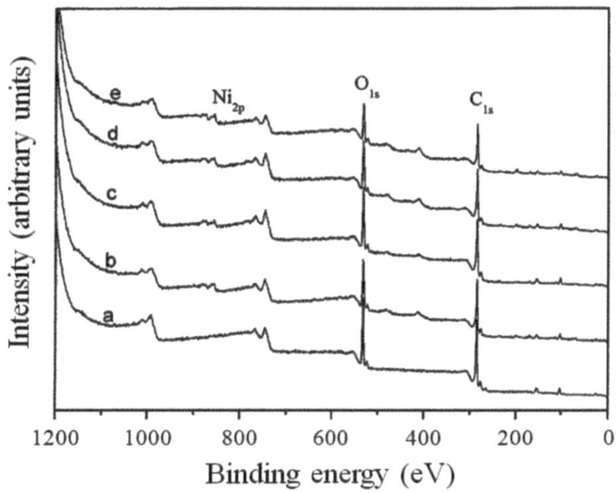

Fig. 7-7 X-ray photoelectron spectroscopic spectra of the nontreated and nickel-plated carbon fibers: (a) 0, (b) 5, (c) 10, (d) 30, and (e) 60 A/m^2.

Often, more than one of these properties is required in a single fiber. High strength and stiffness require almost perfectly oriented polymer molecules and a fully extended polymer chain. Although this has been recognized since the 1930s, the breakthrough in producing a highly oriented aramid fiber did not occur until the late 1960s and early 1970s when it was discovered that nematic solutions of poly(*p*-benzamide) and poly(*p*-phenylene terephthalamide) could be processed into fibers with a highly oriented, extended chain configuration [35].

Commercial high-performance organic fibers first became available with the introduction of a meta-aramid (*m*-aramid) fiber, Nomex, in the second half of the 1960s. Since that time, several classes of fibers have been commercialized. High-strength *p*-aramid fibers were introduced in the early 1970s, followed by liquid crystalline polyesters, high-strength polyethylene, and most recently, the introduction of poly (*p*-phenylene-2, 6-benzobisoxazole). A new fiber M5® has entered late development stages. Table 7-4 summarizes the properties of the existing commercial organic fibers, while Fig. 7-8 compares the properties of these fibers with those of other materials. Both strength and stiffness are expressed as specific values and are corrected for density.

(1) Meta-aramids

The meta-aramid fibers (especially with 100 percent aromatic content) offer very high temperature stability. Homopolymers in this class of materials exhibit a glass transition temperature of 275 °C and a melting temperature of 420 °C, very close to the decomposition temperature. The high aromatic content is also responsible for excellent performance in flame, leading to an intumescent char with adequate mechanical properties

Table 7-4 Mechanical Properties of some Organic Fibers Including Projections for M5®

Fibers	Strength (σ, (GPa)	Failure (ε, %)	Strain Modulus (*E*, GPa)	$(U^*)^{1/3a}$(m/s)
PBO[b]	5.20	3.10	169	813
Spectra 1,000	2.57	3.50	120	801
600-den.[c] KevlarKM2	3.40	3.55	82.6	682
850-den. Kevlar KM2	3.34	3.80	73.7	681
840-den. Kevlar 129	3.24	3.25	99.1	672
1,500-den. Kevlar 29	2.90	3.38	74.4	625
200-den. Kevlar 29	2.97	2.95	91.1	624
1,000-den. Kevlar 29	2.87	3.25	78.8	624
1,140-den. Kevlar 49	3.04	1.20	120	621
Carbon fiber	3.80	1.76	227	593
E-glass	3.50	4.7	74.0	559
Nylon	0.91	N/A	9.57	482
M5 conservation	8.50	2.5	300	940
M5 goal	9.50	2.5	450	1,043
M5 (2001 sample)	3.96	1.4	271	583

[a]*U* is the product of fiber specific toughness and strain wave velocity such that the dimensional analysis indicates that V50 velocity of an armor system scales with* $(U^*)^{1/3}$.
[b]*Poly (p-phenylene-2,6-benzobisoxazole).*
[c]*Denier.*

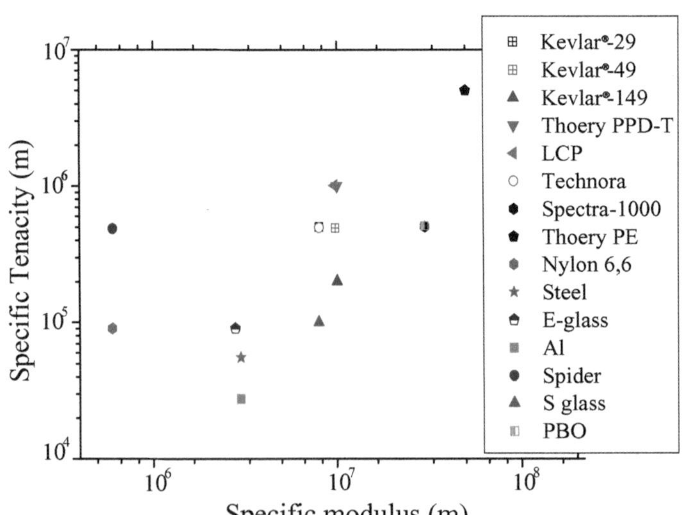

Fig. 7-8 Specific tenacity vs. specific modulus for several advanced organic fibers.

to provide for additional thermal insulation. The thermal stability and fire resistance characteristics of *m*-aramid fibers make them attractive for application such as protective apparel and hot gas filtration.

These fibers are now also being marketed in paper form, making them attractive for structural applications. The *m*-aramid-based paper can be saturated with a matrix (e.g., epoxy, phenolic) resin to form composite boards, or it can be converted into honeycomb structures, which find applications in the aerospace, electrical, and electronic industries. For example, Nomex products are used in protective apparel, hot gas filtration, automotive hoses, electrical insulation, aircraft parts, and sporting goods.

(2) Para-aramids

This class of materials has several products differing in the specific chemistries used. The first commercial product (Kevlar) is based on *p*-(phenylene terephthalamide), which continues to represent the largest global volume. The same chemistry underlies the second-largest-volume product, Twaron. Due to the rigidity of this material it forms anisotropic solutions in its solvent, sulfuric acid. The air-gap spinning process converts the solution into a highly ordered, highly crystalline fiber. This higher rigidity translates into better thermal properties (dimensional stability). Several *p*-aramid copolymers have been evaluated in the past. As copolymers, they are non-crystalline and can be processed by spinning from organic solvents containing salts with orientation, and thus properties, developed by fiber drawing. Technora represents the only firm that has reached a small commercial scale in the West. The Russian scientists had structured their *p*-aramids (SVM, Armos, and Rusar) on a rather expensive monomer diaminophenylbenzimidazole and its copolymers with *p*-phenylenediamine. This technology while producing slightly better mechanical properties, results in a very expensive fiber (monomer cost, very low productivity spinning processes, and very large energy consumption). Many of these products still do not meet Western standards as far as quality and uniformity are concerned.

The p-aramids, such as Kevlar fibers, are used in thermal-resistant clothing, protective vests and helmets, composites, asbestos replacement, reinforcement for tyre and mechanical rubber goods, ropes and cables, and sporting goods.

(3) Liquid crystalline polyesters

The development of liquid crystalline polyesters followed the commercialization of *p*-aramids. Like the *p*-aramids, liquid crystalline polyesters are based on aromatic structures. Finding a balance between processibility and properties has been a constant challenge for this technology. To achieve melt processibility into fibers, various copolymer compositions have been used and materials have to be spun at low molecular weight. To obtain adequate physical properties, a slow heat treatment process has to be applied. While the strength and stiffness of fibers in this class are of the same order of the magnitude as aramids, their thermal properties are somewhat lower. Liquid crystalline polyester fibers are used in applications such as tow ropes, cargo tie-downs, tethers, and cables.

(4) High-strength polyethylene
Ultra-high-molecular-weight polyethylene offers the first demonstration that the fully extended chain configuration and resultant high-performance properties can be achieved in flexible polymers. The shaping process is relatively complex. Gel spinning at low polymer concentration is followed by extraction of the solvent and the critical drawing step. The very high draw ratio achievable in this system is responsible for the development of high strength and stiffness of the fiber [36]. However, the low melting point of this material limits its applicability at elevated temperatures, and its non-polar nature decreases its adhesion to resins. Spectra fibers are used in numerous applications, including ballistic vests, helmets, armored vehicles, sailcloth, fishing lines, marine cordage and lifting slings, and cut-resistant gloves and safety apparel. Ballistic applications have become important for this material since the development of unidirectional composite laminated structures, which allow for diminishing blunt trauma.

(5) Polybenzoxazole
The invention and development of high-performance fibers based on aramids led to an understanding of the principles underlying this area of science and technology. The natural extension was to move to even more rigid polymers. The development of polybenzoxazole and polyphenylene benzobisthiazole compositions by Wolf and others, with the continuing work of Dow Chemical Company and finally Toyobo, culminated in the commercialization of Zylon fiber in the second half of the 1990s. Toyobo increased the capacity of this fiber to 500 megatonnes in 2007.

As for other rigid polymers, the shaping of this fiber is through processing of anisotropic solutions, but this time from polyphosphoric acid. Higher rigidity than aramids and lack of amide bond translate to improved thermal properties and improved mechanical properties (Table 7-5). Expensive ingredients and highly aggressive solvents combined with the

Table 7-5 Physical Properties of some Organic Commercial Fibers

	p-aramid	UHMWPE[a]	PBO[b]
Yarn tenacity, gpd[c]	18–27	30–38	40–42
Density, g/cm^3	1.44	0.97	1.56
Moisture regain, percent	1–7	0	0.6–2
Heat resistance, °C	550	150	650
LOI[d], percent	29	20	68

[a] *Ultrahigh-molecular-weight polyethylene.*
[b] *Poly (p-phenylene-2,6-benzobisoxazole).*
[c] *Grams per denier.*
[d] *Limiting oxygen index.*

extremely high viscosities of these solutions result in significantly higher prices vs. the aramid fibers.

While the advantages previously cited are generically associated with very rigid polymers, some disadvantages are due to specific chemistries. Poorer ultraviolet stability and adhesion to resins limit its use in some applications. More recently, there have been reports of some possible problems with hydrolytic stability. This issue is important enough to warrant careful following of future developments.

(6) M5® fibers

Although the fibers previously discussed are produced at very different volumes, the entire set is considered commercial. The committee has given significant consideration to one fiber, which is in the precommercial stage of development, M5 (poly{2,6-diimidazo[4,5-*b*:4′,5′-*e*]pyridinylene-1,4-(2,5-dihydroxy)phenylene}) [37]. The fiber is being developed by Magellan Systems International, which expects to start up a pilot plant in 2005, with a commercial facility likely in 2007. M5 is a rigid polymer spun

Table 7-6 Properties of M5® Fibers[a]

Fibers	PBO[e] Zylon	M5 1999–2000	M5 April 2002	M5 Targets
Tenacity, GPa	5.5	4	5.3	9.5
Elongation, %	2.5	1.2	1.5	>2
Elastic modulus, GPa	280	330	350	>400
Compressive stress, GPa[b]	0.42	1.6	1.7	2
Compressive strain, %	0.15	0.4	0.5	0.5
σ_{max} compressive, GPa[c]	0.8	3	3	4
Density, g/mL	1.56	1.7	1.7	1.7
Water regain, %	0.6	2.0	2.0	2
Onset of thermal degradation in air, °C	550	530	530	530
LOI[d], $\%O_2$	68	>50		>50
Electrical conduction	—	—		—
Impact resistance, Composites	++	++		+++
Damage tolerance	N/A	++		+++
Weaving properties	+/−	+		+
Stability in UV radiation	—	++		++

[a] *Toyobo data, available at http://www.toyobo.co.jp/*
[b] *Measured in unidirectional composite test bars, three-point bending test, onset of deflection for the organic fiber reinforced composites; catastrophic failure for the carbon composites. M5 composites proved to be able to carry much higher loads than the load at onset of deflection and to absorb much energy at high strains in a mode analogous to the flow behavior in steel being damaged.*
[c] *In a bending test on unidirectional composites, the maximum apparent stress on the outermost fibers (neglecting change of shape, at the compression side, by ductile response at high deformation).*
[d] *Limiting oxygen index. Source: Magellan Systems International.*
[e] *Poly (p-phenylene-2,6-benzobisoxazole).*

from an anisotropic solution; its properties are given in Table 7-6. While the tensile properties realized are similar to those of poly (*p*-phenylene-2,6-benzobisoxazole), the fiber offers potential for significant improvement of compressive properties above those of any of the commercial organic fibers. The compressive strength of the fiber ranges from two to four times that of other organic fibers. This is attributed to the higher level of hydrogen bonding, which improves its structural integrity. Such properties offer potential for combining both structural and ballistic requirements of systems. This is further strengthened by a very good adhesion to resins (imidazole group) and, thus, improved performance in composite applications. The differences between oxazole and imidazole rings result in better ultraviolet stability and potential for improved hydrolytic stability.

The high-performance fibers that are currently available have been in production for decades, and the technology is now mature. Their properties depend sensitively on microstructure and in some cases are approaching theoretical limits. Thus, for commercial fibers only incremental improvements in fiber properties are expected in the near future. The new fiber M5, which is in precommercial development, offers a potential for a unique combination of properties that promise a unique set of applications.

(7) Raw materials for organic fiber development
The development of advanced organic fibers is a complex and expensive endeavor. It usually requires the establishment of a raw materials base in addition to the processes for polymerization and shaping. Of the fibers discussed in this report, only ultrahigh-molecular-weight polyethylene did not require the introduction of a new raw material. The raw materials and their development represent a significant portion of the cost associated with the development of advanced fibers. For example, poly (*p*-phenylene-2, 6-benzobisoxazole) fibers are extremely expensive because of the expensive ingredients required to produce them, despite the fact that the development of these fibers began approximately 20 years ago. The need to develop an ingredients base is also essential to successful commercialization of M5 fibers.

(8) Properties and design needs
It is well established that the thermal properties of organic fibers are controlled by the chemistry and morphology of the fibers and that little can be done to improve these properties within a given chemical composition. Therefore, it is important to choose a material that can maintain its properties over the expected operating temperature range. On the other hand, the mechanical properties of all materials in Table 7-4 can be tailored to meet the needs of specific applications. A good example of such tailoring can be found in the aramid family, where production technology has been developed to alter the balance between the elongation and modulus of fibers to meet the demands of applications that are driven by fatigue (e.g., reinforcement of rubber) as well as polymer matrix

composite applications, where stiffness is a controlling factor. These two extremes cover the range of elongation and modulus described in Table 7-5.

The current estimates of theoretical strength and stiffness for *p*-aramids and polyethylene are shown in Fig. 7-8. The elastic modulus realized in commercially available fibers approaches 80 percent of the theoretical values for these fibers. Thus, a dramatic increase in the stiffness of these organic fibers is not expected. For higher stiffness, one would have to look at the fibers described in Table 7-6. Estimating the theoretical strength of a material is more difficult and less reliable than estimating the theoretical modulus. However, there is general agreement that the strength values attained for organic fibers are further from theoretical values than are the modulus values. For example, the demonstrated values of strength are estimated at 30–50 percent of theoretical expectations. Thus, it is reasonable to expect incremental improvements in the strengths of existing commercial fibers over the next several years.

(9) Properties of organic fibers and their reinforced composites

Table 7-7 lists the experimental surface properties of the Kevlar fibers treated by chemical treatment in aqueous phosphoric acid (H_3PO_4) solution. The results of pH and acid value exhibit that the surface properties of untreated Kevlar fibers show a slight excess of acidic properties. This is probably due to the increase of acidic groups on the fiber surfaces. Both the pH and the acid values of Kevlar fibers treated with H_3PO_4 acidify to those of the untreated fibers, and indicate that the concentration of 10 wt. % H_3PO_4 provides an optimum condition to the surface acidity.

The surface compositions of the fibers for current concentration of H_3PO_4 are also summarized in Table 7-7. As a result, the amounts of surface oxygen are increased with increasing the concentration of H_3PO_4 up to 35 wt.%, whereas carbon and nitrogen are not largely influenced on the current concentration of H_3PO_4. This result suggests that the oxidation of Kevlar fiber surfaces occurs by phosphoric acid solution with optimum concentrations. It then results in increasing the surface acidity or hydroxyl and carboxyl functional groups on the chemical treated Kevlar fibers. From the result of Table 7-7, the chemical treatment in phosphoric

Table 7-7 pH, Acid Value, and Chemical Compositions of the Chemical Oxidation Treated Kevlar Fibers

Concentration of H_3PO_4 (wt.%)	pH	Acid value (mequiv./g)	C_{1s}	O_{1s}	N_{1s}	O_{1s}/C_{1s}
No treatment	7.01	15.9	73.3	24.5	1.3	0.334
1	6.97	30.0	72.5	25.1	1.5	0.346
10	6.95	65.7	70.8	32.4	1.1	0.457
35	6.78	72.5	72.6	32.9	1.6	0.453

acid solution makes an important role in increasing the acidic properties of Kevlar fibers. Consequently, the increase of acidic groups on the fiber surface may expect to promote the surface energy of the fibers as well as the interfacial bonding by establishing secondary or van der Waals forces at the interfaces between fibers and matrix, resulting in increasing mechanical interfacial properties of the composites [38].

Figs. 7-9(a) and (b) show the results for ILSS and K_{IC} of the Kevlar-fibers reinforced composites according to the concentration of phosphoric acid solution. As a result, the effects of fiber surface treatment conditions and the resulting fiber–matrix adhesion on mechanical properties of the

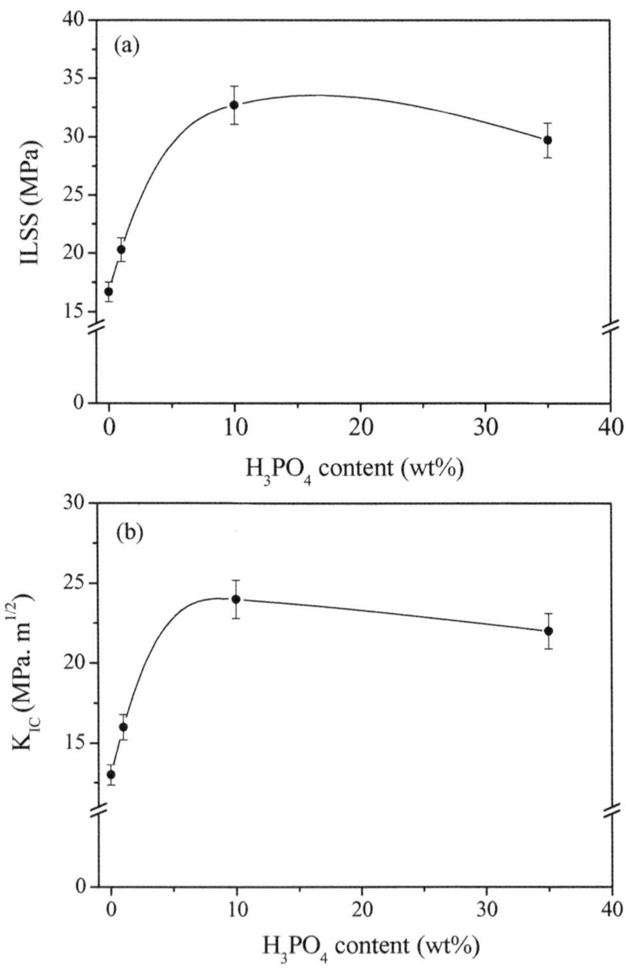

Fig. 7-9 Interlaminar shear strength (ILSS) (a) and K_{IC} (b) of H_3PO_4 treated-Kevlar fibers-reinforced composites as a function of H_3PO_4 concentration.

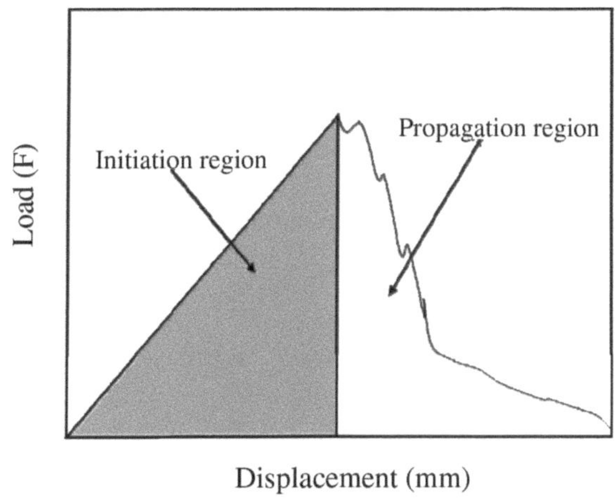

Fig. 7-10 Schematic representation of the load-displacement curve in an impact test.

composites show good relationships. That is, K_{IC} value is to increase on increasing the ILSS value for the degree of adhesion at interfaces. The maximum strength values of ILSS and K_{IC} are obtained at the chemical treatment of 10 wt.% phosphoric acid solution. Therefore, we suggest that additional energy is needed to extend the interfacial crack at this condition, which is attributed to increasing the interfacial adhesion between the fibers and matrix [39]. Also, this result is good in agreement with surface analyses of the Kevlar fibers, as already shown in Table 7-7.

The impact behavior is major mechanical properties in order to evaluate the degree of toughness in composite materials. When the materials are impacted, transferring load is absorbed into the materials and it affects the residual mechanical properties [40].

Fig. 7-10 shows a schematic representation of the load history in an impact test. There is a region of fracture initiation followed by a region of fracture propagation. The total impact energy measured during the test, E_t, is the sum of the initiation energy, E_i, and the propagation energy, E_p. If the material behavior is elastic to failure and the stress in the specimen is adequately described by the simple beam theory, then the initiation energy in a unidirectional composite reinforced with one type of fiber is given by Eq. (7-1) [41].

$$E_i = \frac{Lwt}{18} \cdot \nu_f \cdot \frac{\sigma_f^2}{E_f} \tag{7-1}$$

where L is the span length, w the specimen width, t the specimen thickness, ν_f the fiber volume fraction, σ_f and E_f the fiber stress and the fiber modulus at failure by tension or compression, respectively.

The second portion of the impact energy, E_p, is a more complex way since E_p is dependent on material fracture mechanisms that, in turn, are influenced by specimen geometry as well as by material properties. For specimens having a similar geometry, the relative percentage of energy absorbed in fracture initiation and propagation provides an indication of the ductility of materials. The ductility index (DI) can be defined as the ratio of propagation energy to initiation energy [42, 43].

$$DI = \frac{E_p}{E_i} \tag{7-2}$$

Therefore, brittle materials have low E_p and, as a result, low DI, as deduced in Eq. (7-2).

Fig. 7-11 shows the representative load–displacement curves of the Kevlar fibers-reinforced composites under impact load. The zero displacement indicates the contact of the impact tip onto the sample. And, the samples are deformed after contacting the impact tip and are impacted by the tip. The multiple peaks in Fig. 7-11 are generated because of the sample grip in the impact-testing machine used to prevent slippage of the samples and the first peak load is thus taken to be the maximum load. As a result, the chemical treatment of Kevlar fibers changes the slope of

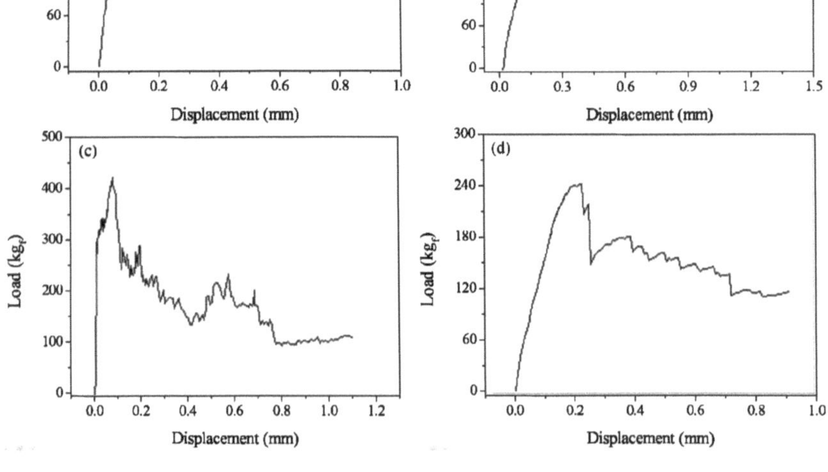

Fig. 7-11 Load-displacement curves of H_3PO_4 treated-Kevlar fibers-reinforced composites as a function of H_3PO_4 concentration after impact test.

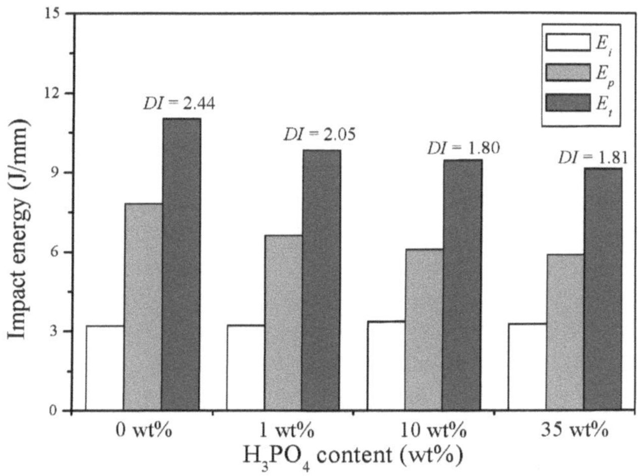

Fig. 7-12 E_i, E_p, and ductility index (DI) of H_3PO_4 treated Kevlar fibers-reinforced composites as a function of H_3PO_4 concentration.

the load–displacement curve to the maximum load. The specimens with H_3PO_4 shows steeper slopes up to the maximum load compared with the untreated specimen. This means that the H_3PO_4 treated composites have a higher impact modulus than the untreated ones. Therefore, H_3PO_4 treated composites show more brittle behaviors than the untreated ones and have a smaller displacement, resulting from the higher impact stiffness [44].

The results of the E_p, E_i, and ductility index of the composites are shown in Fig. 7-12. As a result, the total energy absorbed during the impact test, i.e., both the impact energy and the ductility index of the H_3PO_4 treated Kevlar fibers-reinforced composites are decreased on increasing the concentration of H_3PO_4. This is probably due to the increase of higher impact stiffness, resulting in increasing the brittle behavior of the composites [45, 46]. However, the degree of decrease on impact properties is fairly small compared to the increase of the mechanical interfacial properties, which can compensate for the entire mechanical properties of the composites.

7.2.3. High Performance Matrix Resins

Although the fibers play a dominant role in determining the stiffness and strength of a composite, the choice of the matrix will determine the maximum service temperature, viable processing approaches, and long-term durability. Matrix materials can be divided into two broad categories: thermosetting and thermoplastic. Thermosetting materials are characterized by having a low-viscosity, reactive, starting oligomer that cures (reacts) to form an insoluble, infusible network [47]. The cure temperature and time influence many matrix properties. An incomplete cure affects mechanical properties, swelling behavior in solvents, and moisture susceptibility. On the other hand, an incomplete cure (increased distance between cross-links) may improve flexibility and toughness.

Thermoplastics are linear or slightly branched polymers of relatively high molecular weight. They are of significant interest because they can be remelted, thus easing repair in the field, and because they are melt-processable. The use of thermoplastics also allows other composite processing techniques such as injection molding of short fiber composites and extrusion of long fiber composites. Thermoplastic polymer morphology may either be amorphous (disordered) or semicrystalline (partially ordered). The semicrystalline and even liquid crystalline morphologies can impart superior solvent resistance to the overall matrix resin. The semicrystalline polymers display a melting temperature for the crystalline regions, as well as a glass transition, and processing must be performed in excess of the melting temperature. Amorphous thermoplastics, on the other hand, are processed above the glass transition temperature. A possible limitation for some applications of semicrystalline polymers concerns the morphological changes that may occur during processing or in the service environment, particularly due to the application of heat or exposure to solvent while under stress. These changes in the crystalline structure and/or content may cause changes in the overall composite properties and are highly undesirable. Likewise, the control of morphology during processing is critical to achieving the desired matrix properties such as toughness and chemical resistance.

Various resins that exist today provide a wide range of service temperatures. For example, epoxies, vinyl esters, and polyesters will meet the relatively low-service-temperature requirements (160–180 °F) of the Army's Future Combat System ground, and tactical vehicles as well as Navy ship systems.

Higher-temperature performance can be achieved (250–400 °F) with epoxies, bismaleimides and polyimide resins using traditional prepreg or tow-preg and autoclave, filament winding, and fiber placement process technologies. The formulations of these materials have been developed to enable the use of resin transfer molding processes for smaller-scale components. Prepregs, or preimpregnated fiber assemblies, are commonly used in many applications and are fabricated by spreading an array of fiber tows and impregnating the tows with a thermoset or thermoplastic resin to produce a thin sheet of materials. In this material form, the fibers are continuous and aligned, providing high stiffness and strength in the fiber direction and low matrix-dominated properties in the transverse direction. For structures subjected to multiaxial loadings, the prepreg is laminated to tailor the properties.

Vacuum-assisted resin transfer molding has been identified as an affordable process and is used to fabricate structural armor for ground vehicle hull structures containing integral ceramic composites, as well as large-scale topside ship and hull structures. Desirable resin attributes for these applications include relatively low viscosity at room temperature to enable room-temperature infusion as well as the lowest cure temperature possible to meet hot-wet glass transition temperature requirements. Resins meeting these needs enable

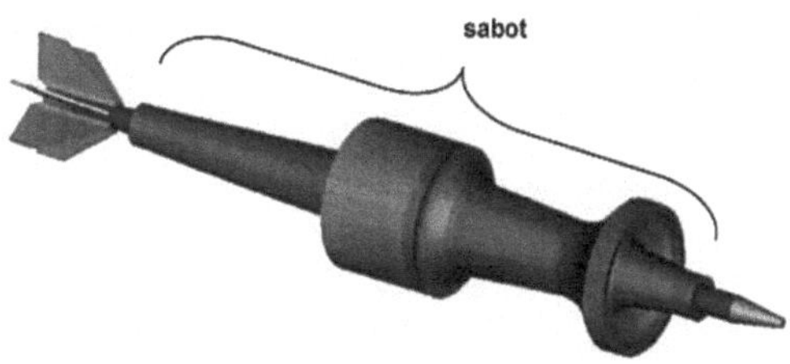

Fig. 7-13 The M829A2 sabot, which stabilizes the trajectory of the munition early in flight, was designed with graphite fiber-reinforced epoxy material that resulted in a 30 percent weight reduction in the sabot component. This weight reduction enabled an increased muzzle velocity of approximately 60 m per second, significantly increasing the range and armor penetration capability of the projectile. The M829A2 accounts for one of the largest uses of composite materials in DOD.

low-temperature tooling materials to be used, providing significant cost savings. However, the performance demands on these composites remain high. For example, structural armor requires resins that have high elongation to failure to survive ballistic impact, but current vacuum-assisted resin transfer molding resins fall short of the service temperature requirement. Higher-cure-temperature resins meet the service temperature requirements, but ballistic performance is degraded. Additional research in resins is needed to balance processing ease and performance for this important U.S. Department of Defense (DOD) application.

The need for more damage-tolerant aerospace structures has led to the development of toughened thermosets and thermoplastic matrices that are resistant to impact damage and delamination growth. High matrix toughness has also been proven to be a key property in ballistic performance of tank munitions such as the M829A2 (toughened thermoset) and the M829A3 (polyetherimide thermoplastic) carbon fiber sabots (see Fig. 7-13). The capability of electron-beam (e-beam) processing for non-autoclave cure of large-scale structures such as rocket motors and fuel tanks has been demonstrated. Improvements in resin toughness and interface optimization for e-beam resins are needed to improve the properties and resistance to microcracking.

The role of the matrix in the long-term durability of a composite is a critical issue. The durability is affected by the state of the resin, which may undergo physical aging or environmental degradation, as well as changes in the interaction with the fiber at the interface. In addition, the stress state within the matrix due to processing, thermal and fatigue cycling, and other mechanical loads is critical to the long-term performance. Microcracking is one of the first damage modes observed in the matrix phase. Microcracking can initiate fiber fracture, interface debonding, and delamination that can limit the lifetime of the component. An even more severe case occurs when microcracks provide pathways for accelerated degradation by the environment, for example, moisture ingress into honeycomb structures followed by freeze-thaw or elevated temperature thermal oxidation that can dramatically reduce mechanical properties. Advances in modeling the failure mechanisms and the development of new microcracking-resistant composites are needed.

7.2.4. Fiber–Matrix Interface

7.2.4.1. Definition of fiber–matrix interface

The interface can be defined as the three-dimensional boundary between the fiber and matrix. It is critical to controlling composite properties because fiber–matrix interaction occurs through the interface. This interaction can occur through three mechanisms: mechanical coupling or micromechanical interlocking of the two materials, physical coupling such as van der Waals or electrostatic interaction, and covalent bonding (by way of a coupling agent) between the fiber and the matrix (see Fig. 7-14 for some examples). These interactions create an interphase region which is a three-dimensional region near the fiber with properties different from either the fiber or the matrix. For example, in thermosets, the interphase can form due to preferential absorption of either the curing agent or the resin at the fiber surface, leading to a region of higher or lower

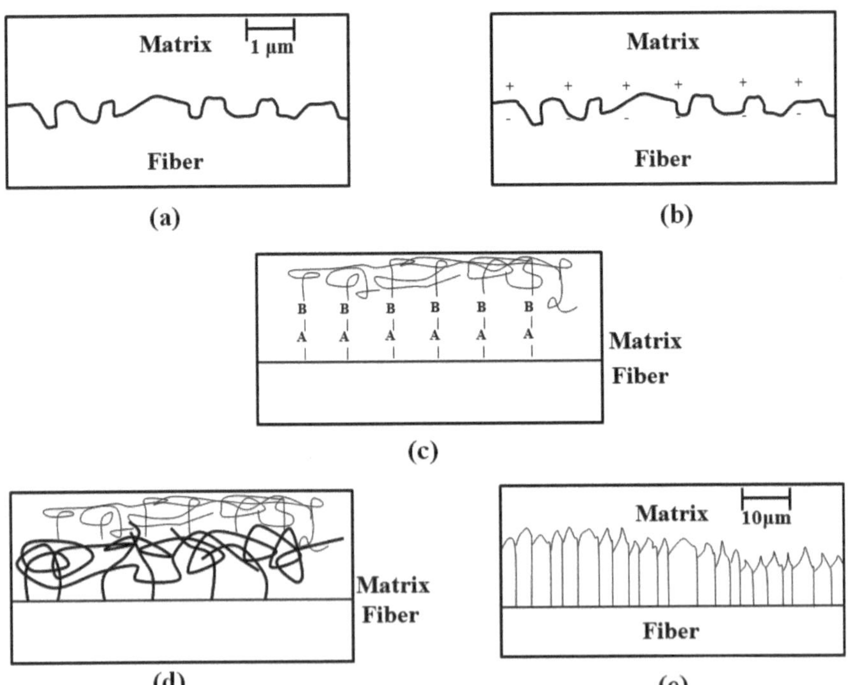

Fig. 7-14 Schematic depictions of various interactions at the fiber-matrix interface: (a) micromechanical interlocking, (b) permanent or induced dipole interactions, (c) chemical bonding, (d) chain entanglement, and (e) transcrystallinity.

cross-linking and in thermoplastics, the interphase can be a region of transcrystallinity. For ease of discussion, the term 'interface' is used to describe both the two-dimensional interface and the three-dimensional interphase.

To control the properties of the interface, surface treatments are often used on the fiber. Surface treatment generally involves surface oxidation of the fiber, either electrolytically or using gas or liquid chemicals. This provides some functionality on the fiber surface by increasing the surface area and the number of reactive groups on the surface and can improve adhesion in terms of mechanical interlocking and physiochemical interactions. The surface can then be coated with different organic and/or inorganic compounds to promote further interaction. A sizing may also be used to protect the fiber from process damage. Some sizings, such as organosilanes, may also be referred to as coupling agents, as these couple the active fiber surface to a non-polar resin, predominantly through increased oxygen functionality. Other types of materials, including oligomer polyimides and polyamides, can be grafted to the surface in order to promote adhesion through other mechanisms, such as transcrystallization.

The chemistry of such treatments is proprietary, but they may perform a variety of functions, as follows [48]:

- Protecting the fiber surface during shipment and handling and subsequent processing,
- Binding the fibers together (which makes processing of prepreg and weaves easier),
- Reducing static electricity,
- Improving the chemical bonding to the matrix.

While it is often assumed that sizing improves the coupling between the fiber and the matrix, several studies have shown that sizing can create a brittle interface region [49, 50]. Thus, sizings can be detrimental to bulk mechanical properties. In addition, sizings and other fiber treatments can lead to variability in the interface properties. This variability in interface properties can originate from several sources, including how evenly the sizing or coating covers the fibers, variability in wetting of the matrix that can create poorly bonded regions, variation in surface roughness of the fibers, and matrix variability.

The interface is crucial in controlling composite properties because load is transferred from the matrix to the fiber through the interface and deformation of the interface region (debonding or yielding) is critical in absorbing energy during failure [51, 52]. In high-performance fiber-reinforced polymer matrix composites, the fiber has a much higher modulus than the matrix. Therefore, when the composite is loaded to a given strain, the fiber will carry more load than the matrix. This load must be transferred from the matrix to the fiber through the interface region. The interface is particularly relevant in applications requiring damage tolerance and durability in aggressive environments [53], because the interface controls damage development. When a fiber breaks, the neighboring fibers are subjected to a stress concentration [54–56]. The location of the next break due to the stress overload is dictated by the interface properties.

Fig. 7-14 shows some different types of interface bonding. Changing interface properties can have a dramatic effect on the overload behavior. For the perfectly elastic case (from a strong interface), the stress concentration factor is large and narrow. This will lead to in-plane fiber failure and prevent energy absorbing mechanisms such as debonding and fiber pull-out. Thus, brittle failure will occur. If the interface is compliant or debonding occurs, the stress concentration factor will decrease significantly and the region of overload in the neighboring fibers will be larger. This type of interface will lead to a brush-like fracture surface and higher ductility and toughness because of the debonding and fiber pull-out that occur. In the organic fibers, the fiber breaks are more diffuse and even the initial stress concentration is lower and more diffuse [57]. However, similar failure mechanisms apply. Thus, a relatively weak interface contributes to high toughness for both organic and carbon fiber composites, but it can result in a significant reduction in strength. This combination has proven to be successful in composite armor [58].

What is clear from prior work is that control over the interface is important not only for protecting the fibers, but for controlling properties. In light of this, some areas of interface development are important for improving composite performance. There are some specialized needs for better sizings for carbon fiber. This includes sizings for vinyl ester resins and, in particular, sizings for carbon-vinyl ester composites of interest to the Navy for topside ship structures. Vinyl ester resins offer significant processing advantage and affordability over bisphenol-based resins. Recent work has shown that thermoplastic sizings of polyhydroxyethers lead to improved processing, improved fiber durability during handling, improved composite properties in a vinyl ester-based resin [59]. Therefore, it is clear, that there are sizing materials that will lead to improved composite properties. Further work in this area is warranted.

Higher-temperature sizings are required that do not degrade in use. As the use of temperatures for composites continues to increase, the stability of the interface at high temperature is required to maintain composite performance. Although some work has been done in this area, namely the use of LaRC PETI-5, a phenylethynyl-terminated imide oligomer, as well as some proprietary resins, this is still an area in need of some focus [60, 61].

The role of the interface in controlling properties is still not fully understood, particularly as it applies to lifetime prediction, and a better understanding is needed. The interface has a significant effect on fatigue behavior, creep, and environmental stability, yet direct inclusion of interface properties into lifetime models is still under development [62]. To complicate matters further, the interface variability is also a relevant parameter and can lead, for example, to enhanced toughness [63]. One opportunity for improving the cost-effectiveness of composites is tighter design criteria, and this requires, in addition to an understanding of matrix and fiber effects, a thorough understanding of the role of interface variability on composite properties.

The vision for the future should be one in which the fiber properties, matrix properties, and interface properties as well as the stochastic aspects are included in lifetime prediction models that have the confidence of composite designers. In a very ideal future, their performance is monitored in service conditions and data are added back into computer-aided design and computer-aided engineering models to validate and improve their predictive properties [64–66].

7.2.4.2. Mechanical interfacial properties of composites

It is generally accepted that good mechanical properties and long durability of the composites largely depend on fiber–matrix interfacial adhesion in the case of the composites, since load stress transfers from one matrix to the other via the fiber [67]. For example, the interlaminar shear strength is improved when the constitutive elements of composites are fabricated in modifications that increase interfacial surface areas and surface functional groups [68, 69].

The analytical expression for critical stress intensity factor (K_{IC}, in mode I fracture) was characterized using a single edge notched beam fracture toughness test in 90 ° three-point bending flexure, as shown schematically in Fig. 7-15.

Also, Mode I interlaminar fracture toughness (G_{IC}) was determined according to ASTM D 5528-94 using the double cantilever beam test, as shown in Fig. 7-16. Specimens were cut to size using a diamond saw with the fiber direction parallel to the length of the sample, 20 mm in width, 3 mm in thickness and 160 mm in length with a 30 mm crack starter film. Aluminum tabs were attached to the end of the specimens containing the insert using a high strength epoxy adhesive and allowed to dry for 48 h. An Instron 1125 Universal Testing Machine was used to perform the tests using a crosshead speed of 1 mm/min and a load vs. displacement curve was produced for each test. From these data, the Mode I critical strain energy release rate (G_{IC}) was calculated using corrected

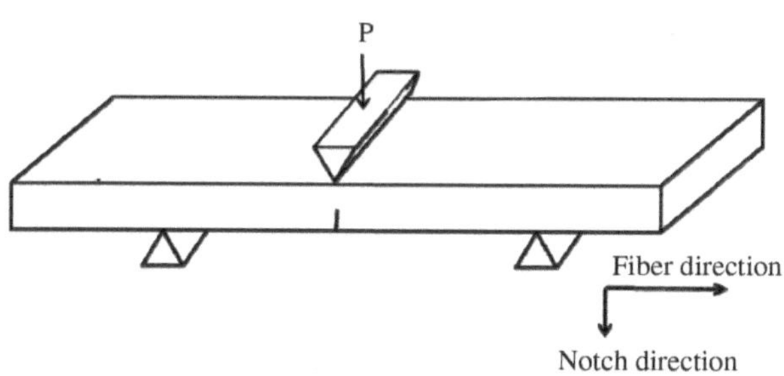

Fig. 7-15 Schematic diagram of single edge notched beam fracture toughness test.

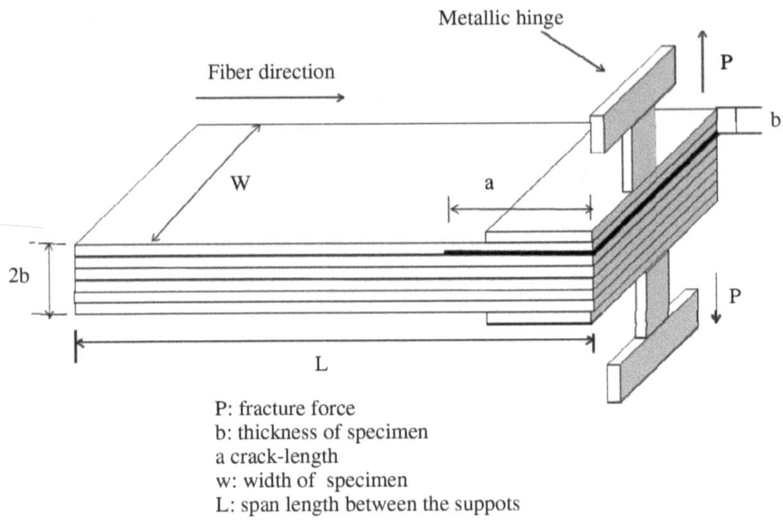

Fig. 7-16 Schematic diagrams of double-cantilever beam specimen.

beam theory according to the Eq. (7-3).

$$G_{IC} = \frac{3P \cdot \delta}{2W(a + |\Delta|)} \tag{7-3}$$

where P is the load, δ the displacement, W the specimen width, a the crack length, and $|\Delta|$ the crack length correction factor which is determined to be the x-axis intercept of the plot of the cube root of the compliance ($C = \delta/P$) vs. the crack length (a). The values of G_{IC} were plotted as a function of crack length (a) to produce a resistance (R) curve.

Fig. 7-17 shows the results of interlaminar shear strength for the carbon fibers-reinforced composites. A good relationship between the characters of oxyfluorinated carbon fiber surfaces and the resulting fiber–matrix adhesions on mechanical interfacial properties of the composites exists in this experimental condition. That is, interlaminar shear strength value increase on increasing the wettability of the fibers for the degree of adhesion at the interfaces due to the oxyfluorination, which can be attributed to the increase of the polarity on the fiber surfaces. And, the maximum strength value of the interlaminar shear strength is obtained at the oxyfluorinated carbon fiber sample condition at 100 °C (CFO-100).

Fig. 7-18 shows the results of fracture toughness (K_{IC}) for the composites, according to the fluorination conditions. Good relationships are shown between the fluorine content and the resulting fracture toughness of the composites. That is, the value of K_{IC} increases on increasing the fluorine content on carbon fiber surfaces, which is corresponding to the work of fracture for the degree of adhesion at interfaces. The maximum strength value of K_{IC} is obtained at the

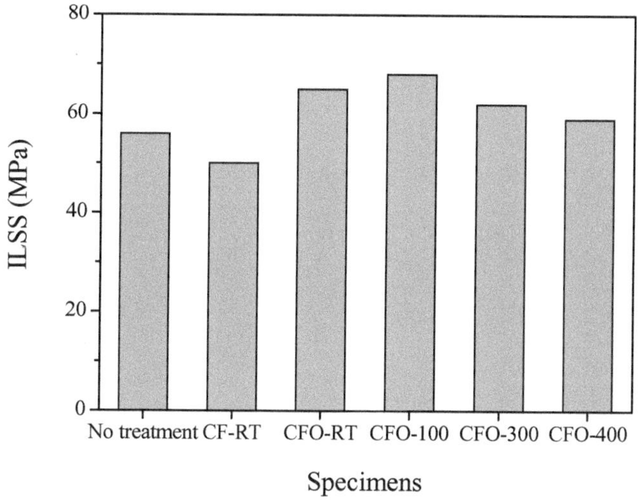

Fig. 7-17 Interlaminar shear strength (ILSS) of oxyfluorinated carbon fibers-reinforced composites.

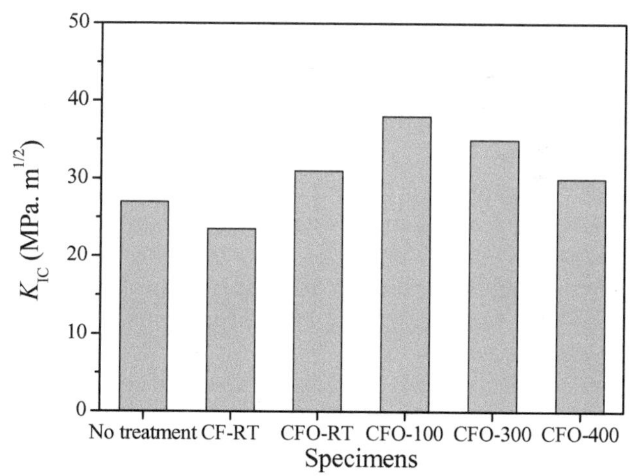

Fig. 7-18 Fracture toughness (K_{IC}) of oxyfluorinated carbon fibers-reinforced composites.

O_{1s}/C_{1s} ratio of 100 °C (CFO-100). Therefore, we suggest that additional energy is needed to extend the interfacial crack at this condition, which is attributed to increasing the interfacial adhesion between the fibers and the matrix [70]. Also, this is also in good agreement with the result of interlaminar shear strength for the composites.

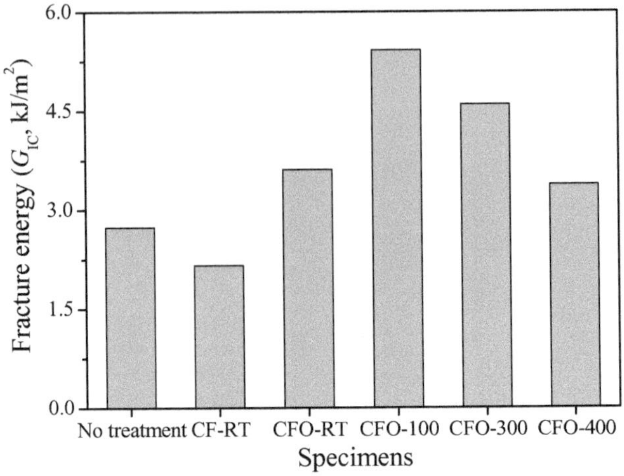

Fig. 7-19 Strain energy release rate (G_{IC}) of oxyfluorinated carbon fibers-reinforced composites.

The specific fracture energy (G_{IC}) of the composites as a function of fluorine concentrations is shown in Fig. 7-19. The composites made by fluorination at 100 °C shows a significant improvement of $G_{IC,}$ studied in any other experimental conditions, which probably results from the increase in fracture surface area through more tortuous path of crack growth or the increase in surface functional groups of the carbon fibers. However, the effects of fiber pull-out, fiber bridging, and fiber fracture are not expected in the present composite system. This fracture mechanism can be due to the following reason: when bend stress (σ) is applied to the specimens with different loads, a shear (τ) or transverse tensile stress (σ_T) can be produced in the fiber–matrix interfaces. When the span length is low, the shear stress is produced in the fiber–matrix interfaces. Thus, the composites show a transition from fiber-dominant fracture to fiber–matrix interface dominated fracture [72–76].

The strain energy release rate (G_{IC}) can be also reduced from the measured load and load point displacement curve (P_a–δ_m curve) (Fig. 7-20). The self-weight of the beam cannot be neglected in the test, and the influence of beam self-weight can be accounted for by introducing an additional equivalent force P_w, which supplies the same amount of energy based on the load–displacement curve (see Fig. 7-20). Therefore the total load P applied on the beam can be represented as $P = P_w + P_a$ where P_a is the applied load, provided that the load and displacement are both applied downward. The total fracture energy Wt absorbed by the beam may be divided into three parts, W_0, W_1, and W_2, as shown in Fig. 7-20. In the physical test, only the $P_a - \delta_m$ curve can be recorded. The area under the measured P_a–δ_m curve is denoted as W_0; the rescaled $P_a - \delta$ curve will include the effect of beam self-weight, and the fracture energy due to

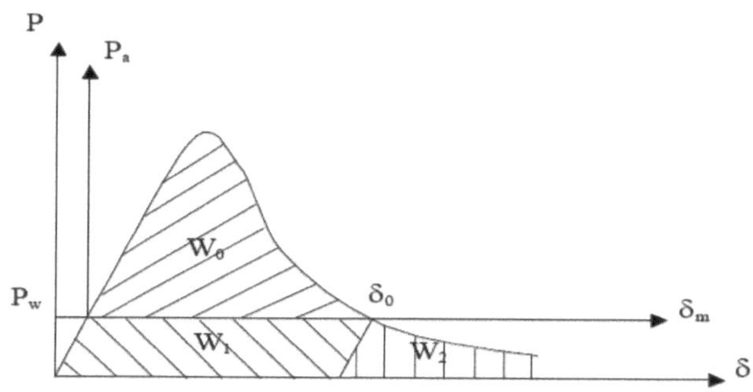

Fig. 7-20 Load-displacement curve for evaluation of the strain energy release rate (G_{IC}).

the beam self-weight (W_1) is approximated as $W_1 = P_w\delta_0$ (where δ_0 is the displacement corresponding to $P_a = 0$). As in the study conducted by Peterson [77] and Swartz [78], W_2 is approximately equal to W_1. Therefore, the total fracture energy of the beam is expressed as

$$Wt = W_0 + 2Pw{\cdot}\delta_0 \tag{7-4}$$

By assuming that energy absorption takes place only in the fracture zone, the strain energy release rate (G_{IC}) of the bonded interface can be calculated by

$$G_{IC} = \frac{W_t}{(b - a_0)t} \tag{7-5}$$

$$= \frac{W_0 + 2P_w\delta_0}{(b - a_0)t} \tag{7-6}$$

where b is the depth of the beam, t is the width of the beam, and a_0 is the length of the notch. Based on this data reduction method, the critical strain energy release rate (G_{IC}) of a notched specimen can be obtained once the load–displacement of the specimen is recorded from the experiment.

7.2.5. Development of Composite System

In the past, research on composites has been viewed as a collection of issues, such as fiber development, matrix studies, interfacial bonding, processing, and finally structural mechanics. Traditionally, matrices have been thought to protect the fibers. However, it is now understood that the matrix will influence the entirety of the process–structure–property relationships for the composite system while leaving the fibers unchanged. Further, the properties and structure of the resin matrix are different when reinforced, and this presents a major challenge. Only recently has the industry matured to the point where the fiber, resin matrix, interface, and surface are routinely seen as a system [79–82].

With the development of computers capable of modeling structures at the molecular scale, the time is approaching when it will be possible to design the optimum composite for a given system performance and cost. This systems perspective will involve modeling the structure and the manufacturing processes at various length scales to relate microstructural behavior to system performance. This modeling will include the placement of the reinforcing phase, the structure of the matrix phase, and even the degree of bonding at the interface. The methodology could be used to predict (and even control) the final structure of the composite. In other words, future engineers will not merely analyze the mechanics of the final product, but will apply a systems perspective and employ advanced modeling techniques to create reinforced structures that best meet the given system requirements. This approach is the key to the reduced cost and accelerated insertion of new materials, into the DOD systems [83–86].

A system approach is also crucial to understanding the stochastic aspects of composite failure, which is essential for improving the design criteria of these systems. Lack of knowledge in this regard can lead to excessive design safety margins that result in increased weight and cost and lower system performance. A better understanding of the effect of constituent variability on composite properties is crucial to taking advantage of fibers, on the market today as well as future fibers, as is the development of micromechanical and continuum-based models that include the stochastic process for the prediction of composite behaviors.

The transition to a systems approach is likely to occur gradually, with full implementation in 10 or more years in the future. While this approach is evolving, the researchers and material suppliers will continue to make incremental improvements in reinforcing fibers, matrix resins, and composite forms and processes.

Composites are expected to play an even greater role in military systems of the future. The Army's Objective Force in the U.S. DOD exemplifies an ongoing transformation to an entirely new future combat system incorporating advanced materials and design concepts for munitions, armaments, and hull structures for manned and unmanned robotic vehicles that will be light enough to be rapidly deployed on C-130 aircraft. Ground vehicles in the Future Combat System, will have to weigh between 10 and 20 tons and have superior mobility, transportability, survivability, and lethality for a variety of missions. The Navy is considering carbon fiber composites for next-generation topside ship structures such as destroyers, aircraft carriers, littoral combat ships, and other high-speed vehicles to satisfy the weight and performance requirements of these systems. These applications represent a possible 100-fold increase in carbon fiber usage.

The future composite applications will also require a new class of multifunctional materials. The Future Combat System of U.S. DOD, next-generation ship structures, and unmanned aircraft are some examples of programs where lightweight structure, ballistics, shock attenuation, radar signature reduction, power generation, and embedded sensors and actuators will be needed. The composites provide the opportunity to integrate these functions into a single material system. Affordable processes and robust design methods will be needed to develop and optimize these multifunctional composite materials.

7.3. CARBON MATRIX COMPOSITES

7.3.1. Introduction

Carbon and graphite are attractive materials for use at elevated temperature in inert atmosphere and ablative environment. The use of monolithic carbon or particulate composites is greatly limited by brittle mechanical behavior, flaw sensitivity, variability in properties, anisotropy and fabrication difficulties associated with large and complex components and structures. Carbon fiber-reinforced carbon matrix composites consist of carbon fibers embedded in a carbonaceous matrix [87–96].

Carbon fiber-reinforced carbon matrix composites are increasingly being considered for aerospace applications. In composite materials, the fibers impart strength, stiffness and fatigue resistance, while the carbon matrix holds the fibers together. A synergistic effect also exists between the fiber and the carbon matrix which results in high fracture toughness and wear resistance. But the most attractive properties of carbon fiber-reinforced carbon matrix composites are their high specific strength and modulus [97].

The aim of these materials is to combine the advantage of fiber-reinforced composites such as high specific strength, stiffness and in-plane toughness with the refractory properties at high temperatures, superior to any other materials (Table 7-8) has resulted in the exploitation of carbon fiber-reinforced carbon matrix composites as structural materials in space vehicle heat shields, rocket nozzles, and aircraft brakes. Additionally, properties such as biocompatibility and chemical inertness have led to new applications in medicine and industry. Carbon fiber-reinforced carbon matrix composites can perhaps lay claim to represent the ultimate development of carbon science.

Table 7-8 Room Temperature Mechanical Properties of Graphite and C/C Composites

	Elastic modulus (GPa)	Tensile strength (MPa)	Compressive strength (MPa)	Fracture energy (kJ/m^2)
Graphite	10–15	40–60	110–200	0.01
Unidirectional C/C* composites	120–150	600–700	500–800	1.4–2.0
3D C/C composites	40–100	200–350	150–200	5–10

C/C: Carbon fiber-reinforced carbon matrix.

Although carbon composites, in the form of polygranular synthetic graphite, were used in the fins of Second World War German rockets, it was not until the advent of carbon fiber technology in the late 1950s that the potential for the development of truly structural components was realized. The advancement of carbon fiber-reinforced carbon matrix composite materials technology was initially very slow, but by the late 1960s it had begun to emerge as a major new genre of engineering materials. During the 1970s, carbon–carbon structures were under extensive development in the USA and Europe, mainly for military use. The original carbon fiber-reinforced carbon matrix composites, to be used in rocket nozzles and re-entry parts, were produced using reinforcements in the form of woven fabrics of low-modulus rayon precursor carbon fibers. The matrix was derived from pyrolysed high-char yield resins such as phenolic and furan. Fiber-reinforced plastic molding techniques were used to fabricate precursor composite structures which were subsequently carbonized.

Since those early days, it is now possible to use the whole variety of available types of carbon fibers with their individual characteristics. The fibers may be combined in a wide variety of woven, knitted, braided, and filament wound forms to provide one-, two-, and multi-directional reinforced composites. Additionally, lower-cost composites employ a wide range of felts, fabric, and short fiber systems. The reinforcement fibers can be combined with any of the forms of carbon, previously described.

The matrix, for example, could be a vapor-deposited carbon, a glassy carbon (resulting from the pyrolysis of a resin), or the coke from the liquid phase pyrolysis of a mesophase pitch. Furthermore, by careful control of the precursor and the heat treatment temperature, the degree of graphitization of the matrix may be varied considerably, thus imparting a wide range of thermomechanical properties to the composites.

Carbon fiber-reinforced carbon matrix composites are not, therefore, a single material, but rather, a family of materials, many combinations of which have yet to be realized and evaluated. To some extent the carbon fiber-reinforced carbon matrix composites resemble the particulate carbon composites, save that the granular filler is replaced by fibers. The characteristics of the precursors and the processes which take place during manufacture are critical in determining the ultimate properties.

There are two key developments required in order to secure a more widespread development of carbon fiber-reinforced carbon matrix composites. Many possible markets are severely limited as a result of the excessive costs of the composites. If the carbon fiber-reinforced carbon matrix composites are to be exploited in cost-sensitive applications such as replacement of asbestos in motor vehicle and passenger train brakes, more efficient and economically viable precursors and processes need to be developed. In very 'high technology' usage, where cost is justified by improved performance (such as aircraft engine parts), prevention of oxidation is the paramount concern. Furthermore, a number of less

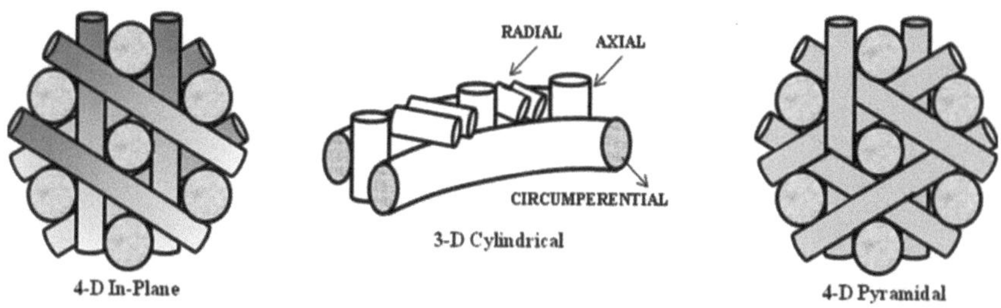

Fig. 7-21 Multi-axial dimensional structures of carbon fiber-reinforced carbon matrix composites.

critical, but still significant problems such as poor matrix mechanical properties, for example, need to be addressed.

7.3.2. Structure of Carbon/Carbon Composites

Carbon/Carbon materials consist of a three-dimensional woven part having carbon fibers in the axial, radial, and circumferential directions, or four-dimensional construction with fiber in *X*, *Y*, and *Z* directions and additional fibers placed in *X*, *Y* planes or in the *XZ* or *YZ* planes, as shown in Fig. 7-21. X-ray techniques are used to determine weave integrity. The woven preform is pressure impregnated with liquid pitch to fill the voids between the yarn, densified at high pressure (up to 15,000 psi), and heat treated at 2550 °C (4622 °F) to produce a rigid multi-dimensional reinforced graphite structure. The impregnation, densification and graphitization cycles are repeated until final desired density is obtained.

7.3.3. Oxidation Behavior and Coating Protection of Carbon/Carbon Composites

Based on their good mechanical performances, potential uses of structural carbon materials have been identified in future manned hypersonic vehicles and in other aircraft applications. However, most of these applications involve extended time periods in oxidizing environments. Unfortunately, carbon reacts rapidly with oxygen at temperatures as low as 500 °C and the composites are subject to oxidation degradation. Less is known about the effects of oxidation on the properties of carbon fiber-reinforced carbon matrix composites, because they vary with the oxidation mechanisms and the microstructural features of the materials [98, 99]. But it is well known that oxidation degrades quickly the properties of carbon materials [100, 101]. Therefore, numerous investigations have been conducted in an attempt to protect carbon based composites against oxidation.

The first protection mechanisms are based on a macroscopic concept with the use of a barrier against oxygen diffusion at high temperature. The main

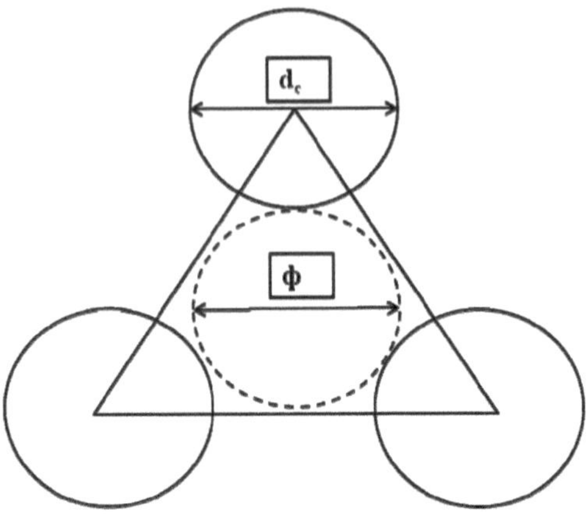

Fig. 7-22 Schematic representation of the oxidation of ceramic-filled carbon fiber-reinforced carbon matrix composites and arrangement of ceramic particles.

problem, when applying this concept, is the thermal mismatch between the coating and carbon substrate which results in cracking of the coating. Thus, other approaches have been developed including: (1) intermediate layers, (2) inhibited matrices, and (3) fiber protections.

7.3.3.1. Oxidation kinetic and mechanism

As shown schematically in Fig. 7-22, the SiC-impregnated carbon fiber-reinforced carbon matrix composites which consist of two phases: a homogenous carbon matrix without any voids and silicon carbide particles dispersed in the matrix uniformly and with a triangular arrangement were used to evaluate oxidation kinetic and mechanism of carbon fiber-reinforced carbon matrix composites.

Thus, during an exposure in an oxidizing atmosphere at high temperature, the carbon matrix is consumed with the formation of CO_2 or/and CO, which results in a mass loss to the composites. Because silicon carbide is assumed to be non-reactive, a porous layer consisting of ceramic particles will be produced with the recession of the carbon matrix. This layer, which covers the outerface of the composite, is called the 'decarbonization layer.' The pore size in the decarbonization layer is also assumed to remain unchanged during the period of oxidation.

(1) Hypotheses

The carbon fiber-reinforced carbon matrix composites used for the model are assumed to possess the following characters: (a) the carbon matrix is homogenous in structure and properties, (b) the ceramic particles are of spherical shape of the same size and are dispersed in the matrix uniformly, (c) the reaction interlayer of carbon is planar, and the ceramic is non-reactive during the oxidation.

(2) Chemical reactions

The oxidation of carbon in pure oxygen results in the formation of gaseous carbon oxides, with an overall mass loss, according to the following reactions:

$$C + O_2 \rightarrow CO_2 \quad (7\text{-}7)$$

$$2C + O_2 \rightarrow CO \quad (7\text{-}8)$$

The values of the intrinsic reaction constant Vi (in kg·m^{-2}·s^{-1}), which is used in the present work for the oxidation of the carbon matrix, are calculated with the following equation: $Vi = 1070 \exp(-123000/RT)$, according to the report of H. Marsh *et al.* [102]. If the reaction order has been taken as 1, then the kinetic law can be expressed as the following equation: $Vi = K_C C_A M_c b$, where K_C is the rate constant (ms^{-1}), C_A is the concentration of oxygen, M_c is the molar mass of carbon, and b is the ratio of gaseous products to every mol of oxygen consumed. The choice of reaction equation will cause some differences in the calculated results. For example, the effective diffusion coefficients of the gaseous products are different, because of Stefen flow, but the difference is small; Fillipuzzi and Naslain [103] and Eckel *et al.* [104] have discussed this issue carefully. The deviation caused by these two systems O_2-CO_2, and O_2-CO is limited to about 5 percent; therefore, in this model only CO_2 is considered as the reaction product.

With the condition of a non-reactive ceramic, as schematically shown in Fig. 7-22, once a porous decarbonized layer is built up at the outer surface of the composites, the process of oxidation at a given temperature may proceed through the following five steps:

(a) Oxygen transports from flow bulk to the outerface of the oxidizing composites;
(b) Oxygen diffuses through the pores to the reaction interface;
(c) Chemical reaction between carbon and oxygen leads to the production of CO, and CO;
(d) CO/CO_2 diffuses through the decarbonized layer to the outerface;
(e) CO/CO_2 transports from the outerface into the bulk flow.

We only take the second, third and fourth steps into account, because the rates of these are considered much slower than the others. Assuming that the overall process of oxidation is in the pseudo-steady state normally adopted for solid gas reactions, the kinetic equation can be derived as follows:

The diffusion rate of oxygen in the decarbonized layer:

$$R_1 = -D_{eff} \times \frac{dC(x)}{dx} \quad (Fick's\ First\ Law) \quad (7\text{-}9)$$

The reaction rate of carbon:

$$R_c = K_c C_{O_2} \tag{7-10}$$

The diffusion rate of CO_2:

$$R_2 = -D_{eff} \times \frac{dC(x)}{dx} \tag{7-11}$$

A pseudo-steady state yields linear concentration gradients of oxygen and dioxide, and hence Eqs. (7-9) and (7-11) can be rewritten as:

$$R_1 = -D_{eff} \times [C_{O_2}(x) - C_{O_2}(0)]/L \tag{7-12}$$

$$R_2 = -D_{eff} \times [C_{CO_2}(0) - C_{CO_2}(x)]/L \tag{7-13}$$

Eqs. (7-10), (7-12), and (7-13) will proceed at the same rate, and taking the reaction of Eq. (7-7) into account, then:

$$R_1 = R_2 = R_c \tag{7-14}$$

Hence, after the elimination of the concentration of $C(x)$,

$$R_c = \frac{C_0}{\frac{1}{k_c} + \frac{L}{D_{eff}}} \tag{7-15}$$

The rate of oxidation is equal to the rate of the burnoff of carbon matrix (in mol per unit time per area) due to oxidation taking place on the planar interface. Therefore:

$$R_c = \frac{1}{S} \times \frac{n(c)}{dt} \tag{7-16}$$

$$= \frac{1}{S} \times \frac{dW}{M_c dt} \tag{7-17}$$

$$= \frac{1}{S} \times \frac{S a d_m dL}{M_c dt} \tag{7-18}$$

$$= \frac{a d_m}{M_c} \times \frac{dL}{dt} \tag{7-19}$$

where S, a, and d_m are the surface area, volume content of carbon, and average density of the composites, respectively.

Substituting Eq. (7-18) into Eq. (7-19), yields:

$$\frac{dL}{dt} = \frac{M_c}{ad_m} \times \frac{C_0}{\frac{1}{k_c} + \frac{L}{D_{eff}}} \tag{7-20}$$

By integrating both sides of the above equation

$$\int_0^L (1/K_c + L/D)dx = \int_0^t \frac{M_c C_0}{ad_m} dt \tag{7-21}$$

The kinetic equation becomes:

$$e\frac{1}{2D_{eff}}L^2 + \frac{1}{K_c}L = \frac{M_c C_0}{ad_m}t \tag{7-22}$$

It is usual to consider the former item on the left-hand side as being under diffusion control and the latter as being under reaction control. The effective diffusion coefficients comprise of two parts: the molecular diffusion regime (diffusion coefficient, D_m) and Knudsen diffusion regime (diffusion coefficient, D_k), so:

$$\frac{1}{D_{eff}} = \frac{1}{D_m} + \frac{1}{D_k} \tag{7-23}$$

For a cylindrical pore, D_k can be calculated as

$$D_k = 1.534d \times \sqrt{\frac{T}{M}} \tag{7-24}$$

where d is the diameter of the pore (m), T absolute temperature (K), and M molecular mass ($kgmol^{-1}$) of oxygen.

The binary diffusion coefficient D_m can be calculated according to a simplified expression using Chapman–Enskog theory [105], the value for the binary mixture of CO_2–O_2 system is

$$D_m = 3.277 \times 10^{-4} \times T^{1.5} \times P^{-1} \quad in\ m^2\ s^{-1} \tag{7-25}$$

C_0 in mol m^{-3} is calculated from the ideal gas law,

$$C_0 = \frac{P_i}{RT} = 1.203 \times 10^{-1} \times \frac{P_i}{T} \tag{7-26}$$

As can be seen from Eq. (7-25), the depth of the decarbonized layer is a function of carbon content, density of the composite, oxygen concentration, reaction time as well as the constants K_c and D_{eff} the former usually has Arrhenius-type relationship with temperature. The reaction constant K_c depends strongly on the graphitization degree of carbon and its chemical properties; therefore, K_c will change greatly as the carbon material is heated at different temperatures; the effective diffusion coefficient is related to the content of ceramic and its size. In the following analysis, some parameters involved such as the content of carbon and the size of the ceramic will be discussed.

The parameters in Eq. (7-26) such as D_m, a, *and* D_{eff} will change with variation of the composition of the ceramic. If we consider the ceramic particles in a triangular arrangement (Fig. 7-22) and assume the pore through which diffusion occurs to be cylindrical, the relation between the diameters of the pores, the content of carbon, and the size of ceramic particles is given by

$$\phi = \left(\frac{1.1}{\sqrt{1-a}} - 1\right) \times d_c \tag{7-27}$$

where a and d_c are the volume content of carbon and the diameter of ceramic particle in the composite. If we take carbon/silicon carbide composite as an example, and the density of carbon is taken as 1.8×10^3 kgm^{-3}, at the same time, the density of the composite changes with the carbon content. By combining Eqs. (7-22)–(7-27), numerical methods can be employed to solve the equation [106].

The calculated variation of oxidized depth vs. time at three different temperatures (Fig. 7-23), and for the composites with three different carbon contents (Fig. 7-24) and three different ceramic particle sizes (Fig. 7-25) are shown. The recession rate of carbon increases with the increase of oxidation temperature. As the temperature decreases by 100 K, the recession rate of carbon decreases by 50 percent. However, recession rates of carbon increase slowly with the increase of carbon content, for example, only an increase of about 18 percent is found as the content of carbon changes from 55 to 75 percent. As the size of ceramic

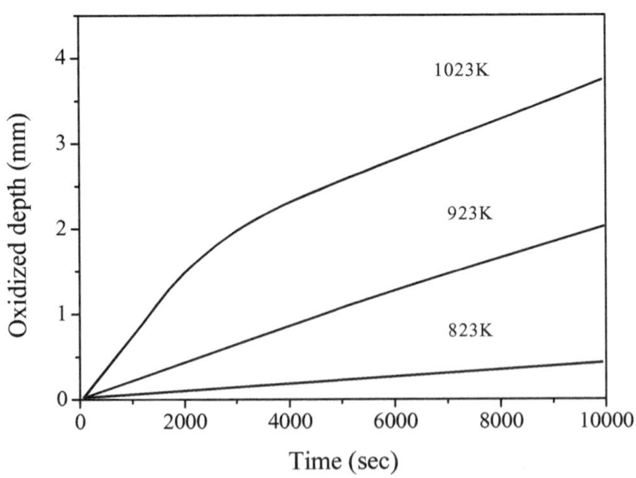

Fig. 7-23 Relationship of oxidized depth of carbon fiber-reinforced carbon matrix composites with duration of time (content of carbon: 65 percent; particle size of ceramic filler: 1 μm).

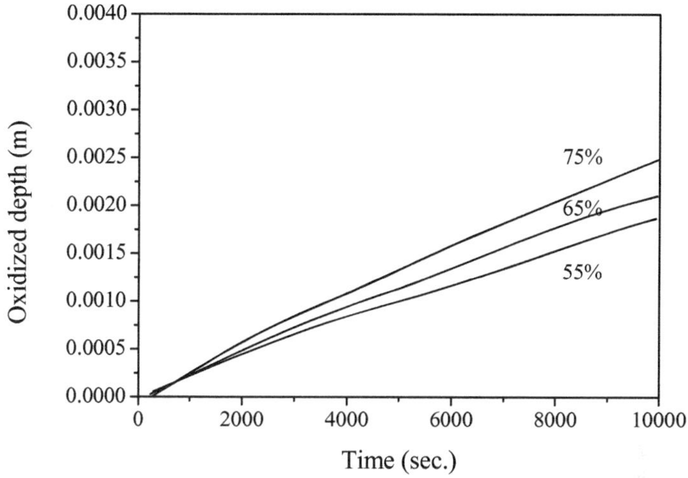

Fig. 7-24 Relationship of oxidized depth of carbon fiber-reinforced carbon matrix composites with duration of time (oxidized temperature: 923 K; particle size of ceramic filler: 1 μm).

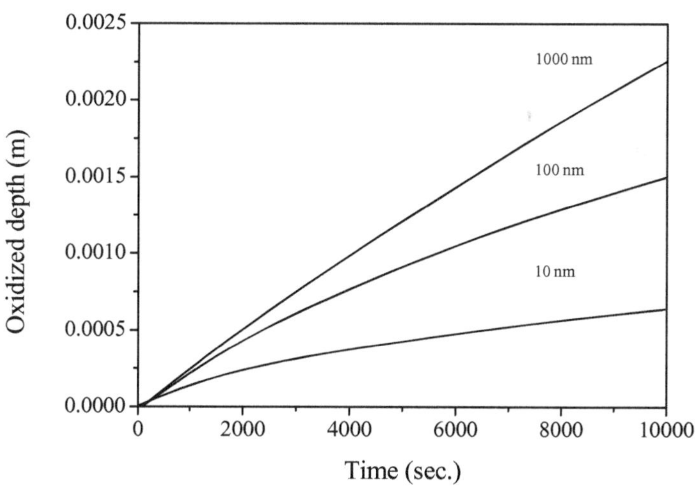

Fig. 7-25 Relationship of oxidized depth of carbon fiber-reinforced carbon matrix composites with duration of time (content of carbon: 65 percent; oxidized temperature: 923 K).

decreases from 1 μm to 10 nm, the recession rate of carbon slows down dramatically. The burn-off of those composites with larger ceramic particles increases quickly with the duration of time at 923 K. With the composites with nanoscaled ceramic particles the oxidized depth increases only slowly, the oxidized depth being about only 0.5 mm after oxidation for 10,000 s.

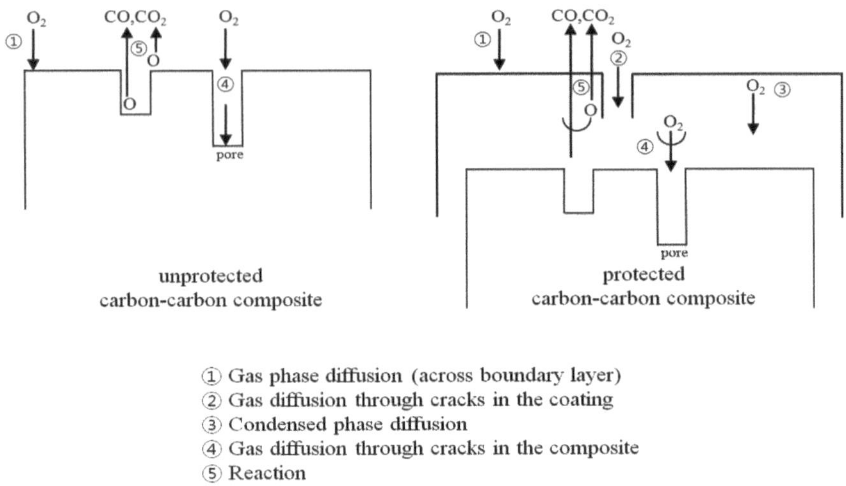

Fig. 7-26 Schematic representation of limiting steps during oxidation.

7.3.3.2. Coating

Oxygen barrier layer

In the presence of a protective coating on the surface, the oxidation of a substrate involves the following steps (Fig. 7-26) [107]:

(a) Gas phase diffusion across the boundary layer: oxygen has to diffuse through across the boundary layer at the surface.

(b) Gas diffusion through cracks in the coating: if the coating is cracked, the oxygen can diffuse through the cracks to the substrate/coating interface. The diffusion coefficient depends on the crack width. For large cracks, the diffusion coefficient is the same as the bulk gas diffusion coefficient. However, for small cracks, where the mean free path of the gas molecules is of the same order as the crack size, the gas molecules collide with the walls. This reduces the effective diffusion coefficient, and transport occurs by a process called Knudsen diffusion.

(c) Condensed phase diffusion: even if the coating is crack-free, oxygen can diffuse through it to oxidize the substrate. The transport mechanism depends on the defect structure of the coating.

(d) It can be gas diffusion through cracks in the composites.

(e) It can also be reaction between carbon and oxygen to form oxidation products.

(f) Desired properties for the coatings on carbon fiber-reinforced carbon matrix composites are thermal expansion compatibility to minimize cracks, low oxygen and carbon diffusion, low vapor pressure, chemical compatibility with carbon, erosion, wear, and impact resistance. The most important

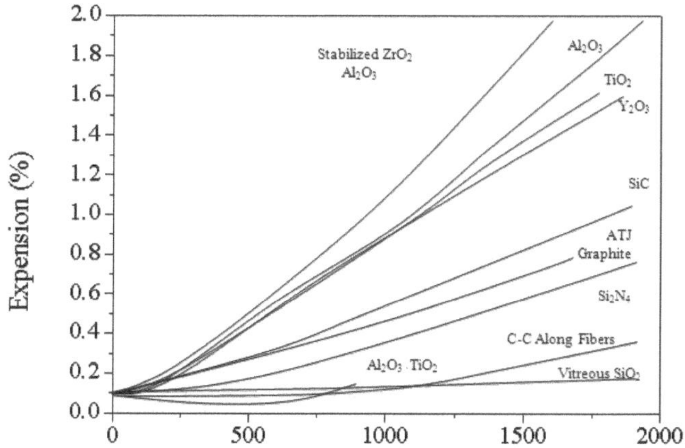

Fig. 7-27 Thermal expansion coefficients of refractory materials.

criterion is a good adequation between the thermal expansion coefficients of the coating and of the composite. Fig. 7-27 compares the thermal expansion coefficients of refractory materials with the expansion of graphite or high performance carbon fiber.

The most suitable ceramics are some oxides (as silica) and silicon carbide or silicon nitride. But in spite of these good thermal expansion compatibilities, it is difficult to form coatings without cracks. Indeed at temperatures below the temperature of the coating application, ($T < 1250$ °C), tension-induced coating cracks exist [108]. Therefore, at low temperature, silicon carbide (nitride) works only for short-term applications. At higher temperatures due to the closure of the cracks and due to the formation of low viscosity silica, the protective properties of this type of coating are improved and this type of protection can be used for medium-term applications.

At very high temperatures or medium temperatures and low oxygen pressures, silicon carbide does not protect anymore [109–121] because of the formation of gaseous silicon monoxide either by direct active oxidation:

$$SiC + O_2 \rightarrow SiO + CO$$

$$Si_3N_4 + 3/2\ O_2 \rightarrow 3SiO + 2N_2$$

or by reaction between silica and silicon carbide (or nitride):

SiC/SiO_2 reaction

$$SiC + 2SiO_2 + 3SiO + CO$$

Si_3N_4/SiO_2 reaction

$$Si_3N_4 + 3SiO_2 \rightarrow 6SiO + 2N_2$$

In this domain (domain III), carbon fiber-reinforced carbon matrix composites must be protected by other ceramic materials such as oxides [122–127] or metal carbides [128–130]. Noble metals [122, 123, 131–134] and borate coatings have also been studied [107, 122, 135–137].

Oxides, metal carbides, and noble metals have numerous advantages but generally they have a too large thermal expansion coefficient (Fig. 7-27). Therefore, it is very difficult to apply crack-free coatings. Other limitations are generally due to fast oxygen diffusion through the oxide phases (for instance ZrO_2, HfO_2 have high oxygen permeability). In addition, oxides can be reduced by carbon (as low as 1200 °C for silica, 1730 °C for ZrO_2, 1750 °C for HfO_2). As a consequence, direct oxide deposit on carbon must be avoided. The oxides are generally used as the upper layer (oxygen barrier layer) in multilayer coatings.

Refractory group metals, such as iridium and ruthenium, present poor adherence to carbon materials, erosion via volatile oxidation, high cost, and short supply. At present, iridium is used in multilayer coating concepts [134].

Boron oxide is an attractive compound for the protection against oxidation. The chemical compatibility of B_2O_3 with carbon, its good wetting property, and its low viscosity (10^4–10^2 dPa in the 600–1100 °C range) allow the formation of thin protective glass films [107, 122, 135–137]. The main limitations of borate glasses are their moisture sensitivities [138] (hydrolysis under ambient conditions) and at higher temperatures the vaporization of boron oxide [135]. Under dry conditions borate glass coatings can be used up to 600 °C and 1,000 °C for long- and short-term applications, respectively.

Many materials are candidate to protect carbon fiber-reinforced carbon matrix composites against oxidation. Unfortunately, because of thermal expansion differences and poor adherence properties, most ceramic materials do not form satisfying coatings on carbon fiber-reinforced carbon matrix composites. These coatings present cracks, and therefore the oxidation occurs beneath the coating leading to its spallation. To minimize the density of cracks, transition layers have been developed. Another way to protect the composites and to limit the effects of cracks is to use glass formers (inhibited matrix) which are able to form a mobile glass which spreads within the cracks.

Transition layers

Carbide conversion layers improve the surface cohesive strength. Most of the transition layers are made by a pack cementation process [147, 148], and sometimes these layers are doped (e.g., zirconium or boron) [131]. This transition layer is often used beneath an outer chemical vapor deposition coating [129, 139–146] and this double protection is efficient up to 1,000 °C and 1600 °C for medium- and short-term applications in a dry atmosphere. Other intermediate layers such as porous chemical vapor deposition silicon layer [149] or graded coatings [150] have been developed but seem less efficient. Physical methods such as ion implantation (e.g., boron) have also been tested [150] in order to reduce the oxidation rate at low temperature.

Inhibited matrix

The oxidation protection concept with the use of a diffusion barrier is not valid when damage probabilities are great (meteorite impact...). For that reason a new approach has been developed by adding in the carbon precursor some glass formers which are able to spread within the cracks [146, 151–155]. The components studied are mainly B_4C, B, CrB_2, ZrB_2, SiB_4, $AIBl_2$, ZrH-Si, ZrB_2-Si, SiC, $MoSi_2$, and phosphates. With only inhibited matrix, no protection occurs before an appreciable fraction of carbon (30 percent) has been gasified. Therefore this type of protection is generally used in addition to other protection systems [129, 139, 143, 144–146]. But the addition of inhibitors can modify the mechanical properties of the composite [156]. Indeed, heat treatment of carbonized composites containing glass forming inhibitors such as boron or zirconium diboride above initial carbonization temperature results in reduced mechanical properties [156].

Coatings on fibers

In order to protect the carbonaceous fibrous reinforcement, several coatings have been developed. Most of them have been applied by chemical vapor deposition [157, 158] and by sol–gel methods [159].

7.3.3.3. Complex systems and multilayer coatings

In many aerospace applications, the desired properties of carbon fiber-reinforced carbon matrix composites are thermal stability, thermal shock resistance, and oxidation resistance under thermal cycling conditions in a one atmosphere air pressure or in a partial reduced pressure of molecular or atomic oxygen, resistance to degradation by moisture, resistance to rain erosion, high emittance, and multi-mission reliability.

In order to reach this objective, many complex systems have been developed including: coatings on fibers, inhibited matrix, transition layer, coating, and outer glass sealant [160]. Most of these oxidation protection systems contain SiC chemical vapor deposition coating on fibers, boron or silicon carbide in the matrix, silicon carbide converted layer, silicon carbide chemical vapor deposition coating, and outer silica (or borosilica) glass [120, 144–146, 149, 160]. Unfortunately, these systems based on silicon carbide are not efficient beyond 1800 °C in a dry atmosphere due to the active oxidation of silicon carbide. Moreover, in a humid atmosphere, this type of protection (which includes boron protection, as glass former in the matrix or component of the outer glass) can only work up to 1,100 °C for medium-term applications. Indeed, moisture absorption and subsequent rapid gas formation during heating lead to the spallation of the coating from the substrate [161]. Therefore in order to adjust their properties to the substrate's composite coatings have been developed.

7.3.3.4. Composite coatings

Composite coatings have been developed using chemical vapor deposition, plasma spraying techniques or powder, and polymer mixtures. Many coatings have been realized by chemical vapor deposition such as Si_3N_4 + TiN [162–166], Si_3N_4 + BN [162, 167], Si_3N_4 + SiC [162], SiC + $TiSi_2$[162, 168], SiC + TiC [169], AlN + BN [170, 171], TiB_2 + TiC [172], AlN + Si_3N_4[173], ZrC + BN [174], et HfC + $HfSi_2$[175]. All these systems offer the promise of tailoring properties such as friction coefficient, hardness, resistance to wear and erosion or optical properties. Only ZrC + BN and HfC + $HfSi_2$ have been developed as protective coatings for carbon fiber-reinforced carbon matrix composites and are used in a multilayer protective concept [170, 175].

Thermal sprayed coatings of HfC and HfB_2 composite systems [176] and low-pressure plasma deposition of SiC-reinforced $MoSi_2$ [177] have also been studied.

Another promising approach is to use an oxide phase to cement particles of refractory materials and to form an impermeable coating [178, 179]. For instance crack-free (but porous) composite coatings composed of yttria-stabilized zirconia and SiC whiskers have been deposited on carbon [178]. Other coatings as high melting point borate (Ce $(BO_2)_3$) with SiC whiskers, zirconium silicate/boron oxide ($ZrSiO_4/B_2O_3$), or hafnium boride/boron oxide (HfB_2/B_2O_3) have been tested to protect carbon fiber-reinforced carbon matrix composites [179]. But all these coatings contain boron oxide which is volatile at high temperature or low pressure and medium temperature and which is moisture sensitive. Therefore, this type of coating is used in a dry atmosphere up to 1200 °C for medium-term applications.

Finally, some authors propose composite graded coatings such as mixed oxides [180]. For instance, the following system: mullite/75% mullite-25% alumina/50% mullite-50% alumina/25% mullite-75% alumina/alumina was deposited on silicon carbide and survived thermal treatments with no observable cracking or spalling [180]. This type of coating protects silicon carbide materials up to 1,100 °C for long-term applications.

7.3.3.5. Protection with the use of an inert gas

Recently a new concept has been developed. It is proposed that the introduction of an inert gas into the boundary layer at the carbon composite surface will provide some protection. Some authors [181] have studied the carbon fiber-reinforced carbon matrix composite oxidation in the presence or absence of protection provided by inert nitrogen introduced into the wall surface boundary layer. They have shown that the carbon mass loss at the tube surface is reduced, by a factor of about 3.3 at 1500 K and by a factor 3 at 2000 K, by the nitrogen gas protective boundary layer flow. This promising new protective concept must be optimized according to other flow conditions, temperatures, and pressures.

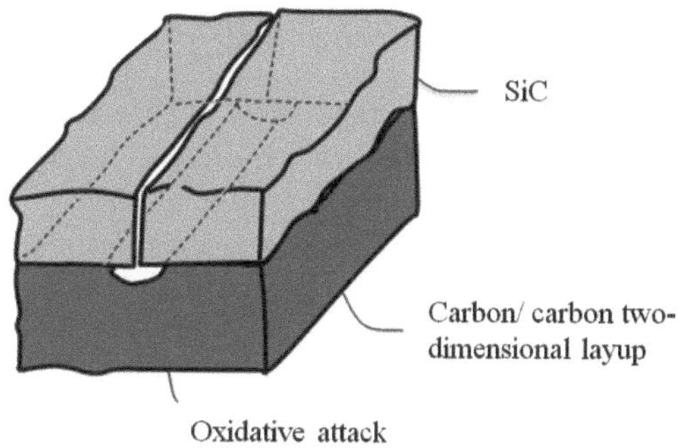

Fig. 7-28 A schematic diagram of SiC-reinforced carbon fiber-reinforced carbon matrix composites.

7.3.3.6. Oxidation through coating cracks

Fig. 7-28 is a schematic of reinforced carbon/carbon with a SiC conversion coating used on the Space Shuttle Orbiter nose cap and wing leading edges. The material is made with a two-dimensional layup of carbon–carbon fabric with repeated applications of a liquid carbon precursor to fill voids [182, 183]. The resultant high strength and light weight make it an ideal aerospace material; however, oxidation is a major concern. An oxidation protection system is based on a SiC conversion coating. Because of the difference in coefficient of thermal expansion of the SiC coating and carbon/carbon substrate, the SiC coating shrinks more than the underlying carbon/carbon on cooldown from the coating application temperature. This leads to vertical cracks in the coating, and these cracks are pathways for oxygen to reach the carbon/carbon substrate [184, 185].

Fig. 7-29 illustrates a typical oxidation void formed because of oxygen diffusion through a coating crack and subsequent reaction. A recent study by

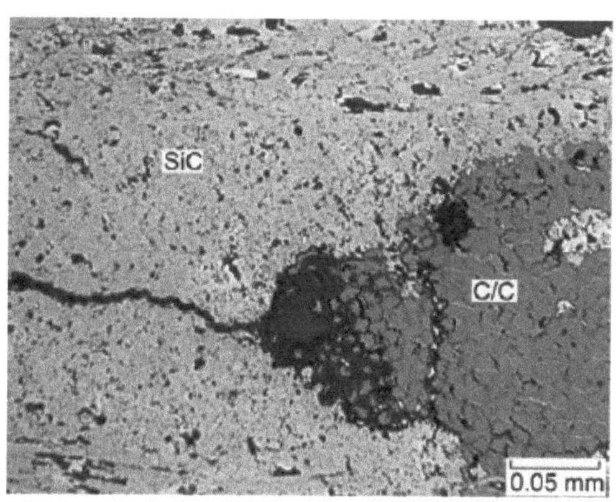

Fig. 7-29 Optical micrograph showing oxidation void in carbon fiber-reinforced carbon matrix composites formed below crack in SiC coating at 1,100 °C and 667 Pa for 1 h.

the authors outlines the microstructural features which distinguish this void from a processing void [186]. In general, an oxidation void is associated with a path for oxygen ingress (i.e., the crack in Fig. 7-29). At higher temperatures (>1,000 °C) these oxidation processes are expected to be diffusion controlled. Diffusion control implies a lack of preferential oxidative attack and hence uniform attack of the carbon/carbon with a plane interface at the void boundary, as illustrated in Fig. 7-28. However, a close examination of the void in Fig. 7-29 indicates some preferential attack along the fiber edges and the resultant 'pointed fibers,' which are characteristic of reaction control [187]. Thus the actual process is a mix of diffusion and reaction control. It will be shown that diffusion control dominates.

There are numerous studies of carbon oxidation through cracks and fissures in the protective coating and associated models found in the literature. Medford [188] developed models of oxidation for carbon fiber-reinforced carbon matrix composites, assuming pathways for oxygen are the cracks and fissures in SiC. His model encompasses the following processes:

(1) Diffusion of oxygen inward through the cracks and/or fissures in the silicon carbide.
(2) Oxidation of SiC crack walls to form SiO_2.
(3) Oxidation of carbon/carbon-matrix, fibers, or both.
(4) Diffusion of CO outward through the cracks/fissures in the silicon carbide.

Medford developed total mass change expressions based on these processes. The expressions were the diffusion equations with modifications for the area of reaction. First he examined the case below 982 °C, where passive oxidation of the silicon carbide is thought to be negligible; then he examined the case above 982 °C, where passive oxidation of the SiC is thought to be significant.

The general process of carbon oxidation in a non-reactive matrix or through a non-reactive coating has been explored by several investigators [189–194]. Carbon oxidizes to CO (g):

$$C(s) + 1/2O_2(g) = CO(g) \tag{7-28}$$

However, thermochemically the equilibrium between O_2, CO, and CO_2, favors the products in the following reaction:

$$1/2O_2(g) + CO(g) = CO_2(g) \tag{7-29}$$

Thus, O_2 (g) and CO (g) cannot coexist in similar amounts—they immediately will react to form CO_2 (g). However, the pairs of species in reaction (2), CO (g) with CO_2 (g) and O_2 (g) with CO_2 (g), may coexist. Thus a two-step oxidation process is required to model carbon oxidation. The two reactions are

(A) At the carbon/gas interface:

$$C(s) + CO_2(g) = 2CO(g) \tag{7-30}$$

Note that CO_2 (g) is the oxidizing species.

(B) At a distance away from the carbon/gas interface:

$$CO(g) + 1/2O_2(g) = CO_2(g) \tag{7-31}$$

In an earlier paper [194], the authors adapted this two-step diffusion-controlled oxidation model to cylindrical pinholes in the conversion coating. The actual oxidation damage was assessed with weight loss measurements and area measurement of oxidation voids below the pinholes, using cross-sections. At 1400 and 1,000 °C hemispherical voids were observed on a macroscopic scale, indicating uniform attack and diffusion control. The two-step model described the process reasonably well at 1,400 and 1,000 °C. However, at 600 °C only minimal oxidation is measured, and reaction control dominated. Hence the model did not fit at this lower temperature.

In the present section, the two-step oxidation model is further extended to describe a half-cylinder trough which forms under a slot or craze crack in the SiC coating. First a series of experiments is performed using machined slots in the coating. This allows testing the model with well-defined oxygen pathway geometry. Then the study is extended to actual through-thickness craze cracks in the coating. Oxidation damage is assessed with cross-sectional views, interrupted or real-time weight loss measurements, and X-ray computed tomography scans. The kinetics of carbon consumption is then compared with the model.

7.3.4. Densification

7.3.4.1. Resin transfer molding of carbon/carbon performs

Carbon/carbon composites are the materials of choice for many demanding high temperature aerospace applications including rocket nozzles, nose cones, and friction materials for commercial and military aircraft. Aerospace structural carbon/carbon composites are often manufactured using woven 2D or 3D C-fiber fabrics, and are sometimes densified using hot isostatic pressing.

The very high cost of these materials has limited their application to aerospace markets where performance and quality requirements take precedence over the cost. Lower cost carbon/carbon composites available today are uniform materials formed by molding chopped C-fiber with phenolic resin. The resulting preforms are then charred to carbon, and densified by adding carbon by chemical vapor deposition. Alternatively, they can be constructed using a non-woven textile technology, wherein oxidized acrylic, or polyacrylonitrile fibers are needled into thick mats, carbonized, and then densified with chemical vapor deposition. These lower cost composites are typically used in applications where structural loads are low to moderate, with the largest commercial application for this type of carbon/carbon composites being in aircraft brake friction materials.

One of the main components in the present approach used to manufacture carbon/carbon composites is the utilization of a resin transfer molding machine

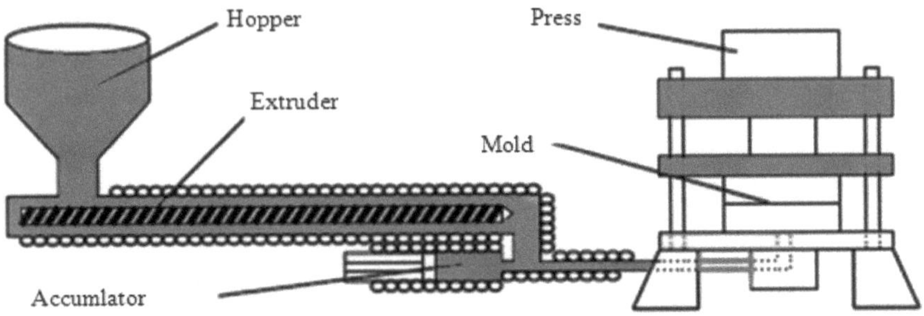

Fig. 7-30 Schematic drawing of the resin transfer molding set-up.

for the infiltration of composite preforms. Resin transfer molding processes are not new in the manufacture of composites where they are used to infiltrate preforms with low viscosity resins. The present resin transfer molding process is special since it uses a high char-yield resin which provides the opportunity for improved carbon yield. The resin transfer molding process thus reduces the number of densification cycles in the subsequent chemical vapor deposition process.

Fig. 7-30 depicts a schematic drawing of the resin transfer molding set-up. The resin is melted in a single screw extruder, and stored under pressure in a hydraulic accumulator. The accumulator is instrumented for position, and pressure, allowing for accurate control of the parameters, volume, pressure, and resin amount of the injection process. Once the part is clamped into the mold, the resin is injected into the part and the clamping force remains, until the resin solidifies in the part. The part is ejected, and the flash is easily removed. The whole cycle takes just a few minutes.

Typical high char-yield resins for resin transfer molding include synthetic mesophase pitches as well as thermally or chemically treated coal pitches and other thermoplastic resins. Before the resin transfer molding, the preform is charred such that it can withstand the temperatures during the resin transfer molding process. Fig. 7-31(b) depicts the preform discussed here after the char stage, as well as with appropriate machining applied. The preform is subsequently preheated and placed in the preheated resin transfer molding mold. Fig. 7-31(c) shows the preform after resin transfer molding.

7.3.4.2. Stabilization

In the resin transfer molding densification, a liquid synthetic mesophase pitch is injected at high pressure into the void spaces of the carbonized, porous carbon/carbon composite, and allowed to solidify. To prevent the expulsion of this pitch during the subsequent high temperature carbonization, the pitch must be stabilized (thermoset) by cross-linking with oxygen. This is accomplished, by

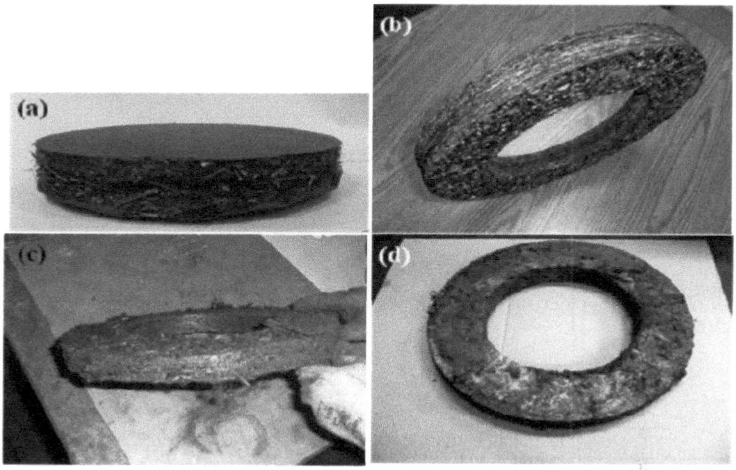

Fig. 7-31 Carbon/Carbon part after: (a) Programmable Powder Preforming Process (P4) process, (b) machining and charring, (c) resin transfer molding, and (d) chemical vapor deposition. The Programmable Powder Preforming Process (P4) technology for composite manufacturing uses a robotic fiber lay-up system which technology was placed in the public domain by Owens Corning for a process of on-line chopping of glass fiber combined with simultaneously deposited powdered epoxy binders.

exposing the solid composite to air at a fixed temperature between 160 °C and 220 °C. The gaseous oxygen enters the composite from an edge surface, diffuses into the pitch, and reacts with the solid. This process depends on the permeability of the oxygen within the pitch, and the kinetics of the oxidation reaction. A model is being developed based on the experimental data that can be used in optimizing the stabilization process. While a low permeability is considered to be a significant limitation in the stabilization process, to our knowledge its value is not known.

As the gaseous oxygen permeates and reacts with the solid, it forms a distributed oxide layer. Such an edge profile of reacted oxygen is shown in Fig. 7-32, as the molecular oxygen to carbon ratio vs. the distance into the composite. Rectangular (4 mm × 4 mm × 2 mm) samples of the carbon/carbon composites produced by resin transfer molding containing an unstabilized mesophase synthetic pitch as carbon matrix were lapped to 0.1 μm, and placed in a horizontal furnace with a continuous dry nitrogen flow until the desired reaction temperature (220 °C) was stabilized. The nitrogen was then replaced by a continuous flow of pure oxygen at 1 atm for a reaction time of 2 h. After the reaction, the samples were flushed with dry nitrogen for 5 h at 150 °C to drive off any moisture, cooled to room temperature, and stored in a dessicator under nitrogen. The solid oxygen concentrations were measured at various depths using the Auger surface spectroscopy with argon ion etching. The depths of the etched regions were measured with atomic force microscopy.

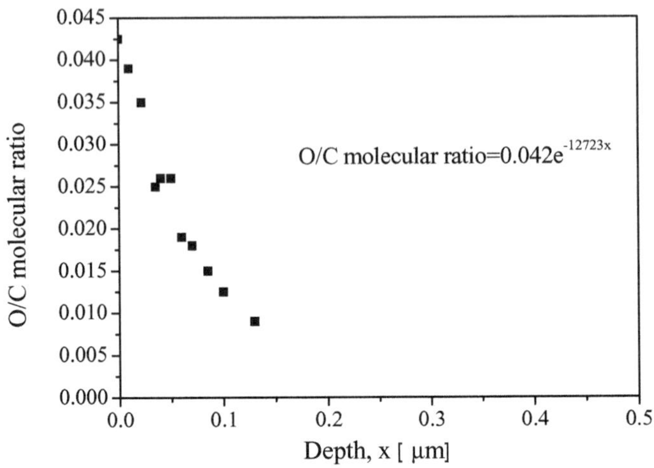

Fig. 7-32 Molecular ratio of oxygen to carbon vs. penetration depth into a C-fiber-untreated C-pitch matrix composite for the oxygen stabilization reaction. Solid points denote experimental data, and the dotted line an exponential curve fit.

The experimental data curve in Fig. 7-32 the result of combining the readings from several locations on a surface—is an average for a single slab. As the Auger oxygen concentration measurements below 1 percent are difficult, an exponential fit, shown by the dotted curve in Fig. 7-32, is used to generate the lower end of the curve. As almost no carbon is lost during the reaction at stabilization temperatures, the molecular carbon distribution is uniform across the solid. The experimental curve of Fig. 7-32 provides a direct observation of the molecular distribution of the reacted oxygen, as well as, the geometry of the cross-linking stabilization of interest to process control. The oxygen distribution demonstrates that while diffusion is an important limiting factor, the reaction is not (as suggested in [195]) truly instantaneous.

A plot relating the density of the part to the process is given in Fig. 7-33. The density of the part after the preforming stage was 0.97 g/cm^3, and progressed steadily toward the target of 1.7 g/cm^3 during the individual process steps. The total time for producing the part was 46 days.

7.3.4.3. Chemical vapor infiltration of carbon/carbon preforms

The reaction of chemical is between two phases, and is on a phase boundary (the gas–fiber interface, so it is a heterogeneous reaction. Because the reaction proceeds on fiber surfaces throughout the preform, on a macroscopic level it looks like a homogeneous reaction, and can be treated like one mathematically. The distinction here is that the average concentrations in the gas phase vary on a much larger length scale than the spacing between the fibers in the preform. Rate-limiting steps are chiefly the first two of the following:

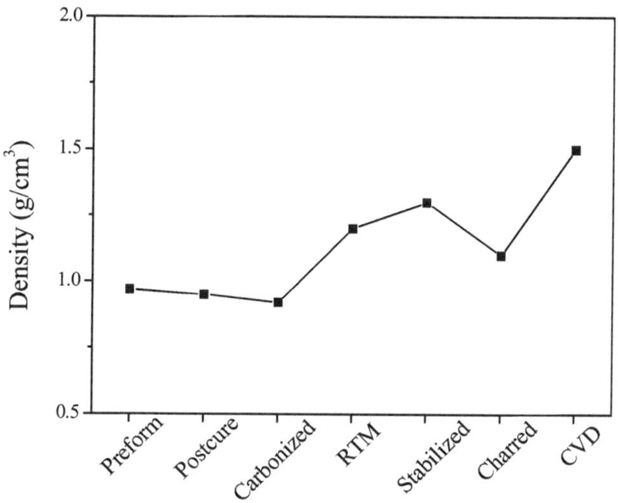

Fig. 7-33 Density increase in the carbon/carbon composite part during processing.

(1) Diffusion of acetylene through a gas boundary layer surrounding the preform.
(2) Diffusion of acetylene through the preform to the reaction sites.
(3) Reaction at the fiber surface forming the graphite matrix.
(4) Diffusion of hydrogen away from reaction sites and out of the preform.

The last of these is not likely to be the slowest step, because small fast-moving hydrogen molecules diffuse more quickly than slower-moving acetylene or other carbon-bearing molecules. The gas boundary layer is possible, but diffusion through the small pores of the preform is likely to be slower than through the open space away from the preform. So likely limiting mechanisms are the second and third: acetylene diffusion to the reaction sites, and reaction at the fiber surface.

7.3.4.4. Coal-tar and petroleum pitches

Because they are thermoplastic, pitches are used mostly for redensification, i.e., further densifying of a carbon/carbon structure that has been ‘rigidized’ by an earlier impregnational densification step (e.g., a resin-impregnated fabric preform) or that has sufficient rigidity from the friction between the elements of the woven structure (e.g., 3D braided preform).

Pitches are unique in passing through a liquid-crystalline transformation at temperatures between about 350 and 550 °C [196]. In this transformation, large lamellar molecules formed by the reactions of thermal cracking and aromatic polymerization are aligned parallel to form an optically anisotropic liquid crystal known as the carbonaceous mesophase [197]. The alignment of the lamellar molecules is the basis for easy thermal graphitizability of the carbonized product. One of the features of a mesophase-based matrix is high bulk density, which is achievable because the matrix density can approach the value for single-crystal graphite, 2.26 g/cm^3.

The topic of pitch impregnation and densification of carbon/carbon composites introduces the subject of densification efficiency, the most meaningful measure of which is volumetric densification efficiency [198]. It is the ratio of the volume of carbon matrix in a process cycle to the volume of porosity available for densification.

For pitches carbonized at atmospheric pressure, coke yields are of the order of 50–60 percent, impregnant densities are −1.35 g/cm^3, and, as we have noted, the densities for pitch-derived matrices are −2.2 g/cm^3. From these values we calculate volumetric densification efficiencies of only 30–40 percent at atmospheric pressure [198]. By resorting to so-called hot isostatic-pressure–impregnation–carbonization, to pressures of about 15,000 psi, carbon yields of pitches can be increased to almost 90 percent [199]. But even with hot isostatic-pressure–impregnation–carbonization, volumetric filling is only 55 percent. Therefore, given a preform with initial porosity of 45 percent, typical for many 3D woven structures, three cycles at maximum densifico/ion efficiency would be required to reduce the porosity to 4 percent. With current hot isostatic-pressure–impregnation–carbonization procedures, however, it is found that at least five cycles at 15,000 psi are required to achieve this same level of porosity. Such reduced efficiency in real systems is the result of forced expulsion of pitch from the preform as a result of the gas-forming pyrolysis reactions accompanying carbonization.

Clearly, one way to increase efficiency, for a given weight-based carbon yield, is to select either an impregnant or a heat treatment temperature that will lower the final matrix density. As will be seen in the next subsection, lower-density carbon matrices can be achieved by using resin precursors that form a glassy-carbon-type structure. But, although this approach fills more of the available space, it does so with a lower-density carbon matrix, which is different in structure from the higher-density graphitic matrix. The trade-offs in properties, particularly mechanical, are not well understood. We will touch on this topic again in the next subsection.

Approaches to improving densification efficiency of the pitch-based matrices without resorting to hot isostatic-pressure–impregnation–carbonization processing include the use of heat-treated and solvent-extracted pitches [200] and partially transformed (to mesophase) pitches [201, 202]. A novel approach, developed by White and Sheaffer [203], is to oxidatively stabilize the mesophase following impregnation and transformation, an approach similar to that employed in mesophase-fiber stabilization. The result is a 'hardened' mesophase that is resistant to the bloating effects of pyrolysis gases but that, upon further heat treatment, yields a dense, graphitic carbon.

The strong orienting effect of the fiber surface on the large lamellar mesophase molecules is an interesting feature of mesophase formation in carbon/carbon composites. This effect was demonstrated by the work of Zimmer and Weitz [204] who used polarized-light microscopy to show that the mesophase molecules near a fiber surface in a close-packed fiber bundle always aligned parallel to the fiber surfaces, even in the presence of strong magnetic fields. Singer and Lewis demonstrated earlier that magnetic fields would orient the mesophase molecules in bulk mesophase [205]. Zimmer and Weitz showed that mesophase would also orient in matrix-rich regions within the fiber bundles, i.e., at points far removed from the fiber surfaces [206]. They calculated a magnetic coherence length of 7 μm, which corresponds roughly to the distance over which the orientation effect acts.

Such localized orientation in the liquid-crystalline state would lead one to expect the final, graphitized matrix also to be well oriented in the immediate vicinity of the fiber. Modulus enhancement in pitch-based carbon/carbon composites has been widely reported, but whether the effect is due to the matrix or to an increase in the fiber modulus, resulting from high-temperature heat-treatment-induced structural changes in the fiber, has not been clarified [207]. The sheath effect is also pronounced in the resin-based carbon matrices, but for different reasons, which we will examine in the next subsection.

Matrix microcracking is characteristic of all carbon/carbon composites, but it is particularly prevalent in graphitic matrices because of the combination of weak shear planes in polycrystalline graphite and the thermal stresses generated during heat treatment [208, 209]. Microcracking also has important effects on the engineering properties of carbon/carbon composite materials—particularly the matrix-dominated properties in the unreinforced directions such as the interlaminar shear strength and perpendicular-to-ply tensile strength in 2D carbon/carbon composite laminates. However, as mentioned above, such microcracking appears to improve in-plane flexural and tensile strength, by way of a Cook–Gordon mechanism [210–213].

7.3.4.5. Thermoset resins

The third, and last, class of carbon/carbon composite impregnant to be discussed is thermoset resins, which is the basis for 'prepreg' fabric and tapes, as noted above; the resin systems can also be used for reimpregnation. In addition to their easy fabricability, thermosets have the advantage of 'charring-in-place' that is, although they soften and deform on heating, they do not fuse or liquefy, and, therefore, no special tools or techniques must be employed to retain the matrix in the composite during pyrolysis.

The thermoset resins are usually highly cross-linked, which makes them resistant to thermal graphitization in bulk form, even to temperatures of 3000 °C [214, 215]. The phenolic resins are currently most commonly used for prepreg operations, whereas furan-based resins are used more for reimpregnating. Both have char yields typically in the 40–50 percent range.

7.3.4.6. Densification efficiency

A number of factors must be considered in the selection of a matrix for carbon/carbon composites. First, the choice of impregnant and the subsequent heat treatment temperature have a significant influence on the fiber–matrix interactions and, hence, on the mechanical properties of the composites. Second is density: Typically, low density glassy-carbon matrices are obtained from the cross-linked resin systems (except within fiber bundles where *in situ* stress graphitization occurs [216]), whereas high-density graphite matrices are obtained from coal-tar and petroleum pitches. The third consideration is densification efficiency, which relates directly to manufacturing cost. The lengthy and complicated processes for fabricating carbon/carbon composites dictate, largely, their very high cost; high cost, in turn, has restricted their use principally to military applications. The non-military applications include aircraft brakes, refractory components in gas-cooled, high-temperature nuclear reactors, and hot-press dies [217]. The third principal consideration, densification efficiency, is the focus of this chapter.

Analytical development

The objective here is to develop a series of relationships between the various factors affecting coal-tar pitch densification of carbon/carbon composites. We begin by defining a unit volume of composites having fractional porosity, θ. The composites are impregnated with a pitch matrix precursor having density ρ_P at the temperature of impregnation, a carbon yield based on weight, y_W, (i.e., $y_{W'}$ = wt. of carbon matrix/wt. of impregnant), and final carbon matrix density ρ_A. Values of y_W and ρ_A correspond to the final heat treatment temperature. We also define Y_I as the overall liquid impregnation efficiency. It is a measure of both the efficiency of the initial impregnation and of keeping the pitch impregnant in place in the composites against the forces of gas evolution during pyrolysis. The volume of carbon matrix ΔV remaining in this unit volume of composite following a complete densification cycle can then be expressed as

$$\Delta V = \frac{y_W \times Y_I \times \rho_P}{\rho_A \times \theta} \tag{7-32}$$

Eq. (7-32), which is the same as that derived by Jortner [218], can be rearranged to give,

$$\frac{\Delta V}{\theta} = \frac{y_W \times Y_I \times \rho_P}{\rho_A} \tag{7-33}$$

which defines $\Delta V/\theta$ ($\cong Y_v$) as a volumetric densitication efficiency: the ratio of the volume of carbon matrix to the volume of porosity available for densification. It is the most useful parameter for measuring efficiency of densification and can be calculated for each step of the process.

Using the above relationships, an upper limit for densification efficiency can be estimated for coal-tar pitches. Taking $y_W = 90$ percent as an upper limit for a number of coal-tar pitches pyrolyzed from 100 to 1000 atm [219, 220], setting $\rho_p = 1.35$ [221] and $\rho_A = 2.2$ g/cm^3, and assuming perfect liquid impregnation efficiency, the maximum densification efficiency per cycle is estimated from Eq. (7-33) to be $Y_v \cong 0.55$. We can express the porosity of carbon/carbon composites after n densification cycles by;

$$\theta_n = \theta_0(1 - Y_V)_n \tag{7-34}$$

where θ_n is the porosity after the nth cycle and θ_0 the initial porosity; this assumes no change in composite dimension with densification.

For the theoretical densification efficiency of 0.55, and an initial porosity of 0.45, which is typical of 3D woven preforms, the porosity following the third cycle is calculated from Eq. (7-34) to be about 0.04. That is, if densifications were completely efficient and there were no composite dimensional change, porosities of the order of 4 percent could be achieved in only three densification cycles.

7.3.5. One-step Manufacturing of Carbon/Carbon Composites with High Density and Oxidative Resistance

Several additives acting as oxidation inhibitors can lower the reactivity of carbon with oxygen. McKee [221] and Savage [222] reported a series of carbon matrix inhibiting materials that were applied to improve oxidation resistance. These materials often had a glass composition, such as boron oxide and other metal phosphates. However, the above-mentioned processes were not good enough to withstand oxygen attack when the temperature exceeds 1,000 °C [223].

Among the non-oxide materials for high-temperature applications, molybdenum disilicide ($MoSi_2$) is a high melting point intermetallic compound noted for its excellent resistance to high-temperature oxidation [224, 225]. It has found engineering applications as high-temperature electric heating elements, for example, $MoSi_2$ heating elements are commercially available for use at temperatures as high as 1900 °C [226]. This interest in $MoSi_2$ is driven by its excellent oxidation resistance, moderate density of 6.24 g/cm^3, and high melting temperature of 2030 °C, which dictates a potential upper use temperature of about 1600 °C. An added advantage when considering the naval environment is $MoSi_2$'s resistance to hot corrosion, which is reportedly an order of magnitude better than that of the best nickel-base superalloy. Furthermore, it exhibits a brittle-to-ductile transition and significant metallic ductility above 900 °C [227–229]. These factors can play important roles in rapid manufacturing processes which avoid catastrophic composite failure during graphitization [230]. Park and Seo reported that the carbon/carbon composites with high oxidative resistance as well as high density were successfully manufactured using the

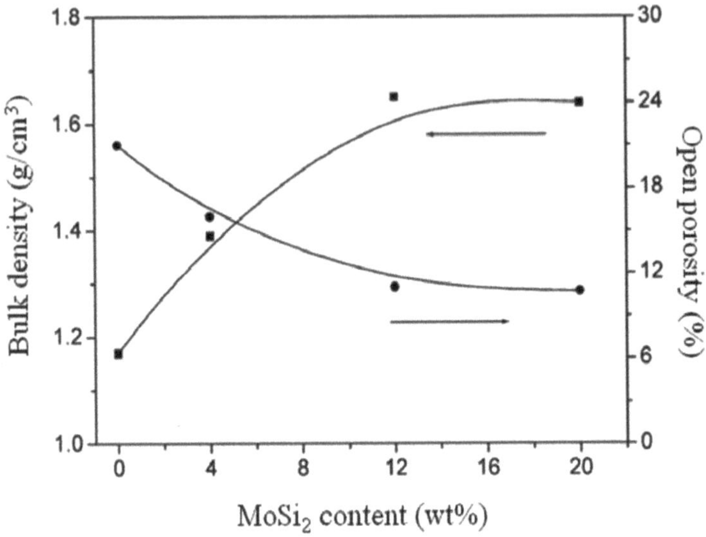

Fig. 7-34 Bulk density and open porosity of carbon/carbon composies as a function of MoSi2 content of the prepreg resin.

prepreg method, and were submitted to carbonization and graphitization (heat treated to 1,100 °C and 2,300 °C), throughout a one-step manufacturing process [231–235].

Fig. 7-34 shows the bulk density (D_b) and open porosity (P_o) of the carbon–carbon composites as a function of $MoSi_2$ content in the prepreg resin. As expected, the composites result in an increasing the D_b and decreasing the P_o with addition of $MoSi_2$ fillers in the composites. And, the composites impregnated with 12 $MoSi_2$ possess the highest density in all composites. This may be attributed to the difference in yield (at 1,000 °C) among the fibers, matrix, and inhibitor. Then, it is possible to consider an inhibition factor (I_f) [236] for determining the efficiency of the inhibition effect. Inhibition factor is defined as the ratio of the oxidation rate of the composites made without $MoSi_2$, r_o, to the rate corresponding to the sample made with $MoSi_2$, r_t:

$$I_f = \frac{r_0}{r_i} \tag{7-35}$$

Both reaction rates, r_o and r_t, are taken at both the same temperature and level of burn-off. Hence the inhibition factor may depend on the level of burn off, the amount of $MoSi_2$ filler, as well as the temperature of oxidation. The values of inhibition factor (I_f) determined at various heat treatment temperatures for different levels of burn-off of the composites are given in Table 7-9 for three inhibitor contents 4, 12, and 20 percent by weight. It is seen that even for a very

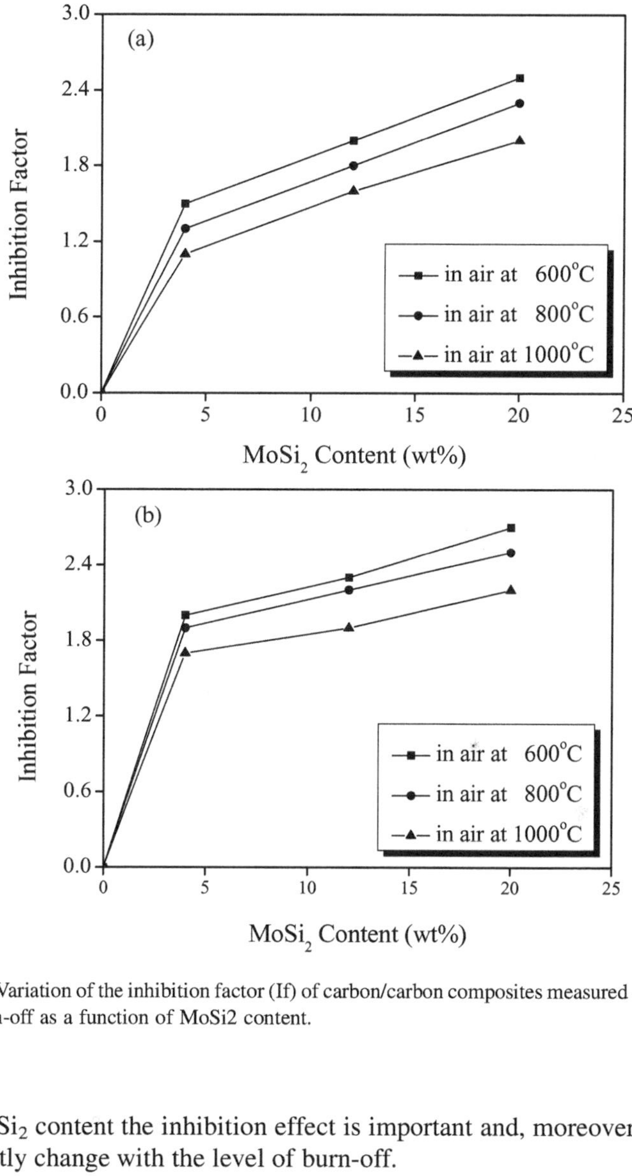

Fig. 7-35 Variation of the inhibition factor (If) of carbon/carbon composites measured at 10 percent level of burn-off as a function of MoSi2 content.

small $MoSi_2$ content the inhibition effect is important and, moreover, does not significantly change with the level of burn-off.

The effect of the amount of $MoSi_2$ on the inhibition factor measured of the carbonized and graphitized composites at different temperatures in a given burn-off level of 10 percent by weight is shown in Figs. 7-35(a) and (b), respectively. For the relatively small presence of $MoSi_2$ in composites, i.e., below 4 percent by weight, I_f increases very sharply with the $MoSi_2$ presence. Thereafter, with the increasing of $MoSi_2$ content, I_f values were increased slowly. These results mean that the additions of $MoSi_2$ are significant effects on oxidation rate of

Table 7-9 Inhibition Factors (I_f) of the Carbonized and Graphitized Composites as Functions of Levels of Burn-off and $MoSi_2$ Content

	Inhibition factor (I_f)								
	Carbonization								
	HTT* (600 °C)			HTT (800°)			HTT (1000°)		
	$MoSi_2$ content (wt %)								
Burn-off (%)	4	12	20	4	12	20	4	12	20
10	1.5	2.0	2.5	1.3	1.8	2.3	1.1	1.6	2.0
20	1.6	1.8	2.2	1.4	1.6	2.0	1.2	1.5	1.8
40	1.5	1.7	2.1	1.3	1.5	1.8	1.1	1.4	1.7
60	1.7	2.1	2.5	1.5	1.9	2.2	1.3	1.8	2.0
80	1.6	2.7	3.4	1.4	2.4	3.1	1.2	2.1	2.9
10	2.0	2.3	2.7	1.8	2.2	2.5	1.5	1.9	2.2
20	2.0	2.1	2.4	1.9	2.0	2.3	1.7	1.9	2.0
40	2.1	2.3	2.5	1.8	2.1	2.0	1.7	2.0	1.9
60	2.0	2.2	2.3	1.9	1.9	2.0	1.8	1.8	1.8
80	1.9	2.1	2.6	1.8	2.0	2.4	1.7	1.9	2.3

**HTT: heat treatment temperature.*

composites in all heat treatment temperatures, though their anti-oxidation rates are somewhat different.

7.3.6. Applications of Carbon/Carbon Composites

Carbon/carbon composites use the strength and modulus of carbon fibers to reinforce a carbon matrix to resist the rigors of extreme environments. At one-tenth the density, carbon/carbon composites offer a high performance, cost effective alternative to refractory metals. Aerospace components commonly fabricated from carbon/carbon composites include rocket motor nozzle throats and exit cones, nosetips/leading edges and thermal protection systems. Reliable performance is the most critical requirement of these components. The carbon/carbon composites have demonstrated reliability and reduced systems costs, especially when multiple components in an assembly can be replaced with a one-piece carbon/carbon composite design.

Commercial applications of carbon/carbon composite materials include furnace fixturing, heatshields, load plates, heating elements, and X-ray targets. Rocket nozzles must withstand an extremely rapid temperature increase in a highly corrosive atmosphere while maintaining a high degree of integrity. In addition to exposure to severe thermal stresses, the nozzle material must perform predictably and uniformly to provide the required accuracy and range.

7.4. METAL MATRIX COMPOSITES

7.4.1. Introduction

Metal composite materials have found application in many areas of daily life for quite some time. Often it is not realized that the application makes use of composite materials. These materials are produced *in situ* from the conventional production and processing of metals. Here, the Dalmatian sword with its meander structure, which results from welding two types of steel by repeated forging, can be mentioned.

Materials like cast iron with graphite or steel with high carbide content, as well as tungsten carbides, consisting of carbides and metallic binders, also belong to this group of composite materials. For many researchers the term metal matrix composites is often equated with the term light metal matrix composites.

Substantial progress in the development of light metal matrix composites has been achieved in recent decades, so that they could be introduced into the most important applications. In traffic engineering, especially in the automotive industry, metal matrix composites have been used commercially in fiber reinforced pistons and aluminum crank cases with strengthened cylinder surfaces as well as particle-strengthened brake disks.

These innovative materials open up unlimited possibilities for modern material science and development; the characteristics of metal matrix composites can be designed into the material, custom-made, dependent on the application. From this potential, the metal matrix composites fulfill all the desired conceptions of the designer. This material group becomes interesting for use as constructional and functional materials, if the property profile of conventional materials either does not reach the increased standards of specific demands, or is the solution of the problem. However, the technology of metal matrix composites is in competition with other modern material technologies, for example powder metallurgy. The advantages of the composite materials are only realized when there is a reasonable cost performance relationship in the component production. The use of a composite material is obligatory if a special property profile can be achieved only by the application of these materials.

The possibility of combining various material systems (metal-ceramic-nonmetal) gives the opportunity for unlimited variation. The properties of these new materials are basically determined by the properties of their single components. Fig. 7-36 shows the allocation of the composite materials into groups of various types of materials [237].

The reinforcement of metals can have many different objectives. The reinforcement of light metals opens up the possibility of application of these materials in areas where weight reduction has first priority. The precondition here is the improvement of the component properties. The development objectives for light metal composite materials are:

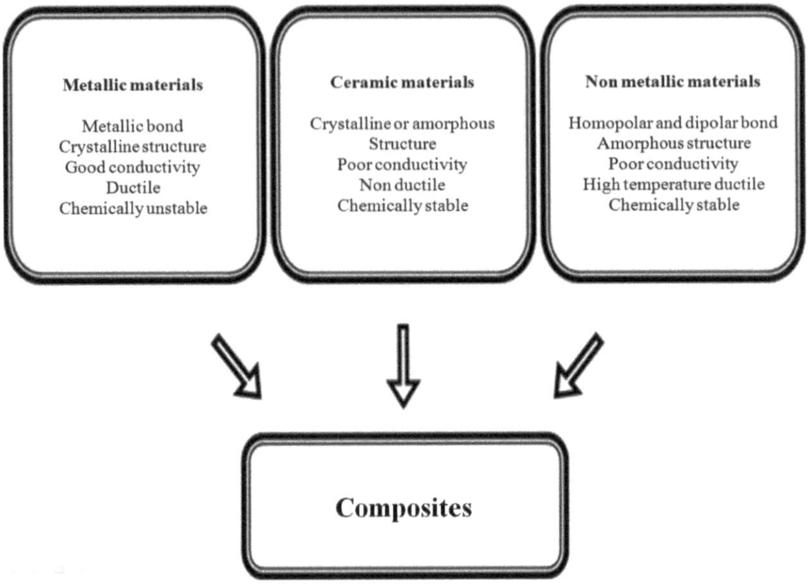

Fig. 7-36 Classification of the composite materials within the group of materials.

- Increase in yield strength and tensile strength at room temperature and above while maintaining the minimum ductility or rather toughness.
- Increase in creep resistance at higher temperatures compared to that of conventional alloys.
- Increase in fatigue strength, especially at higher temperatures.
- Improvement of thermal shock resistance.
- Improvement of corrosion resistance.
- Increase in Young's modulus.
- Reduction of thermal elongation.

To summarize, an improvement in the weight specific properties can result, offering the possibilities of extending the application area, substitution of common materials, and optimization of component properties. With functional materials there is another objective, the precondition of maintaining the appropriate function of the material. Objectives are for example:

- Increase in strength of conducting materials while maintaining the high conductivity.
- Improvement in low temperature creep resistance (reactionless materials).
- Improvement of burnout behavior (switching contact).

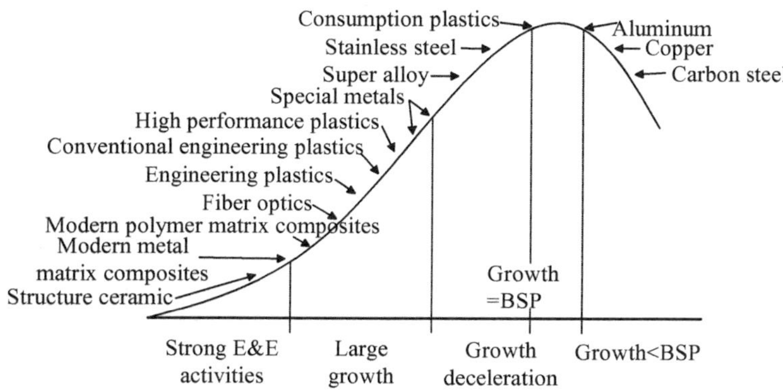

Fig. 7-37 Development curve of the market for modern materials.

- Improvement of wear behavior (sliding contact).
- Increase in operating time of spot welding electrodes by reduction of burn outs.
- Production of layer composite materials for electronic components.
- Production of ductile composite superconductors.
- Production of magnetic materials with special properties.

For other applications different development objectives are given, which differ from those mentioned before. For example, in medical technology, mechanical properties like extreme corrosion resistance and low degradation as well as biocompatibility are expected.

Although increasing development activities have led to system solutions using metal composite materials, the use of especially innovative systems, particularly in the area of light metals, has not been realized. The reason for this is insufficient process stability and reliability, combined with production and processing problems and inadequate economic efficiency. Application areas, like traffic engineering, are quite cost orientated and conservative and the industry is not willing to pay additional costs for the use of such materials. For these entire reasons metal matrix composites are only at the beginning of the evolution curve of modern materials, as shown in Fig. 7-37 [238].

Metal matrix composites can be classified in various ways. One classification is the consideration of type and contribution of reinforcement components in particle-, layer-, fiber- and penetration composite materials (see Fig. 7-38). The fiber composite materials can be further classified into continuous fiber composite materials (multi- and monofilament) and short fibers or, rather, whisker composite materials see Fig. 7-39 [239].

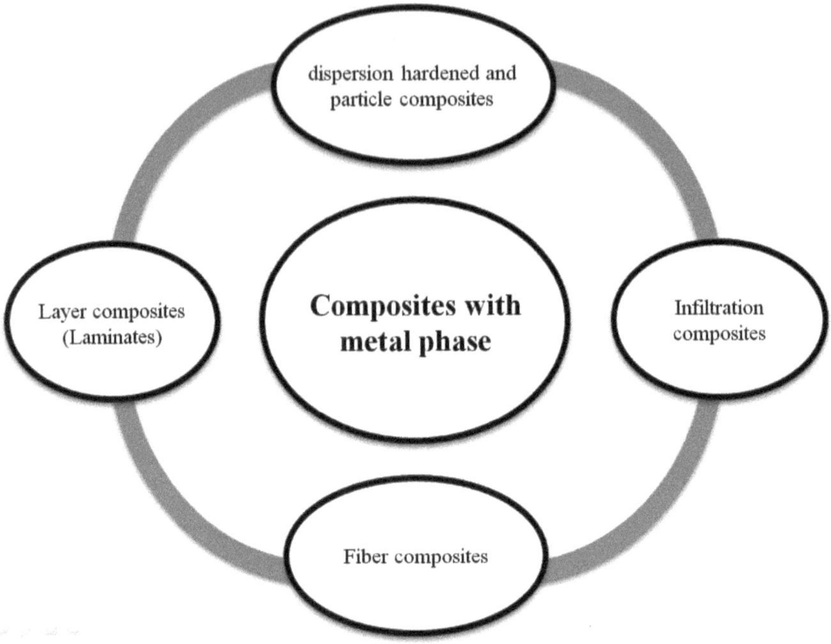

Fig. 7-38 Classification of composite materials with metal matrices.

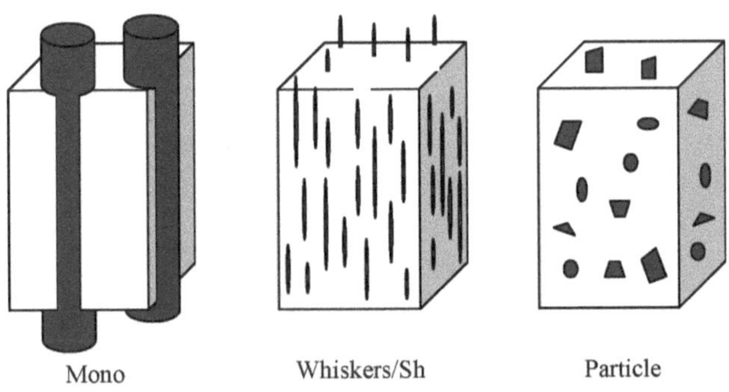

Fig. 7-39 Schematic presentation of three shapes of metal matrix composite materials.

7.4.2. Combination of Materials for Light Metal Matrix Composites

7.4.2.1. Reinforcements

Reinforcements for metal matrix composites have a manifold demand profile, which is determined by production and processing and by the matrix system of the composite material. The following demands are generally applicable [240]: low density, mechanical compatibility (a thermal expansion coefficient which is

Table 7-10 Properties of Typical Discontinuous Reinforcements for Aluminum and Magnesium Reinforcements

Reinforcement	Saffil (Al_2O_3)	SiC particle	Al_2O_3 particle
Crystal structure	δ-Al_2O_3	Hexagonal	Hexagonal
Density (g cm^{-3})	3.3	3.2	3.9
Average diameter (μm)	3.0	Variable	Variable
Length (μm)	ca. 150	—	—
Mohs hardness	7.0	9.7	9.0
Strength (MPa)	2000	—	—
Young's Modulus (GPa)	300	200-300	380

low but adapted to the matrix), chemical compatibility, thermal stability, high Young's modulus, high compression and tensile strength, good processability, and economic efficiency.

These demands can be achieved by using non-metal inorganic reinforcement components only. For metal reinforcement ceramic particles or, rather, fibers or carbon fibers are often used. Due to the high density and the affinity to reaction with the matrix alloy the use of metallic fiber usual fails. Which components are finally used, depends on the selected matrix and on the demand profile of the intended application. In Refs. [240, 241] the information about available particles, short fibers, whiskers, and continuous fibers for the reinforcement of metals is given, including data of manufacturing, processing, and properties. Representative examples are shown in Table 7-10 [244]. The production, processing and type of application of various reinforcements depend on the production technique for the composite materials. A combined application of various reinforcements is also possible (hybrid technique) [239, 242, 243].

Every reinforcement has a typical profile, which is significant for the effect within the composite material and the resulting profile. Table 7-11 gives an

Table 7-11 Property Potential of Different Metal Matrix Composites

MMC[1] type	Properties Strength	Young's modulus	High temperature properties	Wear	Expansion coefficient	Costs
Mineral wool: MMC	*	*	**	**	*	Medium
Discontinuous reinforced MMC	**	**	*	***	**	Low
Long fiber reinforced MMC: C fiber	**	**	**	*	***	High
Other fibers	***	***	***	*	**	High

**low, **mid, ***good.*
[1]MMC: metal matrix composites.

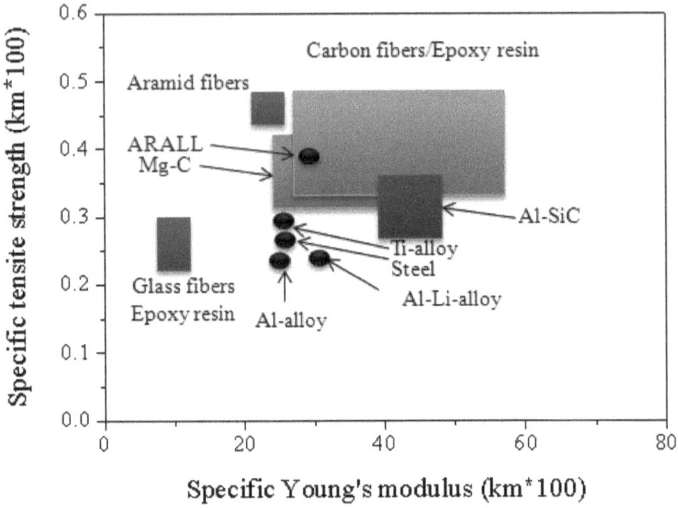

Fig. 7-40 Specific tensile strength and specific Young's modulus of different quasi-isotropic fiber composite materials in comparison to some metal alloys.

overview of possible property profiles of various material groups [238]. Fig. 7-40 shows the specific strength and specific Young's modulus of quasi-isotropic fiber composite materials with various matrixes in comparison to monolithic metals [238]. The group of discontinuous reinforced metals offers the best conditions for reaching development targets; the applied production technologies and reinforcement components, like short fibers, particle and whiskers, are cost effective and the production of units in large item numbers is possible. The relatively high isotropy of the properties in comparison to the long-fiber continuous reinforced light metals and the possibility of processing of composites by forming and cutting production engineering are further advantages.

7.4.2.2. Matrix alloy systems

The selection of suitable matrix alloys is mainly determined by the intended application of the composite material. With the development of light metal composite materials that are mostly easy to process, the conventional light metal alloys are applied as matrix materials. In the area of powder metallurgy special alloys can be applied due to the advantage of fast solidification during the powder production.

Those systems are free from segregation problems that arise in conventional solidification. Also the application of systems with oversaturated or metastable structures is possible. Examples for matrix configurations are given in Refs. [242, 245–251]: conventional cast alloys (G-$AlSi_{12}CuMgNi$, G-$AlSi_9Mg$, G-$AlSi_7$ (A356), AZ91, AE42)), conventional wrought alloys (AlMgSiCu (6061), AlCuSiMn (2014), $AlZnMgCu_{1.5}$ (7075), $TiAl_6V_4$), special alloys (Al-Cu-Mg-Ni-Fe-alloy (2618), Al-Cu-Mg-Li-alloy (8090), AZ91Ca).

For the functional materials non-alloyed or low-alloyed non-ferrous or noble metals are generally used. The reason for this is the demand for the retention of the high conductivity or ductility. A dispersion hardening to reach the required mechanical characteristics at room or higher temperatures is then an optimal solution.

7.4.3. Production and Processing of Metal Matrix Composites

Metal matrix composite materials can be produced by many different techniques. The focus of the selection of suitable process engineering is the desired kind, quantity, and distribution of the reinforcement components (particles and fibers), the matrix alloy and the application. By altering the manufacturing method, the processing and the finishing, as well as by the form of the reinforcement components it is possible to obtain different characteristic profiles, although the same composition and amounts of the components are involved. The production of a suitable precursor material, the processing to a construction unit or a semi-finished material (profile) and the finishing treatment must be separated. For cost effective reasons prototypes, with dimensions close to the final product, and reforming procedures are used, which can minimize the mechanical finishing of the construction units.

Three processing methods have been primarily used to develop metal matrix composites: high-pressure diffusion bonding, casting, and powder-metallurgy techniques. More specifically, the diffusion-bonding and casting methods have been used for continuous-fiber reinforced metal matrix composites. Discontinuously reinforced metal matrix composites have been produced by powder metallurgy and pressure-assist casting processes. Metal matrix composites such as B/Al, Gr/Al, Gr/Mg, and Gr/Cu have been manufactured by diffusion bonding for prototype spacecraft components such as tubes, plates, and panels.

In general the following product engineering types are possible:

(1) Melting metallurgical processes
- Infiltration of short fiber-, particle-, or hybrid preforms by squeeze casting, vacuum infiltration or pressure infiltration [242, 249–251].
- Reaction infiltration of fiber- or particle preforms [252, 253].
- Processing of precursor material by stirring the particles in metallic melts, followed by sand casting, permanent mold casting, or high pressure die casting [245, 246].

(2) Powder metallurgical processes
- Pressing and sintering and/or forging of powder mixtures and composite powders.
- Extrusion or forging of metal-powder particle mixtures [247, 248].
- Extrusion or forging of spraying compatible precursor materials [242, 254, 255].

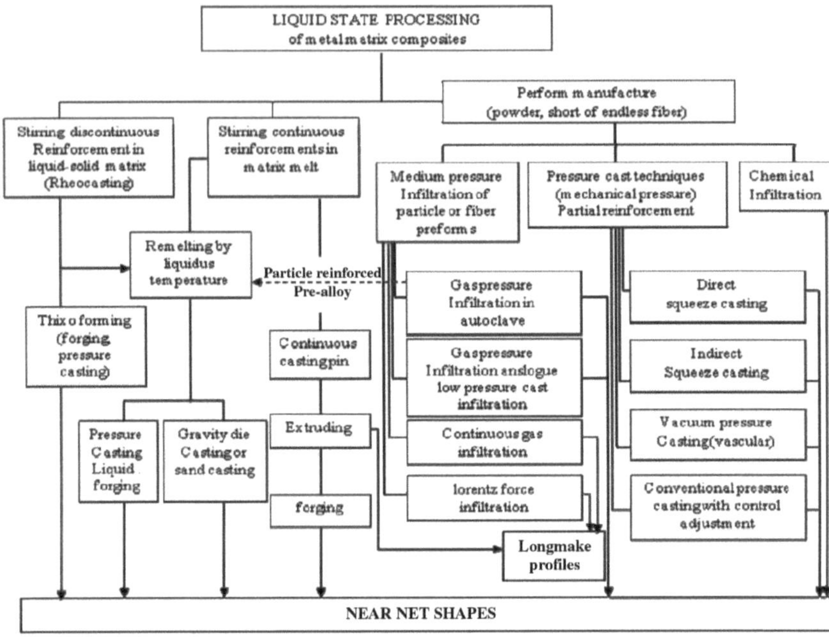

Fig. 7-41 Melting metallurgy of metal matrix composites.

(3) Hot isostatic pressing of powder mixtures and fiber clutches

(4) Further processing of precursor material from the melting metallurgy by thixocasting or-forming, extrusion [256], forging, cold massive forming, or super plastic forming

- Joining and welding of semi-manufactured products.
- Finishing by machining techniques [257].
- Combined deformation of metal wires (group superconductors).

Melting metallurgy for the production of metal matrix composites is at present of greater technical importance than powder metallurgy. It is more economical and has the advantage of being able to use well proven casting processes for the production of metal matrix composites.

Fig. 7-41 shows schematically the possible methods of melting metallurgical production. For melting metallurgical processing of composite materials three procedures are mainly used [252]: compo-casting or melt stirring, gas pressure infiltration, and squeeze casting or pressure casting.

Both the terms compo-casting and melt stirring are used for stirring particles into a light alloy melt. Fig. 7-42 shows the schematic operational sequence of this procedure. The particles are often tending to form agglomerates, which can be only dissolved by intense stirring. However, here gas access into the melt must be absolutely avoided, since this could lead to unwanted porosities or

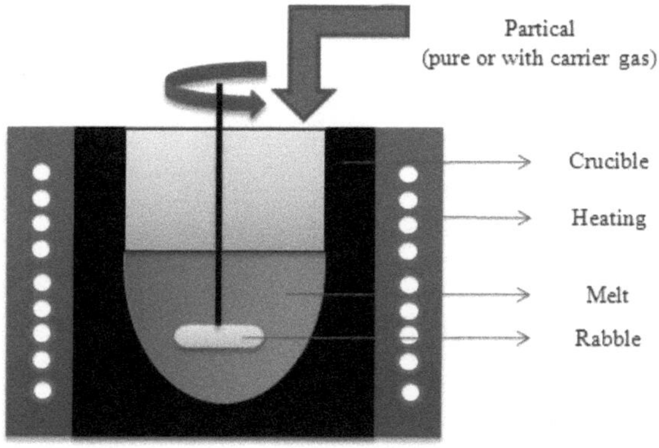

Fig. 7-42 Schematic operational sequence during melt stirring.

reactions. Careful attention must be paid to the dispersion of the reinforcement components, so that the reactivity of the components used is coordinated with the temperature of the melt and the duration of stirring, since reactions with the melt can lead to the dissolution of the reinforcement components. Because of the lower surface to volume ratio of spherical particles, reactivity is usually less critical with stirred particle reinforcement than with fibers. The melt can be cast directly or processed with alternative procedures such as squeeze casting or thixocasting. Melt stirring is used by the Duralcan Company for the production of particle-strengthened aluminum alloys [245, 246]. At the Lanxide Company a similar process is used, with additional reactions between the reinforcement components and the molten matrix being purposefully promoted to obtain qualitatively high-grade composite materials [258]. In the reaction procedures of the Lanxide Company it may be desirable that the reinforcement component reacts completely with the melt to form the component *in situ*, which then transfers the actual reinforcement effect to the second phase in the MMCs.

In gas pressure infiltration the melt infiltrates the preform with a gas applied from the outside. A gas that is inert with respect to the matrix is used. The melting of the matrix and the infiltration take place in a suitable pressure vessel. There are two procedure variants of gas pressure infiltration: in the first variant the warmed up preform is dipped into the melt and then the gas pressure is applied to the surface of the melt, leading to infiltration. The infiltration pressure can thereby be coordinated with the wettability of the preforms, which depends, among other things, on the volume percentage of the reinforcement. The second variant of the gas pressure infiltration procedure reverses the order: the molten bath is pressed to the preform by the applied gas pressure using a standpipe and thereupon infiltrates the bath (see Fig. 7-43). The advantage of this procedure is that there is no development of pores when completely dense parts are present. Since the reaction time is relatively short with these procedures, more reactive materials can be used than, e.g., with the compo-casting. In gas pressure

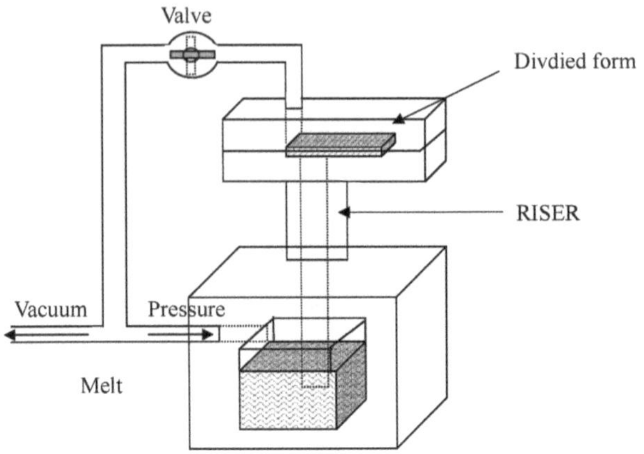

Fig. 7-43 Gas pressure infiltration technique.

infiltration the response times are clearly longer than in squeeze casting, so that the materials must be carefully selected and coordinated, in order to be able to produce the appropriate composite material for the appropriate requirements.

Squeeze casting or pressure casting are the most common manufacturing variants for metal matrix composites. After a slow mold filling the melt solidifies under very high pressure, which leads to a fine-grained structure. In comparison with die-casted parts the squeeze-casted parts do not contain gas inclusions, which permit thermal treatment of the produced parts. One can differentiate between direct and indirect squeeze casting (Fig. 7-44). With direct

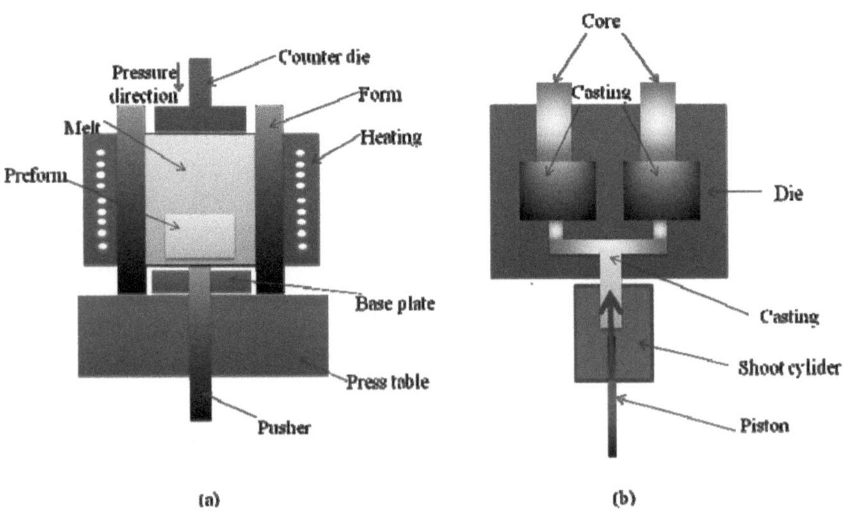

Fig. 7-44 Direct and indirect squeeze casting.

squeeze casting the pressure for the infiltration of the prefabricated preforms is applied directly to the melt. The die is thereby part of the mold, which simplifies the structure of the tools substantially. However, with the direct procedure there is a disadvantage in that the volume of the melt must be determined exactly, since no gate is present and thus the quantity of the melt determines the size of the cast construction unit. A further disadvantage is the appearance of oxidation products, formed in the cast part during dosage. In contrast, in indirect squeeze casting, where the melt is pressed into the form via a gate system, the residues will remain in this gate. The flow rate of the melt through a gate is, due to its larger diameter, substantially less than with die casting, which results in a less turbulent mold filling and gas admission to the melt by turbulences, is avoided.

Both pressure casting processes make the production of composite materials possible, as prefabricated fiber or particle preforms are infiltrated with melt and solidify under pressure. A two-stage process is often used. In the first stage the melt is pressed into the form at low pressure and then at high pressure for the solidification phase. This prevents damage to the preform by too fast infiltration. The squeeze casting permits the use of relatively reactive materials, since the duration of the infiltration and thus the response time, are relatively short. A further advantage is the possibility to manufacture construction units shaped with difficulty and to provide partial reinforcement, to strengthen those areas which are exposed to a higher stress during service.

7.4.4. Mechanism of Reinforcement

The characteristics of metal matrix composite materials are determined by their microstructure and internal interfaces, which are affected by their production and thermal mechanical prehistory. The microstructure covers the structure of the matrix and the reinforced phase. The chemical composition, grain and/or sub-grain size, texture, precipitation behavior, and lattice defects are of importance to the matrix. The second phase is characterized by its volume percentage, its kind, size, distribution, and orientation. Local varying internal tension due to the different thermal expansion behavior of the two phases is an additional influencing factor.

With the knowledge of the characteristics of the components, the volume percentages, the distribution and orientation it might be possible to estimate the characteristics of metallic composite materials. The approximations usually proceed from ideal conditions, i.e., optimal boundary surface formation, ideal distribution (very small number of contacts of the reinforcements among themselves) and no influence of the component on the matrix (comparable structures and precipitation behavior). However, in reality a strong interaction arises between the components involved, so that these models can only indicate the potential of a material. The different micro-, macro-, and meso-scaled models proceed from different conditions and are differently developed. A representation of these models can be seen in Refs. [239, 259]. In the following, simple

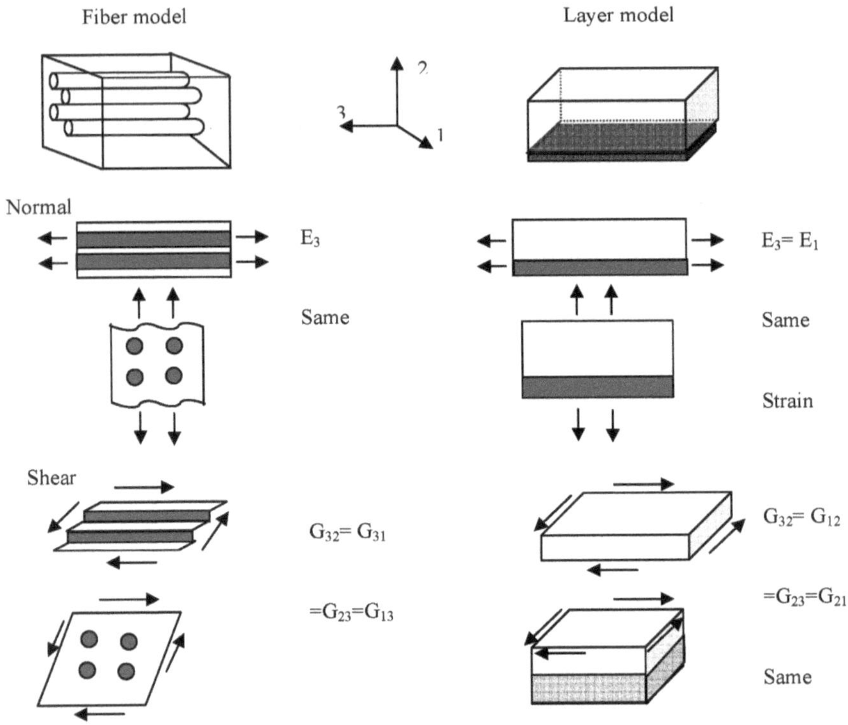

Fig. 7-45 Schematic presentation of elastic constants in composite materials.

models are described, which facilitate our understanding of the effect of the individual components of the composite materials and their form and distribution on the characteristics of the composites.

Simply, we can consider a fiber and/or a plate model. Depending on the load direction, different elastic constants in the metallic composite material result. Fig. 7-45 illustrates the two different models and shows the resulting E and G moduli as a function of the load type. On the basis of these simple considerations an estimate can be made of the attainable strength of the fiber reinforced composite material for the different forms of the fibers.

7.4.5. Influence of Interface

Compared with monolithic materials the microstructure and the interfaces of metal matrix composite materials cannot be considered in isolation, they are mutually related. Chemical interactions and reactions between the matrix and the reinforcement component determine the interface adhesion, modify the characteristics of the composite components, and affect the mechanical characteristics significantly.

In high temperature use of metal matrix composites the microstructure has to remain stable for long service periods. Thermal stability and failure is determined by changes in the microstructure and at the interfaces, e.g., reaction and precipitation processes. Thermal stress of metal matrix composites can take place both isothermally and cyclically. The effects show differences. During a cyclic load of monolithic materials, especially at high temperature gradients and cycle speeds, a high probability of failure by thermal fatigue is to be expected, e.g., short-fiber reinforced aluminum alloys possess good thermal shock stability.

The formation of the interface between the matrix and the reinforcing phase has a substantial influence on the production and characteristics of the metallic composite materials. The adhesion between both the phases is usually determined by the interaction between them. During the production of the molten matrix, e.g., by infiltration, wettability becomes significant.

7.4.5.1. Basics of wettability and infiltration

Basically the wettability of reinforcement with a metal melt can be shown by the edge angle adjustment of a molten droplet on a solid base as the degree of wettability according to Young:

$$\gamma_{SA} - \gamma_{LS} = \gamma_{LA} \cdot \cos\theta \qquad (7\text{-}36)$$

where, γ_{LA} is the surface energy of the liquid phase, γ_{SA} the surface energy of the solid phase, γ_{LS} the interface energy between the liquid and solid phases, and θ is the contact angle.

Table 7-12 Surface and Interface Strains of Selected Metal-ceramic Systems at Different Temperature

Alloy, Ceramic, Systems	Temperature (K)	Yla ($mJ\ m^{-2}$)	Ylsa ($mJ\ m^{-2}$)	Yls ($mJ\ m^{-2}$)	Refs.
Al	953	1050	–	–	37
Mg	943	560	–	–	37
Al_2O_3	0	–	930	–	38
MgO	0	–	1150	–	38
Cu/Al_2O_3	1370	1308	1485	2541	39
	1450	1292	1422	2284	39
Ni/Al_2O_3	1843	1751	1114	2204	39
	2003	1676	988	1598	39
Al/SiC	973	851	2469	2949	40
	1073	840	2414	2773	40
	1173	830	2350	2684	40

Fig. 7-46 shows the contact angle adjustment of a molten droplet on a solid base for different values of the interface energy. At an angle of $>\pi/2$ a non-wettable system is described and for an angle limit of $<\pi/2$ a wettable system. With decreasing angle the wettability improves. In Table 7-12 the surface and interface stresses of the selected metal ceramic systems at different temperatures are summarized. Of special relevance is the system Al/SiC, since it is the basis for the melting metallurgy of particle reinforced aluminum composite materials.

As the contact develops, for example at the beginning of an infiltration, adhesion occurs. The adhesion work W_A for separation is [260]:

$$W_A = \gamma_{SA} - \gamma_{LA} = \gamma_{LS} \tag{7-37}$$

$$W_A = \gamma_{LA} \cdot (1 + \cos\theta) \tag{7-38}$$

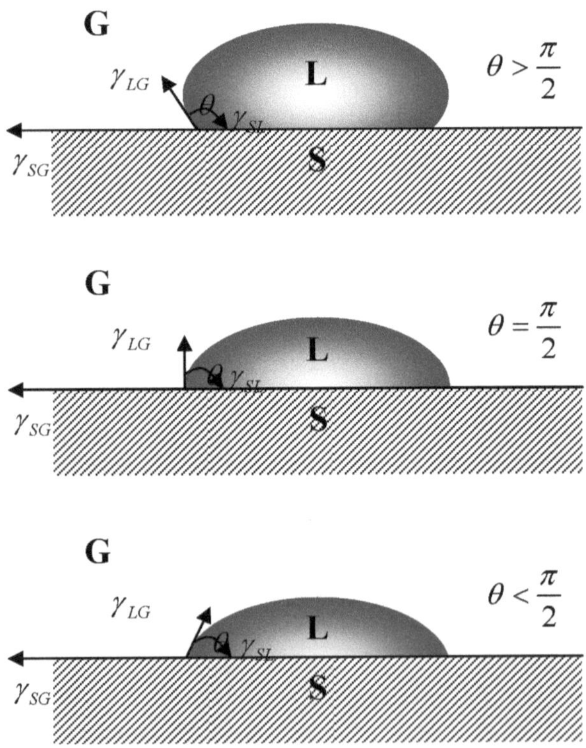

$$\gamma_{SG} = \gamma_{SL} + \gamma_{LG} \cdot \cos\theta$$

Fig. 7-46 Contact angle adjustment of a melt drop on a solid base for various values of the interface energy (after Young).

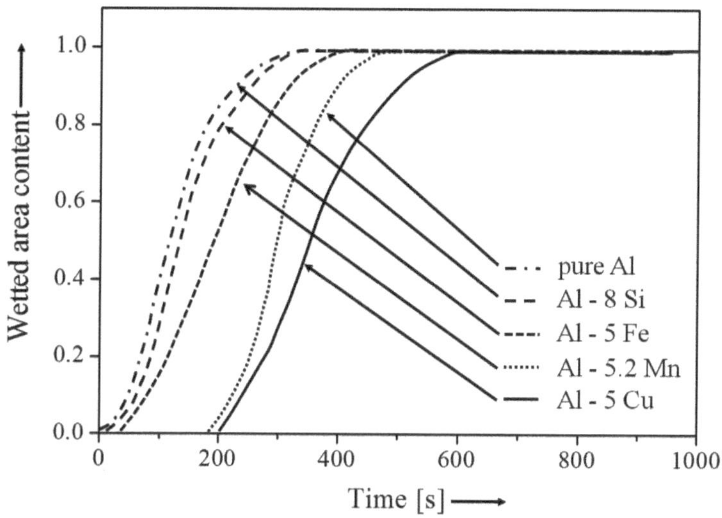

Fig. 7-47 Time dependence of the wetting degree (unit fraction) of silicon carbide plates by aluminum melts at different alloy additions.

In the case of immersion the interface between the solid and the atmosphere disappears, while the interface between the solid and the liquid forms. The immersing work W_I is:

$$W_I = \gamma_{LS} - \gamma_{SA} \quad (7\text{-}39)$$

In the case of spreading the liquid is spread out on a solid surface. During this procedure the solid surface is reduced and a new liquid surface is formed and hence a new solid/liquid interface is formed. The spreading work W_S is:

$$W_S = \gamma_{SA} - \gamma_{LS} - \gamma_{LA} \quad (7\text{-}40)$$

The wetting procedure is kinetic and is dependent on time and temperature. Therefore, the kinetics can be affected by the temperature. Fig. 7-47 shows, as an example of the time dependence of a wetting procedure, the wetting degree (surface fraction) of SiC plates by aluminum alloy melts of different compositions [261].

In Fig. 7-48, the temperature dependence of the wetting angle of one aluminum alloy droplet on a SiC plate is represented. Both figures show the further possibility of the influence of variation of the composition in the appropriate material system [263]. The alloying elements act by changing the surface tension of the melt or by reaction with the reinforcement. On the one hand the composition of the matrix, or rather reinforcement is modifiable and on the other hand there exists the possibility of purposeful influence by applying coatings on the intensifying phase. The role of a reaction at the interface is important, because from it a new system can result and the interface energies can be changed substantially, thus altering the wetting angle.

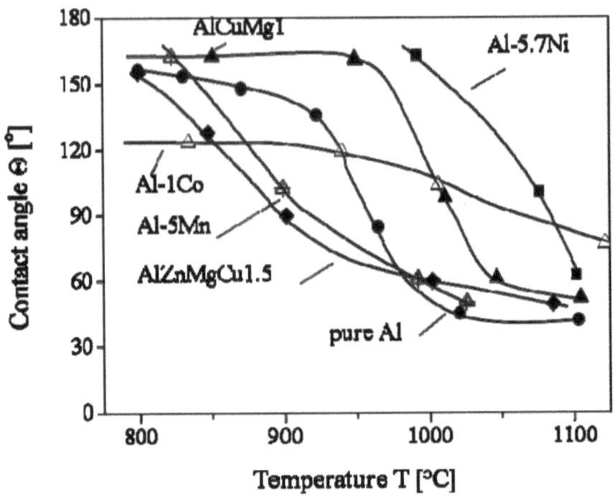

Fig. 7-48 Temperature dependence of the wetting angle of aluminum drops on a SiC plate.

In Eq. (7-36) the change induced by the reaction, for example, of oxide reinforcement $M_{e1}O$ with a matrix alloy part Me_2, has to be taken into consideration [262]:

$$\gamma_{LS} - \gamma_{SA} - = (\gamma_{LS} - \gamma_{SA})_0 - \Delta\gamma r - \Delta G r \tag{7-41}$$

$$Me_2 + M_{e1}O \leftrightarrow M_{e2}O + M_{e1} \tag{7-42}$$

$$\gamma_{LA} \cdot \cos\theta = (\gamma_{SA} - \gamma_{LS})_0 - \Delta\gamma r - \Delta G r \tag{7-43}$$

where $(\gamma_{LS} - \gamma_{SA})_0$ is the wettability without reaction, $\Delta\gamma_r$ the interface tension from the reaction of newly formed interfaces, and ΔG_r the given free energy at the triple line solid/liquid/atmosphere (reaction energy).

Fig. 7-49 shows the influence of pressure-free infiltration through different reactions by using different reactive binder systems and fiber contents such as Mg/Al_2O_3 fibers [253]. With a very reactive SiO_2-containing binder premature infiltration happens at lower temperatures than with an Al_2O_3 binder.

An additional possible influence exists through the change in the surrounding atmosphere or rather the atmosphere in the preform. It is possible, for example, that a preform before infiltration is flushed with gas, which can lead to a change in the oxygen partial pressure. Fig. 7-50 shows the dependence of the wetting angle in the system Al_2O_3/pure aluminum on the temperature for two oxygen partial pressures [264]. For high partial pressures of oxygen high wetting angles occur at low temperatures. Only when starting from temperatures above 1150 K does the value for the wetting angle decrease to that for a low oxygen partial pressure. However, technically this influence is not relevant, because the atmosphere can be changed only with difficulty. An exception is

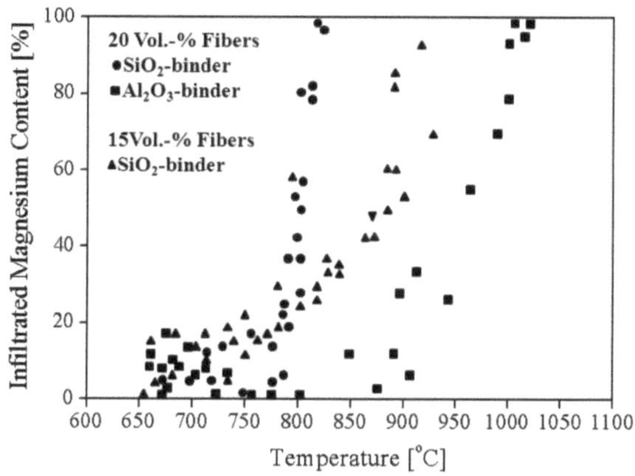

Fig. 7-49 Influence of pressure free infiltration by different reactions under use of different reactive fiber contents on the example of pure magnesium/Al_2O_3 fibers, Dissertation Fritze and.

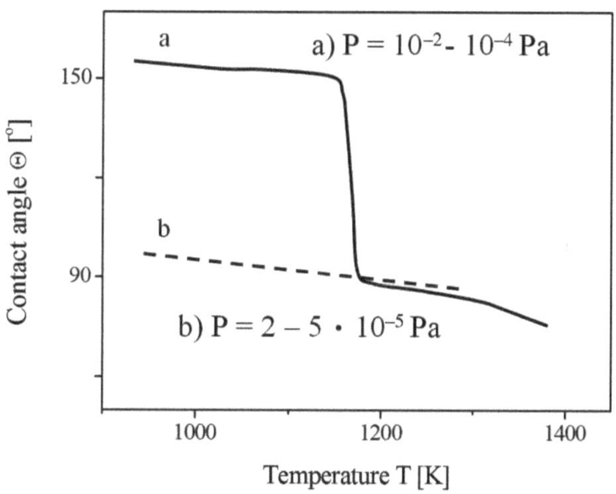

Fig. 7-50 Schematic presentation of the change in the wetting angle with change in the oxygen particle pressure.

the infiltration by the production of a vacuum (gas pressure infiltration); in this case the gas pressures can be modified.

Wetting for the actual infiltration procedure is of substantial importance. This is shown in a simple schematic representation in Fig. 7-51 [265]. In the case of good wetting (small edge angle) a capillary effect occurs (Fig. 7-51(a)). At large edge angles this procedure is inhibited (Fig. 7-51(b)). Additionally this can occur in technical processes by a reaction between the melt and the surrounding

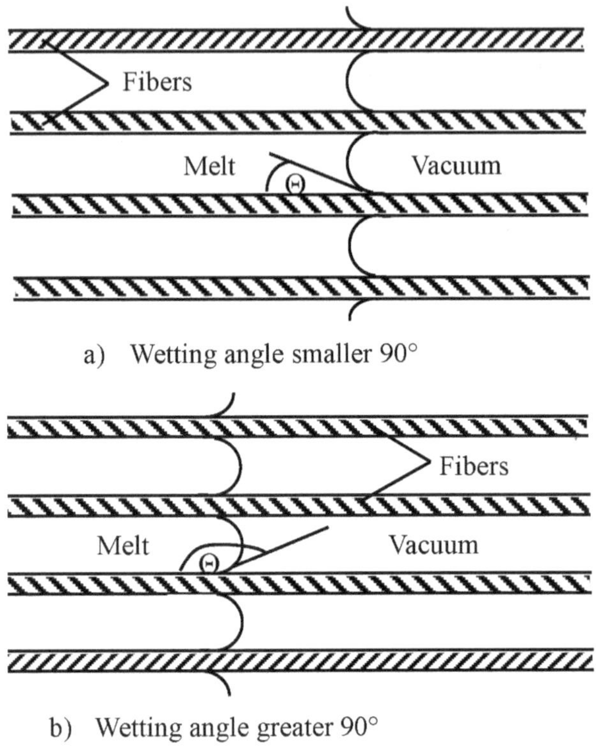

Fig. 7-51 Schematic presentation of an ideal melt infiltration of fiber performs.

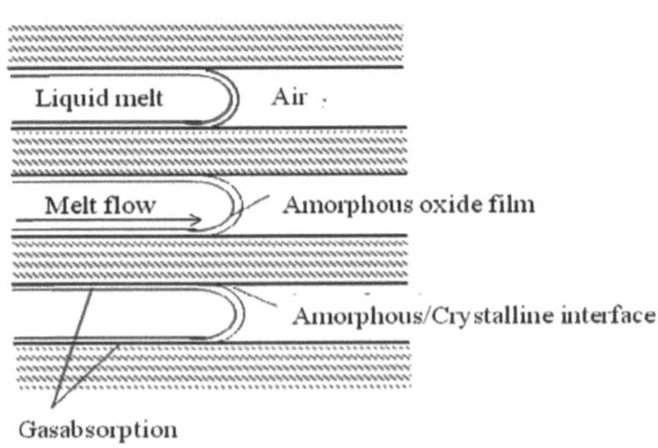

Fig. 7-52 Schematic presentation of the infiltration process of an aluminum oxide preform with molten aluminum.

atmosphere. Then, for example, an oxide film forms, as in the case of magnesium alloys, which affects the wetting behavior by formation of a new interface between the reinforcement and the melt, as clarified in Fig. 7-52 [265]. The statements made above apply only to considerations close to the equilibrium. The influence of wetting on the infiltration during technically relevant processes is thus less, if applied pressure on the melt, or rather the flow rate of the melt in the preform, determines the kinetics of the wetting, for example in the production of a wetted system by high pressure in the melt. However, the wetting nevertheless still has an influence on the adhesion of the components in the composite, which will later be described in more detail.

The actual infiltration process for the production of metallic composite materials consists of several indexing steps: the formation of a contact between the melt and the reinforcement at the surface of a fiber or particle preform, the infiltration with the melt flow through the preform and the solidification procedure. At the beginning of the infiltration a minimum pressure must usually be developed, so that infiltration can follow. Usually a pressure-free spontaneous infiltration is not the rule, and is only possible with thin preforms with reactive systems and with long process times. The resulting pressure as driving force for the infiltration [266] is

$$\Delta P = P_0 - P_a - \Delta P_\gamma \qquad (7\text{-}44)$$

where ΔP is the resulting pressure, the driving force for the reaction, P_0 the pressure in the melt on entering the preform (see Fig. 7-53) [266], P_a the pressure in the melt at the infiltration front (see Fig. 7-54) [266], $\Delta P\gamma$ the pressure decrease in the melt at the infiltration front due to surface influences (effect of wettability). When $P_0 = P_a$ a minimum infiltration pressure ΔP_μ can be defined:

$$\Delta p\mu = \Delta p_a = S_f(\gamma_{LS} - \gamma_{SA}) \qquad (7\text{-}45)$$

where S_f is the surface interface per unit area.

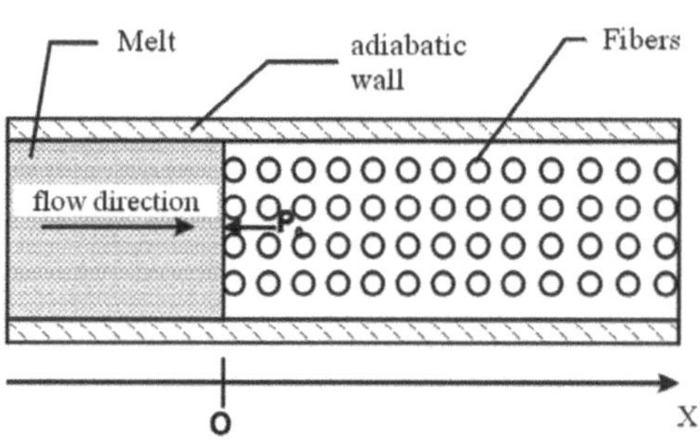

Fig. 7-53 Schematic presentation of an adiabatic, unidirectional infiltration, starting condition.

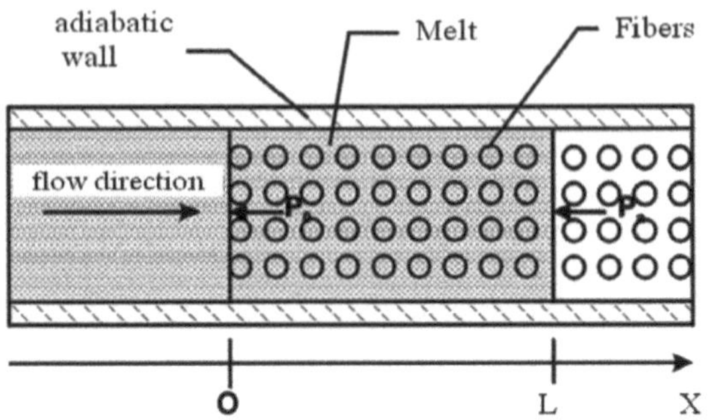

Fig. 7-54 Schematic presentation of an adiabatic, unidirectional infiltration.

Without external applied pressure the effect of induced infiltration by the capillary force can be presented as following [267]:

$$P_\gamma = 2\gamma_{\text{LA}} \cdot \cos\theta / r \tag{7-46}$$

where r is the radius of the capillary.

Using the hydrostatic pressure;

$$h_s = \frac{2Y_{\text{LA}} \cdot \cos\phi}{\rho \cdot g \cdot h} \tag{7-47}$$

where h_s is the rising height, g the gravitational constant, and ρ the density.

As a pressure-free infiltration a preform consisting of many capillaries can be imagined and thus the influence of wettability and structure parameter (surface and pores-or rather capillary diameter) can be seen when introducing the following structure parameter in Eq. (7-40) [267]:

For a spherical particle:

$$S_f = \frac{6V_f}{d_f(1 - V_f)} \tag{7-48}$$

For a long fiber bunch and short fiber perform:

$$S_f = \frac{4V_f}{d_f(1 - V_f)} \tag{7-49}$$

where V_f is the fiber or particle content and d_f the fiber or particle diameter.

Table 7-13 shows the change in the specific surface with increasing fiber portion in the Al_2O_3-preform of Saffil fibers [253, 268]. The specific surface also influences the permeability of a preform. This characteristic is important for the even supply of the preform with the melt and affects the necessary pressure for the infiltration. This is shown in Fig. 7-55 for Saffil preforms with water infiltration [268, 269]. It is noticeable that from fiber contents of 20 vol.% the permeability is significantly reduced. In this context the viscosity of a melt also has an important influence. By variation of the temperature composition

Table 7-13 Specific Surface of Al_2O_3 Preforms

Fiber volume content of Al_2O_3 performs [vol.%]	10	20	24	25
Specific surface: Sf = 106 fiber surfaces (m^2)/pore volume (m^3)	1.26	3.41	4.39	4.58

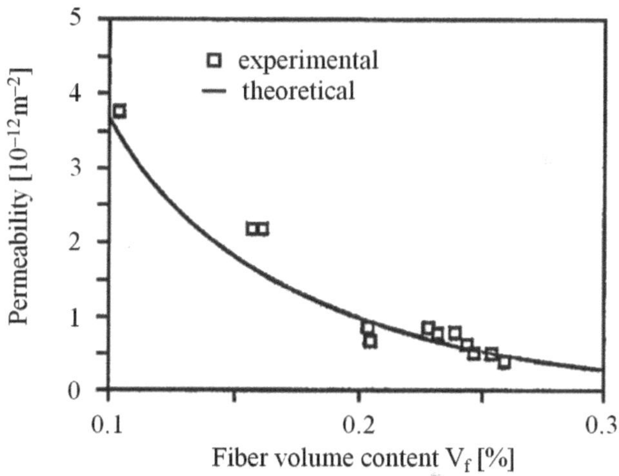

Fig. 7-55 Comparison of permeability of preforms for running water, after Mortensen and calculations of Sangini and Acrivos.

optimization of the infiltration process is controllable. Fig. 7-56 gives information on the change in the viscosity of magnesium and aluminum melts as a function of the temperature for unalloyed systems.

The previous considerations provide only an explanation of the processes and influencing variables on the wetting and infiltration, which are relevant in material systems, for example for stirring particles into melts or infiltrating. For both processes further procedures are of course relevant. Examples are the solidification procedures of the melts. They overlay with the above-mentioned procedures. For poor wetting of particles, for example in the production of particle-strengthened light alloys, a segregation or liquidization of the particles can take place. During the infiltration, solidification procedures can affect the permeability and prevent the complete treatment of the preform. Eq. (7-45) (see Figs. 7-55 and 7-56) [270] assumes a constant heat balance and no partial solidification. In reality there is heat dissipation over the tool and thus directed solidification occurs. Also the given free solidification heat has a substantial influence. The solidification worsens the permeability and influences the flow conditions in the preform. In reality a feed and a solidification

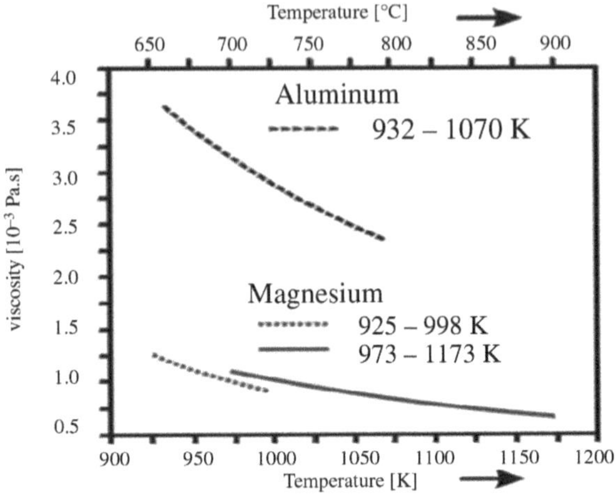

Fig. 7-56 Temperature dependence of the viscosity of magnesium and aluminum melts.

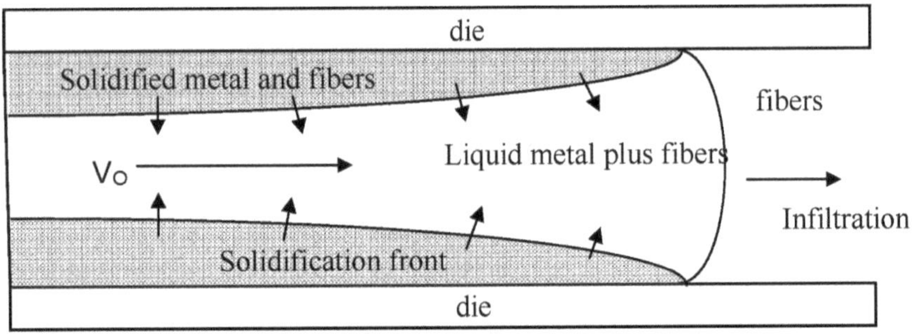

Fig. 7-57 Schematic presentation of solidification with external heat loss during a unidirectional infiltration.

procedure take place during the infiltration as a result of the directed heat dissipation, as represented in Fig. 7-57 [267]. The heat dissipation in the system preform/liquid or solidified melt is essentially determined by the thermal characteristics of the components (specific heat, thermal conductivity). Thus the reinforcements possess essentially higher specific heat values and smaller heat conductivities (exceptions are carbon fibers). In Table 7-14 these characteristic values are summarized for the C and Al_2O_3 fibers. In the case of the applied alloys it is to be noticed that magnesium melts possess a smaller heat capacity than aluminum and therefore are processed at higher temperatures or the preforms have to be at higher temperatures than with aluminum alloys.

Table 7-14 Comparison of Physical Data of C-fibers and Aluminum Oxide Fibers (Saffil)

Fiber	Specific heat ($J\ m^{-3}\ K^{-1}$)	Coefficient of thermal conductivity ($W\ m^{-1}\ K^{-1}$)
Carbon P100	1.988×10^6	520
Carbon T300	1.124×10^6	20.1
Saffil	2.31×10^6	0

7.4.6. Properties of Metal Matrix Composites

Saffil has been actively involved in the development of metal matrix composite applications since 1980, producing grades of Saffil fiber tailored to meet the stringent technical requirements of reinforced aluminum components such as pistons and engine blocks. In this subsection, the properties of the metal matrix composites are based on typical piston alloys which are manufactured by being incorporated with Saffil fiber in the metal matrix.

At 300 °C, the presence of the fiber has improved the ultimate tensile strength (Fig. 7-58(a)) by a factor of 4. It also extends the maximum use limit of the alloy to around 500 °C, a peak temperature which is sometimes approached in the combustion bowl rim of the diesel engine pistons. The modulus has been also increased to values in excess of those predicted by theory for a random planar array of fibers. This indicates a strong fiber matrix bonding and also the effect of modified matrix microstructure. At 300 °C, the composites containing 18 and 24 percent fiber have a higher modulus than the unreinforced alloy at 25 °C (Fig. 7-58(b)).

The presence of the fiber in aluminum alloys generally results in modified microstructure in a reduced grain size, as shown in Fig. 7-59.

Fig. 7-60 shows thermal conductivity of composites containing 5 and 15% v/v Saffil fiber in AA332 alloy. The thermal conductivity is lower than for the unreinforced metal and decreases as the fiber level rises. It is higher in the direction parallel to fiber orientation planes than in the normal direction, but in all cases, is much higher than for Ni-resist iron which is frequently used for piston inserts.

The wear resistance of the alloy is greatly improved by including only a small volume fraction of fiber. Fig. 7-61 shows data for 4.4% v/v addition of RG grade Saffil fiber. Wear of the mating component, in this case Cr steel, might be increased by contact with the reinforced alloy, but can be avoided by using the RG grade fiber with the appropriate alpha alumina level. Additional wear data giving results for a higher fiber loading and also a comparison with Ni-resist are shown in Fig. 7-62.

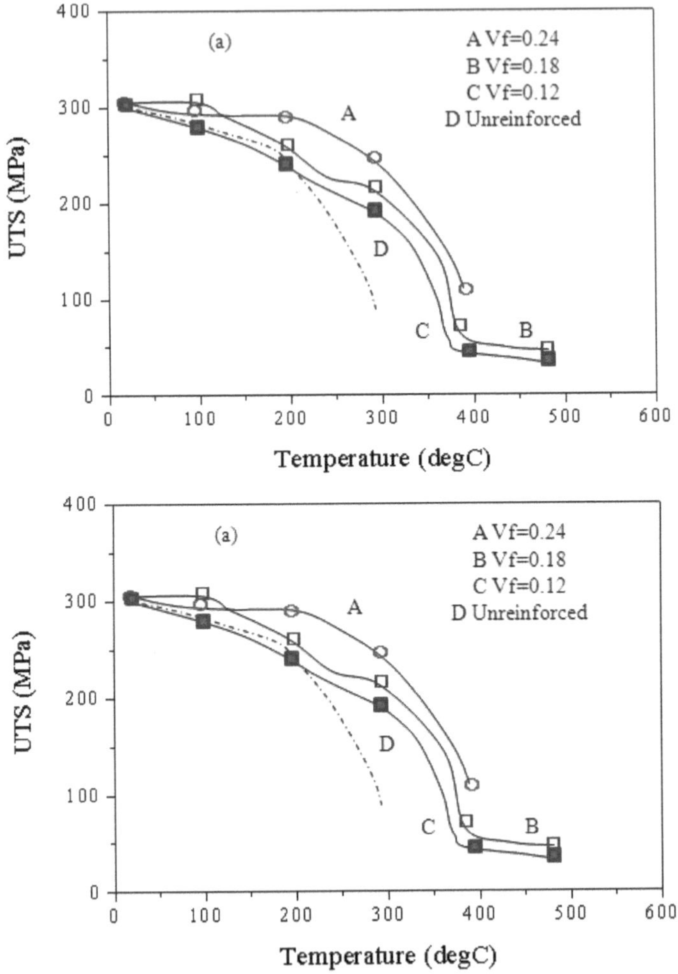

Fig. 7-58 Tensile strength (a) and modulus (b) of composites made from Saffil fiber RF grade in AC8B (T6) alloy.

7.4.7. Possible Applications of Metal Matrix Composites

7.4.7.1. Automobile products

Light alloy composite materials have, in automotive engineering, a high application potential in the engine area (oscillating construction units: valve train, piston rod, piston, and piston pin; covers: cylinder head, crankshaft main bearing; engine block: part-strengthened cylinder blocks), see Table 7-15. An example of the successful use of aluminum composite materials within this range is the partially short fiber reinforced aluminum alloy piston in Fig. 7-63, in which

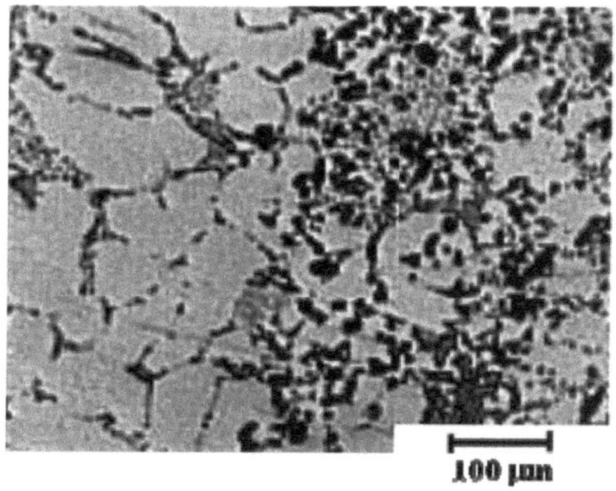

Fig. 7-59 Optical micrograph of interface between reinforced and unreinforced regions of AC8B (T6) alloy showing uniform distribution of Saffil fiber at 0.24 volume fraction.

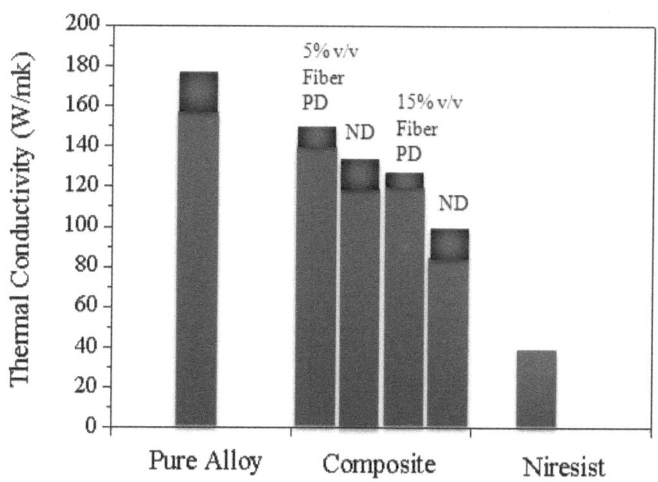

Fig. 7-60 Thermal conductivity of composites containing 5 and 15% v/v Saffil fiber in AA332 alloy measured in Parallel Direction (PD) and Normal Direction (ND) with respect to fiber orientation planes.

the recess range is strengthened by Al_2O_3 short fibers. Comparable construction unit characteristics are attainable only with the application of powder metallurgical aluminum alloys or when using heavy iron pistons. The reason for the application of composite materials is as already described the improved high temperature properties. Potential applications are in the area of undercarriages, e.g., transverse control arms and particle-strengthened brake disks, which can also be applied in the area of railmounted vehicles, e.g., for undergrounds and railway (ICE), see Fig. 7-64. In the following some potential construction units

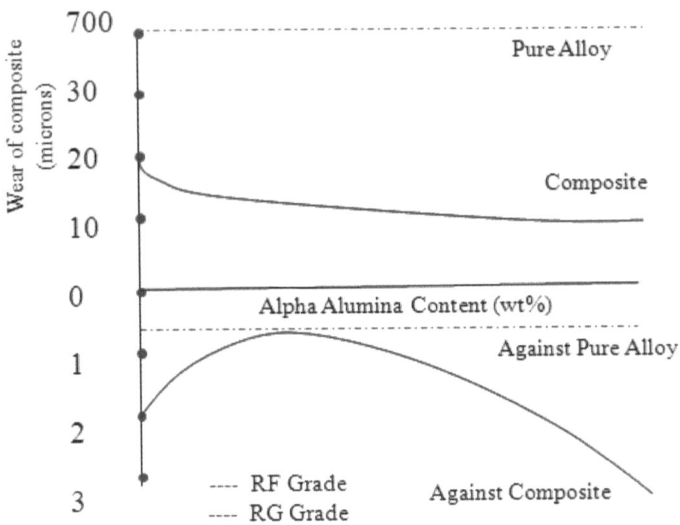

Fig. 7-61 Wear of composites (4.4% v/v Saffil fiber in AC8A (T7) alloy) and mating surface (Cr steel ring).

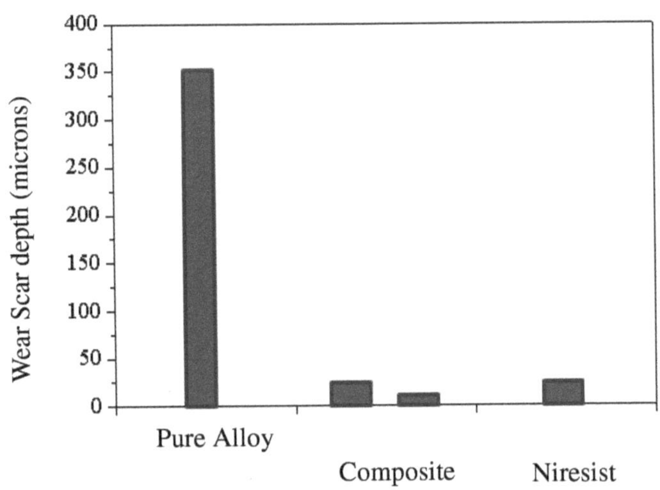

Fig. 7-62 Wear of composites (5 and 15% v/v Saffil fiber RF grade in AA332 (T5) alloy) against modular cast iron at 190 °C.

made out of aluminum matrix composite materials with the data concerning materials, processing, and development targets are presented.

In the aviation industry the high specific strength, the high Young's modulus, the small thermal expansion coefficient, the temperature resistance and the high conductivity of the strengthened light alloys are of interest

Table 7-15 Applications of Metal Matrix Composites

I.	**Drive shaft for people and light load motor vehicles (Fig. 7-65):**		
	Materials:	$AlMg_1SiCu$ + 20 vol.% Al_2O_3	
	Processing:	Extrusion form cast feed materials	
	Development aims:	– High dynamic stability, high Young's modulus (95 GPa) – Low density (2.95 gcm^{-3}) – High fatigue strength (120 MPa for n = 5 × 107, R = –1, RT) – Sufficient toughness (21.5 MPa m1/2) – Substitution of steels	
II.	**Vented passenger car brake disk (Fig. 7-66):**		
	Materials:	$G\text{-}AlSi_{12}Mg$ + 20 vol.% SiCP	
	Processing:	– Sand-or gravity die casting	
	Development aims:	– High wear resistance (better than conventional cast iron brake discs) – Low heat conductivity (factor 4 higher than cast iron) – Substitution of iron materials	
III.	**Longitudinal bracing beam (Stringer) for planes (Fig. 7-67):**		
	Materials:	$AlCu_4Mg_2Zr$ + 15 vol.% SiCP	
	Processing:	Extrusion and forging of casted feed materials	
	Development aims:	– High dynamic stability. high Young's modulus (100 GPa) – Low density (2.8 gcm^{-3}) – High strength (Rm = 540 MPa, $R_{P0.2}$ = 413 MPa, RT) – High fatigue strength (240 MPa for $n = 5 \times 10^7$, $R = -1$, RT) – Sufficient toughness (19.9 MPa $m^{1/2}$)	
IV.	**Disk brake calliper for passenger cars (Fig. 7-68):**		
	Materials:	Aluminum alloy with Nextel ceramic fiber 610	
	Weight reduction:	55% compared to cast iron.	

compared to the polymer materials, e.g., for reinforcements, axle tubes, rotors, housing covers, and structures for electronic devices. A compilation of the potential and realized applications of most different metal matrix composites can be seen in Table 7-16.

7.4.7.2. Space system

From the onset of the space era, both organic-matrix and metal matrix composites, with high specific stiffness and near-zero coefficient of thermal expansion, have been developed for space applications. Of the organic-matrix composites,

Fig. 7-63 Partial short fiber reinforced light metal diesel pistons.

Fig. 7-64 Cast brake disk particle of reinforced aluminum.

Fig. 7-65 Drive shaft particle of reinforced aluminum for passenger cars.

Fig. 7-66 Vented passenger car brake calipers of particle reinforced aluminum.

Fig. 7-67 Longitudinal connecting rod of particle reinforced aluminum.

Fig. 7-68 Disk brake calliper for passenger cars of conventional cast iron.

Table 7-16 Potential and Realistic Technical Applications of Metal Matrix Composites

Applications	Required properties	Material system	Processing technique
Automotive and goods vehicle			
Braking systems, piston rods, frames, piston, piston pins, valve spring cap, brake discs, disc brake calliper, brake pods, carding shaft	High specific strength and stiffness, temperature resistance, low thermal expansion coefficient, wear resistance thermal conductivity	Al-SiC, Al-Al_2O_3, Mg-SiC. Mg-Al_2O_3, discontinuous us reinforcements	Fusion infiltration, extrusion, forging, gravity die casting, die casting, squeeze-casting
Accumulator plate	High stiffness, Creep resistance	PbC, Pb-Al_2O_3	Fusion infiltration
Military and civil air travel			
Axle tubes, reinforcement, blade and gear box casing fan and compressor blades	High specific strength and stiffness, temperature resistance, impact strength, fatigue resistance	Al-B, Al-SiC, Al-C, Ti-SiC, Al- Al_2O_3, Mg- Al_2O_3, Mg-C, continuous and discontinuous reinforcements	Fusion infiltration, hot pressing, diffusion welding and soldering, extrusion, squeeze-casting
Turbine blades	High specific strength and stiffness, temperature resistance, impact strength, fatigue resistance	W super alloys, z, B, Ni, Al, Ni-Ni_3Nb	Fusion infiltration, aligned solidification near net shaped components

(continued)

Table 7-16 *(continued)*

Applications	Required properties	Material system	Processing technique
Aerospace industry			
Frames, reinforcement, aerials, joining elements	High specific strength and stiffness, temperature resistance, low thermal expansion coefficient, thermal conductivity	Al-SiC, Al-B, Mg-C, Al-C, Al- Al_2O_3. continuous and discontinuous reinforcements	Fusion infiltration, extrusion, diffusion welding and joining (special structures)
Energy techniques (electrical components and conducting materials) carbon brushes	High electrical and thermal conductivity, wear resistance	Cu-C	Fusion infiltration, powder metallurgy
Electrical contacts	High electrical conductivity, temperature and corrosion resistance, burn-up resistance	Cu-C, Ag- Al_2O_3	
Ag-C, Ag-SnO3, Ag-Ni	Fusion infiltration, Powder metallurgy, extrusion, pressing		
Super conductor	Superconducting, mechanical strength, ductility	Cu-Nb, Cu-Nb, Sn, Cu-YB_aCO	Extrusion, powder, metallurgy, coating technologies
Other applications			
Spot welding electrodes	Burn-up resistance	Cu-W	Powder metallurgy, Infiltration
Bearings	Load carting capacity, wear resistance	Pb-C, brass-Teflon	Powder metallurgy, Infiltration

graphite/epoxy has been used in space for truss elements, bus panels, antennas, wave guides, and parabolic reflectors in the past 30 years. The metal matrix composites possess high-temperature capability, high thermal conductivity, low coefficient of thermal expansion, and high specific stiffness and strength. Those potential benefits generated optimism for metal matrix composites for critical space system applications in the late 1980s.

Table 7-17 Material Properties of Unidirectional Metal Matrix Composites for Space Applications

Properties	P100/6061 Al	P100/AZ91C Mg	Boron/Al
Reinforcement volume content (%)	42.2	43	50
Density (g/cm^3)	2.5	1.97	2.7
Poisson ratio	0.295	0.3	0.23
Specific heat (J/kg K)	812	795	801
Longitudinal Young's modulus (GPa)	342.5	323.8	235
Ultimate tensile strength (MPa)	905	710	1,100
Thermal conductivity (W/m K)	320	189	—
CTE (10^{-6}/K*)	−0.49	0.54	5.8

Slope of a line joining extreme points of the thermal strain curve.

Table 7-17 lists the typical properties of a few continuous-fiber reinforced metal matrix composites. Generally, measured properties of as-fabricated metal matrix composites are consistent with the analytically predicted properties of each composite. The primary advantage of metal matrix composites over counterpart organic-matrix composites is the maximum operating temperature. For example, B/Al offers useful mechanical properties up to 510 °C, whereas an equivalent B/Epoxy composite is limited to about 190 °C. In addition, metal matrix composites such as Gr/Al, Gr/Mg, and Gr/Cu exhibit higher thermal conductivity because of the significant contribution from the metallic matrix.

7.4.8. Recycling

Of special economic and ecological interest for newly developed materials is the necessity for recirculation of arrears, cycle scrap, and other material from these composites into the material cycle. Since ceramic materials usually occur in the form of particles, short fibers, or continuous fibers for the reinforcement of metallic materials, a material separation of the components with the goal being the reuse of the matrix alloy and the reinforcement is almost impossible. However, with conventional melting treatments in remelting factories the matrix alloy can be recycled without problems.

In melting or powdering the metallurgically manufactured discontinuous short fiber or particle-strengthened light alloys reuse of the composite materials from cycle scrap or splinters can be possible under certain conditions. Particularly for particle strengthened aluminum cast alloys the use of cycle scrap is possible, however splinters on direct remelting cause difficulties due to contamination problems. Different recycling concepts for light alloy composite materials in relation to the alloy composition, the kind of reinforcement, and the production and processing prehistory are also overviewed.

7.5. CERAMIC MATRIX COMPOSITES

7.5.1. Introduction

New materials and processing routes provide opportunities for the production of advanced high performance structures for different applications. Ceramic matrix composites are one of these promising materials [271–275]. By combining different ceramic matrix materials with special suitable fibers, new properties can be created and tailored for interesting technical fields.

7.5.2. Reinforcements

Ceramic reinforcements are available in a variety of forms. Early ceramic composites used discontinuous reinforcements which, when combined with ceramic matrices, could be formed using conventional monolithic ceramic processes. Some of the more common discontinuous reinforcements include whiskers, platelets, and particulates having compositions of Si_3N_4, silicon carbide (SiC), aluminum nitride (AlN), titanium diboride, boron carbide, and boron nitride. Of these, silicon carbide has been the most widely used because of its stability with a broad range of ceramic oxide and non-oxide matrices. The discontinuous oxide ceramic reinforcements are less prevalent because of their incompatibility with many common ceramic matrices.

Modern day composites, which use the fiber in a continuous form to better optimize the structural properties, generally require more costly manufacturing processes. Continuous fiber is available in both monofilament and multifilament tow forms, with multifilament tow fiber being lower in cost on a per-pound basis and, in many cases, easier to process into complex shapes. Some of the more common continuous reinforcements include glass, mullite, alumina, carbon, and silicon carbide. Of these, the silicon carbide fibers have been the most widely used because of their high strength, stiffness, and thermal stability.

In ceramic matrix composites, only those fiber components are used that withstand the relatively high temperatures required for the production of ceramics, without significant damage. Other requirements to be met are long-term high temperature stability, creep resistance, and oxidation stability. The importance of each of these demands depends on the type of application.

Organic, polymeric fiber materials cannot be used in ceramic matrix composites because of their degradation at temperatures below 500 °C. Also conventional glass fibers, with melting or softening points below 700 °C, cannot be used for this purpose.

Possible candidates for the reinforcement of ceramic materials are polycrystalline or amorphous inorganic fibers or carbon fibers. The term ‘ceramic fiber’ summarizes all the non-metallic inorganic fibers (oxide or non-oxide), with the exception of fibers manufactured via solidification of glass melts.

The distinction between ceramic fibers and glass fibers has become more difficult during the last few years, because ceramics produced via new precursor

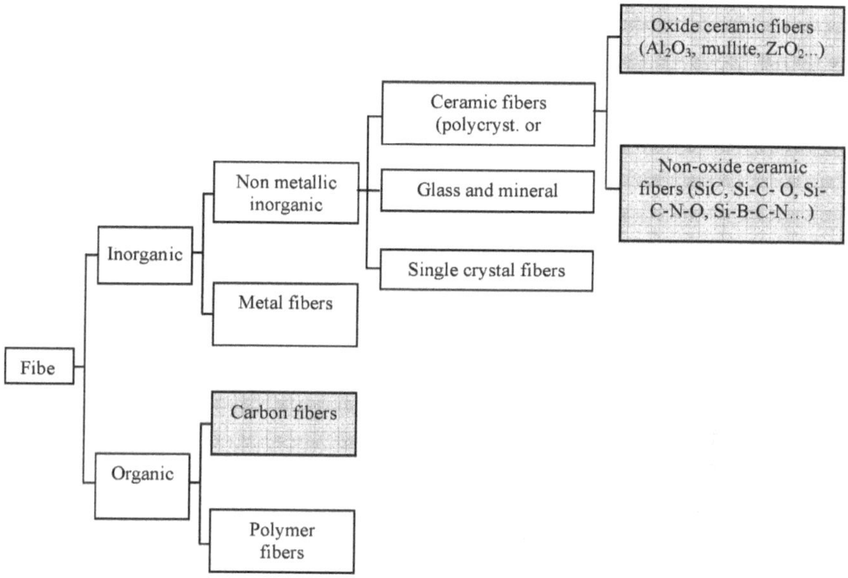

Fig. 7-69 Classification of different fiber types.

or sol–gel routes can also be amorphous (i.e., 'glassy') in structure and the production process can also contain a melt processing step. This means that ceramic fibers can be either polycrystalline, partially crystalline, or amorphous. However, the expression 'glass fibers' should only be applied to fibers that are produced via solidification of typical glass melts based on silicate systems. If these melts are produced by using minerals such as basalt, then the fibers should be called 'mineral fibers'.

Carbon fibers can also be used under certain conditions in ceramic matrix composites. Although these fibers degrade in an oxidizing atmosphere above 450 °C, they are stable under non-oxidizing conditions up to temperatures of 2800 °C. Carbon fibers have a very good cost-performance ratio, if an environmental protection of the composite allows the use of this fiber type. Therefore, an environmental barrier coating is an important field of research in ceramic matrix composites. Fig. 7-69 shows a general classification of fibers, containing ceramic fibers and carbon fibers.

7.5.3. Structure and Properties of Fibers

7.5.3.1. Fiber structure

As shown in Fig. 7-70, the structure of fibers can be considered from different viewpoints, depending on the magnification at which the structure is presented. The example shown is a polymeric fiber. At the molecular level, the displayed fiber is composed of poly (ethylene terephthalate) chain molecules. This

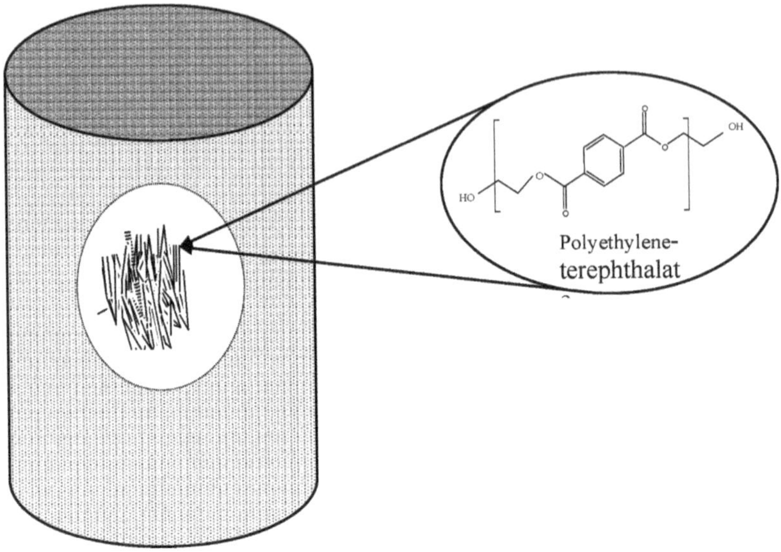

Fig. 7-70 Fiber structure of poly (ethylene terephthalate) fibers.

chemical structure determines the thermal and chemical stability and the theoretical strength of the fibers.

Most fibers also possess supramolecular structures, which are determined by molecular orientation and crystalline and/or amorphous regions (e.g., a two phase crystalline and amorphous structure is indicated in Fig. 7-71). These structures are formed during processing and can be significantly influenced

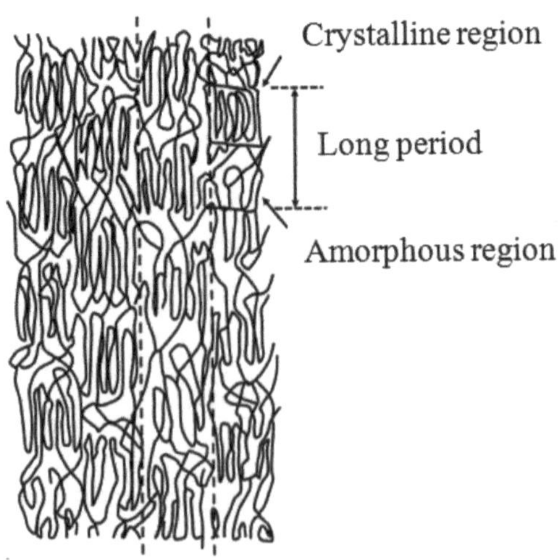

Fig. 7-71 Model of molecular arrangement in fibers.

by the processing conditions. This physical structure has important influence on the thermomechanical properties of fibers.

Finally, the fiber morphology influences the properties at the *macroscopic level*. Important criteria are cross-section, uniformity of the diameter along the fiber, porosity, and structural flaws, as well as surface properties such as roughness and surface energy (which is determined by the chemical structure of the surface). Adhesion to the matrix material and wetting behavior during infiltration steps are strongly influenced by these properties.

7.5.3.2. Structure formation

The structure formation in fibers depends not only on the fiber material itself but also on the processing conditions. Process parameters are controlled in such a way that specific supramolecular and macroscopic structures can be obtained. Important manufacturing processes are melt spinning, dry spinning, and wet spinning, together with modified spinning processes related to these:

- In the melt spinning process, fibers are formed via melts, which are forced through nozzles at high pressure and then solidified by cooling.
- In the dry spinning process, polymer solutions are used, which are also spun through nozzles. In this case, the fiber formation occurs by evaporation of the solvent from the spun solution.
- In the wet spinning process, polymer solutions are also used, but fiber forms by precipitation of the polymer in a liquid precipitation bath.

Important process parameters that influence the structure formation of the fibers are spinning speed, draw ratio, temperature, and other ambient conditions. If specialty fibers are desired, the spinning process is often followed by after treatments, which will determine the final structure of the fibers. Such after treatments include cross-linking of the melt spun ceramic precursors before pyrolysis, annealing, and sintering of oxide-based green fibers in order to achieve ceramic fibers and special surface treatments of carbon fibers. If ceramic or carbon fibers are to be used in ceramic matrix composites, the fibers have to be coated in many cases (e.g., with pyro carbon or boron nitride), which act as interfaces between fiber and matrix.

7.5.3.3. Structure parameters and fiber properties

The physical properties of fibers are determined essentially by three structure parameters: bond type, crystallinity, and molecular orientation [276]. The energy content of different types of chemical bonds is illustrated in Table 7-18 [276]. Covalent and ionic bonds, which can be oriented one-, two-, or three dimensionally within fibers, have the highest energy content and so determine the mechanical strength and modulus of the fibers. Other bond types are of minor importance in high-performance fibers.

Table 7-18 Chemical Bond Types and Bond Energies with Examples of Materials

Bond types	Energy ($kJ\ mol^{-1}$)	Examples	
Ionic	800–15000	NaCl	3D
		ZrO_2	3D
		Al_2O_3	3D
Covalent	200–600 (single bonds)	Aramide	1D
		Graphite (in plane)	2D
		Glass, SiC	3D
Metallic	100–800	Metals	
Hydrogen bonds	20–50	Aramide	
		Aliphat. Polyamide	
		Cellulose	
Dipole-dipole	ca. 2	Polyester	
Van der Waals	ca. 1	Polyolefines	
		Graphite (between planes)	

Different fiber types are shown in Table 7-19, which are distinguished by the structural parameters listed [276]. It is obvious that the fibers possess different architectures and the final properties of the fibers (e.g., modulus and strength) are determined by a combination of these structural parameters. If the bond type and the

Table 7-19 Fiber Structures and Properties

Fiber type	Polyester polyamide	Aramide fibers from LC phase	Carbon	Ceramic (crystalline)	Ceramic (amorphous) glass
Structure	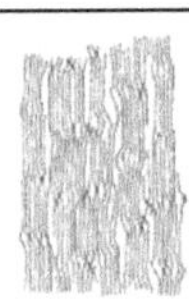	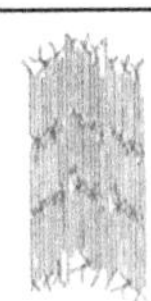	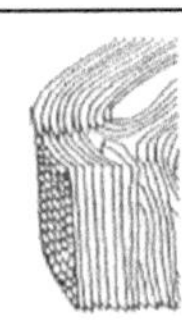	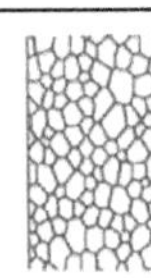	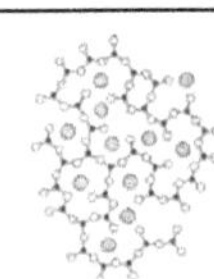
	1D linear 2 phases	1D linear 1 phase	2D layered	3D isotropic	3D isotropic
Bond type	1D covalent, hydrogen bonds (PA), dipole-dipole (PES), van der Waals	1D covalent, hydrogen bonds, van der Waals	2D covalent, van der Waals	3D Covalent/ ionic	3D covalent/ ionic
Crystallinity	Medium	Paracyrstalline	Paracrystalline	Polycrystalline	Amorphous
Orientation	Medium	Very high	High	None	None

spatial orientation of the bonds were the main criterion for good mechanical properties, then ceramic fibers and glass fibers with three dimensional, covalent, or ionic bonds would have far superior properties compared to other fiber types. But since these fibers are isotropic without molecular orientation, they possess lower strength than the carbon fibers. This is because carbon fibers have a structure with two-dimensional covalent bonds, showing pronounced crystallinity and high orientation. Based on this advantageous combination of structural parameters, carbon fibers currently display the highest values for strength and moduli. Depending on the processing conditions, carbon fibers can achieve either modulus as high as 600 GPa, or extremely high strength values (above 7000 GPa).

Polymeric aramide fibers also show high strength values due to their high molecular orientation in the fiber axis, although only one-dimensional covalent bonds and hydrogen bonds are present. Fig. 7-72–7-73 show an overview of mechanical properties of different fiber types. Averages are given because, as mentioned before, properties can differ depending on processing conditions. In reality, no perfect fiber structure can be obtained during processing, which means that the real properties of fibers are usually far below the theoretical property values calculated for a perfect structure. Therefore, one important goal of a fiber spinning and fiber formation process is to reduce structure imperfections to a minimum by optimized process control. Relatively high prices of high-performance fibers are usually caused by highly sophisticated and complex manufacturing processes rather than by more expensive materials used in the fiber production.

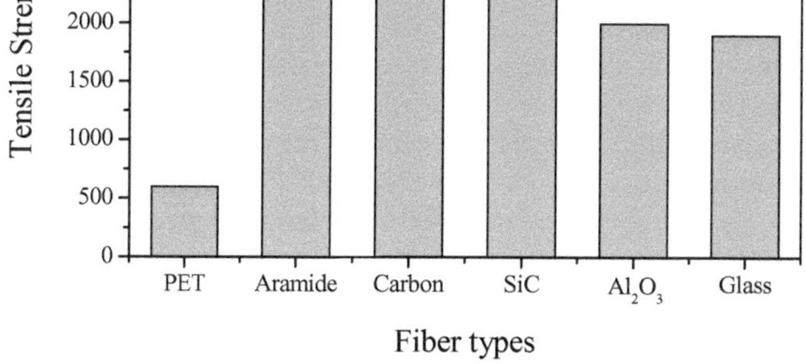

Fig. 7-72 Typical tensile strengths of different fiber types.

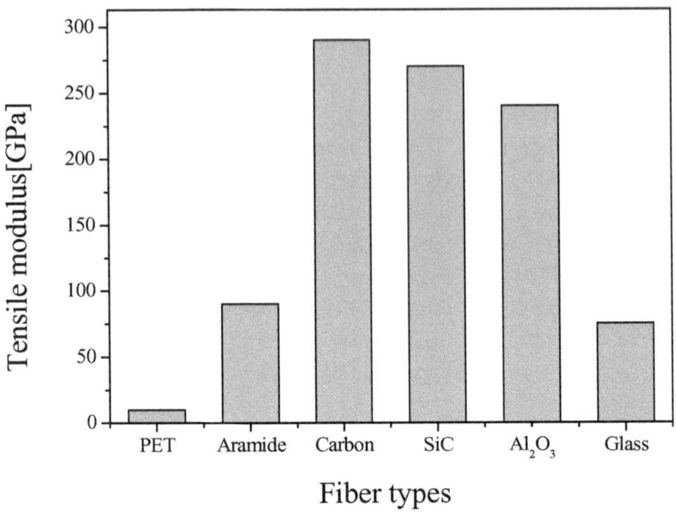

Fig. 7-73 Typical tensile modulus of different fiber types.

7.5.4. Inorganic Fibers

7.5.4.1. Production processes

The manufacturing processes for inorganic fibers can be divided into two categories: one process is called indirect, as the fibers or the non-ceramic precursor fibers are not obtained by a spinning process, but by using other fiber materials. The fiber is soaked with a pre ceramic precursor material, or the precursor material is deposited on the surface. The inorganic fiber is then formed by the pyrolysis of the organic template fibers.

In the second route, called direct process, inorganic precursors (salt solutions, sols, or precursor melts) are directly spun into so-called green fibers, in some cases by using organic polymer additives.

Another important distinction of production processes is based on the fiber length. There exist production processes for endless fibers as well as for short fibers, with fiber lengths from millimeters to some centimeters. Short fibers are usually produced by fibrillation of spinning dopes using fast rotating discs or by air blowing techniques. In addition, a process is conducted in which fabrics made of cellulosic short fibers are saturated with precursors and then pyrolysed and sintered in order to transform the material into a ceramic fiber fabric [277]. Fig. 7-74 shows a schematic depiction of the production of ceramic fibers and Fig. 7-75 shows a dry spinning line for the production of endless, oxide-based green fibers.

(1) Indirect fiber production

a) Chemical vapor deposition process: In the chemical vapor deposition process, ceramic fibers are formed via gas phase deposition of ceramic materials on carrier fibers. The carrier fiber usually forms

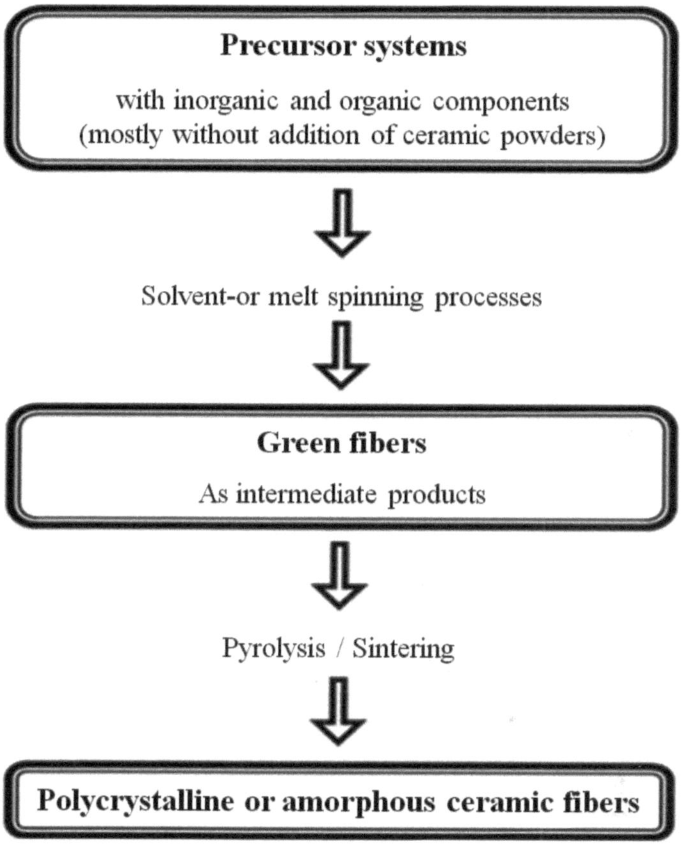

Fig. 7-74 Production routes of ceramic fibers.

the core of the ceramic fiber. Examples of core materials are carbon fibers and tungsten wires.

b) Relic process: In the so-called relic process, absorbent organic fiber materials (mostly cellulose-based) are saturated with salt solutions or sols. Afterwards the organic material is burned off and the salt or sol is converted into a ceramic material at high temperature to obtain ceramic fibers.

(2) Direct fiber production

Direct fiber production processes can be distinguished by the preceramic precursor components used in the spinning dopes:

a) Spinning dopes based on molecularly dispersed precursors: In these processes, soluble salts are used in the spinning dopes, which can be converted into ceramics by a calcination step. Although the salts are dissolved in the form of ions, that is, dispersed at the molecular scale, these processes are often incorrectly summarized under sol–gel processes. In addition to the salt, the spinning dope consists of organic

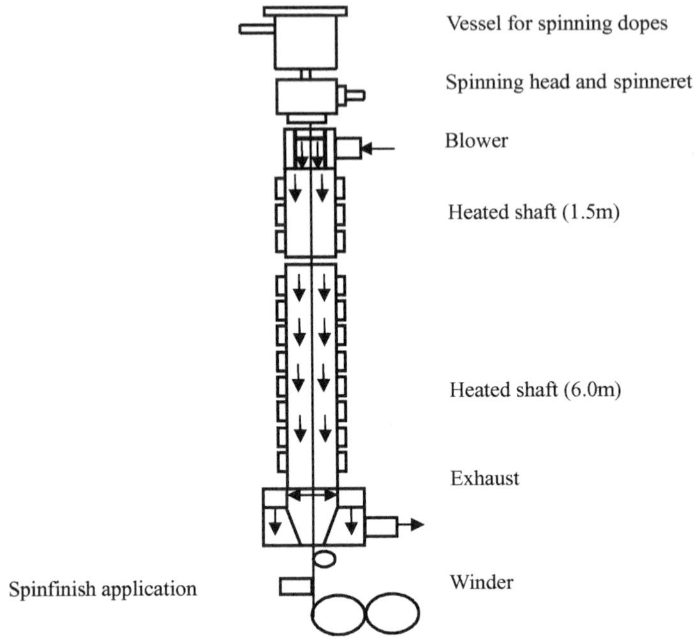

Fig. 7-75 Dry spinning facility at the Institute for textile chemicstry and chemical fibers (ITCF), Denkendorf, for spinning of oxide forming green fibers.

polymers such as poly (ethylenoxide), poly (vinylalcohol), or poly (vinylpyrrolidone), in order to achieve the rheological behavior needed for the spinning process. Solvents are usually water or water/alcohol mixtures. In some cases, nanoscaled ceramic particles are added to the dope to control the structure formation during ceramization.

b) Spinning dopes based on colloidally dispersed precursors: This process is similar to the one mentioned above, but here colloidal inorganic components are used as preceramic precursors (i.e., particle sols). In this case, the term sol–gel process is appropriate. The organic polymers, which have to be added in order to achieve spinnability, are the same as in the above process, with the same solvents also being used.

c) Spinning dopes containing coarse ceramic particles (ceramic powders): Sometimes, coarse ceramic particles are added to salt- or sol-based spinning dopes in order to increase the ceramic yield and to reduce the shrinkage during calcinations and sintering. In this case, the process is called the slurry process.

d) Spinning dopes based on inorganic polymers: In these so-called precursor or precursor polymer processes, the spinning dope consists of either a solution of an inorganic polymer, which can be spun via a dry spinning process or the precursor polymers are fusible and can be spun using a melt spinning process. Here no addition of organic polymer is necessary since the solution or the melt already possess the viscoelastic rheological behavior needed to be spun into fibers. The inorganic polymers usually carry organic functional groups such as methyl or propyl groups, so that during pyrolysis this organic material also has to be burned off. However, these systems have significantly higher ceramic yields compared to the processes described above. The precursors, which are melt spun have to be cross-linked (either chemically or by high energy radiation) before pyrolysis, otherwise the material would remelt and lose its fiber shape if heated above the melting point.

7.5.4.2. Properties of commercial products

(1) Comparison of oxide and non-oxide ceramic fibers

The oxide fibers, currently available commercially, are mostly based on Al_2O_3 or Al_2O_3/SiO_2 ceramics. They possess high values for tensile strength and modulus, and due to their oxidic nature are stable against oxidation at high temperatures. Unfortunately, even the best polycrystalline oxide fibers are prone to creep under load at 1,100 °C. Above this temperature the fibers cannot be used in ceramic matrix composites for long time applications. Also the oxide fibers tend to form larger grains when kept at high temperatures over long time periods. The larger grains tend to grow at the expense of smaller grains because of diffusion processes at grain boundaries, which can lead to brittle fibers.

The commercially available non-oxide ceramic fibers are based on silicon carbide and Si–C–(N)–O materials, which contain undesirable oxygen to a greater or lesser extent and can also contain Ti, Zr, or Al. Non-oxide fibers also exhibit high values for tensile strength and modulus (even higher than oxide fibers) and due to their structure, which is amorphous in many cases, they possess lower creep rates at high temperatures compared to the polycrystalline oxide fibers. Disadvantages of these fibers are their susceptibility to oxidation, which leads to fiber degradation in an oxidizing atmosphere over time. The lower the oxygen content of the fiber itself, the better its oxidation resistance.

The production process, which requires an inert atmosphere in most cases, is complex. Especially in the fabrication of low oxygen fibers (<1 wt.% oxygen), such as the Hi-Nicalon types or Sylramic, sophisticated technologies are needed, which lead to high fiber prices. The limitations of both oxide and non-oxide fiber types have to be known in order to choose the proper material for the intended application. Current research activities are focused on the development of oxide fibers with enhanced creep properties and reduced grain growth rates, as well as on the development of non-oxide fibers with improved oxidation stability and lower production costs [278–286].

(2) Oxide ceramic filament fibers

Table 7-20 gives an overview of the commercially available oxide ceramic filament fibers (i.e., endless fibers). The specifications given are taken from

Table 7-20 Overview About Commercial Oxide Ceramic Filament Fibers

Producer Fiber	Composition (Wt.%)	Diameter (μm)	Density (g/cm^3)	Tensile strength/ modulus (MPa/GPa)	Production technique/ structure	Approx. price
3M Nextel 720	Al_2O_3: 85 SiO_2: 15	10–12	3.4	2,100/260	Sol-Gel/59 vol.% α-Al_2O_3 +41 vol.% Mullite	€790/kg (1,500 den) €600/kg (3,000 den)
3M Nextel 610	Al_2O_3: > 99	10-12	3.9	3,100/380	Sol–Gel/ α-Al_2O_3	€790/kg (1,500 den) €600/kg (3,000 den) €440/kg (10,000 den)
3M Nextel 550	Al_2O_3: 73 SiO_2: 27	10-12	3.03	2,000/193	Sol-Gel/γ-Al_2O_3+$SiO_{2amorph}$	€590/kg (2,000 kg)
3M Nextel 440	Al_2O_3: 70 SiO_2: 28 B_2O_3: 2	10-12	3.05	2,000/190	Sol-Gel/γ-Al_2O_3 + Mullite + $SiO_{2amorph}$	€500/kg (2,000 den)
3M Nextel 312	Al_2O_3: 62.5 SiO_2: 24.5 B_2O_3: 13	10-12	2.7	1,700/150	Sol-Gel/Mullite + amorph or 100% amorph	€260/kg (1,800 den)
Sumitomo Altex	Al_2O_3: 85 SiO_2: 15	10/15	3.3	1,800/210	Polyaluminoxane/γ Al_2O_3	€640–720/kg
Nitivy Nitivy ALF	Al_2O_3: 72 SiO_2: 28	7	2.9	2,000/170	Sol-Gel/γ-Al_2O_3	€390/kg (twised yarn, twists: 10–15)
Mitsui Almax-B	Al_2O_3: 60-80 SiO_2: 40–20	7–10	2.9	Not available	Unknown/δ-Al_2O_3	Price not available

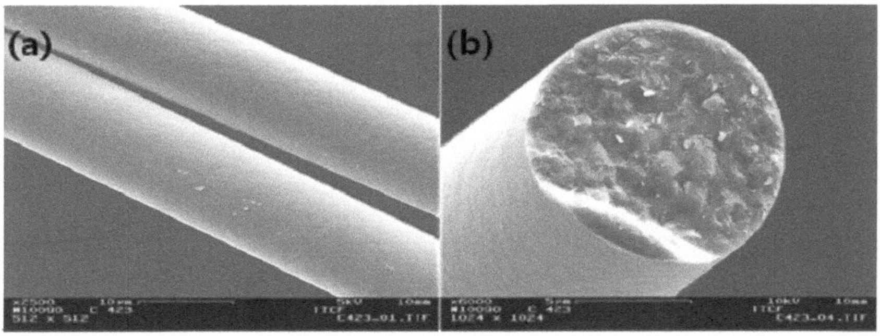

Fig. 7-76 Mullite ceramic fibers (produced at the Institute for textile chemistry and chemical fibers, Denkendorf) as an example for oxide ceramic fibers.

product information of the fiber producers [287–290]. The reported prices are for quantities of over 100 kg of fiber and are usually higher for lower quantities. The denier given for the 3 M fibers is a unit of measure for the linear mass density of fibers. It is defined as the mass in grams per 9,000 m. In the International System of Units, the tex is used instead, which is the mass in grams per 1,000 m. Since the fibers have different densities, the denier does not indicate directly the number of filaments. For Nextel 720, 3000 den corresponds to about 900 filaments; for Nextel 610, 3000 den corresponds to about 800 filaments; for Nextel 550 and 440, 2000 den corresponds to about 700 filaments; and for Nextel 312, 1800 den also corresponds to about 700 filaments (if fiber diameters of 12 μm are assumed). Nevertheless, no responsibility can be taken for the correctness of this information.

Fig. 7-76 shows the structure of a 10 μm mullite-based oxide fiber developed at ITCF, Denkendorf (Germany), as an example of a new oxidic ceramic fiber [291, 292].

(3) Non-oxide ceramic filament fibers

Table 7-21 gives an overview of the commercially available (some only in smaller quantities) non-oxide filament fibers. Prices for the Nicalon fibers were provided by the North American distributor COI Ceramics Inc. Nevertheless, no responsibility can be taken for the correctness of this information. Fig. 7-77 shows an example of an amorphous non-oxide Si–C–N fiber after temperature treatment at 1500 °C for 12 h in air [275]. This fiber was produced at the Institute for textile chemistry and chemical fibers, Denkendorf with a precursor material from the Department of Ceramic Materials Engineering of the University of Bayreuth. Beside the oxidation layer of silicon dioxide at the fiber surface, no further degradation can be observed.

7.5.5. Properties and Applications of Ceramic Matrix Composites

Ceramic matrix composites combine reinforcing ceramic phases with a ceramic matrix to create materials with new and superior properties. In ceramic matrix

Table 7-21 Overview About Commercial Nonoxide Ceramic Filament Fibers

Producer Fiber	Composition (Wt.%)	Diameter (μm)	Density (g/cm^3)	Tensile strength/ modulus (MPa/ GPa)	Production technique/ structure	Approx. price
Nippon Carbon Hi-Nicalon "S"	Si: 68.9 C: 30.9 O: 0.2	12	3.10	2,600/420	Polycarbosilane/β-SiC	€7,000/kg >10 kg
Nippon Carbon Hi-Nicalon	Si: 63.7 C: 35.8 O: 0.5	14	2.74	2800/270	Polycarbosilane/β -SiC + C	€3,250/kg >10 kg
Nippon Carbon Hi-Nicalon NL-200/201	Si: 56.5 C: 31.2 O: 12.3	14	2.55	3000/220	Polycarbosilane/β-SiC+ SiO_2+C	€1,000/kg >10 kg
UBE Industries Tyranno Fiber SA 3	Si: 67.8 C: 31.3 O: 0.3 Al: <2	10/75	3.10	2800/380	Polycarbosilane/β-SiC	€6500/kg >10 kg
UBE Industries Tyranno Fiber ZMI	Si: 56.1 C: 34.2 O: 8.7 Zr: 1.0	11	2.48	3400/200	Polycarbosilane/β-SiC+	€1400/kg >10 kg
UBE Industries Tyranno Fiber LoxM	Si: 55.4 C: 32.4 O: 10.2 Ti: 2.0	11	2.48	3300/187	Polycarbosilane/β-SiC+	€1200/kg > 10 kg
UBE Industries Tyranno Fiber S	Si: 50.4 C: 29.7 O: 17.9 Ti: 2.0	8.5/11	2.35	3300/170	Polycarbosilane/β-SiC+	€1,000/kg > 10 kg

(continued)

Table 7-21 *(continued)*

Producer Fiber	Composition (Wt.%)	Diameter (μm)	Density (g/cm^3)	Tensile strength/ modulus (MPa/ GPa)	Production technique/ structure	Approx. price
COI Ceramics Sytramic-i BN	SiC/BN	10	3.00	3000/4000	Precursor-polymer/SiC/BN and other phases	€10500/kg > 10 kg
COI Ceramics Sytramic	SiC: 96.0 Tib2: 3.0 B4C: 1.0 O: 0.3	10	2.95	2700/310	Precursor-polymer/SiC and other phases	€8500/kg > 10 kg
Specialty Materials SCS-Ultra	SiC on C	140 (with carbon fiber core)	3.0	5865/415	CVD on C-filament/β-SiC on C	€16400/kg
Specialty Materials SCS-9A	SiC on C	78 (with carbon fiber core)	2.8	3450/307	CVD on C-filament/β-SiC on C	€19600/kg
Specialty Materials SCS-6	SiC on C	140 (with carbon fiber core)	3.0	3450/380	CVD on C-filament/β-SiC on C	€4850/kg
Tisics Sigma	SiC on W	100/140 (with carbon fiber core)	3.4	4000/4000	CVD on C-filament/SiC on W	Price not available

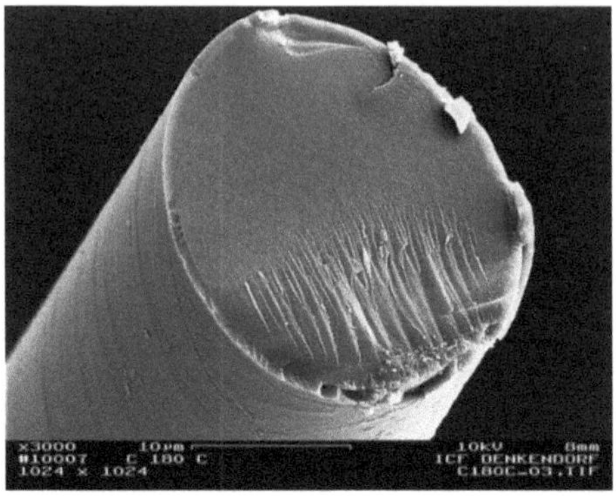

Fig. 7-77 Si–C–N fibers (the Institute for textile chemistry and chemical fibers, Denkendorf), produced from a polycarbosilazane precursor (University of Bayreuth). The depicted fiber was already treated for 12 h at 1500 °C in air.

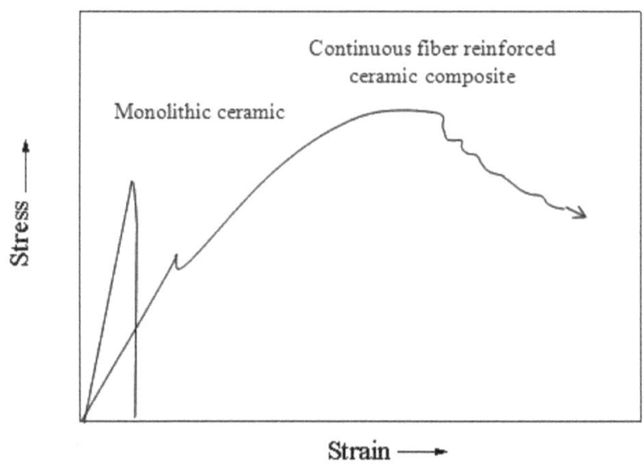

Fig. 7-78 Comparison of failure modes for monolithic ceramics and continuous fiber-reinforced ceramic composites.

composites, the primary goal of the ceramic reinforcement is to provide toughness to an otherwise brittle ceramic matrix (Fig. 7-78). Fillers can also be added to the ceramic matrix during processing to enhance characteristics such as electrical conductivity, thermal conductivity, thermal expansion, and hardness. The desirable characteristics of ceramic matrix composites include high-temperature stability, high thermal shock resistance, high hardness, high corrosion resistance, light weight, non-magnetic and non-conductive properties, and versatility in providing unique engineering solutions. The combination of these characteristics makes ceramic matrix composites attractive alternatives to traditional processing industrial materials such as high alloy steels and refractory metals. For the processing industry, related benefits of

Table 7-22 Representative Ceramic Composite Suppliers

Continuous fiber ceramic composites	Discontinuous reinforced ceramic composites
3M Company	Advanced Refractory Technologies, Inc.
Synterials, Inc.	ALANX Wear Solutions, Inc.
B. F. Goodrich	Cercom
Composite Optics, Inc.	DuPont Lanxide Composites, Inc.
Dow Corning Corporation	Greenleaf Corporation
DuPont Lanxide Composites, Inc.	
McDermott Technology, Inc.	
Techniweave, Inc.	
Textron Systems Corporation	

using ceramic composites include increased energy efficiency, increased productivity, and regulatory compliance. Key barriers to the broad application of ceramic matrix composites include the lack of specifications, databases, attachment concepts, inservice repair methodology, high cost, and scale-up. A representation of suppliers of discontinuous and continuous fiber-reinforced ceramic composites is shown in Table 7-22.

(1) Discontinuous reinforced ceramic composites

The discontinuous reinforced ceramic composites are produced using processes originally developed for monolithic ceramics. Processing methods commonly used including slip casting or injection molding followed by sintering to full density in a high-temperature-capable furnace. The shaping and sintering processes can also be combined using unidirectional hot pressing or hot isotatic pressing. Net or near-net shape processing can be achieved with final machining often limited to satisfying high-tolerance dimensions or surface finishes. An example of the material microstructure showing the matrix and uniformly distributed discontinuous reinforcement is shown in Fig. 7-79.

Discontinuous reinforced ceramics are being used in the mining and abrasives industry, where the use of these ceramics has reduced downtime usually needed for replacing worn parts. The largest application has been chute liners. One product, consisting of 75% SiC + 15%Al_2O_3 + 10% Al, provides excellent resistance to dry erosion, slurry erosion, sliding abrasion, thermal shock, and chemical attack. Similar high-wear applications where discontinuous reinforced ceramics have shown improvements include liners for cyclones, pipes, and pump housings (Fig. 7-80).

Discontinued reinforced ceramics are also being considered for use in a high-pressure (1035-kPa) heat exchanger designed for use in steam reforming of methane. Advantages of ceramic composites for this application

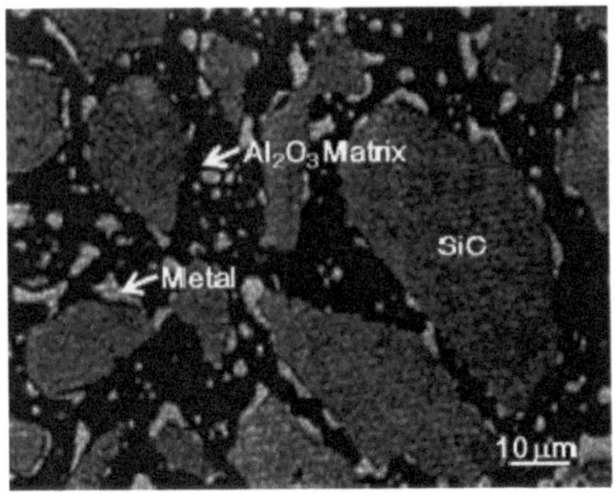

Fig. 7-79 Typical microstructure for a discontinuous reinforced ceramic composite (Silicon carbide/Al2O3-Aluminum).

Fig. 7-80 Examples of discontinuous reinforced industrial wear parts.

include higher temperature capability, reduced fouling, increased energy efficiency, reduced emissions and reduced reformer weight, and size. Of the materials tests, a product composed of $SiC + Al_2O_3 + Al$ proved the least vulnerable to steam attack at elevated temperatures.

(2) Continuous reinforced ceramic composites

Ceramic composites containing continuous fiber reinforcements must be processed by the method of the reinforcement. Typically, processing involves the formation of a fiber preform that contains an interface coating applied by chemical vapor deposition or a particle-filled slurry process followed by impregnation with a second particle-filled slurry mix, pre-ceramic polymer, precursor gases, molten metal, or other raw material that converts to a ceramic matrix when heated. The interface is a very thin layer (<5 mm total thickness), can be multiple layers to achieve the desired result, and is applied to the individual ceramic filaments. The interface serves as protection for the fibers during matrix processing and as a source of debonding during crack propagation in the brittle ceramic matrix. A typical microstructure of ceramic fiber-reinforced ceramic matrix composites shown in Fig. 7-81 illustrates the matrix, fiber, and interface phases.

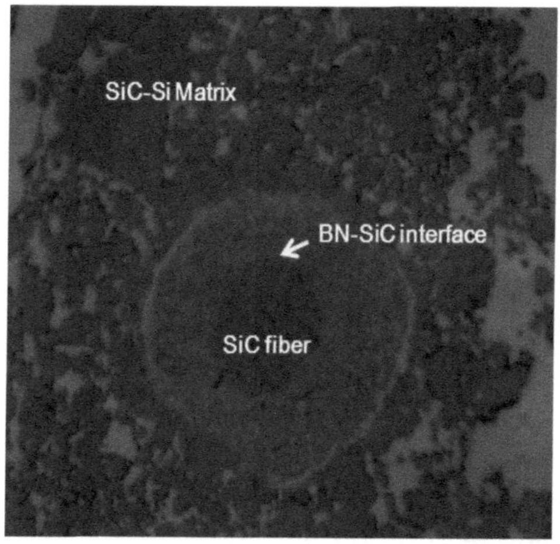

Fig. 7-81 Typical microstructure of silicon carbide/silicon carbide continuous reinforced ceramic composites.

a) (2-1) Silicon carbide matrix continuous reinforced ceramic composites have been successfully demonstrated in a number of applications where a combination of high thermal conductivity, low thermal expansion, light weight, and good corrosion and wear resistance is desired. SiC matrix continuous reinforced ceramic composites—the most mature of the available continuous reinforced ceramic composites—can be fabricated using a variety of processes, fibers, and interface coatings. Fibers widely used for industrial applications where long life is desired include silicon carbide or mullite. Processes available to fabricate the SiC matrix continuous reinforced ceramic composites and the matrix composition formed include chemical vapor infiltration (silicon carbide), polymer infiltration (SiCN, silicon carbide), nitride bonding (Si-SiC-Si_3N_4), and melt infiltration (Si- silicon carbide). The interface coating can either be carbon or boron nitride with a protective overcoat of silicon carbide or Si_3N_4.

The material properties of silicon carbide matrix continuous reinforced ceramic composites are ultimately determined by the combination of fiber, process and interface coating selected with typical values shown in Table 7-23.

Applications for which silicon carbide matrix continuous reinforced ceramic composites are being considered are summarized in Table 7-24 and are shown in Fig. 7-82. The applications are in various stages of development, with gas-fired radiant burner screens and gas turbine engine combustors well along in the product development

Table 7-23 Typical Material Properties for Silicon Carbide Matrix Continuous Reinforced Ceramic Composites

Density	2.1 g/cm^3
Tensile strength	262 MPa
Tensile modulus	96 GPa
Strain to failure	0.4%
Compressive strength	450 MPa
Shear strength	34 MPa
Thermal expansion	2.7×10^{-6}/°C
Thermal conductivity	1.32 W/m°C

Table 7-24 Applications Being Considered for Silicon Sarbide Matrix Continuous Reinforced Ceramic Composites

Application	Conditions imposed on SiC matrix continuous reinforced ceramic composites	Benefits demonstrated or perceived by the customer
Hot-gas recirculating fan	High mechanical stresses, high temperature	Increased life and temperature capability
Heat exchanger	High temperature, high thermal stresses, highly erosive and hot corrosion	Higher temperature capability and life
Gas-fired radiant burner screen	High temperature, high thermal stresses	Increased life and thermal output
Gas-fired radiant burner tubes	High temperature, high thermal stress	Increased life and temperature capability
Gas turbine engine combustion liners and tip shrouds	High temperature, hot corrosion	Reduced emissions and increased efficiency resulting from reduced cooling air, higher temperature capability
Hot-liquid filtration	Harsh chemical environment, high temperature	Increased life, higher temperature capability
Canned motor pump containment shell	Harsh chemical environment	Increased electrical efficiency through reduced eddy current drag
Gas-fired melting immersion burner tube	Harsh molten metal environment, high thermal shock	Reduced contamination, increased efficiency, improved heating uniformity
Furnace pipe hanger	High temperature, high mechanical stresses	Longer life, higher temperature capability, reduced supporting structure

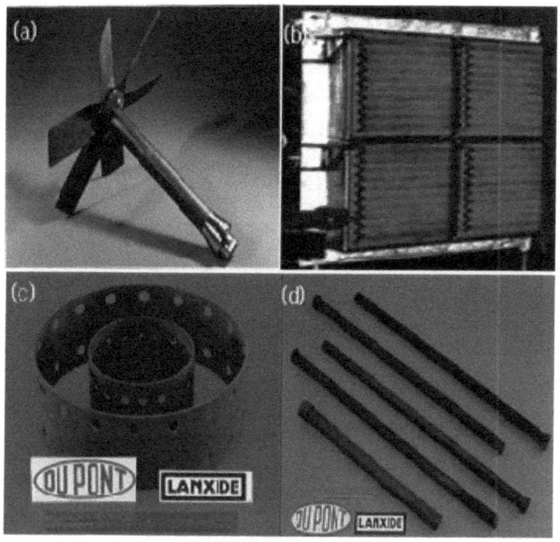

Fig. 7-82 Applications of silicon carbide matrix continuous reinforced ceramic composites: (a) hot gas recirculating fan, (b) radiant burner Screen, (c) gas turbine engine combustor liners, and (d) filters manufactured using chemical vapor infiltration.

cycle and the nearest to commercialization. SiC matrix continuous reinforced ceramic composites have operated in gas turbine engine combustor liners for 1,000 h and radiant burner screens for 10,000 thermal cycles without failure.

b) (2-2) Oxide matrix continuous reinforced ceramic composites

Alumina and mullite matrix continuous reinforced ceramic composites have been successfully demonstrated in applications where a high tolerance to salt corrosion, molten glass or oxidation is required in combination with the high toughness, light weight, and high-thermal-shock resistance characteristic of continuous reinforced ceramic composites. The fibers generally used for industrial applications requiring long life include SiC, mullite, or alumina and mullite matrix continuous reinforced ceramic composites. Processes available to fabricate alumina and mullite matrix continuous reinforced ceramic composites, and the matrix compositions formed, include sol–gel (Al_2O_3, SiO_2), directed metal oxidation (Al_2O_3 + Al), and chemical bonding ($AlPO_4$). The interface coating varies widely and in some cases is not required, resulting in improved thermal stability but reduced interlaminar properties.

The material properties of alumina and mullite continuous reinforced ceramic composites are ultimately determined by the combination of fiber, process, and interface coating sizing selected, with typical values shown in Table 7-25.

Table 7-25 Typical Material Properties for an Alumina Matrix Continuous Reinforced Ceramic Composites

Density	2.82 g/cm^3
Tensile Strength	365 MPa
Tensile Modulus	124 GPa
Strain to Failure	0.35%
Compressive Strength	270 MPa
Thermal Expansion	8.5×10^{-6}/°C
Shear Strength	15.2 MPa

Table 7-26 Applications Being Considered for Alumina and Mullite Matrix Continuous Reinforced Ceramic Composites

Applications	Conditions imposed on SiC matrix continuous reinforced ceramic composites	Benefits demonstrated or perceived by the customer
Thermophotovoltaic burner/emitter	High thermal shock, high temperature, surface combustion	Increased efficiency, multifuel capability, maximum on/off response, increased life
Heat exchanger	High temperature, high thermal stresses, highly erosive and hot corrosion	Higher temperature capability and life
Hot-gas filtration	Harsh chemical environment, high temperature	Increased life, higher temperature capability
Gas turbine engine combustion liners and shrouds	High temperature, hot corrosion	Reduced emissions and increased efficiency resulting from reduced cooling air, higher temperature capability
Burner stabilizer	High corrosion	Longer life

Applications for which alumina or mullite matrix continuous reinforced ceramic composites are being considered are summarized in Table 7-26 and are shown in Fig. 7-83. The applications are in various stages of development, with hot gas filtration being well along in the product development cycle and the nearest to commercialization.

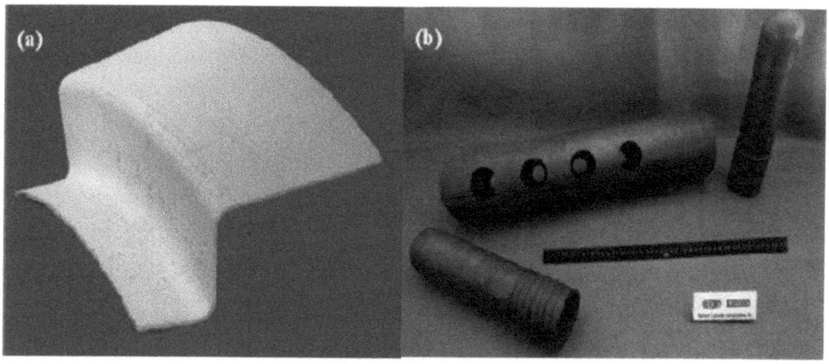

Fig. 7-83 Alumina matrix continuous reinforced ceramic composites: (a) burner stabilizer ring and (b) heat exchanger headers.

7.6. SUMMARY

1. Polymer matrix composites
   Polymer matrix composites are plastics (resins) within which there are embedded fibers. The plastic is known as the matrix, and the fibers orientated within it are known as the reinforcement. The reinforcement tends to be stiffer and stronger than the matrix providing stiffness and strength. The primary reason why polymer matrix composites are chosen for the manufacture of components is due to the weight saving relative to their stiffness and strength. Polymer matrix composites can be used for metal replacement, but there are key property differences which mean that direct substitution should be subject to material selection via materials expert. The factors which determine the direct costs of producing components and structures from polymer matrix composites are: (1) Material costs, (2) Tooling costs, (3) Processing costs, and (4) Finishing costs. Furthermore other additional costs associated with quality control, health and safety and environmental factors need to be considered.
2. Carbon matrix composites
   Carbon/carbon composites are made of fibers in various directions and carbonaceous polymers and hydrocarbons as matrix precursors. Their density and properties depend on the type and the volume fraction of reinforcement matrix precursor used, and end heat treatment temperature. Composites made with thermosetting resins as matrix precursors possess low densities (1.55–1.75 g/cm^3) and well-distributed microporosity whereas those made with pitch as the matrix precursor, after densification exhibit densities of 1.8–2.0 g/cm^3 with some mesopores, and those made by the chemical vapor deposition technique with hydrocarbon gases, possess intermediate densities and matrices with close porosities. At present, many advanced oxidation protection systems have been developed and the main

developed approach is the use of an external diffusion barrier. The protective coatings become more and more elaborated with the use of graded mixed layers. The main components used are based on silicon and boron. With siliceous coatings, oxidation protection is efficient up to 1800 °C in a dry atmosphere (1 bar) for medium-term applications. Also, the carbon/carbon composites made with oxidation inhibitor led to a significantly improved oxidation resistance owing to both the reduction of the crack in composites and the formation of a mobile diffusion barrier for oxygen when compared to those made without oxidation inhibitor. It is possible that protection by an inert gas could be developed with this objective in mind.

3. Metal matrix composites
Metal matrix composites exhibited the particular benefits, such as lower density, increased specific strength and stiffness, increased high-temperature performance limits, and improved wear-abrasion resistance, which are dependent on the properties of the matrix alloy and of the reinforcing phase. The selection of the matrix is empirically based, using readily available alloys; and the major consideration is the nature of the reinforcing phase. A large variety of metal matrix composite materials exist. The reinforcing phase can be fibrous, platelike, or equiaxed (having equal dimensions in all directions); and its size can also vary widely, from about 0.1 to more than 100 μm. The matrices based on most engineering metals have been explored, including aluminum, magnesium, zinc, copper, titanium, nickel, cobalt, iron, and various aluminides. This wide variety of systems has led to an equally wide spectrum of properties for these materials and of processing methods used for their fabrication. Reinforcements used in metal matrix composites fall in five categories: continuous fibers, short fibers, whiskers, equiaxed particles, and interconnected networks.

4. Ceramic matrix composites
Advanced monolithic ceramics are being throughout the industrial processing industry and have demonstrated superior performance to conventional materials at an affordable price. As user confidence continues to grow and as energy savings, increased productivity, and reduced maintenance are confirmed, the need has emerged for advanced ceramics with improved toughness (also known as ceramic composites). The future of ceramic matrix composites is directly depending on progress that would be achieved in the availability of higher performance constituents (fibers, interphases, and tailored matrices) as well as in processing cost reduction.

REFERENCES

[1] R.B. Aronson, Manufacturing Engineering 122 (1999) 52.
[2] E. Fitzer, A. Gkogkidis, M. Heine, High Temperature—High Pressures 16 (1984) 363.
[3] D.D. Edie, R.J. Diefendorf, Carbon Fiber Manufacturing. Carbon-Carbon Materials and Composites Noyes, Park Ridge (1993), p. 19.

[4] N.C. Gallego, D.D. Edie, Carbon 38 (2000) 1003.
[5] L.S. Singer, US Patent 4,005,183 (1977).
[6] I. Mochida, K. Shimizu, Y. Korai, H. Otsuka, Y. Sakai, S. Fujiyama, Carbon 28 (1990) 311.
[7] W. M. Kalback, H. E. Romine, X. M. Bourrat, US Patent 5,259,947 (1991).
[8] D.D. Edie, M.J. Dunham, Carbon 27 (1989) 647.
[9] J.G. Lavin, D.R. Boyington, J. Lahijani, B. Nysten, J.P. Issi, Carbon 31 (1993) 1001.
[10] SRI Consulting, Chemical Economics Handbook, Menlo Park (1999).
[11] D.J. Johnson, Journal of Physics D Applied Physics 20 (1987) 287.
[12] Courtaulds Ltd., British Patent, 144341 (1969).
[13] J.D. Ray, S. Steinsiger, B.A. Cass, US Patent, 16251 (1970).
[14] Morganite Research Development Ltd., British Patent, 493571 (1969).
[15] J.T. Paul, Appl. No. 25208/74, Patent Specification 1433712.
[16] S. Wang, A. Garton, Journal of Applied Polymer Science 45 (1992) 1743.
[17] Z. Wu, C.U. Pittman Jr, S.D. Gardner, Carbon 33 (1995) 597.
[18] A. Fukunaga, S. Ueda, Composites Science and Technology 60 (2000) 249.
[19] L.T. Yuan, S.S. Shyu, J.Y. Lai, Journal of Applied Polymer Science 42 (1991) 2525.
[20] S.J. Park, M.H. Kim, Journal of Materials Science 35 (2000) 1901.
[21] S.R.J. Oliver, N. Bowden, G.M. Whitesides, Journal of Colloid and Interface Science 224 (2000) 425.
[22] Y. Chang, H.J. Ohara, Journal of Fluorine Chemistry 57 (1992) 169.
[23] R. Beydon, G. Bernhart, Y. Segui, Surface and Coatings Technology 126 (2000) 39.
[24] E. Hage, S.F. Costa, L.A. Pessan, Journal of Adhesion Science and Technology 11 (1997) 1491.
[25] S. Abraham, B.C. Pai, K.G. Satyanarayana, V.K. Vaidyan, Journal of Materials Science 25 (1990) 2839.
[26] F. Hoecker, J.K. Kocsis, Journal of Applied Polymer Science 59 (1996) 139.
[27] S.J. Park, Long-Range Force Contributions to Surface Dynamics, Interfacial Forces and Fields: Theory and Applications, Marcel Dekker, New York (1990) pp. 385.
[28] S.J. Park, J.S. Jin, J.R. Lee, Journal of Adhesion Science and Technology 14 (2000) 1677.
[29] S.J. Park, J.R. Lee, Journal of Materials Science 33 (1998) 647.
[30] A. Proctor, P.M.A. Sherwood, Surface and Interface Analysis 4 (1982) 212.
[31] P. Denison, F.R. Jones, J.F. Watts, Surface and Interface Analysis 9 (1986) 431.
[32] P. Denison, F.R. Jones, J.F. Watts, Journal of Physics D Applied Physics 20 (1987) 306.
[33] C. Kozlowski, P.M.A. Sherwood, Carbon 24 (1986) 357.
[34] C.D. Wagner, W.M. Riggs, L.E. Davis, J.F. Moulder, Handbook of X-Ray Photoelectron Spectroscopy. Perkin-Elmer Corp, Norwalk (1979).
[35] V. Gabara, High Performance Fibers 1: Aramid Fibers, Synthetic Fiber Material. Longman, New York (1994), p. 239.
[36] D.C. Prevorsek, High Performance Fibers 2: High Performance Polyethylene Fibers, Synthetic Fiber Materials. Longman, New York (1994), p. 263.
[37] Advanced Lightweight Engineering, M5 Materials Research (1998).
[38] S.J. Park, Long-range Force Contributions to Surface Dynamics. Interfacial Forces and Fields: Theory and ApplicationsMarcel Dekker, New York (1990), p. 400.
[39] P.C. Varelidis, D.G. Papakostopoulos, C.I. Pandazis, C.D. Papaspyrides, Composites Part A 31 (2000) 549.
[40] M.O.W. Richardson, M.J. Wisheart, Composites Part A 27 (1996) 1123.
[41] W.L. Lee, C.F. Lin, Materials Science and Engineering A 308 (2001) 124.

[42] S.J. Park, Y.S. Jang, Journal of Colloid and Interface ScienceJ. Colloid Interface Sci. 237 (2001) 91.
[43] S.J. Park, W.B. Park, J.R. Lee, Journal of Materials Science 35 (2000) 6151.
[44] N. Iwashita, M.V. Swain, J.S. Field, N. Ohta, S. Bitoh, Carbon 39 (2001) 1525.
[45] M. Labronici, H. Ishida, Composite Interfaces 2 (1994) 199.
[46] S.J. Park, M.S. Cho, Carbon 38 (2000) 1053.
[47] A. Knop, L.A. Pilato, Phenolic Resins: Chemistry, Applications, and Performance. Springer, Berlin (1985).
[48] J.K. Kim, Y.W. Mai, Engineered Interfaces in Fiber Reinforced Composites. Elsevier, Oxford (1998).
[49] C. Marston, C. Galiotis, Journal of Materials Science 33 (1998) 5311.
[50] M.S. Amer, M. Koczak, L.S. Schadler, Composites Part A 27 (1996) 861.
[51] M.S. Madhukar, L.T. Drzal, Journal of Composite Materials 26 (1992) 310.
[52] M.S. Madhukar, L.T. Drzal, Journal of Materials Science 28 (1993) 569.
[53] B. Harris, P.W.R. Beaumont, E. Mancunill de Ferran, Journal of Materials Science 6 (1971) 238.
[54] P.W.J. van den Heuvel, T. Peijs, R.J. Young, Composites Science and Technology 58 (1998) 933.
[55] L.S. Schadler, C. Galiotis, International Materials Reviews 40 (1995) 116.
[56] C.D. Wagner, M.S. Amer, L.S. Schadler, Journal of Materials Science 31 (1996) 1165.
[57] K.M. Atallah, C. Galiotis, Composites 24 (1993) 635.
[58] M. Tanoglu, S.H. McKnight, G.R. Palmese, J.W. Gillespie Jr, Composites Science and Technology 61 (2001) 205.
[59] J.J. Lesko, J.S. Riffle, N.S. Broyles, N. Verghese, S.V. Davis, US Patent, 6,020-063 (2000).
[60] D. Cho, L.T. Drzal, Journal of Applied Polymer Science 75 (2000) 1278.
[61] D. Cho, L.T. Drzal, Journal of Applied Polymer Science 76 (2000) 190.
[62] N.S. Broyles, K.N.E. Verghese, S.V. Davis, H. Li, R.M. Davis, J.J. Lesko, J.S. Riffle, Polymer 39 (1998) 3417.
[63] T.U. Marston, A.G. Atkins, D.K. Felbeck, Journal of Materials Science 9 (1974) 447.
[64] Advanced Insertion of Materials (AIM) Program (2004).
[65] G.L. Hahn, K.M. Nelson, C.R. Saff, Accelerated Insertion of Materials–Composites, In Paper Presented at the 34th International Society for the Advancement of Material and Process Engineering (SAMPE) Technical Conference, Baltimore, (2002).
[66] Materials Engineering for Affordable New Systems (MEANS) Program.
[67] P.C. Varelidis, D.G. Papakostopoulos, C.I. Pandazis, C.D. Papaspyrides, Composites Part A 31 (2000) 549.
[68] S.J. Park, J.S. Kim, Carbon 39 (2001) 2011.
[69] L.S. Penn, B. Jutis, Journal of Adhesion 30 (1989) 67.
[70] P. Gopal, L.R. Dharani, F.D. Blum, Polymer Composites 5 (1997) 327.
[71] J.S. Shim, S.J. Park, S.K. Ryu, Carbon 39 (2001) 1635.
[72] I.S. Kim, Materials Research Bulletin 33 (1998) 1069.
[73] S.J. Park, J.B. Donnet, Journal of Colloid and Interface Science 206 (1998) 29.
[74] S.J. Park, J.S. Jin, Journal of Colloid and Interface Science 242 (2001) 174.
[75] Y.V. Basova, H. Hatori, Y. Yamada, K. Miyashita, Electrochemistry Communications 1 (1999) 540.
[76] A. Fjeldly, T. Olsen, J.H. Rysjedal, J.E. Berg, Composites Part A 32 (2001) 373.

[77] P.E. Petersson, Crack Growth and Development of Fracture Zones in Plain Concrete and Similar Materials, In Report TVBM-1006, Lund Institute of Technology, Lund (1981).
[78] S.E. Swartz, S.T. Yap, Materials and Structures 21 (1988) 410.
[79] A.R. Bunsell, Fiber Reinforcements for Composite Materials. Elsevier, New York (1988).
[80] D. Hull, T.W. Clyne, An Introduction to Composite Materials, 2nd edit. Cambridge University Press, Melbourne (1996).
[81] F.T. Wallenberger, Advanced Inorganic Fibers: Processes, Structures, Properties Applications. Kluwer Academic, Dordrecht (2000).
[82] G.L. Hahn, K.M. Nelson, C.R. Saff, Accelerated Insertion of Materials-Composites, 34th International Society for the Advancement of Material and Process Engineering (SAMPE) Technical Conference, Baltimore, (2002).
[83] G. Havskjold, Materials and Manufacturing Directorate, Air Force Research Laboratory, Robust Design Computational System, AFRL-ML-WP-TR-2000-4093 (2000).
[84] National Research Council, Accelerating Technology Transition, National Academies Press, Washington, D.C. (2004).
[85] Advanced Insertion of Materials (AIM) Program.
[86] E. Fitzer, L.M. Manocha, Carbon Reinforcements and Carbon/Carbon Composites, 1st edit. Springer, Berlin (1998).
[87] G. Savage, Carbon-carbon Composites, 1st edit. Chapman & Hall, London (1993).
[88] J.D. Buckley, D.D. Edie, Carbon-Carbon Materials and Composites. Noyes, New Jersey (1993).
[89] C.R. Thomas, Essentials of Carbon-Carbon Composites. The Royal Society of Chemistry, Cambridge (1993).
[90] Y. Waku, N. Nakagawa, T. Wakamoto, H. Ohtsubo, K. Shimizu, Y. Kohtoku, Nature 49 (1997) 389.
[91] M. Schwartz, Composite Materials Handbook, 2nd edit. McGraw-Hill, New York (1992).
[92] M. Granda, J.W. Patrick, A. Walker, E. Casal, J. Bermejo, R. Menendez, Carbon 36 (1998) 943.
[93] G. Rellick, Carbon 28 (1990) 589.
[94] H.T. Fang, J.H. Jeon, J.C. Zhu, Z.D. Yin, Carbon 40 (2002) 2559.
[95] Q.G. Fu, H.J. Li, K.Z. Li, X.H. Shi, Z.B. Hu, M. Huang, Carbon 44 (2006) 1866.
[96] J.D. Buckley, Ceramic Bulletin 67 (1988) 364.
[97] J.X. Zhao, Carbon 23 (1985) 9.
[98] R.H. Knibbs, Proceedings of the 3rd Conference on Industrial Carbon and Graphite. Society Chemical Industry, London (1971), p. 297.
[99] P.A. Thrower, Carbon 20 (1982) 457.
[100] J.L. Wood, Carbon 19 (1981) 61.
[101] H. Marsh, K. Kuo, Introduction to Carbon Science. Butterworths, Newcastle (1989).
[102] L. Fillipuzzi, R. Naslain, Journal of the American Ceramic Society 77 (1988) 467.
[103] A.J. Eckel, J.D. Cawley, T.A. Parthasarathy, Journal of the American Ceramic Society 78 (1995) 972.
[104] W. Geankoplis, Mass Transport Phenomena. Ohio State University Bookstores, Columbus (1987).
[105] W.G. Zhang, H.M. Cheng, T.S. Xie, Z.H. Shen, B.L. Zhou, H. Sano, Y. Uchiyama, K. Kobayashi, Carbon 36 (1998) 991.
[106] K.L. Luthra, Carbon 26 (1988) 217.
[107] S.T. Dimitrijevic, Proceedings of the 20th Biennial Conf. on Carbon, UCSB, Santa Barbara (1991).

[108] L.H. Keys, Proceedings of the Symposium on Properties of High Temperature Alloys. The Electrochemical Society, Princeton (1977), p. 77.
[109] E.A. Gulbransen, Journal of the Electrochemical Society 113 (1966) 1311.
[110] W. Hinze, Journal of the Electrochemical Society 123 (1976) 1066.
[111] T. Narushima, T. Goto, T. Hirai, Active-to-Passive Transition in the Oxidation of CVD-SiC, Proceedings of the International Meeting on Advanced Materials, 4 Materials Research Society, Pittsburgh, (1989).
[112] G.H. Schiroky, GA-A18696 (1986).
[113] S.C. Shingal, Proceedings of the Symposium on Properties of High Temperatures Alloys with Emphasis on Environmental Effects, p. 77 (1976).
[114] R.C. Dickinson, Proceedings of the Oxidation Resistant C/C Composites for Hypersonic Vehicles Application, NASA Conference Publication 2501, Virginia, (1987).
[115] D.E. Rosner, The Journal of Physical Chemistry 74 (1970) 1829.
[116] E. Sheehan, Journal of the American Ceramic Society 65 (1982) C-111.
[117] H. Du, Fundamental Studies on Oxidation of Chemical Vapor Deposited. Silicon Nitride Thesis, Pennsylvania State University, Pennsylvania (1988).
[118] W.C. Tripp, Journal of the American Ceramic Society 59 (1976) 399.
[119] W.L. Vaughn, Journal of the American Ceramic Society 73 (1990) 1540.
[120] A.H. Heuer, Journal of the American Ceramic Society 73 (1990) 2785.
[121] J.E. Sheehan, Carbon 27 (1989) 709.
[122] Y. Grenie, Aerospatiale, NT No. 89188 AQEMIR (1989).
[123] J.R. Striie, Ceramic Bulletin 67 (1988) 369.
[124] Q.G. Fu, H.J. Li, X.H. Shi, K.Z. Li, W. Zhang, M. Huang, Applied Surface Science 253 (2007) 3757.
[125] S.M. Sim, Symposium on Better Ceramics through Chemistly, Palo Alto, (1986).
[126] S.M. Sim, Ultrastructure Processing of Advanced Ceramics, Florida, p. 995, (1983).
[127] W.J. Lackey, Ceramic Engineering and Science Proceedings 9 (1988) 1223.
[128] Q.G. Fu, H.J. Li, K.Z. Li, X.H. Shi, M. Huang, G.D. Sun, Materials Science and Engineering A 445 (2007) 386.
[129] J.R. Dentzer, SAMPE Quarterly 2 (1971).
[130] D.W. McKee, Carbon 26 (1988) 659.
[131] J.M. Criscione, ML-TDR-64-173, Part 11, (1974).
[132] J. Li, R. Luo, Y. Chen, Q. Xiang, C. Lin, P. Ding, N. An, J. Cheng, Applied Surface Science 255 (2008) 1967.
[133] D.B. Fischbach, Proceedings of the 20th Biennal Conference on Carbon, UCSB, Santa Barbara, p. 412, (1991).
[134] D.W. McKee, Carbon 24 (1986) 737.
[135] H.J. Li, G.S. Jiao, K.Z. Li, C. Wang, Materials Science and Engineering A 475 (2008) 279.
[136] M. Inagaki, Carbon 29 (1991) 287.
[137] C.E. Chang, Materials Research Bulletin 6 (1971) 1287.
[138] E.G. Fatzer, US Patent, 3,925,577, (1975).
[139] J.V. Milewski, US Patent, 2,528,823, (1983).
[140] F.S. Galasso, US Patent, 4,425,407, (1984).
[141] R.D. Veltri, US Patent, 4,476,178, (1984).
[142] J.W. Patten, US Patent, 4,500,602, (1985).
[143] C.W. Stroud, NASA TM 81 799, (1980).
[144] G. Rousseau, Brevet Europeen, 0.282.386 A1, (1988).

[145] J. Cullinan, AFOSR-86-0251, (1990).
[146] D.C. Rogers, Proceedings of the 7th National SAMPE Technical Conference, 7, Albuquerque, (1975).
[147] A.H. Klepping, AIAA-86-1499, (1986).
[148] T. Kinjo, Kawasaki Steel Technical Report, 23, 54, (1990).
[149] W. Kowbel, Proceedings of the Third International Symposium on Ceramic Materials and Components for Engineer, Las Vegas (1988).
[150] I. Jawed, Materials Research Bulletin 21 (1986) 1391.
[151] D.W. McKee, Carbon 26 (1988) 659.
[152] D.W. McKee, Carbon 24 (1986) 331.
[153] R.C. Shaffer, US Patent, 4,321,298, (1982).
[154] Vasilos, US Patent, 4,426,600, (1982).
[155] S. Ragan, Proceedings of the 20th Biennial Conference on Carbon, UCSB, Santa Barbara, p. 428, (1991).
[156] S. Miller, D. Papadopoulos, P. Heimann, L. Inghram, L. McCorkle, Composites Science and Technology 67 (2007) 2183.
[157] H. Vincent, Third European Conference on Composite Materials, ECCM, Bordeaux, p. 257, (1989).
[158] Y. Deslandes, Journal of Materials Science Letters 9 (1990) 200.
[159] D.A. Eitman, In Gas Turbine and Aeroengine Congress and Exposition, Toronto (1989).
[160] D.M. Barret, Proceedings of the ASC Symposium on High Temperature Composites, Dayton, p. 282 (1989).
[161] W.J. Lackey, Proceedings of the 10th lnternational Conference on Chemical Vapor Deposition, p. 1008, (1987).
[162] T. Hirai, Proceedings of the 19th Universal Conference on Ceramic Science, p. 329, (1982).
[163] T. Hirai, Journal of Materials Science 17 (1982) 1320.
[164] T. Hirai, Proceedings of the 8th International Conference on Chemical Vapor Deposition', p. 790, (1981).
[165] K. Hiraga, Journal of the American Ceramic Society 66 (1983) 539.
[166] T. Hirai, Proceedings of the 19th Universal Conference on Ceramic Science, p. 347, (1982).
[167] D.P. Stinton, Ceramic Engineering and Science Proceedings 5 (1984) 668.
[168] M. Touanen, Materials Science and Engineering A 147 (1991) 239.
[169] W.Y. Lee, Journal of the American Ceramic Society 74 (1991) 2136.
[170] W.Y. Lee, Journal of the American Ceramic Society 74 (1991) 2649.
[171] K. Bartsch, Journal of Materials Science 26 (1991) 4318.
[172] S. Zirinsky, Journal of the Electrochemical Society 125 (1978) 305.
[173] D.P. Stinton, Ceramic Engineering and Science Proceedings 6 (1985) 707.
[174] B. Bavarian, Proceedings of the 35th lnternational SAMPE Symposium, Book 2, p. 1348, (1990).
[175] E. Lugscheider, High Temperature—High Pressures 21 (1990) 119.
[176] Y.L. Jeng, Scripta Metallurgica et Materialia 28 (1993) 453.
[177] S.M. Sim, Ultrastructure Proceessing of Advanced Ceramics. Wiley, New York (1990) p. 995.
[178] D.W. McKee, Carbon 25 (1987) 551.
[179] M.V. Roode, Ceramic Engineering and Science Proceedings 9 (1988) 1245.
[180] S.S. Sandhu, In Proceedings of the 8th International Conference on Composite Materials, Honolulu, p. 22-F-1, (1991).

[181] N.S. Jacobson, D.J. Roth, R.W. Rauser, D.M. Curry, Oxidation Through Coating Cracks of SiC-Protected Carbon/Carbon, NASA/TM-2007-214834. Surface & Coatings Technology 203 (2008) 372–383.

[182] D. M. Curry, J.W. Latchem, G.B. Whisenhunt, AIAA-1983-483, (1983).

[183] S.W. Yurgartis, M.D. Bush, B.E. Mast, Surface and Coatings Technology 70 (1994) 131.

[184] F.J. Buchanan, J.A. Little, Surface and Coatings Technology 53 (1992) 137.

[185] N.S. Jacobson, D.M. Curry, Carbon 44 (2006) 1142.

[186] W.H. Glime, J.D. Cawley, Carbon 33 (1995) 1053.

[187] J.E. Medford, AIAA-1975-730, (1975).

[188] J. Bernstein, T.B. Koger, Electrochemical Society 135 (1988) 2086.

[189] E.L. Courtright, J.T. Prater, G.R. Holcomb, St. Pierre, R.A. Rapp, Oxidation of Metals 36 (1991) 423.

[190] L. Filipuzzi1, G. Camus, R. Naslain, J. Thebault, Journal of the American Ceramic Society 77 (1994) 459.

[191] L. Filipuzzi, R. Naslain, Journal of the American Ceramic Society 77 (1994) 467.

[192] G.R. Holcomb, Corrosion 52 (1996) 531.

[193] N.S. Jacobson, T.A. Leonhardt, D.M. Curry, R.A. Rapp, Carbon 37 (1999) 411.

[194] W. C. Stevens, R.J. Diefendorf, Thermosetting of Mesophase Carbon II: Theoretical, In Proceedings of the 4th International Carbon Conference, Deutschen Keramischen Gesellschaft, Baden-Baden, p. 40, (1986).

[195] J.D. Brooks, J.H. Taylor, Nature 206 (1965) 697.

[196] J.L. White, Progress in Solid State Chemistry 9 (1975) 59.

[197] G.S. Rellick, Carbon 28 (1990) 589.

[198] L.E. McAllister, W. Lachman, Handbook of Composites. Elsevier, New York (1983), p. 109.

[199] J.E. Zimmer, L.M. Salvador, US Patent, 4,554,024 (1985).

[200] Z. Wang, B. Wu, Q. Gong, H. Song, J. Liang, Materials Letters 62 (2008) 3585.

[201] L.S. Singer, I.C. Lewis, R. Bacon, Proceedings of the JANNAF Rocket Nozzle Technology Subcommittee Meeting, Naval Surface Warfare Center, Silver Spring, (1989).

[202] J.L. White, P.M. Sheaffer, Carbon 27 (1989) 697.

[203] J.E. Zimmer, R.L. Weitz, Carbon 26 (1988) 579.

[204] L.S. Singer, R.T. Lewis, Extended Abstracts, 11th Carbon Conference, 207, (1973).

[205] J. S. Evangelides, Extended Abstracts, 12th Carbon Conference, p. 93, (1975).

[206] Q.F. Zha, Y.S. Guo, Y.Z. Zhang, H.Y. Hou, X.J. Yang, Carbon 45 (2007) 2321.

[207] J. Jortner, Proceeding of Army Symposium on Solid Mechanics, AMMRC MS 76-2, p. 81, (1976).

[208] J. Jortner, Carbon 24 (1986) 603.

[209] J.D. Buch, Structure-Property Relationships in Advanced Carbon Materials, In Report No. TOR-0079(4726-04)-1, the Aerospace Corporation, El Segundo (1979).

[210] L.M. Manocha, O.P. Bahl, Carbon 26 (1988) 13.

[211] L.M. Manocha, E. Yasuda, Y. Tanabe, S. Kimura, Carbon 26 (1988) 333.

[212] R.A. Meyer, S.R. Gyetvay, ACS Symposium Series 303 (1986) 380.

[213] E. Fitzer, W. Huttner, Journal of Physics D Applied Physics 14 (1981) 347.

[214] G. Jiang, S.J. Pickering, G.S. Walker, N. Bowering, K.H. Wong, C.D. Rudd, Thermochimica Acta 454 (2007) 109.

[215] Y. Hishiyama, M. Inagaki, S. Kimura, S. Yamada, Carbon 12 (1974) 249.

[216] F. Fitzer, A. Gkogkidis, ACS Symposium Series 303 (1986) 346.

[217] J. Jortner, Extended Abstracts, 13th Biennial Conference on Carbon, p. 377, (1977).

[218] K.J. Lee, H.Z. Cheng, J.S. Chen, Wear 260 (2006) 99.
[219] N. Nadeau, G.L. Vignoles, C.M. Brauner, Chemical Engineering Science 61 (2006) 7509.
[220] D.W. Mckee, Chemistry and Physics of Carbon. Marcel Dekker, New York (1981).
[221] G. Savage, Carbon-Carbon Composites. Chapman & Hall, London (1993), p. 157.
[222] T.M. Wu, W.C. Wei, S. Hsu, Carbon 29 (1991) 1257.
[223] K. Hagen, T. Katja, S. Christian, H. Waldemar, Journal of the American Ceramic Society 79 (1996) 2429.
[224] J.M. Yang, S.M. Jeng, Journal of Materials Research 6 (1991) 505.
[225] D.A. Berztiss, R.R. Cerchiara, E.A. Gulbransen, F.S. Pettit, G.H. Meier, Materials Science and Engineering A 155 (1992) 165.
[226] D.J. Evans, F.J. Scheltens, J.B. Woodhouse, H.L. Fraser, Philosophical Magazine A 75 (1997) 1.
[227] A.K. Vasudévan, J.J. Petrovic, Materials Science and Engineering A 155 (1992) 1.
[228] R. Xiaobing, O. Kazuhiro, Nature 389 (1997) 579.
[229] Y. Waku, N. Nakagawa, T. Wakamoto, H. Ohtsubo, K. Shimizu, Y. Kohtoku, Nature 389 (1997) 49.
[230] S.J. Park, M.S. Cho, J.R. Lee, P.K. Pak, Carbon 37 (1999) 1685.
[231] S.J. Park, M.S. Cho, J.R. Lee, Journal of Colloid and Interface Science 226 (2000) 60.
[232] S.J. Park, M.K. Seo, Carbon 39 (2001) 1229.
[233] S.J. Park, M.K. Seo, J.R. Lee, Carbon 40 (2002) 835.
[234] S.J. Park, M.K. Seo, J.R. Lee, Carbon 41 (2003) 2991.
[235] P. Ehrburger, P. Baranne, J. Lahaye, Carbon 24 (1986) 495.
[236] G. Ondracek, Werkstoffkunde: Leitfaden fur Studium und Praxis. Expert-Verlag, Wurzburg (1994).
[237] TechTrends, International Reports on Advanced Technologies: Metal Matrix Composites: Technology and Industrial Application, Innovation 128, Paris, (1990).
[238] T.W. Clyne, P.J. Withers, An Introduction to Metal Matrix Composites. Cambridge University Press, Cambridge (1993).
[239] G. Jiang, S.J. Pickering, E.H. Lester, T.A. Turner, K.H. Wong, N.A. Warrior, Composites Science and Technology 69 (2009) 192.
[240] Z. Xu, J. Li, X. Wu, Y. Huang, L. Chen, G. Zhang, Composites Part A 39 (2008) 301.
[241] K.K.C. Ho, A.F. Lee, S. Lamoriniere, A. Bismarck, Composites Part A 39 (2008) 364.
[242] J. Schroder, K.U. Kainer, Materials Science and Engineering A 135 (1991) 33.
[243] K.U. Kainer, Newsletter 82 (1997) 36.
[244] Duralcan Composites for Gravity Castings, Duralcan, USA, San, Diego (1992).
[245] Duralcan Composites for High-Pressure Die Castings Duralcan USA, San Diego (1992).
[246] C.W. Brown, W. Harrigan, J.F. Dolowy, Proceeding of Verbundwerk 90, Demat, Frankfurt, (1990).
[247] Manufacturers of Discontinuously Reinforced Aluminum (DRA), DWA Composite Specialities, Chatsworth (1995).
[248] W. Henning, E. Köhler, Maschinenmarkt 101 (1995) 50.
[249] S. Mielke, N. Seitz Grosche, International Conference on Metal Matrix Composites, The Institute of Metals, London, 4/1, (1987).
[250] S. Tzamtzis, N.S. Barekar, N.H. Babu, J. Patel, B.K. Dhindaw, Z. Fan, Composites Part A 40 (2009) 144.
[251] Lanxide Electronic Components, Lanxide Electronic Components, New York (1995).
[252] S.C. Tjong, Y.W. Mai, Composites Science and Technology 68 (2008) 583.

[253] A.G. Leatham, A. Ogilvy, L. Elias, Proceedings of the International Conference on P/M in Aerospace, Defence and Demanding Applications, MPIF, Princeton, p. 165, (1993).

[254] Cospray Ltd. Banbury, U. K. (1992).

[255] R. Jendrzejewski, K. Van Acker, D. Vanhoyweghen, G. Sliwinski, Applied Surface Science 255 (2009) 5584.

[256] S. Amini, C. Ni, M.W. Barsoum, Composites Science and Technology 69 (2009) 414.

[257] K.K. Chawla, Composite Materials: Science and Engineering. Springer-Verlag, New York (1998).

[258] D.L. McDanels, R.W. Jech, J.W. Weeton, Transactions of Metallurgical Society, AIME 223 (1965) 636.

[259] K.C. Russel, S.Y. Oh, A. Figueredo, MRS Bulletin 16 (1991) 46.

[260] T. Choh, T. Oki, Materials Science and Technology 3 (1987) 378.

[261] A. Mortensen, I. Jin, International Materials Reviews 37 (1992) 101.

[262] R. Warren, C.H. Andersson, Composites 15 (1984) 101.

[263] S.Y. Oh, J.A. Cornie, K.C. Russel, Particulate Welding and Metal: Ceramic Interface Phenomena, In Proceeding of 11th Annual Conference on Composites and Advanced Materials, Ceramic Engineering and Science, Cocoa Beach, (1987).

[264] G.A. Chadwick, Materials Science and Engineering A 135 (1991) 23.

[265] A. Mortensen, L.J. Masur, J.A. Cornie, M.C. Flemings, Metallurgical and Materials Transactions 20 (1989) 2535.

[266] A. Mortensen, J.A. Cornie, Metallurgical and Materials Transactions 18 (1987) 1160.

[267] A. Mortensen, T. Wong, Metallurgical Transactions A 21 (1990) 2257.

[268] A.S. Sangini, A. Acrivos, International Journal of Multiphase Flow 8 (1982) 193.

[269] P. Poddar, V.C. Srivastava, P.K. De, K.L. Sahoo, Materials Science and Engineering A 460 (2007) 357.

[270] W. Krenkel, R. Naslain, H. Schneider, High Temperature Ceramic Matrix CompositesWiley, Weinheim (2001), p. 401.

[271] S.M. Lee, Handbook of Composite Reinforcements. Wiley, Weinheim (1993).

[272] B. Clauß, D. Schawaller, Advanced Sciences and Technologies 50 (2006) 1.

[273] Z. Wang, S. Dong, P. He, L. Gao, H. Zhou, J. Yang, D. Jiang, Journal of the European Ceramic Society 30 (2010) 787.

[274] D. Belitskus, Fiber and Whisker Reinforced Ceramics for Structural Applications. Marcel Dekker, New York (1993).

[275] A.R. Bunsell, Fiber Reinforcements for Composite Materials. Elsevier, New York (1988).

[276] W. Erwied, Foams and Foam Control. Ullmann's Encyclopedia of Industrial ChemistryVCH, Weinheim (1988), p. 465.

[277] M. Shibuya, M. Sakurai, T. Takahashi, Composites Science and Technology 67 (2007) 3338.

[278] A. Morales-Rodriguez, P. Reynaud, G. Fantozzi, J. Adrien, E. Maire, Scripta Materialia 60 (2009) 388.

[279] F. Yang, X. Zhang, J. Han, S. Du, Journal of Alloys and Compounds 472 (2009) 395.

[280] A.R. Bunsell, Journal of Applied Polymer Science Symposium 47 (1991) 87.

[281] F.T. Wallenberger, Advanced Inorganic Fibers: Processes, Structures, Properties Applications. Kluwer Academic, Dordrecht (2000).

[282] A.R. Bunsell, M.H. Berger, Fine Ceramic Fibers. Marcel Dekker, New York (1999).

[283] http://www.mmm.com/ceramics/misc/tech_notebook.html.

[284] http://www.sumitomo-chem.co.jp/english/division/kiso.html.

[285] http://www.nitivy.co.jp/english/nitivy.html.

[286] http://www.mitsui-mmc.co.jp/eindex.html.
[287] M. Schmucker, H. Schneider, T. Mauer, B. Clauß, Journal of the American Ceramic Society 88 (2005) 488.
[288] M. Schmucker, H. Schneider, T. Mauer, B. Clauß, Journal of the European Ceramic Society 25 (2005) 3249.
[289] http://www.coiceramics.com.
[290] http://www.ube.de.
[291] http://www.specmaterials.com/silicarbsite.htm.
[292] http://www.tisics.co.uk/fiber.htm.

Chapter 8

Composite Characterization

8.1. INTRODUCTION

Composite materials make their way into all aspects of modern technological society, but particularly so for applications requiring great strength and light weight such as in the aerospace industry. Because they are hybrid heterogeneous materials, they can be difficult to characterize with any one single methodology. The readers will find in this chapter: (1) major types or methods of characterization techniques that work best for composite materials, including chemical, physical, mechanical properties, and more; (2) advanced characterization methods for surface or morphology analyses such as transmission electron microscopy, scanning probe microscopy, atomic force microscopy, and so on; (3) concise summaries of all major characterization technologies for composite materials.

8.2. EVALUATION OF REINFORCEMENT FIBERS

8.2.1. Introduction

We describe techniques and test methods that are generally used to characterize the chemical, physical, and mechanical properties of reinforcement fibers for application in composite materials. Reinforcements in the form of unidirectional yarns, strands, or tows, and bidirectional fabrics are covered. Sophisticated experimental techniques are generally required for fiber characterization, and test laboratories must be well equipped and experienced for measuring fiber properties. It is also recognized that in many cases the measurement of a fiber property that manifests itself in the reinforced composite can best be accomplished with the composites.

Most reinforcement fibers are surface treated or have a surface treatment (e.g., sizing) applied during their production to improve handleability and/or promote fiber–resin bonding. Surface treatments affect wettability of the fiber during impregnation as well as the dry strength and hydrolytic stability of the fiber–matrix bond during use. Because of the direct relation to composite properties, the effectiveness of any treatment to modify surface chemistry is

Interface Science and Composites, Volume 18
ISSN: 1573-4285, doi: 10.1016/B978-0-12-375049-5.00008-6

generally measured on the composite itself by means of mechanical tests. The amount of sizing and its compositional consistency are significant in quality control of the fiber and the measurement of these parameters is part of the fiber evaluation.

8.2.2. Chemical Techniques

A wide variety of chemical and spectroscopic techniques and test methods are available to characterize the chemical structures and compositions of reinforcement fibers. Carbon fibers are found to range from 90 to 100% carbon. Typically, standard and intermediate modulus polyacrylonitrile carbon fibers are 90–95% carbon, with most of the remaining material being nitrogen. Minor constituents and trace elements can be extremely important when composites containing these fibers are considered for use at elevated temperatures (above 500°F or 260 °C). Organic fibers usually contain significant amounts of hydrogen and one or more additional elements (e.g., oxygen, nitrogen, and sulfur) which can be identified by spectroscopic analysis. Glass fibers contain sulfur dioxide and usually aluminum and iron oxide. Depending on the type of glass, calcium oxide, sodium oxide, and oxides of potassium, boron, barium, titanium, zirconium, sulfur, and arsenic may be found.

8.2.2.1. Elemental analysis

A variety of quantitative wet gravimetric and spectroscopic chemical analysis techniques may be applied to analyze the compositions and trace elements in fibers. The American Society for Testing and Materials (ASTM) test method C 169 may be used to determine the chemical compositions of borosilicate glass fibers [1].

A suitable standardized method for carbon and hydrogen analysis, modified to handle carbon and polymeric fibers, is provided by the ASTM test method D 3178 [2]. Carbon and hydrogen concentrations are determined by burning a weighed quantity of sample in a closed system and fixing the products of combustion in an absorption train after complete oxidation and purification from interfering substances. Carbon and hydrogen concentrations are expressed as percentages of the total dry weight of the fiber. The ASTM test method D 3174 [3] describes a related test in which metallic impurities may be determined by the analysis of ash residue.

Alternatively, a variety of commercial analytical instruments are available which can quickly analyze carbon, hydrogen, nitrogen, silicon, sodium, aluminum, calcium, magnesium, and other elements in reinforcement fibers. X-ray fluorescence, atomic absorption, flame emission, and inductively coupled plasma emission spectroscopic techniques may be employed for elemental analysis. Operating instructions and method details are available from the instrument manufacturers.

Trace metallic constituents are significant in carbon and polymeric fibers because of their possible effect on the rate of fiber oxidation. The presence of metals is usually expressed as parts per million in the original dry fiber and can be determined by analyzing the ash residue. The semi-quantitative determinations are generally made using flame emission spectroscopy. When quantitative values are desired, atomic absorption methods are used. With respect to oxidation of carbon fibers, sodium is usually of most concern because of its tendency to catalyze the oxidation of carbon.

8.2.2.2. Titration

The potential chemical activity of surface groups on fibers may be determined by titration techniques. For example, the relative concentration of hydrolyzable groups introduced during the manufacture or post-treatment of carbon fibers may be determined by measuring the pH. However, titration techniques are typically not used on commercial carbon fibers due to the low levels of surface functionality.

8.2.2.3. Fiber Structure

The X-ray diffraction spectroscopy may be used to characterize the overall structure of crystalline or semi-crystalline fibers. The degree of crystallinity and orientation of crystallites have a direct effect on the modulus and other critical properties of carbon and polymeric fibers. The X-ray powder diffraction using commercial power supplies and diffractometer units is used to characterize the structure of carbon fibers. The fiber is ground into a fine powder and then the X-ray powder diffraction pattern is taken using CuKα radiation. The patterns generally undergo computer analysis to determine the following parameters:

1. average graphite layer spacing: from the 0 0 2 peak position;
2. average crystal size L_c: from the 0 0 2 peak width;
3. average crystal size L_a: from the 1 0 0 peak width;
4. average lattice dimension a-axis: from the 1 0 0 peak position;
5. the ratio of peak area to the diffused area;
6. the 0 0 2 peak area to the total diffraction area;
7. the 1 0 0 peak area to the total diffraction area;
8. the ratio of the 1 0 0 to 0 0 2 peak areas;
9. crystallinity index: from a comparison of the X-ray diffraction of known crystallized and amorphous carbons.

The X-ray scattering of crystalline fibrous materials shows the presence of sharp and diffuse diffraction patterns which are indicative of crystal phases interdispersed with amorphous regions. The concept of the crystallinity index is derived from the fact that a portion of the scattering from a fiber is diffuse and thereby contributes to the so-called amorphous background. Thus, a simple method of estimating crystallinity is obtained by separating the diffraction pattern into crystalline (sharp) and amorphous (diffuse) components. The crystallinity index is a relative measure of crystallinity, and not an absolute numerical result, useful for correlating with physical properties of fibers.

Wide-angle X-ray spectroscopy and infrared spectroscopy (IRS) techniques have also been developed to determine the crystallinity and orientation of molecules in polymeric fibers. Testing and interpretation of results requires specialized equipment, sophisticated computer models, and a high level of technical expertise.

8.2.2.4. Fiber surface chemistry

Fibers generally are given a surface treatment to improve the adhesion between the fibers and resin matrix materials. Gases, plasmas, liquid chemical, or electrolytic treatments are employed to modify fiber surfaces. Introducing surface oxidation is perhaps the most common approach to modifying fiber surfaces.

Fiber surface structure, the modifications which surfaces undergo as a result of the different fiber surface treatments, and the relative importance of these modifications for composite properties are not well understood. This arises because of the small surface areas involved (0.5–1.5 m^2/g) and the very low concentrations of functional groups. If 20% of the surface was covered by one particular species, this would only amount to 1 μmol of chemical groups/g of fiber. Surface characterization should be carried out on fibers which have not been sized. Residual size from solvent desized fiber can interfere with most techniques, while pyrolysis techniques may alter the fiber surface due to oxidation and char products.

The following techniques have been used for characterizing fiber surfaces:

(1) X-ray diffraction – provides information relating to crystallite size and orientation, degree of graphitization, and micropore characteristics.

(2) Electron diffraction – gives crystallite orientation, three-dimensional order, and degree of graphitization. (Better for surfaces since penetration is only 1000.)

(3) Transmission electron microscopy – provides the highest resolution of any of the microscopic techniques routinely available. Ultramicrotomy can be used to prepare specimens, typically about 50 nm thick, for direct transmission electron microscopy analysis of the fiber surfaces. Transmission electron microscopy provides information about surface fine structure and shows fibrils and needle-like pores.

(4) Scanning electron microscopy – gives structural and surface features. Scanning electron microscopy is a useful technique for determining fiber diameters and identifying morphologic characteristics (scales, chips, deposits, pits) on fiber surfaces.

(5) Electron spin resonance spectroscopy – gives crystallite orientation.

(6) X-ray photoelectron spectroscopy or electron spectroscopy for chemical analysis – measures the binding energy of core electrons in atoms excited by low-energy X-rays. Changes in the chemical environment of a surface region 10–15 nm thick (the first few atomic layers) are revealed by slight shifts in the energy of these core electrons giving information on functional group types and concentrations. The surface sensitivity arises because the depth of the electrons is between 1 and 2 nm.

The ratios of total oxygen to total carbon and of oxidized carbon (including hydroxyl, ether, ester, carbonyl, and carboxy functional groups) to total carbon may be determined in carbon fibers using X-ray photoelectron spectroscopy or electron spectroscopy for chemical analysis.

(7) Auger electron spectroscopy – directs high-energy electrons (1–5 keV) onto surfaces to create vacancies in the core levels of atoms. These vacancies represent excited ions which may undergo de-excitation and thereby create Auger electrons. By analyzing the characteristic energies of all the back-scattered Auger electrons in the energy range 0–1 keV, the elemental composition of the first 30 or 40 atomic layers (about 30 nm) is possible and in some cases molecular information can be obtained from the analysis of data.

(8) Ion scattering spectroscopy – uses an ion as a molecular probe to identify elements on the outermost surface layer. Only atomic information can be obtained and sensitivity depends on the atomic element.

(9) Secondary ion mass spectroscopy (MS) – uses a controlled sputtering process with accelerated ions to remove surface atomic layers for direct analysis by MS. Secondary ion mass spectroscopy can be used to identify surface molecules and determine their concentrations.

(10) IRS or Fourier transform IRS – absorption vibrational spectroscopy technique to obtain molecular information about surface composition. IRS yields both qualitative and quantitative information relating to the chemical composition of surface molecules. The quality of the IRS analysis depends on the fiber composition and is directly related to the care taken during sample preparation.

For fibers with diameters between 0.015 and 0.03 mm, no sample preparation is required if an infrared microscope is available to examine the fibers directly. Organic fibers may be pressed (up to 1000 m^2) into a film of fiber grids.

(11) Laser Raman spectroscopy – absorption/vibrational spectroscopic techniques which complement IRS are relatively simple to apply. Little or no sample preparation is necessary. Fibers can be oriented in the path of the incident beam for direct analysis. Fiber sample must be stable to the high-intensity incident light and should not contain species that fluoresce.

(12) Contact angle and wetting measurements – provide an indirect measurement of fiber surface free energy for use in predicting interfacial compatibility and thermodynamic equilibrium with matrix materials. Contact angle and wetting measurement information can be obtained by direct measurement of contact angle, mass pick-up, or surface velocity. Measurement of contact angles on small diameter fibers (<10 μm) is difficult if done optically. If a fiber's dimensions are known, a simple force balance may be used to determine the contact angle by measuring the force induced by immersing the fiber into a liquid of known surface free energy. The apparatus usually employed for this test is the Wilhelmy balance [4].

Contact angles θ also may be measured indirectly by the micro-Wilhelmy technique [5–8]. A single fiber is partially immersed in a liquid and the force exerted on the fiber due to the surface tension of the liquid is measured. The contact angle is determined from the relationship $F = C\gamma LV \cos\theta$, where F is the force measured corrected for buoyancy, C the circumference of the fiber, and γLV the surface tension of the liquid. The results may be used to determine the fiber surface free energy and the contributions of polar and dispersive components to the free energy [7, 8].

(13) Physisorption and chemisorption measurements – adsorption of inert gas or organic molecules can be used to measure fiber surface area. To obtain accurate estimates of surface area, it is important that there is complete monolayer coverage of the surface, the area occupied by the adsorbed gas is known, and significant amounts of the gas are not taken up in micropores. Additional complications arise when the adsorption of organic molecules is used in place of gas adsorption, since it may be necessary to know the orientation of the adsorbed molecules to calculate surface area. Adsorption may also occur only at specific active sites and, if solutions are used, solvent molecules may be adsorbed as well.

The chemical reactivity of fiber surfaces can be determined by oxygen chemisorption and desorption measurements. Topographical changes (e.g., pores, cracks, and fissures) caused by surface treatments often can be readily detected by adsorption measurements. Flow microcalorimetry is a useful technique for directly measuring the heats of adsorption [9].

(14) Thermal desorption measurements – desorption of volatile products from fibers by heat treatment in vacuo. Thermal gravimetric analysis (TGA), gas chromatography (GC), MS, IRS analysis or combinations of pyrolysis GC/MS or TGA/IRS may be used to identify components desorbed from fiber surfaces. Below 150 °C, CO, NH, CH, and various organic molecules are observed, depending on the fiber type.

(15) Chemical identification of functional groups by titrimetric, coulometric, and radiographic techniques.

8.2.2.5. Sizing content and composition

The amount of sizing contained on fibers is expressed as a percentage of the dry sized fiber weight. It is generally determined by extracting the fibers with a heated solvent; then the cleaned fibers are washed, dried, and weighed. The American Society for Testing and Materials test method (ASTM) test method C 613 [10] describes a suitable method utilizing Soxhlet extraction equipment; however, similar extractions using a laboratory hot plate and beaker are also common. The selection of a solvent that quantitatively removes all the sizing but does not dissolve the fiber is essential for accuracy in this determination.

Thermal removal techniques are also utilized and are most practical for the more difficult soluble sizings. Time, temperature, and atmosphere conditions must be predetermined to ensure the sizing is removed without seriously affecting the fiber. The precise amounts of residue from decomposition of the sizing and weight loss of the fibers due to oxidation must also be known from control tests for greatest accuracy. Suppliers of Advanced Composite Materials Association recommended test method standard reference method (SRM) 14-90 'Determination of Sizing Content on Carbon Fibers' describes a pyrolysis technique for carbon fibers.

Sizing compositions and lot-to-lot chemical consistency may be determined by spectroscopic and chromatographic analysis of materials isolated by extracting the fibers with a suitable solvent. Acetone, tetrahydrofuran, and methylene chloride are commonly used solvents for extraction. Liquid and GC and diffuse IRS are used to analyze or 'fingerprint' the chemical compositions of extracts.

8.2.2.6. Moisture content

The moisture content or moisture regains of fibers or textiles may be determined using the common method. Care must be taken when applying the procedure since volatile materials in addition to moisture may be removed. If possible, tests should be performed on fibers that have not been sized. Moisture content is expressed as weight percentage moisture based on the dry weight of the specimen.

8.2.2.7. Thermal stability and oxidative resistance

The susceptibility of fibers and fiber surface to oxidation is measured as weight loss under given conditions of time, temperature, and atmosphere. This is especially important in the evaluation of carbon and organic fibers considered for use in plastics exposed to elevated temperatures since it contributes to the long-term degradation of composite properties. TGA may be used to determine the thermal decomposition temperature T_d of carbon and organic fibers and estimate the relative amounts of volatile, organic additives and inorganic residues.

A standard method for determining the weight loss of carbon fibers is given in ASTM test method D 4102 [11]. Variations in this test method regarding exposure of fibers have been studied and give similar results [12]. In order to minimize variability in test results, proper control of gas flow rates and currents is critical when performing the TGA.

8.2.3. Physical Techniques

The physical properties of fibers of importance in their applications in polymer matrix composites fall into two categories – those inherent in the filament itself (intrinsic) and those derived from the construction of filaments into yarns, tows, or fabrics (extrinsic). The former includes density, diameter, and electrical resistivity; the latter includes yield, cross-sectional area, twist, fabric construction, and areal weight. Density and the derived properties are used in the calculations required for the construction and analysis of the composite products. Density and yield are useful measures of quality assurance [13]. Filament diameter and electrical resistivity are important for the nonstructural aspects of aerospace and aircraft applications.

8.2.3.1. Filament diameter

The average diameter of fibers may be determined by using an indexing microscope fitted with an image-splitting eyepiece or from a photomicrograph of the cross-sectional view of a group of mounted fibers. Since fibers are not always true cylinders, effective diameters may be calculated from the total cross-sectional area of the yarn or tow and dividing by the number of filaments in the bundle. The cross-sectional area may also be estimated from the ratio of mass per unit length to density. For irregular, but characteristically shaped fibers, an area factor may be required in calculating the average fiber diameter.

Optical microscopy can provide information about fiber diameter and variation in diameter with length. The upper limit of resolution of the optical microscope is about one-tenth of a micrometer; hence, features less than 1 μm cannot be well characterized by optical microscopy.

Other techniques, such as scanning electron microscopy, provide much higher resolution than optical microscopy for determining fiber diameter and cross-sectional characteristics. Features of fiber surfaces down to the 5 nm level can be observed. In addition, the large depth of field provided by scanning electron microscopy helps define three-dimensional characteristics on fiber surfaces and defines fiber topography.

8.2.3.2. Density of fibers

8.2.3.2.1. Overview

Fiber density is not only an important quality control parameter in fiber manufacture but also required for determination of the void content of the fibrous composite, as described in ASTM test method D 2734, 'Void Content of Reinforced Plastics' [14]. Fiber density can also be used as a distinguishing parameter to identify a fiber. For example, fiber density results can readily differentiate between E and S-2 glass (E glass is 2.54 g/cm^3 (0.092 lb/in^3); S-2 is 2.485 g/cm^3 (0.090 lb/in^3)).

With few exceptions, the determination of density is accomplished indirectly by measuring the volume and weight of a representative sample of the fiber, and then combining these values to calculate density. The weight measurement is most easily obtained by using a quality analytical balance. To determine volume, however, there are several approaches used. The most common approach uses simple Archimedes methods involving the displacement of liquids of known density. Direct measurement of density can be made by observation of the level to which the test material sinks in a density-graded liquid [15].

Liquids are used almost exclusively in displacement techniques for the determination of volume. However, there are advantages of using a gas medium in place of liquid to determine the volume of fiber. One advantage is the minimization of errors associated with liquid surface tension. The gas displacement approach is often referred to as helium pycnometry. When a gas displacement approach is used, the test specimen volume is determined by measuring pressure changes of a confined amount of a gas that behaves as an ideal gas at room temperature (preferably high-purity helium). Helium pycnometry is not a recognized test method for measuring the volume and density of fibers; yet it has been demonstrated to be a viable technique [16, 17].

ASTM D 3800 [18] deals specifically with obtaining the density of fibers. This standard covers three different liquid displacement procedures: Procedure A, which is very similar to the D 792 liquid displacement method [19]; Procedure B, in which a low-density liquid is slowly mixed with a high-density liquid (containing the fibers) until the fibers become suspended; and Procedure C, which simply refers to D 1505, which is a density gradient method.

8.2.3.2.2. The ASTM test method D 3800 and standard test method for density of high-modulus fibers

The approach taken in ASTM D 3800 is threefold. Procedure A is identical to D 792 except that the immersion fluids recommended have only fibers in mind. The concern is complete fiber wetting and avoiding entrapped microbubbles. Procedure B relies on careful mixing of two liquids of different densities (with the fiber immersed). When the fibers are suspended in the mixed liquid, a hydrometer or liquid pycnometer is used to determine the density of the liquid. The density of the suspended fiber is equal to that of the liquid. Procedure C is D 1505 inserted as a part of D 3800 by reference.

The experimenter needs to be mindful to avoid entrapped bubbles, liquid absorption, and problems involving the fiber sizing coating (if any). Common sense immediately flags roving as a difficult fiber form to wet out; yet complete wetout is required to produce meaningful data. Pay close attention to the interfilament regions. In D 1505 the problem is not as severe because the fibers can be cut and/or spread out prior to insertion. Since the measurement is direct, the size of the fiber sample is irrelevant. Immersing many small fiber fragments allows for direct verification of density variations of the fiber, keeping in mind that small fragments may take hours to sink to their equilibrium density level. It cannot be emphasized enough that complete wetout must be achieved. Use of high-wetting, vacuum-degassed liquids goes a long way to this end. Remember that the fibers are a prime geometry for the nucleation of gas bubbles out of solution. If the liquid is not fully degassed, a bubble-free roving can quickly form new bubbles.

The surface area to volume ratio of composite fibers is extremely high. For cylindrical shapes, S. A./$V = 2/R$, where R the radius, is only several micrometers. For a 0.028-mil (7-μm) fiber, the ratio is 143,000 to 1. It is, therefore, very important to ensure compatibility between the fiber and the liquid. Glass and polyethylene fibers are fairly immune in this regard; however, aramid, for example, is certainly not. The liquid immersion time should be kept to a minimum to avoid liquid diffusion into the fiber.

The mistake is often made of thinking of the fiber by itself, when in reality it is usually coated with an interfacial sizing agent (to promote improved bonding with the matrix resin). It is good practice to research the sizing agent, as it is a completely different material than the fiber (with different absorption and chemical characteristics). Since the sizing is applied to the outer surface of the fiber, even the volume of a thin coat quickly becomes significant. For example, a 0.028-mm (7-μm) diameter carbon fiber with a typical coating of 1 percent sizing agent on a weight basis (with assumed density of 1.2 g/cm^3 (0.043 lb/in^3)) gives a final product which is 98.5% fiber and 1.5% sizing on a volume basis. For precision work, strip the sizing agent off the fiber before measuring fiber density.

8.2.3.3. *Electrical resistivity*

The determination of electrical resistivity is recommended as a control measure for checking processing temperature and to determine compliance with specific resistance specifications, where required. Electrical resistivity is one of the properties dramatically affected by the structural anisotropy of carbon fibers. Measurements can be made on either a single filament or a yarn. The measured value is resistance per given length of fiber as read on an ohm meter or similar device. The contact resistance can be eliminated by obtaining the resistance for two different lengths of fiber and calculating the difference due to the longer length. This difference is then converted to resistance per unit length and then multiplied by the area of the fiber or yarn bundle expressed in consistent units. Resistivity is expressed as ohmcentimeter, ohmmeter, or ohm-inches and refers to the value in the axial direction. Transverse resistivity is seldom reported.

8.2.3.4. *Coefficient of thermal expansion*

Standardized methods for measuring the coefficient of thermal expansion of the fibers are not generally available although good correlations between laboratories making these measurements do exist. The coefficient of thermal expansion is directionally dependent, highly influenced by the anisotropy of fibers. Carbon fibers typically have a negative axial coefficient of thermal expansion and a slightly positive transverse coefficient of thermal expansion. Commercial instruments (e.g., DuPont Model 943 Thermomechanical Analyzer or equivalent) can be used directly or modified to measure the axial coefficient of thermal expansion.

The coefficient of thermal expansion of the fiber can also be derived from measurements made on composites with unidirectional reinforcement. Laser interferometry and dilatometry are the techniques most frequently used. Other techniques, including some applied to the unimpregnated fiber, have also been found satisfactory. When testing the composite, the unidirectional fibers may be oriented parallel or perpendicular to the direction of measurement to obtain the axial or transverse coefficient of thermal expansion. To perform the analysis, the modulus of the fiber, the modulus and coefficient of thermal expansion of the matrix, and the fiber loading must be known. It may be desirable to perform the measurements on composites with different fiber loadings in order to check the results.

8.2.3.5. Thermal conductivity

The thermal conductivity of fibers is generally determined analytically from measurements of axial thermal conductivity on unidirectional reinforced composites. However, some measurements have been made on both fiber bundles and single filaments. These have agreed quite well with values determined from composite measurements [20]. Both types of measurements require considerable operator skill and sophisticated equipment, and are perhaps best left to the thermophysics laboratory. A well-defined relationship between axial thermal conductivity and axial electrical conductivity (or resistivity) has been developed for a wide range of carbon fibers. Since electrical resistivity is relatively easy to measure, reasonable estimates of thermal conductivity can be made for electrical resistivity measurements [21]. Transverse thermal conductivity can be determined for thin composites using a pulsed laser technique to measure thermal diffusivity. The thermal conductivity can then be calculated if the specific heat of the fiber is known.

8.2.3.6. Specific heat

This property is measured in a calorimeter such as described in ASTM test method D 2766 [22]. This also is not a simple measurement and is best left to the experienced laboratory.

8.2.3.7. Thermal transition temperatures

Differential scanning calorimetry, differential thermal analysis, or thermal mechanical analysis instrumentation may be applied to measure the glass transition temperature T_g and, if the fiber is semi-crystalline, its crystalline melting temperature T_m. General procedures for measuring T_g and T_m of organic fibers are given in ASTM test methods D 3417 and D 3418 [23, 24].

8.2.4. Mechanical Testing of Fibers

8.2.4.1. Tensile properties

It is important to note that the fiber stress at specimen failure is test dependent. For example, Table 8-1 shows the difference in fiber tensile stress at failure for typical carbon fibers tested as a filament, an impregnated tow, and a unidirectional laminate. These data reflect the fact that composite tensile strength depends on many factors, including interface characteristics, as well as fiber and matrix properties. These data emphasize the need to define the objective of fiber testing. Thus, for acceptance testing, it is recommended that fiber strength be measured on a material form representative of composite behavior. For carbon fibers, an impregnated tow test is recommended; for boron fibers, single-filament tests are recommended.

8.2.4.1.1. Filament tensile testing

Single-filament tensile properties can be determined using ASTM test method D 3379, tensile strength, and Young's modulus for high-modulus single-filament materials. The following summarizes this method: A random selection of single filaments is made from the material to be tested. The filaments are centerline mounted on special slotted tabs. The tabs are gripped so that the test specimen is aligned axially in the jaws of a constant-speed movable crosshead test machine, and stressed to failure. For this test method, filament cross-sectional areas are determined by planimeter measurements of a representative number of filament cross-sections as displayed on highly magnified photomicrographs. Alternative methods of area determination, such as optical gages, image-splitting microscope, linear weight-density method, etc., may also be used.

Tensile strength and Young's modulus are calculated from the load–elongation records and the cross-sectional area measurements. The specimen is shown in Fig. 8-1 [25].

Table 8-1 Effect of Test Method on Measured Tensile Strength

Test	Nominal Measured Fiber Tensile Strength			
	Typical Carbon Fiber Standard Modulus		Typical Carbon Fiber Intermediate Modulus	
	ksi	MPa	Ksi	MPa
Filament	595	4100	780	5380
Tow	580	4000	790	5450
Laminate	555	3830	665	4590

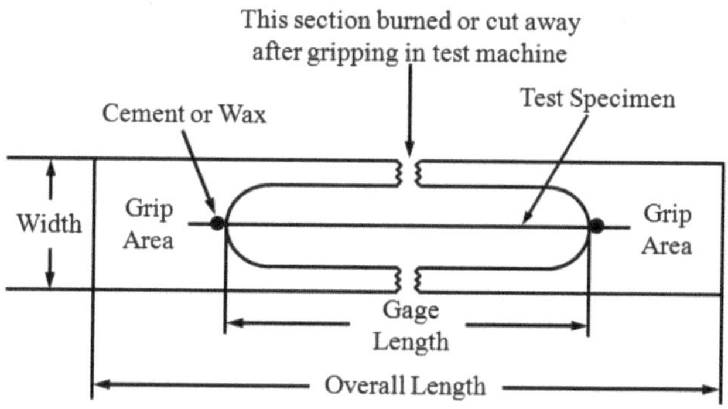

Fig. 8-1 Typical specimen mounting method.

ASTM test method D 4018 Method I test specimens require a special cast-resin end tab and grip design (Figs. 8-2(a) and (b)), to prevent specimen slippage in the grips under high loads [26]. Alternative methods of specimen mounting with end tabs are acceptable, provided that test specimens maintain axial alignment on the test machine centerline, and they do not slip in the grips at high loads. Method II test specimens require no special gripping mechanisms. Standard rubber-faced jaws should be adequate.

8.2.4.1.2. Fiber properties from unidirectional laminate tests

The most general representative procedure for the measurement of composite properties is to combine fiber and resin and test as a cured laminate. It is

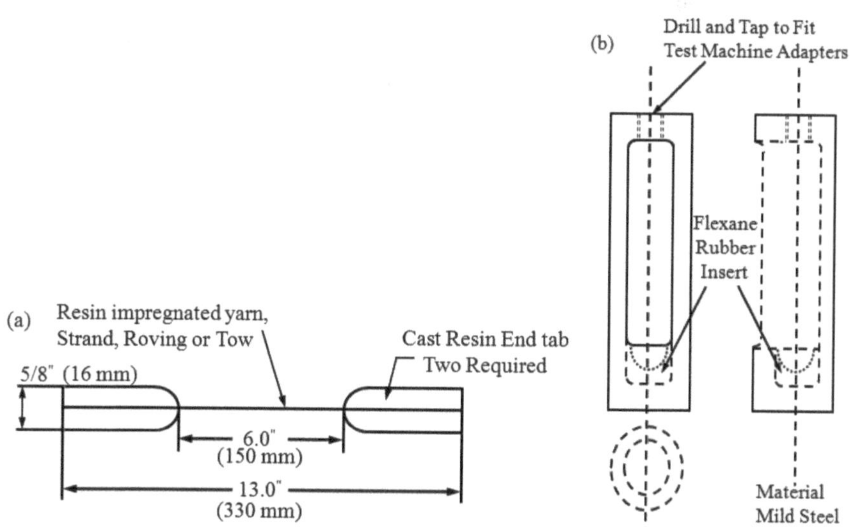

Fig. 8-2 Test specimen with cast-resin tabs (a) and grips for high-load tensile specimen (b).

Table 8-2 Effects of Resin on Laminate Properties

Fiber	Resin	Tensile Strength		Tensile Modulus	
		ksi	MPa	Msi	GPa
AS4	A	527	3630	32.1	221
AS4	B	500	3450	32.7	225
AS4	C	435	3000	32.4	223
AS4	D	432	2980	31.9	220

Tensile strength and modulus data normalized to 100% fiber volume.

important to understand that laminate properties are a function of both fiber and resin. Table 8-2 shows the dependence of measured mechanical properties with various modified epoxy resins. Another factor to consider is the fiber volume fraction of the laminate. A fiber volume of 55%–65% has been found to allow consistent measurement of normalized fiber properties for carbon fiber laminates. Since the objective is to determine fiber properties, the data must be normalized to 100% fiber volume. This is done simply by the following equation:

$$\text{Property}\,(100\%) = \text{Property} \times 100/\text{Fiber Volume} \qquad (8\text{-}1)$$

Laminate testing should be conducted per ASTM test method D 3039 [27].

8.3. EVALUATION OF MATRIX RESINS

8.3.1. Introduction

The function of the matrix in a composite is to hold the fibers in the desired position, and to provide a path for introducing external loads into the fibers. Since the strengths of matrix materials are generally lower than fiber strengths by an order of magnitude or more, it is desirable to orient the fibers within a composite structure so that they will carry the major external loads. Although the success of composites is largely due to this ability, the strength and other properties of matrix materials cannot be ignored. Matrix material properties can significantly affect how a composite will perform, particularly with respect to in-plane compression, in-plane shear, resistance to impact damage, and other interlaminar behavior, and especially when exposed to moisture and elevated temperatures.

A wide range of polymeric resin systems are used as the matrix portion of fiber-reinforced composites. These systems generally fall into two broad categories: thermoplastic materials and thermosetting materials. The thermoplastics are nonreactive solids designed to soften, melt, and intimately infiltrate reinforcement fiber bundles at appropriate processing temperatures and pressures, and to solidify into a desired shape on cooling. Thermosets are reactive materials comprised of organic resins and other constituents required for chemical 'curing.' They may exist in various forms (liquid, solid, film, powder, pellets, etc.) in the uncured state, and may be partially reacted prior to combining with the reinforcing fibers. During composite processing they react irreversibly to form solids. In addition to the organic constituents, thermoset systems may also contain additives such as catalysts, fillers, and processing aids, which may be inorganic or contain metals. Thermoplastic or elastomeric fillers may also be incorporated. Due to their reactive nature, most uncured thermosets must be stored under refrigeration, although some multipart systems designed for component mixing just prior to use may not require cold storage. Both thermoplastics and thermosets can be used to preimpregnate reinforcing fibers to produce prepreg, while processes such as resin transfer molding are generally more suited to thermosets.

This section focuses on methods of testing and characterizing matrix materials and their constituents. Chemical, physical, thermal, and mechanical properties are considered, as well as methods for test specimen preparation and environmental conditioning of test specimens. Tests of thermosets (in both the cured and uncured states) and thermoplastics are addressed.

The properties covered in this chapter will largely be of interest to resin formulators and material suppliers. The composite end user will also find some matrix properties useful, particularly for process cycle development and, to a lesser extent, for initial screening and material selection. A number of matrix properties and tests are also applicable to quality assurance, especially if resins are purchased separately from the reinforcement for use in resin transfer molding or similar processes.

8.3.2. Preparation of Matrix Specimen

Specimens of unreinforced (neat) matrix materials are required for physical and/or mechanical characterization of these polymers in the solid (cured) state. Methods available for specimen preparation are strongly dictated by the type of matrix material being studied. Primary variables include thermoset vs. thermoplastic, viscosity at various processing stages, processing temperature, amount of volatiles evolved, and degree of brittleness in the fabricated state. When working with uncured polymers, personal safety is always a concern, and the appropriate Material Safety Data Sheets should be consulted.

8.3.2.1. Thermoset polymers

Thermoset polymers of interest, that is, those used as matrices in composites, are typically of sufficiently low viscosity at some point during the cure process to flow. Thus, they may be cast into plate forms to provide blanks from which finished specimens can be machined, or molded into even more complex geometries if necessary to create net-dimension specimens directly.

When casting neat (unreinforced) polymers for use as mechanical test specimens, it is critical that voids, inclusions, and similar defects be minimized, in both size and number. Most of the thermoset polymers used as matrices, even those considered to be toughened, tend to be relatively brittle, and thus their ultimate strengths are strongly dictated by critical flaw size.

Inclusions can be present in impure resin as obtained from the supplier, or introduced during the fabrication process (e.g., inadequately cleaned molds, airborne dirt particles, inadequate mixing of components, etc.). Caution also must be exercised when using release agents, to avoid contamination of the polymer.

Defects can be in the form of surface scratches, edge chips, and mold marks. Voids are typically caused by trapped volatiles which evolve during the initial stages of the curing process. The evolution of volatiles can be suppressed, or at least minimized, by subjecting the polymer to pressure during the curing process. However, it is more common to apply a vacuum during the initial stage of the cure cycle, while the polymer is either still in the mixing container or already in the mold. This is done at one or more points in time as the temperature is being elevated, and while the viscosity is at its lowest. Thus, a vacuum oven is useful.

The vacuum can evoke a strong evolution of volatiles, requiring that the container or mold have sufficient volume to contain the frothy polymer until the gas bubbles burst. If a single flat panel is to be fabricated, a simple box mold consisting of five steel plates, viz., a bottom and four sufficiently high sides, held together with screws, works well. This box can be disassembled after cure, for ease of polymer matrix plate removal and easy clean-up. Individual strips of polymer can also be made in this manner, by placing thin steel strips of width equal to the desired polymer matrix specimen width upright on one long edge, spaced apart to the desired polymer specimen thickness.

Since volatiles are being evacuated, the vacuum pump itself should be protected, by the use of a cold trap to condense these vapors before they pass through the pump. If a cavity mold is being used to produce individual specimens of net dimensions, an elastomeric funnel works well to contain the volume of volatiles; the polymer will flow back down into the mold as the bubbles collapse. The funnel can then be left in place during the remainder of the cure. During clean-up, the funnel can be flexed to easily remove the cured polymer residue on it.

The individual specimen cavity molds can be fabricated of metal, usually steel rather than aluminum because of its lower thermal expansion and higher surface hardness. These are typically two-piece split molds, to permit cured specimen removal. Elastomeric molds, themselves easily fabricated by casting around a permanent pattern, are an attractive alternative. The cured mold can be slit along its length to remove it from around the pattern, this slit also permitting it to be later pried open to easily remove the polymer specimen cast in it. In any case, the individual specimen molds are typically ganged together for efficiency. The as-molded specimens are ready for testing with little or no further preparation. At most, and primarily for aesthetic reasons, the mold seam(s) may be lightly sanded off.

If vacuum is not being subsequently used to remove volatiles, the molds can be filled from the bottom to minimize trapped air, but this adds complication and is usually not necessary. Likewise, if the viscosity of the polymer is too high for gravity fill, pressure can be used to force it into the mold. Again, this is not usually necessary considering the composite processing requirements of these polymers as matrix materials anyway.

As an alternative to a box mold for fabricating flat neat matrix plates, the polymer can be cast between two vertically positioned flat plates, the desired cast polymer plate thickness held apart by spacers, and sealed around three edges. The polymer is then poured into the open top edge. The plates may be metal or glass. However, this technique is not always successful. Because of the constraint of the mold at both surfaces of the polymer, and the difficulty of achieving full release, the cast polymer plate may crack due to the stresses induced by differential thermal contraction during cool down from the cure temperature. Also, the polymer, which typically has a higher coefficient of thermal expansion than the mold, may contract away from the mold surfaces, producing a mottled surface. These local depressions are typically very shallow and can be removed by subsequent surface grinding of the cast plate. However, thermal residual strains associated with the formation of these surface irregularities remain (as can be observed under polarized light), and are very difficult to anneal out. Also, the very long path length that any trapped air bubbles or volatiles must travel to reach the free surface makes the production of void-free polymer plates more difficult to achieve.

8.3.2.2. Thermoplastic polymers

Thermoplastic polymers used in composites are typically high-processing-temperature (620–840°F (325–450 °C)) systems and higher-temperature-mold materials must be used. Matrix polymers for use in fabricating neat specimens tend to be available in film or granular forms. Pressure injection or compaction is typically necessary, which is complicated by the fact that the minimum viscosities achievable tend to be higher than for thermosets. Although volatile evolution is usually not an issue when molding thermoplastics since they are typically fully polymerized, trapped air can still be a problem. Thus, the use of vacuum during forming may still be desirable.

These high-temperature thermoplastics tend to be less brittle than the thermoset polymer matrix materials. Thus, cracking of the polymer plate during the molding operation due to differential contraction of plate and mold is less of a problem, but it can still occur.

8.3.2.3. Specimen machining

For both thermosets and thermoplastics, if the neat matrix specimen has been molded to final shape, no additional preparation is needed. Dog-bone cylindrical specimens, typically for use in solid-rod torsion testing, but sometimes used for tension and compression testing, are one such example.

Tension, compression, and Iosipescu shear specimens of thermoset polymers are typically machined from flat plates or strips rather than being molded to net dimensions. Although individual dog-bone flat specimens of commodity thermoplastics are commonly (injection-) molded to final dimensions, high-temperature thermoplastic matrix materials are usually not. Rather, flat rectangular blanks are molded, and dog-bone specimens are machined from them.

The various polymers are relatively easy to machine using abrasive wheels. If desired, the surfaces of as-molded plates can be ground prior to cutting individual specimens from them. The plates are cut into strips and specimen blanks using thin abrasive blades, although sometimes diamond wheels, or even toothed band saw blades, are used. Dog-bone specimens can then be ground to final dimensions. The notches in Iosipescu shear specimens can likewise be ground in, using shaped grinding wheels and multiple passes. Specimens can be stacked together for this operation, mutually supporting each other.

Most polymer matrix specimens will tolerate minor grinding-induced scratches and chipped edges, even though this is never desirable. However, some polymers are extremely sensitive to these surface defects. All surfaces and edges within the specimen gage length must then be carefully smoothed with fine (e.g., down to 600-grit) emery cloth. When working with a new polymer matrix, both as-ground and surface-polished tensile specimens should initially be tested, to determine the polymer's sensitivity to surface defects. Since final polishing adds additional labor cost, it is desirable to only do so when necessary.

8.3.3. Chemical Analysis Techniques

Chemical characterization techniques are listed in Table 8-3. Elemental analysis and functional group analysis provide basic and quantitative information relating to chemical composition. Spectroscopic analysis provides detailed information about molecular structure, conformation, morphology, and physicochemical characteristics of polymers. Chromatographic techniques separate sample components from one another, and thereby simplify compositional characterization and make a more accurate analysis possible. Employing spectroscopic techniques to monitor components separated by gas or liquid chromatography greatly enhances characterization, providing a means to identify and quantitatively analyze even the most minor components.

8.3.3.1. Elemental analysis

Elemental analysis techniques such as ion chromatography, atomic absorption, X-ray fluorescence, or emission spectroscopy can be applied to analyze specific elements, such as boron or fluorine. When necessary, X-ray diffraction may also be used to identify crystalline components, such as fillers, and to determine the relative percent crystallinity for certain resins.

Table 8-3 Techniques for Chemical Characterization

Elemental analysis	-Conventional analytical techniques -X-ray fluorescence -Atomic absorption -Inductively coupled argon plasma -Energy dispersive X-ray microanalysis -Neutron activation analysis
Functional group analysis	-Conventional wet chemical techniques -Potentiometric titration -Coulometry -Radiography
Spectroscopic analysis	-Infrared (pellet, film, dispersion, reflectance), Fourier transform infrared (FTIR), photoacoustic FTIR, internal Reflection (IR), IR microscopy, dichroism -Laser Raman -Nuclear magnetic resonance (NMR) ^{13}C, ^{1}H, ^{15}N; conventional (soluble sample), solid state (machined or molded sample) -Fluorescence, chemiluminescence, phosphorescence -Ultraviolet–visible (UV–vis) -Mass spectroscopy (MS), election impact MS, field desorption MS, laser desorption MS, secondary ion mass spectroscopy (SIMS), chemical ionization MS -Electron spin resonance -Electron spectroscopy for chemical analysis) -X-ray photoelectron -X-ray emission -X-ray scattering (small angle-Saxs) -Small-angle neutron scattering -Dynamic light scattering
Chromatographic analysis	-Gas chromatography (GC) or GC/MS (low-molecular-weight compounds) -Pyrolysis-GC and GC/MS (pyrolysis products) -Headspace GC/MS (volatiles) -Inverse GC (thermodynamic interaction parameters) -Size-exclusion chromatography (SEC), SEC-IR -Liquid chromatography (LC or high-performance liquid chromatography (HPLC)), HPLC–MS, multidimensional/orthogonal LC, microbore LC -Supercritical fluid chromatography (SFC) -Thin-layer chromatography (TLC), 2D TLC

8.3.3.2. Functional group and wet chemical analysis

The analysis of reactive functional groups is particularly important in determining equivalent weights of prepolymers. Titration and wet chemical analysis for specific functional groups are useful techniques for characterizing individual epoxy components but have limited application and may provide misleading results when complex resin formulations are analyzed.

8.3.3.3. Spectroscopic analysis

IRS provides more useful information for identifying polymers and polymer precursors than any other absorption or vibrational spectroscopy technique and is generally available in most laboratories. It yields both qualitative and quantitative information concerning a polymer sample's chemical nature, that is, structural repeat units, end groups and branch units, additives, and impurities [28]. Computerized libraries of spectra for common polymeric materials exist for direct comparison and identification of unknowns. Computer software allows the spectrum of a standard polymer to be subtracted from an unknown to estimate its concentration and perhaps to determine whether another type of polymer is also present in the sample.

IRS is sensitive to changes in the dipole moments of vibrating groups in molecules and, accordingly, yields useful information for the identification of resin components. It provides a fingerprint of the resin composition and is not limited by the solubility of resin components [29, 30]. Indeed, gases, liquids, and solids may be analyzed by IRS. Advances in technology have led to the development of Fourier transform IRS, a computer-supported infrared technique for rapidly scanning and storing infrared spectra. Multiple scans and Fourier transformation of the infrared spectra enhance the signal-to-noise ratio and provide improved spectra for interpretation. In addition, the Fourier transform IRS attenuated total reflection and diffuse reflectance techniques may be applied for quality assurance of thermoset composite materials to assess their state of cure, that is, residual epoxide concentration.

Although not as popular as IRS, laser Raman spectroscopy complements as an identification technique and is relatively simple to apply [31]. As long as the specimen is stable to high-intensity incident light and does not contain species that fluoresce, little or no sample preparation is necessary. Solid specimens need only be cut to fit into the sample holder. Transmission spectra are obtained directly with transparent specimens. For translucent specimens, a hole may be drilled into the specimen for passage of the incident light and transmission spectra obtained by analyzing light scattered perpendicular to the incident beam. The spectrum of a turbid or highly scattering specimen is obtained by analyzing the light reflected from its front surface. Powdered samples are simply tamped into a transparent glass tube and fibers can be oriented in the path of the incident beam for direct analysis.

8.3.3.4. Chromatographic analysis

High-performance liquid chromatography is the more versatile and economically viable quality assurance technique for soluble resin materials [32–39]. It involves the liquid-phase separation and monitoring of separated resin components. Dilute solutions of resin samples are prepared and injected into a liquid mobile phase which is pumped through column(s) packed with a stationary phase to facilitate separation and then into a detector. The detector monitors concentrations of the separated components, and its signal response, recorded as a function of time after injection, provides a 'fingerprint' of the sample's chemical composition. Quantitative information may be obtained if the sample components are known and sufficiently well resolved, and if standards for the components are available. Size-exclusion chromatography, a high-performance liquid chromatography technique, is particularly useful in determining the average molecular weights and molecular weight distributions of thermoplastic resins [39]. Recent advances have resulted in improved and automated high-performance liquid chromatography instrumentation that is relatively low cost and simple to operate and maintain.

A powerful, but technically more demanding, technique for directly analyzing polymers is pyrolysis GC/MS. In this case, the sample only needs to be rendered sufficiently small to fit onto the pyrolysis probe. Not only can the polymer type be identified by comparing the resulting spectrum with standards, but also volatiles and additives can be identified rapidly and quantitatively, and polymer branching and crosslink density can sometimes be measured.

Other chromatographic and spectroscopic techniques have also been considered [31, 40–44]. GC, GC headspace analysis, and GC–MS are useful for analyzing residual solvents and some of the more volatile resin components. Combined thermal analysis-GC–MS can be used to identify volatile reaction products during cure [45, 46].

8.3.3.5. *Molecular weight and molecular weight distribution analysis*

Techniques for evaluating polymer molecular weight, molecular weight distribution, and chain structure are listed in Table 8-4. Size-exclusion chromatography is the most versatile and widely used method for analyzing polymer molecular weight and molecular weight distribution. Once the solubility characteristics of a polymer are known, a suitable solvent can be selected for dilute solution characterization. Tetrahydrofuran is most often the solvent of choice for size-exclusion chromatography; however, toluene, chloroform, dimethylformaldehyde, and *m*-cresol are also used. If the polymer's Mark–Houwink constants, K and a, in the solvent are known, size-exclusion chromatography can be applied to determine the polymer's average molecular weight and molecular weight distribution [47]. If the constants are unknown or the polymer has a complex structure (e.g., branched, a copolymer, or mixture of polymers), size-exclusion chromatography still may be used to estimate the molecular weight distribution and other parameters relating to the structure and composition of the polymer. Although size-exclusion

Table 8-4 Polymer Molecular Weights, Molecular Weight Distribution and Chain Structure

Techniques	Parameters measured	Principle
Size-exclusion chromatography	Molecular weight averages and molecular weight distribution also provides information relating to polymer chain branching, copolymer composition, and polymer shape	Liquid chromatography technique. Separates molecules according to their size in solution and employs various detectors to monitor concentrations and identify sample components. Requires calibration with standard polymers
Light scattering (Rayleigh)	Weight-average molecular weight M_w (g/mol), virial coefficient A_2 (mol cm^3/g^2), radius of gyration $\langle R_g \rangle_z$ (A), polymer structure, anisotropy, polydispersity	Measurement of scattered light intensities from dilute polymer solutions dependent on solute concentration and scattering angle. Requires solubility, isolation, and in some cases fractionation of polymer molecules
Nonaqueous reverse-phase high-performance liquid chromatography and thin-layer chromatography	Molecular weight averages and molecular weight distribution. Requires calibration	Liquid chromatography technique based on equilibrium distribution of polymer molecules between a nonaqueous binary solvent mobile phase and a nonpolar stationary (packing) phase

chromatography indicates the presence of soluble nonpolymeric components, high-performance liquid chromatography is the better technique for characterizing residual monomers, oligomers, and other soluble, low-molecular-weight sample components.

Light scattering, osmometry, and viscometry are also used to analyze polymer molecular weight. Although seldom applied to synthetic polymers, sedimentation is an excellent technique for characterizing the molecular weight of polymers having very large molecular weight. The special techniques tend to be somewhat empirical or have limited utility and therefore are used less often.

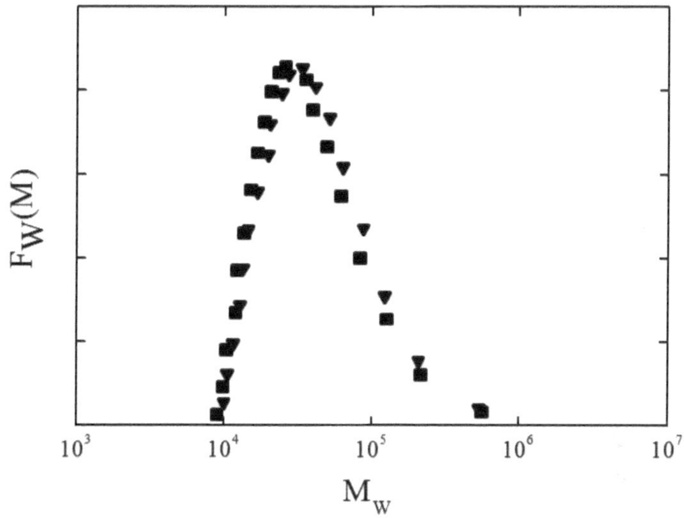

Fig. 8-3 Molecular weight distribution of Kevlar™ in concentrated sulfuric acid using dynamic laser light scattering.

New techniques which show great promise for characterizing polymer chain structure also are listed in Table 8-4. One of the most promising new techniques is dynamic laser light scattering. Unlike size-exclusion chromatography, dynamic light scattering can be applied to any soluble polymer, regardless of temperature or solvent, and does not require polymer standards for calibration. Fig. 8-3 illustrates the molecular weight distribution of poly(1,4-phenyleneter-eph-thalamide) (i.e., Kevlar™) measured by the laser light scattering [34]. As indicated, the polymer's molecular weight distribution can be fully characterized using very little sample and a single solution with concentrated sulfuric acid as the solvent.

Dilute solution viscometry is a simple technique for determining the limiting viscosity number or intrinsic viscosity $[\eta]$ of soluble polymers [48, 49]. The apparatus is inexpensive and simple to assemble and operate. The $[\eta]$ of a polymer depends on its hydrodynamic volume in the solvent and is related to the molecular weight of the polymer.

8.3.4. Thermal and Physical Analysis Techniques

The physical properties of the matrix material will influence the processing method as well as determine the type of application appropriate for the fabricated composite. Thermal analysis methods are used to determine glass transition and crystalline melt temperatures, thermal expansion, thermal decomposition, heat of reaction, and other thermal events in the matrix materials. Rheological methods provide information on the temperature-dependent flow behavior. In addition, the cure-dependent characteristics of thermosetting resins can also be evaluated. Other methods can be employed to determine the morphology and density of the matrix material. The analysis techniques discussed in the following sections are used to determine the physical properties of thermoplastic and thermosetting materials.

8.3.4.1. *Thermal analysis*

Thermal analytical techniques, such as TGA, differential thermal analysis, differential scanning calorimetry, thermal mechanical analysis, dynamic mechanical analysis, and torsional braid analysis, provide useful information relating to the composition and processability of resins.

TGA monitors the weight changes in a sample as a function of temperature. Although primarily used for studying the degradation processes, TGA can also be applied as a quality assurance technique to provide information about the volatiles, resin, fiber, and inorganic residue content of prepreg materials [75]. Since dissimilar materials often degrade and volatilize at different temperatures and rates, compositional differences may be reflected by differences in their TGA thermograms. Thermal oxidative degradation rates determined by TGA are useful for estimating the life cycles of resin materials [50].

Differential scanning calorimetry and differential thermal analysis techniques are frequently employed for characterizing resins and composite materials [51, 52]. Both differential scanning calorimetry and differential thermal analysis monitor enthalpy changes in materials as a function of temperature (differential scanning calorimetry directly and differential thermal analysis indirectly) and thereby provide similar information useful for quality assurance of prepreg materials. Differential thermal analysis measures the temperature difference (ΔT) between the epoxy resin specimen and a reference material, whereas differential scanning calorimetry measures the rate of heat evolution (dH/dt) or enthalpy absorption of the specimen relative to a reference. Differential thermal analysis and differential scanning calorimetry measure thermal changes (1) as a function of time with both the specimen and reference material held at the same temperature (isothermal) or (2) as a function of temperature with both the specimen and reference material heated at the same heating rate (dynamic).

For quality assurance applications, differential thermal analysis and differential scanning calorimetry are usually run in the dynamic mode with the weighed specimen in an aluminum specimen holder and an empty holder used as the reference. Dynamic differential thermal analysis and differential scanning calorimetry measure the glass transition temperature T_g and heat of reaction ΔH of the prepreg resin but do not provide information about chemical composition directly. By monitoring the fraction of heat evolved as a function of temperature or time, information relating to the extent of cure and curing kinetics can be obtained. Differential scanning calorimetry and differential thermal analysis may also be applied to evaluate the melting temperature T and to estimate the degree of crystallinity of thermoplastic resins and composites. Since the average specimen size used in differential scanning calorimetry is only about 10 mg (0.00002 lb), special care must be taken in obtaining representative materials. Multiple specimens' runs are advisable.

Thermal mechanical analysis is used in conjunction with differential thermal analysis and differential scanning calorimetry to study the thermal transition behavior (e.g., T_g) of prepreg resins and cured laminates. Thermal mechanical analysis simulates a linear dilatometer to measure the thermal expansion and contraction of specimens under dynamic or isothermal heating conditions. Adjustable loads are applied via a specially designed probe resting upon the specimen surface. Sensitive displacement devices are employed to monitor the 'nominal' thermal response of a material. Since thermal transition behavior is related to the chemical composition and the extent of cure of a prepreg resin, thermal mechanical analysis can be applied as a quality assurance technique.

As previously discussed, TGA provides an indication of a sample's thermal decomposition temperature T_d and is used to estimate the relative amounts of volatiles, polymers, nonpolymeric additives, and inorganic residues. Differential scanning calorimetry or differential thermal analysis is applied to evaluate the extent of cure and curing characteristics of thermoset resins, to determine a polymer's T_g, and, if the polymer is semi-crystalline, to determine its crystalline melting temperature T_m. Suggested procedures for measuring T_g and T_m are given in the ASTM standards D 3417 and D 3418 [53, 54]. Thermal mechanical analysis can also be used to determine the T_g and to obtain further information about a polymer's heat distortion temperature and thermal expansion coefficient. For pelletized or molded samples, a razor blade or microtome can be used to cut samples to approximately fit the dimensions (thickness and diameter) of the sample holder. If the sample has been cut or is already in film or sheet form with a thickness no greater than 0.015 in (0.04 mm), a punch or cork borer may be used to cut disks of an appropriate size.

Alternatively, a hot stage microscope may be used to observe the heat distortion temperature and onset of flow of powdered samples. Initially the powder particles have sharp, rough edges. As the sample is heated and the heat distortion temperature is approached, the edges first become blurred and then the particles start to agglomerate. Finally, at T_m, for semi-crystalline polymers, or T_g, in the case of glassy polymers, flow occurs and a clear melt or liquid forms. Microscopes equipped with cross-polarizers are useful for defining crystal–crystal transitions and the onset of melting of semi-crystalline polymers.

8.3.4.2. Rheological analysis

The processing characteristics of a thermoplastic or thermosetting resin are dependent on flow behavior, which is characterized by rheological analysis. Methods which measure the temperature-dependent viscosity under constant shear conditions are used to obtain information on flow behavior. These methods include the use of viscometers or capillary rheometers. Since the viscosity of thermosetting materials also depends on the degree of cure, other methods may be used to obtain rheological information during cure.

Dynamic mechanical analysis, torsional braid analysis, and various mechanical spectrometers may be used to measure the rheological response of resins as a function of frequency, temperature, and/or state of cure. Both dynamic mechanical analysis and torsional braid analysis can provide information relating to the storage modulus, loss modulus, complex viscosity, and tan delta of polymers. In addition, information relating to gelation, vitrification, and the T_g of cured thermosetting resins can be obtained [55–57]. Rheological techniques are most often used to optimize processing parameters. However, since rheological properties are related to resin composition and morphology, rheological techniques may also be applied for the quality assurance of resins.

Dynamic dielectric analysis techniques can provide information on the flow behavior and curing characteristics of matrix materials. Dynamic dielectric analysis involves the use of electrical measurements to monitor changes in the dielectric constant, the dissipation factor, capacitance, and/or conductance of the resin during processing as a function of frequency, time, and temperature. Measured electrical parameters are highly responsive to changes in resin viscosity and are often employed to investigate and optimize prepreg processing parameters such as resin flow and gelation time/temperature. Since chemical composition affects the electrical properties and curing behavior of thermosetting resins, dynamic dielectric analysis techniques may also be applied for their quality assurance [57–64]. The ASTM test methods which are applicable for rheological analysis [65–69] include:

1. ASTM D 2393 ‘Viscosity of Epoxy Resins and Related Components.’ Method for measuring the viscosity of the liquid components of an epoxy resin system and/or the mixed formulation.
2. ASTM D 3835 ‘Rheological Properties of Thermoplastics with a Capillary Rheometer.’ Method describes the measurement of the rheological characteristics of thermoplastics at temperatures and shear conditions common to processing equipment.
3. ASTM D 4065 ‘Determining and Reporting Dynamic Mechanical Properties of Plastics.’ Practice for obtaining rheological information by free vibration and resonant or nonresonant forced vibration techniques.
4. ASTM D 4440 ‘Rheological Measurement of Polymer Melts using Dynamic Mechanical Procedures.’ Practice for determining the rheological properties of thermoplastics over a range of temperatures by nonresonant forced vibration techniques.
5. ASTM D 4473 ‘Measuring the Cure Behavior of Thermosetting Resins using Dynamic Mechanical Procedures.’ Practice is intended to provide means for determining the cure behavior of supported and unsupported thermosetting resins over a range of temperatures by free vibration and

resonant and nonresonant forced vibration techniques. Definitions of terms related to dynamic mechanical analysis are provided in ASTM D 4092 [70].

8.3.4.3. Morphology

The morphology of the matrix materials will be dependent on the type of polymer. The formation of a highly cross-linked network in thermosetting materials is controlled by the degree of conversion and the functionality of the components involved in the cure. Their degree of cross-linking is described in terms of the degree of cure which can be determined by thermal analysis and spectroscopic methods.

On a microscopic scale, the semi-crystalline thermoplastics contain regions of three-dimensional order (crystalline) and regions which lack long-range order (amorphous). Typically the crystalline regions consist of spherulites, which are aggregates of lamellar crystals that radiate from a nucleation site. The thermal history of the material, as well as the presence of fibers and/or fillers, will affect the size and number of spherulites and the degree of crystallinity [71, 72]. Differences in the crystalline region may also have an effect on mechanical properties [73, 74].

The analysis of the crystalline region is achieved by a variety of techniques. The size and degree of orientation of crystals can be studied by X-ray diffraction, electron microscopy, and birefringence methods, while a polarizing microscope is typically used for the analysis of spherulites. The degree of crystallinity can be determined by X-ray diffraction, specific volume, and heat of fusion. The specific volume method requires determining the specific volume of the sample as well as completely amorphous and crystalline samples of the material. The heat of fusion method involves radioing the heat of fusion of the sample and a completely crystallized sample of the material. The heats of fusion can be determined using the ASTM D 3417.

The noncrystalline thermoplastics may exhibit different levels of molecular orientation. The liquid crystal polymers may have regions of one- and/or two-dimensional order which can be evaluated by thermal analysis. The amorphous thermoplastics, typically lacking any long-range order, can undergo orientation depending on the processing technique. In general, molecular orientation can produce anisotropic properties in the material. However, the morphologic characteristics of a neat resin sample may be quite different from those found in a fabricated composite.

8.3.4.4. Volatiles content

The volatiles content of resin materials is of interest to prepreg manufacturers, resin transfer molding fabricators, and manufacturers using wet lay-up techniques. Inadequate control of volatiles may affect the handling of prepregs and final laminate quality. The property is most relevant where materials have a solvent incorporated into the formulation. The volatile content represents the solvent(s), resin components, and other constituents that volatilize at the test temperature. The remainder is commonly known as resin solids and filler. The resin solids include the materials that can be dissolved in an organic solvent (usually polymer constituents) and the filler is generally the insoluble inorganic material. The test for volatiles is commonly performed as a quality control check.

Specific methods are material dependent, but they generally involve placing the materials in an air circulating or vacuum oven at a specified temperature for a specified time. Weight loss is measured gravimetrically. The time and temperature are selected such that volatiles completely evaporate and resin solids do not volatilize or degrade. The ASTM D 3530 'Volatiles Content of Carbon-Fiber Prepreg' [75] represents the standard oven exposure test method intended for carbon/epoxy fiber prepregs. The procedure may also be used as a guideline for most thermosetting resins, although the user is cautioned that the temperature may not be appropriate for other resin systems. In the ASTM standard, the time is fixed but the temperature is selected to represent the part process cycle.

Thermogravimetric analysis is sometimes used instead of an oven exposure method. It is an instrumented procedure where a small specimen is placed on a microbalance. The instrument is automated to gradually increase the temperature while continuously monitoring the specimen weight. Thermogravimetric analysis procedure for volatiles content should specify heating rate, gas and flow rate, specimen size, and the temperature range used to define volatile content. While there is no established ASTM procedure for thermogravimetric analysis to determine volatiles in resins, the ASTM E 1131 'Compositional Analysis by Thermogravimetry' [76] gives general guidance. Other techniques such as thermogravimetric analysis–Fourier transform IRS and headspace GC [77] may be used in a research or problem-solving environment for qualitative/quantitative determination of volatiles.

8.3.4.5. Moisture content

Depending on the resin system, moisture may retard cure, create volatiles, or cause other undesirable reactions during processing. For those resin systems affected, control and measurement of the moisture content is required. Automated moisture meters based on the Karl Fischer titration method are commonly used to determine moisture content in most types of resins. A small sample, typically 5 g of liquid resin, is introduced to a cell containing the Karl Fischer reagent and a solvent, typically methanol. A current is passed between two electrodes, creating iodine in a quantitative series of reactions with water and the Karl Fischer reagent. The reaction and test are described in the ASTM D 4672 'Polyurethane Raw Materials.'

An alternate method for determining moisture content uses an instrument that heats a small specimen, typically 10 g of solid. The moisture is volatilized and transported to an electrolytic cell by a nitrogen carrier gas.

The water reacts to form phosphoric acid, which is then quantitatively measured by passing current through the cell. A standard for this test is ASTM D 4019 'Method for Moisture in Plastics by Coulometric Regeneration of Phosphorous Pentoxide' [78].

8.4. EVALUATION OF REINFORCEMENT–MATRIX INTERFACE

8.4.1. Introduction

In our discussion of reinforcements it has been mentioned that sometimes a surface treatment is carried out to achieve the required bonding to the matrix. The load acting on the matrix has to be transferred to the reinforcement via the interface. Thus, the reinforcements must be strongly bonded to the matrix if their high strength and stiffness are to be imparted to the composite. The fracture behavior is also dependent on the strength of the interface. A weak interface results in a low stiffness and strength but high resistance to fracture, whereas a strong interface produces high stiffness and strength but often a low resistance to fracture, that is, brittle behavior. Other properties of a composite, such as resistance to creep, fatigue, and environmental degradation, are also affected by the characteristics of the interface. In these cases the relationship between properties and interface characteristics is generally complex.

The interface is important whether the reinforcement is in the form of continuous fibers, short fibers, or whiskers or particles, although the exact role of the interface may differ with the type of reinforcement. For example, we will see that the stress distribution along the interface is critical in determining the reinforcement efficiency of short fiber composites and the details of failure in the vicinity of the ends of broken continuous fibers.

In some cases a distinct phase, produced by a reaction between the matrix and the reinforcement, exists at the reinforcement–matrix interface. In other instances, the interface can be viewed as a planar region of only a few atoms in thickness across which there is a change in properties from those of the matrix to those of the reinforcement. Thus, at the interface there is usually a discontinuity in chemical nature, crystal and molecular structure, mechanical properties, and other properties. The characteristics of the interface are determined by the discontinuity in properties and are therefore specific to each reinforcement–matrix combination. The surface roughness of the reinforcement has also to be considered.

8.4.2. Wettability

Interfacial bonding is due to adhesion between the reinforcement and the matrix and due to mechanical keying. Clearly, for adhesion to occur during the manufacture of a composite the reinforcement and the matrix must be brought into intimate contact. At some stage in composite manufacture the matrix is often in a condition where it is capable of flowering and its behavior approximates to that of a liquid. A key concept in this context is wettability. Wettability defines the extent to which a liquid will spread over a solid surface. Good wettability means that the liquid (matrix) will flow over the reinforcement covering every 'bump' and 'dip' of the rough surface of the reinforcement and displacing all air.

Wetting will only occur if the viscosity of the matrix is not too high and if wetting results in a decrease in the free energy of the system. Let us consider the latter in more detail by studying a thin film of liquid (matrix) spreading over the solid (reinforcement) surface. All surfaces have an associated energy and the free energy per unit area of the solid–gas, liquid–gas, and solid–liquid interfaces is γ_{SG}, γ_{LG}, and γ_{SL}, respectively. For an increment of area dA covered by the spreading film, extra energy is required for the new areas of solid–liquid and liquid–gas interfaces. This extra energy is ($\gamma_{SL}\,dA + \gamma_{LG}\,dA$), whereas $\gamma_{SG}dA$ is recovered as the solid surface is covered. For the spreading of the liquid to be spontaneous, it has to be energetically favorable and therefore we must have:

$$\gamma_{SL}\,dA + \gamma_{LG}\,dA < \gamma_{SG}\,dA \qquad (8\text{-}2)$$

Dividing by dA gives:

$$\gamma_{SL} + \gamma_{LG} < \gamma_{SG} \qquad (8\text{-}3)$$

From this relationship the spreading coefficient (SC) is defined as:

$$\mathrm{SC} = \gamma_{SG} - (\gamma_{SL} + \gamma_{LG}) \qquad (8\text{-}4)$$

which has to be positive for wetting. It follows from Eq. (8-3) that if γ_{SG} is similar to or less than γ_{LG}, then wetting will not occur. Thus, an epoxy resin with γ_{LG} of about 40 mJ/m^2 might be expected to readily wet alumina ($\gamma_{SG} \sim 1100$ mJ/m^2) but not polyethylene which has a low value of γ_{SG} of about 30 mJ/m^2.

The above analysis demonstrates the significance of the relative values for the surface energies involved in the wetting process but does not yield a parameter that specifies intermediate conditions of wetting. The readers can find more detail explanation about wettability in Chapter 3 (Solid–Liquid Interaction).

8.4.3. Interfacial Bonding

Once the matrix has wet the reinforcement, and is therefore in intimate contact with the reinforcement, bonding will occur. A number of different types of bonds may be formed. Furthermore, for a given system more than one bonding mechanism may be operative at the same time, for example, mechanical and electrostatic bonding, and the bonding mechanism may change during the various production stages or during service, for example, electrostatic bonding changing to reaction bonding. The type of bonding varies from system to system and depends on fine details such as the presence of surface contaminants or of added surface active agents (coupling agents). This section discusses the main features of the different bonding mechanisms.

8.4.3.1. *Mechanical bonding*

A mechanical interlocking or keying of two surfaces, as shown in Fig. 8-4(a), can lead to a reasonable bond. Clearly the rougher the interface, the greater is the interlocking, and hence more effective the mechanical bonding. Also any contraction of the matrix onto the reinforcement is favorable to bonding.

The mechanical bond is most effective when the force is applied parallel to the interface; in other words, the shear strength may be considerable. On the other hand, when the interface is being pulled apart by tensile forces, the strength is likely to be low unless there is a high density of features (designated A in Fig. 8-4(a)) with the re-entrant angles. In most cases a purely

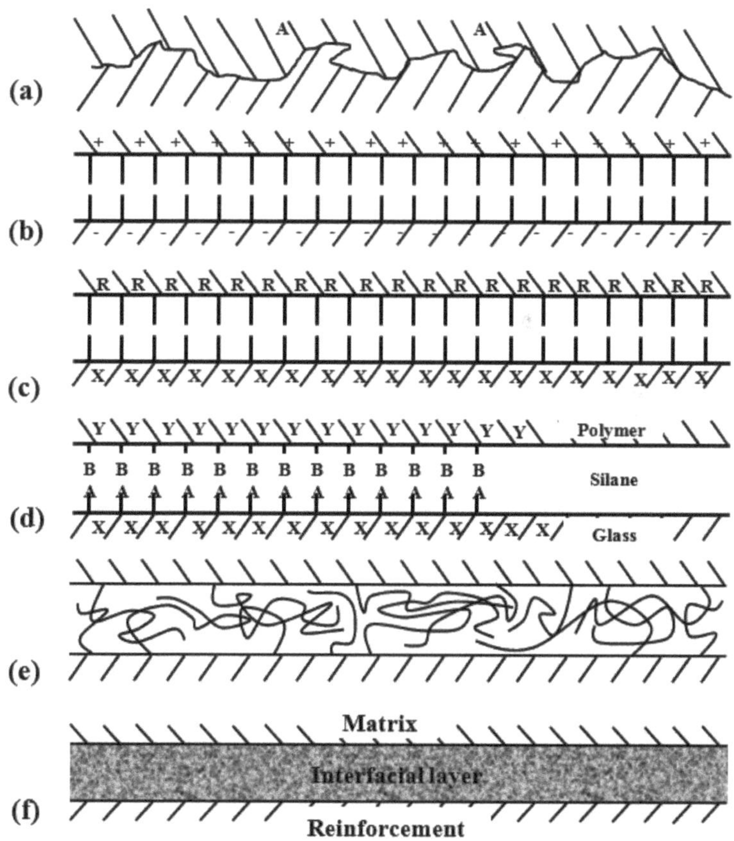

Fig. 8-4 Schematic diagrams of the interfacial bonding mechanisms: (a) mechanical bonding (features designated A have a re-entrant angle); (b) electrostatic bonding; (c) chemical bonding (R and X represent compatible chemical groups); (d) chemical bonding as applied to a silane coupling agent; (e) reaction bonding involving polymers; (f) interfacial layer formed by interdiffusion.

mechanical bond is not encountered and mechanical bonding operates in conjunction with another bonding mechanism.

8.4.3.2. Electrostatic bonding

Bonding occurs between the matrix and the reinforcement when one surface is positively charged and the other negatively charged (Fig. 8-4(b)). This leads to an electrostatic attraction between the components of the composite which will depend on the difference in charge on their surfaces. Electrostatic interactions are short range and are only effective over small distances of the order of atomic dimensions: the reader will therefore appreciate that it is essential that the matrix and reinforcement are in intimate contact and that surface contamination and entrapped gases will decrease the effectiveness of this bonding mechanism.

8.4.3.3. Chemical bonding

In the context of composite science, chemical bonding is the bond formed between chemical groups on the reinforcement surface (marked X in Fig. 8-4(c)) and compatible groups (marked R) in the matrix. Not surprisingly, the strength of the chemical bond depends on the number of bonds per unit area and the type of bond.

It is thought that chemical bonding may account for the success of some coupling agents. For example, *silanes* are commonly employed for coupling the oxide groups on a glass surface to the molecules of a polymer matrix. At one end (A) of the silane molecule a hydrogen bond forms between the oxide (silanol) groups on the glass and the partially hydrolyzed silane, whereas at the other end (B) it reacts with a compatible group in the polymer (Fig. 8-4(d)).

8.4.3.4. Reaction or interdiffusion bonding

The atoms or molecules of the two components of the composite may interdiffuse at the interface to give what is known as reaction or interdiffusion bonding. For interfaces involving polymers this type of bonding can, in simple terms, be considered as due to the intertwining of molecules (Fig. 8-4(e)). Many factors control the strength of this interface including the distance over which the molecules have entwined the extent of the entanglement of the molecules and the number of molecules per unit area of interface.

For systems involving metals and ceramics the interdiffusion of species from the two components can produce an interfacial layer of different composition and structure from either of the components (Fig. 8-4(f)). The interfacial layer will also have different mechanical properties from either the matrix or reinforcement and this consequently greatly affects the characteristics of the interface. In metal matrix composites the interfacial layer is often a brittle intermetallic compound which is a compound that exists at or around a stoichiometric composition such as AB, A_2B, and A_3B, for example, $CuAl_2$.

One of the main reasons why interfacial layers are formed is that ceramic matrix and metal matrix composite production invariably involve high temperatures. Diffusion is rapid at high temperatures as the rate diffusion, or the diffusion coefficient, D_d, increases exponentially with temperature according to the following Arrhenius-type equation:

$$D_d = D_0 \exp(-Q_d/RT), \tag{8-5}$$

where Q_d is the activation energy for diffusion, D_0 the constant, R the gas constant, and T the temperature.

Temperature has a marked effect on the diffusion coefficient; substituting a typical value for Q_d of 250 kJ/mol into Eq. (8-5) it can be shown what D_d is 2×10^{34} greater at 1000 °C than at room temperature. The extent of the interdiffusion x, and hence the thickness of the reaction layer, depends on time, t, as well as temperature and is given by the following approximate relationship:

$$x = (D_d t)^{1/2} \tag{8-6}$$

An interface can change during service. In particular, interfacial layers can form during service at elevated temperature. In addition, previously formed layers may continue to grow and complex multilayer interfaces may develop.

8.4.4. Methods for Measuring Bond Strength

In view of the major role played by the interface in determining some of the most important features of the mechanical behavior of composites, the reader will not be surprised to learn that numerous test methods have been employed to measure interfacial strength. Unfortunately it turns out that there is no simple, reliable method for determining this quantity. Besides the inherent problems associated with the different techniques, the results may not be adhesive, that is, the two components may not have separated at the interface. The failure may take place close to the interface but in the reinforcement or in the matrix, this is a cohesive failure. The situation is further complicated if an interfacial layer of material C has formed between components A and B. In this situation there are two interfaces at which it is possible to have adhesive failure, namely, at A–C and C–B, and three materials (A, B, and C) which could fail cohesively.

It can be argued that, from the practical point of view, it does not matter whether failure during testing is adhesive or cohesive as long as the strength of the 'weak link' at the interface is being measured. This argument may be valid provided it is possible to ensure that the testing conditions reproduce the service conditions, but even then inadequate information on the type and strength of interfaces will undoubtedly limit our knowledge and the development of composites.

8.4.4.1. Single fiber tests

The most direct test involves pulling a partially embedded single fiber out of a block of matrix material (Fig. 8-5(a)). This test, although simple in principle, is difficult to carry out especially for thin brittle fibers. However, if successful,

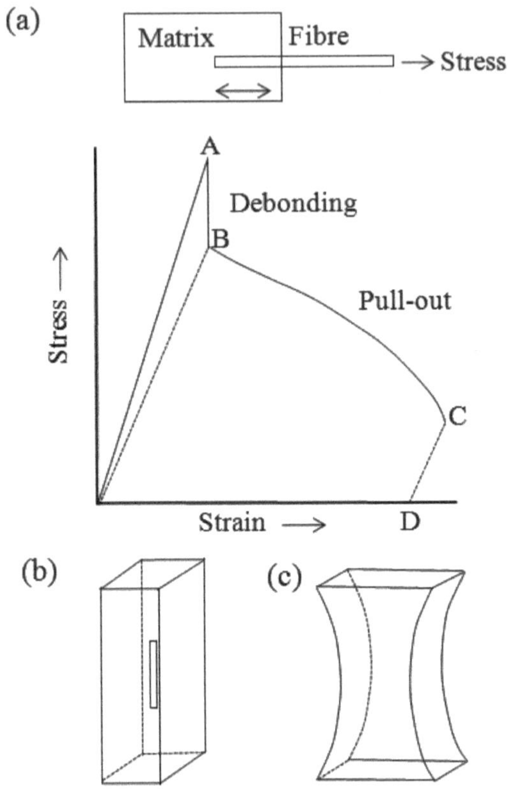

Fig. 8-5 Single fiber tests: (a) fiber pull-out test; (b) compression test for interfacial shear strength; (c) compression test for interfacial tensile strength.

from the resulting tensile stress vs. strain plot the shear strength of the interface and the energies of debonding and pull-out may be obtained.

The interfacial shear strength τ_1 may also be evaluated using a specimen consisting of a block of matrix material with a single, completely embedded short fiber which is accurately aligned longitudinally in the center of the specimen (Fig. 8-5(b)). On testing in compression, shear stresses are set up at the ends of the fibers as a consequence of the difference in elastic properties of the fiber and matrix. The shear stresses eventually lead to debonding at the fiber ends and τ_1 may be evaluated from the compressive stress σ_c at which debonding occurs as:

$$\tau_1 \sim 2.5\sigma_c \tag{8-7}$$

The problem with this test is in determining the onset of debonding and hence σ_c. This is usually done by visual observation which means the test is only really suitable for transparent matrices.

Debonding induced by tensile stresses may be achieved if a curved neck specimen with a single continuous fiber is tested in compression (Fig. 8-5(c)). With this test geometry, transverse tensile stresses, perpendicular to the interface, arise because of the differences in the Poisson's ratio of the fiber and the matrix. At a compressive stress of σ_c, the tensile strength σ_1 of the interface is reached and tensile debonding occurs. σ_1 proportional to σ_c:

$$\sigma_1 = C\sigma_C \tag{8-8}$$

where C is a constant which depends on Poisson's ratio and Young's modulus of the fiber and matrix. As for the previous tests it is difficult to determine the onset of debonding and hence σ_c. Also compression tests require accurate alignment of the fibers.

An alternative method is to carry out a tensile test on a dumbbell-shaped specimen containing a single fiber. Application of the load causes the fiber to fragment; the limiting final fragment length is the critical length, from which the interface shear strength may be calculated.

8.4.4.2. Bulk specimen tests

We can acquire information on the interfacial strength using bulk composite specimens and modifications to conventional mechanical property tests such as torsion, tensile, and three-point bend tests. For present purposes it is inappropriate to discuss all of these and three-point bend and the Iosipescu tests have been selected as commonly used examples of this type of test.

Of these the three-point bend geometry and the variation in the shear and tensile stresses in the loaded specimen is illustrated in Fig. 8-6. At a given load P, the maximum tensile stress σ, which is parallel to the specimen length, lies on a line on the surface below the center loading point (Fig. 8-6(b)) and is given by:

$$\sigma = \frac{3}{2}\frac{(PS)}{(D^2)'} \tag{8-9}$$

where P, D, and S are defined in Fig. 8-6(a); D and P are the height (thickness) and breadth of the specimen, respectively, and S the span between the outer loading points.

Using this testing arrangement with an aligned composite with the fibers perpendicular to the specimen length (Fig. 8-6(d)) enables the tensile strength σ_1 of the interface to be evaluated by the substitution of the failure load into Eq. (8-9).

We have to consider the shear stresses in a specimen if we wish to determine the interfacial shear strength. The shear stress has a maximum at the midplane of the specimen (Fig. 8-6(c)) of:

$$\tau = \frac{3}{4}\frac{P}{BD'} \tag{8-10}$$

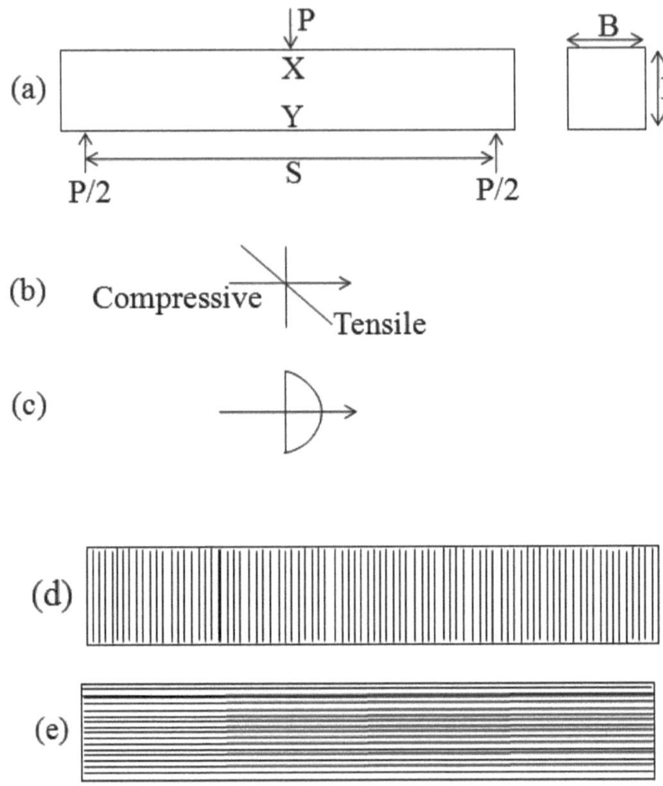

Fig. 8-6 Three-point bend test: (a) specimen geometry; (b) tensile stresses; (c) shear stresses; (d) fiber alignment for determination of interfacial tensile strength; (e) fiber alignment for determination of interfacial shear strength.

Combining the above equation with Eq. (8-9) gives:

$$\frac{\tau}{\sigma} = \frac{D}{2S} \tag{8-11}$$

from which it can be seen that the ratio of the maximum shear stress to the maximum tensile stress depends on S and D, that is, specimen geometry. A short span and a thick specimen increase the ratio and therefore enhance the likelihood of a shear failure. It follows that a short, thick specimen of an aligned fiber composite with the fibers parallel to the length of the specimen (Fig. 8-6(e)) will fail by shear stress at the fiber–matrix interface provided $\tau_1/\hat{\sigma}_{TC}$, where $\hat{\sigma}_{TC}$ is the tensile strength of the composite, is less than $D/2S$. This test configuration is known as a short/beam bend test or the interlaminar shear strength test. Unfortunately both the transverse tensile strength and the shear strength obtained from three-point bend test are found to depend on the volume

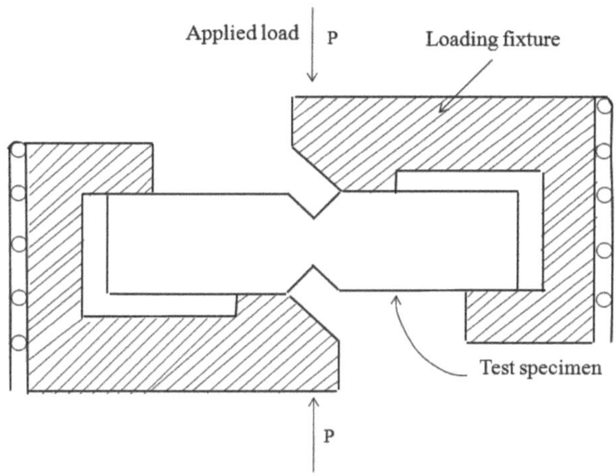

Fig. 8-7 Diagram of Iosipescu testing fixture and specimen.

fraction of fibers and as a result cannot be considered as true values for the bond strengths.

A number of tests use specimens with a double-V-notch and the method attributed to Iosipescu falls into this category. The Iosipescu shear test (Fig. 8-7) aims to achieve a state of pure shear in the mid-length of the specimen through the action of force couples which produce counteraction moments. The principle is that a constant shear force is induced through the mid-section of the specimen and that the moments exactly cancel at the mid-length of the specimen, thus producing a pure shear loading at that location. In practice, however, there is some nonuniformity of the shear stress distribution in the test section with probably small normal compressive stresses. Nevertheless, this test geometry has been found to yield reliable shear strength data.

8.4.4.3. Micro-indentation tests

As the name suggests, this test employs a standard micro-indentation hardness tester. The advantages of the micro-indentation test are that the specimen is small and does not have to receive any special preparation except that one surface has to be polished to a finish suitable for microscopic examination. The indentor is loaded with a force P on to the center of a fiber, whose axis is normal to the surface, and causes the fiber to slide along the fiber–matrix interface. The surface of the fiber is therefore depressed a distance μ below the surrounding matrix surface (Fig. 8-8). For a specimen of reasonable thickness the sliding distance is less than the specimen

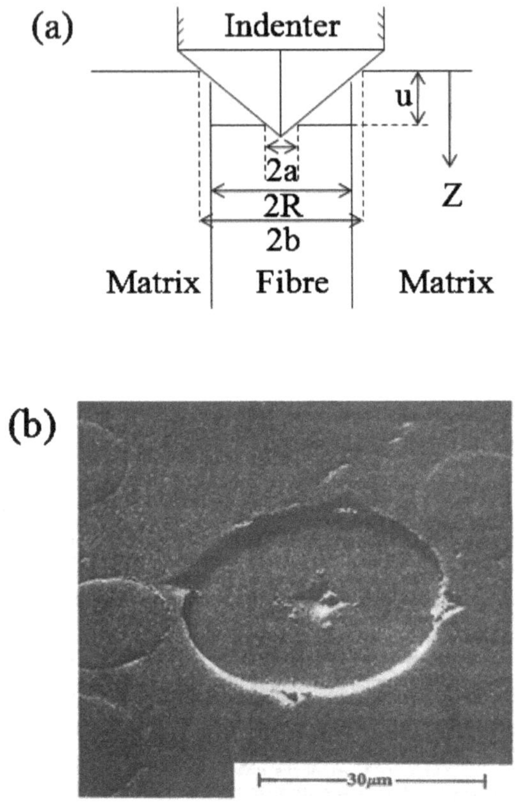

Fig. 8-8 Micro-indentation test: (a) cross-sectional diagram of loaded indentor; (b) scanning electron micrograph of an indented silicon carbide fiber in a lithium aluminosilicate glass–ceramic matrix.

thickness and in these circumstances the interfacial shear strength τ_1 is given by:

$$\tau_1 = P^2/4\pi u R^3 E_f \tag{8-12}$$

where R and E_f are the radius and Young's modulus of the fiber, respectively. For a standard pyramid indentor the depression distance μ may be calculated from:

$$u = (b - a)\cot 74°, \tag{8-13}$$

where a is the half diagonal length of the indentation on the fiber and b the half diagonal length of the indentation on the matrix surrounding the fiber. This method is particularly suitable for ceramic matrix composites and has been shown to be sensitive enough to detect changes in τ_1 with heat treatment time and temperature for glass–ceramic matrix composites (Fig. 8-9).

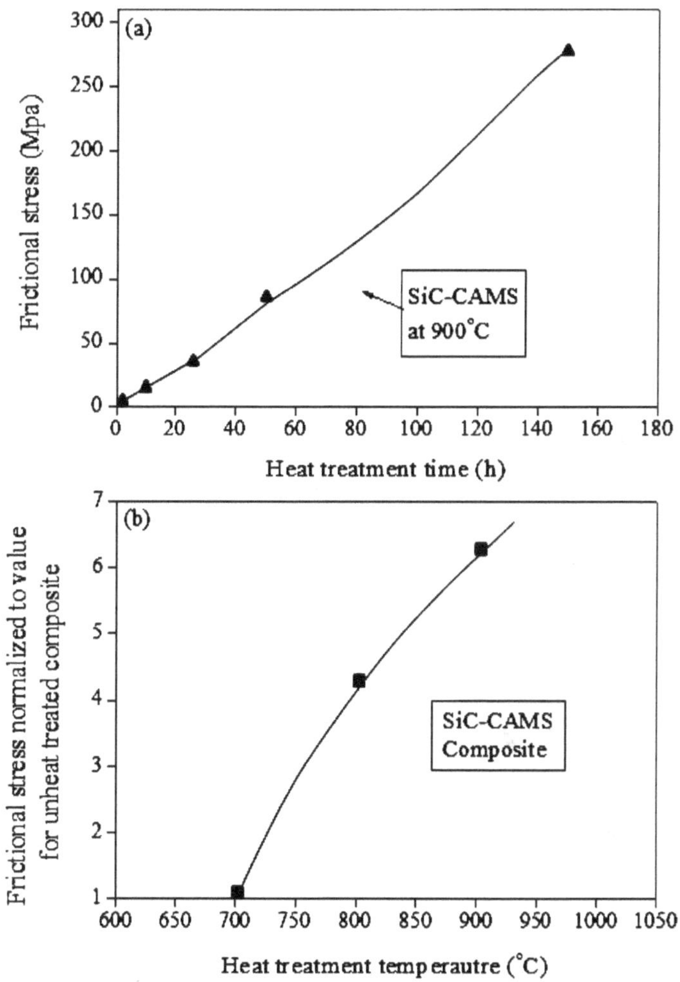

Fig. 8-9 Micro-indentation data for the interfacial shear stress of silicon carbide fiber in 'Silceram' glass–ceramic as a function of heat treatment: (a) time at 900 °C; (b) temperature for 12 h holding.

8.5. EVALUATION OF COMPOSITES

8.5.1. Introduction

Experimental characterization refers to the determination of the material properties through tests conducted on suitably designed specimens. The data obtained from the tests are appropriately reduced to evaluate various material properties that can later be used for analysis and design of practical structures.

Understanding the material response over the entire range of loads is necessary if advanced design procedures are employed for efficient material utilization. In the case of composite materials it may be desirable to begin the design with constituent material properties and arrive at the composite macromechanical properties through micromechanics analyses. However, confidence in the micromechanics analyses is created by necking validity through experiments. Other purposes of the experimental characterization are to check the adequacy of fabrication procedures, assure material uniformity, and as a comparative measure for various candidate materials.

Elastic constants and strengths are the basic mechanical properties of materials. For a unidirectional lamina or composite, there are four independent elastic constants – the elastic moduli in the longitudinal and transverse directions, the shear modulus, and the major Poisson ratio – and five independent strengths, namely, tensile and compressive strengths in the longitudinal and transverse directions and the in-plane shear strength. In the case of a laminate the interlaminar shear strength is also an important property. It is necessary to establish these properties for the minimum characterization of a unidirectional lamina. They are generally established by subjecting suitable material specimens to in-plane loads. However, since composite material structures are often subjected to bending loads, it is desirable to establish flexural properties in addition to the properties just mentioned. It is desirable that all the properties be established for a single ply or lamina of the composite material that is the basic building block for composite structures. However, practical considerations often prevent the construction of single-layer test specimens. Thus, it becomes necessary to conduct tests on multi-layer specimens and use appropriate laminate theory to reduce the results in terms of lamina properties. If the laminates are unidirectional, of course, their behavior simulates the lamina behavior.

In this section, the test procedures commonly employed for evaluating various composite properties are described. The appropriateness of selecting a particular type of specimen for each test is discussed. The methods of reducing the experimental data are also discussed. However, the discussion is limited only to static properties and the details of instrumentation and measuring techniques are omitted. A more detailed discussion can be found in the literature [79–85]. Furthermore, various standard setting bodies, such as the American Society for the Testing of Composites, and these standards should be consulted.

8.5.2. Factors Determining the Properties

The fabrication and properties of composites are strongly influenced by the proportions and properties of the matrix and the reinforcement. The proportions can be expressed either via the weight fraction (w) which is relevant to fabrication or via the volume faction (v), which is commonly used in property calculations. The definitions of w and ν are related simply to the fractions of weight (W) or volume (V) shown as follows:

$$\text{Volume fractions:} \quad \upsilon_f = V_f/V_c \quad \text{and} \quad \upsilon_m = V_m/V_c \tag{8-14}$$

$$\text{Weight fractions:} \quad w_f = W_f/W_c \quad \text{and} \quad w_m = W_m/W_c \tag{8-15}$$

where the subscripts m, f, and c refer to the matrix, fiber (or in the more general case, reinforcement), and composites, respectively.

We note that:

$$\upsilon_f + \upsilon_m = 1 \tag{8-16}$$

and

$$w_f + w_m = 1 \tag{8-17}$$

We can relate weight to volume fractions by introducing the density, ρ, of the composites and their constituents. Now:

$$W_c = W_f + W_m \tag{8-18}$$

which, as $W = \rho V$, becomes:

$$\rho_c V_c = \rho_f V_f + \rho_m V_m \tag{8-19}$$

or

$$\rho_c V_c = \rho_f (V_f/V_c) + \rho_m (V_m/V_c) = \rho_f \upsilon_f + \rho_m \upsilon_m \tag{8-20}$$

It may also be shown that:

$$\frac{1}{\rho_e} = \frac{w_f}{\rho_e} + \frac{w_m}{\rho_e} \tag{8-21}$$

Also we have

$$w_f = \frac{W_f}{W_c} = \frac{(\rho_f V_f)}{(\rho_c V_c)} = \frac{\rho_f}{\rho_c} \upsilon_f, \tag{8-22}$$

and similarly

$$w_m = \frac{W_m}{W_c} = \frac{(\rho_m V_m)}{(\rho_c V_c)} = \frac{\rho_m}{\rho_c} \upsilon_m \tag{8-23}$$

We can convert from weight fraction to volume fraction, and vice versa, provided the densities of the reinforcement (ρ_f) and the matrix (ρ_m) are known.

Eq. (8-14) shows that the density of the composite is given by the volume fraction adjusted sum of the densities of the constituents. This equation not only is applicable to density but also, in certain circumstances, may apply to other properties of composites. A generalized form of the equation is:

$$X_c = X_m v_m + X_f v_f \tag{8-24}$$

where X_c represents an appropriate property of the composite, and, as before, v the volume fraction and the subscripts m and f refer to the matrix and reinforcement, respectively. This equation is known as the law of mixtures.

Most properties of composites are a complex function of a number of parameters as the constituents usually interact in a synergistic way so as to provide properties in the composite that are not fully accounted for by the law of mixtures. The chemical and strength characteristics of the interface between the fibers and the matrix are particularly important in determining the properties of the composites. The interfacial bond strength has to be sufficient for load to be transferred from the unreinforced matrix. On the other hand, if we are also concerned with the toughness of the composites, the interface must not be so strong that it does not fail and allow toughening mechanisms such a debonding and fiber pull-out to take place.

Other parameters which may significantly affect the properties of composites are the shape, size, orientation, and distribution of the reinforcements and various features of the matrix such as the grain size for polycrystalline matrices. These, together with volume fraction, constitute what is called the microstructure of the composites. However, it should be noted that even for properties which are microstructure dependent, and which do not obey the law of mixtures, the volume fraction still plays a major role in determining properties. The volume fraction is generally regarded as the single most important parameter influencing the composite's properties. Also, it is an easily controllable manufacturing variable by which the properties of composites may be altered to suit the application.

A problem encountered during manufacture is maintaining a uniform distribution of the reinforcements. Ideally, the composites should be homogeneous, or uniform, but this is difficult to achieve. Homogeneity is an important characteristic that determines the extent to which a representative volume of the materials may differ in physical and mechanical properties from the average properties of the materials. Nonuniformity of the system should be avoided as much as possible because it reduces those properties that are governed by the weakest part of the composites. For example, failure in nonuniform materials will initiate in an area of lowest strength, thus adversely affecting the overall strength of a component manufactured from those materials.

In components manufactured from continuous fiber–reinforced composites, such as unidirectional or cross-ply laminates, anisotropy may be desirable as it can be arranged for the maximum service stress to be in the direction that has the highest strength. Indeed, a primary advantage of these composites is the ability to control the anisotropy of a component by design and fabrication. It is clear from the preceding discussion that many factors may be important in determining the properties of composites; this should be borne in mind when studying theoretical models as all the factors are seldom accounted for the development of the models.

8.5.3. Principal Coordinate Axes

In the composite tests, the classical beam theory is referred to the text that one wishes to derive from the theory [86–89]. Fig. 8-10 shows the principal coordinate axes for unidirectional lamina. The unidirectional ply is characterized by:

1. E_{11}: longitudinal stiffness (elastic modulus) (E_1).
2. E_{22}: transverse stiffness (E_2).
3. E_{33}: stiffness (E_3).
4. G_{12}: in plane shear modulus = shear modulus in 1–2 plane = G_{21}.
5. G_{23}: shear modulus in 2–3 plane = G_{32}.
6. G_{13}: shear modulus in 1–3 plane = G_{31}.
7. ν_{12}: major Poisson's ratio (load in 1-direction, strain in 2-direction) (please note that this is different from ν_{21}, which is the minor Poisson's ratio).
8. ν_{23}: Poisson's ratio (load in 2-direction, strain in 3-direction) (different from ν_{32}).
9. ν_{13}: Poisson's ratio (load in 1-direction, strain in 3-direction) (different from ν_{31}).
10. σ_{1t}: longitudinal tensile strength.
11. σ_{2t}: transverse tensile strength.

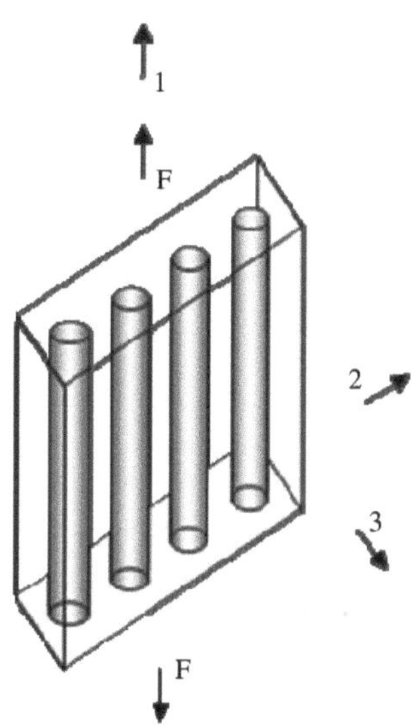

Fig. 8-10 Principal coordinate axes for unidirectional lamina.

12. σ_{3t}: tensile strength in the 3-direction.
13. σ_{1c}: longitudinal compressive strength.
14. σ_{2c}: transverse compressive strength.
15. σ_{3c}: compressive strength in the 3-direction.
16. τ_{12}: in-plane shear failure in 1–2 direction $= \tau_{21}$ (interlaminar shear strength).
17. τ_{23}: in-plane shear failure in 2–3 direction $= \tau_{32}$.
18. τ_{13}: in-plane shear failure in 1–3 direction $= \tau_{31}$.
19. α_1, α_2, and α_3: coefficients of thermal expansion.
20. β, β_2, and β_3: coefficients of moisture expansion.
21. κ_1, κ_2, and κ_3: coefficients of thermal conductivity.

8.5.4. Density

Density of composite materials can be determined by simple using the relationship between volume and mass. A special technique for the determination of the density where pores are taken into account is the determination by using water displacement (Archimedian density).

8.5.4.1. Dry bulk density

A basic method to determine the dry bulk density is using a specimen with regular shape (usually rectanguloid). By measuring the mass and getting the knowledge of the volume the density can be calculated as:

$$\text{Density} = \text{Weight (mass)}/\text{Unit volume} \tag{8-25}$$

8.5.4.2. Density by water displacement (Archimedean density)

This method allows the determination of the density in air compared to its displacement in water or other liquid of known density. Depending on the nature of the specimen (e.g., open or closed cell), the resultant value may deviate from the true mass.

A clean specimen is weighed accurately in air using a laboratory balance. The same specimen is weighed while suspended in water or other liquid of such density that the specimen will sink. Deducting the mass of the suspension wire from the weight in liquid, the volume of the specimen is calculated from the effect of displacement by a liquid of known density (Archimedean principle). This allows the determination of density of specimens with irregular shapes, uneven surfaces, or porosity. Caution must be exercised to ensure that no air is trapped within the specimen. Placing the specimen in a vacuum while submerged in the displacement liquid will usually avoid any error.

8.5.5. Determination of Fiber Content

The mass fraction of the carbon fiber–reinforced composites can be reliably determined by chemical digestion of the resin [90, 91]. The weighed specimen is placed in a Kjeldahl flask in an efficient furnell cupboard utilizing backboard extraction, with the mouth of the flask pointing away from the operator. Concentrated H_2SO_4 is added to the flask, which is then heated to furning. Fifty percent (w/v) H_2O_2 is then added dropwise until the solution becomes colorless. The amount added depends on the resin type. An epoxy novaloc or phenolic resin will use up to some five times that required for an epoxy resin. After digestion, the carbon fiber is filtered into a weighed sintered glass crucible, washed with distilled water until free from acid, dried, and weighed. This process is quite hazardous and a proprietary glass fume extraction/washing system should preferably be attached above the mouths of the Kjeldahl flasks.

8.5.6. Coefficient of Thermal Expansion

8.5.6.1. Dilatometer

The coefficient of thermal expansion is one of the most important properties of metal matrix composites. Since nearly all metal matrix composites are used in various temperature ranges, the measurement of coefficient of thermal expansion as a function of temperature is necessary in order to know the behavior of the material. Several different systems for measurement of coefficient of thermal expansion can be used depending on the temperature conditions. One of the most common used systems is a dilatometer, as shown in Fig. 8-11.

A dilatometer measures the length or the volume changes of the sample, when the sample follows a temperature program and submits a small force

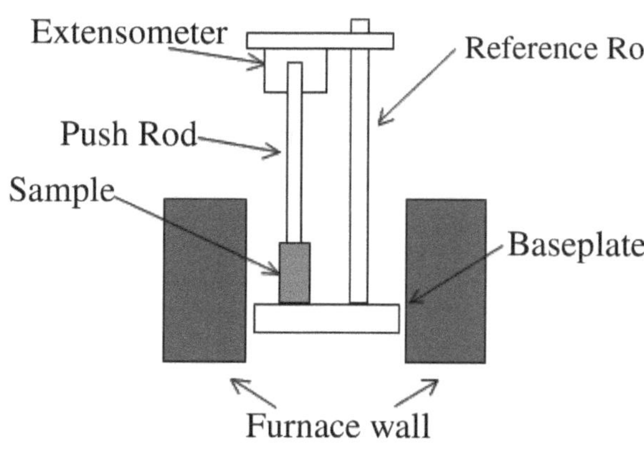

Fig. 8-11 Schematic setup for measurement of coefficient of thermal expansion (dilatometer).

(pushrod contact pressure can be adjusted between 15 and 50 cN). In a pushrod dilatometer the change in length of the sample is detected by an inductive displacement transducer. Calibration and corrections of measurements are done by using various standards and comparison with materials of known expansion. The measurement of the coefficient of thermal expansion can be carried out in the temperature range from approximately −150 up to 1500 °C under various conditions (air, inert gas, or vacuum). The coefficient of thermal expansion can be measured within an error of 3–5%.

8.5.7. Thermal Conductivity

8.5.7.1. Comparative method

The thermal conductivity of metals, alloys, or composites with = 0, 2, …, 100 W/mK can be measured by comparative method with steady-state longitudinal heat flow in a temperature range room temperature up to about 1000 °C. The comparative instrument measures heat flow based on the known thermal properties of standard reference materials. The test specimen is sandwiched between two identical reference samples (Fig. 8-12). This stack is placed between two heating elements controlled at different temperatures. A guard heater is placed around the test stack to ensure a constant heat flux through the stack and no lateral heat flow losses. As heat flows from the hot element to the cold element the temperature gradient (T/L) across the stack is measured with thermocouples. Once the specimen reaches a state of thermal equilibrium, its thermal conductivity is calculated from the following expression:

$$\lambda = \frac{L}{A} \bullet \frac{Q}{\Delta T} \tag{8-26}$$

where Q is the heating power of the heater.

The experimental error is in the range of approximately ±10%. The specimen geometry is cylindrical (diameter 25 or 50 mm, height approximately 10–40 mm).

8.5.8. Specific Heat

8.5.8.1. Differential Scanning Calorimetry

The differential scanning calorimetry is a thermoanalytical method, which allows, by measuring the temperature difference between the sample and a

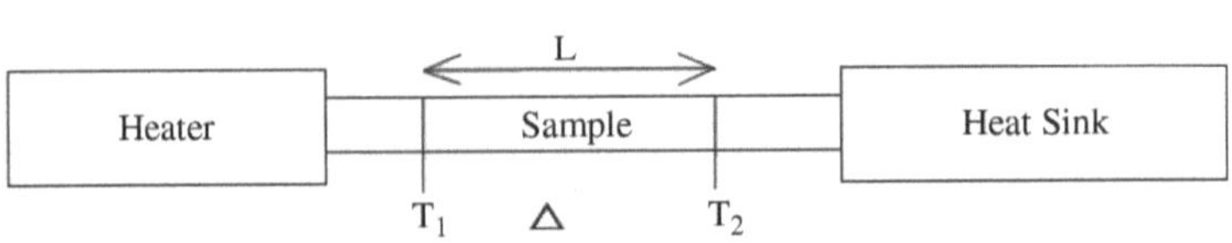

Fig. 8-12 Schematic diagram for measurement of thermal conductivity.

reference, a quantitative measurement of the heat flow difference when the sample and the reference submit the same temperature program. The results are very important for many applications and provide information for the characterization of the materials. Measurements of specific heat can be performed in air, inert gas, or under vacuum. Centipoise (*Cp*) measurements up to 1500 °C with an error <3% are possible. Dimensions of solid samples are a diameter of 5 mm and a thickness of 1 mm. Samples should be placed parallel and the face in contact with the crucible must be polished to have the optimal heat transmission. Powders should be pressed for a better thermal conductivity.

8.5.9. Electrical Resistivity

8.5.9.1. Four-point probe measurements

Measurement of volume and surface resistivity of flat test specimens can be performed under controlled relative humidity conditions at standard temperature. For measurement a specimen is placed in a resistivity test cell constructed to perform volume and surface resistivity measurements. A voltage is applied to one electrode and the current through or across the surface of the specimen is measured. The voltage and current measurements and the fixture geometry are used to compute volume or surface resistivity. Four-point probe measurements can be made on conductive materials using a micro-ohmmeter.

8.5.10. Thermal Cycling

Measurement of thermal cycling is necessary in order to simulate the behavior and the performance of a material during its operation. Therefore, the measurement of coefficient of thermal expansion or thermal conductivity of a material during several temperature cycles can be useful. Temperature ranges can be between room temperature and the proposed temperature of operation. Measuring of coefficient of thermal expansion during several cyclings leads to information about the deformation per cycle or the total resulting deformation during all cycles. The behavior of coefficient of thermal expansion and thermal conductivity during thermal cycling also allows a statement about the interface properties in the materials.

8.5.11. Tensile Modulus

There are many versions of tensile specimen and as a routine test procedure, generally, a simple test bar ($150 \times 10 \times 1.0$ mm) is sufficient with end tabs attached, fabricated from prepreg or aluminum [91]. Preparation of the composite surface by abrasion and subsequent degreasing prior to attaching end pieces is important and the end pieces should be held in a jig during the bonding process. If aluminum is used, then the end pieces should be roughened, or grit blasted and degreased, followed by etching for 10 min at 65 °C in:

22. 30 cm^3 distilled H_2O,
23. 10 cm^3 concentrated H_2SO_4,
24. 1 g $Na_2Cr_2O_7$,

and then rinsed thoroughly with deionized water, air dried, and then must be used within 4 h of surface preparation.

When the two jaws of the test machine are placed in contact, they must be used within 4 h of surface preparation.

When the two jaws of the test machine are placed in contact, they must be directly in line to ensure the load will be applied correctly. A 25-mm gage length extensometer is attached and if it is a single channel, it would be very sensitive to exact alignment, whereas a dual gage model is truly averaging.

It is important to check the extensometer knife edges for damage and to ensure that no additional load is applied by the weight of the extensometer cable, which is best supported with a magnetic clip attached to the metal frame of the testing machine. If wedge action grips are used, make certain that the shear pins are intact.

The degree to which the fiber is off-axis has an important bearing on the Young's modulus (Fig. 8-13) and is derived from:

$$\frac{1}{E_X} = \frac{1}{{}^1E_c}\cos^4\theta\left(\frac{1}{G_{12}} - \frac{2\upsilon_{12}}{{}^1E_c}\right)\sin^2\theta\cos^2\theta + \frac{1}{{}^2E_c}\sin^4\theta \qquad (8\text{-}27)$$

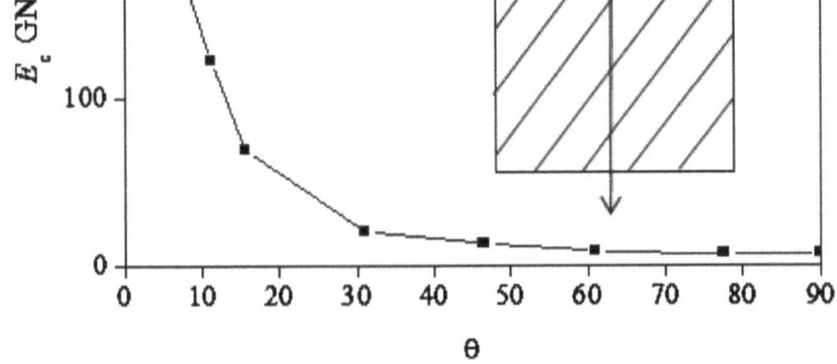

Fig. 8-13 The angular dependence of the modulus of a 50% V_f unidirectional carbon fiber–reinforced composite laminate.

where E_x is the in-plane off-axis modulus and ^1E (longitudinal), ^2E (transverse), G_{12} (shear), and ν_{12} ('major' Poisson's ratio) the four independent elastic or engineering constants defining the elastic behavior of the composite [92–94].

The angular dependence [95] of the modulus of a unidirectional 50% V_f carbon fiber–reinforced composite laminate is shown in Fig. 8-13. The extensometer can be calibrated with a drum micrometer and it is important that the anvils onto which the extensometer is attached are soft metals, such as mild steel, and not chrome plated, since the extensometer would not be able to bite in sufficiently to prevent slippage.

8.5.12. Tensile Strength

To ensure that the tensile specimen will break in the gage length and not at the grips, the specimen is wasted (Fig. 8-14) using a high-speed router cutter. It has been found advantageous to replace the cutter with a small grindstone of the same diameter. The specimen is held in a special template, which the cutter follows exactly as it shapes the tensile specimen. Alternatively, instead of using a router cutter, a 125-mm radius grindstone can be used. This radius was established theoretically by 'Research Assessment Exercise' and adjusted practically, so that the same radius could be used for a compression specimen without the latter failing in a buckling mode. The contour equation (Fig. 8-15) developed by Research Assessment Exercise is:

$$y = y_0 \exp\left(\frac{\tau_{ult}}{\sigma_0} \times \frac{x}{y_0}\right) \tag{8-28}$$

$$\frac{dy}{dx} = \frac{\tau_{ult}}{\sigma_0 y_0} \exp\left(\frac{\tau_{ult}}{\sigma_0} \times \frac{x}{y_0}\right) \tag{8-29}$$

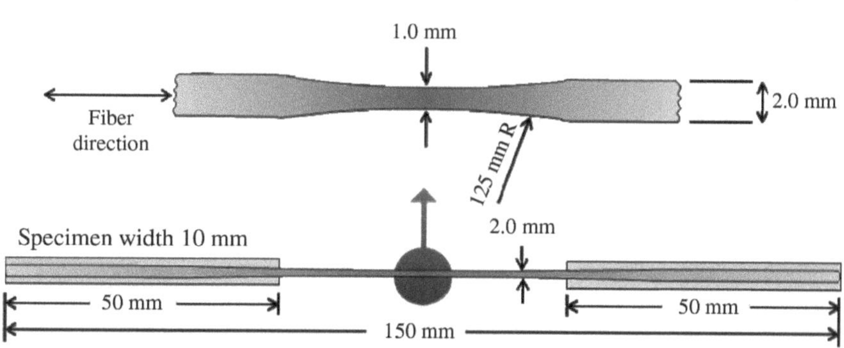

Fig. 8-14 Unidirectional tensile composite specimen.

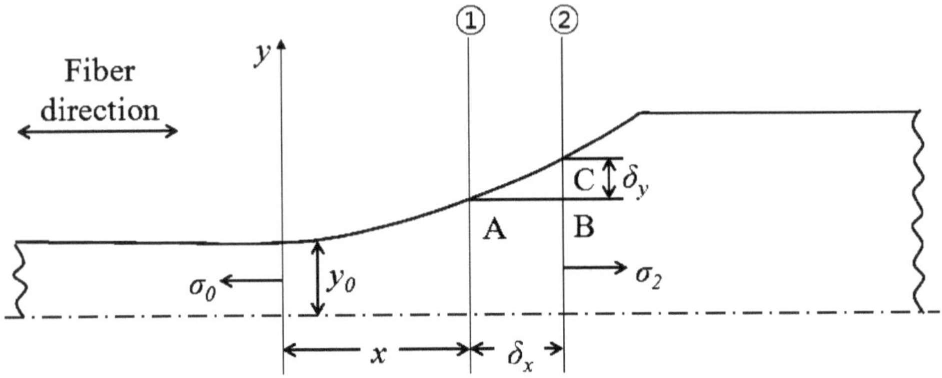

Criterion $\tau_{ult}\delta_x \geq \sigma_2\delta_y$

Contour eq. $y = y_0 \exp\left(\frac{\tau_{ult}}{\sigma_0}\frac{x}{y_0}\right) \quad \left.\frac{d_y}{d_x}\right|_{y=0} = \frac{\tau_{ult}}{\sigma_0}$

Fig. 8-15 Wasting contour of unidirectional carbon fiber–reinforced composite tensile specimen.

$$\frac{d^2y}{dx^2} = \frac{\tau_{\upsilon lt}}{\sigma_0} \times \frac{\tau_{\upsilon lt}}{\sigma_0 y_0} \exp\left(\frac{\tau_{\upsilon lt}}{\sigma_0} \times \frac{x}{y_0}\right) = \frac{\tau_{\upsilon lt}^2}{\sigma_0^2 y_0} \exp\left(\frac{\tau_{\upsilon lt}}{\sigma_0} \times \frac{x}{y_0}\right) \tag{8-30}$$

where $\tau_{\nu lt}$ is the shear strength and σ the tensile strength.

Now the radius of curvature is given by:

$$\frac{1}{R} = \frac{d^2y/dx^2}{\pm\left[1 + (dy/dx)^2\right]^{3/2}} \tag{8-31}$$

Hence:

$$R = \frac{\pm\left[1 + \frac{\tau_{\upsilon lt}^2}{\sigma_0^2}\exp\left(\frac{2\tau_{\upsilon lt}}{\sigma_0} \times \frac{x}{y_0}\right)\right]^{\frac{3}{2}}}{\frac{\tau_{\upsilon lt}^2}{\sigma_0^2}\exp\left(\frac{\tau_{\upsilon lt}}{\sigma_0} \times \frac{x}{y_0}\right)} \tag{8-32}$$

$$\text{at} \quad y = 0, \quad \frac{dy}{dx} = \frac{\tau_{\nu lt}}{\sigma_0} \tag{8-33}$$

Then:

$$R = \frac{y_0\left[1 + (\tau_{\upsilon lt}/\sigma_0)^2\right]^{3/2}}{(\tau_{\upsilon lt}/\sigma_0)^2} \tag{8-34}$$

In the above case, $y_0 = 0.5$ mm. Hence:

$$R = \frac{0.5\left[1 + (\tau_{ult}/\sigma_0)^2\right]^{3/2}}{(\tau_{ult}/\sigma_0)^2} \tag{8-35}$$

Fig. 8-16 shows the mode of failure with different fiber types when using a 125-mm radius necked portion and the evidence does suggest that a radius of 175 mm would better accommodate AU and HTU fibers.

When breaking tensile specimens, the operator should be protected against flying debris with a protective screen, which can be a sliding window attached to the frame of the test machine.

Strain gages can be used for the measurement of strain, or displacement, in mechanical testing and can be a wire wound grid or an etched foil pattern, normally having a gage factor of 2 with a resistance of several hundred ohms.

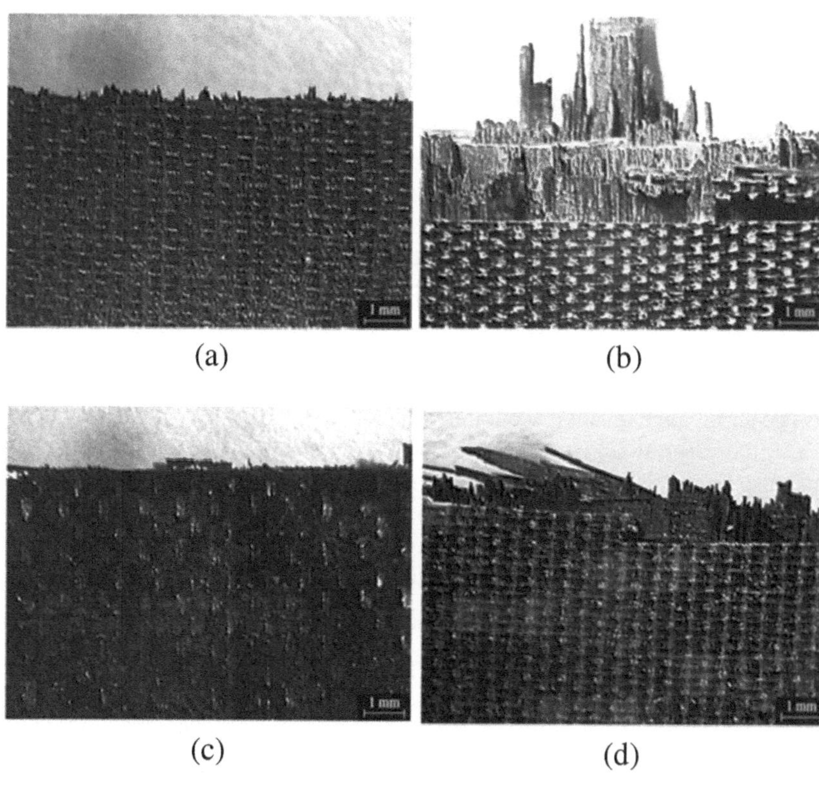

Fig. 8-16 Failure mode types of carbon fiber–reinforced plastic tensile composite specimens: (a) F155/PW, (b) F584/8HS, (c) F584/PW, and (d) F155/8HS.

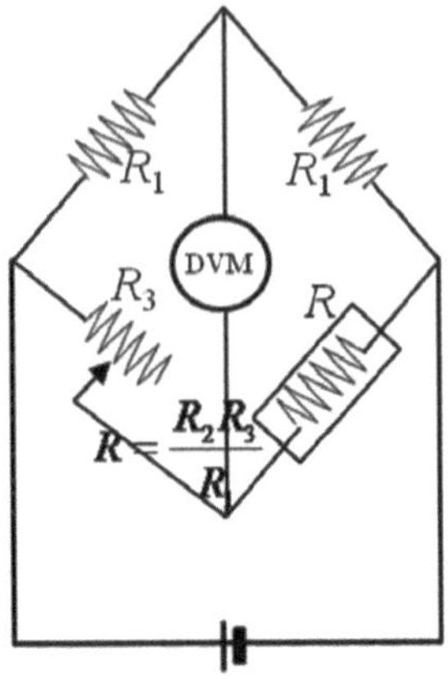

Fig. 8-17 Wheatstone bridge direct current circuit for a strain gage.

High resistances are preferred, since high voltages of the order of 2–4 V can be used with corresponding low current, giving improved hysteresis and zero load stability effects. The gages are used in direct current bridge circuits (Fig. 8-17), or better, in an alternating current bridge circuit, which has many practical advantages.

Attaching a strain gage t to the specimen is an acquired art and the correct surface preparation of the gage and specimen, as recommended by the gage supplier, is paramount. If connections have to be made, it is better to undertake these before the gage is installed, requiring a fine soldering iron and a steady hand, although it is better to use gages with the lead wires already attached.

The strain gage must be accurately aligned and even a 2° misalignment can give a 15% error. If dummy gages are used for temperature compensation, they must also be precisely aligned.

If the gage factor is to be checked on sample gages, then it can be calibrated most conveniently mounted on a flexural specimen (Fig. 8-18) against an accurate extensometer. Such a procedure will also check the proficiency of the bonding method:

$$e = \frac{dt}{R}c = dt/L^2 \quad \text{and} \quad e = \frac{e't}{(t+2h)} \tag{8-36}$$

where e is the actual surface strain, d the deflection through a distance of L, t the thickness of beam, e' the apparent strain measured, and h the height of strain gage grid above beam surface.

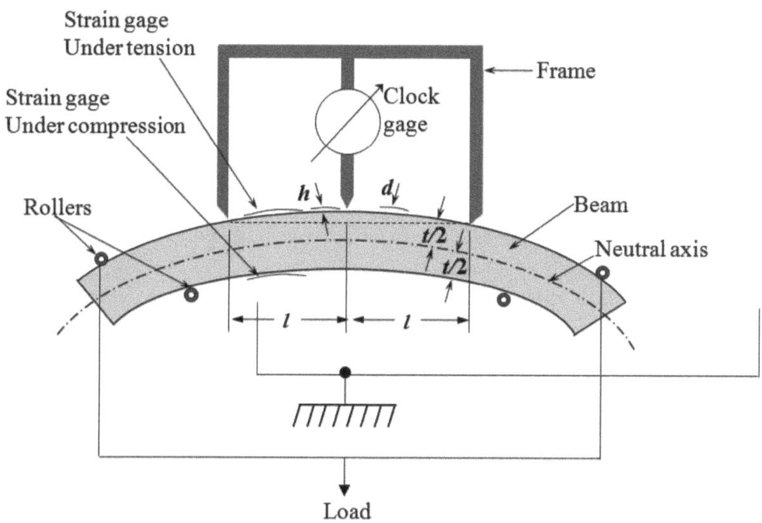

Fig. 8-18 Test jig for calibrating strain gages.

8.5.13. Shear Strength

8.5.13.1. Interlaminar shear strength

The interlaminar shear strength is the interfacial shear stress or shear strength of the matrix materials and is measured with the three-point bend test (ASTM D 2344), which is ideal for outline testing. Fig. 8-19 shows a diagrammatic view of the test and shows how shear, tensile, and compressive forces are involved. Shear failure will take places at the midplane in the form of delamination, while Fig. 8-20 shows how a parabolic shear stress distribution occurs [96–99].

The shear strength (strictly for a homogeneous specimen) is derived from:

$$\tau \cong \frac{3P}{4bh} \tag{8-37}$$

where P is the force at failure, b the width, and h the thickness of specimen.

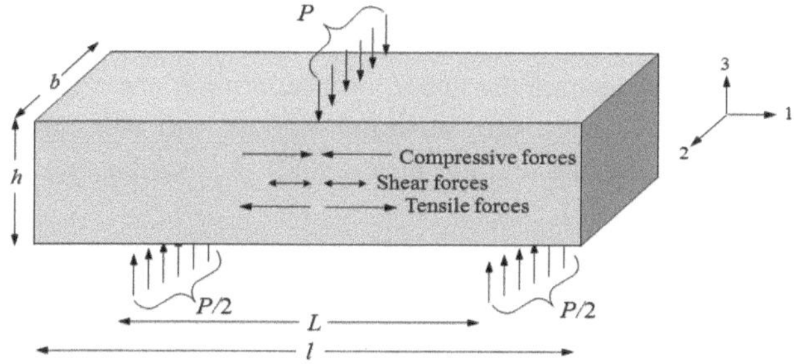

Fig. 8-19 Short beam shear test showing test configuration.

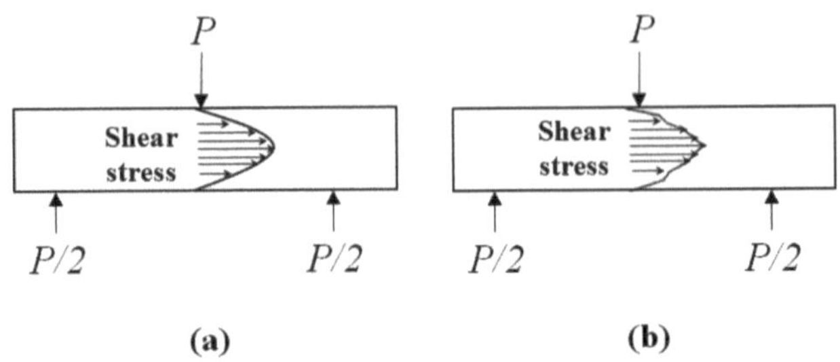

Fig. 8-20 Effect of shear stress distribution in (a) homogeneous specimen and (b) laminated specimen.

If the beam is too long compared to its depth, then a flexural failure (tensile or compressive) can occur at the outer plies of the beam. To ensure that an interlaminar shear failure does occur, the span to depth ratio must satisfy:

$$\frac{2L}{h} \prec \frac{F_1}{\tau} \tag{8-38}$$

where L is the beam span and F_1 the flexural strength of the beam.

The span to depth ratio has been chosen to ensure a shear failure. Although the physical validity and accuracy is questionable, useful results can be achieved. However, attention is drawn to the following points:

1. Use a resin system that ensures a shear-type failure.
2. Ensure that the test jig has been correctly lined up and the loading nose is at right angles to the longitudinal direction of the specimen.
3. When comparing interlaminar shear strength values, be certain to compare like for like, as a 4:1 span/depth ratio will give higher interlaminar shear strength values than a 5:1.
4. Undercure of the resin matrix can also give higher interlaminar shear strength values.
5. Do not apply the law of mixtures to correct for fiber content, although the volume fraction of the fibers does affect the interlaminar shear strength.
6. Record the mode of failure. With a valid mode of failure, the calculated shear stress can be termed the apparent interlaminar shear strength.
7. The radii chosen for the loading nose and supports will affect the results and putting rubber under the loading nose avoids surface damage, improving the test reliability [100].
8. Browning et al. have described a four-point shear test [100].

8.5.13.2. *In-plane shear tests*

The full characterization of a unidirectional composite under an in-plane shear load parallel to the fiber direction will entail the measurement of shear modulus

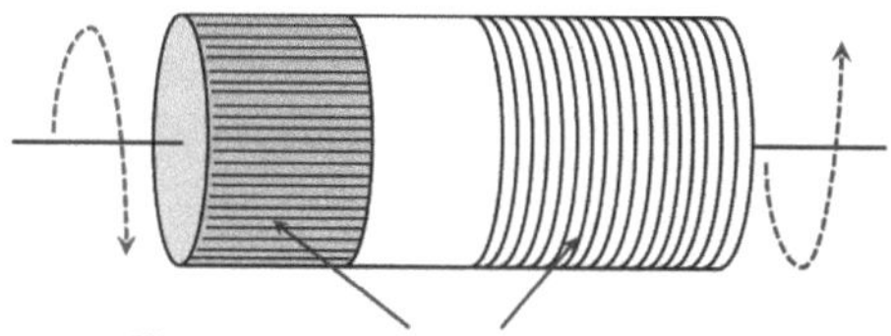

Fig. 8-21 Tube subjected to axial torsion.

G_{12}, shear strength (subscript '6'), and ultimate shear strain γn_6 [101]. These can be determined by:

(1) The torsion test: The torsion test is generally considered to be the most accurate shear test, and Fig. 8-21 [102, 103] shows a thin-walled tube subject to axial tension, which gives a state of pure shear and produces reliable measurements for both the strength and modulus, but it is a costly procedure:

$$(\tau_6)_{\max} = \frac{2Tr_0}{\pi(r_0^4 - r_i^4)} \tag{8-39}$$

$$(\tau_6)_{\max} = \varphi r_0 = (\varepsilon_x)_\theta = 45^\circ = 2(\varepsilon_x)_\theta = 45^\circ \tag{8-40}$$

where r_i and r_o are inner and outer radii, φ the angle of twist per unit length, and $(\varepsilon_x)_\theta = 45^\circ$ and -45° the surface strains at 45° and -45° with tube axis, respectively.

As an approximation for a thin-walled tube:

$$(r_6)_{\max} \cong \frac{T}{2\pi \bar{r}^2 h} \tag{8-41}$$

where the mean radius $\bar{r} = (1/2)(r_0 + r_i)$ and for a solid rod, $r_i = 0$. The shear strain can be measured from the angle of twist or the strains at 45° and/or -45° using strain gages.

(2) Two- or three-rail shear test: A more practical method is to use the two- or three-rail shear test (ASTM Test Method D 4255) (Fig. 8-22), although Lockwood has questioned the reproducibility [104] of the test. The ASTM is not strictly a standard, but is referred to as a guide.

The average shear stress applied to the specimen is:

$$\tau_6 = \frac{P}{2lh} \tag{8-42}$$

where P is the load, l the specimen length along the rails, and h the specimen thickness.

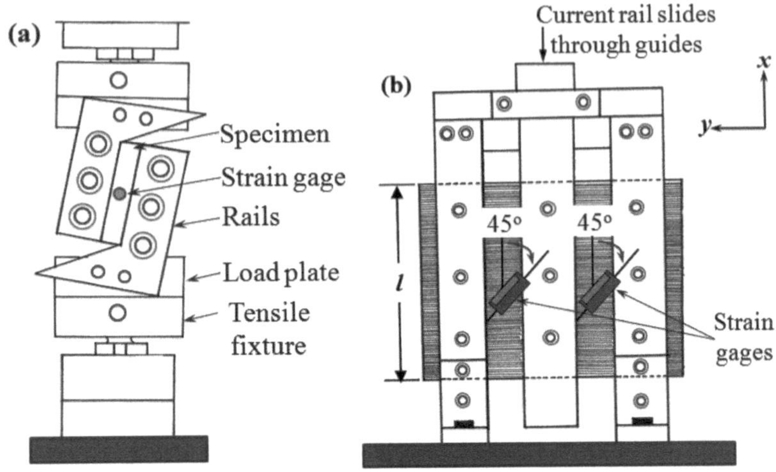

Fig. 8-22 ASTM two-rail (a) and (b) three-rail shear tests.

The two-rail shear test is only suitable for the determination of shear moduli. The three-rail shear does give a better approximation to pure shear as the compressive load is applied parallel to the clamped edges of the specimen.

(3) The double V-notch shear (Iosipescu test): The Iosipescu test (Figs. 8-23 and 8-24) has two opposing V-notches to impose pure shear in the middle of the specimen (ASTM Test Method D 5379) and gives results that compare well with the torsion tube method [103, 105]. The average shear stress is:

$$\tau_6 = \frac{P}{lh} \tag{8-43}$$

where l is the specimen height at notch location and h the specimen thickness.

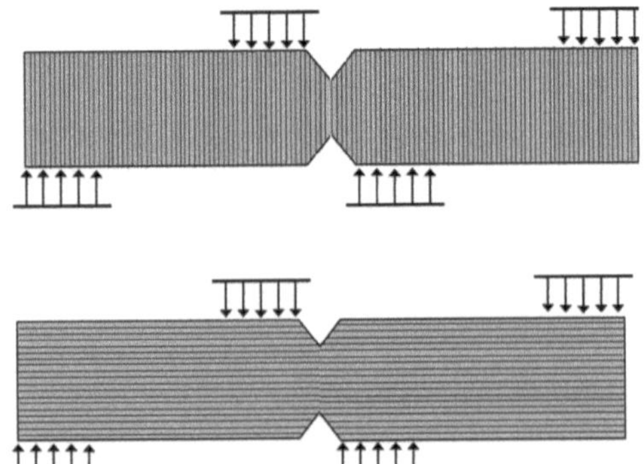

Fig. 8-23 Iosipescu shear test configuration.

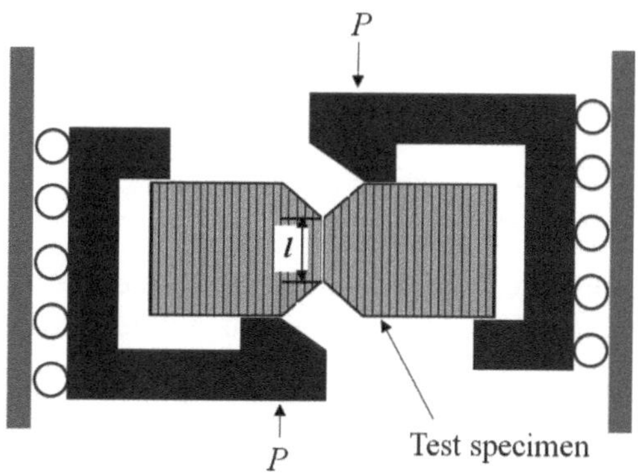

Fig. 8-24 Iosipescu shear test jig.

(4) Tension coupon test: Fig. 8-25 shows the $[\pm 45]_{ns}$ tension coupon test (ASTM D 3518), where the shear strain is measured at ±45° to the axis using strain gages where:

$$\gamma_{xy} = \varepsilon_{11} - \varepsilon_{22} \tag{8-44}$$

Output from the strain gages is in volts given by:

$$e = \frac{4V_0}{KV_{in}G} \tag{8-45}$$

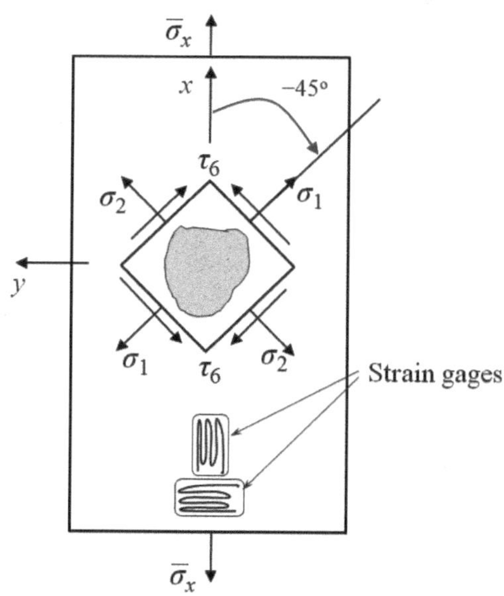

Fig. 8-25 $[\pm 45]_{ns}$ angle ply shear test.

where V_0 is the output voltage, K the gage factor, V_{in} the bridge voltage, and G the shear strain.

And shear modulus G_{12} is given by:

$$G_{12} = \frac{P}{2bh(\varepsilon_{11} - \varepsilon_{22})} \quad \text{(shear modulus in the 1–2 plane)} \tag{8-46}$$

where P is the tensile force, b the specimen width, h the specimen thickness, ε_{11} the axial strain, and ε_{22} the transverse strain but will require derivation of axial (ε_{11}) and transverse strains (ε_{22}).

An equation for modulus can be expressed as:

$$G_{12} = \frac{E_x}{2(1 + \upsilon_{xy})} \tag{8-47}$$

where E and E_x are the effective Young's modulus and Poisson's ratio of the specimen, respectively.

(5) The 10° off-axis test: This test [106] is reputed to minimize the effect of longitudinal (σ_1) and transverse (σ_2) tensile stresses on the shear response (Fig. 8-26). A two-gage rosette is placed on either side of the specimen, with the two gages of the rosette at ±54° to the fiber direction. The algebraic difference of the strain readings of gages A and B gives the in-plane shear strain:

$$\gamma_6 = \gamma_A - \gamma_B \tag{8-48}$$

This test is more accurate than the $[\pm 54°]_n$ coupon test, but less simple. When subjected to a uniaxial tensile load, three types of stress are experienced –

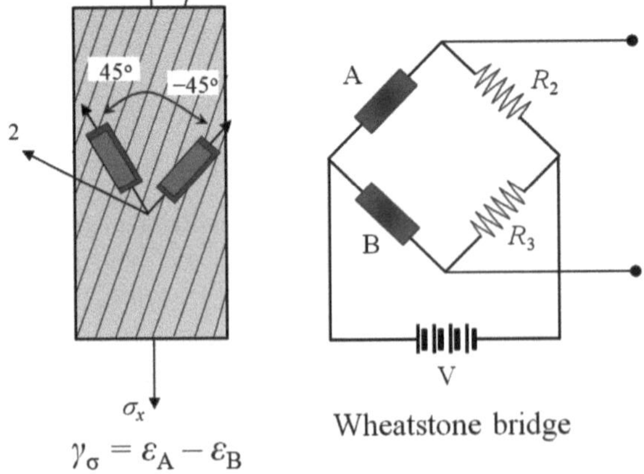

Fig. 8-26 Schematic diagram of 10° off-axis shear test.

longitudinal, transverse, and in-plane shear – on the 10° plane. The intralaminar shear strain approaches a maximum value at an angle of about 10° between the fiber direction and the applied load, when the shear stress is the major component contributing to fracture. The intralaminar shear strength depends on the shear strengths of polymer matrix and the interfacial bond.

Type I carbon fiber in epoxy resin system shows three modes of failure – longitudinal fracture, shear failure, and transverse failure – and the strength at any given angle is determined by the weakest of these failure modes.

8.5.14. Flexural Strength and Modulus

Flexural testing is a simple vend test involving no end tabs but, unfortunately, the test conditions do vary widely with the chosen test method, so results must be treated with caution, although the test is useful as a quality control procedure. There are three relevant standards – ASTM D 790M, BSI 2782 Method 1005, and CRAG – and all of them cover a three-point test procedure (Fig. 8-27(a)), while the ASTM also covers a four-point procedure (Fig. 8-27(b)) [107–110].

Modulus measurements carried out at higher span/depth ratios are less influenced by the penetration of the loading nose and supports into the specimen. A better technique is to measure the deflection at different span/depth ratios for the same load.

For flexural strength:

$$\text{ASTM (three-point): } s = \left(\frac{3PS_2}{2wh^2}\right)\left(1 + 6\left(\frac{\delta}{S_2}\right)^2 - 4\left(\frac{\delta h}{S_2^2}\right)\right) \tag{8-49}$$

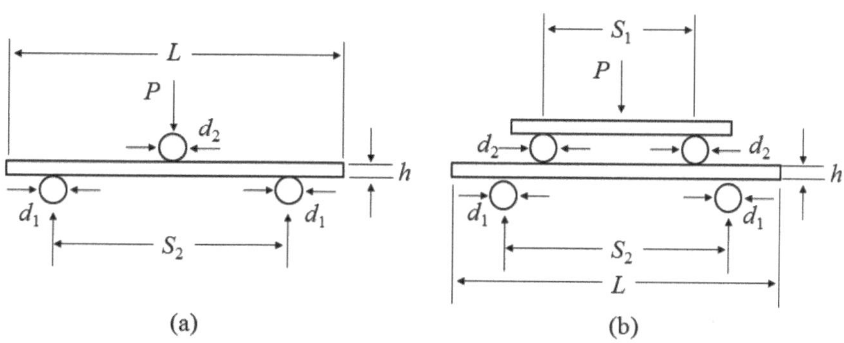

Fig. 8-27 A three-point flexure bending (a) and a four-point flexure bending (b) arrangement.

$$\text{ASTM (four-point): } s = \left(\frac{3PS_2}{2\pi h^2}\right)\left(1 + 4.7\left(\frac{\delta}{S_2}\right)^2 - 7.04\left(\frac{\delta h}{S_2^2}\right)\right) \tag{8-50}$$

$$\text{BSI(3-point): } s = \left(\frac{3PS_2}{2\pi h^2}\right)\left(1 + 4\left(\frac{\delta}{S_2}\right)^2\right) \tag{8-51}$$

$$\text{CRAG(3-point): } s = \frac{3PS_2}{2\pi h^2} \tag{8-52}$$

where S is the stress in the outer fibers at mid-span (N/m^2), P the load at break (N), S_1 the load span (m), S_2 the support span (m), w the width of beam (m), h the thickness of beam (m), and δ the deflection at center of span (m).

And for flexural modulus:

$$\text{ASTM, BSI, and CRAG (3-point flexure): } E_{\mathrm{B}} = \frac{S_2^3 m}{4wh^3} \tag{8-53}$$

$$\text{ASTM (4-point flexure): } E_{\mathrm{B}} = \frac{0.21S_2^3 m}{wh^3} \tag{8-54}$$

where E_B is the tangent modulus of elasticity (N/m^2) and m the slope of tangent to initial portion of the load–deflection curve (N/m).

Using the variable span method with three-point loading:

$$\delta = \frac{PL^3}{4wh^3E} + C.P \tag{8-55}$$

where P is the load applied and C the softness constant due to machine softness, deformation of sample at load points, and shear in the sample.

Now the equation can be written in the following form:

$$y = mx + c \tag{8-56}$$

where

$$x = L^3, \qquad C = C.P, \qquad m = \frac{P}{4wh^3E} \tag{8-57}$$

Hence, m can be obtained by the method of least squares and the value for the modulus can be determined.

Table 8-5 gives the dimensions and test conditions for the ASTM D 790, BSI 2782 Method 1005, and CRAG test standards based on an adopted laminate thickness of 2 mm. The typical values for modulus and strength are given for Civa Geigy's 913/X-ray absorption spectroscopy system [111].

Table 8-5 Dimensions and Test Conditions for Flexure Tests Based on a Laminate Thickness of 2 mm

Standard	w (mm)	h (mm)	S/h	L (mm)	d1 (mm)	d2 (mm)	Test speed (mm·min^{-1})	Typical value for UD composites E (GPa)	σ (GPa)
ASTM 3pt	25	2	60	150	6.4	6.4	12	116.7	1.56
4pt,1/3 span pts	25	2	20[a]/60[b]	150	6.4	6.4	13.3	132.4	1.61
BSI 3pt	15	2	16	50	4	10	1.0	73.7	1.46
CRAG 3pt	10	2	40[c]	100	10	25	5.0	114.9	1.72

[a]*load roller S_1/h ($^1/_3$ span points may also be 5.3, 10.7 and 13.3 mm; $^1/_4$ span points may also be 8, 16 and 20 mm;*
[b]*support roller S_2/h ($^1/_3$ span points and $^1/_4$ span points may be also 16, 32 and 40 mm;*
[c]*may also be 25 min for 90° and 0/90° carbon.*

8.5.15. Uniaxial Compressive Strength and Modulus

A uniaxial compressive load can be introduced into a specimen by [112–123]:

1. Direct loading of the specimen end, but this is not suitable for high-strength composites due to the specimen ends failing by crushing, which is termed brooming.
2. Shear loading of the specimen end.
3. The mixed shear and direct loading method.

The method of shear loading of the specimen end was used in the original Celanese test method [115], where the tabbed specimen is held in conical grips, held in tapered sleeves accommodated in an outer cylinder, which replaced the conical grips with tapered grips that fitted into matching pockets in a steel block, aligned by pillar guides fitted with linear bearings.

Adaptations of this version of a compressive test were made by Wyoming University [124]. Purslow and Collings, later modified by the port of Royal Academy of Engineering (RAE) [125], used a mixed shear and end loading method, where a specimen is bonded into slots machined in Al end blocks and the value did depend on the type of adhesive used. A similar test modified with an additional alignment device was adopted as ASTM D 695, which in its original form was introduced for testing unreinforced plastics [126, 127].

The RAE method was modified by Birmingham University [128], where a wasted specimen was clamped at the ends in steel cubes instead of bonding and the method was further improved by Imperial College [126] using a four-pillar die set to obtain good alignment.

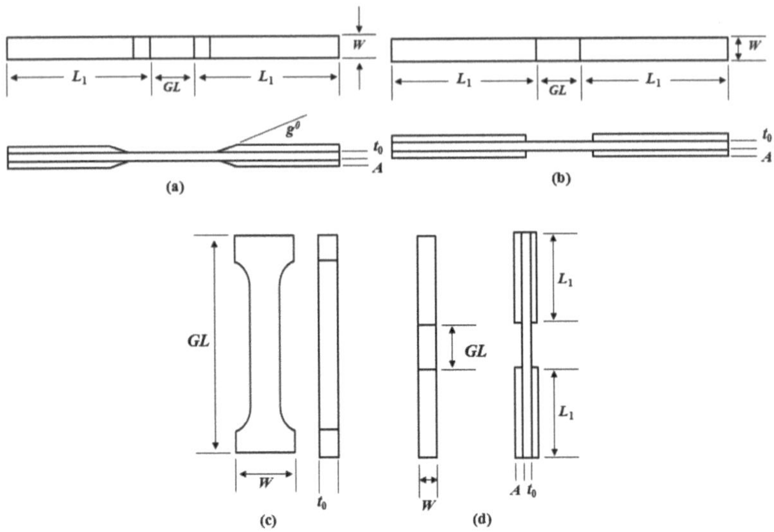

Fig. 8-28 Details of compression test specimens: (a) ASTM D 3410; (b) CRAG; (c) ASTM D 695; (d) ASTM D 695 (modified).

The specimens are shown in Fig. 8-28, while the dimensions and test conditions are given in Table 8-6 for the three main compression test specimens used for carbon/epoxy composites. The correct alignment of fiber in the specimen, specimen in the jig, and the jig in the test machine are paramount. Compression testing is difficult to undertake because of premature failure due to crushing or buckling and is not used for routine testing. Kumar and Helminiak

Table 8-6 Dimensions and Test Conditions for Compression Testing of Carbon Fiber/Epoxy Composites

	Celanese test ASTM D 3410	ASTM D 695M		CRAG
		Standard	Modified	
t_0	2	2	2	2
G.L.	12.7	80	5	10
L_1	63.5	—	35	50
W	6.35	19/12	10	10
A	1	—	2	1
End tag materials	Steel	—	Carbon/epoxy	Aluminum
Strain measurement	Strain gauge	Extensometer	—	Strain gauge
Test speed (mm/min)	1.3	1	1	1.26

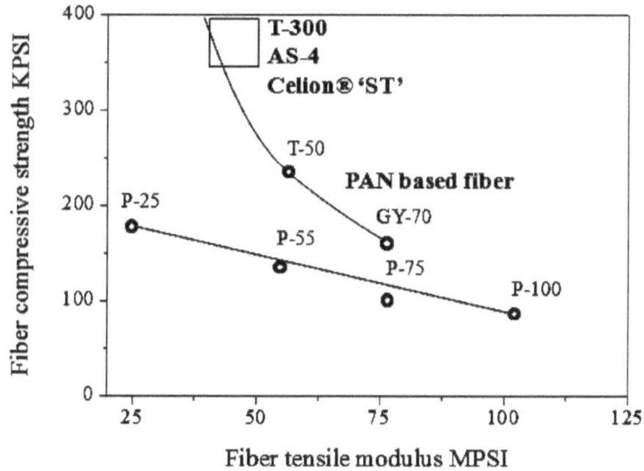

Fig. 8-29 Relation between compressive strength and tensile modulus in carbon fibers.

[129] have shown a relationship (Fig. 8-29) between compressive strength and tensile modulus in carbon fibers.

8.5.16. Fatigue

Fatigue is determined under cyclic loading, when a crack is initiated and then grows to a critical size, with failure occurring at stresses below the value obtained for normal static tests. In general, any static test adopted for fatigue testing is much more demanding. The shape of the test specimen is chosen to fail in gage length, in a manner akin to the failure of a structural component. Fatigue tests can be undertaken in the tensile, flexural, compression, and shear modes, generally using a servo-hydraulic test machine. Optical microscopy, ultrasonics, X-ray, and infrared thermography are used for detecting the onset and study of fatigue failure.

When designing with carbon fiber–reinforced plastic, strain levels are kept low (0.3%–0.4%) due to possible impact damage and the notch sensitivity of carbon fiber–reinforced plastic in compression, so fatigue is not seen as a major problem in aircraft design, except for such components as helicopter blades. A typical tensile fatigue curve is shown in Fig. 8-30 for unidirectional carbon fiber–reinforced plastic, where peak tensile stress is plotted against the cycles to failure (*S–N* plot). The effect of fatigue on different carbon fiber types is shown in Fig. 8-31 [130–137].

Carbon fibers are not sensitive to fatigue loading, but *S–N* curves are determined mainly by the strain in the matrix. The weaker fibers fail first due to the presence of flaws introducing a stress concentration in the matrix and at the fiber/resin interface, thereby inducing microcracks, which further develop in fatigue, making such fibers ineffective. The tougher resin composites do have increased fatigue sensitivity (Fig. 8-32).

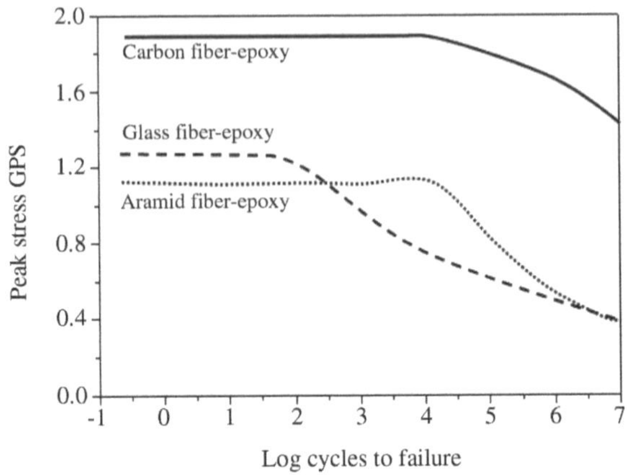

Fig. 8-30 *S–N* diagrams for representative unidirectional composite materials.

8.5.17. Creep

The application of a force to a material over a sustained period of time induces creep and if this force eventually leads to cracking, fracture, or rupture, the process is termed stress cracking, static fatigue, or creep rupture and if the environment has speeded up the process, it is termed environmental stress cracking [138, 139].

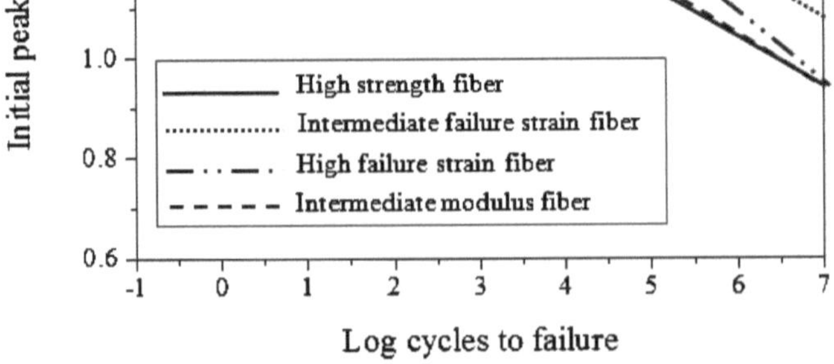

Fig. 8-31 *S–N* fatigue data for unidirectional carbon fiber–reinforced plastic showing the limited effect of different fibers with the same epoxy resin.

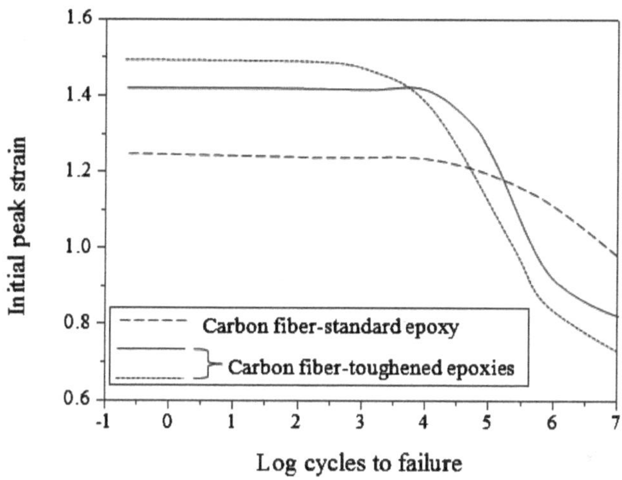

Fig. 8-32 *S–N* data for unidirectional carbon fiber–reinforced plastic with the same fiber type in different matrices.

The creep of a composite material will depend on the reinforcement and the matrix. Carbon fibers are not much affected by creep, although there will be some loss of strength under sustained loading, but the matrix will exhibit creep, a thermoplastic more so than a thermosetting resin. However, if the fiber is well aligned in a unidirectional composite, it will have good resistance to creep along the fiber axis, but would display creep in torsion and flexure. Plastics do tend to recover when the stress is removed, as long as no damage has occurred. It is normal to plot strain against log time (Fig. 8-33) where the initial strain develops

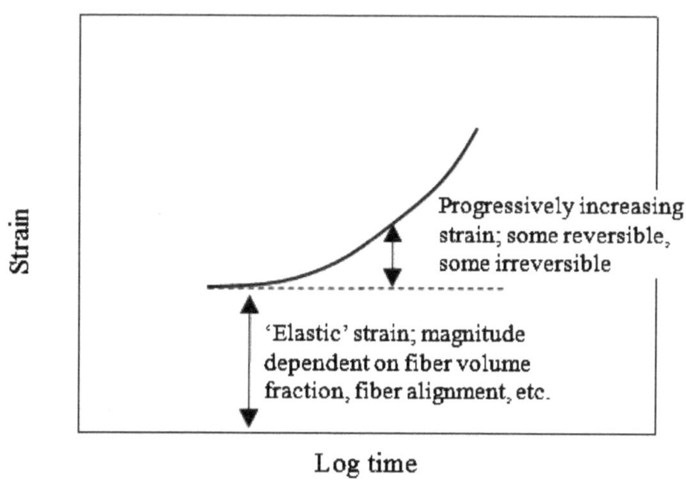

Fig. 8-33 The creep response of a composite specimen to an applied stress.

rapidly, reflecting the modulus, whereas the time-dependent portion of the curve is associated with a viscoelastic response in the matrix eventually leading to a run-away creep, terminating in creep rupture. An idealized creep equation would be:

$$\varepsilon(t) = \sigma C \int_0^\infty f(\tau)(l - e^{-l/\tau})d\tau \qquad (8\text{-}58)$$

where C is a coefficient with the dimensions of elastic compliance, τ the retardant time of a single Voigt element, $f(\tau)$ the distribution function of retardation times, and σ the applied stress.

8.5.18. Impact Behaviors

The impact behavior is major mechanical property in order to evaluate the degree of toughness in composite materials. When the materials are impacted, transferring load is absorbed into the materials and it affects the residual mechanical properties [140–143].

A composite derives its toughness from the ability of the reinforcing fibers to blunt incipient cracks together with the work necessary to either fracture them or pull them out of the matrix. Glass fibers are tougher than carbon, while aramid fibers are tougher than glass and toughened resins can contribute to toughness.

Impact tests are carried out at a fast rate of loading to promote a brittle failure. Izod and Charpy tests are used for isotropic and laminated materials and are not suitable for unidirectional materials, or out-of-plane measurements with laminated materials, since notches are not always effective. Preferred impact test methods are:

(1) Out-of-plane impact testing onto a thin shell structure (e.g., to simulate dropping a tool, or the impact of stones, onto an aircraft structure). There are many variants of this test – the specimen can be simply supported or clamped; the projectile can be dropped or fired (1–10 m/s) and may take the form of a sphere, a cylinder with a hemispherical end, or a dart. The energy and/or the test speed can be varied by adjusting the mass or drop height. The test can be instrumented to record the force, energy absorbed, and the deflection.

(2) Transient tensile and compressive in-plane loads to thin shell structures (e.g., an unexpected service load at a high rate).

(3) Sustained in-plane compressive loads to thin- and thick-walled structures (e.g., crushing under crash conditions).

Therefore, the low-velocity impact test is performed with a drop-weight impact tester instead of Izod and Charpy tests [144, 145].

8.5.19. Fracture Toughness

The subject of fracture mechanics is concerned with the study of fracture of materials by crack propagation. The fracture mechanics approach has been found to be very useful in predicting the strength of isotopic and homogeneous materials in the presence of flaws or cracks and also in developing techniques for quality control and in-service inspection. Fracture mechanics, applied to composite materials, is more complicated because self-similar crack growth, which occurs in homogeneous materials, usually does not occur in composite materials. Fracture toughness tests cannot be well interpreted for oriented composites such as unidirectional continuous fiber–reinforced composites, but they are still used. They can be used for random-fiber composites that are isotropic in the plane of the composite [146–149]. Commonly employed fracture toughness tests are described in this section.

The fracture toughness of materials is measured by using notched specimens in tension and bend tests. In the case of tension specimens a notch is provided on its one edge or at both the edges. The single-edge-notched and double-edge-notched specimens are shown in Fig. 8-34(a) and (b). Tension tests are performed on single-edge-notched or double-edge-notched specimens until failure occurs. Load–elongation data are recorded. The stress–intensity factor can be calculated from the following relationship:

$$K_I = \sigma Y \sqrt{c} \tag{8-59}$$

where σ is the applied stress, Y the K-calibration factor, and c the initial crack length. The values of Y used in calculating K_I are often taken the same as the ones used for isotropic materials [150].

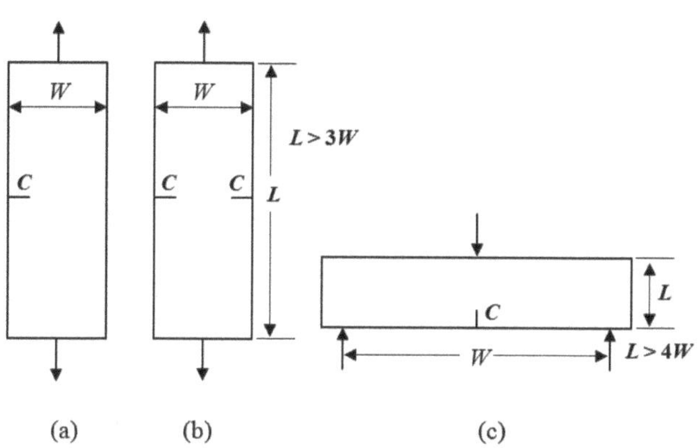

Fig. 8-34 Fracture toughness testing sepcimens: (a) single-edge notched, (b) double-edge notched, and (c) notch bend test specimen.

For the single-edge-notched specimens, Y is [151]:

$$Y = 1.99 - 0.41\left(\frac{c}{w}\right) + 18.70\left(\frac{c}{w}\right)^2 - 38.48\left(\frac{c}{w}\right)^3 + 53.85\left(\frac{c}{w}\right)^4 \quad (8\text{-}60)$$

and for double-edge-notched specimens [152]:

$$Y = 1.98 + 0.36\left(\frac{2c}{w}\right) - 2.12\left(\frac{2c}{w}\right)^2 + 3.42\left(\frac{2c}{w}\right)^3 \quad (8\text{-}61)$$

where w is the specimen width.

Fracture toughness tests in bending are referred to as notched bend tests. This technique involves loading a cracked specimen in three-point bending with the crack located at the midspan on the tension surface of the beam as shown in Fig. 8-34(c). The stress–intensity factor K_I can be calculated by Eq. (8-59), where σ is taken as the maximum bending stress given by Eq. (8-59) and the calibration factor Y is given by:

$$Y = 1.93 - 3.07\left(\frac{c}{w}\right) + 14.53\left(\frac{c}{w}\right)^2 - 25.11\left(\frac{c}{w}\right)^3 + 25.80\left(\frac{c}{w}\right)^4 \quad (8\text{-}62)$$

A detailed discussion on fracture toughness testing is presented in Brown and Srawly [153–631].

8.6. RELATIONSHIP BETWEEN SURFACE AND MECHANICAL INTERFACIAL PROPERTIES IN COMPOSITES

8.6.1. Surface Free Energy and Work of Adhesions

The surface science is related to the adhesion force, supposing that the strength of an adhesive bond is proportional to the work of adhesion, W_A [162, 163]. When the surface free energy of a solid can be determined on the basis of the contact angle (θ) measurements using the geometric mean, W_A according to Fowkes' proposition [164] based on the solid–liquid droplet–air system can be described by the following equation:

$$W_A = \gamma_L(1 + \cos\theta) \quad (8\text{-}63)$$

where W_A is the work of adhesion between the reinforcements and the matrices.

Also, the work of adhesion is simply equal to the sum of dispersive and polar interactions, according to Owens and Wendt's geometric approach [165, 166]:

$$W_A = W_A^L + W_A^{SP} \quad (8\text{-}64)$$

The readers can find more detail explanation about work of adhesion in Chapter 4.

8.6.2. Surface Free Energy Analysis using a Linear Fit Method

In the early methods, Owens and Wendt [165] and Kaelble [169] also introduced the concept of a linear first-order function for the determination of the surface free energy and its London dispersive component and γ_S^{SP} of a solid. The contact angles for several nonpolar and polar testing liquids are measured, [167–169] such as:

$$\frac{\gamma_L(1+\cos\theta)}{2\sqrt{\gamma_L^L}} = \sqrt{\gamma_S^{SP}}\frac{\sqrt{\gamma_L^{SP}}}{\sqrt{\gamma_L^L}} + \sqrt{\gamma_S^L} \tag{8-65}$$

In the above equation, we then believe that the plot of $\left(\frac{\gamma_L(1+\cos\theta)}{2\sqrt{\gamma_L^L}}\right)$ against $\left(\frac{\sqrt{\gamma_L^{SP}}}{\sqrt{\gamma_L^L}}\right)$ should give a straight line with intercept $\sqrt{\gamma_S^L}$ and slope $\sqrt{\gamma_S^{SP}}$ in a simple way to obtain surface energetics of a solid, γ_S^L and γ_S^{SP}.

8.6.3. Surface Free Energy and Fractural Properties

Fracture is defined as the separation of materials into two or more parts by an applied stress, which can be classified as either brittle or ductile. For highly crystalline carbons brittle fracture may occur after little or no plastic deformation and some plastic deformation may also happen for materials possessing a degree of ductility.

The process of brittle fracture is characterized by the rapid propagation of a crack under applied stresses, which are several orders of magnitude lower than that of the theoretical stress required to break the intermolecular bonds in the solid. The explanation of this process was provided by Griffith [170] who proposed there are many microcracks in brittle materials that concentrate any applied stress at their tips. Therefore, the maximum stress at the tip is given by the following equation for elliptical crack:

$$\sigma_m = \frac{2\sigma}{(c/p)^{1/2}} \tag{8-66}$$

where σ is the applied stress, c the half the length of an interfacial crack or the full length of a surface crack, and ρ the radius of curvature at the tip.

The tip geometry may approach atomic sharpness in brittle materials and the cohesive strength can be exceeded at the tip of the crack with such stress concentrations under a fairly low applied tensile stress. For the crack to propagate through the materials, the release of the stored elastic energy must be at least equal to the energy of formation of the two new surfaces produced by the separation process.

Consider a surface crack of depth, c, and unit width; then the strain energy released by this crack can be found by integrating the strain field per unit volume from infinity to the surface of the crack, where E is Young's modulus.

This gives:

$$G_E = \frac{\sigma^2 \pi c^2}{2E} \tag{8-67}$$

The energy required to form the two new crack surfaces will be given by:

$$G_S = 2\gamma{\cdot}c \tag{8-68}$$

where γ is the surface energy per unit area.

8.6.4. Mechanical Approach

The micromechanical analysis of an interface requires a complete knowledge of the mechanical properties of the materials involved. In the case of carbon fiber–reinforced polymer matrix composites, the matrix material can be characterized mechanically with considerable accuracy.

A small single fiber is embedded in a matrix and the debonding load for the pull-out of the fiber from the matrix is measured and calculated for the interfacial shear strength (τ) using the following equation [171, 172]:

$$\tau = \frac{F}{2\pi{\cdot}\gamma_f{\cdot}L} \tag{8-69}$$

where F is the debonding force (N), γ_f the fiber radius (μm), and L the embedded length (mm).

According to Desarmot and Favre's proposition [173] on the basis of Greszczuk's geometric method [174], interfacial shear strength (τ) measured from the pull-out tests may allow for the determination of the maximum interfacial shear strength, $\tau_{\max}$, derived from the maximum debonding load, $F_{\max}$, at $L = 0$, using the following equation:

$$\tau_\mathrm{i} = \left(\frac{F_i}{2\pi{\cdot}\gamma_f}\right) \times \alpha \times [\coth(\alpha{\cdot}L) \times \cosh(\alpha{\cdot}L) - \sinh(\alpha{\cdot}L)] \tag{8-70}$$

$$\alpha = \left(\frac{2G_m}{b_i{\cdot}\gamma_f{\cdot}E_f}\right)^{1/2} \quad in \quad b_i = \gamma_f \times \ln\left(\frac{R}{2\gamma_f}\right) \tag{8-71}$$

where b_i is the effective thickness of the interface (mm) and R the radius of the matrix (mm). The parameters E_f, G_m, and r_f are the Young's modulus of the fiber (GPa), the shear modulus of the matrix in Greszczuk experiments (GPa), and the fiber radius (μm).

The average parameters used in the experimental condition are listed in Table 8-7.

The quality of the fiber/matrix interface was also assessed by fragmentation tests. These tests consist of the stretching of a single fiber, embedded on a matrix

Table 8-7 Average Parameters used in this Experimental Condition

Items	Values
G_m (GPa)	1.30
E_f (GPa)	10.35
γ_f (mm)	3.50
R (mm)	22.098
b_i (mm)	0.028
α (mm^{-1})	50.449

material so that, as the applied strain increases, the embedded fiber breaks at the points where the ultimate fiber strength is reached [175]. The process continues until the remaining fiber fragments are so small that the shear stress transfer along their surface is insufficient to build up the tensile strength necessary for another failure. The interpretation of the test results was based on the Kelly–Tyson approach [176]. This approach considers that there is a maximum shear strength value, τ_i, which can be sustained by the interface, characteristic of each particular interface. Above τ_i, decohesion (debonding), or yielding of the matrix, occurs at the fiber/matrix interface. To each τ_i value corresponds a critical fiber length, l_c, or more accurately, a fiber length distribution ranging from $l_c/2$ to l_c, since any fiber with length *just* above l_c will still break. Considering the distribution of the fragment lengths to be nearly symmetrical, their average should be:

$$\bar{l} = \frac{3}{4} l_c \tag{8-72}$$

The average interfacial shear strength, $\bar{\tau}$, can be determined by applying the equilibrium condition between the tensile force acting on the fiber and the shear forces transferred through the interface [177]:

$$\bar{\tau} = \frac{d}{2l_c} \sigma(l_c) \tag{8-73}$$

For a given fiber/matrix system, the smaller the fiber fragments obtained in the fragmentation test, the higher the interfacial strength.

The total work of fracture, W_f, of single-edge-notched bending specimens are calculated using the following equation:

$$W_f = \frac{A_t}{2(b-a)d} \tag{8-74}$$

where A_t is the total area under the load–deflection curve.

The energy absorbed corresponding to load P can be considered to be the work at crack initiation (W_i) and which is calculated as follows:

$$W_i = \frac{A_p}{2(b-a)d} \tag{8-75}$$

where A_p represents the area under the load–deflection curve up to the load P.

8.6.5. Energetic Approach

The interface is characterized by its fracture toughness G_i and debonding is due to crack propagation along the interface. In the pull-out configuration, the Penn and Lee's fracture criterion [178] considered the strain energy stored in the fiber free length to give the energy balance equation:

$$2\pi\gamma_f G_i = \frac{dU_t}{da} + \frac{dU_L}{da} \tag{8-76}$$

where a is the fiber free length, U_t the strain energy stored in the embedded fiber length plus the strain energy stored in the surrounding matrix, and U_L the strain energy stored in the fiber free length.

Then, the debonding force becomes:

$$F_d = 2\pi\gamma_f\sqrt{\gamma_f \cdot G_i \cdot E_f}\,\tanh(\alpha L) \tag{8-77}$$

where α is a constant given by the following expression:

$$\alpha = \sqrt{\frac{2\mu_m}{\gamma_f^2 \cdot E_f \cdot \ln(R/2\gamma_f)}} \tag{8-78}$$

8.6.6. Weibull Distribution

Both the tensile and pull-out strength of carbon fibers are usually assessed by single-filament tensile tests [179]. The experimental data generated by these tests have high scatter, mainly due to the presence of flaws along the fibers. Thus, the interpretation of the data must be done statistically. Several statistical distributions have been used to describe tensile data, the more flexible so far being the Weibull distribution [180]:

$$F(\sigma;\sigma_0,m) = 1 - \exp\left(-\left(\frac{\sigma}{\sigma_0}\right)^m\right) \tag{8-79}$$

where σ is the load factor, σ_0 the Weibull scale, and m the shape parameters.

The flaw-induced nature of fiber failure leads to a length dependence of its tensile strength; the longer the fiber length, the larger the number of flaws that are present and the higher the probability of occurrence of a severe flaw. The Weibull distribution has to be adapted to account for that dependence.

It also assumes that the strength distribution of each independent link is described by a simple Weibull distribution, characterized by identical parameters. This leads to a fiber survival probability equal to the product of the survival probability of each link. On the basis of this consideration the Weibull cumulative distribution function $F(\sigma;\ \sigma_0,\ m)$ and the corresponding mean Weibull strength ($\bar{\sigma}$), adapted to account for the gauge length dependence of the fibers (L), may be described by the following equations:

$$F(\sigma;\sigma_0,m) = 1 - \exp\left[-L\left(\frac{\sigma}{\sigma_0}\right)^m\right] \tag{8-80}$$

$$\bar{\sigma} = \sigma_0 \cdot L^{-1/m}\Gamma\left(1 + \frac{1}{m}\right) \tag{8-81}$$

where Γ is the gamma function.

The parameter estimate for the two-parameter Weibull distribution is performed for the strength data obtained at all gauge lengths simultaneously. In this work, the maximum likelihood theory is used to determine the Weibull parameters. The method used is that described in Stoner's thesis [181], as well as the calculation programs therein. In this way, for each type of fiber, a single set of parameters is obtained that fit all the gauge lengths tested [182]. The estimate of the tensile strength at any gauge length needed for the interfacial shear strength calculations is performed simply by substituting the calculated parameters in Eq. (8-81), for the specified fiber length.

Also, the load factor (σ) and the gauge length (L) taken as part of the σ_0 parameter are redefined with the unit of stress (MPa) such that the Weibull distribution function becomes:

$$P = 1 - \exp[-(\sigma/\sigma_0)^m] \tag{8-82}$$

By taking twice the logarithms of the survival probability $(1 - P)$, we also obtained the following equation:

$$LnLn[1/(1 - P)] = m \cdot Ln(\sigma) + Ln[(1/\sigma_0)^m] \tag{8-83}$$

The parameters m and σ_0 defining the Weibull distribution are usually determined either graphically or numerically by the least squares method. Therefore, we can assign a failure probability P_i to each value of σ_i after ranking all the measured values in ascending order (i taking value from 1 to n which corresponds to the number of measurements of the sample tested) according to one of the principal probabilities used:

$$P = i/(n + 1) \tag{8-84}$$

8.6.7. Experimental Results of Composites

8.6.7.1. Single fiber tensile strength

Fig. 8-35 presents the ultimate fiber properties of surface oxy-fluorinated carbon fibers. The tensile strength of the oxy-fluorinated single carbon fiber is increased by increasing the fluorine and oxygen concentration on carbon fiber surfaces, whereas the strength of all the carbon fibers is decreased by increasing the gauge length of the specimens. It seems that the fiber surfaces are modified by the oxy-fluorination process, regardless of their treatment conditions, and the minor differences are usual compared with surface analyses when the single fiber tensile test is used for characterization. An increase in the number of specimens tested is not expected to lead to more detailed information about the surface treatment effects on the fiber tensile strength, since the single fiber strength can vary as a result of inevitable imperfect process [183].

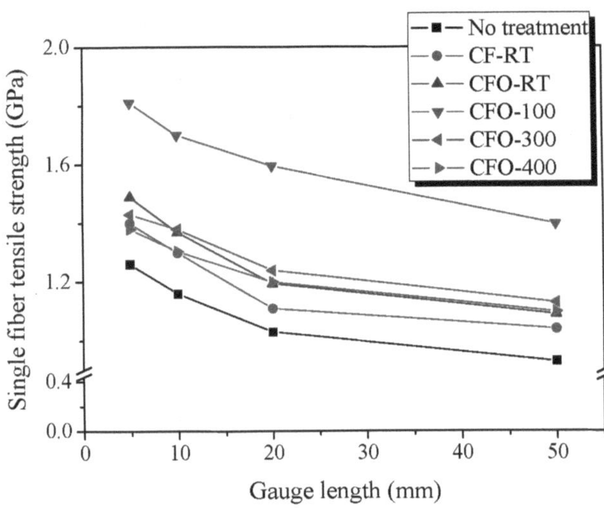

Fig. 8-35 Single fiber tensile strength of oxy-fluorinated carbon fibers as a function of gauge length (F_2/O_2 = 50/50 wt.%).

8.6.7.2. Weibull Distribution Parameter

The probability P_i to each value σ_i can be calculated after achieving the tensile and pull-out tests on carbon fibers using Eq. (8-84). The different cumulative probability distributions are shown in Fig. 8-36. For an initial failure probability at 6% the fracture strength is increased from 5.3 MPa for the as-received state to 6.9 MPa after oxy-fluorination at 100 °C (CFO-100). And the distributions for entire oxy-fluorinated carbon fibers are comparably closer to each other with the exception of a few high extreme values. Also, the Weibull parameter, m, and the scale parameter, σ_0, of all the fluorinated carbon fibers after pull-out tests

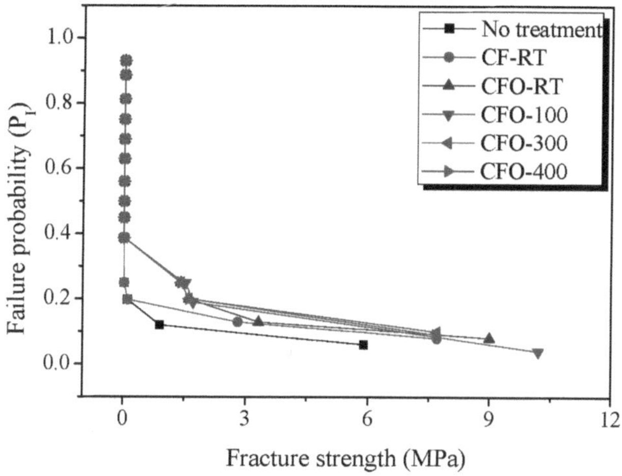

Fig. 8-36 Fracture strength vs. failure probability plots of oxy-fluorinated carbon fibers as a function of fluorination temperature (F_2/O_2 = 50/50 wt.%).

exhibit the similar behaviors, compared with as-received carbon fibers, as listed in Table 8-8. This means that the interfacial shear strength (τ) or failure strength of the oxy-fluorinated single-filament carbon fiber is dependent on the failure mechanism of the nature of carbon fibers in an epoxy matrix system.

Table 8-8 Maximum Interfacial Shear Strength (τ_{max}), Interfacial Toughness (G_i), and Weibull Distribution Parameters (m and σ_0) of All Fluorinated Carbon Fibers in an Epoxy Cure System

Specimens	τ_{max} (MPa)	G_i (kJ)	Weibull parameter (m)	Scale parameter (σ_0)
No treatment	5.28	9.25	2.66	1.43×10^{-2}
CF-RT	4.83	8.52	2.57	1.41×10^{-2}
CFO-RT	6.68	11.56	2.39	1.26×10^{-2}
CFO-100	6.85	14.42	2.41	1.22×10^{-2}
CFO-300	6.12	10.59	2.46	1.33×10^{-2}
CFO-400	5.98	10.35	2.49	1.37×10^{-2}

8.6.7.3. Pull-out behaviors and apparent shear strength

The tensile load on the fiber has been recorded as a function of the displacement of the upper crosshead. Typical experimental pull-out curves are shown in Fig. 8-37. First, consider the Type 1 pull-out curve depicted in Fig. 8-37(a). At the start of the test, the system is elastically loaded; nevertheless, the curve does not keep a linear shape and some jerks occasionally occur as the load is increased.

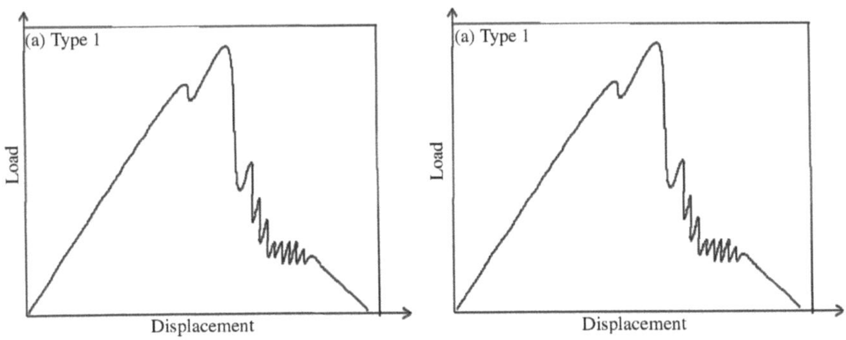

Fig. 8-37 Typical pull-out curves of the fibers.

These observations indicate that debonding is neither complete nor sudden. When the load maximum is reached, debonding is complete and load abruptly falls. The fiber free length elastic relaxation leads to partial fiber pull-out [184].

The second part of the curve corresponds to the fiber pull-out with a pronounced stick–slip phenomenon. Type 2 pull-out curve also depicted in Fig. 8-37(b) is only observed for large embedded lengths. In this case, the curve shows a sequence of rise and fall of the load. Debond and pull-out processes can no more be distinguished so that the debonding load cannot be determined. Differentiation between these two kinds of behavior will be made on plots giving the debonding load vs. embedded length, and in case of the latter and the reported load will be the recorded maximum load.

Fig. 8-38 shows the load–displacement curve of the carbon fibers (oxy-fluorination at 100 °C, CFO-100) after pull-out test. As expected, the oxy-fluorinated carbon fibers exhibit a typical Type I curve because of the natural characteristic of

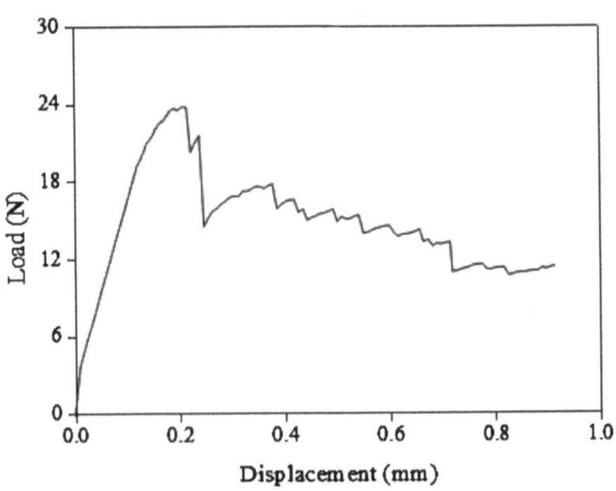

Fig. 8-38 Typical load–displacement curve of oxy-fluorinated carbon fiber sample, CFO-100 after pull-out test.

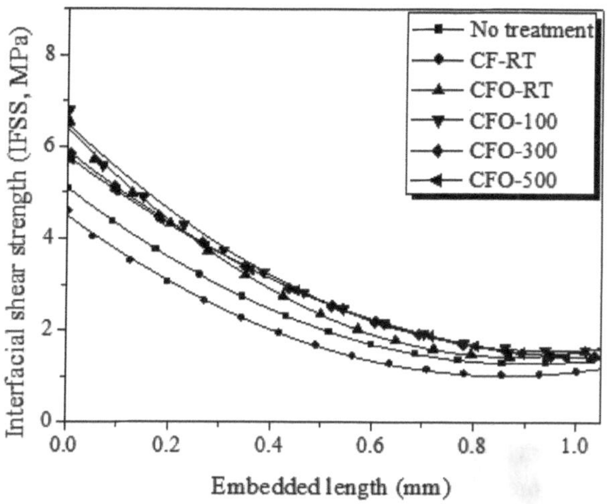

Fig. 8-39 Dependence of interfacial shear strength on embedded length (L_e) of oxy-fluorinated single carbon fiber in an epoxy cure system (F_2/O_2 = 50/50 wt%).

carbon fibers in the epoxy resin matrix system. Therefore, we take a maximum load to calculate apparent shear strength of the oxy-fluorinated carbon fibers in an epoxy matrix resins.

Fig. 8-39 shows the results of the measured interfacial shear strength between the carbon fibers and the epoxy matrix as obtained by single fiber pull-out tests. The slightly straight or curved lines in the figure should be taken only as an indicator of the trend, and therefore they do not represent any data fitting. From these trends, it is possible to characterize the different fracture behavior based on the dependence of the measured interfacial shear strength as a function of the embedded fiber length.

According to the theoretical analysis [185], the independence of interfacial shear strength on embedded length indicates that the failure behavior is mainly due to the ductile fracture during the pull-out process. On the other hand, a decrease in the interfacial shear strength with increasing embedded fiber length would represent mainly brittle failure behavior. That is, the measured shear strength represents that the occurrence of the pull-out is not sudden and catastrophic but only shear yielding failure, which is the characteristic behavior of ductile fracture.

8.7. EVALUATION OF LAMINATED COMPOSITES

8.7.1. Introduction

The bonding state at interfaces between two adjacent laminates clearly plays a critical role in determining the mechanical behavior of composite laminates. In most analytical and numerical work on composite materials, a perfect interface which implies the continuity of both displacements and tractions across this idealized interlaminar interface has been assumed. However, from a more rigorous physical point of view, the existence of a perfect interfacial bond in a real laminated composite seems impossible, since such an idealized interface cannot be delaminated. An appropriate model of composite laminates with imperfect interfaces should therefore be adopted as a more realistic approach to the study of laminates. Literature indicates the interfaces in the analysis of composite laminates fall into two categories: two-dimensional interfaces with zero thicknesses and three-dimensional interfaces. Desai et al. [186] proposed the existence of a thin solid interface, referred to as a thin-layer interface between the structural and geological materials. The shear stiffness for the thin-layer was found from special laboratory tests and the normal stiffness was assumed to be composed of participations of the thin-layer interface and the adjoining solids. Hohberg and Schweiger [187] discussed the penalty behavior of thin-layer elements and explained certain locking phenomena observed in thin solids and layer elements. Mal et al. [188] studied the feasibility of using their developed ultrasonic technique to determine certain macroscopic properties of the interface zones of composite laminates. The thickness and elastic properties of the interlaminar interface zone in a cross-ply graphite–epoxy laminate were estimated by the proposed approach. Allix and Ladevèze [189] used a two-dimensional entity to model interlaminar interfaces for the prediction of delamination initiation and growth in the case of static loadings. Schellekens and de Borst [190] carried out eigenmode analyses of the element stiffness matrices to assess the impact of the applied integration scheme on the stress predictions of two- and three-dimensional interface elements. It was demonstrated that large stress gradients over the element and coupling of the individual node-sets of the interface element might result in an oscillatory type of response. Pandey and Sun [191] developed a nonlinear viscous interlaminar interface model to describe the effect of interlaminar slip on the deformation of thermoplastic composite laminates subjected to processing conditions. Cheng et al. [192] introduced a spring-layer to model imperfectly bonded interfaces of multilayered composites. By modeling the composite laminates as a set of monolayers separated with zero-thickness adhesive interfaces, Bui et al. [193] investigated the effects of imperfect interlaminar interfaces on the overall mechanical behavior of composite laminates. Recently, Wang et al. [194] monitored interlaminar thermal damage in continuous carbon fiber polymer–matrix composites in real time during thermal cycling by measurement of the contact electrical resistivity of the interlaminar interface.

In this section, an alternative approach is presented to characterize the mechanical behavior of composite laminate interfaces. The theoretical analysis is mainly based on the generalized composite laminate theory. The interface model proposed in this chapter belongs to the category of three-dimensional ones, but the distinguishing feature of the model herein lies in the convenience of deducing the mechanical parameters for the interfaces using data obtained via convenient standard measurements, thus avoiding cumbersome laboratory tests.

8.7.2. Analysis of Laminated Composites

During laminated composite manufacturing, an adhesive (e.g., epoxy resin) is usually applied between unidirectional composite prepregs before the composite system is cured under pressure. A third phase thus forms at the interface between the two components (the lamina and the adhesive) where a certain amount of polymeric chain interpenetration occurs [195]. The molecular entanglements play a major role in the transfer of stresses. So we define the thin entity between the lamina and the adhesive as the interface, or more appropriately [196], the interphase. An example of the interphase is shown in Fig. 8-40(a).

The interphase between the lamina and the adhesive results from polymeric intermolecular interactions [196]. The mechanical characteristics of this interphase are different from those of either the bulk adhesive or the lamina. The effect of the interphase on the mechanical properties of the laminated composites is to be incorporated into the conventional composite laminate theory hereafter in order to more accurately represent the composite laminates.

Between two adjacent composite laminae, all the components including the two interphase layers and the pure adhesive between them can be treated as an equivalent layer termed herein the bonding layer, as shown in Fig. 8-40(b), whose thickness can be determined by nondestructive and mechanical testing [197]. The properties of the bonding layer in fact vary along its thickness dimension gradually from those of the lamina to those of the adhesive. However, for practice we ignore the gradient and focus on the average or the equivalent properties.

In order to determine the mechanical properties of this bonding layer, the classical composite laminate theory is generalized here by considering the laminated composite as a system consisting of laminae and the bonding layers. The assumptions for the classical laminate theory are still adopted in

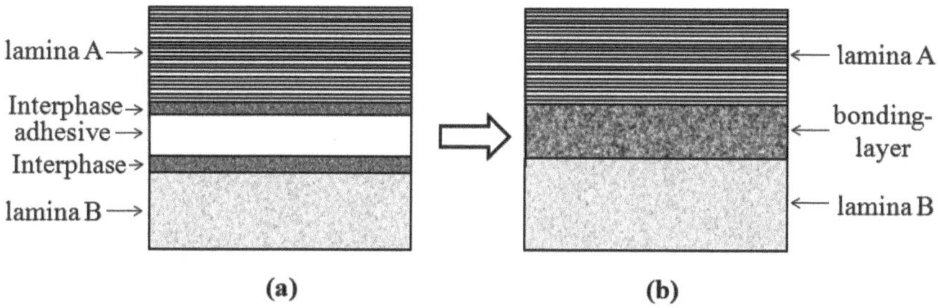

Fig. 8-40 Sketch of interphases (a) and bonding layers (b).

this chapter, for example, piecewise homogeneous material, linearly elastic deformation for a lamina and a bonding layer, and perfect connection between a lamina and a bonding layer.

For a laminate composite made up of m_k ($k = 1, 2, \ldots, n$) laminae with in-plane stiffness, Q_{ij}^k in a reference coordinate system, it is reasonable to assume that the bonding layers between the composite laminae are identically homogeneous materials with equivalent in-plane stiffness, Q_{ij}^{b}. According to the composite laminate theory [198], the overall in-plane stiffness of the composite system, Q_{ij}^s, can be derived as:

$$Q_{ij}^s = \frac{m_1 h}{h^s} Q_{ij}^1 + \frac{m_2 h}{h^s} Q_{ij}^2 + \cdots + \frac{m_n h}{h^s} Q_{ij}^n + \frac{h^b}{h^s} Q_{ij}^b \tag{8-85}$$

where the superscripts s and b denote the laminated composite system and the bonding layers, respectively. So h^s is the total thickness of the composite system, h^b the total thickness of the bonding layers in the composite system, and h the thickness of each lamina.

Because

$$\frac{(m_1 + m_2 + \cdots + m_n)h}{h^s} + \frac{h^b}{h^s} = 1 \tag{8-86}$$

Substituting Eq. (8-85) into the above equation leads to:

$$Q_{ij}^s = \left(1 - \frac{h^b}{h^s}\right) \frac{m_1 Q_{ij}^1 + m_2 Q_{ij}^2 + \cdots + m_n Q_{ij}^n}{m_1 + m_2 + \cdots m_n} + \frac{h^b}{h^s} Q_{ij}^b \tag{8-87}$$

That is:

$$Q_{ij}^s - \frac{m_1 Q_{ij}^1 + m_2 Q_{ij}^2 + \cdots + m_n Q_{ij}^n}{m_1 + m_2 + \cdots m_n} = \frac{h^b}{h^s} \left(Q_{ij}^b - \frac{m_1 Q_{ij}^1 + m_2 Q_{ij}^2 + \cdots + m_n Q_{ij}^n}{m_1 + m_2 + \cdots m_n} \right) \tag{8-88}$$

Here $\frac{m_1 Q_{ij}^1 + m_2 Q_{ij}^2 + \cdots + m_n Q_{ij}^n}{m_1 + m_2 + \cdots m_n}$ is the average stiffness property of the laminae in relation to the reference coordinates, and $\frac{h^b}{h^s}$ the thickness ratio of the total bonding layers relative to the composite system. If Eq. (8-88) is divided by $\frac{m_1 Q_{ij}^1 + m_2 Q_{ij}^2 + \cdots + m_n Q_{ij}^n}{m_1 + m_2 + \cdots m_n}$, then the normalized relationship between the mechanical properties of the bonding layers and the whole composite system is obtained as follows:

$$\frac{Q_{ij}^s}{\frac{m_1 Q_{ij}^1 + m_2 Q_{ij}^2 + \cdots + m_n Q_{ij}^n}{m_1 + m_2 + \cdots m_n}} - 1 = \frac{h^b}{h^s} \left(\frac{Q_{ij}^b}{\frac{m_1 Q_{ij}^1 + m_2 Q_{ij}^2 + \cdots + m_n Q_{ij}^n}{m_1 + m_2 + \cdots m_n}} \right) - 1 \tag{8-89}$$

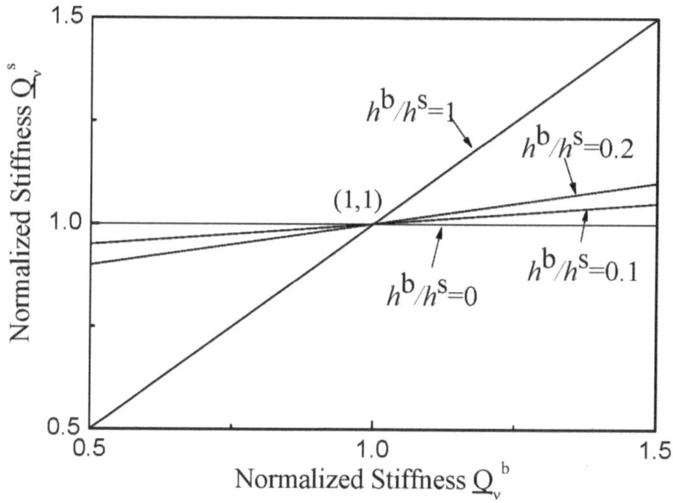

Fig. 8-41 Relationship between the normalized stiffness of the composite system and the normalized stiffness of the bonding layers.

wherein Q_{ij}^b and Q_{ij}^s are normalized from being divided by the average stiffness properties of the laminae in the composite system, which are thus denoted by Q_{ij}^b and Q_{ij}^s, respectively. So Eq. (8-89) can be rewritten as:

$$Q_{ij}^s - 1 = \frac{h^b}{h^s}(Q_{ij}^b - 1) \tag{8-90}$$

The relationship between the two normalized parameters, Q_{ij}^b and Q_{ij}^s, is plotted in Fig. 8-41. It shows apparently that the normalized stiffness, Q_{ij}^s, of the composite system is in direct proportion to the normalized stiffness Q_{ij}^b of the bonding layers. The proportionality factor is the thickness ratio, $\frac{h^b}{h^s}$. All lines cross at the point (1, 1) as dictated by Eq. (8-90). There are two extreme cases, where the thickness ratio, $\frac{h^b}{h^s}$, approaches to 0 and 1, respectively. In fact, only the areas bounded by the two extreme lines are the allowable composite systems. Therefore, in the area on the right side of the point (1, 1), the two normalized stiffness parameters Q_{ij}^b and Q_{ij}^s are both greater than 1, indicating that stiffer bonding layers of Q_{ij}^b reinforce the composite system. On the left side of point (1, 1), however, the two normalized stiffness parameters are less than 1, meaning softer bonding layers yield a softer composite system. Note all these are independent of the properties of the constituent laminae. Furthermore, when the thickness ratio, $\frac{h^b}{h^s}$, is unity, this special case represents a system composed of bonding layers only. When the thickness ratio, $\frac{h^b}{h^s}$, is zero, that is, the thickness of the bonding layers vanishes, so that the normalized stiffness of the composite system remaining a constant 1, that is, equal to that of the average stiffness properties of the laminae. This represents exactly the laminated composite without considering the effects of the bonding layers on the composite system according to the classical laminate theory, revealing that the generalized composite laminate theory developed in this chapter indeed depicts various cases for the effects of the bonding layers on the laminate system.

8.7.3. Numerical Illustration

In this section, a numerical example is given to demonstrate how to get the mechanical properties of the bonding layers by this generalized composite laminate theory using the simple testing data of the composite. The data used here are excerpted from Ref. [199] for a $[0_2, 90_2]_{2S}$ carbon/epoxy laminate composite system; the system tensile modulus is 73.39 GPa and the Poisson's ratio is 0.04, which were determined through more than six individual tests on the laminated composite. The mechanical properties of a single carbon/epoxy ply for the composite are as follows: $E_{11} = 142$ GPa, $E_{22} = 9.0$ GPa, and $\upsilon_{12} = 0.32$. From Eq. (8-89), one can easily write:

$$\frac{Q_{ij}^b}{\frac{Q_{ij}^{(0)}+Q_{ij}^{(90)}}{2}} - 1 = \frac{\frac{Q_{ij}^s}{\frac{Q_{ij}^{(0)}+Q_{ij}^{(90)}}{2}} - 1}{\frac{h^b}{h^s}} \tag{8-91}$$

Given the mechanical properties of the laminate composite system and each lamina with fibers oriented along 0° or 90° axis, the relationship between the stiffness Q_{ij}^b of the bonding layers and their total thickness ratio $\frac{h^b}{h^s}$ can be obtained. Then, from the stiffness Q_{ij}^b it is ready to get the mechanical characteristics bonding layers sandwiched between the composite laminate:

$$E_{11}^b = \frac{Q_{11}^b Q_{22}^b - Q_{21}^b Q_{12}^b}{Q_{22}^b} \tag{8-92}$$

$$E_{22}^b = \frac{Q_{11}^b Q_{22}^b - Q_{21}^b Q_{12}^b}{Q_{11}^b} \tag{8-93}$$

$$G_{12}^b = Q_{66}^b \tag{8-94}$$

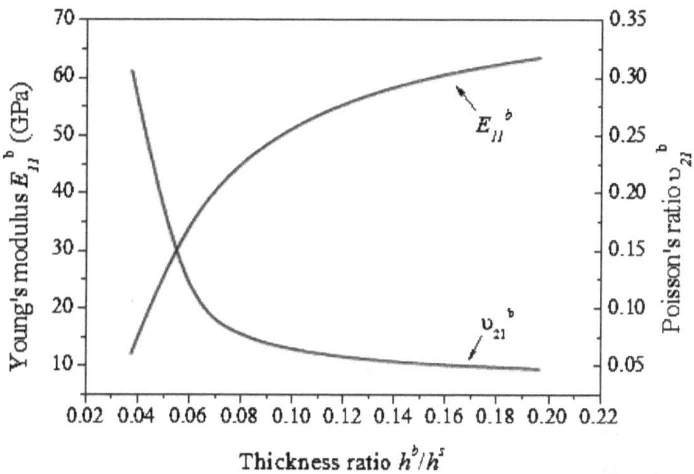

Fig. 8-42 Mechanical properties vs. thickness ratio for bonding layers.

$$v_{21}^{b} = \frac{Q_{21}^{b}}{Q_{22}^{b}} \tag{8-95}$$

$$v_{12}^{b} = \frac{Q_{12}^{b}}{Q_{11}^{b}} \tag{8-96}$$

The above equations show that the bonding layers do not necessarily behave as isotropic materials as expected for a layer of resin. For the specific example presented here, the overall bonding layers are in fact an orthotropic material. Moreover, E_{11}^{b} is equal to E_{22}^{b} and v_{21}^{b} is equal to v_{12}^{b}, that is, the overall bonding layers behave the same along the two principal axes, the so-called balanced system. The properties of the bonding layers are now the functions of the thicknesses only and the Young's modulus E_{11}^{b} and the Poisson's ratio v_{21}^{b} with the variation of the thickness ratio $\frac{h^b}{h^s}$ are shown in Fig. 8-42. For a given thickness ratio, $\frac{h^b}{h^s}$, the corresponding mechanical properties of the overall bonding layers can thus be determined.

8.8. NONDESTRUCTIVE TESTING OF COMPOSITES

8.8.1. Introduction

Fiber-reinforced composite materials continue to be used in a large number of applications ranging from aerospace systems to automotive, industrial, and consumer products. In many other cases fiber-reinforced composite materials are being developed and used to replace metal components, particularly those used in corrosive environments [200].

Because of their increasing utilization in structural applications, the nondestructive evaluation of fiber-reinforced composites continues to receive considerable attention for research and development. Due to the heterogeneous nature of composites, the form of defects is often very different from those typically found in metallic materials, and the fracture mechanisms are more complex.

Composite materials represent acoustically and thermally heterogeneous materials in which a variety of defects with different dimensions may be formed. Typical defects of composite materials include fiber breaks, microcracks, delaminations, foreign objects, or contaminants, impact damage, and porosity or voids. Fiber/matrix distribution, fiber waviness, and fiber orientation are also important microstructural properties of composites. Many of these defects or microstructural variations cause changes in the acoustic attenuation and the speed of sound in composite materials. Consequently, ultrasonic testing is currently one of the most frequently used methods for the nondestructive evaluation of fiber-reinforced composite materials.

8.8.2. Techniques for Evaluating of Properties and Defects of Composites

8.8.2.1. Typical defects of composites

8.8.2.1.1. Porosity

Porosity is a perennial problem in composite manufacturing, especially for laminated structures, such as carbon fiber–reinforced polymer composites. The term porosity generally refers to the voids caused by the trapped air or the volatile gas that is released during the cure process. The size and shape and distribution of the voids depend on the structure of the composites.

The presence of porosity has been shown to have a strong correlation with certain mechanical properties [201–203]. It has generally been found that the interlaminar shear strength decreases by approximately 7% per 1% of voids up to a total void content of about 4% [204]. Other mechanical properties are also affected. The 2% porosity has commonly become the nominal acceptance threshold for many composite components. Research programs in nondestructive evaluation have therefore concentrated on providing techniques that could detect and measure porosity in the range of 1%–5% (v/v). A nondestructive evaluation technique that could determine the level of porosity in a composite laminate independent of other variables, such as pore morphology and the fiber and matrix materials, would be ideal. However, to date no single nondestructive evaluation method has been able to provide this universal porosity meter.

8.8.2.1.2. Fiber/matrix distribution

Fiber and matrix distribution is critical to the performance of composite structures. Knowledge of the size and location of matrix-rich pockets and of matrix-starved regions provides important information that can be used in finite element codes to calculate performance influences. Steiner [205] evaluated 64-ply graphite–epoxy composite samples using ultrasonic pulse-echo C-scan imaging [202]. The fiber and matrix components react differently to ultrasonic energy. While the fibers reflect the energy, the matrix tends to absorb some of the energy. Although the differences may be subtle, advanced image enhancing techniques aid in the determination of matrix-rich or matrix-starved areas in the scanned specimen.

8.8.2.1.3. Fiber volume fraction

Strengths of composites are determined largely by the interaction between the fiber and the matrix. Since the matrix distributes the load onto and between the fibers, it is important to know the respective volume amounts (volume fraction) of the fiber and matrix. It has been shown that fiber volume fraction in carbon fiber–reinforced composites can be nondestructively measured using cold neutron prompt gamma activation analysis. Fiber volume determination is accomplished in this technique by measuring and comparing the hydrogen and carbon content in a given sample. Fiber volume measurement using prompt gamma activation analysis can potentially be applied to any composites where the composition of the matrix and reinforcement differ.

8.8.2.1.4. Fiber orientation

The orientation of the fibers must be known and carefully controlled to take full advantage of the flow molding process for fiber-reinforced composite materials. A small variation in the ply orientation will cause a reduction in the stiffness of a composite laminate. Ultrasonic polar scans have been used to nondestructively determine the fiber orientation of carbon fiber–carbon matrix composites [206]. Fig. 8-43 shows a schematic diagram of an ultrasonic polar scan where the amplitude of the transmitted beam is determined for many different angles.

8.8.2.1.5. Fiber waviness

Fiber waviness is a defect that is often introduced during the manufacturing process. All of the nondestructive evaluation techniques based on ultrasonic and X-ray radiography have major limitations in obtaining quantitative information when used to inspect polymer matrix composites for fiber waviness.

8.8.2.1.6. Delaminations

Fiber-reinforced polymer matrix composites are widely used as the critical structural components for applications in many different industries. In most of these applications, delamination between the fiber plies is one of the most serious types of flaws that can occur. Delamination can affect the structural integrity of the composites by reducing the compressive strength and the

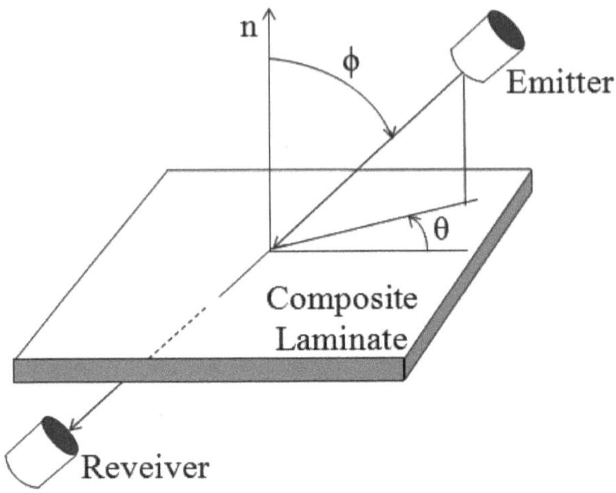

Fig. 8-43 Target spot is impinged at constant distance from all possible angles (ϕ, θ) in a polar scan.

mechanical stiffness. In the worst case, a delamination may propagate and cause fracture of the composite materials. Therefore, the nondestructive evaluation of delaminations is very important in both the manufacture and in-service sustainment of these materials.

A number of non-destructive evaluation techniques have been widely used to detect delaminations. Techniques that are well established include visual inspection, tap testing, resonance methods, ultrasonic testing, thermography, eddy current testing, and X-ray radiography. However, the inspection of composites is a difficult task due to their multilayered structure, anisotropy, and heterogeneity, and inspection with current techniques can still be a difficult task.

Inspection of composite structures for naval applications has been reported in literature [207] where the detection of a variety of defect conditions in composite materials is being investigated by Navy. Fig. 8-44 shows two examples in which a significant amount of composite materials were used in naval structures [208] (see Section 4.6.6).

8.8.2.1.7. Impact damage

Nondestructive evaluation of impact damage in fiber-reinforced composite materials is an important issue for designers and maintainers of components or systems that utilize these materials. It is not uncommon for composite structures to have substantial subsurface damage from foreign object impact without exhibiting any visible surface damage. With the ever-increasing use of composites for primary and secondary aircraft structures, a major concern for the designer is the knowledge of the effect of impact damage on the residual structural performance of these materials. The capability of nondestructive

Fig. 8-44 Composite materials in naval structures: (a) USS Arleigh Burke (DDG 51) with composite forward director's room; (b) USS Radford (DD 968) underway with advanced enclosed mast sensor system.

evaluation methodology to locate and characterize impact damage, combined with the ability of the structural engineer to predict the residual strength of damaged structures, is of major importance in terms of having an effective damage tolerant design and providing tools for in-service inspection.

Several non-destructive evaluation techniques provide for a rapid, whole field or wide-angle view of composite structures to locate potential impact damage. These techniques include various types of thermography, electronic speckle pattern interferometry, and laser shearography. There are also nondestructive evaluation techniques that are slower and have a smaller field of view but have a greater capability for quantifying the impact damage. These techniques include various types of ultrasonics and thermography.

8.8.2.2. Nondestructive evaluation

In a research program to investigate the failure of fiber-reinforced composite materials it was thought that better insight into the failure mechanism would be gained by employing ultrasonic scanning and acoustic emission techniques on the test specimens. These techniques are basically non-destructive evaluation methods that are usually used to detect damage in structures of composites during regular inspection routines [209–213].

Also, in order to encourage the expanded application of composite materials, the use of appropriate non-destructive evaluation approaches is critical to effective process control and the assurance of high-quality products and reliable performance in service [214–218].

According to the ASTM Guideline of Standard Terminology for Nondestructive Examinations [219], nondestructive evaluation or nondestructive testing is defined as the development and application of technical methods to examine materials or components in ways that do not impair future usefulness and serviceability in order to detect, locate, measure, and evaluate flaws; to assess integrity, properties, and composition; and to measure geometric characteristics. Through nondestructive evaluation, the defects and/or flaws in a material can be determined as false, relevant, or nonrelevant indications. This defect information is then evaluated to determine if the component meets

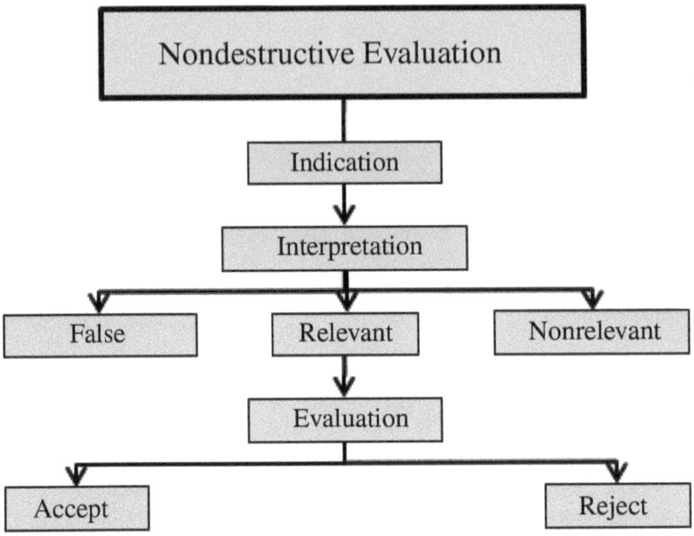

Fig. 8-45 A schematic description of the procedure for nondestructive evaluation.

specified acceptance criteria. Fig. 8-45 presents a schematic of a typical nondestructive evaluation procedure [220]. Fig. 8-46 shows the linear dimensions of microstructures and defects as well as the resolution capabilities of various nondestructive evaluation techniques [220].

The most important issue in selecting a nondestructive evaluation technique involves the number and types of flaws contained within materials. Generally, composite materials have some degree of porosity and dislocations; fibers– or particulates–matrix composites contain several types of defects, including interlaminar porosity and processing-induced voids. To consider when and how various nondestructive evaluation techniques can be best applied to the examination of composite materials, this section overviews the following approaches: ultrasonic testing, radiography, X-ray computed tomography, and acoustic emission. Table 8-9 provides a comparison of these nondestructive evaluation techniques as applied to the realm of advanced composite materials.

8.8.2.2.1. Ultrasonic testing

In recent years, it has been proven that indirect assessment of the microstructure by ultrasonic velocity or attenuation measurements is possible [221–225]. Therefore, establishing a relationship between the microstructural parameters and ultrasonic evaluation results may be useful for fast and nondestructive quality control of the materials. Generally, ultrasonic testing can be used to detect flaws; determine the size, shape, and location of defects; and identify

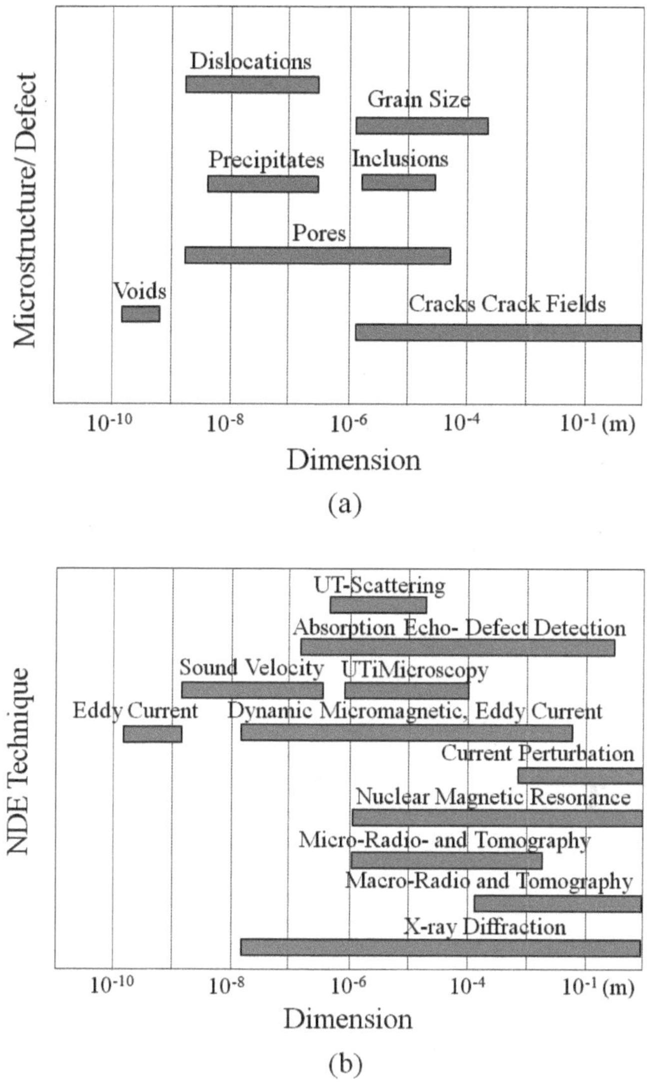

Fig. 8-46 The (a) linear dimensions of microstructures and defects and (b) resolution capabilities of nondestructive evaluation techniques.

discontinuities of materials. Also, the determination of ultrasonic velocities can be used to measure the modulus of elasticity or Young's modulus of materials [226–228].

Sound that possesses frequencies so high that it cannot be heard is called ultrasound (the frequency range is typically greater than 20 kHz). In ultrasonic

Table 8-9 Key Nondestructive Evaluation Techniques for Analyzing Advanced Composite Materials

Characteristics	Ultrasonics	X-ray computed tomography	Radiography	Acoustic emission
Principles	Sonic transmission	X-ray transmission	X-ray, gamma-ray, and neutron transmission of penetrating radiation	Stress wave emission
Variables	Scattering, attenuation, and velocity	Absorption and attenuation coefficients	Absorption and attenuation coefficients	Amplitude, counts, and number of events
Advantages	Suitable for thick materials; relatively quick testing time	Creates cross-sectional view of the entire transmitted thickness	Extensive available database	Real-time monitoring
Limitations	Requires water immersion or acoustic coupling	Expensive; limited specimen size; radiation hazard	Expensive; depth of defect not indicated; radiation hazard	Requires a prehistory of stresses for flaw detection
Detectable flaws	Voids, delaminations, porosity, and inclusions	Voids, delaminations, porosity, and inclusions	Voids, delaminations, porosity, and inclusions	Delaminations and inclusions

testing, beams of high-frequency sound waves are introduced into materials so as to detect both surface and internal flaws. The sound waves travel through the materials (with some attendant loss of energy) and are deflected at interfaces and/or defects. The deflected beam can be displayed and analyzed to assess the presence of flaws or discontinuities. Most ultrasonic inspections are performed at frequencies between 0.1 and 25 MHz.

When transmitting an ultrasonic beam through materials, there will be energy loss by absorption and dispersion through the thickness [229]. In fiber-reinforced composites the presence of voids, cracks, and delamination will cause extra energy losses which result in a reduction of the amplitude of the received signal. By scanning the materials with an ultrasonic beam and simultaneously recording the value of the signal amplitude, an absorption pattern is obtained. This technique is usually called ultrasonic C-scan and it has been used for some time as a qualitative test [230]. However, some researchers have shown that it can also be used to give a quantitative assessment of composite void content.

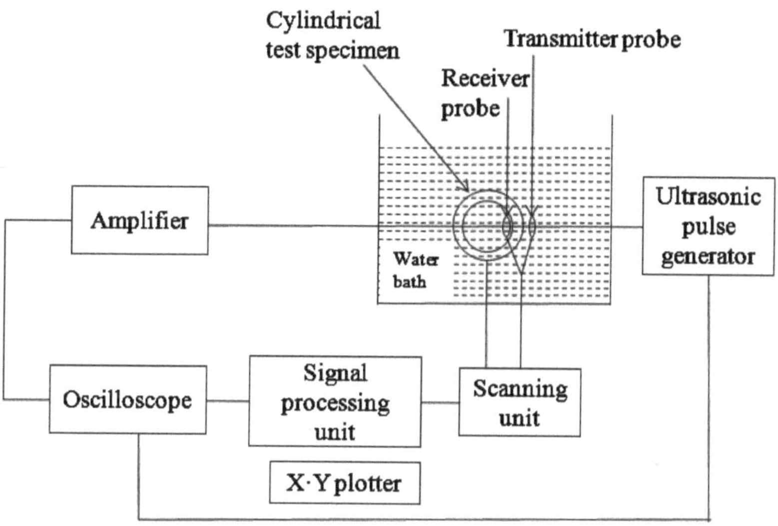

Fig. 8-47 Schematic diagram of the ultrasonic scanning test arrangement.

The test setup is shown schematically in Fig. 8-47. The scanning unit rotates the cylindrical specimen and moves the transmitter–receiver system along the whole surface of the tube. The scanning is performed in a water tank as the energy loss in the water is constant, and therefore any variation in the received pulse is due to losses in the specimen. The measurements were carried out using the ultrasonic testing equipment. Pulse-echo technique (using a single pulser/receiver transducer) was applied using longitudinal ultrasonic waves excited by ~MHz straight-beam probes.

Fig. 8-48 presents a pulse-echo mode setup with a pulser/receiver transducer and the through-transmission ultrasonic setup using a pair of focused transducers [231]. The scan is performed in an immersion tank. In Fig. 8-48(a), the pulser/receiver transducer is used to generate ultrasonic sound waves and

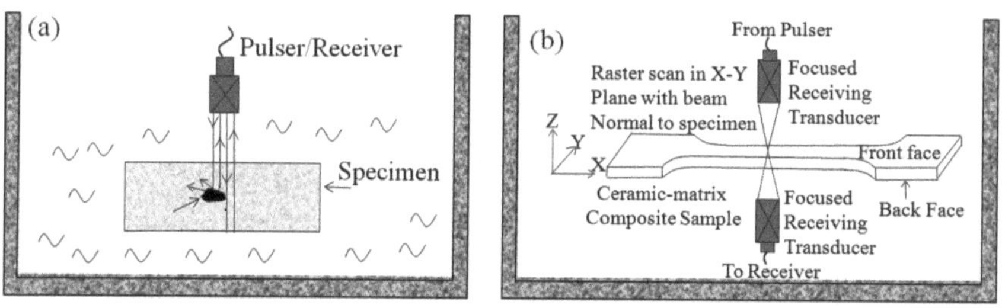

Fig. 8-48 Setups for ultrasonic scanning: (a) pulse-echo mode with a pulser/receiver transducer and (b) through-transmission using a pair of focused transducers in an immersion tank.

Fig. 8-49 Ultrasonic C-scans of the (a) $[0/90]_{4S}$ cross-ply panel and (d) $[0]_{12}$ unidirectional panel from the Nicalon/calcium aluminosilicate (CAS) composites with scanning electron microscopic cross-sectional views of the microstructures in (b) Section A, (c) Section B, (e) Section C, and (f) Section D.

receive the reflected beams. The transducer obtains the traveling sound-wave signals, which are displayed on the oscilloscope with amplitudes against the traveling time. Finally, a computer gathers the amplitudes and forms a scan image. In the through-transmission ultrasonic mode, the transmitting transducer (pulser) and receiving transducer (receiver) are aligned, and the beam path is kept perpendicular to the test specimen during the scan (Fig. 8-48(b)). The two transducers are ganged using a yoke arrangement that maintains the alignment of the focused beam.

A constant pressing force on the probes was applied to maintain a constant thickness of oil film at the probe-specimen interface. After ultrasonic measurements, both scanning electron microscope and optical microscope were usually used to characterize the microstructural parameters of the composites.

Fig. 8-49 illustrates ultrasonic amplitude measurements derived from continuous Nicalon fiber–reinforced calcium aluminosilicate (CAS) glass–ceramic matrix composites—designated Nicalon/CAS. Two types of materials were prepared: a $[0/90]_{4S}$ cross-ply composite panel and a $[0]_{12}$ unidirectional composite panels. All panels had dimensions of 152.4 mm × 152.4 mm × 3 mm. For the $[0/90]_{4S}$ panel, the Nicalon/CAS composites were prepregged into unidirectional plies with a 40% fiber volume and hot-pressed into a $[0/90]_{4S}$ laminate panel. For the $[0]_{12}$ unidirectional composite, the Nicalon/Corning Glass Works composites were supplied in a 12-ply, unidirectionally reinforced

($[0]_{12}$) panel. Fabricated using 500-count fiber tows, this material was produced by hot pressing at about 1350 °C using prepregged unidirectional plies.

In the case of ultrasonic testing for the $[0/90]_{4S}$ cross-ply panel (Fig. 8-49(a)), there are variations of ultrasonic amplitude, particularly along a horizontal line and a vertical line through the center of the panel. For the $[0]_{12}$ unidirectional panel (Fig. 8-49(d)), the ultrasonic transmitted amplitudes revealed two cracks in addition to the long crack across the width (horizontal line). As a whole, the upper and lower sections of the panel contained more defects than the center section.

Analysis via scanning electron microscopy was performed to relate the different amplitude distributions on the panel edges. As shown in Fig. 8-49(a), Section A had relatively greater ultrasonic amplitude values than Section B; hence, the processing-induced porosity was found not in Section A (Fig. 8-49 (b)) but in Section B (Fig. 8-49(c)). In the case of the $[0]_{12}$ unidirectional composite panel, Sections C (lower amplitude values, as shown in Fig. 8-49 (d)) and D (higher amplitude values, again as shown in Fig. 8-49(d)) were selected for scanning electron microscopy investigation. In Section C (Fig. 8-49(e)), severe porosity is visible at the fiber–matrix interface; Section D (Fig. 8-49(f)) reveals good microstructural features. Clearly, the suite of figures that comprise Fig. 8-49 demonstrates that lower ultrasonic amplitudes correspond to the presence of porosity. This occurs because such defects scatter and deflect the ultrasonic testing waves.

Modulus measurements were performed using a dry-coupling ultrasonic technique and specimens machined from woven Nicalon fabric–reinforced silicon carbide composite disks (Fig. 8-50). Longitudinal moduli along the in-plane (fiber–fabric plane) direction (X_1 or X_2 axis) and through-thickness (X_3) direction were determined. Note that along the through-thickness direction, the specimen contained approximately 20 fiber fabric plies. Ultrasonic measurements were performed using longitudinal waves. The longitudinal modulus (C) is related to the longitudinal velocity (V_L) by:

$$C = \rho V_L^2 \tag{8-97}$$

where ρ is the density of the composites.

The efforts in nondestructive evaluation of materials have also been extended to aluminum (Al) metal matrix composites [232–238]. Mott and Liaw [233] have shown that the ultrasonic methods are promising techniques for differentiating the regions that have varying mechanical properties due to inhomogeneous distribution of silicon carbide (SiC) whiskers in a 7090 aluminum–25% silicon carbide composite sample. Similarly, Lu and Liaw [237] developed a theoretical model to establish the relationship between ultrasonic velocity and the particle orientation distribution. In extruded aluminum–silicon carbide composites comprising the 0–30 vol.% particulate phase, the numerically calculated velocities were in agreement with the measured velocities

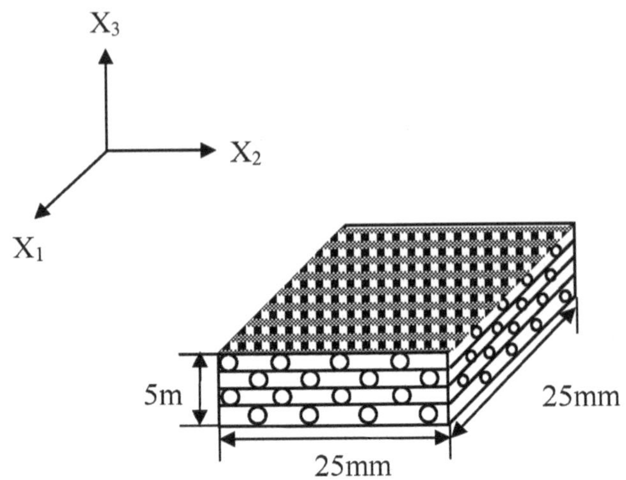

Fig. 8-50 An illustration of specimen geometry for the woven Nicalon/SiC composites used in modulus measurements.

indicating that randomly oriented particles have led to a higher velocity in the extrusion plane with respect to the ones in transverse direction. Later, Jeong [238] developed multiple nondestructive evaluation techniques for determining the constituent volume fractions of silicon carbide particulate–reinforced 7091 aluminum–alloy matrix composites.

The principal characteristics of ultrasonic waves are particle displacement, particle velocity, and sound pressure. Propagation of ultrasonic waves in solids is of high complexity because a solid body tries to maintain not only its volume but also its shape. This results in an increased variety of possible wave types.

For the silicon carbide–reinforced aluminum matrix composites, the reinforcement is much stiffer than the matrix material, while their densities are comparable. Thus, ultrasound would travel faster in the silicon carbide phase than in the aluminum matrix. This would account for that the composites with higher fraction of silicon carbide would also yield a higher velocity.

8.8.2.2.2. Acoustic emission testing

Acoustic emission can be a very powerful nondestructive evaluation technique for the in situ monitoring of damage evolution during mechanical testing. When a material is subjected to stress, it experiences plastic deformation, the formation of flaws, or fracture; these conditions produce small stresses or ultrasonic waves in the material and acoustic emissions are generated.

Acoustic emission is a transient elastic wave generated by the rapid release of energy within materials. Although acoustic emission can be audible

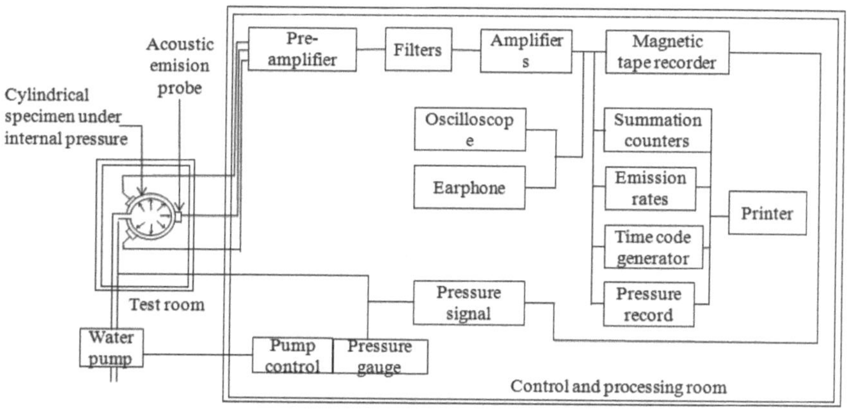

Fig. 8-51 Schematic diagram of the acoustic emission test system.

(during failure of timber and rock, e.g.), most emissions are either too low in amplitude or too high in frequency to be detected by the unaided ear, and therefore require the use of proper detecting equipment. A schematic diagram of the test equipment is shown in Fig. 8-51. To eliminate the effect of any unwanted external noise, the specimens were tested in a special insulated test room.

Three cylindrical specimens made of glass fiber–reinforced araldite wound at ±45° were tested. A typical stress/strain curve is shown in Fig. 8-52. Specimens (1) and (2) were pressurized to above the linear limit, whereas specimen (3) was tested to complete failure.

Four acoustic emission probes were mounted onto each specimen, three round the circumference at the gauge section of the tube and the fourth on the top of the specimen. The probes were distinguished from each other by the color of their wires, that is, red, yellow, or blue. The total emission counts emanating from the internal pressurization test of specimen (2) as detected by the blue probe are shown in Fig. 8-53. The other probes detected a similar pattern of acoustic emission events but with less counts according to their position from the damaged area. The probe on the top of the specimen picked up the hydraulic noise of the water used to pressurize the specimen. Probes on specimen (1) detected a similar pattern of total counts.

On specimen (3), which was tested to complete failure, the acoustic emission counts have similar pattern to that of specimen (2) but the energy released in the form of a stress wave was much higher. Fig. 8-54 shows the total emission counts as detected by the red probe which was mounted on the failed area. Here again, the other probes picked lower counts but of similar pattern. The emission rate was generally steady with increasing load (note that the total counts in Fig. 8-54 are plotted vs. time) until the fibers fractured producing audible acoustic emission.

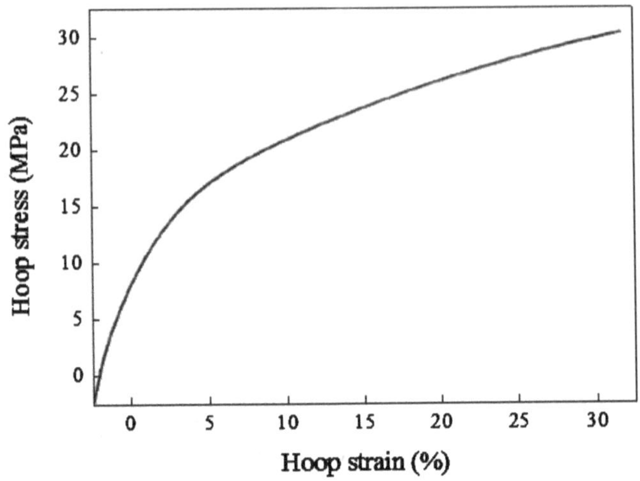

Fig. 8-52 Stress/strain curve of the test specimens.

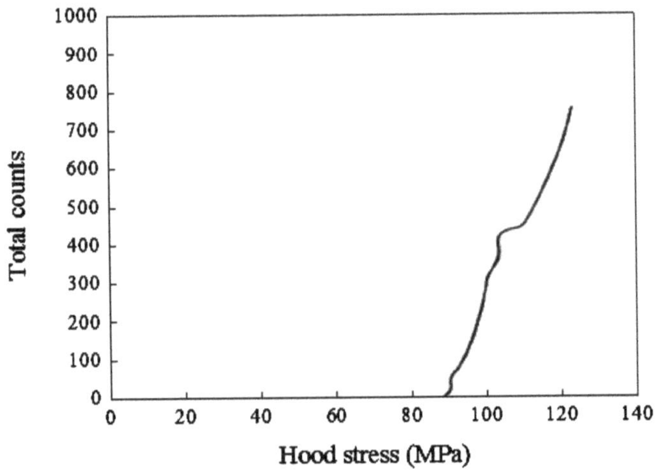

Fig. 8-53 Total acoustic emission counts of specimen (2) vs. hoop stress.

8.8.2.2.3. Radiography

There are two primary applications of radiography: determining inherent material properties (e.g., composition, particle size, and density) and evaluating manufacturing properties (e.g., existing flaws and complexities). In radiography, radiation is passed through the subject part or material, and a detector

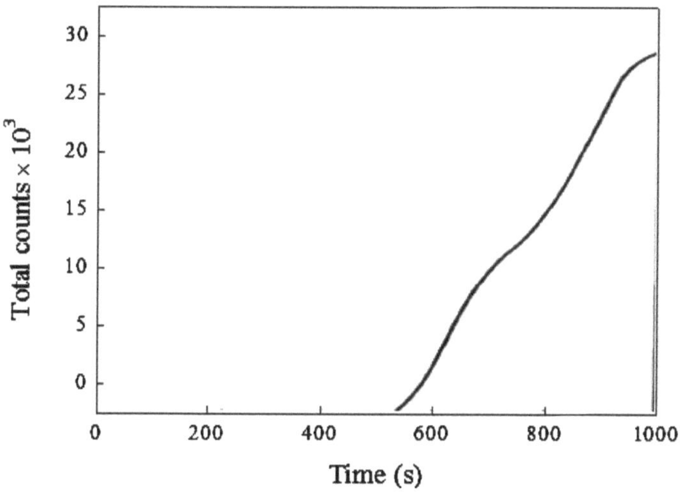

Fig. 8-54 Total acoustic emission counts of specimen (3) vs. time.

senses variations in the intensity of the radiation exiting the sample, thereby enabling a profile of the internal defect distribution [239–242]. The resulting two-dimensional radiation pattern can be visualized using photographic films or fluorescent screens. Using this technique, a wide range of material thickness levels can be assessed, as can complex shapes that would be difficult to scan using an ultrasonic technique.

In general, radiographic nondestructive evaluation can be classified according to the type of radiation employed: X-ray, gamma, or neutron.

Nondestructive evaluation using X-ray or gamma rays includes capturing and processing a shadow photograph. Specifically, the varying amounts of transmitted radiations cause ionization in the emulsion surface of the film, forming a latent image. The radiographic sensitivity of the resulting photograph is limited by the quality of the shadow image, by the scatter of radiation in the test object, by the test environment, and by the nature of the film emulsion.

In neutron radiography, the image is essentially a two-dimensional shadow display representing the intensity distribution of thermal neutrons passing through an object. As compared to X-ray or gamma-ray radiography, the attenuation characteristics of neutron energy are different. Here, the total neutron cross-section is the criterion for utilizing neutron radiography, whereas density and atomic number (linear absorption coefficient) are the main parameters in X-ray and gamma-ray radiography. Generally, neutron radiography complements conventional X-ray and gamma-ray radiography as it can detect flaws and material conditions that other methods cannot effectively assess.

Although radiography has numerous advantages, it is not without certain drawbacks. For example, access to opposite sides of the test object is required, and some structures are not amenable to study (e.g., components in radioactive or high-temperature environments). To maximize the likelihood of detection, cracks must be oriented nearly parallel to the beam. Delaminations are almost always undetectable. Finally, the use of radiography is a health risk, and the technique is expensive.

8.8.2.2.4. X-ray computed tomography

X-ray computed tomography provides a cross-sectional view of an object's interior and is well suited to characterizing a material's integrity [243, 244]. In essence, X-ray computed tomography is an advanced form of X-ray radiography. Conventional radiography provides a two-dimensional presentation of a three-dimensional object as the image plane is approximately normal to the X-ray beam. X-ray computed tomography creates a digital representation of a thin slice parallel to the X-ray beam. Typical slice thickness ranges from 0.025 to 3 mm, with pixel sizes (picture elements) ranging from 0.025 to 1 mm [241, 245, 246].

Fig. 8-55 illustrates the X-ray computed tomography setup and reconstruction procedure. As shown in Fig. 8-55(a), X-ray computed tomography system typically contains a tungsten X-ray radiation source and a cadmium/tungstate radiation detector as well as a precision manipulator to scan cross-sectional slices from different angles. The X-ray source, which can be operated at 420 kV and 3 mA, is collimated to form a thin fan beam. The fraction of the attenuated X-ray beam is directly related to the density and thickness of the test object, the composition of the material, and the energy of the X-ray beam. To obtain a full set of imaging data, the test subject, X-ray source, or detector array moves and a sequence of measurements is made from multiple incremental angles (θ). The data-acquisition system reads the signal from each detector in the array, converts the measurements into numerical values, and transfers the data to a computer for digital reconstruction. The reconstruction process (Fig. 8-55(b)) uses an algorithm to identify the point-by-point distribution of the X-ray densities in the two-dimensional image of the cross-sectional slice [247]. In the example depicted in Fig. 8-55(b), the incremental rotation angle equals 18°; this means that 10 different cross-sectional views were collected in order to construct an X-ray computed tomography image.

8.8.2.2.5. Photomicrographic Study

To further examine the failure caused to the matrix which was detected by acoustic emission and ultrasonic scanning, a photomicrographic study was usually used by a scanning electron microscopy [248, 249].

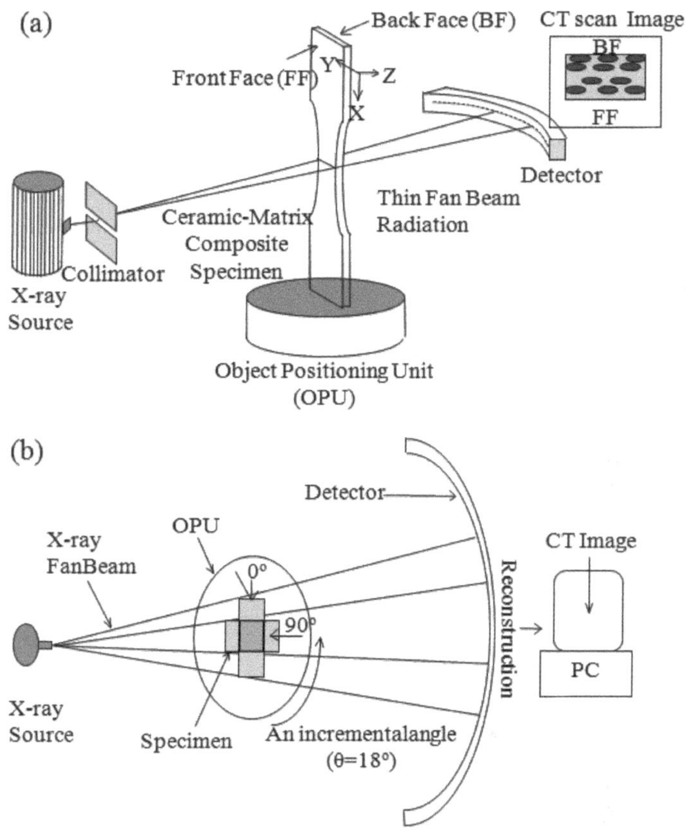

Fig. 8-55 The (a) experimental setup for obtaining the X-ray computed tomography-scan image and (b) reconstruction procedure.

8.9. SUMMARY

The present discussion on experimental characterization of composites has only a limited objective of familiarizing a reader with the most common type of tests that are performed on composite materials. The tests described herein are only static tests conducted to evaluate static properties. The discussion on static tests is also by no means complete. There are many static tests that are performed to evaluate properties needed in design under certain circumstances. Composite materials are relatively new and more complex compared to homogeneous and isotropic metallic materials. Their characterization poses many problems that are not present with isotropic materials. Experience with these materials is also limited. The major nondestructive evaluation methods currently in use for inspecting composite materials are ultrasonics, acoustic emission, X-ray computed tomography, X-ray radiography, etc. However, the nondestructive evaluation capability for inspecting parts with complex curvature is reduced because of the difficulty of the inspection. Inspection of composite joints/fasteners/adhesive bonds is also a more difficult inspection problem.

REFERENCES

[1] ASTM C 169, ASTM Book of Standards. ASTM, Philadelphia, 15.02 (1984).
[2] ASTM D 3178, ASTM Book of Standards. ASTM, Philadelphia, 05.05 (1984).
[3] ASTM D 3174, ASTM Book of Standards. ASTM, Philadelphia, 05.05 (1984).
[4] A.W. Adamson, Physical Chemistry of Surfaces. Wiley, New York (1980).
[5] G.E. Hammer, L.T. Drzal, Applied Surface Science 4 (1980) 340.
[6] P.J. Dynes, D.H. Kaeble, Journal of Adhesion 6 (1974) 195.
[7] D.H. Kaeble, Journal of Adhesion 6 (1974) 239.
[8] L.T. Drzal, J.A. Mescher, D.L. Hall, Carbon 17 (1979) 375.
[9] B. Rand, R. Robinson, Carbon 15 (1977) 311.
[10] ASTM C 613-67, ASTM Book of Standards. ASTM, Philadelphia, 15.03 (1988).
[11] ASTM D 4102-82, ASTM Book of Standards. ASTM, Philadelphia, 15.03 (1988).
[12] B.H. Eckstein, Weight Loss of Carbon Fibers in Air, 18th International SAMPE Technical Conference, Texas (1986).
[13] ASTM D 3800-79, ASTM Book of Standards. ASTM, Philadelphia, 15.03 (1988).
[14] ASTM D 2734, Annual Book of ASTM Standards. ASTM, West Conshohocken (1988).
[15] ASTM D 1505, Annual Book of ASTM Standards. ASTM, West Conshohocken (1988).
[16] S.R. Ghiorse, MIL Handbook 17 Testing Working Group, Santa Fe (1996).
[17] S.R. Ghiorse, J. Tiffany, Evaluation of Gas Pycnometry as a Density Measurement Method. In Proceedings of the 34th MIL-HDBK-17 PMC Coordination Group, Schaumberg (1996).
[18] ASTM D 3800, Annual Book of ASTM Standards. ASTM, West Conshohocken (1990).
[19] ASTM D 792, Annual Book of ASTM Standards. ASTM, West Conshohocken (1990).
[20] L. Piraux, B. Nysten, A. Haquenne, J.P. Issi, Solid State Communications 50 (1984) 697.
[21] I.L. Kalnin, Thermal Conductivity of High Modulus Carbon Fibers, ASTM STP 580 Composite Reliability (1975).
[22] ASTM D 2766, ASTM Book of Standards. ASTM, Philadelphia, 5.02 (1984).
[23] ASTM D 3417-83, ASTM Book of Standards. ASTM, Philadelphia, 8.03 (1987).
[24] ASTM D 3418-82, ASTM Book of Standards. ASTM, Philadelphia, 8.03 (1987).
[25] ASTM D 3379-75, Annual Book of ASTM Standards. American Society for Testing and Materials, Philadelphia, 15.03 (1990).
[26] ASTM D 4018-81, Annual Book of ASTM Standards. American Society for Testing and Materials, Philadelphia, 15.03 (1990).
[27] ASTM D 3039-76, Annual Book of ASTM Standards. American Society for Testing and Materials, Philadelphia, 15.03 (1990).
[28] J. Schultz, L. Lavielle, C. Martin, Journal of Adhesion 23 (1987) 45.
[29] C.A. May, D.W. Hadad, C.E. Browning, Polymer Engineering & Science 19 (1979) 545.
[30] A. Garton, Polymer Preprints 25 (1984) 163.
[31] H.W. Siesler, K. Holland-Moritz, Infrared and Raman Spectroscopy of Polymers. Marcel Dekker, New York (1980).

[32] J.L. Koenig, Quality Control and Nondestructive Evaluation Techniques for Composites Part II: Physiochemical Characterization Techniques. A State-of-the-Art Review. U.S. Army Aviation R & D Command, AVRADCOM TR 83-F-6 (1983).
[33] G.L. Hagnauer, Polymer Composites 1 (1980) 81.
[34] G.L. Hagnauer, D.A. Dunn, 12th National SAMPE Technical Conference, Vol. 12, pp 648 (1980).
[35] G.L. Hagnauer, Industrial and Engineering Chemistry Research 23 (1981) 128.
[36] G.L. Hagnauer, D.A. Dunn, Industrial and Engineering Chemistry Product Research and Development 21 (1982) 68.
[37] G.L. Hagnauer, D.A. Dunn, Quality Assurance of Epoxy Resin Prepregs. In Plastics in a World Economy, ANTEC 84, Society of Plastics Engineers 42nd Annual Technical Conference and Exhibition, pp 330 (1984).
[38] S.A. Mestan, C.E.M. Morris, Journal of Macromolecular Science—Reviews in Macromolecular Chemistry & Physics 24 (1984) 117.
[39] ASTM Test Method D 5296, in Annual Book of ASTM Standards. American Society for Testing and Materials, West Conshohocken, 8.03 (1990).
[40] M. Kowalsha, A. Wirsen, National SAMPE Symposium and Exhibition Proceeding 25 (1980) 389.
[41] C.S. Lu, J.L. Koenig, Journal of the American Chemical Society 32 (1972) 112.
[42] C.F. Poranski Jr., W.B. Moniz, D.L. Birkle, J.T. Kopfle, S. A. Sojka, Carbon-13 and Proton NMR Spectra for Characterizing Thermosetting Polymer Systems I. Epoxy Resins and Curing Agents. Naval Research Laboratory, NRL Report 8092 (1977).
[43] J.A. Happe, R.J. Morgan, C.M. Walkup, Composite Technology Review 6 (1984) 77.
[44] G.A. Morse, Journal of Polymer Science—Polymer Chemistry 22 (1984) 3611.
[45] A.B. Hunter, Analysis of Reaction Products of Polyimide by High Pressure DSC with GC/MS. In Proceedings of 12th North American Thermal Conference, Williamsburg, p. 527 (1983).
[46] J.S. Chen, A.B. Hunter, Polymer Preprints 22 (1981) 253.
[47] ASTM Test Method D 2857, Annual Book of ASTM Standards. American Society for Testing and Materials, West Conshohocken, 8.02 (1990).
[48] M. Naoki, I.H. Park, S.L. Bunder, B. Chu, Journal of Polymer Science—Polymer Physics 23 (1985) 2567.
[49] V. Bellenger, E. Fontaine, A. Fleishmann, J. Saporito, J. Verdu, Polymer Degradation and Stability 9 (1984) 195.
[50] J.H. Flynn, L.A. Wall, Journal of Research of the National Bureau of Standards A 70 (1966) 487.
[51] J.F. Carpenter, Assessment of Composite Starting Materials: Physiochemical Quality Control of Prepregs. In AIAA/ASME Symposium on Aircraft Composites, San Diego (1977).
[52] N.S. Schnieder, J.F. Sprouse, G.L. Hagnauer, J.K. Gillham, Polymer Engineering & Science 19 (1979) 304.
[53] ASTM Test Method D 3417, Annual Book of ASTM Standards. American Society for Testing and Materials, Philadelphia, 8.02 (1987).
[54] ASTM Test Method D 3418, Annual Book of ASTM Standards. American Society for Testing and Materials, Philadelphia, 8.03 (1987).
[55] J.K. Gillham, Polymer Engineering & Science 16 (1976) 353.
[56] J.B. Enns, J.K. Gillham, Journal of Applied Polymer Science 28 (1983) 2567.
[57] W.X. Zukas, W.J. MacKnight, N.S. Schneider, ACS Symposium Series, Journal of the American Chemical Society 227 (1983) 223.

[58] D. Crozier, F.W. Tervet, SAMPE Journal 18 (1982) 12.
[59] C.A. May, M.R. Dusi, J.S. Fritzen, D.K. Hadad, M.G. Maximovich, K.G. Thrasher, A. Wereta Jr, ACS Symposium Series, Journal of the American Chemical Society 227 (1983) 1.
[60] R.J. Hinrichs, ACS Symposium Series, Journal of the American Chemical Society 227 (1983) 187.
[61] M.B. Roller, Rheology of Curing Thermosets: Critique and Review. In Plastics in a World Economy, ANTEC '84, Society of Plastics Engineers 42nd Annual Technical Conference and Exhibition (1984).
[62] W.E. Baumgartner, T. Ricker, SAMPE Journal 19 (1983) 6.
[63] S.D. Senturia, N.F. Sheppard Jr, H.L. Lee, S.B. Marshall, SAMPE Journal 19 (1983) 22.
[64] B.B. Chowdhury, American Laboratory., 49.
[65] ASTM Test Method D 2393, Annual Book of ASTM Standards. American Society for Testing and Materials, West Conshohocken, 8.02 (1990).
[66] ASTM Test Methods D 3835, Annual Book of ASTM Standards. American Society for Testing and Materials, West Conshohocken, 8.02 (1990).
[67] ASTM Practice D 4065, Annual Book of ASTM Standards. American Society for Testing and Materials, West Conshohocken, 8.02 (1990).
[68] ASTM Practice D 4440, Annual Book of ASTM Standards. American Society for Testing and Materials, West Conshohocken, 8.03 (1990).
[69] ASTM Practice D 4473, Annual Book of ASTM Standards. American Society for Testing and Materials, West Conshohocken, 8.03 (1990).
[70] ASTM Terminology D 4092, Annual Book of ASTM Standards. American Society for Testing and Materials, West Conshohocken, 8.02 (1990).
[71] P. Cebe, S.D. Hong, Polymer 27 (1986) 1183.
[72] D.J. Blundell, B.N. Osborn, Polymer 24 (1983) 953.
[73] Y. Lee, R.S. Porter, Polymer Engineering Science 26 (1986) 633.
[74] J.C. Seferis, Polymer Composites 7 (1986) 158.
[75] ASTM Test Method D 3530, Annual Book of ASTM Standards. American Society for Testing and Materials, West Conshohocken, 15.03 (1988).
[76] ASTM Test Method E 1131, Annual Book of ASTM Standards. American Society for Testing and Materials, West Conshohocken, 14.02 (1990).
[77] ASTM Test Method D 4526, Annual Book of ASTM Standards. American Society for Testing and Materials, West Conshohocken, 8.03 (1990).
[78] ASTM Test Method D 4672, Annual Book of ASTM Standards. American Society for Testing and Materials, West Conshohocken (1990).
[79] C.W. Bert, Experimental Characterization of Composites. In Structural Design and Analysis—Part II. Academic, New York (1975).
[80] M.E. Waddoups, Characterization and Design of Composites Materials. In Composite Material Workshop. Technomic Publishing Company, Stamford (1968).
[81] G. Epstein, Testing of Reinforced Plastics. In Handbook of Fiberglass and Advanced Plastics Composites. Van Nostrand, Princeton, New Jersey (1969).
[82] G.C. Grimes, M.E. Bronstad, Testing Methods for Advanced Composites. In Handbook of Fiberglass and Advanced Plastics Composites. Van Nostrand, Princeton, New Jersey (1969).
[83] J.E. Ashton, J.C. Halpin, P.H. Petit, Primer on Composite Materials: Analysis. Technomic Publishing Company, Stanford (1969).
[84] S. Dastin, G. Lubin, J. Munyak, A. Slobodzinski, Mechanical Properties and Test Techniques for Reinforced Plastic Laminates. In Composite Materials: Testing and Design, ASTM STP 460. American Society for Testing and Materials, Philadelphia (1969).

[85] E.A. Rothman, G.E. Molter, Characterization of the Mechanical Properties of a Unidirection Carbon Fiber Reinforced Epoxy Matrix Composite. In Composite Materials: Testing and Design, ASTM STP 460. American Society for Testing and Materials, Philadelphia (1969).
[86] J. Case, Strength of Materials. Edward Arnold, London (1957).
[87] J. Marin, Mechanical Behaviour of Engineering Materials. Prentice Hall, New Jersey (1962).
[88] A. Morley, Strength of Materials. McGraw-Hill, New York (1916).
[89] R.J. Roark, Formulas for Stress Strain. McGraw-Hill, Washington, D.C. (1965).
[90] W.M. Haynes, T.L. Tolbert, Journal of Computer Material 3 (1969) 709.
[91] I. Krucinska, Composites Science and Technology 41 (1991) 287.
[92] D. Hull, D.W. Clyne, An Introduction to Composite Materials. Cambridge University Press, Cambridge (1996).
[93] R.M. Jones, Mechanics of Composite Materials. McGraw-Hill, Washington, D.C. (1973).
[94] S.W. Tsai, H.T. Hahn, Introduction to Composite Materials. Technomic, Westport (1980).
[95] P.W. Manders, I.M. Kowalski, The Effect of Small Angular Fiber Misalignments and Tabbing Techniques on the Tensile Strength of Carbon Fiber Composites. In 32nd International SAMPE Symposium, pp 985, Anaheim (1987).
[96] N.R. Sottos, J.M. Hodgkinson, F.L. Matthews, A Practical Comparison of Standard Test Methods using Carbon Fiber Reinforced Epoxy. In Proceeding of ICCM-VI and ECCM-II, pp 10. Elsevier, London (1987).
[97] W.C. Cui, M.R. Wisnom, M. Jones, Computer Engineering 4 (1994) 299.
[98] J. Harding, L. Dong, Composites Science and Technology 51 (1994) 347.
[99] K. Liu, M.R. Piggott, Composites 26 (1995) 841.
[100] C.E. Browning, F.L. Abrams, J.M. Whitney, A Four Point Shear Tests for Graphite/Epoxy Composites. In Composite Materials: Quality Assurance and Processing, ASTM STP 797. American Society for Testing and Materials, Philadelphia (1983).
[101] S. Lee, M. Munroe, In-plane Shear Properties of Graphite/Epoxy Composites for Aerospace Applications: Evaluation of Test Methods by the Decision Analysis Method, Aeronautical Note NAE-AS22, NRC No. 23778. Ottawa University, Ottawa (1984).
[102] Y. Sawada, A. Shindo, Carbon 30 (1992) 619.
[103] S.R. Swanson, M. Merrick, G.R. Toombes, Composites 16 (1985) 8220.
[104] P.A. Lockwood, Composite Technology Review 3 (1981) 83.
[105] W.R. Broughton, M. Kumosa, D. Hull, Composites Science and Technology 38 (1990) 299.
[106] C.C. Chamis, J.H. Sinclair, Experimental Mechanics 17 (1977) 339.
[107] M.R. Wisnom, Journal of Composite Materials 26 (1992) 1173.
[108] M. Buggy, A. Carew, Journal of Materials Science 29 (1994) 1925.
[109] M.R. Wisnom, Composite Structures 18 (1991) 47.
[110] M. Wisnom, The Flexural Strength of Unidirectional Carbon Fiber-epoxy. In 5th European Conference on Composite Materials, ECCM, Bordeaux, pp 165 (1992).
[111] N.R. Sottos, J.M. Hodgkinson, F.L. Matthews, A Practical Comparison of Standard Test Methods using Carbon Fiber Reinforced Epoxy. In Proceedings of 6th International Conference on Composite Materials, London, p. 1310 (1987).
[112] M.J. Shuart, Compression Testing of Graphite/Epoxy, ASTM STP 734, p. 152. American Society for Testing Materials (1981).
[113] N.R. Adsit, Compression Testing of Graphite/Epoxy. In: Compression Testing of Homogeneous Materials and Composites, ASTM STP 808, p. 175. American Society for Testing and Materials (1983).

[114] R.K. Clark, W.D. Lisagor, Compression Testing of Graphite/Epoxy Composite Materials. In Test Methods and Design Allowables for Fibrous Composites, ASTM STP 734, p. 34 (1981).
[115] G.C. Grimes, Experimental Study of Compression–Compression Fatigue of Graphite/Epoxy Composites. In ASTM STP 734, pp 281. American Society for Testing and Materials (1981).
[116] R.M. Lamothe, J. Numes, Evaluation of Fixturing for Compression Testing of Metal Matrix and Polymer/Epoxy Composites. In Compression Testing of Homogeneous Materials and Composites, ASTM STP 808, pp 241. American Society for Testing and Materials (1983).
[117] M. Nakatani, M. Shioya, J. Yamashita, Carbon 37 (1999) 601.
[118] T. Ohsawa, M. Miwa, M. Kawade, E. Tsushima, Journal of Applied Polymer Science 39 (1990) 1733.
[119] A.H. Shinohara, T. Sato, F. Saito, T. Tomioka, Y. Arai, Journal of Materials Science 28 (1993) 6611.
[120] C. Sourtis, Composites Science and Technology 42 (1991) 373.
[121] M.R. Wisnom, Composites 21 (1990) 403.
[122] D.H. Woolstencroft, A.R. Curtis, R.I. Haresceugh, Composites 12 (1981) 3456.
[123] D.H. Woolstencroft, A.R. Curtis, R.I. Haresceugh, Composites 16 (1985) 8220.
[124] J.S. Berg, D.F. Adams, Journal of Computer Science and Technology 11 (1989) 41.
[125] K.F. Port, The Compressive Strength of cfrp. Royal Aircraft Establishment. In Technical Report 82083, Farnborough (1982).
[126] J.G. Hãberle, F.L. Matthews, Studies on Compressive Failure in Unidirectional cfrp using an Improved Test Method. In Proceedings of 4th European Conference on Composite Materials, ECCM, Stuttgart (1980).
[127] K.N. Hofer, P.N. Rao, Journal of Testing Evaluation 5 (1977) 278.
[128] A.J. Barker, Composites 18 (1987) 217.
[129] S. Kumar, T.E. Helminiak, Materials Research Symposium Proceeding 134 (1989) 363.
[130] M. Buggy, G. Dillon, Composites 22 (1991) 191.
[131] A.R. Bunsell, A. Somer, Plastics Rubber and Composites Processing and Applications 18 (1992) 263.
[132] M.D. Gilchrist, A.J. Kinloch, F.L. Matthews, Composites Science and Technology 59 (1999) 179.
[133] A. Haque, P.K. Raju, Materials Evaluation 56 (1998) 765.
[134] S.A. Hitchen, S.L. Ogin, P.A. Smith, Composites 26 (1995) 303.
[135] P.E. Irving, C. Thiagrarajan, Smart Materials and Structures 7 (1998) 456.
[136] J.M. Lifschitz, Journal of Computer Science and Technology 10 (1988) 100.
[137] C.T. Lin, P.W. Kao, F.S. Yang, Composites 22 (1991) 135.
[138] M.E. Nichols, S.S. Wang, P.H. Geil, Journal of Macromolecular Science Part B Physics B29 (1990) 303.
[139] J.B. Sturgeon, Creep of Fiber Reinforced Thermosetting Resins. In Creep of Engineering Materials. Mechanical Engineering Publications (1978).
[140] W.E. Cantwell, J. Morton, Composites 21 (1991) 347.
[141] G. Dorey, S.M. Bishop, Impact Performance of Carbon Fiber/PEEK Composites. In International Conference on Impact Testing and Performance of Polymeric Materials. Surrey University (1985).
[142] M.W. Wardle, G.E. Zahr, Instrumented Impact Testing of Aramid and Reinforced Composite Materials. In Instrumented Impact Testing of Plastics and Composite Materials, ASTM STP 936, p. 219 (1987).
[143] S.J. Park, Y.S. Jang, Journal of Colloid and Interface Science 237 (2001) 91.
[144] I.S. Kim, Materials Research Bulletin 33 (1998) 1069.
[145] V.K. Srivastava, K. Maile, A. Klenk, Materials Science and Engineering: A 271 (1999) 38.

[146] S.K. Gaggar, L.J. Broutman, Materials Science Engineering 21 (1975) 1975.
[147] S.K. Gaggar, L.J. Broutman, Journal of Computational Materials 9 (1975) 216.
[148] S.K. Gaggar, L.J. Broutman, Fiber Science and Technology 9 (1976) 205.
[149] S.K. Gaggar, L.J. Broutman, Fracture Toughness of Raman Glass Fiber Epoxy Composites: An Experimental Investigation. In Flow Growth and Fracture, ASTM STP 631. American Society for Testing and Materials, Philadelphia (1977).
[150] J.F. Mandell, S.S. Wang, F.J. McGarry, Journal of Computational Materials 9 (1975) 266.
[151] W.F. Brown, J.E. Strawly, Plane Strain Crack Toughness Testing of High Strength Metallic Materials. In ASTM STP 410. American Society for Testing and Materials, Philadelphia (1966).
[152] D.L. Bowie, Rectangular Tensile Sheet with Symmetric Edge Cracks. American Society of Mechanical Engineers, 64-APM-3 (1964).
[153] A.A. Aliyu, I.M. Daniel, Effects of Strain Rate on Delamination Fracture Toughness of Graphite/Epoxy. In Delamination and Debonding of Materials, ASTM STP 876. American Society for Testing and Materials, Philadelphia (1985).
[154] C.R. Corletto, W.L. Bradley, Mode II Delamination Fracture Toughness of Unidirectional Graphite/Epoxy Composites. In Composite Materials: Fatigue and Fracture, ASTM STP 1012. American Society for Testing and Materials. Philadelphia (1989).
[155] I.M. Daniel, I. Shareef, A.A. Aliyu, Rate Effects on Delamination Fracture Toughness of a Toughened Graphite/Epoxy. In Toughened Composites, ASTM STP 937. American Society for Testing and Materials, Philadelphia (1987).
[156] J.W. Gillespie, L.A. Carlsson, A.J. Smiley, Composites Science and Technology 28 (1987) 1.
[157] S.A. Hitchen, A.L. Ogin, P.A. Smith, C. Soutis, Composites 25 (1994) 407.
[158] R.L. Ramkumar, J.D. Whitcomb, Characterization of Mode I and Mixed-mode Delamination Growth in T300/5208 Graphite/Epoxy. In Delamination and Debonding of Materials, ASTM STP 876. American Society for Testing and Materials, Philadelphia (1985).
[159] D.J. Wilkins, J.R. Eisenmann, R.A. Camin, W.S. Margolis, R.A. Benson, Characterizing Delamination Growth in Graphite/Epoxy. In Damage in Composite Materials: Basic Mechanisms, Accumulation, Tolerance Characterization, ASTM STP 775. American Society for Testing and Materials, Philadelphia (1982).
[160] S.J. Park, M.K. Seo, T.J. Ma, J.R. Lee, Journal of Colloid and Interface Science 252 (2002) 249.
[161] S.J. Park, M.K. Seo, J.R. Lee, Journal of Colloid and Interface Science 268 (2003) 127.
[162] S.J. Park, Interfacial Forces and Fields: Theory and Applications. Marcel Dekker, New York (1999).
[163] M.S. James, C.F. Paul, Composites Part A 27 (1996) 339.
[164] F.M. Fowkes, Journal of Physics Chemistry 67 (1963) 2538.
[165] D.K. Owens, R.C. Wendt, Journal of Applied Polymer Science 13 (1969) 1741.
[166] S.J. Park, M. Brendlé, Journal of Colloid and Interface Science 188 (1997) 336.
[167] S. Wu, Polymerization Interface and Adhesion. Marcel Dekker, New York (1982).
[168] S.J. Park, W.B. Park, J.R. Lee, Polymer Journal 31 (1999) 28.
[169] D.H. Kaelble, Journal of Adhesion 2 (1970) 66.
[170] A.A. Griffith, Philosophical Transactions of the Royal Society A221 (1920) 163.
[171] S. Meretz, W. Auersch, C. Marotzke, E. Schulz, A. Hampe, Composites Science and Technology 48 (1993) 285.
[172] Y.C. Kim, S.J. Park, Journal of Applied Polymer Science 74 (1999) 15.
[173] G. Desarmot, J.P. Favre, Composites Science and Technology 42 (1991) 151.
[174] L.B. Greszczuk, Interface in Composites ASTM STP 452 (1969) 42.

[175] P.J. Herrera-Franco, L.T. Drzal, Composites 23 (1992) 2.
[176] A. Kelly, W.R. Tyson, Journal of the Mechanics and Physics of Solids 13 (1965) 329.
[177] W.A. Fraser, F.H. Achker, A.D. Di Benedetto, A Computer Modelled, Single Filament Technique for Measuring Coupling and Sizing Agent Effects in Fiber Reinforced Composites. In Proceedings 30th Annual Technical Conference on Reinforcing Plastics 22-A. The Society of the Plastics Industry, Washington, D.C. (1975).
[178] L.S. Penn, S.M. Lee, Fiber Science and Technology 17 (1982) 91.
[179] A. Hampe, G. Klinka, S. Meretz, E. Schulz, Composites 26 (1995) 40.
[180] W. Weibull, Journal of Applied Mechanics 18 (1951) 293.
[181] E.G. Stoner, The Effect of Shape on the Tensile Strength of Pitch-based Carbon Fibers. Clemson University, Clemson SC (1991).
[182] S.J. Park, Y.S. Jang, Journal of Colloid and Interface Science 237 (2001) 91.
[183] A. Bismarck, M.E. Kumru, B. Song, J. Springer, E. Moos, J. Karger-Kocsis, Composites Part A 30 (1999) 1351.
[184] L.S. Penn, S.M. Lee, Journal of Composite Technology and Research 11 (1989) 23.
[185] C.Y. Yue, H.C. Looi, International Journal of Adhesion and Adhesives 21 (2001) 309.
[186] C.S. Desai, M.M. Zaman, J.G. Lightner, H.J. Siriwardane, International Journal for Numerical and Analytical Methods in Geomechanics 8 (1984) 19.
[187] J.H. Hohberg, H.F. Schweiger, On the Penalty Behaviour of Thin-layer Elements. In Numerical Models in Geomechanics: Proceedings of the Fourth International Symposium on Numerical Models in Geomechanics, Swansea (1992).
[188] A.K. Mal, M.R. Karim, Y. Bar-Cohen, Materials Science, Engineering A 126 (1990) 155.
[189] O. Allix, P. Ladevèze, Composite Structures 22 (1992) 235.
[190] J.C.J. Schellekens, R. de Borst, International Journal for Numerical Methods in Engineering 36 (1993) 43.
[191] R.K. Pandey, C.T. Sun, Mechanical Composite and Material Structure 3 (1996) 183.
[192] Z.Q. Cheng, A.K. Jemah, F.W. Williams, American Society of Mechanical Engineers Journal of Applied Mechanics 63 (1996) 1019.
[193] V.Q. Bui, E. Marechal, H. Nguyen-Dang, Composites Science and Technology 59 (1999) 2269.
[194] S. Wang, Z. Mei, D.D.L. Chung, International Journal of Adhesion and Adhesives 21 (2001) 465.
[195] R.W. Richards, S.K. Peace, Polymer Surfaces and Interfaces III. John Wiley & Sons, Chichester, New York (1999).
[196] A. Ghatak, K. Vorvolakos, H. She, D.L. Malotky, M.K. Chaudhury, Journal of Physics Chemistry B 104 (2000) 4018.
[197] C.S. Desai, Mechanics of Materials And Interfaces: The Distributed State Concept, pp 425. CRC Press, Boca Raton (2001).
[198] B.D. Agarwal, L.J. Broutman, Analysis and performance of fiber composites, 2nd edition. John Wiley & Sons, New York (1990).
[199] A.S. Chen, F.L. Matthews, G.D. Sims, Composites Science and Technology 58 (1998) 613.
[200] M.J. Mathews, S.R. Swanson, Composites Science and Technology 67 (2007) 1489.
[201] K. Hsu, M. Uhl, B. Lucht, H. Jeong, D.K. Hsu, Mechanical Strength Degradation of Graphite Fiber Reinforced Thermoset Composites Due to Porosity. In Review of Progress in QNDE, p. 107 (1988).
[202] D. Hsu, Ultrasonic Nondestructive Evaluation of Void Content in CFRP. In Proceedings of the 46th Annual Technical Conference, p. 1273. Society of Plastics Engineers (1988).

[203] D.K. Hsu, Ultrasonic Methods for Evaluating Porosity Content and Mechanical Strength of CFRP Laminates. In ASNT Spring Conference: Fourth Annual Research Symposium, NDE of Composites and Composite Materials, pp 223. American Society for Nondestructive Testing (ASNT) (1995).
[204] N.C.W. Judd, W.W. Wright, SAMPE Journal 10. (1978) .
[205] K.V. Steiner, Defect Classifications in Composites using Ultrasonic Nondestructive Evaluation Techniques. In Damage Detection in Composite Materials, ASTM STP 1128. American Society for Testing and Materials (1992).
[206] J. Degrieck, N.F. Declercq, O. Roy, Insight 45 (2003) 196.
[207] K.G. Lipetzky, B.G. Bandos, SAMPE Journal 39 (2003) 66.
[208] E.T. Camponeschi, R. Crane, B. Lipetzky, B. Bandos, Materials Evaluation., 752.
[209] D. Hsu, Ultrasonic Nondestructive Evaluation of Void Content in CFRP. In Proceedings of the 46th Annual Technical Conference, pp 1273. Society of Plastics Engineers (1988).
[210] K.I. McRae, C.A. Zala, Review of Progress in Quantitative Nondestructive Evaluation 14A (1995) 703.
[211] K.V. Steiner, Defect Classifications in Composites using Ultrasonic Nondestructive Evaluation Techniques. In Damage Detection in Composite Materials, STP 1128, p. 72. American Society for Testing and Materials (1992).
[212] D.J. Dorsey, R. Hebner, W.S. Charlton, Journal of Computational Materials 38 (2004) 1505.
[213] T. Monnier, Journal of Intelligent Material Systems and Structures 17 (2006) 411.
[214] L. Cartz, Nondestructive Testing. ASM, Materials Park, OH (1995).
[215] M.M. Schwartz, Composite Materials. Prentice-Hall, New Jersey (1997).
[216] D.E. Bray, D. McBride, Nondestructive Testing Techniques. John Wiley & Sons, New York (1992).
[217] P.E. Mix, Introduction to Nondestructive Testing. John Wiley & Sons, New York (1987).
[218] D.W. Richerson, Modern Ceramic Engineering, 2nd edition. Marcel Dekker, New York (1992).
[219] Standard Terminology for Nondestructive Examinations. ASTM E1316-96, Philadelphia (1996).
[220] J.P. Monchalin, R. Heon, J.F. Bussiere, B. Farahbakhsh, Nondestructive Characterization of Materials. Plenum Press, New York (1987).
[221] C.H. Gü, B. Ogel, Material Characteristics 47 (2001) 227.
[222] J.P. Panakkal, H. Willems, W. Arnold, Journal of Materials Science 25 (1990) 1397.
[223] L.P. Martin, D. Doden, M. Rosen, Journal of the American Ceramic Society 79 (1996) 1281.
[224] K.K. Phani, S. Mukherjee, D. Basu, Journal of the American Ceramic Society 79 (1996) 3331.
[225] L.P. Martin, D. Nagle, M. Rosen, Materials Science and Engineering A 246 (1998) 151.
[226] L. Adler, J.H. Rose, C. Mobley, Journal of Applied Physics 59 (1986) 336.
[227] B. Champagne, L. Piche, Powder Metallurgy International 18 (1986) 261.
[228] C.H. Schilling, V.J. Garcia, R.M. Smith, R.A. Roberts, Journal of the American Ceramic Society 81 (1998) 2629.
[229] W.H.M. van Dreumel, NDT International 11 (1978) 235.
[230] D.E.W. Stone, B. Clarke, Non-Destruction Testing 8 (1975) 137.
[231] J. Kim, P.K. Liaw, Materials Science Engineering A 409 (2005) 302.
[232] H.M. Ledbetter, S.K. Datta, Journal of Acoustical Society of America 79 (1986) 239.
[233] G. Mott, P.K. Liaw, Metallurgical and Materials Transactions 19A (1988) 2233.
[234] P.K. Rohatgi, S. Raman, B.S. Majumdar, A. Banerjee, Materials Science and Engineering A 123 (1990) 89.

[235] P.K. Liaw, R.E. Shannon, W.G. Clark, W.C. Harrigan, H. Jeong, D.K. Hsu, Journal of the Minerals Metals & Materials Society 44 (1992) 36.
[236] P.K. Liaw, R. Pitchumani, D.K. Hsu, H. Jeong, S.C. Yao, Journal of Engineering for Gas Turbines and Power 116 (1994) 647.
[237] Y. Lu, P.K. Liaw, Journal of Composite Materials 29 (1995) 1096.
[238] H. Jeong, Research in Nondestructive Evaluation 9 (1998) 41.
[239] Y. Nir-El, V. Katz, J. Haddad, D. Eliezer, Nondestructive Testing and Evaluation 11 (1994) 149.
[240] W.J. Lewis, B.J. Lewis, K.M. Jaansalu, D.R. Black, C.J. Smith, Nondestructive Testing and Evaluation 11 (1994) 155.
[241] S. Ekinci, A.G. Bilge, British Journal of Non-destructive Testing 33 (1991) 450.
[242] ASNT Handbook. In Nondestructive Testing Handbook, 2nd edition ASNT, Columbus, OH (1985).
[243] P.C. Copley, J.W. Eberhand, G.A. Mohr, Journal of Minerals Metallurgy and Materials 46 (1994) 14.
[244] ASTM Handbook. In Nondestructive Evaluation and Quality Control, pp 358. ASTM International, Materials Park, OH (1992).
[245] J.C. Elliott, V. Katz, Y. Haddad, D. Eliezer, Journal of Minerals, Metallurgy and Materials 46 (1994) 11.
[246] A. Abdul-Aziz, Materials Evaluation 64 (2006) 20.
[247] Standard Guide for Computed Tomography Imaging. ASTM E1441-95, Philadelphia (1995).
[248] M.N. Nahas, Composites 16 (1985) 148.
[249] R. Prakash, Non-destructive Testing and Fatigue Behaviour of Carbon-fiber Reinforced Polymer. Cranfield Institute of Technology, Cranfield (1975).

Chapter 9

Modeling of Fiber–Matrix Interface in Composite Materials

9.1. INTRODUCTION

The properties of the fiber–matrix interface are of great importance for the macroscopic mechanical properties of composite materials. With a strong interface, more efficient stress transfer between fibers is assured. Fiber–matrix debonding would then be suppressed, which otherwise would induce premature accumulation of damage leading to failure. This is of particular importance under fatigue conditions [1–4]. A thorough investigation of the influence of interfacial shear strength on various static strength properties of a fiber-reinforced polymer matrix has shown that in particular off-axis strengths with notable shear stresses could be significantly improved by a more effective interface [5].

Interfacial adhesion phenomena is usually analyzed in either a mechanics framework, or by its chemistry. The former concerns notions from fracture mechanics, interfacial stresses and strains, interlocking, slip-stick frictional sliding, keying, etc., whereas the chemistry approach deals with the composition of the polymer and the fiber surface and the interaction between them, i.e., covalent, hydrogen, electrostatic, and van der Waals bonding in decreasing order of efficiency.

In the literature, the two-dimensional 'interphases' or finite-thickness 'interphases' are considered when analyzing the interaction between fibers and matrix in composites. On the molecular level, an interphase can be envisaged, with a chemical or physical bond to the residual groups on the fiber surface, and an effective functional gradient in the radial direction with properties different from those of the bulk matrix, e.g., by interdiffusion of polymer networks of the size and the matrix.

Interface Science and Composites, Volume 18
ISSN: 1573-4285, doi: 10.1016/B978-0-12-375049-5.00009-8

The goal of the manipulation of the interfacial chemistry is to achieve a more efficient interfacial stress transfer capability, which would result in improved mechanical properties of the composites. There are many techniques to characterize the stress transfer capability of a fiber–matrix interface, e.g., the pull-out test [6], the single-fiber composite tensile test [7], the transverse debond test [8], interlaminar shear strength in short-beam shear specimens [9], etc. The transferability of the results of different test methods is still relatively limited, although the ranking order in interfacial efficiency usually seems to agree. The choice of test method should therefore reflect the stress state to which the interface is subjected in the composite application, and preferably also convey a material property. Since the propagation of interfacial cracks in shear of embedded fibers aligned in the load direction is an important contributor to fatigue damage degradation and failure of composite laminates [10, 11].

9.2. EVALUATION OF FIBER–MATRIX INTERFACIAL SHEAR STRENGTH AND FRACTURE TOUGHNESS

On the basis of microscopical analyses of the fiber distribution and longitudinal shear deformation in unidirectional fiber composites, a simple approach is presented for characterizing the interfacial shear strength and fracture toughness [10].

9.2.1. Microscopical Geometric Analysis of Fiber Distributions in Unidirectional Composites

Assuming the fibers are distributed hexagonally, through a simple geometrical analysis the following relationship is obtained:

$$S = 2\left[\left(\frac{\pi}{2\sqrt{3}v_f}\right)^{1/2} - 1\right]^r \tag{9-1}$$

where v_f is the volume fraction of the fiber, s the distance between two fibers, and r the fiber radius.

For square distribution of the fibers, Eq. (9-1) changes to

$$S = 2\left[\left(\frac{\pi}{4v_f}\right)^{1/2} - 1\right]^r \tag{9-2}$$

The distance between the centers of two fibers is $S = 2r + s$.

Because the fiber modulus is much larger than that of the matrix, i.e., the modulus of glass fiber is larger than that of epoxy by an order of magnitude, when the composites undergo a longitudinal shear deformation [12], the fiber modulus can be assumed to be infinitely large and the shear deformation is all

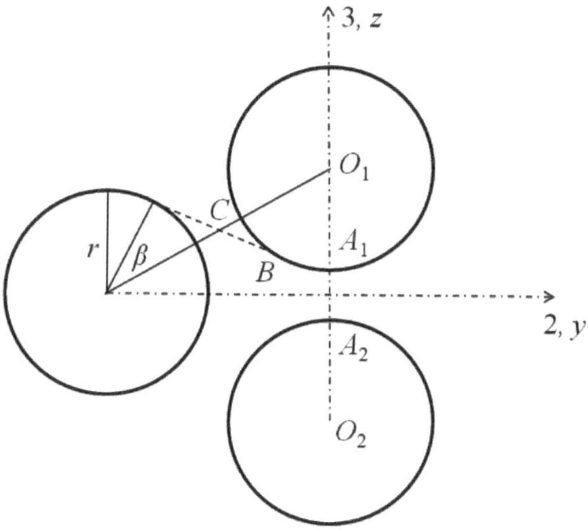

Fig. 9-1 Hexagonal fiber distribution and interfacial crack.

caused by the matrix. Thus, the average shear strain is given as

$$\gamma_{av} = \frac{\Delta u}{s + 2r} \tag{9-3}$$

where Δu is the difference between the longitudinal displacement of the centers of the two adjacent fibers in the through-thickness direction, i.e., $\Delta u = u_{02} - u_{01}$.

In the matrix, the maximum shear strain takes place at the location where the separation distance of the two fibers is the shortest, as shown by $\bar{A}_1\bar{A}_2$ in Fig. 9-1. Thus the maximum shear strain in the matrix is given as:

$$\gamma_{max} = \frac{\Delta u}{s} \tag{9-4}$$

Consequently, the relationship between the average shear strain of the composite and the maximum strain in matrix is derived as:

$$\frac{\gamma_{av}}{\gamma_{max}} = \frac{s}{s + 2r} \tag{9-5}$$

9.2.2. Measurement of Interfacial Shear Strength

The method of measuring the interlaminar shear strength by the short-beam bending test has been well documented. The interlaminar shear strength in the short-beam bending test is given as [13–15]

$$ILSS = \frac{3p}{4bh} \tag{9-6}$$

where p is bending load, b the width, and h the thickness of the beam.

The interlaminar shear strength in Eq. (9-6) is that in the middle surface of the beam, and it is the apparent average stress. According to Eq. (9-3), we have

$$ILSS = G_{13c} \frac{\Delta u}{s + 2r} \tag{9-7}$$

where G_{13c} is the principal (longitudinal-transverse) shear modulus of the composites, and subscripts 1 and 3 refer to fiber direction and transverse direction, respectively (Fig. 9-1).

The maximum shear strength in the matrix is

$$\tau_{\max} = G_m \frac{\Delta u}{s} \tag{9-8}$$

Consequently,

$$\tau_{\max} = ILSS \cdot \frac{G_m}{G_{13c}} \times \frac{s + 2r}{s} \tag{9-9}$$

Thus, after measurement of the interlaminar shear strength from Eq. (9-9), the shear strength of the interface (or matrix) can be obtained. If the shear strength of the matrix is larger than that of the interface, Eq. (9-9) gives the interfacial shear strength, otherwise it is matrix strength.

The shear modulus, G_{13c}, of unidirectional fiber composites can be determined by the role of mixtures. However, the following formula yields better predictions [16]:

$$G_{13c} = \frac{G_m}{1 - \sqrt{v_f(1 - G_m/G_{13f})}} \tag{9-10}$$

Taking glass-fiber-reinforced low-modulus epoxy as an example, the matrix modulus is, $G_m = 0.77$ GPa, the modulus of glass fiber, $G_{13f} = 30$ GPa, and we assume that $v_f = 0.6$. Thus, Eq. (9-9) gives $\tau_{max} = 1.31$ interlaminar shear strength. Compared with the experimental results of Ref. [17], this proved to be a good prediction.

9.2.3. Measurement of Interfacial Fracture Toughness

The end-notched-beam flexure test is a commonly used method for the evaluation of interlaminar shear (mode II) fracture toughness, G_{II}^c. The formula is given as [18]

$$G_{II}^c = \frac{9a^2p_c^2}{b^2h^3E_{11}} \tag{9-11}$$

where E_{11} is longitudinal modulus, p_c the fracture load of the end-notched-beam flexure test, a the crack length, b the width, and h the half of the thickness of the beam.

Considering the shear effect the expression is modified to be [18]

$$G_{II}^c = \frac{9a^2p_c^2}{b^2h^3E_{11}}\left[1+0.2\frac{E_{11}}{G_{13}}\left(\frac{h}{a}\right)^2\right] \quad (9\text{-}12)$$

The critical stress intensity factor is given as:

$$K_{II}^c = \tau_c(\pi a)^{1/2}H \quad (9\text{-}13)$$

where τ_c is the apparent fracture shear stress determined by Eq. (9-6) and

$$H\left(\frac{a}{2}\right)^{1/2} = 0.368 - 0.0717\left(\frac{e}{d}\right) + 0.0045\left(\frac{e}{d}\right)^2 + 0.545\left(\frac{a}{e}\right) + 0.490\left(\frac{e}{d}\right)\left(\frac{a}{e}\right) - 0.0209\left(\frac{a}{e}\right)\left(\frac{e}{d}\right)^2 \quad (9\text{-}14)$$

where the parameters in the above expression are shown in Fig. 9-2.

From fracture mechanics, we have:

$$G_{II}^c = \frac{S_{11}}{\sqrt{2}}\left[\left(\frac{S_{33}}{S_{11}}\right)^{1/2} + \frac{2S_{13}+S_{55}}{2S_{11}}\right]^{1/2} \times (K_{II}^c)^2 \quad (9\text{-}15)$$

where S_{ij} is compliance tensor of the composites, and 1 and 3 indicate longitudinal and through-thickness directions, respectively.

For unidirectional fiber composites, the fracture surface of the interlaminar crack is microscopically uneven. The interfacial crack surface of the fiber/matrix is a partial cylindrical (or circle arc) surface and they are linked by matrix cracks. The lengths of the matrix crack and of the interfacial crack are determined by the difference of the matrix strength and interfacial strength [19, 20]. Higher strength leads to shorter crack length. The simplest situation is that the matrix crack is a tangent to interfacial cracks of the adjacent fibers (see Fig. 9-1 for the hexagonal distribution). From the geometry of Fig. 9-1, the lengths of the two cracks can be calculated. The length of a half interfacial crack is:

$$l_i = r\left(\frac{\pi}{3} - \beta\right) \quad (9\text{-}16)$$

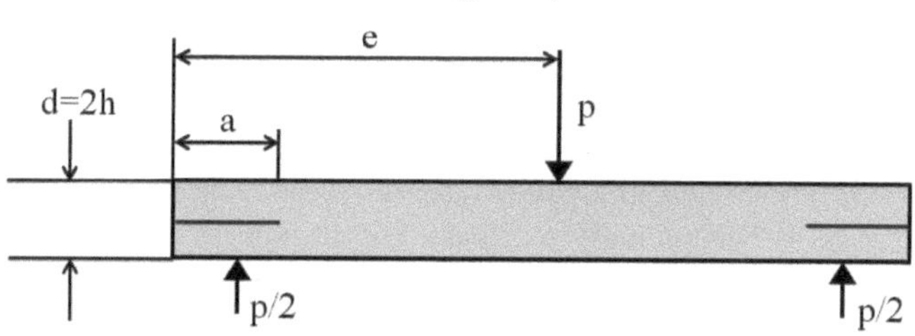

Fig. 9-2 Schematic showing end-notched-beam flexure specimen.

where the angle β is determined by:

$$\cos^{-1}\beta = \frac{r}{r + s/2} \tag{9-17}$$

The length of a half matrix crack l_m is:

$$l_m = \mathrm{r}\tan(\beta) \tag{9-18}$$

For $v_f = 0.6$, it is obtained that $l_i = 0.43r$ and $l_m = 0.71r$.

The interlaminar fracture toughness of composite, G_{II}^c, consists of two parts: contributions of matrix crack and interfacial crack. According to the rule of mixtures, we have:

$$G_{II}^c = \frac{1}{l_i + l_m}(l_i G_{II}^i + l_m G_{II}^m) \tag{9-19}$$

where G_{II}^i is the critical strain-energy release rate (fracture toughness) for interfacial fracture. G_{II}^m is critical strain-energy release rate of the matrix.

After G_{II}^c is measured by the end-notched-beam flexure test and G_{II}^m is known from handbooks of materials properties. G_{II}^i can be calculated from Eq. (9-19). As an example, a T300 carbon fiber-reinforced XU292 resin composites is considered. From the handbook, it is known that the fracture toughness of the resin XU292 is $G_{II}^m = 210.0\ \mathrm{J/m^2}$ (1.2 in.-lb/in.). The measured data give $G_{II}^c = 1.48\ \mathrm{kJ/m^2}$; thus, Eq. (9-19) gives $G_{II}^i = 3.57\ \mathrm{kJ/m^2}$. It is easy to see that the interfacial debonding crack absorbs more energy than the matrix crack. This is why fiber composites possess much larger fracture toughness than matrix. The above discussion is given for the case of a hexagonal distribution of the fibers. For the square fiber distribution, longer interfacial cracks may be expected if the interfacial strength is smaller than that of the matrix. In the limit, it is assumed that $l_i = (\pi/2)r$ and $l_m = (s/2) = 0.288r$, and we obtain $G_{\mathrm{IIi}} = 1.71\ \mathrm{kJ/m^2}$, which is approximately equal to G_{IIc}.

Furthermore, this method can also be used to evaluate the Mode I fracture toughness for interfacial cracking by using delamination fracture tests. For example, one can use the double-cantilever beam test to measure mode I fracture toughness G_{Ic} of composites. From Eq. (9-19), the interfacial fracture toughness can be estimated. If one carries out mixed-mode bending tests, then the mixed-mode fracture toughness can also be obtained.

9.3. INTERPRETATION OF SINGLE FIBER PULL-OUT TEST

The load versus extension trace generated by the single-fiber pull-out test is thoroughly interpreted and mathematically modeled. The single-fiber pull-out test is employed experimentally to model the failure of composite materials [21, 22].

9.3.1. Early Observations of Single-Fiber Pull-out Test

One of the earliest descriptions of the single-fiber pull-out test, which detailed the initial debonding, crack propagation, completion, and fiber pullout, was given by Kelly in 1970 [23]. It was recognized that if the embedded length of the fiber was greater than the length of catastrophic interfacial failure, complete debonding would not occur at once.

This same type of load *versus* displacement curve was also reported by Takaku and Arridge in 1979 [24]. In their interpretation of the load-extension trace they stated that the peak of the maximum load correlated to 'the wire debonding along the full embedded length from the matrix, i.e., crack initiation through to failure.' They did not, however, account for the time required for the debonding to occur and the initiation and steady state or completion loads were not separated.

Although early observations of the single-fiber pull-out test included the onset of crack initiation as well as the propagation of this crack along the length of the interface, quantitative distinctions between the events were not made. Instead, the maximum load the system was able to withstand was used as the basis of comparison between sample sets. As efforts progressed, however, more details were recognized and more appropriate experimental interpretations presented.

9.3.2. Calculation of Single-Fiber Pull-out Test

A single-fiber pull-out technique has been used to determine the interfacial bond strength of carbon fibers embedded in epoxy resin. A method is presented whereby the maximum interfacial bond shear stress may be evaluated from the simple measurement of the pull-out force and fiber dimensions [25].

The study of load transfer between a fiber and a matrix started with the work of Cox [26], who analyzed the stress transfer around a single discontinuous fiber embedded in a matrix. This has come to be called the shear-lag theory.

Recent work by Galiotis *et al.* [27] has experimentally confirmed these predictions. The basic assumption of the shear lag theory is that all the tensile loads are carried by the fiber and all the shear loads are carried by the matrix, although in practice the matrix surrounding the fiber also sees some tensile stresses. However, the major drawback for all these tests is that they only produce a value for the pull-out force from which the mean interracial shear stress, τ, for that length of embedded fiber may be calculated using the following formula:

$$\tau = \frac{F}{\pi D l} \tag{9-20}$$

The load transfer into the fiber from the matrix occurs over a length of fiber commonly called the ineffective length (Fig. 9-3) [26], which is defined when some proportion of the remote fiber load is reached. If the shear strength is

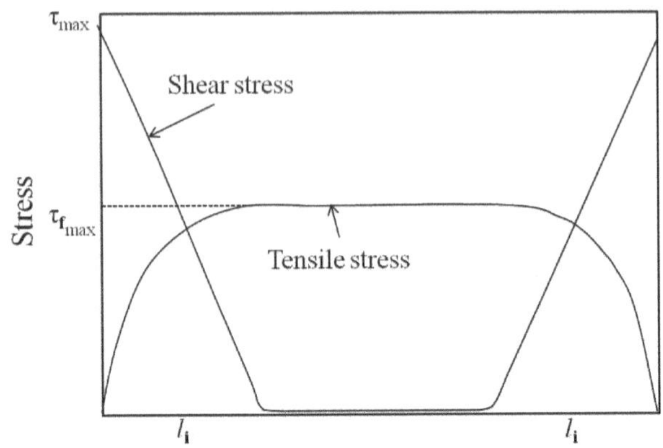

Fig. 9-3 Stress distributions for a short embedded fiber.

calculated from Eq. (9-20) for a fiber whose embedded length is greater than this ineffective length, then the shear strength will be underestimated. This is because a length of fiber is included over which little stress transfer is taking place. So the smaller the embedded length the greater the estimate of the shear strength. Extending this argument the maximum shear stress will be obtained when the embedded length tends to zero, and embedded fiber length becomes impossibly small to measure directly. However, as will be shown, it is not necessary to go to such lengths to obtain the maximum shear stress of the interface.

Greszczuk [28] has taken the shear-lag equations and applied them to the case of a single fiber, one end of which is embedded in resin (Fig. 9-4). The result of his analysis is an expression for the maximum shear strength

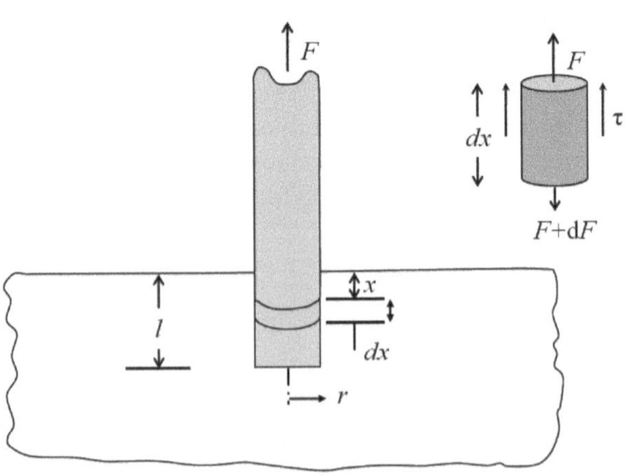

Fig. 9-4 Model for matrix stresses around a partially embedded fiber.

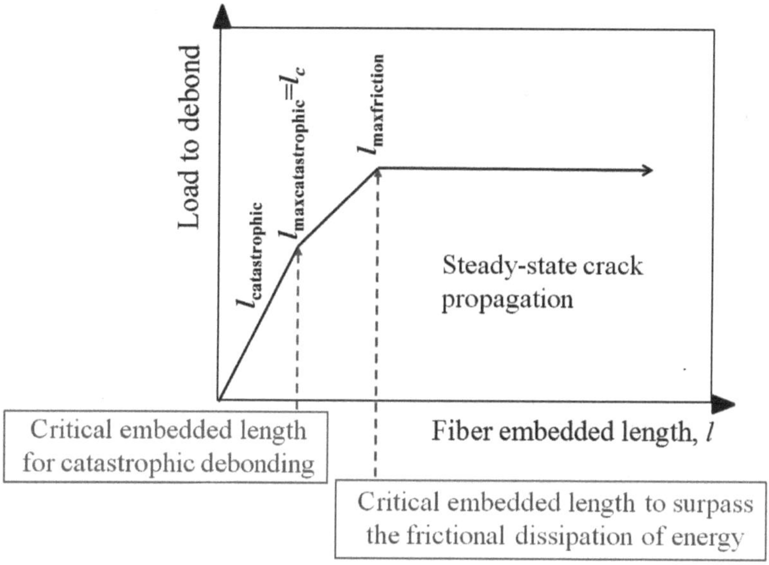

Fig. 9-5 Theoretical embedded fiber length *versus* load required to debond.

9.3.7. Stress Transfer from Matrix to Fibers

The term critical fiber length l, has been used to describe many phenomena. It has been applied to the weighted average fiber length resulting from fiber fragmentation [44–48] to the primary frictional zone associated with debonding and pullout and to the bimodal relationship between embedded length and load to debond [49–51]. The common denominator of this descriptor in these applications is the tensile properties of the fiber relative to those of the matrix, and the normal forces exerted by the resin on the fiber. The following is a brief discussion of the critical fiber length value relating to fiber pullout.

Here, the critical fiber length is the maximum embedded length that will catastrophically fail, $l = l_{maxcatastrophic}$. The primary frictional zone that results during debonding and subsequent pullout is an event related to the Poisson's contraction of the fiber under a given load.

Removal of the fiber from the surrounding matrix is realized when the load required for the radial contraction is less than the load required for steady-state debonding, which, in turn, is less than the ultimate strength of the fiber. The relationship between fiber length up to l_c, and associated stress, and the relationship between fiber embedded length and load required to debond, is the same. This is the result of the load-carrying ability of a fiber with length less than l_c. The heavily documented embedding length *versus* load required to debond has been reported as a bimodal relationship based on catastrophic interfacial failure. Given the inherent scatter in the data, it was unclear to many if the relationship between debonding force and embedded length was a linear increase then a flat plateau, or a linear increase followed by another linear increase with a modestly different slope. As will be elaborated upon in this article, the initially rising portion of the curve does reflect catastrophic interfacial failure of embedded fiber lengths less than l_c. For fiber lengths greater than l_c, the role of friction is seen as a second region displaying a new positive slope. Only when the various embedded fiber lengths are all greater than l_c whether there will be an initial positive slope followed by a plateau. In this case, the first region is the result of failure plus friction and the plateau the result of steady-state primary frictional zone.

9.4. INTERPRETATION OF SINGLE-FIBER PUSH-OUT TEST

The fiber push-out test is one of the more convenient techniques for obtaining interfacial properties [52]. This method affords a simple screening test and makes it possible to probe into individual fibers within a bulk composite fabricated by conventional processes. In this test, an indenter is used to apply load to the fiber end. During the test, the applied load and the indenter tip displacement are continuously monitored, providing a force–displacement curve. Data can be analyzed using both stress-based and energy-based schemes. The stress-based criterion is straightforward. By measuring the applied force F, the fiber length H (the thickness of the specimen), and the fiber diameter D, the average applied interfacial shear stress τ may be obtained from Eq. (9-32).

$$\tau = \frac{F}{\pi DH} \tag{9-32}$$

where F is the maximum applied force required for debonding and pushing a fiber out of the specimen and τ gives the interfacial shear strength τ_s. when F is the applied force to cause frictional sliding of a fiber in the matrix (after interfacial debonding), τ becomes the frictional shear stress τ_f.

Alternatively, energy-based schemes have been developed to convert force/displacement data from pushout tests to interface properties [53–58]. These theories are based on the shear lag concept in which load transfers by shear from an infinite matrix to a fiber. Among these schemes, the model of Kerans and Parthasarathy [57] represents a recent refinement of earlier solutions and is described as follows.

The equation for the fiber displacement [58] is modified here by using positive values for applied load, debond load, and displacement, giving

$$\delta = \frac{1 - 2\nu_f k}{2\mu k \pi r E_f}\left[P_a - P_d + P_r - (P^* - P_r)\ln\left(\frac{P^* + P_a}{P^* + P_d - P_r}\right)\right] \tag{9-33}$$

where δ is the fiber displacement, μ the friction coefficient, ν_f the Poisson's ratio of the fiber, r the radius of the fiber, E_f the Young's modulus of the fiber, P_a the applied load, P_r the residual axial load in the fiber, caused by the mismatch in the coefficient of thermal expansion, P^* the maximum value of applied force due to friction for pull-out tests (it has no equivalent physical significance in pushout tests), P_d a critical force in the fiber at the crack tip and is a function only of the interface toughness (mode II) Γ_i, and k is given by

$$k = \frac{E_m v_f}{E_f(1 + v_m) + E_m(1 - v_f)} \tag{9-34}$$

P^* is assumed to be independent of the roughness, so P^* and P_r may be written as

$$P^* = C_1 \sigma_N \tag{9-35}$$

$$P = C_2 \sigma_N \tag{9-36}$$

where σ_N is the normal stress on the fiber surface, and where

$$C_1 = -\frac{\pi r^2}{k} \tag{9-37}$$

$$C_2 = -\frac{\pi r^2 v_f}{k}\left(\frac{1 + 2k}{1 - 2v_f k}\right) \tag{9-38}$$

From Ref. [56], we have

$$P_d = C_4 \sqrt{\Gamma_i} \tag{9-39}$$

where

$$C_4 = \left(\frac{4\pi^2 r^3 E_f}{1 - 2v_f k}\right)^{1/2} \tag{9-40}$$

Thus, Eq. (9-29) may be changed as

$$\delta = \frac{C_3}{\mu}\left[P_a - C_4\sqrt{\Gamma_i} + C_2\sigma_N - (C_1 - C_2)\sigma_N \ln\left(\frac{C_1\sigma_N + P_a}{C_1\sigma_N + C_4\sqrt{\Gamma_i} - C_2\sigma_N}\right)\right] \tag{9-41}$$

where $C_3 = (1-2\nu_f k)/(2k\pi r E_f)$.

In the present section, three-parameter (μ, Γ_i, σ_N) fitting was performed by applying Eq. (9-41) to push-out measurements for E-glass/epoxy composites.

Single-fiber push-out tests were conducted using scanning electron microscopy equipped with a screw-driven testing machine that included a 1000 g load cell. The schematic of the test apparatus is shown in Fig. 9-6.

For push-out tests, the measurement of average bond strength would appear to be straightforward. However, the average bond strength may not be the true

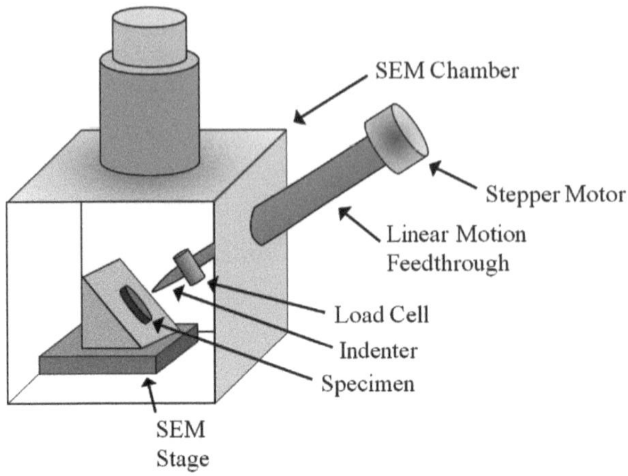

Fig. 9-6 Schematic diagram of single-fiber push-out test set-up.

bond strength because debonding often occurs progressively as a fiber is pushed into the matrix before the load drop. The push-out procedure is shown schematically in Fig. 9-7. Debonding is likely to initiate near the top surface [59] and proceed along the length of the fiber until the process becomes unstable. Mechanical constraint may be exerted by nearby fibers, unlike the fiber fragmentation test, contributing to different debonding behavior. Thus, accurate measurement of interface strength requires a more rigorous analysis.

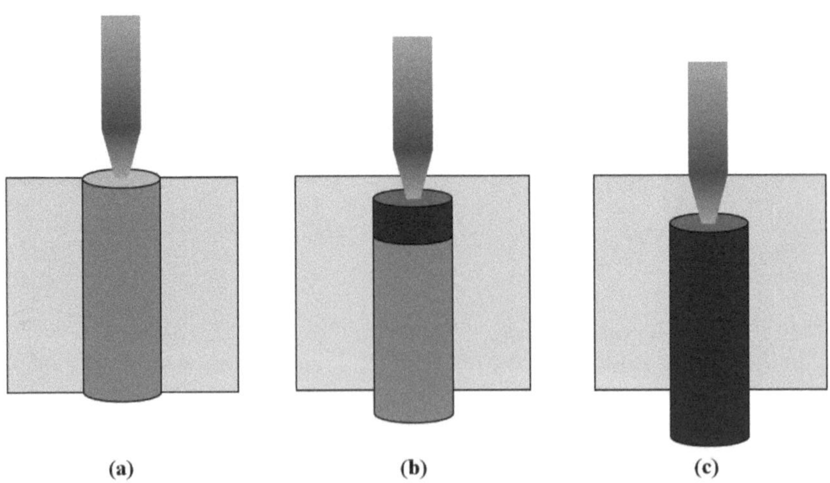

Fig. 9-7 Schematic representation for progressive debonding in pushout tests: (a) no debonding occurs at low load level; (b) debonding is in progress with the increase of applied load; (c) fiber is pushed out of the matrix with full debonding.

9.5. INTERPRETATION OF SINGLE-FIBER FRAGMENTATION TEST

The interface shear strength τ is traditionally calculated from fragmentation test data via the Kelly–Tyson equation [60]:

$$\tau = \frac{3}{8}\frac{\sigma_f(l_{sat})d_f}{l_{sat}} \tag{9-42}$$

where $\sigma_f(l_{sat})$ the strength of a fiber segment of length equal to the saturation length and d_f the fiber diameter.

The saturation length l_{sat} is measured at the saturation limit, which is reached at the end of the fragmentation test when no more fiber breaks occur. Note that in the original model of Kelly and Tyson [60], the strength of the fiber was assumed to be deterministic and constant, without size effects, which is experimentally incorrect. To correctly calculate σ_f (l_{sat}), one must either: (i) perform a large number of tests with single fibers at various lengths, followed by extrapolation to the saturation length (which usually is too short to be directly tested) or (ii) perform a continuously monitored fragmentation test, measure the average fiber strength at progressively decreasing average fragment lengths by either acoustic or optical means, followed by extrapolation to the saturation length from the nonsaturating (linear) portion of the fragment length *versus* fiber strength plot [61, 62]. The latter procedure was adopted here. Specifically, logarithmic plots of the decreasing fragment length against increasing fiber stress (corrected for the presence of prestress and thermal stress) were prepared. The slope of the linear portion of the plot gives the Weibull shape parameter ρ, and the value of the fiber fragment strength at the saturation length may then be obtained from [63, 64]

$$\langle\sigma_f(l_{sat})\rangle = \langle\sigma_f(l_1)\rangle\left(\frac{l_{sat}}{l_1}\right)^{-1/\rho} \tag{9-43}$$

where $\langle\sigma_f(l_{sat})\rangle$ is the average fiber stress at the saturation length l_{sat} and $\langle\sigma_f(l_1)\rangle$ the average stress of a fiber fragment of length l_1.

Using Eq. (9-43), the interfacial strength is obtained from Eq. (9-42), the interfacial toughness is proposed by Zhou *et al.* [65], and is expressed as follows:

$$\Gamma_i = \frac{r_f\sigma_{f\infty}^2}{4E_f}\left[1-\frac{(1-t)^2}{3}+\frac{t(4-t)}{\beta L_d}-\frac{2SE_f r_f^2}{L_d}\times\left(\frac{\beta t^2}{4}+\frac{(1-t)^2}{L_d}\right)\right]-\frac{r_f\Gamma_f}{2L_d} \tag{9-44}$$

where

$$S = \frac{1}{4G_f}+\frac{1}{G_m}\ln(r_m/r_f) \tag{9-45}$$

and

$$t = 1 - \frac{\Psi L_d}{r_f} \quad (9\text{-}46)$$

Here, L_d is the length of the debond (counting on both sides), E_f the fiber Young's modulus, r_f the fiber radius, Γ_f the fiber fracture energy, $\sigma_{f\infty}$ the far-field fiber stress, r_m the effective radius of the matrix, G_f and G_m are the fiber and matrix shear modulus, respectively, ψ the interfacial friction factor (which is different from the friction coefficient in physical meaning), and β the Nayfeh's [65, 66] shear-lag parameter.

Analysis of single fiber fragmentation test data relies on the Kelly–Tyson approach, a statistics-based scheme. The accuracy of the results obtained from this scheme strongly depends on the fragmentation history of the fiber in the tests [67–69]. While longer test specimens provide more accurate measurements, practical limitations restrict specimens to a few centimeters and shear stress values obtained by this method are not absolute values. Furthermore, the energy-based model proposed by Zhou *et al.* is also not completely accurate. Their model involves two parameters (the friction factor Ψ and the ratio of r_m/r_f), which currently cannot be measured. The numerical values used for these two parameters represent the best estimates obtained by the method described in Refs [64–66]. Comparisons with interface properties reported in the literature suggest that the Zhou–Nairn–Wagner analysis tends to overestimate interface energy, which may exaggerate the differences in interface energies measured by the two techniques.

9.6. FIBER–MATRIX ADHESION FROM SINGLE-FIBER COMPOSITE TEST

The local strain $\varepsilon(x)$ along an embedded fiber can be measured during tensile testing [70]. By simple force equilibrium, the interfacial shear stress τ can be determined as

$$\tau = \frac{E_f r}{2} \cdot \frac{d\varepsilon}{dx} \quad (9\text{-}47)$$

where E_f is the fiber Young's modulus and r is the fiber radius.

More elaborate models have been developed to account for the damage process at the interface, including, e.g., debonding with frictional sliding, matrix plasticity, Poisson effects, etc. (see Ref. [71] for a review), most of which require the characterization of several parameters each pertinent to a separate deformation mechanism. In cases such as the present, where the details of the inelastic interfacial deformation mechanisms are unknown, Eq. (9-43) may be used to give an effective maximum shear stress τ_{max} as a physical measure of the efficiency of the interface by taking the derivative of the strain profile measured close to a fiber end at high far-field strains. This method has the advantage to give a more direct measure of the interfacial shear strength compared to that based on average fragment length at saturation in a single-fiber composite. Loose fiber ends molded in the polymer may be analyzed with micro-Raman spectroscopy without interference from damage caused by the disruptive motion when the relatively elevated elastic energy in the carbon fiber is released as the fiber fragments. The interfacial shear strength determined from fiber fragmentation in a single-fiber composite test is shown to give considerably underestimated values compared to measures determined more directly with micro-Raman spectroscopy [72].

The fiber stress profile, $\varepsilon(x)$, is experimentally measured, and some scatter in the measured values is expected from, e.g., the measurement equipment and inhomogeneities in the tested materials. The derivative in Eq. (9-45) is therefore infelicitous, since any variation in the measurement inaccuracy would then be amplified. This variation is commonly suppressed by fitting the stress profile with a smooth function, typically a polynomial of a lower order. However, the choice of this function is arbitrary and user dependent. Fitting the maximum shear stress, $\tau_{\max}$, directly to the fiber strain profile, $\varepsilon(x)$, without any derivative would be more suitable. This method is schematically outlined in Fig. 9-8. We employ Cox' shear-lag solution with the constraint of a maximum constant shear stress close to the fiber end [73]:

$$\varepsilon(x) = \frac{2\tau_{\max}}{E_f r}, \quad \text{for } 0 \leq x \leq x_y \tag{9-48}$$

and

$$\varepsilon(x) = \frac{2x_y\tau_{\max}}{E_f r} + \left(\varepsilon_0 - \frac{2x_y\tau_{\max}}{E_f r}\right)[1 - \exp\{-\beta(x - x_y)\}], \quad \text{for } x > x_y \tag{9-49}$$

where

$$\beta = \frac{1}{8r}\sqrt{\frac{E_m}{(1+\nu_m)E_f \ln(R/r)}} \tag{9-50}$$

Here, R is the radial distance from the fiber axis where the matrix is deemed unaffected by the presence of the fiber. The matrix Young's modulus and Poisson ratio are denoted E_m and ν_m, respectively. The far-field fiber strain ε_0 is the asymptotic strain in the fiber far away from the fiber end.

Some residual fiber strain may develop after curing from chemical shrinkage of the matrix and/or from mismatch in thermal expansion coefficients of the fiber and matrix materials if cure were carried out at an elevated temperature. The far-field fiber strain is related to the applied specimen strain ε_a,

$$\varepsilon_0 = \varepsilon_a + \varepsilon_{res} \tag{9-51}$$

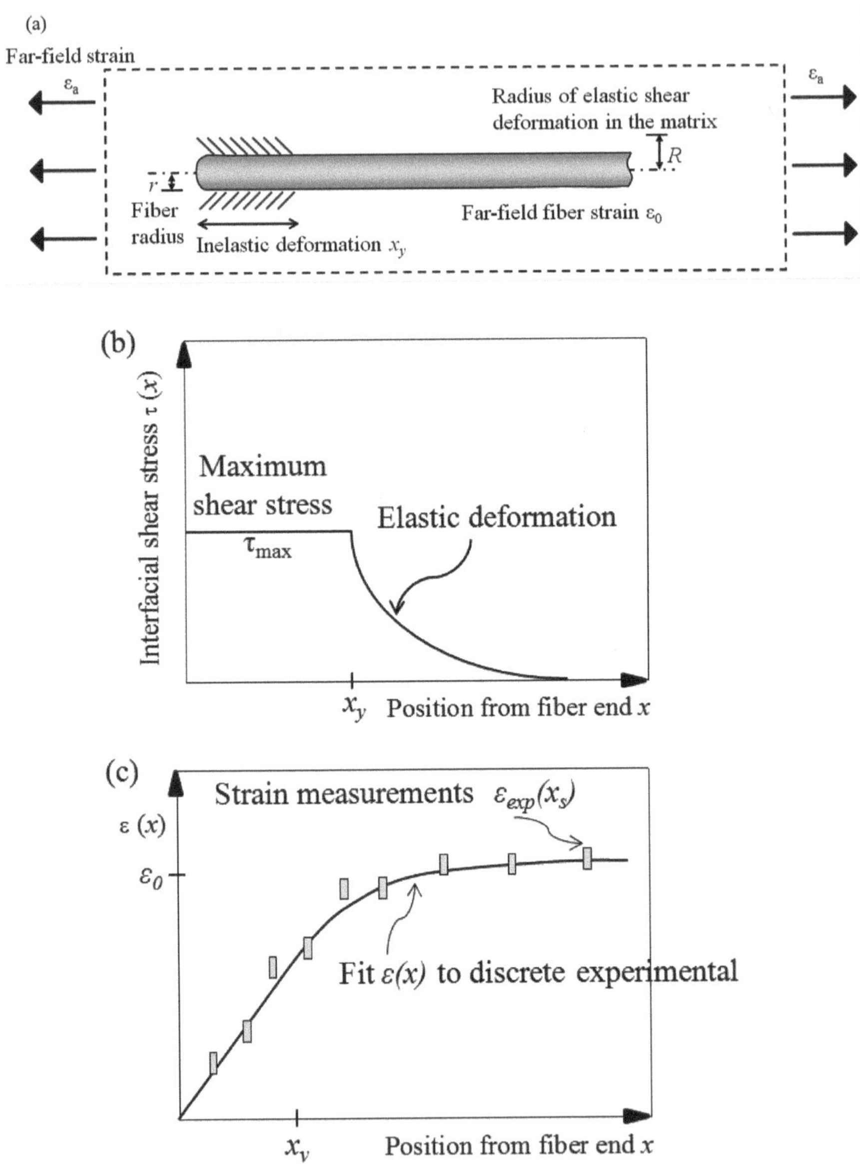

Fig. 9-8 Schematic illustrations of (a) deformation in a single-fiber composite specimen, (b) interfacial shear stress distribution, and (c) fiber strain distribution fitted to experimental data.

where ε_{res} is the far-field residual strain of the fiber, which is typically compressive, and can be measured with micro-Raman spectroscopy far away from the fiber end in the unloaded specimen using the fiber as an *in situ* strain gauge.

It is assumed that all residual deformations are elastic, which implies that the strain is zero at the fiber end and then decreases toward ε_{res} in the unloaded case according to the shear-lag model. This corresponds to the special case when $x_y = 0$ and $\varepsilon_0 = \varepsilon_{res}$ in Eq. (9-49).

As load is applied to the single-fiber composites, the interfacial shear stress close to the fiber end eventually attains a maximum value and inelastic deformation takes place in this zone, as depicted in Fig. 9-8(a). Prior to debonding, this value is constant in the inelastic zone, whereas the shear stress in the intact zone is described by the elastic solution, as shown in Fig. 9-8(b). These assumptions would allow us to fit the strain profiles given by Eqs. (9-46) and (9-47) to the experimental data of the fiber strain, which is shown in Fig. 9-8(c). This would include both zones—the one with inelastic interfacial deformation the other with elastic deformation. To reduce the number of fitting parameters, some constraints can be imposed on the shear lag parameter β and the length of the inelastic zone x_y.

The shear lag parameter β in Eq. (9-50) should be independent of the interfacial strength, since it reflects shear stress transferability under purely elastic conditions. However, a noticeable difference in the value of β has been found for sized and unsized carbon fibers, with otherwise identical composite composition [74]. This mismatch can be attributed to the difference in stiffness between the matrix and the size layer, in particular, for small radii. Due to this variability, the value of β or R is commonly found by fitting it to experimental data for each new material combination. However, a micromechanical model in which β or R is expressed in terms of constituent elastic properties and the fiber radius would be preferable. Such models have been developed by Cherepanov [75], based on the stress intensity factor at the fiber breaks, and by Budiansky *et al.* [76] for multifiber composites. Aksel *et al.* [77] have proposed.

$$\frac{R}{r} = \sqrt{1 + \frac{E_f}{(c-1)E_m}} \tag{9-52}$$

This is based on force equilibrium within a concentric fiber–matrix cylinder of radius R. The stress concentration factor in the matrix is $c = \sigma_m(x = 0)/\sigma_m(x = \infty)$ on average in the cylinder cross section. Different R/r ratios have been proposed based on experimental results under elastic conditions, but the values have a wide range of 3–15 [78]. To avoid fitting R/r to the experimental data for every new material, it has been suggested to assume that $c = 2$ [79], which is in concert with several experimental results, where the size thickness and inelastic deformations are deemed to be small. For instance, the stress concentration factor c can be found to be 1.95 for unsized carbon fiber/epoxy system [80], 1.92 for another unsized carbon fiber/epoxy system [81], and 2.22 for a desized α-alumina fiber/epoxy system [82]. In the present study, it is assumed that $c = 2$, which reduces the number of unknown fitting parameters.

The number of fitting parameters can be further reduced by considering the relation between the length of the inelastic damage zone and the maximum interfacial shear stress. Continuity of the shear stress at $x = x_y$ in Eqs. (9-48) and (9-49) implies

$$x_y = \frac{E_f \varepsilon_0 r}{2\tau_{\max}} - \frac{1}{\beta} \tag{9-53}$$

which can be reinserted into Eq. (9-49).

Given the elastic properties of the constituents and the far-field fiber strain, the only fitting parameters are the maximum interfacial shear stress $\tau_{\max}$ (which is the sought property), and ε_0, which correlates with the applied strain according to Eq. (9-51). By fitting the two-zone strain profile of the fiber expressed in Eqs. (9-48) and (9-49), $\varepsilon(x)$, to an experimental strain profile at discrete points at a fiber end, $\varepsilon_{\exp}(x_i)$, a measure of the interfacial shear strength is obtained (Fig. 9-8(c)). This is done by numerically finding the least-squares sum with respect to $\tau_{\max}$:

$$\tau_{\max}^{\min} \sum_{i=1}^{n} \{\varepsilon_{\exp}(x_i) - \varepsilon(\tau_{\max} \cdot x_i)\}^2 \tag{9-54}$$

where n is the number of discrete points x_i at which the strain was measured.

The advantage of this methodology compared with straightforward use of Eq. (9-47) is that it is less sensitive to inaccuracy in measurement data, and that it also takes the elastic deformations of the matrix into account. Furthermore, it gives average interfacial shear strength along the entire inelastic region ($0 \le x \le x_y$) in contrast to the maximum local shear strength in Eq. (9-47), which may arise from an exceptionally strong part of the interface that does not reflect the average interfacial behavior [83].

9.7. MICROMECHANICAL MODELING OF MICROBOND TEST

Test methods using single-fiber model composites, such as pull-out, microbond, and pushout tests, are helpful in extracting the debonding process of the interface from the overall damage process of a composite. In particular, the microbond test [84, 85] is a simple fiber pull-out technique for evaluating shear failure of the fiber–matrix interface. The microbond test involves pulling a fiber out of a bead of matrix (i.e., droplet) through a knife-edge (as shown in Fig. 9-9), accompanied by the debonding at the fiber–matrix interface, and then it evaluates the interfacial fracture properties based on the resulting force required in the pullout. This technique is quite useful for small-diameter fibers, such as carbon fibers, because the fiber pullout can be completed without fiber breakage. However, it has frequently been pointed out that the microbond test fails to quantify the 'true' interfacial properties representing the interfacial fracture, partly because the interfacial properties obtained from the microbond test using conventional data reduction techniques (e.g., the averaged interfacial shear

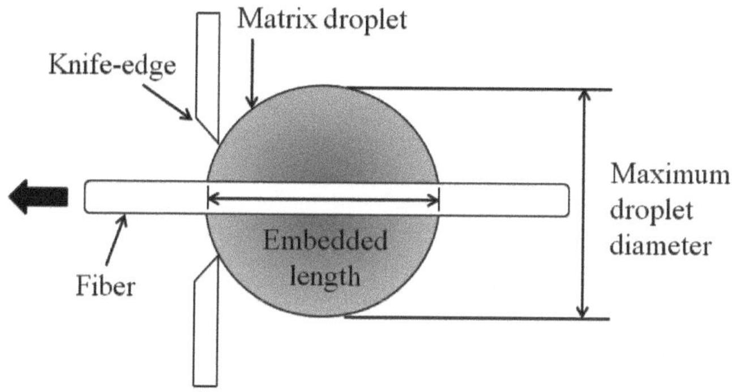

Fig. 9-9 Schematic of the microbond test configuration.

strength) are not consistent with those determined using other test methods [86]. In addition, some researchers [87] have reported significant data scatter as a limiting aspect of the microbond test. This inconvenience is caused by the fact that the micromechanical modeling for the microbond test has not been well developed and the overall damage process during the test is not well understood. This issue must be solved by the contributions of a detailed numerical analysis.

Recent researches [88, 89] have contributed to finite-element modeling of the fiber pull-out and push-out tests. These models address the interfacial fracture during the tests, utilizing interface elements embedding the damage process zone onto the fiber–matrix interface. This framework had been frequently applied to the fracture at the fiber–matrix interface in composites, first in Ref. [90] and then extended in Ref. [91], etc. In most interface elements, the model of the damage process zone was constructed based on the conventional fracture mechanics. (This type of model was referred to as an embedded process zone model in Refs Hutchinson [91] and Scheer [92]). Overall, these approaches to interfacial fracture are also effective for the modeling of single-fiber test methods. In addition, the predicted results of the debonding process during the tests can be utilized to obtain interfacial fracture properties (strength and toughness) of the composites through their fitting to the experimental data.

The present section has focused on achieving a micromechanical understanding of the microbond test, which involves pulling a fiber out of a bead of matrix (i.e., droplet) through a knife-edge, in order to quantify the interfacial fracture properties of fiber-reinforced composites [93].

Fig. 9-10(a) presents a typical force history obtained by the experiment. The fiber-end force increases almost proportionally to the applied displacement until it reaches a maximum force and then drops. The pull-out force F_d, measured as the maximum force applied to the fiber end, is then plotted as a function of the embedded length l of the fiber in the droplet in Fig. 9-10(b). The lines evaluated

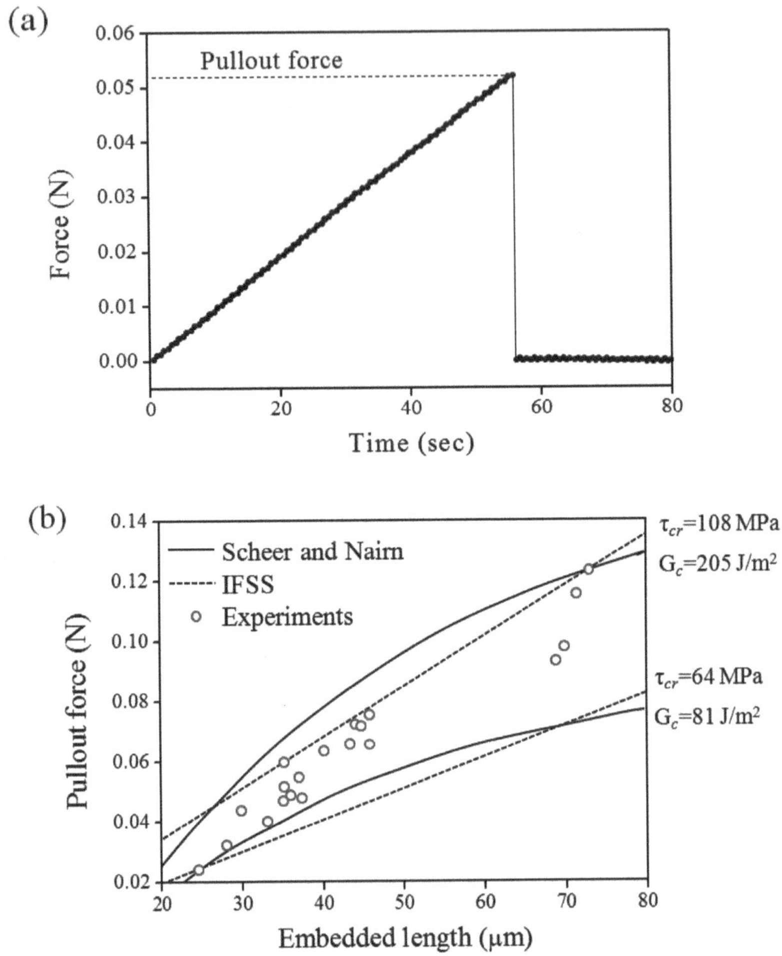

Fig. 9-10 Microbond tests for T800S carbon fiber and epoxy resin system. (a) Typical fiber-end force history during the fiber pullout. (b) Pull-out force as a function of the embedded length.

by the interfacial shear strength criterion and by the analytical model based on the energy-based criterion [94] are also included in Fig. 9-10:

$$F_d = 2\pi r_f l \cdot \tau_{cr} \text{ (interfacial shear strength)} \tag{9-55}$$

$$F_d = F_d(l, G_c \Delta T) \text{ (Energy − based criterion)} \tag{9-56}$$

where τ_{cr} is the interfacial shear strength, G_c is the critical energy release rate, r_f is the fiber radius, and ΔT is the temperature change from the stress-free temperature (−80K).

The experimental results have a large scatter in the pull-out force, and the interfacial parameters estimated by fitting with the experimental data vary excessively. Interfacial shear strength varies from 64 to 108 MPa (average: 92.3 MPa), and G_c varies from 81 to 205 J/m^2 (average: 131.5 J/m^2).

This issue is inherent in small-diameter fibers in attempts to obtain a quantitative fit with experiments. Here, the typical fiber breaking strain is about 2.0% according to the data sheet of the supplier. Therefore, when fiber pullout can be completed without fiber breakage, the pull-out force is at most 0.12 N for the present material system, according to the fiber breaking strain and Hooke's law. Then, the fiber embedded length where pull-out data are available is at most 75 lm, as seen in Fig. 9-10. In contrast, in the previous literature where existing theoretical models have worked well [94], the target material system provides a wide range of embedded lengths in the experiments. For example, Scheer *et al.* [92] demonstrate a quantitative fit with experiments for pulling E-glass fibers (21 μm diameter) out of epoxy droplets, where the maximum embedded length is about 0.45 mm. In their case, the interfacial property may be determined by use of a least-squares regression, because of the wide range of available experimental data.

According to the microbond test results for carbon fiber and epoxy resin system, matrix cracking occurred during the fiber pullout, in addition to the debonding at the fiber–matrix interface. Therefore, in evaluating the fracture properties of the fiber–matrix interface, we should pay attention to the coupling effects of matrix failure and interfacial debonding on the test results.

9.8. INTERPHASE EFFECT ON FIBER-REINFORCED POLYMER COMPOSITES

9.8.1. Introduction

Composites have good dissipative properties, and are hence employed to reduce the mechanical vibration of machine and structural systems [95–97]. The study of mechanical behavior of polymer composites is often restricted to a two-phase system. However, with the development of fiber coating technology, the effect of the interphase zone, generally called simply the interphase, is of practical importance for the prediction of overall behavior of the composites. The effect of the interphase on mechanical properties of composites may be investigated by studying two extremes—perfect bonding of fiber, interphase and matrix; and debonding.

A mathematical model has been developed for the prediction of elastic moduli of three-phase fiber-reinforced composites. This model considers composite materials consisting of three phases, namely, fiber, interphase, and matrix, and incorporates the effect of fiber packing. In the present section, the interphase is assumed to be homogeneous and isotropic materials. The elastic moduli E_{11}, E_{22}, G_{12}, and G_{23} have been evaluated by application of a correspondence principle using the developed model.

The influence of the interphase on the effective material properties of three-phase composites has been studied by Kari *et al.* [95] using a numerical homogenization technique based on finite-element methods and obtained from the relevant elastic moduli.

9.8.2. Three-Phase Bridging Model

A two-phase bridging model [98] for unidirectional fiber-reinforced composite has been extended to incorporate the effect of the interphase. In the two-phase model, elastic stresses in the matrix material are correlated with the fiber by bridging matrix. The bridging matrix represents the load sharing capacity of one constituent phase in the composite with respect to the other phase. On the same basis, stresses in matrix, interphase, and fiber are correlated by Eqs. (9-57) and (9-58):

$$\{d\sigma_i^m\} = [A_{ij}^{ml}]\{d\sigma_j^l\} \tag{9-57}$$

$$\{d\sigma_i^l\} = [A_{ij}^{lf}]\{d\sigma_j^f\} \tag{9-58}$$

where $\{\mathrm{d}\sigma_i\} = \{\mathrm{d}\sigma_{11}, \mathrm{d}\sigma_{22}, \mathrm{d}\sigma_{33}, \mathrm{d}\sigma_{13}, \mathrm{d}\sigma_{12}, \mathrm{d}\sigma_{23}\}^T$ is the incremental stress tensor; $[A_{ij}]$ is the bridging matrix; and superscripts f, I, and m represent the fiber, interphase, and matrix, respectively. A quantity without a suffix represents the composite; subscripts i, j represent the contracted tensor notation.

Constituent material properties used as input for bridging model are is given in Table 9-1. A concentric cylinder model for unidirectional three-phase fiber-reinforced composite is shown in Fig. 9-11 in the form of a representative volume element. It consists of fiber of radius r_f bounded by an interface region of outer radius r_I embedded into matrix of outer radius r_m. Furthermore, the condition of perfect fiber–interphase–matrix bonding and variable state of stress is assumed within the representative volume element. Formulation has been done for the elastic deformation.

Table 9-1 **Constituent Material Properties Used as Input for Bridging Model**

Properties	E-glass fiber	Epoxy matrix	Hard interphase	Soft interphase
E (GPa)	72.4	2.76	37.58	0.5
G (GPa)	30.2	1.02	15.61	0.178
Y	0.2	0.35	0.204	0.4
H	0.0018	0.015	0.0084	0.0084

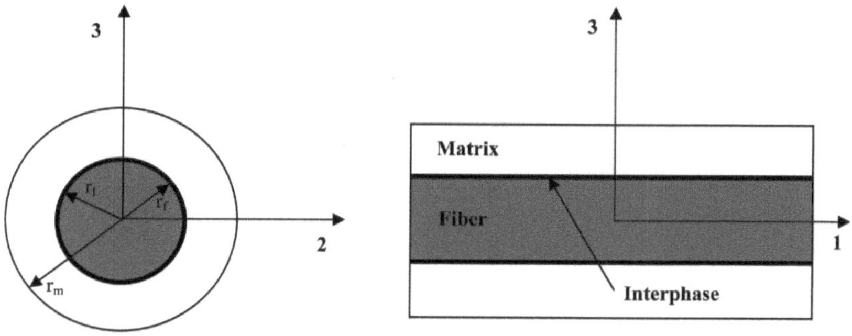

Fig. 9-11 Concentric cylinder model for fiber-reinforced composites.

Let V' denote the volume of representative volume element of unidirectional fiber-reinforced composites, as shown in Fig. 9-11. V_f, V_I, and V_m are the volume of fiber, interphase, and matrix in the representative volume element, respectively.

The volume average stress $\bar{\sigma}_i$ of the composites is defined as

$$\overline{\sigma_i} = 1/V' \int \sigma_i dv = 1/V' \left[\int_{v_f'} \sigma_i dv + \int_{v_i'} \sigma_i dv + \int_{v_m'} \sigma_i dv \right] \tag{9-59}$$

$$\overline{\sigma_i} = \left[\left(\frac{V_f'}{V'} \right) \left(\frac{1}{V_f'} \int_{v_f'} \sigma_i dv \right) + \left(\frac{V_I'}{V'} \right) \left(\frac{1}{V_I'} \int_{v_i'} \sigma_i dv \right) + \left(\frac{V_m'}{V'} \right) \left(\frac{1}{V_m'} \int_{v_m'} \sigma_i dv \right) \right] \tag{9-60}$$

$$\overline{\sigma_i} = V_f \overline{\sigma_i^f} + V_I \overline{\sigma_i^I} + V_m \sigma \tag{9-61}$$

$$\overline{\sigma_{\mathrm{i}}^f} = \left[\frac{1}{V_f'} \int_{v_f'} \sigma_i dv \right] 60 \tag{9-62}$$

$$\overline{\sigma_{\mathrm{i}}^I} = \left[\frac{1}{V_I'} \int_{v_I'} \sigma_i dv \right] \tag{9-63}$$

$$\overline{\sigma_i^m} = \left[\frac{1}{V_m'} \int_{v_m'} \sigma_i dv \right] \tag{9-64}$$

$$V_f = V_f'/V' \tag{9-65}$$

$$V_I = V_I'/V' \tag{9-66}$$

$$V_m = V_m'/V' \tag{9-67}$$

where V_f, V_I, and V_m represent the volume fractions of fiber, interphase, and matrix, respectively, and $\bar{\sigma}_i^f, \bar{\sigma}_i^I, \bar{\sigma}_i^m$ are the volume average constituent stresses.

Similarly, the volume averaged strain is given as

$$\overline{\varepsilon_i} = V_f\overline{\varepsilon_i^f} + V_I\overline{\varepsilon_i^I} + V_m\overline{\varepsilon_i^m} \tag{9-68}$$

In further discussion, the volume averaged stresses and strains will be represented by omitting the overbar. Volume averages stresses and strains in representative volume element of lamina in incremental form are given by Eqs. (9-69) and (9-70):

$$\{d\sigma\} = V_f\{d\sigma^f\} + V_I\{d\sigma^I\} + V_m\{d\sigma^m\} \tag{9-69}$$

and

$$\{d\varepsilon\} = V_f\{d\varepsilon^f\} + V_I\{d\varepsilon^I\} + V_m\{d\varepsilon^m\} \tag{9-70}$$

Constitutive equations correlating the averaged stresses and strains in different phases of the representative volume element are expressed as:

$$\{d\varepsilon^f\} = [S^f]\{d\sigma^f\} \tag{9-71}$$

$$\{d\varepsilon^I\} = [S^I]\{d\sigma^I\} \tag{9-72}$$

$$\{d\varepsilon^m\} = [S^m]\{d\sigma^m\} \tag{9-73}$$

Substituting Eqs. (9-71), (9-72), and (9-73) into Eq. (9-70) and after simple algebraic manipulation we get

$$\{d\varepsilon\} = [s_{ij}][V_f + V_I[A_{ij}^{lf}] + V_m[A_{ij}^{ml}][A_{ij}^{lf}]]\{d\sigma^f\} \tag{9-74}$$

$$S_{ij} = [V_f S_{ij}^f + V_I S_{ij}^l A_{ij}^{lf} + V_m S_{ij}^m A_{ij}^{ml} A_{ij}^{lf}][V_f[I] + V_I S_{ij}^{lf} + V_m S_{ij}^{lf}]^{-1} \tag{9-75}$$

The overall compliance matrix of the lamina is given by Eq. (9-76):

$$S_{ij} = \begin{bmatrix} [S_{ij}]_\sigma & 0 \\ 0 & [S_{ij}]_\tau \end{bmatrix} \tag{9-76}$$

where $[S_{ij}]_\sigma$ and $[S_{ij}]_\tau$ are normal and shear compliance matrices:

$$[S_{ij}]_\sigma = \begin{bmatrix} 1/E_{11} & -\nu_{12}/E_{11} & -\nu_{12}/E_{11} \\ & 1/E_{22} & -\nu_{23}/E_{22} \\ symmetry & & 1/E_{22} \end{bmatrix}, \tag{9-77}$$

$$[S_{ij}]_\tau = \begin{bmatrix} 1/G_{23} & 0 & 0 \\ 0 & 1/G_{12} & 0 \\ 0 & 0 & 1/G_{12} \end{bmatrix} \tag{9-78}$$

where $[S_{ij}]_\sigma$ and $[S_{ij}]_\tau$ are the normal and the shear compliance matrix.

Material parameters E_{22},G_{23}, and ν_{23} are related to one other as in Eq. (9-79):

$$G_{23} = \frac{E_{22}}{2(1+\nu_{23})} \tag{9-79}$$

Thus, only five independent elastic moduli exist and so only five independent elements in the bridging matrix are required. The general form of the bridging matrix is given by Eq. (9-80):

$$[A_{ij}] = \begin{bmatrix} a_{11} & a_{12} & a_{13} & 0 & 0 & 0 \\ a_{21} & a_{22} & a_{23} & 0 & 0 & 0 \\ a_{31} & a_{32} & a_{33} & 0 & 0 & 0 \\ 0 & 0 & 0 & a_{44} & 0 & 0 \\ 0 & 0 & 0 & 0 & a_{55} & 0 \\ 0 & 0 & 0 & 0 & 0 & a_{66} \end{bmatrix} \tag{9-80}$$

We take a_{11}, a_{22}, a_{12}, a_{23}, and a_{66} to be independent.

The bridging matrix between fiber and interphase is represented as

$$[A_{ij}^{\mathrm{fI}}] = \begin{bmatrix} a_{11}^{\mathrm{fI}} & a_{12}^{\mathrm{fI}} & a_{13}^{\mathrm{fI}} & 0 & 0 & 0 \\ a_{21}^{\mathrm{fI}} & a_{22}^{\mathrm{fI}} & a_{23}^{\mathrm{fI}} & 0 & 0 & 0 \\ a_{31}^{\mathrm{fI}} & a_{32}^{\mathrm{fI}} & a_{33}^{\mathrm{fI}} & 0 & 0 & 0 \\ 0 & 0 & 0 & a_{44}^{\mathrm{fI}} & 0 & 0 \\ 0 & 0 & 0 & 0 & a_{55}^{\mathrm{fI}} & 0 \\ 0 & 0 & 0 & 0 & 0 & a_{66}^{\mathrm{fI}} \end{bmatrix} \tag{9-81}$$

The matrix between the interphase and the polymer matrix is represented as

$$[A_{ij}^{Im}] = \begin{bmatrix} a_{11}^{\mathrm{Im}} & a_{12}^{\mathrm{Im}} & a_{13}^{\mathrm{Im}} & 0 & 0 & 0 \\ a_{21}^{\mathrm{Im}} & a_{22}^{\mathrm{Im}} & a_{23}^{\mathrm{Im}} & 0 & 0 & 0 \\ a_{31}^{\mathrm{Im}} & a_{32}^{\mathrm{Im}} & a_{33}^{\mathrm{Im}} & 0 & 0 & 0 \\ 0 & 0 & 0 & a_{44}^{\mathrm{Im}} & 0 & 0 \\ 0 & 0 & 0 & 0 & a_{55}^{\mathrm{Im}} & 0 \\ 0 & 0 & 0 & 0 & 0 & a_{66}^{\mathrm{Im}} \end{bmatrix} \tag{9-82}$$

The independent elements a_{11}, a_{22}, a_{12}, and a_{23} depend on the elastic properties of the matrix, interphase, and fiber as well as on fiber packing. Owing to the axisymmetry of concentric cylinders,

$$a_{33} = a_{22} \tag{9-83}$$

$$a_{13}^{fl} = a_{12}^{fl} = \frac{(S_{12}^{f} - S_{12}^{l})(a_{11}^{fl} - a_{22}^{fl})}{(S_{11}^{f} - S_{11}^{l})} \tag{9-84}$$

$$a_{13}^{lm} = a_{12}^{lm} = \frac{(S_{12}^{l} - S_{12}^{m})(a_{11}^{lm} - a_{22}^{lm})}{(S_{11}^{l} - S_{11}^{m})} \tag{9-85}$$

$$a_{11}^{fl} = E^{1}/E^{f}, \qquad a_{11}^{\mathrm{lm}} = E^{m}/E^{l} \tag{9-86}$$

$$a_{22}^{fl} = \beta + (1-\beta)E^{l}/E^{f} \tag{9-87}$$

$$a_{22}^{lm} = \beta + (1-\beta)E^{m}/E^{l} \tag{9-88}$$

$$a_{55}^{fl} = a_{66}^{fl} = \alpha + (1-\alpha)G^{l}/G^{f} \tag{9-89}$$

$$a_{55}^{lm} = a_{66}^{lm} = \alpha + (1-\alpha)G^{m}/G^{l} \tag{9-90}$$

where α and β are the longitudinal and transverse fiber packing factors.

By using the bridging compliance matrices $[S_{ij}]$, four elastic moduli E_{11}, E_{22}, G_{12}, and G_{23} of the unidirectional fiber composites are derived:

$$E_{11} = \frac{V_f + V_i a_{11fi} + V_m a_{11fi} a_{11im}}{V_f S_{11f} + V_i S_{11i} a_{11fi} + V_m S_{11m} a_{11fi} a_{1im}} \tag{9-91}$$

$$E_{22} = \frac{(V_f + V_i a_{22fi} a_{22im})(V_f + V_i a_{11fi} + V_m a_{11fi} a_{11im})}{V_f^2 S_{22f} + a_{22fi} a_{11fi}(V_i^2 S_{22i} + V_m^2 S_{22m} a_{11fi} a_{22im}) - V_f V_i A - V_i V_m B - V_m V_f C} \tag{9-92}$$

$$A = a_{12fi}(S_{12f} - S_{12i}) - S_{22f} a_{11fi} - S_{22i} a_{22fi} \tag{9-93}$$

$$B = a_{11fi}(S_{12i} - S_{12m})(a_{11fi} a_{12im} - a_{12fi} a_{11im} + a_{22im} a_{12fi}) \tag{9-94}$$

$$C = a_{12fi} a_{22im}(S_{12f} - S_{12m}) + a_{11fi}(S_{12f} a_{12im} + S_{22f} a_{11im} - S_{12m} a_{12im}) \tag{9-95}$$

$$G_{12} = \frac{G_f G_i G_m (V_f + V_i a_{66fi} + V_m a_{66fi} a_{66im})}{G_f G_m V_i a_{66fi} + G_f G_i V_m a_{66im} + G_i G_m V_f} \tag{9-96}$$

$$G_{23} = \frac{V_f + V_i a_{44fi} + V_m a_{44fi} a_{44im}}{V_f S_{44f} + V_i S_{44i} a_{44fi} + V_4 S_{44m} a_{44fi} a_{44im}} \tag{9-97}$$

Furthermore, using the correspondence principle in which E_f, E_I, E_m become complex with the loss modulus associated represented by the imaginary part, the compliance matrix would also be complex. As a result, composite moduli in complex form are obtained as given in Eq. (9-98), which represents the storage moduli (E'_{ij}) and loss moduli (E''_{ij}) components as real and imaginary parts, respectively, and using these loss factors η_{11}, η_{22}, η_{12}, and η_{23} are predicted:

$$E_{ij} = E'_{ij} + jE''_{ij}, \quad \eta_{ij} = \frac{E''_{ij}}{E'_{ij}} \tag{9-98}$$

9.8.3. Finite-Element Model

The 3D finite-element model with varying values of fiber volume fraction and interphase volume fraction for various loading with single cell square array packing is analyzed. Fig. 9-12 shows 3D finite-element models for transverse

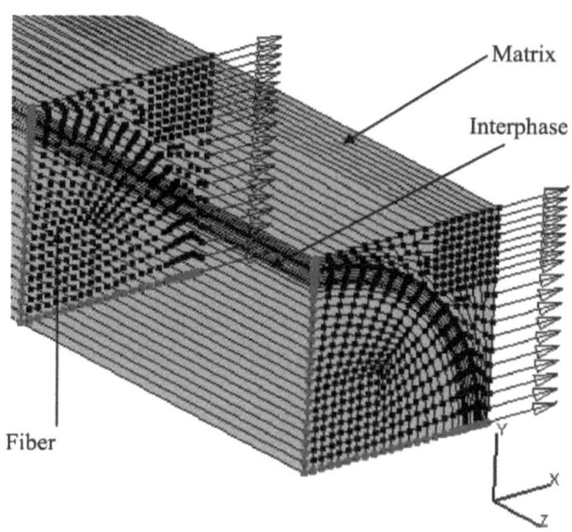

Fig. 9-12 Finite-element model of three-phase fiber-reinforced composites. Number of elements $N_f = 2560$, $N_i = 2561$–3840, and $N_m = 3841$–5960. $V_f = 0.6$, $V_i = 0.1$.

loading modeled in NISA version 15 finite element software. The model consist of 1–2560 eight-noded hexahedron elements in fiber, 2561–3840 eight-noded hexahedron elements in the interphase, and 3840–5960 eight-noded hexahedron elements in the polymer matrix.

By applying the strain energy concept, the unit cell [99] is analyzed to evaluate the damping capacity of the composite from those of its constituents. The loss factor is expressed as:

$$\eta = \frac{\sum_{i=1}^{n} \eta_i W_i}{\sum_{i=1}^{n} W_i} \tag{9-99}$$

The loss factor for the three-phase model considering the effect of the interphase is expressed as:

$$\eta = \frac{\eta_f W_f + \eta_l W_l + \eta_m W_m}{W_f + W_l + W_m} \tag{9-100}$$

The strain energy stored in the composite under loading is written as:

$$W_C = \frac{1}{2}\int_V \sigma_{ij}\varepsilon_{ij} dV = \frac{1}{2}\int_V \sigma_{ij}^T S_{ij}\sigma_{ij} dV \tag{9-101}$$

The strain energy stored in fiber, interphase, and matrix in a unit cell is written as

$$W_f = \frac{1}{2}\sum \left\{\sigma_{ij}^T\right\}_f \{S_{ij}\}_f \{\sigma_{ij}\}_f dV_f \tag{9-102}$$

$$W_l = \frac{1}{2}\sum \left\{\sigma_{ij}^T\right\}_l \{S_{ij}\}_l \{\sigma_{ij}\}_l dV_l \tag{9-103}$$

$$W_m = \frac{1}{2}\sum \left\{\sigma_{ij}^{\mathrm{T}}\right\}_m \{S_{ij}\}_m \{\sigma_{ij}\}_m dV_m \tag{9-104}$$

Equating expression W_c as the sum of the strain energy stored in the components, the overall elastic moduli can be evaluated.

Elastic moduli and loss factors are predicted using a three-phase bridging model developed for various values of the fiber volume fraction. The effect of fiber packing factors is also evaluated on the elastic moduli and loss factors. Figs. 9-13 and 9-14 demonstrate the variation of normal elastic moduli E_{11} and E_{22} with respect to fiber volume fraction V_f. Longitudinal elastic modulus E_{11} shows insensitivity to the fiber packing factors α and β. In this case, fiber, interphase, and matrix are stressed in the longitudinal direction and not much interaction takes place between fiber and matrix. Therefore, arrangement of fiber and matrix does not have any role in taking the load. E_{11} as predicted [100] by all models, bridging and concentric cylinder assembly method [99] and Mori-Tanaka, is practically the same and varies linearly with fiber

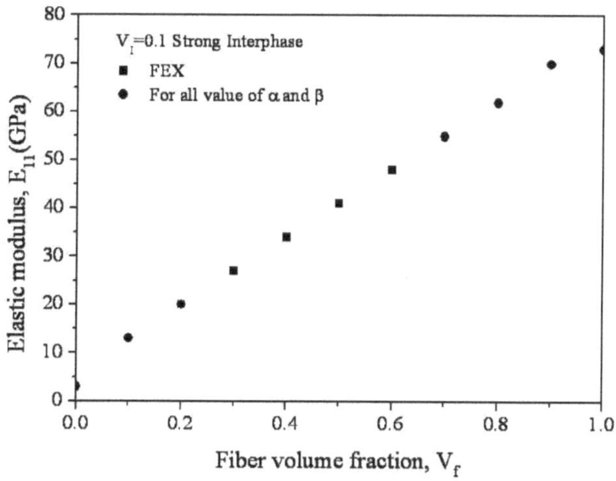

Fig. 9-13 Variation of elastic modulus, E_{11}, with respect to fiber volume fraction.

volume fraction. Fig. 9-14 indicates dependence of transverse elastic modulus E_{22} on transverse packing factor β. With increase in β, the value of E_{22} decreases. Fiber packing effect is more pronounced at higher value of fiber volume fraction. At $V_f = 0.4$, transverse elastic modulus E_{22} decreases by 36% and at $V_f = 0.6$, E_{22} decrease by 42% for β increase from 0.5 to 1.0. Fig. 9-14 also shows the comparison with the concentric cylinder assembly

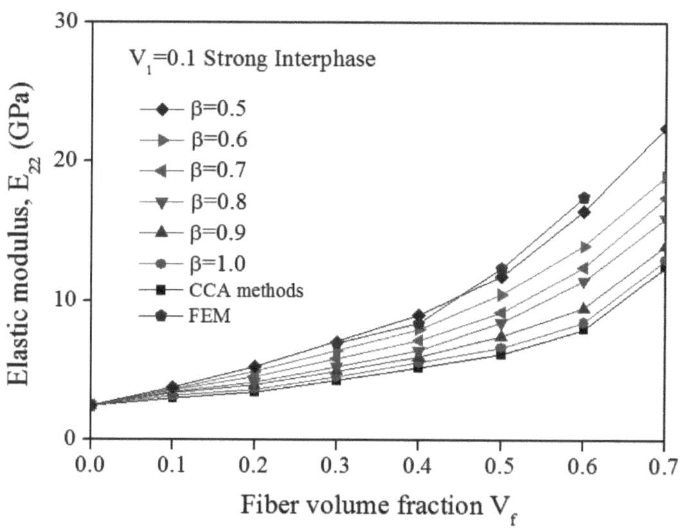

Fig. 9-14 Variation of elastic modulus, E_{22}, with respect to fiber volume fraction.

Table 9-2 Transverse Elastic Moduli E_{22} for Unidirectional Fiber-Reinforced Composites

V_f	β	α	E_i (GPa)	Bridging model, E_{22} (GPa)	Finite-element model wacker, E_{22} (GPa)
0.5	0.51	0.5	6	13.46	13.71
0.5	0.51	0.5	8	14.68	14.68
0.5	0.51	0.5	12	16.10	15.91

method. For $\beta = 1$, bridging model results match with those obtained with the concentric cylinder assembly method. Finite-element model results are in good agreement with bridging model results for lower values of fiber volume fractions.

The bridging model results predicted for transverse elastic moduli are compared with finite-element results reported by Wacker *et al.* [101] for the following input data: $E_f = 84$ GPa, $\nu_f = 0.22$, $E_m = 4$ GPa, $\nu_m = \nu_i = 0.34$, $r_f = 8.5$ μm, and interphase thickness = 1 μm. The results of the bridging model for transverse bridging factor $\beta = 0.51$ show very close matching to the finite-element model results, as seen from Table 9-2.

Figs 9-15 and 9-16 show the effect of fiber volume fraction V_f on shear moduli G_{12} and G_{23}, respectively. In Fig. 9-15, various curves represent

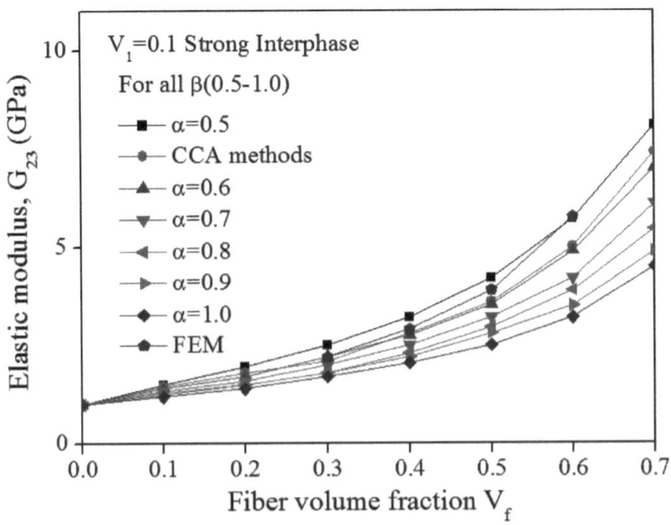

Fig. 9-15 Variation of elastic modulus, G_{12}, with respect to fiber volume fraction.

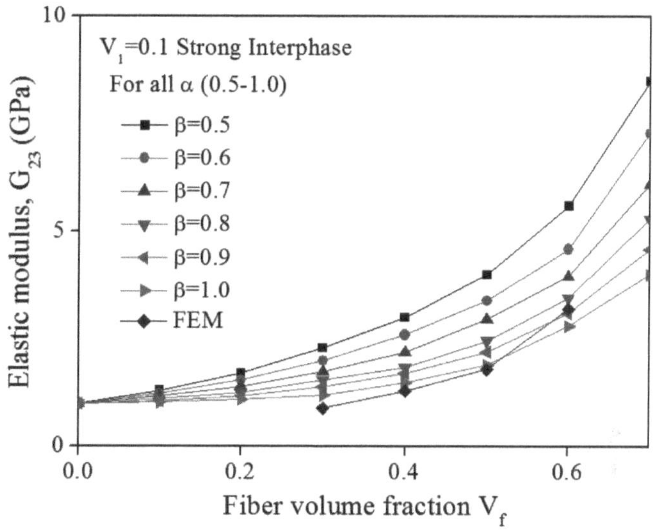

Fig. 9-16 Variation of elastic modulus, G_{23}, with respect to fiber volume fraction.

variation of G_{12} for fiber packing factor $0.5 < \alpha < 1.0$. With the increase in the value of α, G_{12} decreases rapidly for lower range of α. Increase in longitudinal shear modulus is more for lower values of α when fiber volume fraction is large. The results for G_{12} indicate that the transverse packing factor β has no effect on its prediction. Loss in longitudinal shear modulus with increase in α is regular; i.e., similar curves are obtained for all value of α. Finite element model results are closer to bridging model results for $\alpha = 0.6$ for lower value of fiber volume fraction. Fig. 9-16 shows a similar trend as for G_{12} but with different magnitudes. In relation to different values of transverse packing factor β, finite-element model results have similar trend with less magnitude. As is evident here, the magnitude of G_{23} is independent of longitudinal packing factor α.

9.9. SUMMARY

A simple relationship between the fiber strength and the interface strength has been obtained. This relationship makes it possible to establish what the strength properties of the interface should be to obtain an overall composite failure instead of a fiber–matrix interface failure. A mathematical expression also has been obtained that relates the interlaminar shear stress, interfacial shear stress, fracture toughness, and physical and mechanical properties of the composites.

A test technique based on shear-lag theory appears quite attractive for the determination of both the shear strength of the interface and its shear modulus. The techniques are based on pull-out, pushout, fragmentation, and debonding tests with fibers that are embedded in a matrix to various lengths. The extension of the shear-lag theory to composites that consist of discontinuous fibers surrounded by continuous fibers shows that there is a significant variation in the interface shear stress along the length as well as around the circumference of a discontinuous fiber.

In principle, the above methods cannot produce exact experimental values. However, it is simple and able to obtain reasonable estimation, especially, for the case that differences between the interfacial strength and fracture toughness and those of the matrix are not too large. The methods discussed here have the potential for providing more meaningful information about the influence of fiber treatment and resin modification on composite mechanical or mechanical interfacial properties than bulk specimen tests. This is currently the subject of further investigations.

REFERENCES

[1] S. Blassiau, A. Thionnet, A.R. Bunsell, Composite Structures 83 (2008) 312.
[2] S. Behzadi, P.T. Curtis, F.R. Jones, Composite Science Technology 69 (2009) 2421.
[3] K.K. Kratmann, M.P.F. Sutcliffe, L.T. Lilleheden, R. Pyrz, O.T. Thomsen, Composite Science Technology 69 (2009) 228.
[4] E.K. Gamstedt, L.A. Berglund, T. Peijs, Composite Science Technology 59 (1999) 759.
[5] L.T. Drzal, M. Madhukar, Journal of Materials Science 28 (1993) 569.
[6] M.R. Piggott, P.S. Chua, D. Andison, Polymer Composites 6 (1985) 242.
[7] C.Y. Hui, D. Shia, L.A. Berglund, Composite Science Technology 59 (1999) 2037.
[8] H. Zhang, M.L. Ericson, J. Varna, L.A. Berglund, Composites: Part A 28 (1997) 309.
[9] J. Harvey, C. Kozlowski, P.M.A. Sherwood, Journal of Materials Science 22 (1987) 1585.
[10] S.Y. Zhang, Composite Science Technology 60 (2000) 145.
[11] E.K. Gamstedt, Journal of Applied Polymer Science 76 (2000) 457.
[12] A.R. Maligno, N.A. Warrior, A.C. Long, Composite Science Technology 70 (2010) 36.
[13] M.K. Seo, S.J. Park, Composites Interfaces 13 (2006) 249.
[14] M.K. Seo, S.J. Park, Macromolecular Research 17 (2009) 430.
[15] S.J. Park, M.K. Seo, K.Y. Rhee, Materials Science and Engineering A 356 (2003) 219.
[16] C.C. Chamis Simplified Composite Micromechanics Equations for Strength, Fracture Toughness and Environmental Effects. In 39th Annual Conference on Reinforced Plastics and Composites Institute, Session 11-D, pp. 1–16. The Society of the Plastics Industry Inc. (1984).
[17] D. Hull, An Introduction to Composite Materials. Cambridge University Press, Cambridge (1981).
[18] J.E. Grady Fracture Toughness Testing of Polymer Matrix Composites. In Handbook of Ceramics and Composites. Marcel Dekker, New York (1992).
[19] Y. Li, Q. Lin, L. Chen, X. Zhou, Composite Science Technology 69 (2009) 1919.
[20] A.G. Salvi, A.M. Waas, A. Caliskan, Composites: Part A 40 (2009) 19.
[21] J.R. Reeder, D.H. Crews Jr., AIAA Journal 28 (1990) 1270.
[22] C. DiFrancis, T.C. Ward, R.O. Claus, Composites: Part A 27A (1996) 591.
[23] A. Kelly, Proceedings of the Royal Society A: Mathematical. Physical & Engineering Sciences A319 (1970) 95.
[24] A. Takaku, R.G.C. Arridge, Journal of Physics D 6 (1973) 2038.
[25] M.J. Pitkethly, J.B. Doble, Composites: Part A 21 (1990) 389.
[26] H.L. Cox, Journal of Applied Physics 3 (1952) 72.

[27] C. Galiotis, R.J. Young, P.H.J. Yeung, D.N. Batchelder, Journal of Materials Science 19 (1984) 3640.
[28] L.B. Greszczuk, Theoretical Studies of the Mechanics of the Fiber–Matrix Interface in Composites. In Interfaces in Composites, ASTM STP 452, American Society for Testing and Materials, Philadelphia (1969).
[29] W.K. Tredway, K.M. Prewo, C.G. Pantano, Carbon 27 (1989) 717.
[30] J. Bowling, G.W. Groves, Journal of Materials Science 14 (1979) 431.
[31] Outwater, J.O., Murphy, M.C., On the Fracture Energy of Unidirectional Laminates. In Paper Presented at 24th Annual Technical Conference on Reinforced Plastics/Composites Division, The Society of the Plastics Industry (1969).
[32] P. Lawrence, Journal of Materials Science 7 (1972) 1.
[33] H. Emadipour, P. Chiang, J.L. Koenig, Res Mechanica 5 (1989) 165.
[34] A.T. DiBenedetto, L. Nicolais, L. Ambrosio, J. Groeger, Stress Transfer and Fracture in Single Fiber/Epoxy Composites, Composite Interfaces, London, Elsevier (1986).
[35] I. Palley, D. Stevens, Journal Adhesion Science and Technology 3 (1989) 141.
[36] J.W. Hutchinson, H.M. Jensen, Mechanics of Materials 9 (1990) 139.
[37] C.K.Y. Leung, V.C. Li, Composites: Part A 21 (1990) 305.
[38] B. Miller, U. Gaur, D. Hirt, Composite Science Technology 42 (1991) 207.
[39] M.R. Piggott, Interfaces Properties and their Influence of Fiber-reinforced Polymers, Composite Applications, The Role of Matrix, Fiber, and Interface, VCH Publishers, New York (1992).
[40] L.S. Sigl, A.G. Evans, Mechanics of Materials 8 (1989) 1.
[41] J.K. Morrison, S.P. Shah, Y.S. Jeng, Engineering Mechanics 114 (1988) 271.
[42] L.S. Penn, S.M. Lee, Journal of Composites Technology and Research 11 (1989) 23.
[43] M. Romanowicz, Computational Materials Science 47 (2009) 225.
[44] A. Ghavami, A. Abedian, M. Mondali, Materials Design 31 (2010) 2616.
[45] W.D. Bascom, R.M. Jensen, Journal of Adhesion 19 (1986) 219.
[46] A.T. DiBenedetto, L. Nicolais, L. Ambrosio, J. Groeger, Stress Transfer and Fracture in Single Fiber/epoxy Composites, in: H. Ishida., J. L. Koenig (Eds.), Composite Interfaces, Elsevier, London (1986).
[47] S.M. Lee, S. Holguin, Journal of Adhesion 31 (1990) 91.
[48] V. Rao, L.T. Drzal, Polymer Composites 12 (1991) 48.
[49] C.D. DiFrancia, R.O. Claus, J.W. Hellgeth, T.C. Ward Characterization of Polyimide-Coated Optical Fibers for Embedding in Epoxy Resin for the Development of Smart Composite Structures. In Proceeding of Adhesion Society Fourteenth Annual Meeting, Florida (1991).
[50] M.R. Piggott, Composite Science Technology 30 (1987) 295.
[51] F. Rezaei, R. Yunus, N.A. Ibrahim, Materials Design 30 (2009) 260.
[52] H. Chen, W. Hu, Y. Zhong, G. Gottstein, Materials Science and Engineering A 460 (2007) 624.
[53] Y.C. Gao, Y.W. Mai, B. Cotterell, Journal of Applied Mathematics and Physics 37 (1988) 550.
[54] D.B. Marshall, W.C. Oliver, Materials Science and Engineering A 126 (1987) 95.
[55] C.H. Hsueh, Materials Science and Engineering A 123 (1990) 1.
[56] C. Liang, J.W. Hutchinson, Mechanics of Materials 14 (1993) 207.
[57] R.J. Kerans, T.A. Parthasarathy, Journal of American Chemical Society 74 (1991) 1585.
[58] T.A. Parthasarathy, D.B. Marshall, R.J. Kerans, Acta Metallurgica et Materialia 42 (1994) 3773.
[59] B. Luethi, R. Reber, J. Mayer, E. Wintermantel, J. Janczak-Rusch, L. Rohr, Composites: Part A 29A (1998) 1553.
[60] A. Kelly, W.R. Tyson, Journal of Mechanics and Physics of Solids 13 (1965) 329.

[61] B. Yavin, H.E. Gallis, J. Scherf, A. Eitan, H.D. Wagner, Polymer Composites 12 (1991) 436.
[62] B. Yavin, H.D. Wagner, Advanced Composites Letters 2 (1993) 47.
[63] H.D. Wagner, J.R. Wood, G. Marom, Advanced Composites Letters 2 (1993) 176.
[64] X.F. Zhou, H.D. Wagner, Composite Science Technology 60 (2000) 367.
[65] X.F. Zhou, H.D. Wagner, J.A. Nairn, Composites: Part A 30A (1999) 1387.
[66] X.F. Zhoua, H.D. Wagnerb, S.R. Nutt, Composites: Part A 32 (2001) 1543.
[67] X. Wang, B. Zhang, S. Du, Y. Wu, X. Sun, Materials Design 31 (2010) 2464.
[68] F.A. Ramirez, L.A. Carlsson, B.A. Acha, Composites: Part A 40 (2009) 679.
[69] M. Nishikawa, T. Okabe, N. Takeda, Materials Science and Engineering A 480 (2008) 549.
[70] B. Luethi, R. Reber, J. Mayer, E. Wintermantel, J. Janczak-Rusch, L. Rohr, Composites: Part A 29A (1998) 1553.
[71] D. Tripathi, F.R. Jones, Journal of Materials Science 33 (1998) 1.
[72] A. Paipetis, C. Galiotis, Composites: Part A 27A (1996) 755.
[73] A. Kelly, N.H. Macmillan, Strong Solids. Oxford University Press, Oxford (1986).
[74] C. Galiotis, A. Paipetis, Journal of Materials Science 33 (1998) 1137.
[75] G.P. Cherepanov, Mechanics of Brittle Failure. McGraw-Hill, New York (1979).
[76] B. Budiansky, J.W. Hutchinson, A.G. Evans, Journal of Mechanics and Physics of Solids 34 (1986) 167.
[77] B. Aksel, C.Y. Hui, D.C. Lagoudas, Internation Journal of Solids and Structures 27 (1991) 833.
[78] C. Galiotis, Composite Science Technology 42 (1991) 125.
[79] X.F. Zhou, J.A. Nairn, H.D. Wagner, Composites: Part A 30A (1999) 1387.
[80] Z.F. Li, D.T. Grubb, Journal of Materials Science 29 (1994) 189.
[81] R.B. Yallee, R.J. Young, Composite Science Technology 58 (1998) 1907.
[82] M. Nishikawa, T. Okabe, K. Hemmi, N. Takeda, Internation Journal of Solids and Structures 45 (2008) 4098.
[83] M. Nishikawa, T. Okabe, N. Takeda, Materials Science and Engineering A 480 (2008) 549.
[84] B. Miller, P. Muri, L. Rebenfeld, Composite Science Technology 28 (1987) 17.
[85] P.J. Herrera-Franco, L.T. Drzal, Composites: Part A 23 (1992) 2.
[86] P. Zinck, H.D. Wagner, L. Salmon, J.F. Gerard, Polymer 42 (2001) 5401.
[87] X. Bi, Z. Li, P.H. Guebelle, J. Lambros, Mechanics of Materials 34 (2002) 433.
[88] J.H. Tsai, A. Patra, R. Wetherhold, Composites: Part A 36 (2005) 827.
[89] S.R. Nutt, A. Needleman, Scripta Metallurgica et Materialia 21 (1987) 705.
[90] V. Tvergaard, Materials Science and Engineering A 125 (1990) 203.
[91] J.W. Hutchinson, A.G. Evans, Acta Materialia 48 (2000) 125.
[92] R.J. Scheer, J.A. Nairn, Journal of Adhesion 53 (1995) 45.
[93] D. Yang, Y. Sheng, J. Ye, Y. Tan, Computational Materials Science 49 (2010) 253.
[94] P. Kumar, R. Chandra, S.P. Singh, Composites Interfaces 17 (2010) 15.
[95] H. Berger, S. Kari, U. Gabbert, R. Rodriguez-Ramos, J. Bravo-Castillero, R. Guinovart-Diaz, Materials Science and Engineering A 412 (2005) 53.
[96] A.R. Maligno, N.A. Warrior, A.C. Long, Composite Science Technology 70 (2010) 36.
[97] S.H. Lee, S. Wang, G.M. Pharr, H. Xu, Composites: Part A 38 (2007) 1517.
[98] E.J. Barbero, T.M. Damiania, J. Trovil1ion, Internation Journal of Solids and Structures 42 (2005) 2489.
[99] I.C. Finegan, R.F. Gibson, Composite Science Technology 60 (2000) 1077.
[100] Z.M. Haung, Composites: Part A 32 (2001) 143.
[101] G. Wacker, A.K. Bledzki, A. Chate, Composites: Part A 29A (1998) 619.

Chapter 10

Comprehension of Nanocomposites

10.1. INTRODUCTION

Tremendous developments have been made [1–11] in many aspects of composites research and technology during the two decades since the publication of Kelly's paper [12]. Recent advances in producing nanostructured materials with novel material properties have stimulated research to create multifunctional macroscopic engineering materials by designing structures at the nanometer scale. Motivated by the recent enthusiasm in nanotechnology, development of nanocomposites is one of the rapidly evolving areas of composites research.

Nanotechnology can be broadly defined as, 'The creation, processing, characterization, and utilization of materials, devices, and systems with dimensions on the order of 0.1–100 nm, exhibiting novel and significantly enhanced physical, chemical, and biological properties, functions, phenomena, and processes due to their nanoscale size.' Current interests in nanotechnology encompass nano-biotechnology, nano-systems, nanoelectronics, and nano-structured materials, of which nanocomposites are a significant part.

Through nanotechnology, it is envisioned that nanostructured materials will be developed using a bottom-up approach. 'More materials and products will be made from the bottom-up, that is, by building them from atoms, molecules, and the nanoscale powders, fibers and other small structural components made from them. This differs from all previous manufacturing, in which raw materials ... get pressed, cut, molded and otherwise coerced into parts and products' [13].

Fillers play important roles in modifying the desirable properties of polymers and reducing the cost of their composites. In conventional polymer composites, many inorganic filers with dimensions in the micrometer range, for example, calcium carbonate, glass beads, and talc have been used extensively to enhance the mechanical properties of polymers. Such properties can indeed be tailored by changing the volume fraction, shape, and size of the filler particles [14–16]. A further improvement of the mechanical properties can be achieved by using filler materials with a larger aspect ratio such as short glass fibers [17–19]. It is logical

Interface Science and Composites, Volume 18
ISSN: 1573-4285, doi: 10.1016/B978-0-12-375049-5.00010-4

to anticipate that the dispersion of fillers with dimensions in the nanometer level having very large aspect ratio and stiffness in a polymer matrix could lead to even higher mechanical performances. These fillers include layered silicates and carbon nanotubes. Carbon nanotubes have a substantially larger aspect ratio (~1000) in comparison with layered silicates (~200) [20, 21]. Moreover, flexible carbon nanotubes also possess very high strain to failure [22, 23]. Rigid inorganic nanoparticles with a smaller aspect ratio are also promising reinforcing and/or toughening materials for the polymers.

Polymer nanocomposites are an area of substantial scientific interest and of emerging industrial practice. When hybrid combinations of natural fillers and polymers were presented to the public for the first time in the 1990 s [24–27], a huge market opportunity was forecast [28]. In 2006 the global revenues achieved in reality were estimated at only 33.7 million $ with a forecast of 144.6 million $ in 2013 [29]. Consequently European Plastics News concluded more recently that 'nanocomposites are still looking to break into mainstream volume markets' [30]. Nevertheless the long-term perspective of nanocomposites is estimated at 9 billion $ in 2025 [31].

A morphological characteristic that is of fundamental importance in the understanding of the structure-property relationship of nanocomposites is the surface area/volume ratio of the reinforcement materials. This section discusses nanocomposites based upon the three categories of reinforcement materials: particles (silica, metal, and other organic and inorganic particles), layered materials (graphite, layered silicate, and other layered minerals), and fibrous materials (nanofibers and nanotubes). As illustrated in Fig. 10-1, the change in particle diameter, layer thickness, or fibrous material diameter from micrometer to nanometer, changes the ratio by three orders in magnitude. At this scale, there is often distinct size dependence of the material properties. In addition, with the

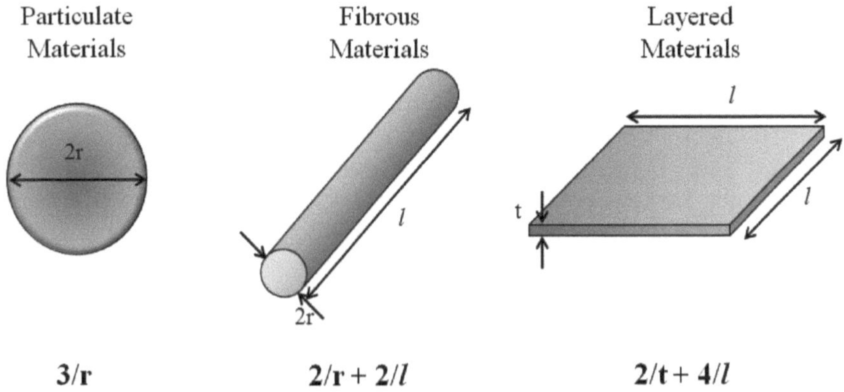

Fig. 10-1 Surface area/volume relations for varying reinforcement geometries.

drastic increase in interfacial area, the properties of the composite become dominated more by the properties of the interface or interphase.

In this chapter, we address the state of knowledge in processing, characterization, and analysis/modeling of nanocomposites with a particular emphasis on identifying fundamental structure/property relationships and compare the properties and performance of nanocomposites with traditional fiber composites.

10.2. TYPES OF NANOCOMPOSITES

10.2.1. Nanoparticle-Reinforced Composites

Particulate composites reinforced with micron-sized particles of various materials are perhaps the most widely utilized composites in everyday materials. Particles are typically added to enhance the matrix elastic modulus and yield strength. By scaling the particle size down to the nanometer scale, it has been shown that novel material properties can be obtained. A few systems are reviewed below for illustrating the resulting modification in matrix properties.

Micron-scale particles typically scatter light making otherwise transparent matrix materials appear opaque. Naganuma and Kagawa [32] showed in their study of SiO_2-epoxy composites that decreasing the particle size resulted in significantly improved transmittance of visible light. Singh *et al.* [33] studied the variation of fracture toughness of polyester resin due to the addition of aluminum particles of 20, 3.5, and 100 nm in diameter. Fig. 10-2 shows that the initial enhancement in fracture toughness is followed by decreases at higher

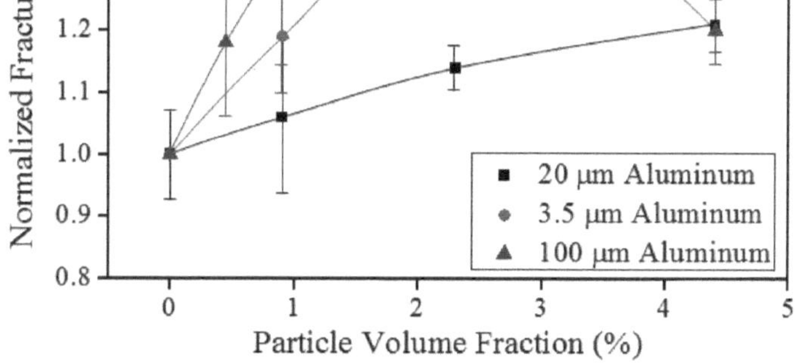

Fig. 10-2 Normalized fracture toughness with respect to volume fraction for various sized particles.

particle volume fraction. This phenomenon is attributed to the agglomeration of nanoparticles at higher particle volume content.

Lopez *et al.* [34] examined the elastic modulus and strength of vinyl ester composites with the addition of 1, 2, and 3 wt.% of alumina particles in the sizes of 40 nm, 1 μm, and 3 μm. For all the particle sizes, the composite modulus increases monotonically with particle weight fraction. However, the strengths of composites are all below the strength of neat resin due to nonuniform particle size distribution and particle aggregation. The work of Thompson *et al.* [35] on metal oxide-polyimide nanocomposite films also noted similar difficulties in processing. Their study utilized antimony tin oxide (11 ~ 29 nm), indium tin oxide (17 ~ 30 nm), and yttrium oxide (11 ~ 44 nm) in two space-durable polyimides: TOR-NC and LaRC TMCP-2. The nanoscale additives resulted in higher stiffness, comparable or lower strengths and elongation, and lower dynamic stiffness (storage modulus). The dispersion of metal oxides on a nanometer scale was not achieved. Ash *et al.* [36] studied the mechanical behavior of alumina particulate-poly(methyl methacrylate) composites. They concluded that when a weak particle/matrix interface exists, the mode of yielding for glassy, amorphous polymers changes from cavitational to shear, which leads to a brittle-to-ductile transition. This behavior is attributed to increased polymer chain mobility, due to the presence of smaller particles, and also the capability to relieve tri-axial stress because of poorly bonded larger particles. An extensive review of the structure-property relationships in nanoparticle/semicrystalline thermoplastic composites has been made by Karger-Kocsis and Zhang [37].

10.2.2. Nanoplatelet-Reinforced Composites

Two types of nanoplatelet-reinforced composites are reviewed: clay and graphite. In their bulk state, both clay and graphite exist as layered materials. In order to utilize these materials most efficiently, the layers must be separated and dispersed throughout the matrix phase. The morphology of clay/polymer nanocomposites is illustrated in Fig. 10-3 [38]. In the conventional miscible state, the interlayer spacing in a clay particle is at its minimum. When polymer resin is inserted into the gallery between the adjacent layers, the spacing expands, and it is known as the intercalated state. When the layers are fully separated, the clay is considered to be exfoliated.

Regarding the other layered materials, the exfoliated graphite or graphene sheet has about the same thickness as exfoliated clay. Table 10-1 shows their high tensile modulus, tensile strength, thermal conductivity, and low electrical resistivity, comparing to clay platelets. The low electrical resistivity of exfoliated graphite facilitates the conductivity of polymer composites when a threshold percolation weight content of the conductive phase is reached. Fig. 10-4 shows the results of Fukushima and Drzal [39] on the resistivity of epoxy matrix (EPON 828) composites with the addition of vapor grown carbon fiber, carbon

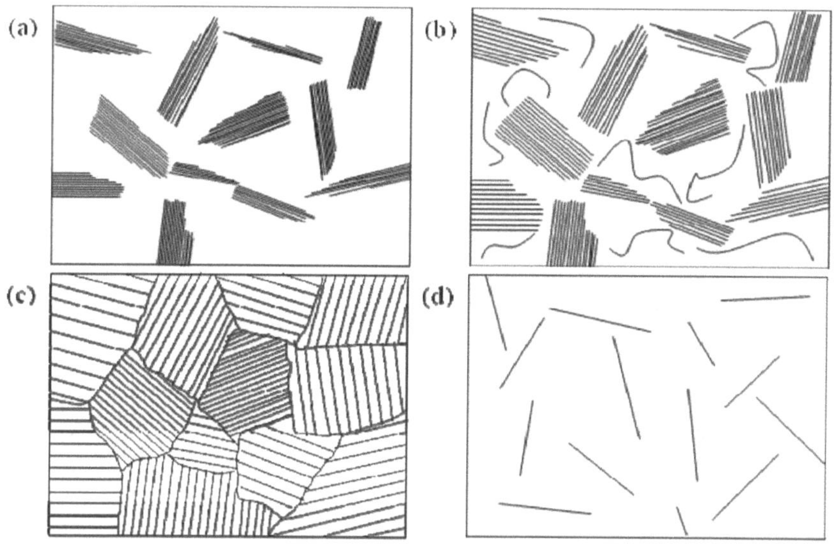

Fig. 10-3 Morphologies of polymer–clay nanocomposites: (a) conventional miscible, (b) partially intercalated and exfoliated, (c) fully intercalated and dispersed, and (d) fully exfoliated and dispersed.

Table 10-1 Properties of Exfoliated Graphite Platelets as Compared to Exfoliated Clay Platelets

Properties	Graphene sheet	Clay platelet
Physical structure	~1 nm × 100 nm	~1 nm × 1,000 nm
Tensile modulus (GPa)	1,000	170
Tensile strength (GPa)	10–20	1
Resistivity (X cm)	$\sim 50 \times 10^{-6}$	10^{10}–10^{16}
Thermal conductivity (W/m K)	3,000	0.67
Coefficient of thermal efficiency (K^{-1})	$\sim 1 \times 10^{-6}$	8–16×10^{-6}
Density (g/cm^3)	2.0	2.5–3.0
D-spacing (nm)	0.34	1.85

black, polyacrylonitrile carbon fiber, exfoliated graphite, and milled graphite. The graphite nanoplatelets are treated in O_2 plasma for initiating radical polymerization. The percolation threshold for exfoliated graphite is around 1 wt.%. Zheng and coworkers [40, 41] have also reported reduced percolation thresholds in exfoliated graphite nanoplatelet–thermoplastic composites.

Song *et al.* [42] have modeled the von Mises stress distribution in graphite/polyacrylonitrile nanocomposites with respect to the level of exfoliation. They

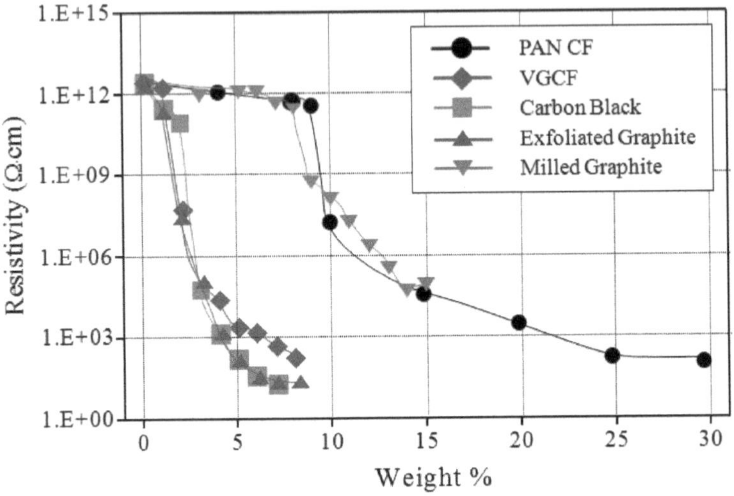

Fig. 10-4 Comparison of various carbon reinforcement materials on the composite bulk resistivity.

demonstrated the reduction in the magnitude of stress concentration with layer thickness at the tip of graphite layers, aligned with the axial tensile load. Such conclusion is similar to those obtained for short fiber composites where the dispersion of a fiber bundle has the benefit of reducing the stress concentration in the matrix material at the bundle end and thus, the chance of matrix cracking or plastic deformation.

The greatly enhanced electrical conductivity of polymeric materials with the addition of small amount of graphite platelets has found many practical applications. These include electromagnetic interference shielding and heat management of electronic and computer devices or equipment, electrostatic paint for automobiles, and polymer sheath of electric cables [43]. Nanoclay/polypropylene composites are being used as functional parts in automobiles. For instance, General Motors is using about 660,000 lbs of nanocomposite material per year.

10.2.3. Nanofibers-Reinforced Composites

Vapor grown carbon nanofibers have been used to reinforce a variety of polymers, including polypropylene, polycarbonate, nylon, poly(ether sulfone), poly(ethylene terephthalate), poly(phenylene sulfide), acrylonitrile–butadiene–styrene, and epoxy. Carbon nanofibers are known to have wide-ranging morphologies, from structures with a disordered bamboo-like structure [44] (Fig. 10-5(a)) to highly graphitized 'cup stacked' structures [45, 46] (Fig. 10-5 (b) and (c)), where the conical shells of the nanofiber are nested within each other. Carbon nanofibers typically have diameters on the order of 50–200 nm.

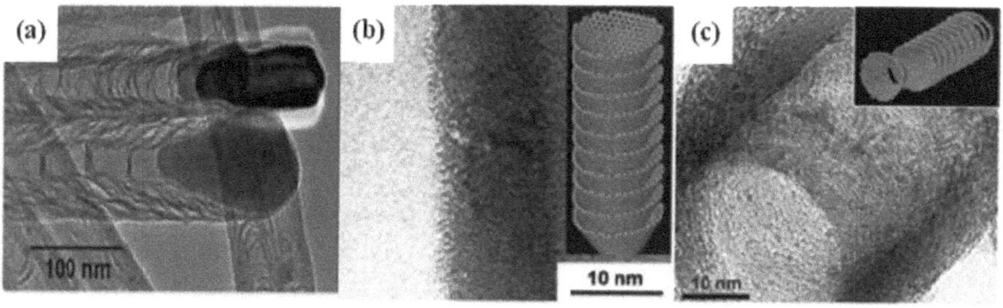

Fig. 10-5 Transmission electron microscopy of the nanoscale structure of carbon nanofibers showing: (a) disordered bamboo-like structures, (b) highly graphitized sidewall of a cup-stacked (molecular models inset) nanofibers showing the shell tilt angle, and (c) a nesting of the stacked layers.

Wei and Srivastava [47] have modeled the mechanical properties of carbon nanofibers with varying morphology using continuum elastic theory and molecular dynamics simulations. The axial Young's modulus of the nanofiber is particularly sensitive to the shell tilt angle, where fibers that have small tilt angles from the axial direction show much higher Young's modulus than fibers with large tilt angles. The wide-ranging morphology of the carbon nanofibers and their associated properties results in a broad range of scatter for experimental results on processing and characterization of nanofiber composites.

Many of the key challenges associated with the processing, characterization, and modeling of carbon nanofiber composites, such as dispersion and adhesion, are similar to those for nanotube-reinforced composites and are discussed in the following section.

10.2.4. Carbon Nanotube-Reinforced Composites

10.2.4.1. Introduction

Since the documented discovery of carbon nanotubes in 1991 by Iijima [48] and the realization of their unique physical properties, including mechanical, thermal, and electrical, many investigators have endeavored to fabricate advanced carbon nanotube composite materials that exhibit one or more of these properties [49–52]. For example, as conductive filler in polymers, carbon nanotubes are quite effective compared to traditional carbon black microparticles, primarily due to their large aspect ratios [53].

The electrical percolation threshold was recently reported at 0.0025 wt.% carbon nanotubes and conductivity at 2 S/m at 1.0 wt.% carbon nanotubes in epoxy matrices [54]. Similarly, carbon nanotubes possess one of the highest thermal conductivities known [55], which suggests their use in composites for thermal management [49].

As reviewed by Thostenson *et al.* [56], the morphology of a carbon nanotube is defined by the orientation and magnitude of the chiral vector in a graphene

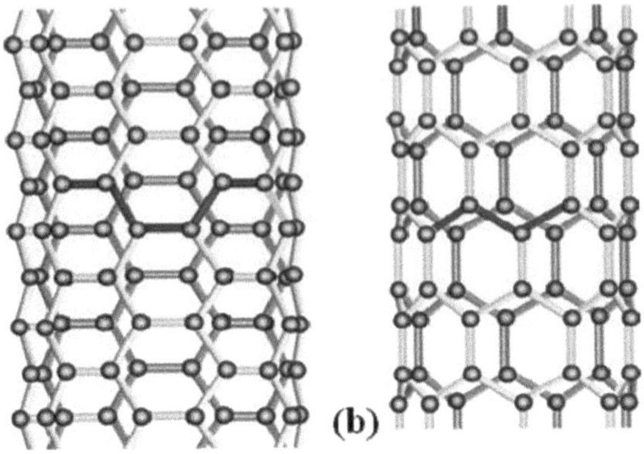

Fig. 10-6 Atomic structures of (a) armchair and (b) zigzag carbon nanotubes.

sheet, which is 'wrapped up' to form the single-walled carbon nanotubes. The two limiting configurations are armchair and zigzag nanotubes (Fig. 10-6). Since their observation over a decade ago [57], numerous investigators have reported remarkable physical and mechanical properties of carbon nanotube. The outstanding thermal and electric properties combined with their high specific stiffness and strength, and very large aspect ratios have stimulated the development of nanotube-reinforced composites for both structural and functional applications [58, 60, 61].

10.2.4.2. Properties of Carbon Nanotube–Polymer Composites

The carbon nanotubes can be thought of as the ultimate carbon fibers with break strengths reported as high as 200 GPa, and elastic moduli in the 1 TPa range [56, 57]. This, coupled with approximately 500 times more surface area per gram (based on equivalent volume fraction of typical carbon fiber) and aspect ratios of around 10^3, has spurred a great deal of interest in using carbon nanotubes as a reinforcing phase for polymer matrices.

Both single-walled carbon nanotubes and multi-walled carbon nanotubes have been utilized for reinforcing thermoset polymers (epoxy, polyimide, and phenolic), as well as thermoplastic polymers (polypropylene, polystyrene, poly (methyl methacrylate), nylon 12, and polyether ether ketone).

Thostenson and Chou [62, 63] have characterized the nanotube structure and elastic properties of a model composite system of aligned multi-walled carbon nanotubes embedded in a polystyrene matrix. Fig. 10-7(a) shows a transmission electron microscopy of the as-processed 5 wt.% nanocomposite film showing large scale dispersion and alignment of carbon nanotubes in the polymer matrix. The arrow in Fig. 10-7(a) indicates the direction of alignment taken as the principal material direction with a nanotube orientation of 0°. The grey lines perpendicular to the arrow in the transmission electron microscopy are artifacts

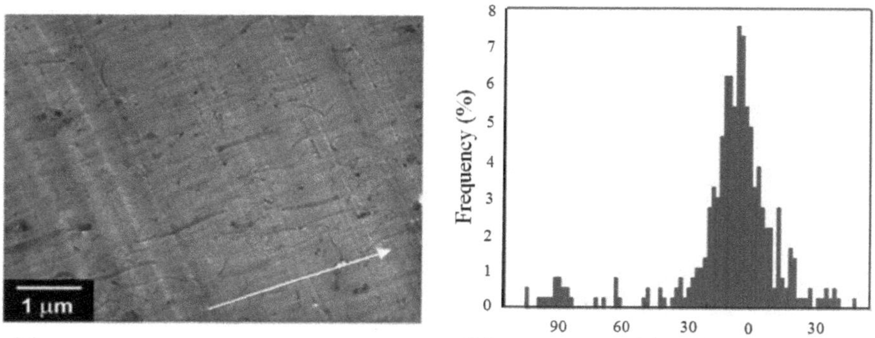

(a) (b)

Fig. 10-7 (a) Transmission electron microscopy of process-induced orientation in nanocomposite ribbons [63] and (b) image analysis of orientation distribution.

from microtome cutting process. Fig. 10-7(b) shows the distribution of nanotube alignment from the image analysis. Based on the data, the standard deviation of nanotube alignment from the principal material direction is less than ± 15°.

The axial elastic properties of the multi-walled carbon nanotubes/polystyrene system have been modeled by Thostenson and Chou [62] using a 'micromechanics' approach through defining an equivalent effective fiber property for the diameter-sensitive carbon nanotube elastic modulus. To accurately model the elastic properties of the composites, the contribution to the overall elastic modulus of each nanotube diameter, and the volume fraction that tubes of a specific diameter occupy within the composite have been taken into account. Using the knowledge of the tube diameter distribution functions as well as the nanotube density and volume distribution as functions of tube diameter, the micromechanics approach identifies the correlation between axial Young's modulus, and the diameter, volume fraction, and length of nanotubes of the aligned nanocomposite model system.

In the processing of nanocomposites, carbon nanotubes need to be separated from bundles and dispersed uniformly in a polymer matrix for maximizing their contact surface area with the matrix. Modification of nanotube surfaces, for example, the creation of covalent chemical bonds between nanotubes and the polymer matrix, enhances their interactions, and gives rise to higher interfacial shear strength than van der Waals bonds [58, 64]. The two major approaches to functionalization are chemical methods and physical methods such as irradiation with electrons or ions.

The chemical methods involve the attachment of chemical bonds to either the nanotube ends or sidewalls. First, nanotube caps, which are more reactive because of their high degree of curvature, can react with a strong acid. The open ends can be stabilized by carboxylic acid and hydroxide groups. Sinnott [65] has provided an in-depth review of the chemical functionalization of carbon nanotubes, where the chemical bonds are used to tailor the interaction of the nanotube with the other entities, such as a solvent, a polymer matrix, or other nanotubes. Hu *et al.* [66] and Ni *et al.* [67] approached the modification of carbon nanotube/polystyrene composites through polyatomic ion beam deposition. Molecular dynamic simulations have demonstrated the modification of a composite of (10, 10) single-walled carbon nanotubes–polystyrene through the deposition of a beam of $C_3F_5^+$ polyatomic ions. Covalent cross-links were induced between otherwise unfunctionalized single-walled carbon nanotubes and polystyrene matrix resins.

Although the length scales of reinforcements in nanocomposites, i.e., diameter of nanoparticles, diameter and length of nanofibers and nanotubes, as well as thickness of nanoplatelets are smaller than those of traditional composites by about three orders in magnitude, there are considerable differences and similarities in the properties and performance of nanocomposites and traditional fiber composites. It is highly desirable to present some key properties of nanocomposites in contrast to the same properties in traditional composites; only a few examples are selected below.

It is well known in short-fiber composites that the presence of fiber-ends induces stress concentrations in the matrix materials when the composites are subjected to loading. The nature of stress concentration and the associated singularities have been studied in terms of the fiber bundle-end shape and bundle aspect ratio [68].

In the case of uniaxial compression of continuous fiber composite, it is well known that fiber defects or fiber misalignment may activate fiber bending and subsequent fiber buckling. Fig. 10-8(a) shows slip band formation in a composite with Al_2O_3 fiber in an aluminum matrix [69]. The activation and subsequent observation of nanotube buckling in a composite under compressive loading are much more difficult because of the small size and alignment of nanotubes. Thostenson and Chou [70] have succeeded in such experiments. The multiple buckling of individual multiwalled carbon nanotubes in polymer composites can be seen in Fig. 10-8(b). An atomistic modeling of elastic buckling of carbon nanotubes has been performed by Li and Chou [71].

Several key mechanisms of energy dissipation have been identified in the fracture of short as well as continuous fiber-reinforced composites. These are

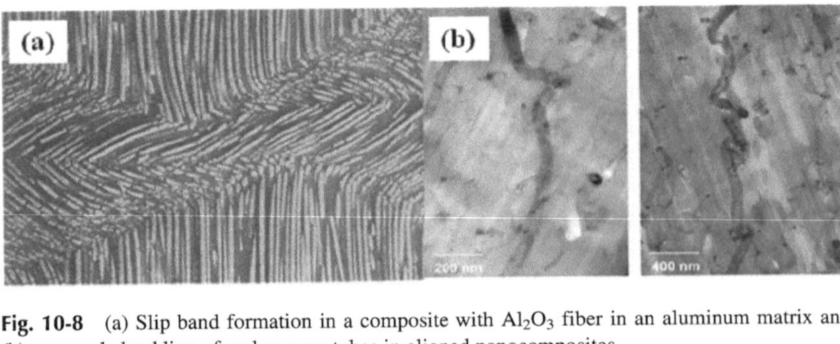

Fig. 10-8 (a) Slip band formation in a composite with Al_2O_3 fiber in an aluminum matrix and (b) nanoscale buckling of carbon nanotubes in aligned nanocomposites.

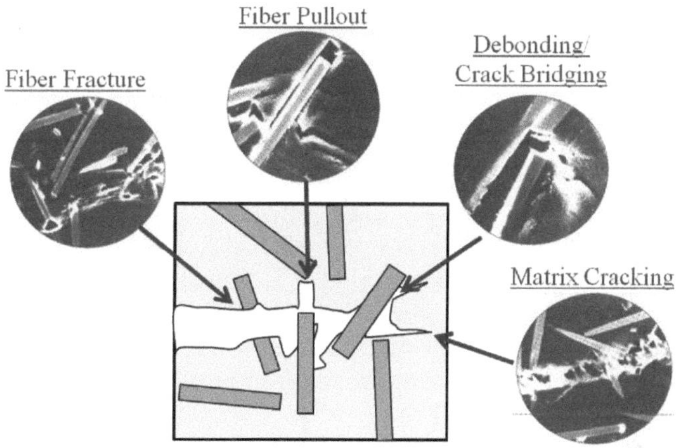

Fig. 10-9 Key mechanisms of energy dissipation have been identified in the fracture of short as well as continuous fiber-reinforced composites.

fiber fracture, fiber pullout, fiber/matrix debonding/crack bridging, and matrix cracking. Fig. 10-9 shows schematically these mechanisms operating at a crack tip [72] as well as the micrographs demonstrating the individual failure modes. It is interesting to note that all these failure modes have also been observed in nanotube reinforced polymer composites as demonstrated by Thostenson and Chou [63] in Fig. 10-10.

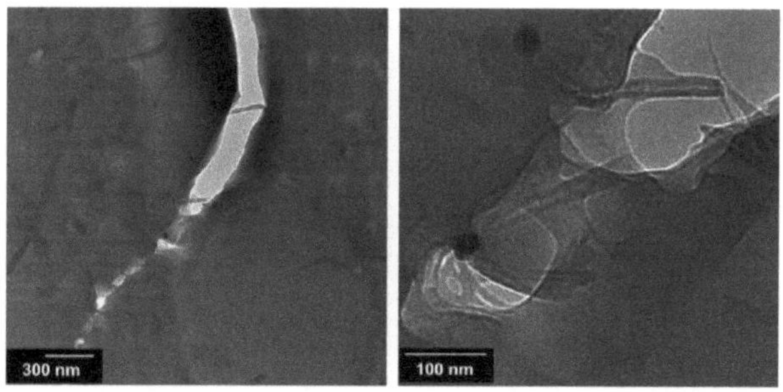

Fig. 10-10 Fracture mechanisms in carbon nanotube-reinforced composites.

One of the reasons of the recent enthusiasm toward carbon nanotubes as reinforcements for composite materials is their reported high elastic modulus and strength comparing to those of existing continuous fibers [73–78]. The addition of carbon nanotubes has also been shown to significantly reduce the resistivity and percolation threshold in both polymer matrix materials, such as epoxy [79, 80] and poly(butylene terephthalate) [81] as well as ceramic matrix materials, such as silicon carbide [82] and aluminum oxide [83]. Some recent mechanical property results for carbon nanotube/polymer composites are summarized in Table 10-2.

Table 10-2 Selected Mechanical Properties of Carbon Nanotubes–Polymer Composites

Materials	Matrix	Conc. (%)	Loading	Modulus	Yield stress	Strength	Toughness	Max. strain	ILSS[11]	Refs
MWCNT[1]	None	100	Tensile[a]	910 GPa		150 GPa				[9]
MWCNT	None	100	Tensile[b]	270–950 GPa		11–200 GPa		12%		[39]
MWCNT	PAN[4]	1.8 vol.	Tensile	+36%	+46%	+31%	+80%			[4]
SWCNT[2]	PVA[5]	60 wt.	Tensile	80 GPa		1.8 GPa	570 J/g	~100%		[16]
MWCNT	PS[6]	5 wt.	DMA[10] (25 °C)	+49% (aligned) + 10% (random)						[17]
MWCNT	PE-B[c]		AFM[c]						47 MPa	[22]
MWCNT	UHMWPE[7]	1 wt.	Tensile	+25%	+48%	+25%	+150%	+60 to +140%		[28]
SWCNT	PAN	4 wt.	AFM[d]	~+100%						[14]
CNF[3]	PMMA[8]	5 wt.	Tensile	+50%		+200%[e]		– 37.5%		[34]
MWNT	PMEMA[9]	1 wt.	DMA	+200%						[33]

[1] *MWCNT: multi-walled carbon nanotubes.*
[2] *SWCNT: single-walled carbon nanotubes.*
[3] *CNF: carbon nanofibers.*
[4] *PAN: polyacrylonitrile.*
[5] *PVA: polyvinylalcohol.*
[6] *PS: polystyrene.*
[7] *UHMWPE: ultra-high molecular weight polyethylene.*
[8] *PMMA: poly(methyl methacrylate).*
[9] *PEMA: poly(ethyl methacrylate).*
[10] *DMA: dynamic mechanical analysis.*
[11] *ILSS: interlaminar shear strength.*
[a] *In situ transmission electron microscopic tensile testing of individual multi-walled carbon nanotubes via a micro-fabricated device.*
[b] *In situ scanning electron microscopic tensile testing of individual multi-walled carbon nanotubes via dual atomic force microscopic tips.*
[c] *In situ atomic force microscopic pull-out of individual multi-walled carbon nanotubes from polyethylene-B film.*
[d] *Moduli measured using an atomic force microscopic nanoindentation technique.*
[e] *Compressive strength via loop test.*
(+/-) indicates an enhancement/diminishment from the neat matrix. Recently reported mechanical property enhancements for various carbon nanotubes/polymer composites are presented with additional information on the method used to collect the data. Experimentally measured values of mechanical properties of multi-walled carbon nanotubes are also presented.

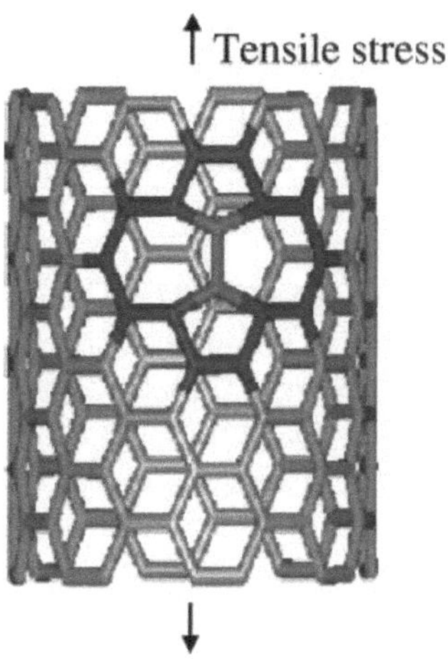

Fig. 10-11 A Stone–Wales defect. A representation of a Stone–Wales defect in the graphitic structure of carbon nanotubes under tensile stress shows the rearrangement of the hexagonal graphitic structure into two pairs of heptagons and pentagons.

Recent reports indicate that tensile failure of multi-walled carbon nanotubes that carry all of the load within the outer shell is initiated by the formation of a Stone–Wales defect in which 2 C–C bonds are broken and 2 new C–C bonds are formed such that a pair of pentagons and a pair of heptagons results (Fig. 10-11) [84, 85]. Increasing numbers of these dislocations leads to necking and ultimately, failure of the carbon nanotubes [85]. Often, however, in carbon nanotubes/polymer composites a reduction in strength is observed suggesting that the carbon nanotubes may promote crystalline defects in the matrix or, considering their size, act as defects themselves. This suggests a critical volume fraction of carbon nanotubes is required for hindering matrix strain allowing strengthening to occur as typically observed in composites.

10.2.4.3. Interfaces of Carbon Nanotube–Polymer Composites

Increasing attention is being focused on the carbon nanotube surface, namely the interface between the carbon nanotubes and surrounding polymer matrices [86]. From micromechanics, it is through shear stress build-up at this interface that stress is transferred from the matrix to the carbon nanotubes. Numerous researchers have attributed lower-than-predicted carbon nanotubes–polymer composite properties to a lack of interfacial bonding [87–89]. If one considers the surface of carbon nanotubes, essentially an exposed graphene sheet, it is not surprising that interfacial traction is a concern. It is the weak inter-planar interaction of graphite that provides its solid lubricant quality, and resistance to matrix adhesion. This is exaggerated by the chemically inert nature of graphene structures. A June 2002 publication on carbon nanotube composites noted 'In depth study on the stress transfer mechanism of the nanotube composites with different chemical and geometrical properties, matrix environments and loading conditions are essential' [90]. Since then, some interesting published results have described progress on addressing this issue.

The pull-out force necessary to remove a given length of an individual multi-walled carbon nanotubes embedded in polyethylene–butane using an atomic force microscopy was measured [91]. The calculated average interfacial shear strength using a Kelly–Tyson approach of 47 MPa; typical values for carbon fiber/epoxy are in the 30–80 MPa range depending on fiber modulus and surface treatment [92] was sufficiently high to suggest that covalent bonding between defects in the outer shell of the multi-walled carbon nanotubes and the polymer was occurring. It also suggested that the polymer chains close to the interface behaved differently than the bulk, a logical result when considering the multi-walled carbon nanotube outer diameter is of similar magnitude to the radius of gyration for the polymer. In a report corroborating an interfacial region of nonbulk polymer, a 'sheathing layer' of polycarbonate on pulled out multiwalled carbon nanotubes was imaged, which gave further evidence of significant interfacial interaction between the multi-walled carbon nanotubes and the polymer matrices [93].

It was also found that chemical functionalization of the multi-walled carbon nanotubes augmented the diameter of the polymer sheath suggesting chemical augmentation of interfacial bonding. Acid oxidation of multi-walled carbon nanotubes was again reported to attach carboxylic groups on the surface, which were then reacted with epoxide-terminated molecules up to 12 wt.% by thermogravimetric analysis [94]. These types of functionalized tubes could enhance reinforcement of epoxy resins. Carboxylated tubes have also been reported to augment the cure rate of epoxy resins at lower temperatures [95]. Unfortunately, direct and indirect measurement of polymer–carbon nanotube interfacial shear strength suggesting good bonding exists conflict with other reports which often cite clean pullout of carbon nanotubes and poor interfacial bonding [87, 88, 96]. The magnitude of carbon nanotube strength (>10 times that of typical carbon fiber) may preclude embedded carbon nanotube tensile failure in large numbers resulting in the dominant failure mode to be carbon nanotube pullout. Order of magnitude increases in interfacial shear strength may be required for the most efficient strengthening of polymers with carbon nanotubes. Optimizing the polymer–carbon nanotube interface for nanoscale mechanical reinforcement remains unclear, but the evidence available indicates that chemical means can be effective, and this is likely to be a major focus in the near term.

Theoretical treatments of carbon nanotube pullout were recently reported [96, 97]. Including one in which a (10, 10) single-walled carbon nanotube pulled out from a polyethylene matrix was modeled via molecular dynamics simulations, and quite a low effective viscosity of 0.2 cP was found for interfacial sliding [96]. A force of approximately 0.1 nN was required for pullout to initiate for a single tube.

10.2.5. Graphene-Based Composite Materials

10.2.5.1. Introduction

Graphene, as two-dimensional (2D) nanomaterials of sp^2-hybridized carbon discovered by Professor Andre Geim's research group at the University of Manchester, has attracted increasing interest due to its great potential for applications in electronic devices and composite materials [98–100]. It consists of a hexagonal array of sp^2 bonded carbon atoms, just like those found in bulk graphite. 2D materials display very interesting properties, and are fundamentally different from the three-dimensional (3D) materials we encounter everyday. The discovery of 2D materials means that scientists now have access to materials of all dimensionalities, including 0D (quantum dots, atoms) and 1D (nanowires, carbon nanotubes). Moreover, with high surface area and good chemical stability, graphene can be used as gas absorbant [101], ultracapacitor materials [102], or support materials for developing novel heterogeneous catalysts with enhanced activity (Fig. 10-12) [103].

Fig. 10-12 The surface of a graphite crystal. Graphite is a layered material, and can be separated into sheets, forming structures like folds and pleats.

Some synthesis routes to graphene started to be reported in 2001, using heat treatments, at 1600 °C of diamond nanoparticles over highly ordered pyrolytic graphite [104]. However, extensive studies on graphene started only after Novoselov *et al.* [105] were able to isolate graphene sheets using a simple micromechanical exfoliation method placing scotch-tape over commercial highly ordered pyrolytic graphite (as shown in Fig. 10-13). The latter technique allowed scientists all over the world to isolate single- and double-layered graphene flakes. Subsequently, other methods to obtain graphene flakes were reported, including its epitaxial growth over SiC substrates [106], and chemical vapor deposition growth over thin metal layers [107].

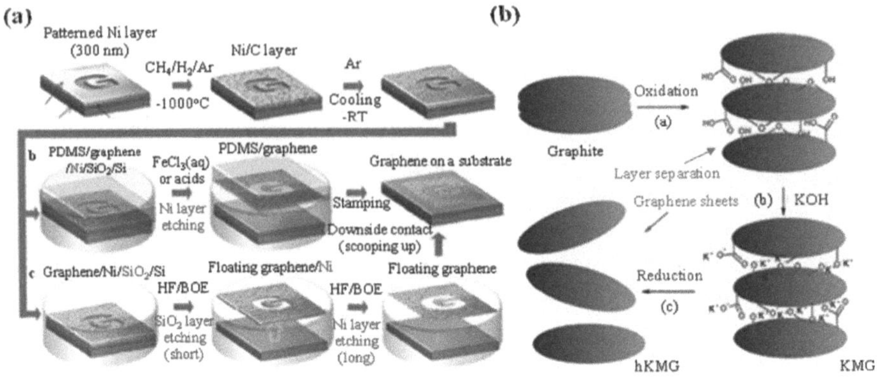

Fig. 10-13 Synthesis method of graphene: (a) chemical vapor deposition method and (b) chemically exfoliation method.

The topological nature of a certain type of grain boundary in polycrystalline graphene may be used to tailor the electrical transport properties [108, 109]. For a long time since graphene, a single atomic layer of graphite, was first isolated using adhesive tape, it has been an open question whether these micrometer-sized specimens of an exotic 2D material could be used in practical applications. More recently, however, several research groups have demonstrated the scaling up of graphene production to the centimeter [110, 111] or even meter scale [112]. The trouble is that when the lateral size increases, the hexagonal graphene lattice breaks inevitably into many domains of single-crystal regions, with grain boundaries between adjacent domains resulting in a disruption of the crystallographic structure. Understanding electron transport across such extended defects is of critical importance if graphene is to be used for such large-scale technological applications as touch-screen panels and solar-cell electrodes. Writing in *Nature Materials*, Oleg Yazyev and Steven Louie now present a theoretical investigation into the electronic structures of various types of grain boundary that can form between two single-crystal graphene domains [113]. These authors find that the electronic momentum mismatch across certain types of boundary can block the transport through the boundary completely for electrons with energies up to ~1 eV, opening up the possibility of device applications based on the structural engineering of graphene boundaries.

For a polycrystalline graphene boundary, the 'topological' nature of the defects at the boundary plays an important role. In the 2D atomic honeycomb arrangement of carbon in graphene, there are several ways of breaking the translational symmetry, thus producing topologically distinct defects in a way that each atom is still chemically bonded to its three nearest neighbors [114]. For example, Fig. 10-14(a, left) shows a five-membered carbon ring embedded in the carbon honeycomb network, which is known as a positively charged

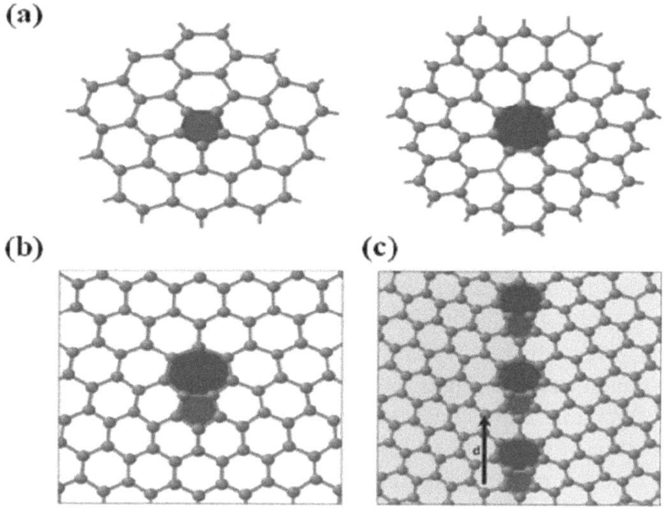

Fig. 10-14 Topological disorders in graphene. (a) Left: positively charged disclination. Right: negatively charged disclination. (b) Edge dislocation where combining the positive and negative disclinations results in the addition of an atomic row. (c) Grain boundary formed by a periodic array of edge dislocations with a periodicity defined by the vector **d**.

disclination. The likelihood of such a defect occurring is extremely low because this defect introduces a positive curvature in the graphene lattice that costs a large amount of energy to form. Likewise, a seven-membered ring defect, a negatively charged disclination, is unlikely as it introduces a negative curvature in the lattice Fig. 10-14(b, right). However, a combination of five- and seven-membered rings can be embedded in the honeycomb lattice with only a small additional energy, as a new row of atoms only needs to be introduced along a certain direction (Fig. 10-14(b)), forming an edge dislocation. In single-walled carbon nanotubes, the presence of a dislocation changes the chiral atomic structure of the nanotubes [115]. In graphene, an isolated dislocation only produces a local disturbance, which slowly dies out away from the dislocation core. However, when the dislocations form a concerted array, their presence can cause a global effect. Fig. 10-14(c) shows a periodic arrangement of dislocations along one direction that makes a clear grain boundary between the differently oriented left and right single-crystal graphene domains. The structure of such a grain boundary line can be described by a vector **d** running along the boundary, whose magnitude quantifies the periodicity of the grain boundary.

Using the periodic translational invariance along this idealized grain boundary, Yazyev and Louie employ a simple and very powerful analytical argument based on the 'zone' folding method. This method involves folding the 2D band structure along the 1D band that follows the grain boundary. The folded 2D graphene band is then projected onto the 1D mini-band, reflecting the periodic structure imposed by the periodicity of the grain boundary. A similar zone-folding approach has been successfully applied to nanotubes to explain the chirality dependence of metallic and semi-conducting nanotubes [116].

In graphene this approach results in two types of folded 1D band structure, depending on the 'chiral' nature of **d**, which is determined by the atomic lattice vectors in each domain. Matching the relative chiral nature of each graphene domain across the boundary thus leads to two different categories of the grain boundary: type I, where both sides of the grain boundary have similar chirality, and type II, where the relative chirality is different.

As in nanotubes, the chiral nature of each domain of graphene is closely related to the momentum conservation along the grain boundary. Considering the momentum of conducting channels flowing across the boundary, one can speculate that the momentum-conserving type I grain boundary is relatively conducting, whereas the non-momentum-conserving type II grain boundary is nonconducting due to the formation of a transport gap. Yazyev and Louie show this explicitly via both an analytical and a numerical first-principles calculation in their paper. The sensitivity of conduction through a graphene boundary to its atomic structure means a careful study of this grain structure is necessary to understand transport in large-area polycrystalline graphene samples. Moreover, if type II boundaries can be engineered in these samples, the tunable energy gaps that form at the local boundary region could be utilized in future graphene electronics.

Although this first important step has been made, a complete understanding of the electronic structure of graphene grain boundaries is still far away. In realistic graphene grain boundaries, the periodic arrangement of dislocations assumed in the authors' work may not be achieved. Randomly oriented grain-boundary segments with complicated dislocation arrangements may be more common features. Any type of disorder breaking the translational periodicity would then result in electronic transmission within the transport gap of the perfectly reflecting type II grain boundaries. The magnitude of this induced transmission would depend on the concentration of such disorder, and may eventually lead to finite conduction within the transport gap of pristine grain-boundary structures. Further experimental and theoretical studies with realistic atomic-scale defects will surely help clarify what happens in realistic samples.

10.2.5.2. Properties of Graphene

As mentioned above, graphene is considered a 2D carbon nanofiller with a one-atom-thick planar sheet of sp^2 bonded carbon atoms that are densely packed in a honeycomb crystal lattice. It is regarded as the 'thinnest material in the universe' with tremendous application potential [117]. Graphene is predicted to have remarkable properties, such as high thermal conductivity, superior mechanical properties, and excellent electronic transport properties [118]. These intrinsic properties of graphene have generated enormous interest for its possible implementation in a myriad of devices [119]. These include future generations of high speed and radio frequency logic devices, thermally and electrically conducting reinforced nanocomposites, ultra-thin carbon films, electronic circuits, sensors, and transparent and flexible electrodes for displays and solar cells.

The crystal structure of graphene can be thought of as two equivalent carbon triangular sublattices. Due to symmetry considerations, the hopping of electrons between the sub-lattices leads to the formation of two energy bands, which intersect at the K point (Fig. 10-15(a)) [120]. Near these crossing points, the electron energy is linearly dependent on the wave vector. This linear dispersion results in massless excitons, which are described by the Dirac equation. Dirac fermions (electrons or holes) exhibit very different and unusual properties

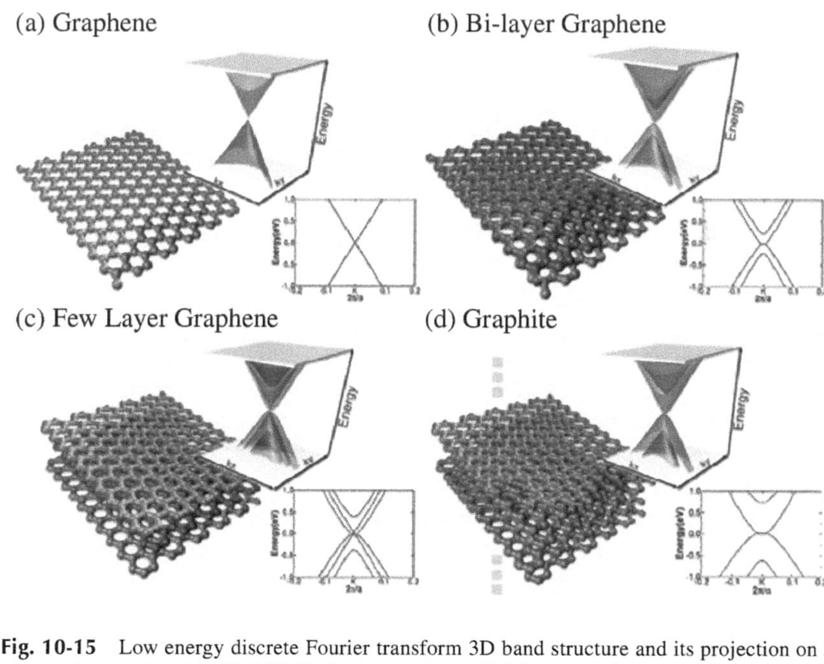

Fig. 10-15 Low energy discrete Fourier transform 3D band structure and its projection on kx close to k point K (π/a*[2/3,0.00.0]) for (a) graphene, (b) bilayer graphene, (c) trilayer graphene, and (d) graphite. Part (a) shows the characteristic Dirac point of graphene. The Dirac point (i.e., relativistic fermion characteristic) is lost in bilayer graphene (b), but appears again in trilayer graphene (c); (d) shows the 3D graphite structure that displays a semimetallic band structure with parabolic-like bands. The Fermi level has been set at zero in all cases.

compared to ordinary electrons, thus leading to new phenomena. Other outstanding properties of graphene include insensitivity to external electrostatic potentials (Klein paradox), jittery motion of the wave function under confining potentials, and large mean free paths. For a comprehensive review of the properties of graphene [121].

The electronic properties of graphene change with the number of layers and by the relative position of atoms in adjacent layers (stacking order). For bilayer graphene, the stacking order can be either AA, with each atom on top of another atom; or AB, where a set of atoms in the second layer sits on top of the empty center of a hexagon in the first layer. As the number of layers increase, the stacking order can become more complicated. For graphite, there are three common types of stacking: (i) AB or Bernal stacking, (ii) ABC or rhombohedral stacking, and (iii) no discernible stacking order or turbostatic stacking. The most stable stacking is AB, thus it has been studied more than other graphene-based stacks. However, the other stacking orders are certainly possible, especially in few-layer graphenes, and remain to be studied in detail. In fact, it has been recently found that bilayer graphene often exhibits AA stacking [122].

Thermal conductivity and mechanical stiffness of the graphene sheets may rival the remarkable in-plane values for graphite ($\sim$3000 $Wm^{-1} K^{-1}$ and 1060 GPa, respectively); their fracture strength should be comparable to that of carbon nanotubes for similar types of defects [123–125]; and recent studies have shown that individual graphene sheets have extraordinary electronic transport properties [126–130]. One possible route to harnessing these properties for applications would be to incorporate graphene sheets in composite materials. The manufacturing of such composites requires not only that graphene sheets be produced on a sufficient scale but that they also be incorporated, and homogeneously distributed, into various matrices.

The limit of thickness at which graphene can still be considered as such, can be determined by the rapid change in the electronic structure as the number of layers increase. The first AB stacks of graphene show very different electronic spectra (Fig. 10-15). The bilayer (Fig. 10-15(b)) shows parabolic bands (thus no Dirac electrons), which touch at the Fermi level. Under the presence of an electric field, the gap on bilayer graphene can be opened, which is of interest in technological applications [131]. The stacking order or disorder has been found to dramatically change the electronic properties of multilayer graphene, introducing Dirac fermions due to symmetry breaking, even in graphite crystals to form graphene stacks [132].

Table 10-3 gives a comparative chart on the mechanical, thermal, and electrical properties of graphene with carbon nanotube, steel, plastic, rubber, and fiber. The tensile strength of graphene is similar or slightly higher than carbon nanotube, but much higher than steel, Kevlar, high-density polyethylene, and natural rubber. The thermal conductivity of graphene is higher than all these

Table 10-3 Properties of Graphene, Carbon Nanotube, Nanosized Steel, and Polymers

Materials	Tensile strength	Thermal conductivity (W/m k) at room temperature	Electrical conductivity (S/m)	Refs
Graphene	130 ± 10 GPa	$(4.84 \pm 0.44) \times 10^3$ to $(5.30 \pm 0.48) \times 10^3$	7,200	[134]
Carbon nanotube	60–150 GPa	3,500	3,000–4,000	[135]
Nanosized steel	1,769 GPa	5–6	1.35×10^6	[136]
Plastic (high-density polyethylene)	18–20 GPa	0.46–0.52	Insulator	[137]
Rubber (nature rubber)	20–30 GPa	0.13–0.142	Insulator	[138]
Fiber (Kevlar)	3,620 MPa	0.04	Insulator	[139]

materials. The electrical conductivity of graphene is also higher than these materials except for steel [133].

10.2.5.3. Surface Treatment of Graphene

Pristine graphene materials are unsuitable for intercalation by large species, such as polymer chains, because graphene as a bulk material has a pronounced tendency to agglomerate in a polymer matrix [140]. It is likely that oxidation followed by chemical functionalization is an essential factor to facilitate the dispersion of graphene and stabilize graphene without agglomeration [140]. The functional groups attached to graphene can be small molecules or polymer chains. The chemical functionalization of graphene is a particularly attractive target because it can improve the solubility and processability as well as enhance the interactions with organic polymers. Considerable works have been carried out on the amination, esterification, isocyanate modification, and polymer wrapping as routes for the functionalization of graphene. The electrochemical modification of graphene using ionic liquids has also been reported. The general approaches for preparing organo-modified graphene areas as follows. The general approaches for preparing organo-modified graphene areas shown in Fig. 10-16 [141].

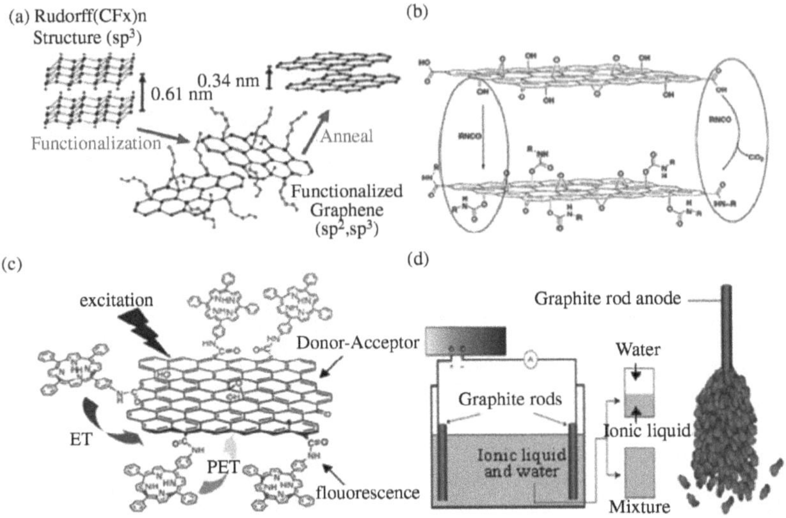

Fig. 10-16 (a) Production of functionalized graphene layers from graphite fluoride and alkyl lithium reagent. (b) Schematic showing the modification of graphene via the covalent attachment of a porphyrin ring on to the graphene oxide surfaces. (c) Schematic showing the modification of graphene oxide using organic isocyanate in dimethy formaldehyde medium. (d) Experimental set-up (left) diagram for the electrochemical modification of graphene by ionic liquid sand exfoliation of the graphite anode (right) to graphene.

10.2.5.4. Graphene–Polymer Nanocomposites

The development of a nanolevel dispersion of graphene particles in a polymer matrix has opened a new and interesting area in materials science in recent years. These nanohybrid materials show considerable improvement in properties that cannot normally be achieved using conventional composites or virgin polymers. The extent of the improvement is related directly to the degree of dispersion of the nanofillers in the polymer matrix. The most important aspect of these nanocomposites is that all these improvements are obtained at very low filler loadings in the polymer matrix [141]. Different types of nano graphite forms, such as expanded graphite and exfoliated graphite, have also been used to produce conducting nanocomposites with improved physicochemical properties.

The superior properties of graphene compared to polymer are also reflected in graphene/polymer nanocomposites which show superior mechanical, thermal, gas barrier, electrical, and flame retardant properties compared to the neat polymer. It was also reported that the improvement in mechanical and electrical properties of graphene-based polymer nanocomposites are much better in comparison to that of clay or other carbon filler-based polymer nanocomposites.

There are many studies on expanded and exfoliated graphite composites based on a range of polymers, including epoxy, poly(methyl methacrylate), polypropylene, linear low density polyethylene, high-density polyethylene, polystyrene, poly(phenylene sulfide), Nylon, polyaniline, phenylethynyl-terminated polyimide, and silicone rubber.

Although carbon nanotubes show comparable mechanical properties compared to graphene, still graphene is better nanofiller than carbon nanotubes in certain aspects, such as thermal and electrical conductivity. However, the improvement in the physicochemical properties of the nanocomposites depends on the distribution of graphene layers in the polymer matrix as well as interfacial bonding between the graphene layers and polymer matrix. Interfacial bonding between graphene and the host polymer dictates the final properties of the graphene-reinforced polymer nanocomposites. Pristine graphene is not compatible with organic polymers and does not form homogeneous composites. In contrast, graphene oxide sheets are heavily oxygenated graphene (bearing hydroxyl, epoxide, diol, ketone, and carboxyl functional groups) that can alter the van der Waals interactions significantly and be more compatible with organic polymers [142].

There are also some additional carbonyl and carboxyl groups located at the edge of the sheets, which makes graphene oxide sheets strongly hydrophilic, allowing them to readily swell and disperse in water [143]. For this reason, graphene oxide has attracted considerable attention as nanofiller for polymer nanocomposites. However, graphene oxide sheets can only be dispersed in aqueous media, which are incompatible with most organic polymers. In addition, unlike graphene, graphene oxide is electrically insulating, which makes it unsuitable for the synthesis of conducting nanocomposites. The surface modification of graphene is an essential step for obtaining a molecular level dispersion of individual graphene in a polymer matrix. The electrical conductivity of the resulting nanocomposites can be increased by the chemical reduction of graphene oxide, presumably by restoring the graphitic network of sp^2 bonds [141, 144, 145].

Graphite, inexpensive and available in large quantity, unfortunately does not readily exfoliate to yield individual graphene sheets. Ruoff *et al.* [125] present a general approach for the preparation of graphene–polymer composites via complete exfoliation of graphite [146] and molecular-level dispersion of individual, chemically modified graphene sheets within polymer hosts. Polystyrene–graphene composites formed by this route exhibits a percolation threshold [147] of ~0.1 vol% for room-temperature electrical conductivity, the lowest reported value for any carbon-based composites except for those involving carbon nanotubes [148]; at only 1 vol%, these composites have a conductivity of ~0.1 S/m, sufficient for many electrical applications [149].

Graphite oxide is a layered material produced by the oxidation of graphite (Fig. 10-17(a)) [98]. In contrast to pristine graphite, the graphene derived sheets in graphite oxide (graphene oxide sheets) are heavily oxygenated, bearing hydroxyl and epoxide functional groups on their basal planes, in addition to carbonyl and

Fig. 10-17 Process flow of graphene–polymer composite fabrication.

carboxyl groups located at the sheet edges [150, 151]. The presence of these functional groups makes graphene oxide sheets strongly hydrophilic, which allows graphite oxide to readily swell and disperse in water [124, 152, 153].

In fact, at present, exfoliation of graphite oxide is the only way to produce stable suspensions of quasi-2D carbon sheets, making this a strategic starting point for large-scale synthesis of graphene sheets. As such, graphite oxide has recently attracted attention as a filler for polymer nanocomposites [154–156].

Nanocomposites have been heralded [157] as a 'radical alternative to conventional filled polymers or polymer blends,' especially in the fields of transportation and electronics. Unfortunately, owing to their hydrophilic nature, graphene oxide sheets can only be dispersed in aqueous media that are incompatible with most organic polymers. In addition, graphite oxide is electrically insulating, unlike graphite, which limits its usefulness for the synthesis of conductive nanocomposites.

It has been demonstrated [143, 152, 158, 159], however, that the electrical conductivity of graphite oxide can be significantly increased by chemical reduction, presumably owing to the restoration of a graphitic network of sp^2 bonds. But reduction of exfoliated graphene oxide nanoplatelets in water results in their irreversible coagulation [143], which then makes dispersion within a polymer matrix at the individual sheet level impossible.

The exfoliation behavior of graphite oxide can be altered by changing the surface properties of graphene oxide sheets by way of chemical functionalization.

A number of chemically modified graphite oxides were prepared by treating graphite oxide with organic isocyanates [160]. The isocyanate treatment reduces the hydrophilic character of graphene oxide sheets by forming amide and carbamate ester bonds to the carboxyl and hydroxyl groups of graphite oxide, respectively. As a result, such isocyanate-derivatized graphite oxides no longer exfoliate in water but readily form stable dispersions in polar aprotic solvents (such as *N,N*-dimethylformamide), consisting of completely exfoliated, functionalized individual graphene oxide sheets with thickness ~1 nm, as determined by atomic force microscopy (Fig. 10-17(c)). These dispersions of isocyanate-derivatized graphite oxide allow graphene oxide sheets to be intimately mixed with many organic polymers, facilitating synthesis of graphene–polymer composites.

The quality of nanofiller dispersion in the polymer matrix directly correlates with its effectiveness for improving mechanical, electrical, thermal, impermeability, and other properties. The properties of the composites are also intimately linked to the aspect ratio and surface-to-volume ratio of the filler. The potential properties of graphene sheet-based composites thus appear promising owing to the extremely high aspect ratios of the sheets as determined from scanning electron microscopic images (Fig. 10-17(g)), in which the average lateral dimension was estimated to be ~1 μm, similar to the values found by atomic force microscopy (Fig. 10-17(b) and (c)). However, in contrast to the typically flat sheets seen by atomic force microscopy when deposited onto atomically flat substrates, the sheets in the composites are crumpled and wrinkled, and at times folded.

Scanning electron microscopic image analysis was used to estimate the apparent graphene sheet surface area at different concentrations (Fig. 10-18 (a)–(d)) [98]. It revealed a surface area lower by roughly a factor of two than the calculated values for each concentration. The nanoscale corrugation of the graphene sheets may contribute to the discrepancy.

At present, research in nanocomposite materials using carbon based nanofillers is dominated by carbon nanotubes. The electrical properties of graphene-based composites compare well with the best values reported in the literature for nanotube–polymer composites. The percolation threshold of around 0.1 vol% is almost the same as was achieved for single-walled carbon nanotubes–polystyrene composites made by latex technology [161], and just twice and four times higher than those measured for single-walled carbon nanotubes–polyimide [145, 161] and poly(phenyleneethynylene)-functionalized single-walled carbon nanotubes–polystyrene composites [162], respectively. Absolute conductivities of graphene–polystyrene composites are also essentially the same as the values reported for single-walled carbon nanotubes–filled polystyrene composites [161]. Furthermore, graphene sheets have higher surface-to-volume ratios than single-walled carbon nanotubes owing to the inaccessibility of the inner nanotube surface to polymer molecules. This makes graphene sheets potentially more favorable for altering all matrix properties such as the mechanical, rheological, permeability properties, and degradation stability. Also, single-walled

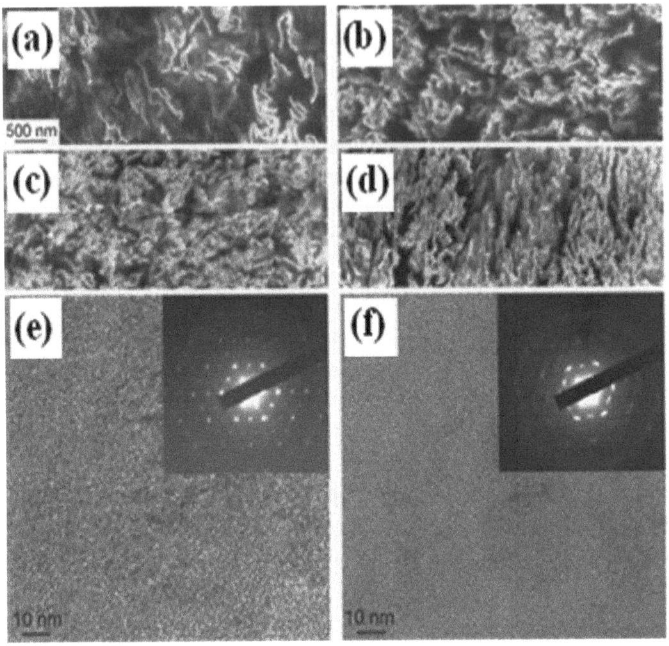

Fig. 10-18 Scanning electron microscopic and transmission electron microscopic images of graphene–polystyrene composites.

carbon nanotubes are still much more expensive than graphite, which is a commodity material with about 1,000,000 ton used annually, and which is sold for a few US dollars per kilogram [163]. In contrast, graphite oxide can be readily prepared from a vast array of graphites, and its industrial production has also recently become a topic of interest [164].

The conductive nature of the graphene filler motivated the study of the percolation behavior through electrical measurements. All measurements and subsequent analysis revealed minor anisotropy, with the in-plane conductivity, $\sigma_{\parallel}$, always slightly higher than the transverse conductivity, $\sigma_{\perp}$. A rapid increase in the direct current electrical conductivity of composite materials takes place when the conductive filler forms an infinite network of connected paths through the insulating matrix.

When the filler particles are rigid bodies, the conductivity of such media is typically described with a bond percolation model [144]. The conductivity of the composites, σ_c, above the percolation threshold is then treated with a power law [165]:

$$\sigma_c = \sigma_f \left[\frac{\phi - \phi_c}{1 - \phi_c}\right]^t \tag{10-1}$$

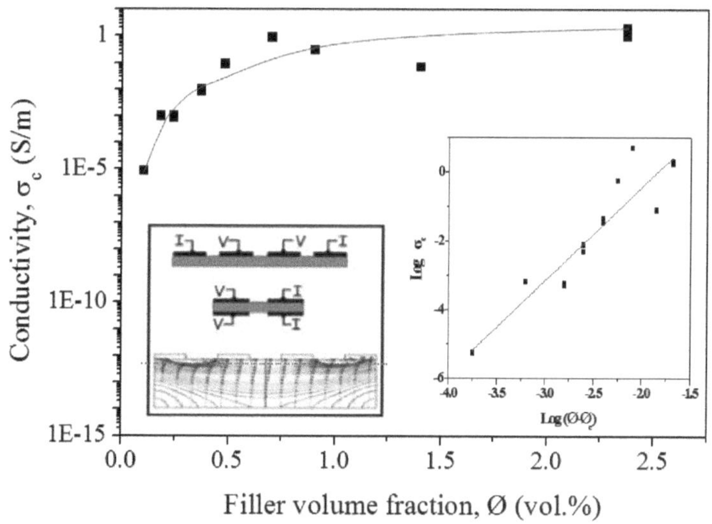

Fig. 10-19 Electrical conductivity of the polystyrene–graphene composites as a function of filler volume fraction.

where σ_f is the conductivity of the filler, ϕ the filler volume fraction, ϕ_c the percolation threshold (the onset of the transition), and t is the 'universal critical exponent.'

Percolation in polystyrene–graphene composites occurs when the filler concentration, ϕ_c, is near 0.1 vol% (Fig. 10-19) [112, 141]. This percolation threshold is about three times lower than reported for any other 2D filler [166]. Such a low percolation threshold for a 3D isotropic case is evidently due to the extremely high aspect ratio of the graphene sheets and their excellent homogeneous dispersion in these composites.

Kuila *et al.* examined the level of thermal expansion for graphene oxide sheet-incorporated epoxy composites, which were prepared by *in situ* process [167]. The epoxy resin showed very poor thermal conductivity but the inclusion of graphene sheets resulted in significant improvements. The addition of 1 wt.% graphene oxide to the epoxy resins had a similar effect on improving the thermal conductivity to that of filling with 1 wt.% of single-walled carbon nanotubes. A 5 wt.% graphene oxide-filled epoxy resin showed a thermal conductivity of ~1 W/m K, which is 4 times higher than that of the neat epoxy resin. These results are consistent with the values reported in the literature [168].

In general, polystyrene/isocyanate-modified graphene composites were prepared using a solution blending method with dimethyl formaldehyde as a solvent. Reduction of the composites was accomplished by dimethyl hydrazine at 80 °C for 24 h. The composites were coagulated by the dropwise addition of dimethyl formaldehyde solution into a large volume of vigorously stirred

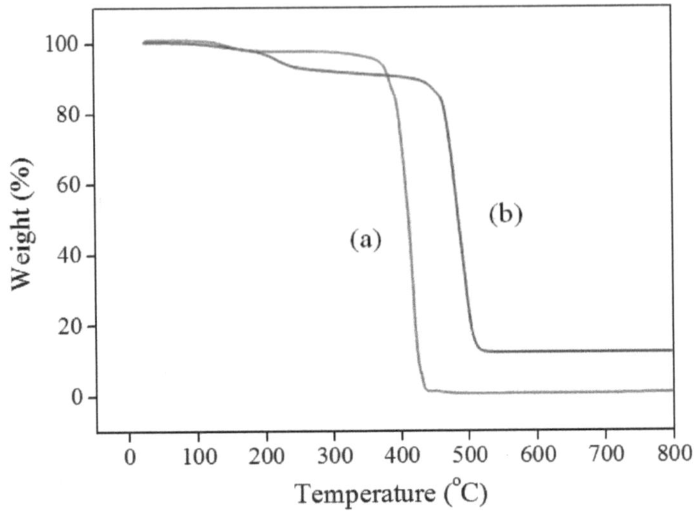

Fig. 10-20 Thermogravimetric analysis (heating rate 10 °C/min, N_2 flow 100 mL/min) curves of (a) polystyrene and (b) polystyrene–graphene nanoparticle composites. The composites show higher thermal stability compared to pristine polystyrene.

methanol (10:1 with respect to the volume of dimethyl formaldehyde used). The composites appeared to be filled almost completely with graphene sheets at a filler loading of only 2.4 vol% due to the large surface area of modified graphene.

Fig. 10-20 shows the thermal stability of the polystyrene–graphene nanoparticle composites and pure polystyrene. The second step degradation temperature of the polystyrene–graphene nanoparticle composites was ~50 °C higher than that of neat polystyrene. This suggests a strong interaction between the polymer matrix and graphene nanoparticle at the interface, which could decrease the mobility of the polymer chains near the interface and increase the thermal stability of the nanocomposites.

Chhowalla *et al.* [169] prepared polystyrene–functionalized graphene sheet composite thin films using a solution blending method. The composite thin films were semiconducting in nature and exhibited an ambipolar field effect. The composite thin films were electrically conducting with a conductivity ranging from 1 to 24 S/m, which is good in agreement with the values reported for bulk composites. The conductivity of the thin films decreased with decreasing temperature to approximately 50 K and increased slightly with further decreases in temperature. Such an anomalous behavior was also observed in the case of graphitic flakes and reduced graphene oxide thin films. It is likely that electron–phonon scattering limits the carrier transport at temperatures <50 K. However, the composite materials become increasingly p-type at higher temperatures. The carrier mobility is limited by the disorder of each functionalized

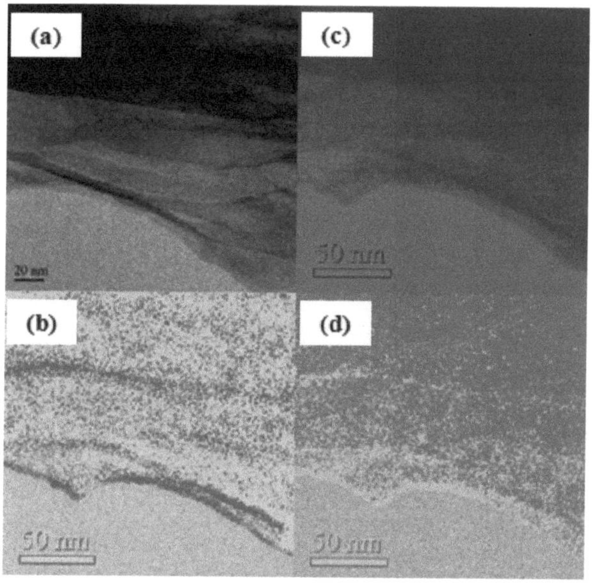

Fig. 10-21 Electron energy loss spectroscopic mapping images of the polyaniline–graphene composite paper (GPCP) – 900 s: (a) zero-energy loss bright field image, (b) nitrogen element mapping, (c) carbon element mapping, and (d) oxygen element mapping.

graphene sheet and sheet-to-sheet junction. Further optimization of the material parameters and reduction conditions are expected to result in better performing devices, rivaling their organic counterparts.

Freestanding and flexible polyaniline/graphene composite paper was prepared by the *in situ* anodic electro polymerization of aniline on graphene paper [141, 170]. The morphological characterization of polyaniline/graphene composite paper was performed using electron energy loss spectroscopy. Fig. 10-21 (a)–(d) shows the distribution of nitrogen, carbon, and oxygen over the entire area of the composite sheet which indicated the presence of homogeneous polyaniline films on an individual graphene sheet. This suggests that the distribution of polyaniline on individual 2D graphene is homogeneous but inhomogeneous in the 3D structure of polyaniline/graphene composite paper.

Fig. 10-22 (a)–(d) [141, 170] shows the cyclic voltammetry results of graphene paper and polyaniline/graphene composite paper. In the case of graphene paper, only one pair of redox peaks was observed due to the transition between quinone/hydroquinone groups. On the other hand, two couples of redox peaks were observed from polyaniline/graphene composite paper indicating the presence of pseudocapacitive polyaniline. The Nyquist plot displayed in Fig. 10-22(c) shows that polyaniline/graphene composite paper has higher redox resistance than graphene paper. This was attributed to the agglomeration of the lower conducting polyaniline. However, the cycling

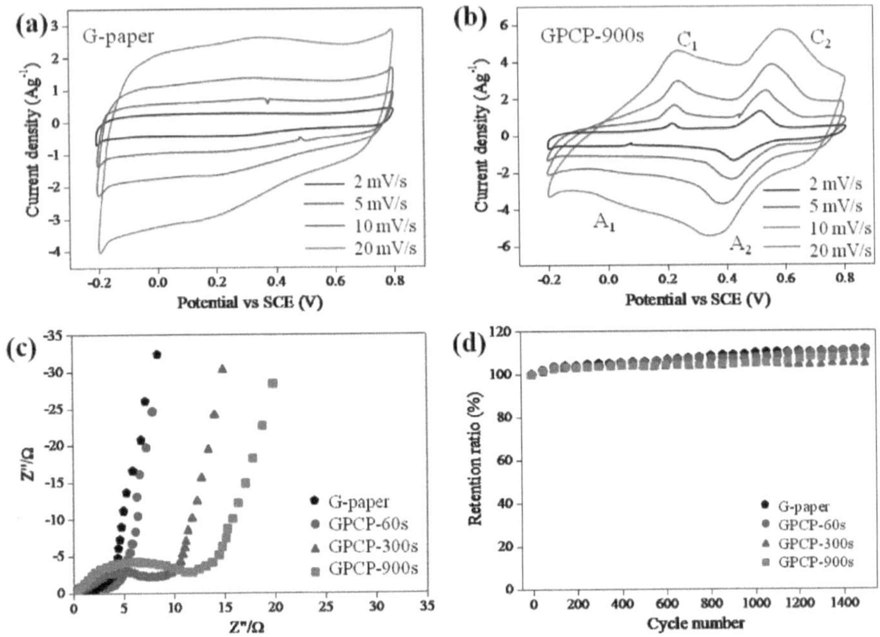

Fig. 10-22 Electrochemical properties of the graphene paper and polyaniline/graphene composite paper (a and b). Cyclic voltammograms recorded from 2 to 20 mV/s in 1 M H_2SO_4. (c) Nyquist plots of the graphene paper (GP) and polyaniline/graphene composite paper (GPCP) – 60 s/300 s/900 s. (d) Cycling stability measured at 50 mV/s.

stability of polyaniline/graphene composite paper was much better than graphene paper (Fig. 10-22 (d)).

10.3. PROCESSING OF NANOCOMPOSITES

10.3.1. Introduction

To maximize the advantage of carbon nanotubes as effective reinforcement for high strength polymer composites, the carbon nanotubes should not form aggregates, and must be well dispersed to enhance the interfacial interaction with the matrix. Several processing methods available for fabricating carbon nanotube/polymer composites based on either thermoplastic or thermosetting matrices have been described in past review articles [171–174]. They mainly include solution mixing, *in situ* polymerization, melt blending, and chemical modification processes. Although inherently different processing routes have been attempted, they all address important issues that affect composite properties, such as exfoliation of carbon nanotube bundles and ropes, homogeneous dispersion of the individual tubes into the matrix, alignment and interfacial bonding.

The effective utilization of carbon nanotube materials in composite applications depends strongly on their ability to be dispersed individually and homogeneously within a matrix. An optimized interfacial interaction between the carbon nanotube sidewalls and the matrix should result in an efficient load transfer to the 'hard' component of the composite. All processing methods known thus far for carbon nanotube-based composites are described below. Due to the huge volume of related literature, only the seminal works will be discussed for each processing method.

10.3.2. Solution Processing of Carbon Nanotube and Polymer

The most common method for preparing carbon nanotube–polymer composites has been to mix both components into a certain solvent and evaporate the latter to form a composite film. The general protocol for all solution processing methods includes the dispersion of carbon nanotube powder in a liquid medium by vigorous stirring and/or sonication, mixing the carbon nanotube dispersion with a polymer solution and controlled evaporation of the solvent with or without vacuum conditions. In general, the most efficient dispersion of tubes during the first step is achieved by bath or tip sonication.

Regarding thermosetting epoxy matrices, an early example of solution-based composite formation was described by Ajayan *et al.* [175]. In this work, multi-walled carbon nanotubes produced by arc discharge were dispersed in ethanol by sonication and mechanically mixed with a mixture of epoxy monomer and curing agent. After evaporation of the volatile solvent, the carbon nanotube–epoxy mixture was poured into capsular molds and then cured. Electron microscopy images showed that slicing the composite material caused partial alignment of the carbon nanotubes on the cut surface.

Concerning the fabrication of carbon nanotube-based composites with thermoplastic matrices, Jin *et al.* [176] have reported a similar method consisting of solution mixing, casting, and drying. The chosen polymer, polyhydroxyaminoether was dissolved in the arc discharge-grown carbon nanotubes in chloroform under sonication. The suspension was then poured into a Teflon mold and dried under ambient conditions for a certain time. Mechanical stretching of thin strips of the composite at 100 °C was found to cause orientation of tubes in the axis of tension. Slight processing variations were used in most subsequent studies. Shaffer and Windle [177] dispersed chemically oxidized catalytic multi-walled carbon nanotubes in water. The material was carefully blended with solutions of poly(vinyl alcohol) in water to give composite dispersions, which could be drop-cast to form films with up to 60 wt.% nanotubes.

In an alternative processing method, Zhou and coworkers [178] utilized a roll-casting technique to prepare composite films of poly(ethylenoxide) embedded with fluorinated carbon nanotubes. The roll-cast system comprises two opposing parallel rollers with a gap distance that can be adjusted. A suspension of the carbon nanotube/polymer mixtures was slowly dropped onto one roller when rotating, while a solid film was formed after evaporation of the solvent. The resulting blends showed enhanced mechanical properties due to the efficient load transfer between the matrix and the functionalized carbon nanotube sidewalls (see below). In order to succeed rapid evaporation of solvent during the suspension casting process, Safadi *et al.* [179] have used the spin coating technique. The authors elucidated the basic relationships between simple-casting and spin-casting conditions, concluding that the processing method employed had no effect on the mechanical/electrical properties of the composites.

Since the casting/evaporation process results in the reagglomeration of carbon nanotubes within the composite film, Winey and co-workers [180] proposed an alternative approach for fabricating composites with individually dispersed carbon nanotubes, the so called coagulation. After the solution mixing step, the carbon nanotube/poly(methyl methacrylate) suspension was dripped into a large excess of bad solvent (water) in order to induce instant precipitation of the polymer chains. The precipitating chains entrapped the carbon nanostructures and prevented them from bundling again. After filtration and drying in vacuum, nanocomposites were obtained with homogeneously distributed carbon nanotubes.

In order to periodically functionalize carbon nanotube sidewalls using controlled polymer crystallization process, the hot coagulation method was developed for the fabrication of carbon nanotube/polyethylene composites [181, 182]. First, the polymer was dissolved in an organic solvent at temperatures approaching its melting point. Subsequently, the carbon nanotube materials were suspended in the same medium using ultrasonication in a hot bath and the hot polymer solution was added to the nanotube suspension. After further sonication, the mixture was brought to the crystallization temperature T_c at which point the polymer crystallized on the carbon nanotube surface, forming shish-kebab structures.

10.3.3. Bulk Mixing

Milling is a mechanical process that leads to local generation of high pressure as a result of collisions throughout the grinding media. Concerning applications in carbon nanotube nanotechnology, this method has been used to shorten the lengths of carbon nanostructures [183]. A solid state mechanochemical pulverization process, namely pan milling, was used to prepare a carbon nanotube–polypropylene composite powder [184]. This powder was subsequently melt-mixed with a twin-roll masticator to obtain a homogeneous composite. The length of the carbon nanotubes was reduced from a few micrometers to $\sim$500 nm.

Similarly, high energy ball milling was utilized to incorporate carbon nanotubes into polymer matrices [185]. In this way, a satisfactory level of dispersion of carbon nanotubes into the polymer matrix was obtained, resulting in an improvement of the physical properties of the samples. In addition, a mechanical–chemical high-speed vibration milling technique was applied for solubilizing carbon nanotubes in various media due to the formation of noncovalent-type complexes with species, such as cyclodextrins [186].

10.3.4. Melt Mixing

Due to the fact that thermoplastic semicrystalline polymers soften when heated above their melting point, melt processing has been a very valuable technique for the fabrication of carbon nanotube-based composites. In addition, the method is suitable for polymers that cannot be processed with solution techniques due to their insolubility in common solvents.

In general, melt processing involves the blending of polymer melt with carbon nanotube materials by application of intense shear forces. Depending on the final morphology/shape of the composites, the bulk samples can then be processed by several techniques, such as extrusion [187]. An early study on the melt mixing of single-walled carbon nanotubes and poly(methyl methacrylate) matrix was carried out by Winey and coworkers [188]. Carbon nanotubes and poly(methyl methacrylate) were first blended in dimethylformamide and the resulting suspension was casted onto Teflon dishes and dried. The resultant films were broken up in small pieces and hot pressed to form a new film. This was then broken up and hot pressed, a process that was repeated as many as 25 times. The authors observed that the carbon nanotube dispersion improved with each melting step. The final composite film was then extruded through a melt spinning apparatus with a spinneret hole of 600 μm diameter to form a fiber. In the final step, the composite fibers were melt spun to achieve draw ratios between 20 and 3600. Due to the alignment of tubes along the fiber axis, the elastic modulus and yield strength of the composite fibers increased with nanotube loading and draw ratio (see below).

Using a different protocol for bulk composites, Goh and co-workers [189] have melt-blended multi-walled carbon nanotubes and poly(methyl methacrylate) in a laboratory mixing molder at a speed of 120 rpm (blending temperature ~200 °C). The mixed samples were then compressed under pressure at 210 °C using a hydraulic press to yield composite films. Electron microscopy images showed that the nanotubes are well dispersed in the polymer matrix. Concerning the upscaled fabrication of composites, Potschke *et al.* [190] have fabricated 1 kg batches of multi-walled carbon nanotube/polycarbonate composites by melt extrusion and compression molding. Rheology measurements of the composites at 260 °C showed that viscosity increases significantly with increasing carbon nanotube concentration, especially for carbon nanotube weight fractions above 2%.

10.3.5. *In Situ* Polymerization

Since 1999, *in situ* polymerization of vinyl monomers in the presence of carbon nanotube materials has been extensively studied for the preparation of functional composites. The main advantage of this method is that it produces polymer-grafted tubes, mixed with free polymer chains. Moreover, due to the small size of monomeric molecules, the homogeneity of the resulting composite adducts is much higher than mixing carbon nanotubes and polymer chains in solution. In this sense, the method allows the preparation of composites with high carbon nanotube weight fraction.

Initially, *in situ* radical polymerization was applied for the synthesis of poly (methyl methacrylate) composites by Jia *et al.* [191]. In this work, the reaction was performed using radical initiator 2,2'-azobis isobutyronitrile. The authors suggested that π-bonds of the carbon nanotube graphitic network were opened by the radical fragments of initiator and therefore the carbon nanostructures could participate in poly(methyl methacrylate) polymerization by acting as efficient radical scavengers. If all reactants were mixed simultaneously, the growth of polymer chains was inhibited since many of the initiator molecules would be consumed by the carbon nanotubes. By using an improved *in situ* process [191, 192], in which carbon nanotube materials is added a little time after the mixing of initiator and monomer, longer polymer chains were obtained in the resulting composite, giving rise to better mechanical properties of the composites (see below).

Almost simultaneously with the work of Jia in 1999, two independent groups have studied the *in situ* radical polymerization of pyrrole [193] and phenylacetylene [194] in the presence of carbon nanotube materials. Spectroscopic characterization of the composites indicated no significant chemical attachment of polymer chains onto the graphitic sidewalls, but rather a helical wrapping mode [195]. Such a non-covalent interaction gave rise to attenuated fluorescence quantum yield in the carbon nanotube/polymer blend. Concerning the processability, the resulting composites exhibited enhanced dispersability in various organic solvents, whereas the electrical, magnetic, and thermal properties of the carbon nanotubes were modified by the conducting polymer chains [193].

As an alternative approach, efficient doping of carbon nanotube sidewalls by conducting polymers has been studied by *in situ* electrochemical means [195, 196]. Microscopy images showed a remarkably uniform polymer coating on individual carbon nanotubes, which paves the way for controlled modification of the outer surface of carbon nanotubes with selected functionalities. Moreover, the observation of joining individual carbon nanotubes by the conducting polymer chains in the composite film was suggested to have great potential in applications such as carbon nanotube-based optoelectronic devices.

Concerning the preparation of anisotropic carbon nanotube–polymer composites, Kimura *et al.* [197] have mixed styrene monomer with nanotubes and subjected the suspension to a constant magnetic field of 10 T. By polymerizing the mixture, the nanotubes were found to be kept aligned within the polymer matrix. Analogous experiments with an epoxy thermoset as a matrix showed that the thermal and electrical properties of the composites were significantly enhanced by magnetic alignment during processing [198]. In an analogous work, Bauhofer and co-workers [199] dispersed carbon nanotubes in an epoxy system based on a bisphenol-A resin and an amine hardener. The application of both alternating current and direct current electric fields during nanocomposite curing was used to induce the formation of aligned conductive nanotube networks between the electrodes. The network structure formed in alternating current fields was found to be more uniform and more aligned compared to that in direct current fields.

Concerning the fabrication of carbon nanotube-based silicone elastomer composites, Liu *et al.* [200] dispersed the graphitic nanostructures into a viscous mixture of monomers by mechanical grinding of the components. Most probably, the thermal energy given by the grinding process caused the efficient condensation of the precursor substances. The thermal conductivities were found to increase with the carbon nanotube amount. There was a 65% enhancement with 3.8 wt.% carbon nanotube loading.

Similarly, Gojny *et al.* [201] studied the dispersion of carbon nanotubes into viscous epoxy monomers via a common shear mixing technique, which is, calendering. In a first step, the nanoparticles were manually mixed into the resin while in the following step they were homogeneously dispersed using a three-roll calender. A first primary dispersion of the agglomerates could be achieved by the knead-vortex between the rolls, while the final exfoliation and dispersion of the carbon nanotubes occurs in the area between the rolls. The suspension was collected and then mixed with the curing agent to obtain homogeneous composites.

10.4. CHARACTERIZATION OF NANOCOMPOSITES

Characterization tools are crucial to comprehend the basic physical and chemical properties of polymer nanocomposites. For structural applications, it facilitates the study of emerging materials by giving information on some intrinsic properties [202–206]. Various techniques for characterization have been used extensively in polymer nanocomposite research. The commonly used powerful techniques are wide-angle X-ray diffraction, small-angle X-ray scattering, scanning electron microscopy, and transmission electron microscopy [207–213]. The scanning electron microscopy provides images of surface features associated with a sample. However, there are two other microscopies, scanning probe microscopy and scanning tunneling microscopy, which are indispensable in nanotube research. The scanning probe microscopy uses the interaction between a sharp tip and a surface to obtain an image. In scanning tunneling microscopy, a sharp conducting tip is held sufficiently close to a surface (typically about 0.5 nm), such that electrons can 'tunnel' across the gap. This method provides surface structural and electronic information at atomic level. The invention of the scanning tunneling microscopy inspired the development of other 'scanning probe' microscopes, such as the atomic force microscopy. The atomic force microscopy uses a sharp tip to scan across the sample. Raman spectroscopy has also proved a useful probe of carbon-based material properties.

Due to the easiness and availability, wide-angle X-ray diffraction is most commonly used to probe the nanocomposite structure and occasionally to study the kinetics of the polymer melt intercalation. In layered silicate nanocomposite systems, a fully exfoliated system is characterized by the absence of intensity peaks in wide-angle X-ray diffraction pattern, e.g., in the range $1.5° \leq 2\theta \leq 10°$, which corresponds to a *d*-spacing of at least 6 nm. Therefore, a wide-angle X-ray diffraction pattern concerning the mechanism of nanocomposite formation and their structure are tentative issues for making any conclusion. On the other hand, transmission electron microscopy allows a qualitative understanding of the internal structure, spatial distribution of the various phases, and views of the defective structure through direct visualization, in some cases of individual atoms. Therefore, transmission electron microscopy complements wide-angle X-ray diffraction data. Small-angle X-ray scattering is typically used to observe structures on the order of 10Å or larger, in the range of 0° or 0.5–5°. The transmission electron microscopy, atomic force microscopy, and scanning electron microscopy, are also required to characterize nanoparticle, carbon nanofiber dispersion, or distribution. However, X-ray diffraction has found relatively limited success in carbon nanotube research. For thermal characterization and to study the cure behavior (typically for thermoset resin systems) of polymer nanocomposites, the commonly used techniques are differential scanning calorimeter, thermogravimetric analysis, thermomechanical analysis, Fourier-transform infrared, dynamic modulus analysis, rheometer, etc.

10.5. SUMMARY

The potential of nanocomposites in various sectors of research and application is promising and attracting increasing investment from Governments and business in many parts of the world. While there are some niche applications where nanotechnology has penetrated the market, the major impact will be at least a decade away.

At present, it is difficult to predict which, if any, market sector would not be able to benefit from this technology. Thus, it may well be that polymeric nanocomposites in the mid- and longer-term will pervade all aspects of life, similar to the way plastics did in the last century. Clearly a diverse range of sectors such as aerospace, automotive, packaging (particularly food but also solar cells), electrical and electronic goods, household goods, etc. will profit substantially from a new range of materials which would be offered by this technology. In the short term (<5 years), the commercial impact may include inkjet markets, nanoparticles in cosmetics, and automotive applications such as body moldings, engine covers and catalytic converts, batteries, computer chips. In the mid-term (10 years), memory devices, biosensors for diagnostics, advances in lighting are all possible. The time-scale for automotive, aerospace, bio-nanotechnology is a long-term prospect (>15 years) as these are risk-averse sectors and thereby for large-scale production it is necessary to carry out strict testing and validation procedures.

An additional area not discussed in detail in this chapter involves the importance of property control, which includes both dispersion and alignment. These issues have been discussed in many of the papers. Obtaining the optimum properties for nanocomposites will usually require excellent dispersion of the nanoparticles. The tendency for nanoparticles (including platelets and fibers of nanoscale dimensions) to coalesce into macrosize agglomerates can seriously impact the achievable properties.

REFERENCES

[1] T.J. Pinnavaia, G.W. Beall, Polymer–Clay Nanocomposites. John Wiley & Sons, New York (2000).
[2] S. Yariv, H. Cross, Organo-Clay Complexes and Interactions. Marcel Dekker, New York (2002).
[3] S.J. Park, M.K. Seo, Roles of Interfaces on Physicochemical Properties of Carbon Nanotubes/Epoxy Matrix Composites, in Advanced Research in Polymer ScienceTransworld Research Network, India (2006).
[4] M.K. Seo, S.J. Park, Journal for Nanoscience and Nanotechnology 9 (2009) 7186.
[5] C. Harrats, G. Groeninckx, Macromolecular Rapid Communications 29 (2008) 14.
[6] M.K. Seo, J.R. Lee, S.J. Park, Materials Science and Engineering A 404 (2005) 79.
[7] Y.L. Loo, R.A. Register, A.J. Ryan, Macromolecules 35 (2002) 2365.
[8] M. Alexandre, P. Dubois, Materials Science and Engineering R: Reports 28 (2000) 1.
[9] J. Wen, G.L. Wilkes, Chemistry of Materials 8 (1996) 1667.
[10] G.L. Wilkes, B. Orler, H. Huang, Polymer Preprints 26 (1985) 300.
[11] J.E. Mark, C.Y. Jiang, M.Y. Tang, Macromolecules 17 (1984) 2613.
[12] A. Kelly, Composites Science and Technology 23 (1985) 171.
[13] National Science and Technology Council, Nanotechnology: Shaping the World Atom by Atom. Committee on Technology, Interagency Working Group on Nanoscience, Engineering and Technology, Washington, District of Columbia (1999).
[14] Y. Long, R.A. Shanks, Journal of Applied Polymer Science 61 (1996) 1877.
[15] Z. Bartczak, A.S. Argon, R.E. Cohen, M. Weinberg, Polymer 40 (1999) 2347.
[16] R.K.D. Misra, P. Nerikar, K. Bertrand, D. Murphy, Materials Science and Engineering A 384 (2004) 284.
[17] H. Unal, A. Mimarogluamd, M. Alkan, Polymer International 53 (2004) 56.
[18] A. Takahara, T. Magome, T. Kajiyama, Journal of Polymer Science Part B: Polymer Physics 32 (1994) 839.
[19] A.T. Dibenedetto, Materials Science and Engineering A 302 (2001) 74.
[20] P. Potschke, A.R. Bhattacharyya, A. Janke, Polymer 44 (2003) 8061.
[21] P.C. LeBaron, T.J. Pinnavaia, Chemistry of Materials 13 (2001) 3760.
[22] M.F. Yu, O. Lourie, M.J. Dryer, K. Molor, T.F. Kelly, R.S. Ruoff, Science 287 (2000) 637.
[23] M.F. Yu, B.S. Files, S. Arepalli, R.S. Ruoff, Physical Review Letters 84 (2000) 5552.
[24] P. Rudolf, Polymer Degradation and Stability 95 (2010) 369.
[25] A. Usuki, M. Kawasumi, Y. Kojima, Y. Fukushima, A. Okada, T. Kurauchi, Journal of Materials Research 8 (1993) 1179.
[26] Y. Kojima, A. Usuki, M. Kawasumi, Y. Fukushima, A. Okada, T. Kurauchi, Journal of Materials Research 8 (1993) 1185.
[27] E. Giannelis, Advanced Materials 8 (1996) 29.
[28] R. Pfaendner, Polymer Degradation and Stability 95 (2010) 369.

[29] www.nanosconference.com.
[30] D. Vink, Eur. Plast. News 36 (2008).
[31] A. Michael, P. Dubois, Materials Science and Engineering A 28 (2000) 1.
[32] T. Naganuma, Y. Kagawa, Composites Science and Technology 62 (2002) 1187.
[33] R.P. Singh, M. Zhang, D. Chan, Journal of Materials Science 37 (2002) 781.
[34] Lopez, L., Song, B.M.K., Hahn, H.T., The Effect of Particle Size in Alumina Nanocomposites, in Proceedings of the 14th International Conference on Composite Materials (ICCM-14), San Diego (2003).
[35] C.M. Thompson, H.M. Herring, T.S. Gates, J.W. Connell, Composites Science and Technology 63 (2003) 1591.
[36] B.J. Ash, R.W. Siegel, L.S. Schadler, Macromolecules 37 (2004) 1358.
[37] J. Karger-Kocsis, Z. Zhang, Structure–Property Relationships in Nanoparticle/Semi-crystalline Thermoplastic Composites. Mechanical Properties of Polymers Based on Nanostructure and MorphologyMarcel Dekker, New York (2004).
[38] J.J. Luo, I.M. Daniel, Composites Science and Technology 63 (2003) 1607.
[39] H. Fukushima, L.T. Drzal, Graphite Nanocomposites: Structural and Electrical Properties, in Proceedings of the 14th International Conference on Composite Materials (ICCM-14), San Diego (2003).
[40] W. Zheng, S.C. Wong, Composites Science and Technology 63 (2003) 225.
[41] W. Zheng, X.H. Lu, S.C. Wong, Journal of Applied Polymer Science 91 (2004) 2781.
[42] Song, J.H., Huh, H., Hahn, H.T., Stress Evaluation of Nanocomposites with Nanoplatelets, in Proceedings of the 14th International Conference on Composite Materials (ICCM-14), San Diego (2003).
[43] J. Tsai, C.T. Sun, Journal of Composite Materials 38 (2004) 567.
[44] V.I. Merkulov, D.H. Lowndes, Y.Y. Wei, G. Eres, E. Voelkl, Applied Physics Letters 76 (2000) 3555.
[45] M. Endo, Y.A. Kim, T. Hayashi, Y. Fukai, K. Oshida, M. Terrones, T. Yanagisawa, S. Higaki, M.S. Dresselhaus, Applied Physics Letters 80 (2002) 1267.
[46] M. Endo, Y.A. Kim, M. Ezaka, K. Osada, T. Yanagisawa, T. Hayashi, M. Terrones, M.S. Dresselhaus, Nano Letters 3 (2003) 723.
[47] C.Y. Wei, D. Srivastava, Applied Physics Letters 85 (2004) 2208.
[48] S. Iijima, Nature 354 (1999) 56.
[49] M.J. Biercuk, M.C. Llaguno, M. Radosvljevic, J.K. Hyun, A.T. Johnson, Applied Physics Letters 80 (2002) 15.
[50] Z. Ounaies, C. Park, K.E. Wise, E.J. Siochi, J.S. Harrison, Composites Science and Technology 63 (2003) 1637.
[51] M.C. Weisenberger, E.A. Grulke, D. Jacques, T. Rantell, R. Andrews, Journal of Nanoscience and Nanotechnology 3 (2003) 535.
[52] R. Andrews, M.C. Weisenberger, Current Opinion in Solid State and Material Science 8 (2004) 31.
[53] C.Y. Wei, D. Srivastava, Applied Physics Letters 85 (2004) 2208.
[54] J.K.W. Sandler, J.E. Kirk, I.A. Kinloch, M.S.P. Shaffer, A.H. Windle, Polymer 44 (2003) 5893.
[55] P. Kim, L. Shi, A. Majumdar, P.L. McEuen, Physical Review Letters 87 (2001) 2155021.
[56] K.T. Lau, D. Hui, Composites: Part B 33 (2002) 263.
[57] B.G. Demczyk, Y.M. Wang, J. Cumings, M. Hetman, W. Han, A. Zettl A., R.O. Ritchie, Materials Science and Engineering A 334 (2002) 173.

[58] E.T. Thostenson, Z.F. Ren, T.W. Chou, Composites Science and Technology 61 (2001) 1899.
[59] D. Sanchez-Portal, E. Artacho, J.M. Soler, A. Rubio, P. Ordejon, Physical Review B 59 (1999) 12678.
[60] R.A. Vaia, H.D. Wagner, Materials Today 7 (2004) 33.
[61] W.A. Curtin, B.W. Sheldon, Materials Today 7 (2004) 44.
[62] E.T. Thostenson, T.W. Chou, Journal of Physics D: Applied Physics 36 (2003) 573.
[63] E.T. Thostenson, T.W. Chou, Journal of Physics D: Applied Physics 35 (2002) L77.
[64] H.D. Wagner, R.A. Vaia, Materials Today 7 (2004) 38.
[65] S.B. Sinnott, Journal of Nanoscience and Nanotechnology 2 (2002) 113.
[66] Y. Hu, I. Jang, S.B. Sinnott, Composites Science and Technology 63 (2003) 1663.
[67] B. Ni, R. Andrews, D. Jacques, D. Qian, M.B.J. Wijesundara, Y.S. Choi, L. Hanley, S.B. Sinnott, Journal of Physical Chemistry B 105 (2001) 12719.
[68] T.W. Chou, Microstructural Design of Fiber Composites. Cambridge University Press, Cambridge (1992).
[69] H.R. Shetty, T.W. Chou, Metallurgical Transactions A 16 (1985) 853.
[70] E.T. Thostenson, T.W. Chou, Carbon 42 (2004) 3015.
[71] C.Y. Li, T.W. Chou, Mechanics of Materials 36 (2004) 1047.
[72] C.Y. Li, T.W. Chou, Composites Science and Technology 63 (2003) 1517.
[73] A. Shindo, Polyacrylonitrile (PAN)-based Carbon Fibers. Comprehensive Composite Materials Elsevier, Amsterdam (2000), p. 1.
[74] M.F. Yu, B.S. Files, S. Arepalli, R.S. Ruoff, Physical Review Letters 84 (2000) 5552.
[75] E. Hernandez, C. Goze, P. Bernier, A. Rubio, Physical Review Letters 80 (1998) 4502.
[76] C.Y. Li, T.W. Chou, International Journal of Solids and Structures 40 (2003) 2487.
[77] P. Zhang, Y. Huang, P.H. Geubelle, P.A. Klein, K.C. Hwang, International Journal of Solids and Structures 39 (2002) 3893.
[78] B.I. Yakobson, M.P. Campbell, C.J. Brabec, J. Bernholc, Computational Materials Science 8 (1997) 341.
[79] C.A. Martin, J.K.W. Sandler, M.S.P. Shaffer, M.K. Schwarz, W. Bauhofer, K. Schulte, M.S.P. Shaffe, Composites Science and Technology 64 (2004) 2309.
[80] J. Sandler, M.S.P. Shaffer, T. Prasse, W. Bauhofer, K. Schulte, A.H. Windle, Polymer 40 (1999) 5967.
[81] A. Nogales, G. Broza, Z. Roslaniec, K. Schulte, I. Sics, B.S. Hsiao, T.A. Ezquerra, Macromolecules 37 (2004) 7669.
[82] S. Cui, R. Canet, A. Derre, M. Couzi, P. Delhaes, Carbon 41 (2003) 797.
[83] G.D. Zhan, J.D. Kuntz, J.E. Garay, A.K. Mukherjee, Applied Physics Letters 83 (2003) 1228.
[84] C. Wei, K. Cho, D. Srivastava, Applied Physics Letters 82 (2003) 2512.
[85] C. Wei, K. Cho, D. Srivastava, Physical Review B 67 (2003) 1154071.
[86] R. Andrews, M.C. Weisenberger, Current Opinion in Solid State and Material Science 8 (2004) 31.
[87] T. Wang, C.H. Lei, A.B. Dalton, C. Creton, Y. Lin, K.A.S. Fernando, Y.P. Sun, M. Manea, J.M. Asua, J.L. Keddie, Advanced Materials 18 (2006) 2730.
[88] J. Bai, Carbon 41 (2003) 1309.
[89] C.W. Nan, Z. Shi, Y. Lin, Chemical Physics Letters 375 (2003) 666.
[90] K.T. Lau, D. Hui, Composites: Part B 33 (2002) 263.
[91] A.H. Barber, S.R. Cohen, H.D. Wagner, Applied Physics Letters 82 (2003) 4140.
[92] J.B. Donnet, T.K. Wang, J.C.M. Peng, S. Rebouillat, Carbon Fibers, 3rd edit. Marcel Dekker, New York (1998), p. 185.

[93] W. Ding, A. Eitan, F.T. Fisher, X.H. Chen, D.A. Dikin, R. Andrews, L.C. Brinson, L.S. Schadler, R.S. Ruoff, Nano Letters 3 (2003) 1593.
[94] A. Eitan, K. Jiang, D. Dukes, R. Andrews, L.S. Schadler, Chemistry of Materials 15 (2003) 3198.
[95] J. Bae, J. Jang, S.H. Yoon, Macromolecular Chemistry and Physics 203 (2002) 2196.
[96] S.J.V. Frankl, V.M. Harik, Surface Science 525 (2003) L103.
[97] K.T. Lau, Chemical Physics Letters 370 (2003) 399.
[98] S. Stankovich, D.A. Dikin, G.H.B. Dommett, K.M. Kohlhaas, E.J. Zimney, E.A. Stach, R.D. Piner, S.T. Nguyen, R.S. Ruoff, Nature 442 (2006) 282.
[99] K.S. Novoselov, A.K. Geim, S.V. Morozov, D. Jiang, Y. Zhang, S.V. Dubonos, I.V. Grigorieva, A.A. Firsov, Science 306 (2004) 666.
[100] A. Nogales, G. Broza, Z. Roslaniec, K. Schulte, I. Sics, B.S. Hsiao, A. Sanz, M.C. Garcia-Gutierrez, D.R. Rueda, C. Domingo, T.A. Ezquerra, Macromolecules 37 (2004) 7669.
[101] M.D. Stoller, S.J. Park, Y.W. Zhu, J.H. An, R.S. Ruoff, Nano Letters 8 (2008) 3498.
[102] A. Nogales, G. Broza, Z. Roslaniec, K. Schulte, I. Sics, B.S. Hsiao, A. Sanz, M.C. Garcia-Gutierrez, D.R. Rueda, C. Domingo, T.A. Ezquerra, Macromolecules 37 (2004) 7669.
[103] E. Yoo, T. Okata, T. Akita, M. Kohyama, J. Nakamura, I. Honma, Nano Letters 9 (2009) 2255.
[104] A.M. Affoune, B.L.V. Prasad, H. Sato, T. Enoki, Y. Kaburagi, Y. Hishiyama, Chemical Physics Letters 348 (2001) 17.
[105] K.S. Novoselov, A.K. Geim, S.V. Morozov, D. Jiang, Y. Zhang, S.V. Dubonos, Science 306 (2004) 666.
[106] C. Berger, Z. Song, X. Li, X. Wu, N. Brown, C. Naud, D. Mayou, T. Li, J. Hass, A.N. Marchenkov, E.H. Conrad, P.N. First, W.A. de Heer, Science 312 (2006) 254.
[107] A. Reina, X. Jia, J. Ho, D. Nezich, H. Son, V. Bulovic, M.S. Dresselhaus, J. Kong, Nano Letters 9 (2009) 30.
[108] K. Philip, Nature Materials 9 (2010) 792.
[109] R. Ramasubramaniam, J. Chen, H. Liu, Applied Physics Letters 83 (2003) 2928.
[110] A. Reina, X.T. Jia, J. Ho, D. Nezich, H.B. Son, V. Bulovic, Nano Letters 9 (2009) 30.
[111] X. Li, W. Cai, J. An, S. Kim, J. Nah, D. Yang, R. Piner, A. Velamakanni, I. Jung, E. Tutuc, S.K. Banerjee, L. Colombo, R.S. Ruoff, Science 324 (2009) 1312.
[112] S. Bae, H. Kim, Y. Lee, X. Xu, J.S. Park, Y. Zheng, J. Balakrishnan, T. Lei, H.R. Kim, Y.I. Song, Y.J. Kim, K.S. Kim, B. Özyilmaz, J.H. Ahn, B.H. Hong, S. Iijima, Nature Nanotechnology 5 (2010) 574.
[113] O.V. Yazyev, S.G. Louie, Nature Materials 9 (2010) 806.
[114] O.V. Yazyev, S.G. Louie, Physical Review B 81 (2010) 195420.
[115] L. Chico, V.H. Crespi, L.X. Benedict, S.G. Louie, M.L. Cohen, Physical Review Letters 76 (1996) 971.
[116] M.S. Dresselhaus, G. Dresselhaus, P.C. Eklund, Science of Fullerenes and Carbon Nanotubes. Academic Press, New York (1996).
[117] Y. Si, T. Samulski, Nano Letters 8 (2008) 1679.
[118] X. Li, X. Wang, L. Zhang, S. Lee, H. Dai, Science 319 (2008) 1229.
[119] M.J. Allen, V.C. Tung, R.B. Kaner, Chemical Reviews 110 (2010) 132.
[120] R. Saito, G. Dresselhaus, M.S. Dresselhaus, Physical Properties of Carbon Nanotubes. Imperial College Press, London (1998).
[121] A.H. Castro Neto, F. Guinea, N.M.R. Peres, K.S. Novoselov, A.K. Geim, Reviews of Modern Physics 81 (2009) 109.

[122] Z. Liu, K. Suenaga, P.J.F. Harris, S. Iijima, Physical Review Letters 102 (2009) 015501.
[123] M.S. Dresselhaus, G. Dresselhaus, Advances in Physics 51 (2002) 1.
[124] M. Hirata, T. Gotou, S. Horiuchi, M. Fujiwara, M. Ohba, Carbon 42 (2004) 2929.
[125] M.F. Yu, O. Lourie, K. Moloni, T.F. Kelly, R.S. Ruoff, Science 287 (2000) 637.
[126] S.V. Morozov, K.S. Novoselov, A.K. Geim, Uspekhi Fizicheskikh Nauk 178 (2008) 776.
[127] K.S. Novoselov, A.K. Geim, S.V. Morozov, D. Jiang, M.I. Katsnelson, I.V. Grigorieva, S.V. Dubonos, A.A. Firsov, Nature 438 (2005) 197.
[128] Y. Zhang, Y.W. Tan, H.L. Stormer, P. Kim, Nature 438 (2005) 201.
[129] Y. Zhang, J.P. Small, M.E.S. Amori, P. Kim, Physical Review Letters 94 (2005) 176803.
[130] C. Berger, Z. Song, T. Li, X. Li, A.Y. Ogbazghi, R. Feng, Z. Dai, A.N. Marchenkov, E.H. Conrad, P.N. First, W.A. de Heer, Journal of Physical Chemistry B 108 (2004) 19912.
[131] Y. Zhang, T. Tang, C. Girit, Z. Hao, M.C. Martin, A. Zettl, M.F. Crommie, Y.R. Shen, F. Wang, Nature 459 (2009) 820.
[132] S. Latil, V. Meunier, L. Henrard, Physical Review B 76 (2007) 201402.
[133] G. Ventura, V. Martelli, Cryogenics 49 (2009) 735.
[134] J. Liang, Y. Huang, L. Zhang, Y. Wang, Y. Ma, T. Guo, Advanced Functional Materials 19 (2009) 2297.
[135] S.M. Yuen, C.C.M. Ma, C.L. Chiang, J.A. Chang, S.W. Huang, S.C. Chen, Composites: Part A 38 (2007) 2527.
[136] D. Shin, F. Gitzhofer, C. Moreau, Journal of Materials Science 42 (2007) 5915.
[137] Y.C. Li, G.H. Chen, Polymer Engineering & Science 47 (2007) 882.
[138] Y. Sun, Y. Luo, D. Jia, Journal of Applied Polymer Science 107 (2008) 2786.
[139] G. Ventura, V. Martelli, Cryogenics 49 (2009) 735.
[140] Y. Geng, S.J. Wang, J.K. Kim, Journal of Colloid and Interface Science 336 (2009) 592.
[141] T. Kuila, S. Bhada, D. Yao, N.H. Kim, S. Bose, J.H. Lee, Progress in Polymer Science 35 (2010) 1350.
[142] H. Becerril, J. Mao, Z. Liu, M. Stoltenberg, Z. Bao, Y. Chen, ACS Nano 2 (2008) 463.
[143] C. Nethravathi, J.T. Rajamathi, N. Ravishankar, C. Shivakumara, M. Rajamathi, Langmuir 24 (2008) 8240.
[144] I. Jung, D.A. Dikin, R.D. Piner, R.S. Ruoff, Nano Letters 8 (2008) 4283.
[145] B.S. Kong, H.W. Yoo, H.T. Jung, Langmuir 25 (2009) 11008.
[146] S. Stankovich, R.D. Piner, X. Chen, N. Wu, S.T. Nguyen, R.S. Ruoff, Journal of Materials Chemistry 16 (2006) 155.
[147] D. Stauffer, A. Aharnoy, Introduction to Percolation Theory. Taylor & Francis, London, (1991).
[148] Z. Ounaies, C. Park, K.E. Wise, E.J. Siochi, J.S. Harrison, Composites Science and Technology 63 (2003) 1637.
[149] D.D.L. Chung, Journal of Materials Science 39 (2004) 2645.
[150] H. He, J. Klinowski, M. Forster, A. Lerf, Chemical Physics Letters 287 (1998) 53.
[151] A. Lerf, H. He, M. Forster, J. Klinowski, Journal of Physical Chemistry B 102 (1998) 4477.
[152] M. Hirata, T. Gotou, M. Ohba, Carbon 43 (2005) 503.
[153] T. Szabo, A. Szeri, I. Dekany, Carbon 43 (2005) 87.
[154] N.I. Kovtyukhova, P.J. Ollivier, B.R. Martin, T.E. Mallouk, S.A. Chizhik, E.V. Buzaneva, A.D. Gorchinskiy, Chemistry of Materials 11 (1999) 771.
[155] N.A. Kotov, I. Dekany, J.H. Fendler, Advanced Materials 8 (1996) 637.
[156] T. Cassagneau, F. Guerin, J.H. Fendler, Langmuir 16 (2000) 7318.

[157] R.A. Vaia, H.D. Wagner, Materials Today 7 (2004) 32.
[158] X.S. Du, M. Xiao, Y.Z. Meng, A.S. Hay, Synthetic Metals 143 (2004) 129.
[159] P. Liu, K. Gong, Carbon 37 (1999) 706.
[160] S. Stankovich, R.D. Piner, S.T. Nguyen, R.S. Ruoff, Carbon 44 (2006) 3342.
[161] D.S. McLachlan, C. Chiteme, C. Park, K.E. Wise, S.E. Lowther, P.T. Lillehei, E.J. Siochi, J.S. Harrison, Journal Polymer Science Part B: Polymer Physics 43 (2005) 3273.
[162] G. Chen, W. Weng, D. Wu, C. Wu, European Polymer Journal 39 (2003) 2329.
[163] N. Grossiord, J. Loos, C.E. Koning, Journal of Materials Chemistry 15 (2005) 2349.
[164] R. Ramasubramaniam, J. Chen, H. Liu, Applied Physics Letters 83 (2003) 2928.
[165] D.W. Olson, USGS Mineral Commodity Summaries (2006).
[166] M. Hirata, T. Gotou, M. Ohba, Carbon 43 (2005) 503.
[167] T. Kuila, S.K. Srivastava, A.K. Bhowmick, Journal of Applied Polymer Science 111 (2009) 635.
[168] A. Yu, P. Ramesh, M.E. Itkis, B. Elena, R.C. Haddon, Journal of Physical Chemistry C 111 (2007) 7565.
[169] G. Eda, M. Chhowalla, Nano Letters 9 (2009) 814.
[170] L. Peponi, A. Tercjak, R. Verdejo, M.A. Lopez-Manchado, I. Mondragon, J.M. Kenny, Journal of Physical Chemistry 113 (2009) 17973.
[171] H.B. Zhang, W.G. Zheng, Q. Yan, Y. Yang, J.W. Wang, Z.H. Lu, G.Y. Ji, Z.Z. Yu, Polymer 51 (2010) 1191.
[172] J.N. Coleman, U. Khan, W.J. Blau, Y.K. Gun'ko, Carbon 44 (2006) 1624.
[173] O. Breuer, U. Sundararaj, Polymer Composites 25 (2004) 630.
[174] M. Moniruzzaman, K.I. Winey, Macromolecules 39 (2006) 5194.
[175] P.M. Ajayan, O. Stephan, C. Colliex, D. Trauth, Science 265 (1994) 1212.
[176] L. Jin, C. Bower, O. Zhou, Applied Physics Letters 73 (1998) 1197.
[177] M.S.P. Shaffer, A.H. Windle, Advanced Materials 11 (1999) 937.
[178] H. Geng, R. Rosen, B. Zheng, H. Shimoda, L. Fleming, J. Liu, O. Zhou, Advanced Materials 14 (2002) 1387.
[179] B. Safadi, R. Andrews, E.A. Grulke, Journal of Applied Polymer Science 84 (2002) 2660.
[180] F. Du, J.E. Fischer, K.I. Winey, Journal Polymer Science Part B: Polymer Physics 41 (2003) 3333.
[181] C.Y. Li, L. Li, W. Cai, S.L. Kodjie, K.K. Tenneti, Advanced Materials 17 (2005) 1198.
[182] R. Hagenmueller, J.E. Fischer, K.I. Winey, Macromolecules 39 (2006) 2964.
[183] N. Pierard, A. Fonseca, Z. Konya, I. Willems, G. van Tendeloo, J.B. Nagy, Chemical Physics Letters 1 (2001) 335.
[184] H. Xia, Q. Wang, K. Li, G.H. Hu, Journal of Applied Polymer Science 93 (2004) 378.
[185] S. Ghose, K.A. Watson, K.J. Sun, J.M. Criss, E.J. Siochi, J.W. Connell, Composites Science and Technology 66 (2006) 1995.
[186] A. Ikeda, K. Hayashi, T. Konishi, J.I. Kikuchi, Chemical Communications 1334 (2004).
[187] C.A. Cooper, D. Ravich, D. Lips, J. Mayer, H.D. Wagner, Composites Science and Technology 62 (2002) 1105.
[188] R. Haggenmueller, H.H. Gommans, A.G. Rinzler, J.E. Fischer, I. Winey, Chemical Physics Letters 330 (2000) 219.
[189] Z. Jin, K.P. Pramoda, G. Xu, S.H. Goh, Chemical Physics Letters 337 (2001) 43.
[190] P. Potschke, T.D. Fornes, D.R. Paul, Polymer 43 (2002) 3247.
[191] Z. Jia, Z. Wang, C. Xu, J. Liang, B. Wei, D. Wu, S. Zhu, Materials Science and Engineering A 271 (1999) 395.

[192] C. Velasco-Santos, A.L. Martinez-Hernandez, F.T. Fisher, R. Ruoff, V.M. Castano, Chemistry of Materials 15 (2003) 4470.
[193] J. Fan, M. Wan, D. Zhu, B. Chang, Z. Pan, S. Xie, Journal of Applied Polymer Science 74 (1999) 2605.
[194] B.Z. Tang, H. Xu, Macromolecules 32 (1999) 2569.
[195] C. Downs, J. Nugent, P.M. Ajayan, D.J. Duquette, K.S.V. Santhanam, Advanced Materials 11 (1999) 1028.
[196] G.Z. Chen, M.S.P. Shaffer, D. Coleby, G. Dixon, W. Zhou, D.J. Fray, A.H. Windle, Advanced Materials 12 (2000) 522.
[197] T. Kimura, H. Ago, M. Tobita, S. Ohshima, M. Kyotani, M. Yumura, Advanced Materials 14 (2002) 1380.
[198] E.S. Choi, J.S. Brooks, D.L. Eaton, M.S. Al-Haik, M.Y. Hussaini, H. Garmestani, D. Li, K. Dahmen, Journal of Applied Physics 94 (2003) 6034.
[199] C.A. Martin, J. Sandler, A.H. Windle, M. Schwarz, W. Bauhofer, K. Schulte, M.S.P. Shaffer, Polymer 46 (2005) 877.
[200] C.H. Liu, H. Huang, Y. Wu, S.S. Fan, Applied Physics Letters 84 (2004) 4248.
[201] F.H. Gojny, M. Wichmann, U. Kopke, B. Fiedler, K. Schulte, Composites Science and Technology 64 (2004) 2363.
[202] R.A. Vaia, H.D. Wagner, Materials Today 7 (2004) 32.
[203] M. Bhattacharya, A.K. Bhowmick, Polymer 49 (2008) 4808.
[204] S.S. Ray, M. Okamoto, Progress in Polymer Science 28 (2003) 1539.
[205] Y. Mai, Z. Yu, Polymer Nanocomposites. Woodhead Pub, Cambridge (2006).
[206] F. Hussain, M. Hojjati, J. Comp. Mater. 40 (2006) 1511.
[207] J.F. Wu, M.Q. Xu, G.C. Zhao, Electrochemistry Communications 12 (2010) 175.
[208] A.B. Morgan, J.W. Gilman, Journal of Applied Polymer Science 87 (2003) 1329.
[209] D.M. Lincoln, R.A. Vaia, Z.G. Wang, B.S. Hsiao, R. Krishnamoorti, Polymer 42 (2001) 9975.
[210] D.M. Lincoln, R.A. Vaia, Z.G. Wang, B.S. Hsiao, Polymer 42 (2001) 1621.
[211] R.A. Vaia, W. Liu, H. Koerner, Journal Polymer Science Part B: Polymer Physics 41 (2003) 3214.
[212] R.S. Justice, D.W. Schaefer, R.A. Vaia, D.W. Tomlin, T.J. Bunning, Polymer 46 (2005) 4465.
[213] D.W. Schaefer, R.S. Justice, Macromolecules 40 (2007) 8501.

Index

A

B

D

E

M

Q

R

S

X

Y

Z

Kolibri 360 GmbH
Ferdinand-Jühlke-Straße 7
99095 Erfurt, Deutschland
produktsicherheit@kolibri360.de